INSTRUCTORS SOLUTIONS MANUAL

by Julie Levandosky
Framingham State College

to accompany

Differential Equations:

An Introduction to Modern Methods and Applications

by

James R. Brannan
Clemson University
and

William E. Boyce
Rensselaer Polytechnic Institute

John Wiley & Sons, Inc.

PUBLISHER   Laurie Rosatone
ACQUISITIONS EDITOR   David Dietz
FREELANCE DEVELOPMENT EDITOR   Anne Scanlan-Rohrer
SENIOR PRODUCTION EDITOR   Ken Santor
MARKETING MANAGER   Jaclyn Elkins
MARKETING ASSISTANT   Chelsee Pengal

This book was set in LaTeX by Julie Levandosky and Aptara Inc. and printed and bound by Lightning Source, Inc. The cover was printed by Lightning Source.

To order books or for customer service please, call 1-800-CALL WILEY (225-5945).

ISBN 978-0-470-12554-3

Table of Contents

Chapter 1

Section 1.1

1.

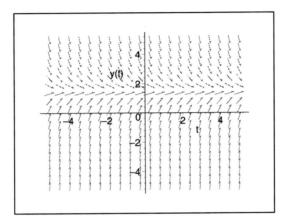

For $y > 3/2$, the slopes are negative, and, therefore the solutions decrease. For $y < 3/2$, the slopes are positive, and, therefore, the solutions increase. As a result, $y \to 3/2$ as $t \to \infty$

2.

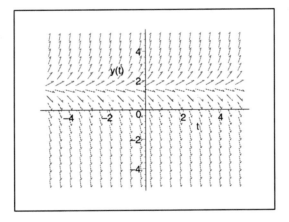

For $y > 3/2$, the slopes are positive, and, therefore the solutions increase. For $y < 3/2$, the slopes are negative, and, therefore, the solutions decrease. As a result, y diverges from $3/2$ as $t \to \infty$

3.

For $y > -3/2$, the slopes are positive, and, therefore the solutions increase. For $y < -3/2$, the slopes are negative, and, therefore, the solutions decrease. As a result, y diverges from $-3/2$ as $t \to \infty$

4.

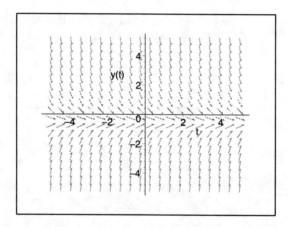

For $y > -1/2$, the slopes are negative, and, therefore the solutions decrease. For $y < -1/2$, the slopes are positive, and, therefore, the solutions increase. As a result, $y \to -1/2$ as $t \to \infty$

5.

For $y > -1/2$, the slopes are positive, and, therefore the solutions increase. For $y < -1/2$, the slopes are negative, and, therefore, the solutions decrease. As a result, y diverges from $-1/2$ as $t \to \infty$

6.

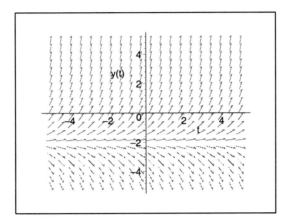

For $y > -2$, the slopes are positive, and, therefore the solutions increase. For $y < -2$, the slopes are negative, and, therefore, the solutions decrease. As a result, y diverges from -2 as $t \to \infty$

7. For the solutions to satisfy $y \to 3$ as $t \to \infty$, we need $y' < 0$ for $y > 3$ and $y' > 0$ for $y < 3$. The equation $y' = 3 - y$ satisfies these conditions.

8. For the solutions to satisfy $y \to 2/3$ as $t \to \infty$, we need $y' < 0$ for $y > 2/3$ and $y' > 0$ for $y < 2/3$. The equation $y' = 2 - 3y$ satisfies these conditions.

9. For the solutions to satisfy y diverges from 2, we need $y' > 0$ for $y > 2$ and $y' < 0$ for $y < 2$. The equation $y' = y - 2$ satisfies these conditions.

10. For the solutions to satisfy y diverges from 1/3, we need $y' > 0$ for $y > 1/3$ and $y' < 0$ for $y < 1/3$. The equation $y' = 3y - 1$ satisfies these conditions.

11.

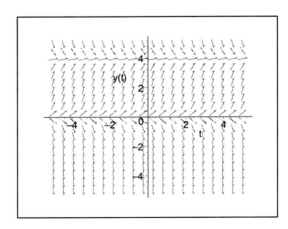

$y = 0$ and $y = 4$ are equilibrium solutions; $y \to 4$ if initial value is positive; y diverges from 0 if initial value is negative.

12.

$y = 0$ and $y = 5$ are equilibrium solutions; y diverges from 5 if initial value is greater than 5; $y \to 0$ if initial value is less than 5.

13.

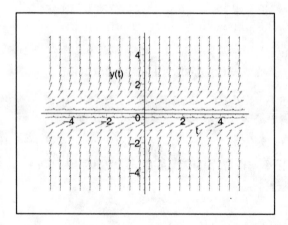

$y = 0$ is equilibrium solution; $y \to 0$ if initial value is negative; y diverges from 0 if initial value is positive.

14.

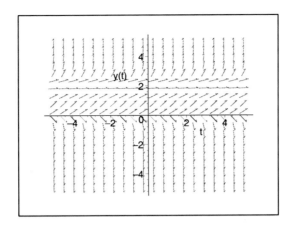

$y = 0$ and $y = 2$ are equilibrium solutions; y diverges from 0 if initial value is negative; $y \to 2$ if initial value is between 0 and 2; y diverges from 2 if initial value is greater than 2.

15. (j)

16. (c)

17. (g)

18. (b)

19. (h)

20. (e)

21.

(a) Let $q(t)$ denote the amount of chemical in the pond at time t. The chemical q will be measured in grams and the time t will be measured in hours. The rate at which the chemical is entering the pond is given by 300 gallons/hour $\cdot .01$ grams/gal $= 300 \cdot 10^{-2}$. The rate at which the chemical leaves the pond is given by 300 gallons/hour $\cdot q/1,000,000$ grams/gal $= 300 \cdot q 10^{-6}$. Therefore, the differential equation is given by $dq/dt = 300(10^{-2} - q 10^{-6})$.

(b) As $t \to \infty$, $10^{-2} - q 10^{-6} \to 0$. Therefore, $q \to 10^4$ g. The limiting amount does not depend on the amount that was present initially.

22. The surface area of a spherical raindrop of radius r is given by $S = 4\pi r^2$. The volume of a spherical raindrop is given by $V = 4\pi r^3/3$. Therefore, we see that the surface area $S = cV^{2/3}$ for some constant c. If the raindrop evaporates at a rate proportional to its surface area, then $dV/dt = -kV^{2/3}$ for some $k > 0$.

23. The difference between the temperature of the object and the ambient temperature is $u - 70$. Since the difference is decreasing if $u > 70$ (and increasing if $u < 70$) and the rate constant is 0.05, the corresponding differential equation is given by $du/dt = -0.05(u - 70)$ where u is measured in degrees Fahrenheit and t is measured in minutes.

24.

(a) Let $q(t)$ be the total amount of the drug (in milligrams) in the body at a given time t (measured in hours). The drug enters the body at the rate of 5 mg/cm^3 $\cdot 100$ cm^3/hr

= 500 mg/hr, and the drug leaves the body at the rate of $0.4q$ mg/hr. Therefore, the governing differential equation si given by $dq/dt = 500 - 0.4q$.

(b) If $q > 1250$, then $q' > 0$. If $q < 1250$, then $q' > 0$. Therefore, $q \to 1250$.

25.

(a) Following the discussion in the text, the equation is given by $mv' = mg - kv^2$.

(b) After a long time, $v' \to 0$. Therefore, $mg - kv^2 \to 0$, or $v \to \sqrt{mg/k}$

(c) We need to solve the equation $\sqrt{.025 \cdot 9.8/k} = 35$. Solving this equation, we see that $k = 0.0002$ kg/m

26.

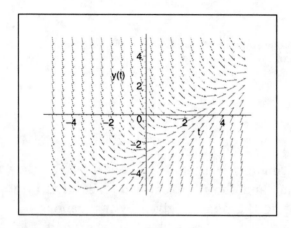

y is asymptotic to $t - 3$ as $t \to \infty$

27.

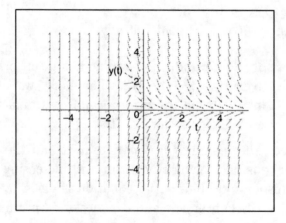

$y \to 0$ as $t \to \infty$

28.

6

$y \to \infty, 0,$ or $-\infty$ depending on the initial value of y

29.

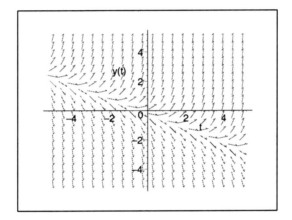

$y \to \infty$ or $-\infty$ depending whether the initial value lies above or below the line $y = -t/2$.

30.

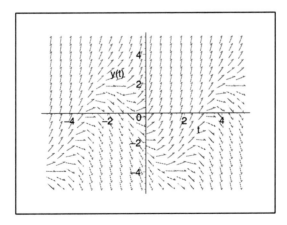

$y \to \infty$ or $-\infty$ or y oscillates depending whether the initial value of y lies above or below the sinusoidal curve.

31.

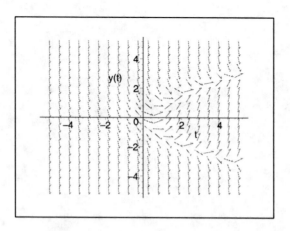

$y \to -\infty$ or is asymptotic to $\sqrt{2t-1}$ depending on the initial value of y

32.

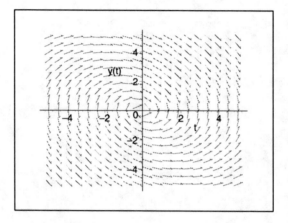

$y \to 0$ and then fails to exist after some $t_f \geq 0$

33.

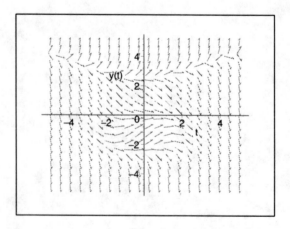

$y \to \infty$ or $-\infty$ depending on the initial value of y

Section 1.2

1.

(a) Rewrite the equation as

$$\frac{dy}{5 - y} = dt$$

and then integrate both sides. Doing so, we see that $-\ln|5 - y| = t + c$. Applying the exponential function, we have $5 - y = ce^{-t}$. Substituting in our initial condition $y(0) = y_0$, we have $5 - y_0 = c$. Therefore, our solution is $y(t) = 5 + (y_0 - 5)e^{-t}$.

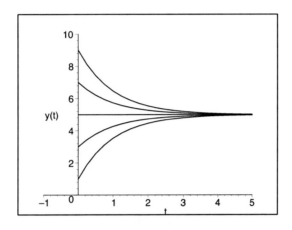

(b) Rewrite the equation as

$$\frac{dy}{5 - 2y} = dt$$

and then integrate both sides. Doing so, we see that $\ln|5 - 2y| = -2t + c$. Applying the exponential function, we have $5 - 2y = ce^{-2t}$. Substituting in our initial condition $y(0) = y_0$, we have $5 - 2y_0 = c$. Therefore, our solution is $y(t) = (5/2) + [y_0 - (5/2)]e^{-2t}$

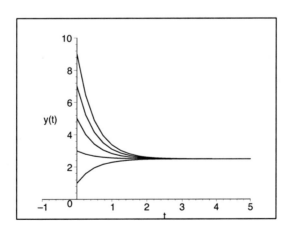

(c) Rewrite the equation as

$$\frac{dy}{10 - 2y} = dt$$

and then integrate both sides. Doing so, we see that $\ln|10 - 2y| = -2t + c$. Applying the exponential function, we have $10 - 2y = ce^{-2t}$. Substituting in our initial condition $y(0) = y_0$, we have $10 - 2y_0 = c$. Therefore, our solution is $y(t) = 5 + [y_0 - 5]e^{-2t}$

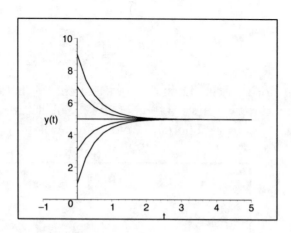

The equilibrium solution is $y = 5$ in (a) and (c), $y = 5/2$ in (b). The solution approaches equilibrium faster in (b) and (c) than in (a).

2.

(a) Rewrite the equation as

$$\frac{dy}{y - 5} = dt$$

and then integrate both sides. Doing so, we see that $\ln |y - 5| = t + c$. Applying the exponential function, we have $y - 5 = ce^t$. Substituting in our initial condition $y(0) = y_0$, we have $y_0 - 5 = c$. Therefore, our solution is $y(t) = 5 + [y_0 - 5]e^t$

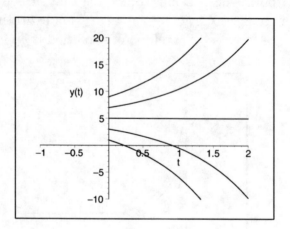

(b) Rewrite the equation as

$$\frac{dy}{2y - 5} = dt$$

and then integrate both sides. Doing so, we see that $\ln |2y - 5| = 2t + c$. Applying the exponential function, we have $2y - 5 = ce^{2t}$. Substituting in our initial condition $y(0) = y_0$, we have $2y_0 - 5 = c$. Therefore, our solution is $y(t) = (5/2) + [y_0 - (5/2)]e^{2t}$

10

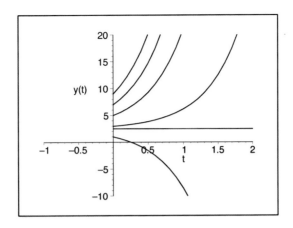

(c) Rewrite the equation as

$$\frac{dy}{2y - 10} = dt$$

and then integrate both sides. Doing so, we see that $\ln|2y - 10| = 2t + c$. Applying the exponential function, we have $2y - 10 = ce^{2t}$. Substituting in our initial condition $y(0) = y_0$, we have $2y_0 - 10 = c$. Therefore, our solution is $y(t) = 5 + [y_0 - 5]e^{2t}$

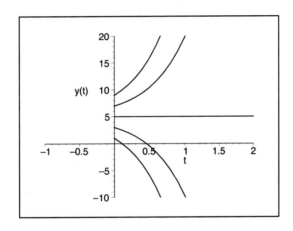

The equilibrium solution is $y = 5$ in (a) and (c), $y = 5/2$ in (b); solution diverges from equilibrium faster in (b) and (c) than in (a).

3.

(a) Rewrite the equation as

$$\frac{dy}{b - ay} = dt$$

and then integrate both sides. Doing so, we see that $\ln|b - ay| = -at + c$. Applying the exponential function, we have $b - ay = ce^{-at}$, or $y = ce^{-at} + (b/a)$

(b) Below we show solution curves for various initial conditions under the cases $a = 1, b = 1$, $a = 5, b = 1$, $a = 1, b = 5$ and $a = 5, b = 5$, respectively.

11

(c) (i) As a increases, the equilibrium is lower and is approached more rapidly. (ii) As b increases, the equilibrium is higher. (iii) As a and b increase, but a/b remains the same, the equilibrium remains the same and is approached more rapidly.

4.

(a) The equilibrium solution occurs when $dy/dt = ay - b = 0$. Therefore, the equilibrium solution is $y_e = b/a$

(b) If $Y(t) = y - y_e$, then $Y'(t) = y' - y_e' = y' = ay - b = a(Y + y_e) - b = aY + ay_e - b = aY + a(b/a) - b = 0 = aY$. Therefore, Y satisfies the equation $Y' = aY$.

5. The solution of the homogeneous problem is $y = ce^{-2t}$. Therefore, we assume the solution will have the form $y = ce^{-2t} + At + B$. Substituting a function of this form into the differential equation leads to the equation

$$2At + A + 2B = t - 3.$$

Equating like coefficients, we see that $A = 1/2$ and $B = -7/4$. Therefore, the general solution is

$$y = ce^{-2t} + \frac{1}{2}t - \frac{7}{4}.$$

6. The solution of the homogeneous problem is $y = ce^{3t}$. Therefore, we assume the solution will have the form $y = ce^{3t} + Ae^{-t}$. Substituting a function of this form into the differential equation leads to the equation

$$-4Ae^{-t} = e^{-t}.$$

Equating like coefficients, we see that $A = -1/4$. Therefore, the general solution is

$$y = ce^{3t} - \frac{1}{4}e^{-t}.$$

7. The solution of the homogeneous problem is $y = ce^{-t}$. Therefore, we assume the solution will have the form $y = ce^{-t} + A\cos(2t) + B\sin(2t)$. Substituting a function of this form into the differential equation leads to the equation

$$[-2A + B]\sin(2t) + [2B + A]\cos(2t) = 3\cos(2t).$$

Solving the two equations, $-2A + B = 0$ and $2B + A = 3$, we see that $A = 3/5$ and $B = 6/5$. Therefore, the general solution is

$$y = ce^{-t} + \frac{3}{5}\cos(2t) + \frac{6}{5}\sin(2t).$$

8. The solution of the homogeneous problem is $y = ce^{2t}$. Therefore, we assume the solution will have the form $y = ce^{2t} + A\cos(t) + B\sin(t)$. Substituting a function of this form into the differential equation leads to the equation

$$[-A - 2B]\sin(t) + [B - 2A]\cos(t) = 2\sin(t).$$

Solving the system of equations $-A - 2B = 2$ and $B - 2A = 0$, we see that $A = -2/5$ and $B = -4/5$. Therefore, the general solution is

$$y = ce^{2t} - \frac{2}{5}\cos(t) - \frac{4}{5}\sin(t).$$

9. The solution of the homogeneous problem is $y = ce^{-2t}$. Therefore, we assume the solution will have the form $y = ce^{-2t} + At + B + C\cos(t) - D\sin(t)$. Substituting a function of this form into the differential equation leads to the equation

$$2At + [A + 2B] + [C + 2D]\cos(t) + [2C - D]\sin(t) = 2t + 3\sin(t).$$

Equating like coefficients, we see that $A = 1$, $B = -1/2$, $C = 6/5$ and $D = -3/5$. Therefore, the general solution is

$$y = ce^{-2t} + t - \frac{1}{2} + \frac{6}{5}\sin(t) - \frac{3}{5}\cos(t).$$

10. The solution of the homogeneous problem is $y = ce^{2t}$. Therefore, we assume the solution will have the form $y = ce^{2t} + Ae^t + Bt^2 + Ct + D$. Substituting a function of this form into the differential equation leads to the equation

$$-Ae^t - 2Bt^2 + [2B - 2C]t + [C - 2D] = 3e^t + t^2 + 1.$$

Equating like coefficients, we see that $A = -3$, $B = -1/2$, $C = -1/2$ and $D = -3/4$. Therefore, the general solution is

$$y = ce^{2t} - 3e^t - \frac{1}{2}t^2 - \frac{1}{2}t - \frac{3}{4}.$$

11.

(a) The general solution is $p(t) = 900 + ce^{t/2}$. Plugging in for the initial condition, we have $p(t) = 900 + (p_0 - 900)e^{t/2}$. With $p_0 = 850$, the solution is $p(t) = 900 - 50e^{t/2}$. To find the time when the population becomes extinct, we need to find the time T when $p(T) = 0$. Therefore, $900 = 50e^{T/2}$, which implies $e^{T/2} = 18$, and, therefore, $T = 2\ln 18 \cong 5.78$ months.

(b) Using the general solution, $p(t) = 900 + (p_0 - 900)e^{t/2}$, we see that the population will become extinct at the time T when $900 = (900 - p_0)e^{T/2}$. That is, $T = 2\ln[900/(900 - p_0)]$ months

(c) Using the general solution, $p(t) = 900 + (p_0 - 900)e^{t/2}$, we see that the population after 1 year (12 months) will be $p(6) = 900 + (p_0 - 900)e^6$. If we want to know the initial population which will lead to extinction after 1 year, we set $p(6) = 0$ and solve for p_0. Doing so, we have $(900 - p_0)e^6 = 900$ which implies $p_0 = 900(1 - e^{-6}) \cong 897.8$

12.

(a) The general solution is $p(t) = p_0 e^{rt}$, where t is measured in days. If the population doubles in 30 days, then $p(30) = 2p_0 = p_0 e^{30r}$. Therefore, $r = (\ln 2)/30$ day^{-1}.

(b) If the population doubles in N days, then $p(N) = 2p_0 = p_0 e^{Nr}$. Therefore, $r = (\ln 2)/N$ day^{-1}

13.

(a) The solution is given by $v(t) = 35(1 - e^{-0.28t})$. The limiting velocity is 35 m/sec. Therefore, we want to find the time T when $v(T) = .98 \cdot 35 = 34.3$ m/sec. Plugging this value into our equation for v, we have $34.3 = 35(1 - e^{-0.28T})$, or $e^{-0.28T} = .02$ which implies $T = (\ln 50)/0.28 \cong 13.97$ sec

(b) To find the position, we integrate the velocity function above. For $v(t) = 35(1 - e^{-0.28t})$, the height is given by $s(t) = \int v(t) = 35t + 125te^{-0.28t}\, dt + C$. Assuming, $s(0) = 0$, we see that $c = -125$. Therefore, $s(t) = 35t + 125e^{-0.28t} - 125$. When $T = 13.97$ seconds, we see that the distance traveled is approximately 366.5 m.

14.

(a) Assuming no air resistance, Newton's Second Law can be expressed as

$$m\frac{dv}{dt} = mg$$

where g is the gravitational constant. Dividing the above equation by m and assuming that the initial velocity is zero, we see that our initial value problem is $dv/dt = 9.8$, $v(0) = 0$

(b) We are assuming the object is released from a height of 300 meters above the ground. The height at a later time t satisfies $ds/dt = v = 9.8t$. Taking the point of release as the origin and integrating the above equation for s, we have $s(t) = 4.9t^2$. We need to find the time T when $s(T) = 300$. That is, $4.9T^2 = 300$. Solving this equation, we have $T = \sqrt{300/4.9} \cong 7.82$ sec

(c) Using the equation $v = 9.8t$, we see that when $T \cong 7.82$ seconds, $v \cong 76.68$ m/sec

15.

(a) If we are assuming that the drag force is proportional to the square of the velocity, equation (1) becomes

$$m\frac{dv}{dt} = mg - \gamma v^2.$$

Plugging in $m = 0.025$, $g = 9.8$, the equation can be written as

$$\frac{dv}{dt} = 9.8 - \frac{\gamma}{.025}v^2.$$

If the limiting velocity is 35 m/sec, then $\gamma(35)^2 = 9.8 \cdot .025$ which implies $\gamma = 0.0002$. Therefore,

$$\frac{dv}{dt} = 9.8 - 0.008v^2,$$

or

$$\frac{dv}{dt} = [(35)^2 - v^2]/125.$$

(b) The equation can be rewritten as

$$\frac{dv}{(35)^2 - v^2} = \frac{dt}{125}.$$

Integrating both sides, we have

$$\ln \left| \frac{v + 35}{v - 35} \right| = \frac{70}{125} t + c.$$

Plugging in the initial condition $v(0) = 0$, we have $c = 0$. Applying the exponential function to both sides of the equation, we have

$$v + 35 = e^{70t/125}(35 - v).$$

Solving this equation for v, we have

$$v(t) = 35 \left[\frac{e^{70t/125} - 1}{e^{70t/125} + 1} \right]$$

or

$$v(t) = 35 \left[\frac{e^{35t/125}\left(e^{35t/125} - e^{-35t/125}\right)}{e^{35t/125}\left(e^{35t/125} + e^{-35t/125}\right)} \right] = 35 \tanh(7t/25)$$

(c) Below we show the graphs of $v(t)$ above (the top curve) and the solution to the problem in example 2 (the bottom curve)

(d) The quadratic force leads to the falling object attaining its limiting velocity sooner.

16

(e) The distance $x(t) = \int v(t)\, dt = \int 35 \tanh(7t/25)\, dt = 125 \ln \cosh(7t/25)$.

(f) Plugging 300 in for $x(T)$ in the answer to part (d), we have $300 = 125 \ln \cosh(7T/25)$. Therefore, $T = (25/7)\text{arccosh}(e^{12/5}) \cong 11.04$ sec

16.

(a) The general solution of the equation is $Q(t) = ce^{-rt}$. Given that $Q(0) = 100$, we have $c = 100$. Assuming that $Q(1) = 82.04$, we have $82.04 = 100e^{-r}$. Solving this equation for r, we have $r = -\ln(82.04/100) = .19796$ per week or $r = 0.02828$ per day.

(b) Using the form of the general solution and r found above, we have $Q(t) = 100e^{-0.02828t}$

(c) Let T be the time it takes the isotope to decay to half of its original amount. From part (b), we conclude that $.5 = e^{-0.2828T}$ which implies that $T = -\ln(0.5)/0.2828 \cong 24.5$ days

17. The general solution of the differential equation is $Q(t) = Q_0 e^{-rt}$ where $Q_0 = Q(0)$. Let τ be the half-life. Plugging τ into the equation for Q, we have $0.5Q_0 = Q_0 e^{-r\tau}$. Therefore, $0.5 = e^{-r\tau}$ which implies $\tau = -\ln(0.5)/r = \ln(2)/r$. Therefore, we conclude that $r\tau = \ln 2$.

18. The differential equation for radium-226 is $dQ/dt = -rQ$. The solution of this equation is $Q(t) = Q_0 e^{-rt}$. Using the result from exercise 17 and the fact that the half-life is 1620 years, we conclude that the decay rate $r = \ln(2)/\tau = \ln(2)/1620$. Let T be the time it takes for the isotope to decay to 3/4 of its original amount. Then

$$\frac{3}{4}Q_0 = Q_0 e^{-\ln(2)T/1620}$$

which implies $T = -1620 \ln(3/4)/\ln(2) \cong 672.4$ years.

19.

(a) We rewrite the equation as
$$\frac{du}{u - T} = -k.$$
Integrating both sides, we have $\ln|u - T| = -kt + c$. Applying the exponential function to both sides of the equation and plugging in the initial condition $u(0) = u_0$, we arrive at the general solution $u(t) = T + (u_0 - T)e^{-kt}$

(b) Since T is a constant, we see that if u satisfies the equation $du/dt = -k(u - T)$, then $d(u-T)/dt = du/dt = -k(u-T)$. Then using the result from exercise 17 above, we know that the relationship between the decay rate k and the time τ when the temperature difference is reduced by half satisfies the relationship $k\tau = \ln 2$.

20. Based on exercise 19 above, the differential equation for the temperature in the room is given by
$$\frac{du}{dt} = -.15(u - 10)$$

with an initial condition of $u(0) = 70$. As shown in exercise 19 above, the solution is given by $u(t) = 10 + 60e^{-0.15t}$. We need to find the time t such that $u(t) = 32$. That is, $22 = 60e^{-0.15t}$. Solving this equation for t, we have $t = -\ln(22/60)/0.15 \cong 6.69$ hours.

21.

(a) The solution of the differential equation with $q(0) = 0$ is $q(t) = CV(1 - e^{-t/RC})$. Below we show a sketch in the case when $C = V = R = 1$.

(b) As $t \to \infty$, the exponential term vanishes. Therefore, the limiting value is $q_L = CV$

(c) If the battery is removed, then $V = 0$. Therefore, our differential equation is

$$R\frac{dq}{dt} + \frac{q}{C} = 0.$$

Also, we are assuming that $q(t_1) = q_L = CV$. Solving the differential equation, we have $q = ce^{-t/RC}$. Using the initial condition $q(t_1) = CV$, we have $q(t_1) = ce^{-t_1/RC} = CV$. Therefore, $c = CVe^{t_1/RC}$. We conclude that $q(t) = CV\exp[-(t - t_1)/RC]$ Below we show a graph of the solution taking $C = V = R = 1$ and $t_1 = 5$.

22.

18

(a) The accumulation rate of the chemical is $(0.01)(300)$ grams per hour. At any given time t, the concentration of the chemical in the pond is $Q(t)/10^6$ grams per gallon. Therefore, the chemical leaves the pond at the rate of $300Q(t)/10^6$ grams per hour. Therefore, the equation for Q is given by $Q' = 3(1 - 10^{-4}Q)$. Since initially there are no chemicals in the pond, $Q(0) = 0$.

(b) Rewrite the equation as
$$\frac{dQ}{10000 - Q} = 0.0003dt.$$
Integrating both sides of this equation, we have $\ln|10000 - Q| = -0.0003t + C$. Applying the exponential function to both sides of this equation, we have $10000 - Q = ce^{-0.0003t}$. Assuming $Q(0) = 0$, we see that $c = 10000$. Therefore, $Q(t) = 10000(1 - e^{-0.0003t})$ where t is measured in hours. Since 1 year is 8760 hours, we see that the amount of chemical in the pond after 1 year is $Q(8760) = 10000(1 - e^{-0.0003t}) \cong 9277.77$ grams.

(c) With the accumulation rate now equal to zero, the equation becomes $dQ/dt = -0.0003Q(t)$ grams/hour. Resetting the time variable, we assign the new initial value as $Q(0) = 9277.77$ grams.

(d) The solution of the differential equation is $Q(t) = 9277.77e^{-0.0003t}$ after t hrs. Therefore, after 1 year $Q(8760) \cong 670.07$ g

(e) Letting T be the amount of time after the source is removed, we obtain the equation $10 = 9277.77e^{-0.0003t}$. Solving this equation, we have $T - \ln(10/9277.77)/0.0003 \cong 2.60$ years

(f)

23.

(a) We are assuming that no dye is entering the pool. The rate at which the dye is leaving the pool is given by $200 \cdot (q/60,000)$ g/min $= q/300$ g/min. Since initially, there are 5 kg of the dye in the pool, the initial value problem is $q' = -q/300$, $q(0) = 5000$ g

(b) The solution of this initial value problem is $q(t) = 5000e^{-t/300}$ where g is in grams and t is in minutes.

(c) In 4 hours (240 minutes), the amount of dye in the pool will be $q(240) \cong 2246.6$ grams. Since there is $60,000$ gallons of water in the pool, the concentration will be $2246.6/60,000 \cong 0.0374$ grams/gallon. So, no, the pool will not be reduced to the desired level within 4 hours.

(d) Let T be the time that it takes to reduce the concentration level of the dye to 0.02 grams/gallon. At that time, the amount of dye in the pool needs to be 1200 grams (as $1200/60000 = 0.02$). Plugging $q(T) = 1200$ into our equation for q, we have $1200 = 5000e^{-T/300}$. Solving this equation, we have $T = 300\ln(25/6) \cong 7.136$ hr

(e) Let r be the necessary flow rate. As in part (a), if the water leaves the pool at the rate of r gallons/minute, then the initial value problem will be $q' = -rq/60,000$, $q(0) = 5000$. The solution of this initial value problem is given by $q(t) = 5000e^{-rt/60,000}$. We need to find the decay rate r such that when $t = 240$ minutes, the amount of dye $q = 1200$ grams. That is, we need to solve the equation $1200 = 5000e^{-240r/60,000}$. Solving this equation, we have $r = 250\ln(25/6) \cong 256.78$ gal/min

Section 1.3

1. The Euler formula is given by $y_{n+1} = y_n + h(3 + t_n - y_n) = (1-h)y_n + h(3 + t_n)$.

(a) 1.2, 1.39, 1.571, 1.7439

(b) 1.1975, 1.38549, 1.56491, 1.73658

(c) 1.19631, 1.38335, 1.56200, 1.73308

(d) The differential equation is linear with solution $y(t) = 2 + t - e^{-t}$.

   1.19516, 1.38127, 1.55918, 1.72968

2. The Euler formula is given by $y_{n+1} = y_n + h(2y_n - 1) = (1 + 2h)y_n - h$.

(a) 1.1, 1.22, 1.364, 1.5368

(b) 1.105, 1.23205, 1.38578, 1.57179

(c) 1.10775, 1.23873, 1.39793, 1.59144

(d) The differential equation is linear with solution $y(t) = (1 + e^{2t})/2$.

   1.1107, 1.24591, 1.41106, 1.61277

3. The Euler formula is given by $y_{n+1} = y_n + h(0.5 - t_n + 2y_n) = (1 + 2h)y_n + h(0.5 - t_n)$.

(a) 1.25, 1.54, 1.878, 2.2736

(b) 1.26, 1.5641, 1.92156, 2.34359

(c) 1.26551, 1.57746, 1.94586, 2.38287

(d) The differential equation is linear with solution $y(t) = 0.5t + e^{2t}$.

    1.2714, 1.59182, 1.97212, 2.42554

4. The Euler formula is given by $y_{n+1} = y_n + h(3\cos(t_n) - 2y_n) = (1 - 2h)y_n + 3h\cos(t_n)$.

(a) 0.3, 0.538501, 0.724821, 0.866458

(b) 0.284813, 0.513339, 0.693451, 0.831571

(c) 0.277920, 0.501813, 0.678949, 0.815302

(d) The differential equation is linear with solution $y(t) = (6\cos(t) + 3\sin(t) - 6e^{-2t})/5$.

    0.271428, 0.490897, 0.665142, 0.799729

5.

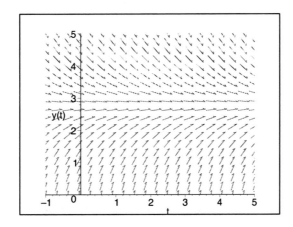

The solutions converge for $y \geq 0$. Solutions are undefined for $y < 0$.

6.

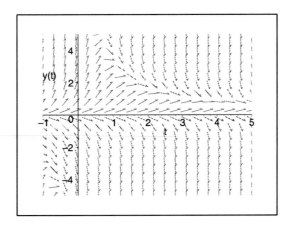

The solutions converge for $y \geq 0$. They diverge for $y < 0$.

7.

All solutions converge.

8.

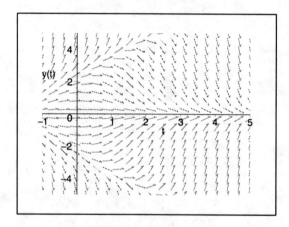

The solutions converge for $|y(0)| < 2.37$ (approximately). They diverge otherwise.

9.

All solutions diverge.

10.

All solutions diverge.

11. The Euler formula is $y_{n+1} = y_n - 3h\sqrt{y_n} + 5h$. The initial value is $y_0 = 2$.

(a) 2.30800, 2.49006, 2.60023, 2.66773, 2.70939, 2.73521

(b) 2.30167, 2.48263, 2.59352, 2.66227, 2.70519, 2.73209

(c) 2.29864, 2.47903, 2.59024, 2.65958, 2.70310, 2.73053

(d) 2.29686, 2.47691, 2.58830, 2.65798, 2.70185, 2.72959

12. The Euler formula is $y_{n+1} = (1 + 3h)y_n - ht_n y_n^2$. The initial value is $(t_0, y_0) = (0, 0.5)$.

(a) 1.70308, 3.06605, 2.44030, 1.77204, 1.37348, 1.11925

(b) 1.79548, 3.06051, 2.43292, 1.77807, 1.37795, 1.12191

(c) 1.84579, 3.05769, 2.42905, 1.78074, 1.38017, 1.12328

(d) 1.87734, 3.05607, 2.42672, 1.78224, 1.38150, 1.12411

13. The Euler formula is $y_{n+1} = y_n + h\frac{(4 - t_n y_n)}{(1 + y_n^2)}$ with $(t_0, y_0) = (0, -2)$.

(a) $-1.48849, -0.412339, 1.04687, 1.43176, 1.54438, 1.51971$

(b) $-1.46909, -0.287883, 1.05351, 1.42003, 1.53000, 1.50549$

(c) $-1.45865, -0.217545, 1.05715, 1.41486, 1.52334, 1.49879$

(d) $-1.45212, -0.173376, 1.05941, 1.41197, 1.51949, 1.49490$

14. The Euler formula is $y_{n+1} = (1 - ht_n)y_n + hy_n^3/10$, with $(t_0, y_0) = (0, 1)$.

(a) 0.950517, 0.687550, 0.369188, 0.145990, 0.0421429, 0.00872877

(b) 0.938298, 0.672145, 0.362640, 0.147659, 0.0454100, 0.0104931

(c) 0.932253, 0.664778, 0.359567, 0.148416, 0.0469514, 0.0113722

(d) 0.928649, 0.660463, 0.357783, 0.148848, 0.0478492, 0.0118978

15. The Euler formula is $y_{n+1} = y_n + h3t_n^2/(3y_n^2 - 4)$ with initial value $(t_0, y_0) = (1, 0)$.

(a) -0.166134, -0.410872, -0.804660, 4.15867

(b) -0.174652, -0.434238, -0.889140, -3.09810

(c) Since the line tangent to the solution is parallel to the $y-$axis when $y \cong \pm 1.155$, Euler's formula can be off by quite a bit. As the slope tends to ∞, using that slope as an approximation to the change in the function can cause a large error in the approximation.

16. The Euler formula is $y_{n+1} = y_n + h(t_n^2 + y_n^2)$ with $(t_0, y_0) = (0, 1)$. A reasonable estimate for y at $t = 0.8$ is between 5.5 and 6. No reliable estimate is possible at $t = 1$ from the specified data.

17. The Euler formula is $y_{n+1} = y_n + h(y_n^2 + 2t_n y_n)/(3 + t_n^2)$ with $(t_0, y_0) = (1, 2)$. A reasonable estimate for y at $t = 2.5$ is between 18 and 19. No reliable estimate is possible at $t = 3$ from the specified data.

18.

(a) See the direction field in exercise 8 above.

(b) The Euler formula is $y_{n+1} = y_n + h(-t_n y_n + 0.1y_n^3)$. For $y_0 < 2.37$, the solutions seem to converge, while the solutions seem to diverge if $y_0 > 2.38$. We conclude that $2.37 < \alpha_0 < 2.38$

19.

(a)

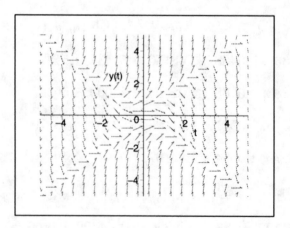

(b) The Euler formula is $y_{n+1} = y_n + h(y_n^2 - t_n^2)$. For $y_0 < 0.67$, the solutions seem to converge, while the solutions seem to diverge if $y_0 > 0.68$. Therefore, we conclude that $0.67 < \alpha_0 < 0.68$

Section 1.4

1. The differential equation is second order, since the highest derivative in the equation is of order two. The equation is linear since the left hand side is a linear function of y and its derivatives and the right hand side is just a function of t.

2. The differential equation is second order since the highest derivative in the equation is of order two. The equation is nonlinear because of the term $y^2 d^2y/dt^2$.

3. The differential equation is fourth order since the highest derivative in the equation is of order four. The equation is linear since the left hand side is a linear function of y and its derivatives and the right hand side does not depend on y.

4. The differential equation is first order since the only derivative in the equation is of order one. The equation is nonlinear because of the y^2 term.

5. The differential equation is second order since the highest derivative in the equation is of order two. The equation is nonlinear because of the term $\sin(t + y)$ which is not a linear function of y.

6. The differential equation is third order since the highest derivative in the equation is of order three. The equation is linear because the left hand side is a linear function of y and its derivatives, and the right hand side is only a function of t.

7. $y_1(t) = e^t \implies y_1' = e^t \implies y_1'' = e^t$. Therefore, $y_1'' - y_1 = 0$. Also, $y_2 = \cosh t \implies y_2' = \sinh t \implies y_2'' = \cosh t$. Therefore, $y_2'' - y_2 = 0$.

8. $y_1 = e^{-3t} \implies y_1' = -3e^{-3t} \implies y_1'' = 9e^{-3t}$. Therefore, $y_1'' + 2y_1' - 3y_1 = (9 - 6 - 3)y_1 = 0$. Also, $y_2 = e^t \implies y_2' = y_2'' = e^t$. Therefore, $y_1'' + 2y_1' - 3y_1 = (1 + 2 - 3)e^t = 0$.

9. $y = 3t + t^2 \implies y' = 3 + 2t$. Therefore, $ty' - y = t(3 + 2t) - (3t + t^2) = t^2$.

10. $y_1 = t/3 \implies y_1' = 1/3 \implies y_1'' = y_1''' = y_1'''' = 0$. Therefore, $y_1''' + 4y_1''' + 3y = t$. Also, $y_2 = e^{-t} + t/3 \implies y_2' = -e^{-t} + 1/3 \implies y_2'' = e^{-t} \implies y_2''' = -e^{-t} \implies y_2'''' = e^{-t}$. Therefore, $y_2'''' + 4y_2''' + 3y = e^{-t} - 4e^{-t} + 3(e^{-t} + t/3) = t$.

11. $y_1 = t^{1/2} \implies y_1' = t^{-1/2}/2 \implies y_1'' = -t^{-3/2}/4$. Therefore, $2t^2 y_1'' + 3t y_1' - y_1 = 2t^2(-t^{-3/2}/4) + 3t(t^{-1/2}/2) - t^{1/2} = (-1/2 + 3/2 - 1)t^{1/2} = 0$. Also, $y_2 = t^{-1} \implies y_2' = -t^{-2} \implies y_2'' = 2t^{-3}$. Therefore, $2t^2 y_2'' + 3t y_2' - y_2 = 2t^2(2t^{-3}) + 3t(-t^{-2}) - t^{-1} = (4 - 3 - 1)t^{-1} = 0$.

12. $y_1 = t^{-2} \implies y_1' = -2t^{-3} \implies y_1'' = 6t^{-4}$. Therefore, $t^2 y_1'' + 5t y_1' + 4y_1 = t^2(6t^{-4}) + 5t(-2t^{-3}) + 4t^{-2} = (6 - 10 + 4)t^{-2} = 0$. Also, $y_2 = t^{-2}\ln t \implies y_2' = t^{-3} - 2t^{-3}\ln t \implies y_2'' = -5t^{-4} + 6t^{-4}\ln t$. Therefore, $t^2 y_2'' + 5t y_2' + 4y_2 = t^2(-5t^{-4} + 6t^{-4}\ln t) + 5t(t^{-3} - 2t^{-3}\ln t) + 4(t^{-2}\ln t) = (-5 + 5)t^{-2} + (6 - 10 + 4)t^{-2}\ln t = 0$.

13. $y = (\cos t)\ln\cos t + t\sin t \implies y' = -(\sin t)\ln\cos t + t\cos t \implies y'' = -(\cos t)\ln\cos t - t\sin t + \sec t$. Therefore, $y'' + y = -(\cos t)\ln\cos t - t\sin t + \sec t + (\cos t)\ln\cos t + t\sin t = \sec t$.

14. $y = e^{t^2}\int_0^t e^{-s^2}\,ds + e^{t^2} \implies y' = 2te^{t^2}\int_0^t e^{-s^2}\,ds + 1 + 2te^{t^2}$. Therefore, $y' - 2ty = 2te^{t^2}\int_0^t e^{-s^2}\,ds + 1 + 2te^{t^2} - 2t(e^{t^2}\int_0^t e^{-s^2}\,ds + e^{t^2}) = 1$.

15. Let $y = e^{rt}$. Then $y' = re^{rt}$. Substituting these terms into the differential equation, we have $y' + 2y = re^{rt} + 2e^{rt} = (r + 2)e^{rt} = 0$. This equation implies $r = -2$.

16. Let $y = e^{rt}$. Then $y' = re^{rt}$ and $y'' = r^2 e^{rt}$. Substituting these terms into the differential equation, we have $y'' - y = (r^2 - 1)e^{rt} = 0$. This equation implies $r = \pm 1$.

17. Let $y = e^{rt}$. Then $y' = re^{rt}$ and $y'' = r^2 e^{rt}$. Substituting these terms into the differential equation, we have $y'' + y' - 6y = (r^2 + r - 6)e^{rt} = 0$. In order for r to satisfy this equation, we need $r^2 + r - 6 = 0$. That is, we need $r = 2, -3$.

18. Let $y = e^{rt}$. Then $y' = re^{rt}$, $y'' = r^2 e^{rt}$ and $y''' = r^3 e^{rt}$. Substituting these terms into the differential equation, we have $y''' - 3y'' + 2y' = (r^3 - 3r^2 + 2r)e^{rt} = 0$. In order for r to satisfy this equation, we need $r^3 - 3r^2 + 2r = 0$. That is, we need $r = 0, 1, 2$.

19. Let $y = t^r$. Then $y' = rt^{r-1}$ and $y'' = r(r-1)t^{r-2}$. Substituting these terms into the differential equation, we have $t^2 y'' + 4ty' + 2y = t^2(r(r-1)t^{r-2}) + 4t(rt^{r-1}) + 2t^r = (r(r-1) + 4r + 2)t^r = 0$. In order for r to satisfy this equation, we need $r(r-1) + 4r + 2 = 0$. Simplifying this expression, we need $r^2 + 3r + 2 = 0$. The solutions of this equation are $r = -1, -2$.

20. Let $y = t^r$. Then $y' = rt^{r-1}$ and $y'' = r(r-1)t^{r-2}$. Substituting these terms into the differential equation, we have $t^2 y'' - 4ty' + 4y = t^2(r(r-1)t^{r-2}) - 4t(rt^{r-1}) + 4t^r = (r(r-1) - 4r + 4)t^r = 0$. In order for r to satisfy this equation, we need $r(r-1) - 4r + 4 = 0$. Simplifying this expression, we need $r^2 - 5r + 4 = 0$. The solutions of this equation are $r = 1, 4$.

21.

(a) Consider Figure 1.4.1 in the text. There are two main forces acting on the mass: (1) the tension in the rod and (2) gravity. The tension, T, acts on the mass along the direction of the rod. By extending a line below and to the right of the mass at an angle θ with the vertical, we see that there is a force of magnitude $mg \cos \theta$ acting on the mass in that direction. Then extending a line below the mass and to the left, making an angle of $\pi - \theta$, we see the force acting on the mass in the tangential direction is $mg \sin \theta$.

(b) Newton's Second Law states that $\sum \mathbf{F} = m\mathbf{a}$. In the tangential direction, the equation of motion may be expressed as $\sum F_\theta = ma_\theta$. The tangential acceleration, a_θ is the linear acceleration along the path. That is, $a_\theta = Ld^2\theta/dt^2$. The only force acting in the tangential direction is the gravitational force in the tangential direction which is given by $-mg \sin \theta$. Therefore, $-mg \sin \theta = mLd^2\theta/dt^2$.

(c) Rearranging terms, we have

$$\frac{d^2\theta}{dt^2} + \frac{g}{L} \sin \theta.$$

22.

(a) The kinetic energy of a particle of mass m is given by $T = \frac{1}{2}mv^2$ where v is its speed. A particle in motion on a circle of radius L has speed $L(d\theta/dt)$ where $d\theta/dt$ is its angular speed. Therefore,

$$T = \frac{1}{2}mL^2 \left(\frac{d\theta}{dt}\right)^2.$$

26

(b) The potential energy of a particle is given by $V = mgh$ where h is the height above some point. Here we measure h as the height above the pendulum's lowest position. Since $\dfrac{L-h}{L} = \cos\theta$, we have $h = L(1 - \cos\theta)$. Therefore,

$$V = mgL(1 - \cos\theta).$$

(c) Since the total energy is conserved. We know that for $E = T + V$, $dE/dt = 0$. Here,

$$E = \frac{1}{2}mL^2\left(\frac{d\theta}{dt}\right)^2 + mgL(1 - \cos\theta)$$

implies

$$\frac{dE}{dt} = mL^2\frac{d\theta}{dt}\frac{d^2\theta}{dt^2} + mgL\sin\theta\frac{d\theta}{dt} = 0.$$

Simplifying this equation, we conclude that

$$\frac{d^2\theta}{dt^2} + \frac{g}{L}\sin\theta = 0.$$

23.

(a) Angular momentum is the moment about a certain point of linear momentum, which is given by

$$mv = mL\frac{d\theta}{dt}.$$

The moment about a pivot point is given by

$$M_p = mL^2\frac{d\theta}{dt}.$$

(b) The moment of the gravitational force is

$$M_g = -mg \cdot L\sin\theta.$$

Then $dM_p/dt = M_g$ implies

$$mL^2\frac{d^2\theta}{dt^2} = -mgL\sin\theta.$$

Rewriting this equation, we have

$$\frac{d^2\theta}{dt^2} + \frac{g}{L}\sin\theta = 0.$$

Chapter 2

Section 2.1

1.

(a)

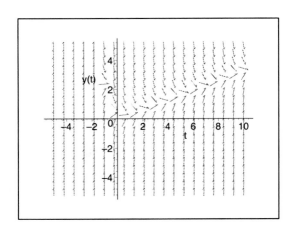

(b) All solutions seem to converge to an increasing function as $t \to \infty$.

(c) The integrating factor is $\mu(t) = e^{3t}$. Then

$$e^{3t}y' + 3e^{3t}y = e^{3t}(t + e^{-2t}) \implies (e^{3t}y)' = te^{3t} + e^t$$
$$\implies e^{3t}y = \int (te^{3t} + e^t)\, dt = \frac{1}{3}te^{3t} - \frac{1}{9}e^{3t} + e^t + c$$
$$\implies y = \frac{t}{3} - \frac{1}{9} + e^{-2t} + ce^{-3t}.$$

We conclude that y is asymptotic to $t/3 - 1/9$ as $t \to \infty$.

2.

(a)

(b) All slopes eventually become positive, so all solutions will eventually increase without bound.

(c) The integrating factor is $\mu(t) = e^{-2t}$. Then

$$e^{-2t}y' - 2e^{-2t}y = e^{-2t}(t^2 e^{2t}) \implies (e^{-2t}y)' = t^2$$

$$\implies e^{-2t}y = \int t^2 \, dt = \frac{t^3}{3} + c$$

$$\implies y = \frac{t^3}{3}e^{2t} + ce^{2t}.$$

We conclude that y increases exponentially as $t \to \infty$.

3.

(a)

(b) All solutions appear to converge to the function $y(t) = 1$.

(c) The integrating factor is $\mu(t) = e^t$. Therefore,

$$e^t y' + e^t y = t + e^t \implies (e^t y)' = t + e^t$$

$$\implies e^t y = \int (t + e^t) \, dt = \frac{t^2}{2} + e^t + c$$

$$\implies y = \frac{t^2}{2}e^{-t} + 1 + ce^{-t}.$$

Therefore, we conclude that $y \to 1$ as $t \to \infty$.

4.

(a)

(b) The solutions eventually become oscillatory.

(c) The integrating factor is $\mu(t) = t$. Therefore,

$$ty' + y = 3t\cos(2t) \implies (ty)' = 3t\cos(2t)$$

$$\implies ty = \int 3t\cos(2t)\,dt = \frac{3}{4}\cos(2t) + \frac{3}{2}t\sin(2t) + c$$

$$\implies y = +\frac{3\cos 2t}{4t} + \frac{3\sin 2t}{2} + \frac{c}{t}.$$

We conclude that y is asymptotic to $(3\sin 2t)/2$ as $t \to \infty$.

5.

(a)

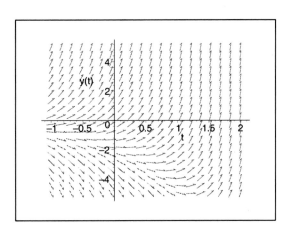

(b) All slopes eventually become positive so all solutions eventually increase without bound.

(c) The integrating factor is $\mu(t) = e^{-2t}$. Therefore,

$$e^{-2t}y' - 2e^{-2t}y = 3e^{-t} \implies (e^{-2t}y)' = 3e^{-t}$$

$$\implies e^{-2t}y = \int 3e^{-t}\,dt = -3e^{-t} + c$$

$$\implies y = -3e^t + ce^{2t}.$$

We conclude that y increases exponentially as $t \to \infty$.

3

6.

(a)

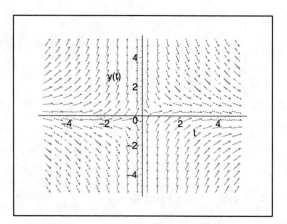

(b) For $t > 0$, all solutions seem to eventually converge to the function $y = 0$.

(c) The integrating factor is $\mu(t) = t^2$. Therefore,

$$t^2 y' + 2ty = t\sin(t) \implies (t^2 y)' = t\sin(t)$$
$$\implies t^2 y = \int t\sin(t)\, dt = \sin(t) - t\cos(t) + c$$
$$\implies y = \frac{\sin t - t\cos t + c}{t^2}.$$

We conclude that $y \to 0$ as $t \to \infty$.

7.

(a)

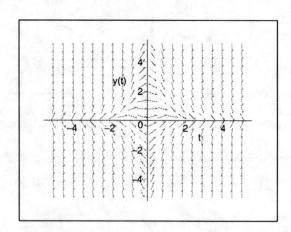

(b) For $t > 0$, all solutions seem to eventually converge to the function $y = 0$.

(c) The integrating factor is $\mu(t) = e^{t^2}$. Therefore, using the techniques shown above, we see that $y(t) = t^2 e^{-t^2} + ce^{-t^2}$. We conclude that $y \to 0$ as $t \to \infty$.

8.

(a)

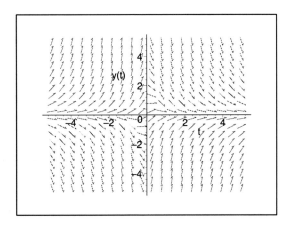

(b) For $t > 0$, all solutions seem to eventually converge to the function $y = 0$.

(c) The integrating factor is $\mu(t) = (1 + t^2)^2$. Then

$$(1 + t^2)^2 y' + 4t(1 + t^2)y = \frac{1}{1 + t^2}$$
$$\implies ((1 + t^2)^2 y) = \int \frac{1}{1 + t^2} \, dt$$
$$\implies y = (\tan^{-1}(t) + c)/(1 + t^2)^2.$$

We conclude that $y \to 0$ as $t \to \infty$.

9.

(a)

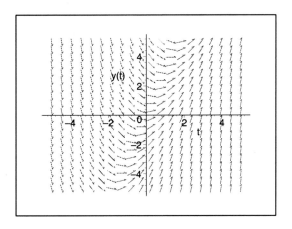

(b) All slopes eventually become positive. Therefore, all solutions will increase without bound.

(c) The integrating factor is $\mu(t) = e^{t/2}$. Therefore,

$$2e^{t/2}y' + e^{t/2}y = 3te^{t/2} \qquad \implies 2e^{t/2}y = \int 3te^{t/2}\, dt = 6te^{t/2} - 12e^{t/2} + c$$

$$\implies y = 3t - 6 + ce^{-t/2}.$$

We conclude that $y \to 3t - 6$ as $t \to \infty$.

10.

(a)

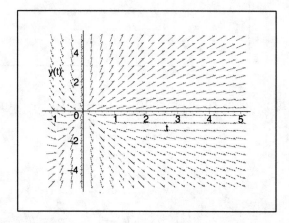

(b) For $y > 0$, the slopes are all positive, and, therefore, the corresponding solutions increase without bound. For $y < 0$ almost all solutions have negative slope and therefore decrease without bound.

(c) By dividing the equation by t, we see that the integrating factor is $\mu(t) = 1/t$. Therefore,

$$y'/t - y/t^2 = te^{-t} \implies (y/t)' = te^{-t}$$

$$\implies \frac{y}{t} = \int te^{-t}\, dt = -te^{-t} - e^{-t} + c$$

$$\implies y = -t^2 e^{-t} - te^{-t} + ct.$$

We conclude that $y \to \infty$ if $c > 0$, $y \to -\infty$ if $c < 0$ and $y \to 0$ if $c = 0$.

11.

(a)

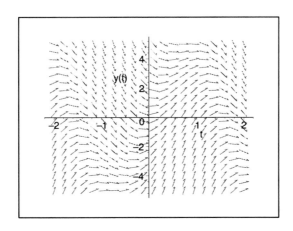

(b) The solution appears to be oscillatory.

(c) The integrating factor is $\mu(t) = e^t$. Therefore,

$$e^t y' + e^t y = 5e^t \sin(2t) \implies (e^t y)' = 5e^t \sin(2t)$$

$$\implies e^t y = \int 5e^t \sin(2t)\, dt = -2e^t \cos(2t) + e^t \sin(2t) + c \implies y = -2\cos(2t) + \sin(2t) + ce^{-t}.$$

We conclude that $y \to \sin(2t) - 2\cos(2t)$ as $t \to \infty$.

12.

(a)

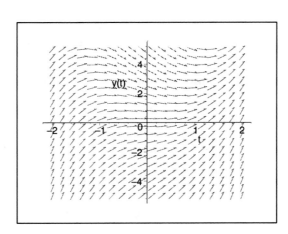

(b) All slopes are eventually positive. Therefore, all solutions increase without bound.

(c) The integrating factor is $\mu(t) = e^{t/}$. Therefore,

$$2e^{t/2} y' + e^{t/2} y = 3t^2 e^{t/2} \implies (2e^{t/2} y)' = 3t^2 e^{t/2}$$

$$\implies 2e^{t/2} y = \int 3t^2 e^{t/2}\, dt = 6t^2 e^{t/2} - 24te^{t/2} + 48e^{t/2} + c$$

$$\implies y = 3t^2 - 12t + 24 + ce^{-t/2}.$$

We conclude that y is asymptotic to $3t^2 - 12t + 24$ as $t \to \infty$.

13. The integrating factor is $\mu(t) = e^{-t}$. Therefore,

$$(e^{-t}y)' = 2te^t \implies y = e^t \int 2te^t \, dt = 2te^{2t} - 2e^{2t} + ce^t.$$

The initial condition $y(0) = 1$ implies $-2 + c = 1$. Therefore, $c = 3$ and $y = 3e^t + 2(t-1)e^{2t}$

14. The integrating factor is $\mu(t) = e^{2t}$. Therefore,

$$(e^{2t}y)' = t \implies y = e^{-2t} \int t \, dt = \frac{t^2}{2}e^{-2t} + ce^{-2t}.$$

The initial condition $y(1) = 0$ implies $e^{-2t}/2 + ce^{-2t} = 0$. Therefore, $c = -1/2$, and $y = (t^2 - 1)e^{-2t}/2$

15. Dividing the equation by t, we see that the integrating factor is $\mu(t) = t^2$. Therefore,

$$(t^2y)' = t^3 - t^2 + t \implies y = t^{-2} \int (t^3 - t^2 + t) \, dt = \left(\frac{t^2}{4} - \frac{t}{3} + \frac{1}{2} + \frac{c}{t^2} \right).$$

The initial condition $y(1) = 1/2$ implies $c = 1/12$, and $y = (3t^4 - 4t^3 + 6t^2 + 1)/12t^2$.

16. The integrating factor is $\mu(t) = t^2$. Therefore,

$$(t^2y)' = \cos(t) \implies y = t^{-2} \int \cos(t) \, dt = t^{-2}(\sin(t) + c).$$

The initial condition $y(\pi) = 0$ implies $c = 0$ and $y = (\sin t)/t^2$

17. The integrating factor is $\mu(t) = e^{-2t}$. Therefore,

$$(e^{-2t}y)' = 1 \implies y = e^{2t} \int 1 \, dt = e^{2t}(t + c).$$

The initial condition $y(0) = 2$ implies $c = 2$ and $y = (t + 2)e^{2t}$.

18. After dividing by t, we see that the integrating factor is $\mu(t) = t^2$. Therefore,

$$(t^2y)' = 1 \implies y = t^{-2} \int t\sin(t) \, dt = t^{-2}(\sin(t) - t\cos(t) + c).$$

The initial condition $y(\pi/2) = 1$ implies $c = (\pi^2/4) - 1$ and $y = t^{-2}[(\pi^2/4) - 1 - t\cos t + \sin t]$.

19. After dividing by t^3, we see that the integrating factor is $\mu(t) = t^4$. Therefore,

$$(t^4y)' = te^{-t} \implies y = t^{-4} \int te^{-t} \, dt = t^{-4}(-te^{-t} - e^{-t} + c).$$

The initial condition $y(-1) = 0$ implies $c = 0$ and $y = -(1 + t)e^{-t}/t^4, \quad t \neq 0$

20. After dividing by t, we see that the integrating factor is $\mu(t) = te^t$. Therefore,

$$(te^ty)' = te^t \implies y = t^{-1}e^{-t} \int te^t \, dt = t^{-1}e^{-t}(te^t - e^t + c) = t^{-1}(t - 1 + ce^{-t}).$$

The initial condition $y(\ln 2) = 1$ implies $c = 2$ and $y = (t - 1 + 2e^{-t})/t, \quad t \neq 0$

21.

(a)

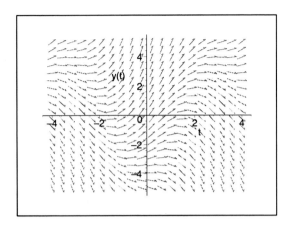

The solutions appear to diverge from an oscillatory solution. It appears that $a_0 \approx -1$. For $a > -1$, the solutions increase without bound. For $a < -1$, the solutions decrease without bound.

(b) The integrating factor is $\mu(t) = e^{-t/2}$. From this, we conclude that the general solution is $y(t) = (8\sin(t) - 4\cos(t))/5 + ce^{t/2}$. The solution will be sinusoidal as long as $c = 0$. The initial condition for the sinusoidal behavior is $y(0) = (8\sin(0) - 4\cos(0))/5 = -4/5$. Therefore, $a_0 = -4/5$.

(c) y oscillates for $a = a_0$

22.

(a)

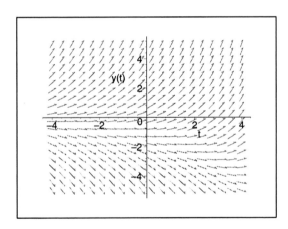

All solutions eventually increase or decrease without bound. The value a_0 appears to be approximately $a_0 = -3$.

(b) The integrating factor is $\mu(t) = e^{-t/2}$, and the general solution is $y(t) = -3e^{t/3} + ce^{t/2}$. The initial condition $y(0) = a$ implies $y = -3e^{t/3} + (a + 3)e^{t/2}$. The solution will behave like $(a + 3)e^{t/2}$. Therefore, $a_0 = -3$.

(c) $y \to -\infty$ for $a = a_0$

23.

(a)

Solutions eventually increase or decrease without bound, depending on the initial value a_0. It appears that $a_0 \approx -1/8$.

(b) Dividing the equation by 3, we see that the integrating factor is $\mu(t) = e^{-2t/3}$. Therefore, the solution is $y = [(2 + a(3\pi + 4))e^{2t/3} - 2e^{-\pi t/2}]/(3\pi + 4)$. The solution will eventually behave like $(2 + a(3\pi + 4))e^{2t/3}/(3\pi + 4)$. Therefore, $a_0 = -2/(3\pi + 4)$.

(c) $y \to 0$ for $a = a_0$

24.

(a)

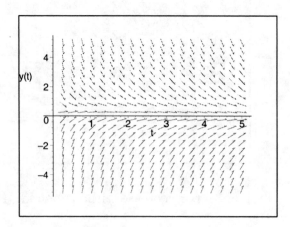

It appears that $a_0 \approx .4$. As $t \to 0$, solutions increase without bound if $y > a_0$ and decrease without bound if $y < a_0$.

(b) The integrating factor is $\mu(t) = te^t$. The general solution is $y = te^{-t} + ce^{-t}/t$. The initial condition $y(1) = a$ implies $y = te^{-t} + (ea - 1)e^{-t}/t$. As $t \to 0$, the solution will behave like $(ea - 1)e^{-t}/t$. From this, we see that $a_0 = 1/e$.

(c) $y \to 0$ as $t \to 0$ for $a = a_0$

25.

(a)

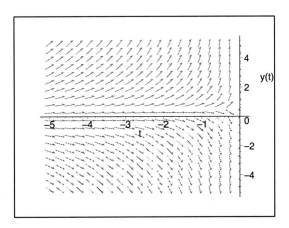

It appears that $a_0 \approx .4$. That is, as $t \to 0$, for $y(-\pi/2) > a_0$, solutions will increase without bound, while solutions will decrease without bound for $y(-\pi/2) < a_0$.

(b) After dividing by t, we see that the integrating factor is t^2, and the solution is $y = -\cos t/t^2 + \pi^2 a/4t^2$. Since $\lim_{t \to 0} \cos(t) = 1$, solutions will increase without bound if $a > 4/\pi^2$ and decrease without bound if $a < 4/\pi^2$. Therefore, $a_0 = 4/\pi^2$.

(c) For $a_0 = 4/\pi^2$, $y = (1 - \cos(t))/t^2 \to 1/2$ as $t \to 0$.

26.

(a)

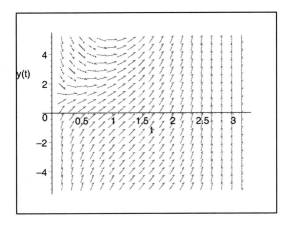

It appears that $a_0 \approx 2$. For $y(1) > a_0$, the solution will increase without bound as $t \to 0$, while the solution will decrease without bound if $y(t) < a_0$.

(b) After dividing by $\sin(t)$, we see that the integrating factor is $\mu(t) = \sin(t)$. As a result, we see that the solution is given by $y = (e^t + c)\sin(t)$. Applying our initial condition, we see that our solution is $y = (e^t - e + a\sin 1)/\sin t$. The solution will increase if $1 - e + a\sin 1 > 0$ and decrease if $1 - e + a\sin 1 < 0$. Therefore, we conclude that $a_0 = (e-1)/\sin 1$

(c) If $a_0 = (e-1)\sin(1)$, then $y = (e^t - 1)/\sin(t)$. As $t \to 0$, $y \to 1$.

27. The integrating factor is $\mu(t) = e^{t/2}$. Therefore, the general solution is $y(t) = [4\cos(t) + 8\sin(t)]/5 + ce^{-t/2}$. Using our initial condition, we have $y(t) = [4\cos(t) + 8\sin(t) - 9e^{t/2}]/5$. Differentiating, we have

$$y' = [-4\sin(t) + 8\cos(t) + 4.5e^{-t/2}]/5$$
$$y'' = [-4\cos(t) - 8\sin(t) - 2.25e^{t/2}]/5.$$

Setting $y' = 0$, the first solution is $t_1 = 1.3643$, which gives the location of the first stationary point. Since $y''(t_1) < 0$, the first stationary point is a local maximum. The coordinates of the point are $(1.3643, .82008)$.

28. The integrating factor is $\mu(t) = e^{2t/3}$. The general solution of the differential equation is $y(t) = (21 - 6t)/8 + ce^{-2t/3}$. Using the initial condition, we have $y(t) = (21 - 6t)/8 + (y_0 - 21/8)e^{-2t/3}$. Therefore, $y'(t) = -3/4 - (2y_0 - 21/4)e^{-2t/3}/3$. Setting $y'(t) = 0$, the solution is $t_1 = \frac{3}{2}\ln[(21 - 8y_0)/9]$. Substituting into the solution, the respective value at the stationary point is $y(t_1) = \frac{3}{2} + \frac{9}{4}\ln 3 - \frac{9}{8}\ln(21 - 8y_0)$. Setting this result equal to zero, we obtain the required initial value $y_0 = (21 - 9e^{4/3})/8 = -1.643$.

29.

(a) The integrating factor is $\mu(t) = e^{t/4}$. The general solution is

$$y(t) = 12 + [8\cos(2t) + 64\sin(2t)]/65 + ce^{-t/4}.$$

Applying the initial condition $y(0) = 0$, we arrive at the specific solution

$$y(t) = 12 + [8\cos(2t) + 64\sin(2t) - 788e^{-t/4}]/65.$$

For large values of t, the solution oscillates about the line $y = 12$.

(b) To find the value of t for which the solution first intersects the line $y = 12$, we need to solve the equation $8\cos(2t) + 64\sin(2t) - 788e^{-t/4} = 0$. The time t is approximately 10.519.

30. The integrating factor is $\mu(t) = e^{-t}$. The general solution is $y(t) = -1 - \frac{3}{2}\cos(t) - \frac{3}{2}\sin(t) + ce^t$. In order for the solution to remain finite as $t \to \infty$, we need $c = 0$. Therefore, y_0 must satisfy $y_0 = -1 - 3/2 = -5/2$.

31. The integrating factor is $\mu(t) = e^{-3t/2}$ and the general solution of the equation is $y(t) = -2t - 4/3 - 4e^t + ce^{3t/2}$. The initial condition implies $y(t) = -2t - 4/3 - 4e^t + (y_0 + 16/3)e^{3t/2}$. The solution will behave like $(y_0 + 16/3)e^{3t/2}$ (for $y_0 \neq -16/3$). For $y_0 > -16/3$, the solutions will increase without bound, while for $y_0 < -16/3$, the solutions will decrease without bound. If $y_0 = -16/3$, the solution will decrease without bound as the solution will be $-2t - 4/3 - 4e^t$.

32. By equation (41), we see that the general solution is given by

$$y = e^{-t^2/4} \int_0^t e^{s^2/4} \, ds + ce^{-t^2/4}.$$

Applying L'Hospital's rule,

$$\lim_{t\to\infty} \frac{\int_0^t e^{s^2/4} \, ds}{e^{t^2/4}} = \lim_{t\to\infty} \frac{e^{t^2/4}}{(t/2)e^{t^2/4}} = 0.$$

Therefore, $y \to 0$ as $t \to \infty$.

33. The integrating factor is $\mu(t) = e^{at}$. First consider the case $a \neq \lambda$. Multiplying the equation by e^{at}, we have

$$(e^{at}y)' = be^{(a-\lambda)t} \implies y = e^{-at} \int be^{(a-\lambda)t} = e^{-at}\left(\frac{b}{a-\lambda}e^{(a-\lambda)t} + c\right) = \frac{b}{a-\lambda}e^{-\lambda t} + ce^{-at}.$$

Since a, λ are assumed to be positive, we see that $y \to 0$ as $t \to \infty$. Now if $a = \lambda$ above, then we have

$$(e^{at}y)' = b \implies y = e^{-at}(bt + c)$$

and similarly $y \to 0$ as $t \to \infty$.

34. We notice that $y(t) = ce^{-t} + 3$ approaches 3 as $t \to \infty$. We just need to find a first-order linear differential equation having that solution. We notice that if $y(t) = f + g$, then $y' + y = f' + f + g' + g$. Here, let $f = ce^{-t}$ and $g(t) = 3$. Then $f' + f = 0$ and $g' + g = 3$. Therefore, $y(t) = ce^{-t} + 3$ satisfies the equation $y' + y = 3$. That is, the equation $y' + y = 3$ has the desired properties.

35. We notice that $y(t) = ce^{-t} + 3 - t$ approaches $3 - t$ as $t \to \infty$. We just need to find a first-order linear differential equation having that solution. We notice that if $y(t) = f + g$, then $y' + y = f' + f + g' + g$. Here, let $f = ce^{-t}$ and $g(t) = 3 - t$. Then $f' + f = 0$ and $g' + g = -1 + 3 - t = -2 - t$. Therefore, $y(t) = ce^{-t} + 3 - t$ satisfies the equation $y' + y = -2 - t$. That is, the equation $y' + y = -2 - t$ has the desired properties.

36. We notice that $y(t) = ce^{-t} + 2t - 5$ approaches $2t - 5$ as $t \to \infty$. We just need to find a first-order linear differential equation having that solution. We notice that if $y(t) = f + g$, then $y' + y = f' + f + g' + g$. Here, let $f = ce^{-t}$ and $g(t) = 2t - 5$. Then $f' + f = 0$ and $g' + g = 2 + 2t - 5 = 2t - 3$. Therefore, $y(t) = ce^{-t} + 2t - 5 -$ satisfies the equation $y' + y = 2t - 3$. That is, the equation $y' + y = 2t - 3$ has the desired properties.

37. We notice that $y(t) = ce^{-t} + 4 - t^2$ approaches $4 - t^2$ as $t \to \infty$. We just need to find a first-order linear differential equation having that solution. We notice that if $y(t) = f + g$, then $y' + y = f' + f + g' + g$. Here, let $f = ce^{-t}$ and $g(t) = 4 - t^2$. Then $f' + f = 0$ and

$g' + g = -2t + 4 - t^2 = 4 - 2t - t^2$. Therefore, $y(t) = ce^{-t} + 2t - 5-$ satisfies the equation $y' + y = 4 - 2t - t^2$. That is, the equation $y' + y = 4 - 2t - t^2$ has the desired properties.

38. Multiplying the equation by $e^{a(t-t_0)}$, we have

$$e^{a(t-t_0)}y' + ae^{a(t-t_0)}y = e^{a(t-t_0)}g(t)$$
$$\implies (e^{a(t-t_0)}y)' = e^{a(t-t_0)}g(t)$$
$$\implies y(t) = \int_{t_0}^{t} e^{-a(t-s)}g(s)\, ds + e^{-a(t-t_0)}y_0.$$

Assuming $g(t) \to g_0$ as $t \to \infty$,

$$\int_{t_0}^{t} e^{-a(t-s)}g(s)\, ds \to \int_{t_0}^{t} e^{-a(t-s)}g_0\, ds = \frac{g_0}{a} - \frac{e^{-a(t-t_0)}}{a}g_0 \to \frac{g_0}{a} \quad \text{as } t \to \infty$$

For an example, let $g(t) = e^{-t} + 1$. Assume $a \neq 1$. By undetermined coefficients, we look for a solution of the form $y = ce^{-at} + Ae^{-t} + B$. Substituting a function of this form into the differential equation leads to the equation

$$[-A + aA]e^{-t} + aB = e^{-t} + 1 \implies -A + aA = 1 \text{ and } aB = 1.$$

Therefore, $A = 1/(a-1)$, $B = 1/a$ and $y = ce^{-at} + \frac{1}{a-1}e^{-t} + 1/a$. The initial condition $y(0) = y_0$ implies $y(t) = (y_0 - \frac{1}{a-1} - \frac{1}{a})e^{-at} + \frac{1}{a-1}e^{-t} + 1/a \to 1/a$ as $t \to \infty$.

39.

(a) The integrating factor is $e^{\int p(t)\, dt}$. Multiplying by the integrating factor, we have

$$e^{\int p(t)\, dt}y' + e^{\int p(t)\, dt}p(t)y = 0.$$

Therefore,

$$\left(e^{\int p(t)\, dt}y\right)' = 0$$

which implies

$$y(t) = Ae^{-\int p(t)\, dt}$$

is the general solution.

(b) Let $y = A(t)e^{-\int p(t)\, dt}$. Then in order for y to satisfy the desired equation, we need

$$A'(t)e^{-\int p(t)\, dt} - A(t)p(t)e^{-\int p(t)\, dt} + A(t)p(t)e^{-\int p(t)\, dt} = g(t).$$

That is, we need

$$A'(t) = g(t)e^{\int p(t)\, dt}.$$

(c) From equation (iv), we see that

$$A(t) = \int_{0}^{t} g(\tau)e^{\int p(\tau)\, d\tau}\, d\tau + C.$$

Therefore,

$$y(t) = e^{-\int p(t)\, dt}\left(\int_{0}^{t} g(\tau)e^{\int p(\tau)\, d\tau}\, d\tau + C\right).$$

14

40. Here, $p(t) = -2$ and $g(t) = t^2 e^{2t}$. The general solution is given by

$$y(t) = e^{-\int p(t)\, dt} \left(\int_0^t g(\tau) e^{\int p(\tau)\, d\tau}\, d\tau + C \right)$$

$$= e^{\int 2\, dt} \left(\int_0^t \tau^2 e^{2\tau} e^{\int -2\, d\tau}\, d\tau + C \right)$$

$$= e^{2t} \left(\int_0^t \tau^2\, d\tau + C \right)$$

$$= e^{2t} \left(\frac{t^3}{3} + c \right).$$

41. Here, $p(t) = 1/t$ and $g(t) = 3\cos(2t)$. The general solution is given by

$$y(t) = e^{-\int p(t)\, dt} \left(\int_0^t g(\tau) e^{\int p(\tau)\, d\tau}\, d\tau + C \right)$$

$$= e^{-\int \frac{1}{t}\, dt} \left(\int_0^t 3\cos(2\tau) e^{\int \frac{1}{\tau}\, d\tau}\, d\tau + C \right)$$

$$= \frac{1}{t} \left(\int_0^t 3\tau \cos(2\tau)\, d\tau + C \right)$$

$$= \frac{1}{t} \left(\frac{3}{4}\cos(2t) + \frac{3}{2}t\sin(2t) + C \right).$$

42. Here, $p(t) = 2/t$ and $g(t) = \sin(t)/t$. The general solution is given by

$$y(t) = e^{-\int p(t)\, dt} \left(\int_0^t g(\tau) e^{\int p(\tau)\, d\tau}\, d\tau + C \right)$$

$$= e^{-\int \frac{2}{t}\, dt} \left(\int_0^t \frac{\sin(\tau)}{\tau} e^{\int \frac{2}{\tau}\, d\tau}\, d\tau + C \right)$$

$$= \frac{1}{t^2} \left(\int_0^t \frac{\sin(\tau)}{\tau} \tau^2\, d\tau + C \right)$$

$$= \frac{1}{t^2} \left(\int_0^t \tau \sin(\tau)\, d\tau + C \right)$$

$$= \frac{1}{t^2} \left(\sin(t) - t\cos(t) + C \right).$$

43. Here, $p(t) = 1/2$ and $g(t) = 3t^2/2$. The general solution is given by

$$y(t) = e^{-\int p(t)\,dt}\left(\int_0^t g(\tau)e^{\int p(\tau)\,d\tau}\,d\tau + C\right)$$

$$= e^{-\int \frac{1}{2}\,dt}\left(\int_0^t \frac{3t^2}{2}e^{\int \frac{1}{2}\,d\tau}\,d\tau + C\right)$$

$$= e^{-t/2}\left(\int_0^t \frac{3\tau^2}{2}e^{\tau/2}\,d\tau + C\right)$$

$$= e^{-t/2}\left(3t^2 e^{t/2} - 12te^{t/2} + 24e^{t/2} + C\right)$$

$$= et^2 - 12t + 24 + ce^{-t/2}.$$

Section 2.2

1. Rewriting as $y\,dy = x^2\,dx$, then integrating both sides, we have $y^2/2 = x^3/3 + C$, or $3y^2 - 2x^3 = c$; $\quad y \neq 0$

2. Rewriting as $y\,dy = [x^2/(1 + x^3)]\,dx$, then integrating both sides, we have $y^2/2 = \ln|1 + x^3|/3 + C$, or $3y^2 - 2\ln|1 + x^3| = c$; $\quad x \neq -1, y \neq 0$

3. Rewriting as $y^{-2}\,dy = -\sin(x)\,dx$, then integrating both sides, we have $-y^{-1} = \cos(x) + C$, or $y^{-1} + \cos x = c$ if $y \neq 0$;. Also, we have $y = 0$ everywhere

4. Rewriting as $(3 + 2y)\,dy = (3x^2 - 1)\,dx$, then integrating both sides, we have $3y + y^2 - x^3 + x + C$ as long as $y \neq -3/2$.

5. Rewriting as $\sec^2(2y)\,dy = \cos^2(x)\,dx$, then integrating both sides, we have $\tan(2y)/2 = x/2 + \sin(2x)/4 + C$, or $2\tan 2y - 2x - \sin 2x = C$ as long as $\cos 2y \neq 0$. Also, if $y = \pm(2n + 1)\pi/4$ for any integer n, then $y' = 0 = \cos(2y)$

6. Rewriting as $(1 - y^2)^{-1/2}\,dy = dx/x$, then integrating both sides, we have $\sin^{-1}(y) = \ln|x| + C$. Therefore, $y = \sin[\ln|x| + c]$ as long as $x \neq 0$ and $|y| < 1$;. We also notice that if $y = \pm 1$, then $xy' = 0 = (1 - y^2)^{1/2}$ is a solution.

7. Rewriting as $(y + e^y)\,dy = (x - e^{-x})\,dx$, then integrating both sides, we have $y^2/2 + e^y = x^2/2 + e^{-x} + C$, or $y^2 - x^2 + 2(e^y - e^{-x}) = C$ as long as $y + e^y \neq 0$.

8. Rewriting as $(1 + y^2)\,dy = x^2\,dx$, then integrating both sides, we have $y + y^3/3 = x^3/3 + C$, or $3y + y^3 - x^3 = c$;.

9.

(a) Rewriting as $y^{-2}\,dy = (1 - 2x)\,dx$, then integrating both sides, we have $-y^{-1} = x - x^2 + C$. The initial condition, $y(0) = -1/6$ implies $C = 6$. Therefore, $y = 1/(x^2 - x - 6)$.

(b)

16

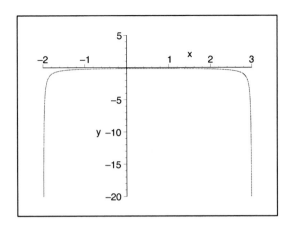

(c) $-2 < x < 3$

10.

(a) Rewriting as $y\,dy = (1 - 2x)dx$, then integrating both sides, we have $y^2/2 = x - x^2 + C$. Therefore, $y = \pm\sqrt{2x - 2x^2 + 4}$. The initial condition, $y(1) = -2$ implies $C = 2$ and $y = -\sqrt{2x - 2x^2 + 4}$.

(b)

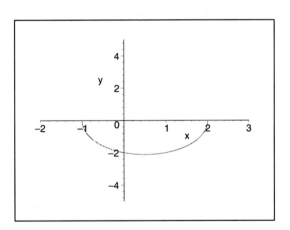

(c) $-1 < x < 2$

11.

(a) Rewriting as $xe^x\,dx = -y\,dy$, then integrating both sides, we have $xe^x - e^x = -y^2/2 + C$. The initial condition, $y(0) = 1$ implies $C = -1/2$. Therefore, $y = [2(1 - x)e^x - 1]^{1/2}$.

(b)

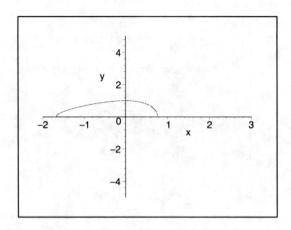

(c) $-1.68 < x < 0.77$ approximately

12.

(a) Rewriting as $r^{-2}dr = \theta^{-1}d\theta$, then integrating both sides, we have $-r^{-1} = \ln\theta + C$. The initial condition, $r(1) = 2$ implies $C = -1/2$. Therefore, $r = 2/(1 - 2\ln\theta)$.

(b)

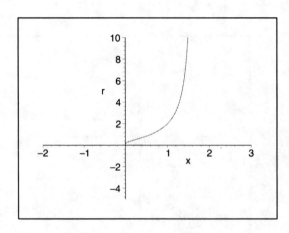

(c) $0 < \theta < \sqrt{e}$

13.

(a) Rewriting as $ydy = 2x/(1+x^2)dx$, then integrating both sides, we have $y^2/2 = \ln(1+x^2) + C$. The initial condition, $y(0) = -2$ implies $C = 2$. Therefore, $y = -[2\ln(1+x^2)+4]^{1/2}$.

(b)

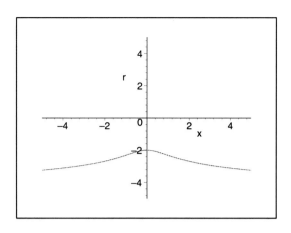

(c) $-\infty < x < \infty$

14.

(a) Rewriting as $y^{-3}dy = x(1+x^2)^{-1/2}dx$, then integrating both sides, we have $-y^{-2}/2 = \sqrt{1+x^2} + C$. The initial condition, $y(0) = 1$ implies $C = -3/2$. Therefore, $y = [3 - 2\sqrt{1+x^2}]^{-1/2}$.

(b)

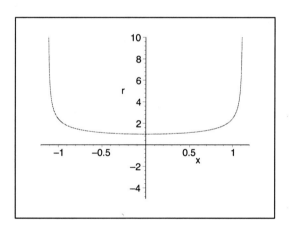

(c) $|x| < \dfrac{1}{2}\sqrt{5}$

15.

(a) Rewriting as $(1 + 2y)dy = 2xdx$, then integrating both sides, we have $y + y^2 = x^2 + C$. The initial condition, $y(2) = 0$ implies $C = -4$. Therefore, $y^2 + y = x^2 - 4$. Completing the square, we have $(y + 1/2)^2 = x^2 - 15/4$, and, therefore, $y = -\dfrac{1}{2} + \dfrac{1}{2}\sqrt{4x^2 - 15}$.

(b)

19

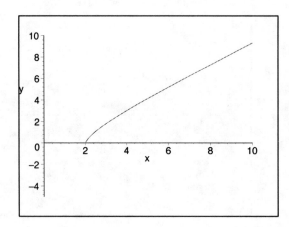

(c) $x > \dfrac{1}{2}\sqrt{15}$

16.

(a) Rewriting as $4y^3 dy = x(x^2+1)dx$, then integrating both sides, we have $y^4 = (x^2+1)^2/4 + C$. The initial condition, $y(0) = -1/\sqrt{2}$ implies $C = 0$. Therefore, $y = -\sqrt{(x^2+1)/2}$.

(b)

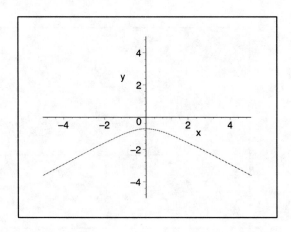

(c) $-\infty < x < \infty$

17.

(a) Rewriting as $(2y - 5)dy = (3x^2 - e^x)dx$, then integrating both sides, we have $y^2 - 5y = x^3 - e^x + C$. The initial condition, $y(0) = 1$ implies $C = -3$. Completing the square, we have $(y - 5/2)^2 = x^3 - e^x + 13/4$. Therefore, $y = 5/2 - \sqrt{x^3 - e^x + 13/4}$.

(b)

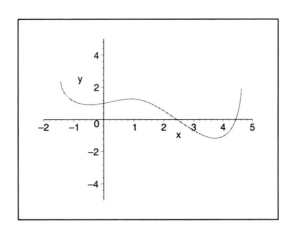

(c) $-1.4445 < x < 4.6297$ approximately

18.

(a) Rewriting as $(3 + 4y)dy = (e^{-x} - e^x)dx$, then integrating both sides, we have $3y + 2y^2 = -(e^x + e^{-x}) + C$. The initial condition, $y(0) = 1$ implies $C = 7$. Completing the square, we have $(y + 3/4)^2 = -(e^x + e^{-x})/2 + 65/16$. Therefore, $y = -\frac{3}{4} + \frac{1}{4}\sqrt{65 - 8e^x - 8e^{-x}}$.

(b)

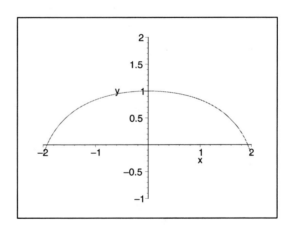

(c) $|x| < 2.0794$ approximately

19.

(a) Rewriting as $\cos(3y)dy = -\sin(2x)dx$, then integrating both sides, we have $\sin(3y)/3 = \cos(2x)/2 + C$. The initial condition, $y(\pi/2) = \pi/3$ implies $C = 1/2$. Therefore, $y = [\pi - \arcsin(3\cos^2 x)]/3$.

(b)

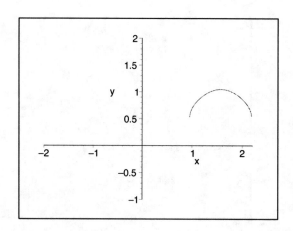

(c) $|x - \pi/2| < 0.6155$ approximately

20.

(a) Rewriting as $y^2 dy = \arcsin(x)/\sqrt{1 - x^2}\, dx$, then integrating both sides, we have $y^3/3 = (\arcsin(x))^2/2 + C$. The initial condition, $y(0) = 1/$ implies $C = 0$. Therefore, $y = \left[\frac{3}{2}(\arcsin x)^2\right]^{1/3}$.

(b)

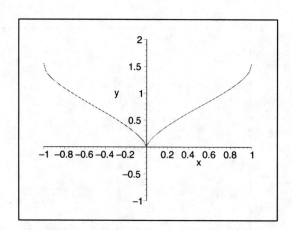

(c) $-1 < x < 1$

21. Rewriting the equation as $(3y^2 - 6y)dy = (1 + 3x^2)dx$ and integrating both sides, we have $y^3 - 3y^2 = x + x^3 + C$. The initial condition, $y(0) = 1$ implies $c = -2$. Therefore, $y^3 - 3y^2 - x - x^3 + 2 = 0$. When $3y^2 - 6y = 0$, the integral curve will have a vertical tangent. In particular, when $y = 0, 2$. From our solution, we see that $y = 0$ implies $x = 1$ and $y = 2$ implies $x = -1$. Therefore, the solution is defined for $-1 < x < 1$.

22. Rewriting the equation as $(3y^2 - 4)dy = 3x^2 dx$ and integrating both sides, we have $y^3 - 4y = x^3 + C$. The initial condition $y(1) = 0$ implies $C = -1$. Therefore, $y^3 - 4y - x^3 = -1$. When $3y^2 - 4 = 0$, the integral curve will have a vertical tangent. In particular, when $y = \pm 2/\sqrt{3}$. At these values for y, we have $x = -1.276, 1.598$. Therefore, the solution is defined for $-1.276 < x < 1.598$

23. Rewriting the equation as $y^{-2}dy = (2+x)dx$ and integrating both sides, we have $-y^{-1} = 2x + x^2/2 + C$. The initial condition $y(0) = 1$ implies $C = -1$. Therefore, $y = -1/(x^2/2 + 2x - 1)$. To find where the function attains it minimum value, we look where $y' = 0$. We see that $y' = 0$ implies $y = 0$ or $x = -2$. But, as seen by the solution formula, y is never zero. Further, it can be verified that $y''(-2) > 0$, and, therefore, the function attains a minimum at $x = -2$.

24. Rewriting the equation as $(3+2y)dy = (2 - e^x)dx$ and integrating both sides, we have $3y + y^2 = 2x - e^x + C$. By the initial condition $y(0) = 0$, we have $C = 1$. Completing the square, it follows that $y = -3/2 + \sqrt{2x - e^x + 13/4}$. The solution is defined if $2x - e^x + 13/4 \geq 0$, that is, $-1.5 \leq x \leq 2$ (approximately). In that interval, $y = 0$ for $x = \ln 2$. It can be verified that $y''(\ln 2) < 0$, and, therefore, the function attains its maximum value at $x = \ln 2$.

25. Rewriting the equation as $(3+2y)dy = 2\cos(2x)dx$ and integrating both sides, we have $3y + y^2 = \sin(2x) + C$. By the initial condition $y(0) = -1$, we have $C = -2$. Completing the square, it follows that $y = -3/2 + \sqrt{\sin(2x) + 1/4}$. The solution is defined for $\sin(2x) + 1/4 \geq 0$. That is, $-0.126 \leq x \leq 1.44$. To find where the solution attains its maximum value, we need to check where $y' = 0$. We see that $y' = 0$ when $2\cos(2x) = 0$. In the interval of definition above, that occurs when $2x = \pi/2$, or $x = \pi/4$.

26. Rewriting this equation as $(1+y^2)^{-1}dy = 2(1+x)dx$ and integrating both sides, we have $\tan^{-1}(y) = 2x + x^2 + C$. The initial condition implies $C = 0$. Therefore, the solution is $y = \tan(x^2 + 2x)$. The solution is defined as long as $-\pi/2 < 2x + x^2 < \pi/2$. We note that $2x + x^2 \geq -1$. Further, $2x + x^2 = \pi/2$ for $x = -2.6$ and 0.6. Therefore, the solution is valid in the interval $-2.6 < x < 0.6$. We see that $y' = 0$ when $x = -1$. Furthermore, it can be verified that $y''(x) > 0$ for all x in the interval of definition. Therefore, y attains a global minimum at $x = -1$.

27.

(a) First, we rewrite the equation as $dy/[y(4-y)] = tdt/3$. Then, using partial fractions, we write

$$\frac{1/4}{y}dy + \frac{1/4}{4-y}dy = \frac{t}{3}dt.$$

Integrating both sides, we have

$$\frac{1}{4}\ln|y| - \frac{1}{4}\ln|4-y| = \frac{t^2}{6} + C$$

$$\implies \ln\left|\frac{y}{y-4}\right| = \frac{2}{3}t^2 + C$$

$$\implies \left|\frac{y}{y-4}\right| = Ce^{2t^2/3}.$$

From the equation, we see that $y_0 = 0 \implies C = 0 \implies y(t) = 0$ for all t. Otherwise, $y(t) > 0$ for all t or $y(t) < 0$ for all t. Therefore, if $y_0 > 0$ and $|y/(y-4)| = Ce^{2t^2/3} \to \infty$, we must have $y \to 4$. On the other hand, if $y_0 < 0$, then $y \to -\infty$ as $t \to \infty$. (In particular, $y \to -\infty$ in finite time.)

23

(b) For $y_0 = 0.5$, we want to find the time T when the solution first reaches the value 3.98. Using the fact that $|y/(y-4)| = Ce^{2t^2/3}$ combined with the initial condition, we have $C = 1/7$. From this equation, we now need to find T such that $|3.98/.02| = e^{2T^2/3}/7$. Solving this equation, we have $T = 3.29527$.

28.

(a) Rewriting the equation as $y^{-1}(4-y)^{-1}dy = t(1+t)^{-1}dt$ and integrating both sides, we have $\ln|y| - \ln|y-4| = 4t - 4\ln|1+t| + C$. Therefore, $|y/(y-4)| = Ce^{4t}/(1+t)^4 \to \infty$ as $t \to \infty$ which implies $y \to 4$.

(b) The initial condition $y(0) = 2$ implies $C = 1$. Therefore, $y/(y-4) = -e^{4t}/(1+t)^4$. Now we need to find T such that $3.99/-.01 = -e^{4T}/(1+T)^4$. Solving this equation, we have $T = 2.84367$.

(c) Using our results from part (b), we note that $y/(y-4) = y_0/(y_0-4)e^{4t}/(1+t)^4$. We want to find the range of initial values y_0 such that $3.99 < y < 4.01$ at time $t = 2$. Substituting $t = 2$ into the equation above, we have $y_0/(y_0-4) = (3/e^2)^4 y(2)/(y(2)-4)$. Since the function $y/(y-4)$ is monotone, we need only find the values y_0 satisfying $y_0/(y_0-4) = -399(3/e^2)^4$ and $y_0/(y_0-4) = 401(3/e^2)^4$. The solutions are $y_0 = 3.6622$ and $y_0 = 4.4042$. Therefore, we need $3.6622 < y_0 < 4.4042$.

29. We can rewrite the equation as

$$\left(\frac{cy+d}{ay+b}\right) dy = dx \implies \frac{cy}{ay+b} + \frac{d}{ay+b}dy = dx \implies \frac{c}{a} - \frac{bc}{a^2y+ab} + \frac{d}{ay+b}dy = dx.$$

Then integrating both sides, we have

$$\frac{c}{a}y - \frac{bc}{a^2}\ln|a^2y + ab| + \frac{d}{a}\ln|ay+b| = x + C.$$

Simplifying, we have

$$\frac{c}{a}y - \frac{bc}{a^2}\ln|a| - \frac{bc}{a^2}\ln|ay+b| + \frac{d}{a}\ln|ay+b| = x + C$$

$$\implies \frac{c}{a}y + \left(\frac{ad-bc}{a^2}\right)\ln|ay+b| = x + C.$$

Note, in this calculation, since $\frac{bc}{a^2}\ln|a|$ is just a constant, we included it with the arbitrary constant C. This solution will exist as long as $a \neq 0$ and $ay + b \neq 0$.

30.

(a) Factoring an x out of each term in the numerator and denominator of the right-hand side, we have

$$\frac{dy}{dx} = \frac{x((y/x) - 4)}{x(1 - (y/x))} = \frac{(y/x) - 4}{1 - (y/x)},$$

as claimed.

24

(b) Letting $v = y/x$, we have $y = xv$, which implies that $dy/dx = v + x \cdot dv/dx$.

(c) Therefore,

$$v + x \cdot \frac{dv}{dx} = \frac{v - 4}{1 - v}$$

which implies that

$$x \cdot \frac{dv}{dx} = \frac{v - 4 - v(1 - v)}{(1 - v)} = \frac{v^2 - 4}{1 - v}.$$

(d) To solve the equation above, we rewrite as

$$\frac{1 - v}{v^2 - 4} dv = \frac{dx}{x}.$$

Integrating both sides of this equation, we have

$$-\frac{1}{4} \ln|v - 2| - \frac{3}{4} \ln|v + 2| = \ln|x| + C.$$

Applying the exponential function to both sides of the equation, we have

$$|v - 2|^{-1/4}|v + 2|^{-3/4} = C|x|.$$

(e) Replacing v with y/x, we have

$$\left|\frac{y}{x} - 2\right|^{-1/4}\left|\frac{y}{x} + 2\right|^{-3/4} = C|x| \implies |x||y-2x|^{-1/4}|y+2x|^{-3/4} = C|x| \implies |y+2x|^3|y-2x| = C.$$

(f)

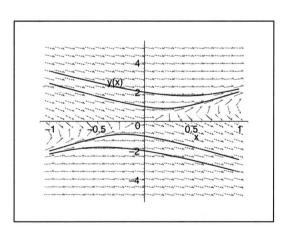

31.

(a)

$$\frac{dy}{dx} = 1 + (y/x) + (y/x)^2.$$

Therefore, the equation is homogeneous.

(b) The substitution $v = y/x$ results in the equation

$$v + x\frac{dv}{dx} = 1 + v + v^2 \implies x\frac{dv}{dx} = 1 + v^2.$$

This equation can be rewritten as

$$\frac{dv}{1 + v^2} = \frac{dx}{x}$$

which has solution $\arctan(v) = \ln|x| + c$. Rewriting back in terms of y, we have $\arctan(y/x) - \ln|x| = c$.

(c)

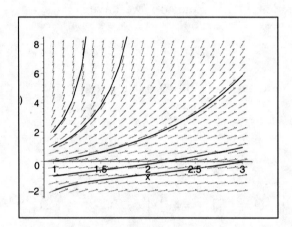

32.

(a)

$$\frac{dy}{dx} = (y/x)^{-1} + \frac{3}{2}(y/x).$$

Therefore, the equation is homogeneous.

(b) The substitution $v = y/x$ results in the equation

$$v + x\frac{dv}{dx} = \frac{x^2 + 3x^2v^2}{2x^2v} \implies \frac{dv}{dx} = \frac{1 + v^2}{2xv}.$$

The solution of this separable equation is $v^2 + 1 = cx$. Rewriting back in terms of y, we have $x^2 + y^2 - cx^3 = 0$.

(c)

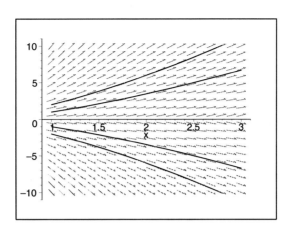

33.

(a)

$$\frac{dy}{dx} = \frac{4(y/x) - 3}{2 - (y/x)}.$$

Therefore, the equation is homogeneous.

(b) The substitution $v = y/x$ results in the equation

$$v + x\frac{dv}{dx} = \frac{4v - 3}{2 - v} \implies x\frac{dv}{dx} = \frac{v^2 + 2v - 3}{2 - v}.$$

This equation can be rewritten as

$$\frac{2 - v}{v^2 + 2v - 3}dv = \frac{dx}{x}.$$

Integrating both sides and simplifying, we arrive at the solution $|v + 3|^{-5/4}|v - 1|^{1/4} = |x| + c$. Rewriting back in terms of y, we have $|y - x| = c|y + 3x|^5$. We also have the solution $y = -3x$.

(c)

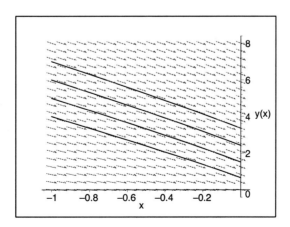

27

34.

(a)
$$\frac{dy}{dx} = -2 - \frac{y}{x}\left[2 + \frac{y}{x}\right]^{-1}.$$

Therefore, the equation is homogeneous.

(b) The substitution $v = y/x$ results in the equation

$$v + x\frac{dv}{dx} = -2 - \frac{v}{2+v} \implies \frac{dv}{dx} = -\frac{v^2 + 5v + 4}{x(2+v)}.$$

This equation is separable with solution $(v+4)^2|v+1| = C/x^3$. Rewriting back in terms of y, we have $|y + x|(y + 4x)^2 = c$.

(c)

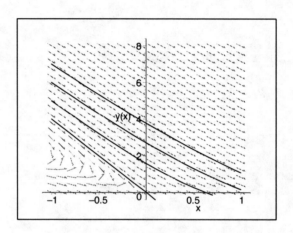

35.

(a)
$$\frac{dy}{dx} = \frac{1 + 3(y/x)}{1 - (y/x)}.$$

Therefore, the equation is homogeneous.

(b) The substitution $v = y/x$ results in the equation

$$v + x\frac{dv}{dx} = \frac{1 + 3v}{1 - v} \implies x\frac{dv}{dx} = \frac{v^2 + 2v + 1}{1 - v}.$$

This equation can be rewritten as

$$\frac{1 - v}{v^2 + 2v + 1}dv = \frac{dx}{x}$$

which has solution $-\frac{2}{v+1} - \ln|v + 1| = \ln|x| + c$. Rewriting back in terms of y, we have $2x/(x + y) + \ln|x + y| = c$. We also have the solution $y = -x$.

28

(c)

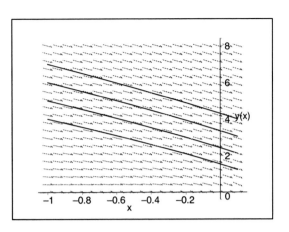

36.

(a)
$$\frac{dy}{dx} = 1 + 3(y/x) + (y/x)^2.$$

Therefore, the equation is homogeneous.

(b) The substitution $v = y/x$ results in the equation

$$v + x\frac{dv}{dx} = 1 + 3v + v^2 \implies x\frac{dv}{dx} = 1 + 2v + v^2.$$

This equation can be rewritten as

$$\frac{dv}{1 + 2v + v^2} = \frac{dx}{x}$$

which has solution $-1/(v + 1) = \ln|x| + c$. Rewriting back in terms of y, we have $x/(x + y) + \ln|x| = c$. We also have the solution $y = -x$.

(c)

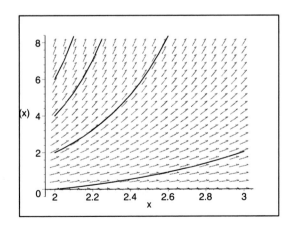

29

37.

(a)
$$\frac{dy}{dx} = \frac{1}{2}(y/x)^{-1} - \frac{3}{2}(y/x).$$

Therefore, the equation is homogeneous.

(b) The substitution $v = y/x$ results in the equation

$$v + x\frac{dv}{dx} = 1 + \frac{1}{2v} - \frac{3}{2}v \implies x\frac{dv}{dx} = \frac{1 - 5v^2}{2v}.$$

This equation can be rewritten as

$$\frac{2v}{1 - 5v^2}dv = \frac{dx}{x}$$

which has solution $-\frac{1}{5}\ln|1 - 5v^2| = \ln|x| + c$. Applying the exponential function, we arrive at the solution $1 - 5v^2 = c/|x|^5$. Rewriting back in terms of y, we have $|x|^3(x^2 - 5y^2) = c$

(c)

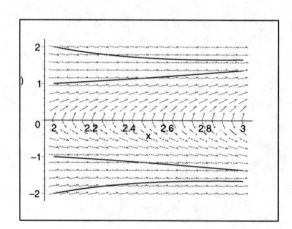

38.

(a)
$$\frac{dy}{dx} = \frac{3}{2}(y/x) - \frac{1}{2}(y/x)^{-1}.$$

Therefore, the equation is homogeneous.

(b) The substitution $v = y/x$ results in the equation

$$v + x\frac{dv}{dx} = \frac{3}{2}v - \frac{1}{2}v^{-1} \implies x\frac{dv}{dx} = \frac{v^2 - 1}{2v}.$$

30

This equation can be rewritten as

$$\frac{2v}{v^2 - 1}dv = \frac{dx}{x}$$

which has solution $\ln|v^2 - 1| = \ln|x| + c$. Applying the exponential function, we have $v^2 - 1 = C|x|$. Rewriting back in terms of y, we have $c|x|^3 = (y^2 - x^2)$

(c)

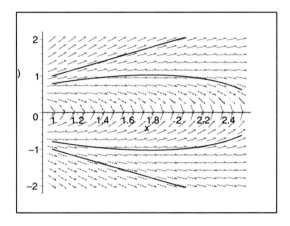

Section 2.3

1. Let $Q(t)$ be the quantity of dye in the tank. We know that

$$\frac{dQ}{dt} = \text{rate in} - \text{rate out}.$$

Here, fresh water is flowing in. Therefore, no dye is coming in. The dye is flowing out at the rate of $(Q/200)g/l \cdot 2l/min = Q/100$ l/min. Therefore,

$$\frac{dQ}{dt} = -\frac{Q}{100}.$$

The solution of this equation is $Q(t) = Ce^{-t/100}$. Since $Q(0) = 200$ grams, $C = 200$. We need to find the time T when the amount of dye present is 1% of what it is initially. That is, we need to find the time T when $Q(T) = 2$ grams. Solving the equation $2 = 200e^{-t/100}$, we conclude that $T = 100\ln(100)$ minutes.

2. Let $Q(t)$ be the quantity of salt in the tank. We know that

$$\frac{dQ}{dt} = \text{rate in} - \text{rate out}.$$

Here, water containing γ g/liter of salt is flowing in at a rate of 2 liters/minute. The salt is flowing out at the rate of $(Q/120)g/l \cdot 2l/min = Q/60$ l/min. Therefore,

$$\frac{dQ}{dt} = 2\gamma - \frac{Q}{60}.$$

31

The solution of this equation is $Q(t) = 120\gamma + Ce^{-t/60}$. Since $Q(0) = 0$ grams, $C = -120\gamma$. Therefore, $Q(t) = 120\gamma[1 - e^{-t/60}]$. As $t \to \infty$, $Q(t) \to 120\gamma$.

3. Let $Q(t)$ be the quantity of salt in the tank. We know that

$$\frac{dQ}{dt} = \text{rate in} - \text{rate out}.$$

Here, water containing 1/2 lb/gallon of salt is flowing in at a rate of 2 gal/minute. The salt is flowing out at the rate of $(Q/100)lb/gal \cdot 2gal/min = Q/50$ gal/min. Therefore,

$$\frac{dQ}{dt} = 1 - \frac{Q}{50}.$$

The solution of this equation is $Q(t) = 50 + Ce^{-t/50}$. Since $Q(0) = 0$ grams, $C = -50$. Therefore, $Q(t) = 50[1 - e^{-t/50}]$ for $0 \le t \le 10$ minutes. After 10 minutes, the amount of salt in the tank is $Q(10) = 50[1 - e^{-1/5}] \approx 9.06$ lbs. Starting at that time (and resetting the time variable), the new equation for dQ/dt is given by

$$\frac{dQ}{dt} = -\frac{Q}{50},$$

since fresh water is being added. The solution of this equation is $Q(t) = Ce^{-t/50}$. Since we are now starting with 9.06 lbs of salt, $Q(0) = 9.06 = C$. Therefore, $Q(t) = 9.06e^{-t/50}$. After 10 minutes, $Q(10) = 9.06e^{-1/5} \cong 7.42$ lbs.

4. Let $Q(t)$ be the quantity of salt in the tank. We know that

$$\frac{dQ}{dt} = \text{rate in} - \text{rate out}.$$

Here, water containing 1 lb/gallon of salt is flowing in at a rate of 3 gal/minute. The salt is flowing out at the rate of $(Q/(200+t))lb/gal \cdot 2gal/min = 2Q/(200+t)$ lb/min. Therefore,

$$\frac{dQ}{dt} = 3 - \frac{2Q}{200+t}.$$

This is a linear equation with integrating factor $\mu(t) = (200+t)^2$. The solution of this equation is $Q(t) = 200 + t + C(200 + t)^{-2}$. Since $Q(0) = 100$ lbs, $C = -4,000,000$. Therefore, $Q(t) = 200 + t - 4,000,000/(200+t)^2$. Since the tank has a net gain of 1 gallon of water every minute, the tank will reach its capacity after 300 minutes. When $t = 300$, we see that $Q(300) = 484$ lbs. Therefore, the concentration of salt when it is on the point of overflowing is 121/125 lbs/gallon. The concentration of salt is given by $Q(t)/(200+t)$ (since t gallons of water are added every t minutes). Using the equation for Q above, we see that if the tank had infinite capacity, the concentration would approach 1 as $t \to \infty$.

5.

(a) Let $Q(t)$ be the quantity of salt in the tank. We know that

$$\frac{dQ}{dt} = \text{rate in} - \text{rate out}.$$

Here, water containing $\frac{1}{4}\left(1 + \frac{1}{2}\sin t\right)$ oz/gallon of salt is flowing in at a rate of 2 gal/minute. The salt is flowing out at the rate of $Q/100 oz/gal \cdot 2gal/min = Q/50$ oz/min. Therefore,

$$\frac{dQ}{dt} = \frac{1}{2} + \frac{1}{4}\sin t - \frac{Q}{50}.$$

This is a linear equation with integrating factor $\mu(t) = e^{t/50}$. The solution of this equation is $Q(t) = (12.5\sin t - 625\cos t + 63150 e^{-t/50})/2501 + C$. The initial condition, $Q(0) = 50$ oz implies $C = 25$. Therefore, $Q(t) = 25 + (12.5\sin t - 625\cos t + 63150 e^{-t/50})/2501$.

(b)

(c) The amount of salt approaches a steady state, which is an oscillation of amplitude $1/4$ about a level of 25 oz.

6.

(a) Using the Principle of Conservation of Energy, we know that the kinetic energy of a particle after it has fallen from a height h is equal to its potential energy at a height t. Therefore, $mv^2/2 = mgh$. Solving this equation for v, we have $v = \sqrt{2gh}$.

(b) The volumetric outflow rate is (outflow cross-sectional area) × (outflow velocity): $\alpha a\sqrt{2gh}$. The volume of water in the tank is

$$V(h) = \int_0^h A(u)\,du$$

where $A(u)$ is the cross-sectional area of the tank at height u. By the chain rule,

$$\frac{dV}{dt} = \frac{dV}{dh} \cdot \frac{dh}{dt} = A(h)\frac{dh}{dt}.$$

Therefore,

$$\frac{dV}{dt} = A(h)\frac{dh}{dt} = -\alpha a\sqrt{2gh}.$$

33

(c) The cross-sectional area of the cylinder is $A(h) = \pi(1)^2 = \pi$. The outflow cross-sectional area is $a = \pi(.1)^2 = .01\pi$. From part (a), we take $\alpha = 0.6$ for water. Then by part (b), we have

$$\pi \frac{dh}{dt} = -0.006\pi\sqrt{2gh}.$$

This is a separable equation with solution $h(t) = 0.000018gt^2 - 0.006\sqrt{2gh(0)}t + h(0)$. Setting $h(0) = 3$ and $g = 9.8$, we have $h(t) = 0.0001764t^2 - 0.046t + 3$. Then $h(t) = 0$ implies $t \approx 130.4$ seconds.

7.

(a) The equation for S is

$$\frac{dS}{dt} = rS$$

with an initial condition $S(0) = S_0$. The solution of the equation is $S(t) = S_0 e^{rt}$. We want to find the time T such that $S(T) = 2S_0$. Our equation becomes $2S_0 = S_0 e^{rT}$. Dividing by S_0 and applying the logarithmic function to our equation, we have $rT = \ln(2)$. That is, $T = \ln(2)/r$.

(b) If $r = .07$, then $T = \ln(2)/.07 \cong 9.90$ years.

(c) By part (a), we also know that $r = \ln(2)T$ where T is the doubling time. If we want the investment to double in $T = 8$ years, then we need $r = \ln(2)/8 \cong 8.66\%$.

8.

(a) The equation for S is given by

$$\frac{dS}{dt} = rS + k.$$

This is a linear equation with solution $S(t) = \dfrac{k}{r}[e^{rt} - 1]$.

(b) Using the function in part (a), we need to find k so that $S(40) = 1,000,000$ assuming $r = .075$. That is, we need to solve

$$1,000,000 = \frac{k}{.075}[e^{.075 \cdot 40} - 1].$$

The solution of this equation is $k \cong 3930$.

(c) Now we assume that $k = 2000$ and want to find r. Our equation becomes

$$1,000,000 = \frac{2000}{r}[e^{40r} - 1].$$

The solution of this equation is approximately 9.77%.

9.

34

(a) Let $S(t)$ be the balance due on the loan at time t. To determine the maximum amount the buyer can afford to borrow, we will assume that the buyer will pay \$800 per month. Then

$$\frac{dS}{dt} = .09S - 12(800).$$

The solution is given by equation (18), $S(t) = S_0 e^{.09t} - 106,667(e^{.09t} - 1)$. If the term of the mortgage is 20 years, then $S(20) = 0$. Therefore, $0 = S_0 e^{.09(20)} - 106,667(e^{.09(20)} - 1)$ which implies $S_0 \approx \$89,035$.

(b) Since the homeowner pays \$800 per month for 20 years, he ends up paying a total of \$192,000 for the house. Since the house loan was \$89,035, the rest of the amount was interest payments. Therefore, the amount of interest was approximately \$102,965.

10.

(a) Let $S(t)$ be the balance due on the loan at time t. Taking into account that t is measured in years, we rewrite the monthly payment as $800(1 + t/10)$ where t is now in years. The equation for S is given by

$$\frac{dS}{dt} = .09S - 12(800)(1 + t/10).$$

This is a linear equation. Its solution is $S(t) = 225185 + 10667t + ce^{.09t}$. The initial condition $S(0) = 100,000$ implies $c = -125185$. Therefore, the particular solution is $S(t) = 225185 + 10667t - 125185e^{.09t}$. To find when the loan will be paid, we just need to solve $S(t) = 0$. Solving this equation, we conclude that the loan will be paid off in 11.28 years (135.36 months).

(b) From part (a), we know the general solution is given by $S(t) = 225185 + 10667t + ce^{.09t}$. Now we want to find c such that $S(20) = 0$. The solution of this equation is $c = -72486$. Therefore, the solution of the equation will be $S(t) = 225185 + 10667 - 72846e^{.09t}$. Therefore, $S(0) = 225185 - 72846 = 152699$.

11.

(a) If S_0 is the initial balance, then the balance after one month is

$$\begin{aligned} S_1 &= \text{initial balance} + \text{interest - monthly payment} \\ &= S_0 + rS_0 - k. \end{aligned}$$

Similarly,

$$\begin{aligned} S_2 &= S_1 + rS_1 - k \\ &= (1 + r)S_1 - k. \end{aligned}$$

In general,

$$S_n = (1 + r)S_{n-1} - k.$$

35

(b) $R = 1 + r$ implies $S_n = RS_{n-1} - k$. Therefore,

$$S_1 = RS_0 - k$$
$$S_2 = RS_1 - k = R[RS_0 - k] - k = R^2 S_0 - (R+1)k$$
$$S_3 = RS_2 - k = R[R^2 S_0 - (R+1)k] - k = R^3 S_0 - (R^2 + R + 1)k.$$

(c) We check the base case, $n = 1$. We see that

$$S_1 = RS_0 - k = RS_0 - \left(\frac{R-1}{R-1}\right)k,$$

which implies that that the condition is satisfied for $n = 1$. We assume that

$$S_n = R^n S_0 - \frac{R^n - 1}{R - 1}k$$

to show that

$$S_{n+1} = R^{n+1} S_0 - \frac{R^{n+1} - 1}{R - 1}k.$$

We see that

$$S_{n+1} = RS_n - k$$
$$= R\left[R^n S_0 - \frac{R^n - 1}{R - 1}k\right] - k$$
$$= R^{n+1} S_0 - \left(\frac{R^{n+1} - R}{R - 1}\right)k - k$$
$$= R^{n+1} S_0 - \left(\frac{R^{n+1} - R}{R - 1}\right)k - \left(\frac{R - 1}{R - 1}\right)k$$
$$= R^{n+1} S_0 - \left(\frac{R^{n+1} - R + R - 1}{R - 1}\right)k$$
$$= R^{n+1} S_0 - \left(\frac{R^{n+1} - 1}{R - 1}\right)k.$$

(d) We are assuming that $S_0 = 20,000$ and $r = .08/12$. We need to find k such that $S_{48} = 0$. Our equation becomes

$$S_{48} = R^{48} S_0 - \left(\frac{R^{48} - 1}{R - 1}\right)k = 0.$$

Therefore,

$$\left(\frac{(1 + .08/12)^{48} - 1}{.08/12}\right)k = \left(1 + \frac{.08}{12}\right)^{48} \cdot 20,000,$$

which implies $k \approx 488.26$, which is very close to the result in example 2.

12.

(a) The general solution is $Q(t) = Q_0 e^{-rt}$. If the half-life is 5730, then $Q_0/2 = Q_0 e^{-5730r}$ implies $-5730r = \ln(1/2)$. Therefore, $r = 1.2097 \times 10^{-4}$ per year.

(b) Therefore, $Q(t) = Q_0 e^{-1.2097 \times 10^{-4} t}$.

(c) Given that $Q(T) = Q_0/5$, we have the equation $1/5 = e^{-1.2097 \times 10^{-4} T}$. Solving for T, we have $T = 13,304.65$ years.

13. Let $P(t)$ be the population of mosquitoes at any time t, measured in days. Then

$$\frac{dP}{dt} = rP - 20,000.$$

The solution of this linear equation is $P(t) = P_0 e^{rt} - \frac{20,000}{r}(e^{rt} - 1)$. In the absence of predators, the equation is $dP_1/dt = rP_1$. The solution of this equation is $P_1(t) = P_0 e^{rt}$. Since the population doubles after 7 days, we see that $2P_0 = P_0 e^{7r}$. Therefore, $r = \ln(2)/7 = .09902$ per day. Therefore, the population of mosquitoes at any time t is given by $P(t) = 200,000 e^{.099t} - 201,997(e^{.099t} - 1) = 201,997 - 1997 e^{.099t}$.

14.

(a) The solution of this separable equation is given by $y(t) = \exp[2/10 + t/10 - 2\cos(t)/10]$. The doubling-time is found by solving the equation $2 = \exp[2/10 + t/10 - 2\cos(t)/10]$. The solution of this equation is given by $\tau \approx 2.9632$.

(b) The differential equation will be $dy/dt = y/10$ with solution $y(t) = y(0)e^{t/10}$. The doubling time is found by setting $y(t) = 2y(0)$. In this case, the doubling time is $\tau \approx 6.9315$.

(c) Consider the differential equation $dy/dt = (0.5 + \sin(2\pi t))y/5$. This equation is separable with solution $y(t) = \exp[(1 + \pi t - \cos(2\pi t))/(10\pi)]$. The doubling time is found by setting $y(t) = 2$. The solution is given by 6.9167.

(d)

15.

(a)

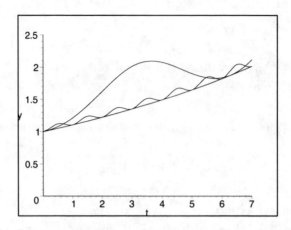

(b) Based on the graph, we estimate that $y_c \approx 0.83$.

(c) We sketch the graphs below for $k = 1/10$ and $k = 3/10$, respectively. Based on these graphs, we estimate that $y_c(1/10) \approx .41$ and $y_c(3/10) \approx 1.24$.

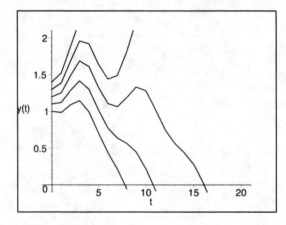

(d) From our results from above, we conclude that y_c is a linear function of k.

38

16. Let $T(t)$ be the temperature of the coffee at time t. The governing equation is given by

$$\frac{dT}{dt} = -k(T - 70).$$

This is a linear equation with solution $T(t) = 70 + ce^{-kt}$. The initial condition $T(0) = 200$ implies $c = 130$. Therefore, $T(t) = 70 + 130e^{-kt}$. Using the fact that $T(1) = 190$, we see that $190 = 70 + 130e^{-k}$ which implies $k = -\ln(12/13) \cong .08$ per minute. To find when the temperature reaches 150 degrees, we just need to solve $T(t) = 70 + 130e^{-0.08t} = 150$. The solution of this equation is $t = -\ln(80/130)/.08 \cong 6.07$ minutes.

17.

(a) The solution of this separable equation is given by

$$u^3 = \frac{u_0^3}{3\alpha u_0^3 t + 1}.$$

Since $u_0 = 2000$, the specific solution is

$$u(t) = \frac{2000}{(6t/125 + 1)^{1/3}}.$$

(b)

(c) We look for τ so that $u(\tau) = 600$. The solution of this equation is $t \approx 750.77$ seconds.

18.

(a) The integrating factor is $\mu(t) = e^{kt}$. Then $u = e^{-kt} \int ke^{kt}(T_0 + T_1 \cos(\omega t)) = ce^{-kt} + T_0 + kT_1(k \cos \omega t + \omega \sin \omega t)/(k^2 + \omega^2)$. Since $e^{-kt} \to 0$ as $t \to \infty$, we see that the steady state is $S(t) = T_0 + kT_1(k \cos(\omega t) + \omega \sin(\omega t))/(k^2 + \omega^2)$.

(b)

The amplitude R of the oscillatory part of S(t) is approximately 9 degrees Fahrenheit. The time lag τ between maxima is approximately 3.5 seconds.

(c) From above, the oscillatory part of $S(t)$ is given by

$$kT_1 \frac{k\cos(\omega t) + \omega \sin(\omega t)}{k^2 + \omega^2} = \frac{kT_1}{\sqrt{k^2 + \omega^2}}(\cos(\omega t)\cos(\omega \tau) + \sin(\omega t)\sin(\omega \tau))$$

for τ such that $\cos(\omega \tau) = k/\sqrt{k^2 + \omega^2}$ and $\sin(\omega \tau) = \omega/\sqrt{k^2 + \omega^2}$. That is, $\tau = \frac{1}{\omega}\arctan(\omega/k)$. Further, letting $R = kT_1/\sqrt{k^2 + \omega^2}$, we can write the oscillatory part of $S(t)$ as

$$R[\cos(\omega t)\cos(\omega \tau) + \sin(\omega t)\sin(\omega \tau)] = R\cos(\omega(t - \tau)).$$

Below we show graphs of R and τ versus k.

40

19.

(a) The differential equation for Q is

$$\frac{dQ}{dt} = kr + P - \frac{Q(t)}{V}r.$$

Therefore,

$$V\frac{dc}{dt} = kr + P - c(t)r.$$

The solution of this equation is $c(t) = k + P/r + (c_0 - k - P/r)e^{-rt/V}$. As $t \to \infty$, $c(t) \to k + P/r$.

(b) In this case, we will have $c(t) = c_0 e^{-rt/V}$. The reduction times are $T_{50} = \ln(2)V/r$ and $T_{10} = \ln(10)V/r$.

(c) Using the results from part (b), we have: Superior, $T = 431$ years; Michigan, $T = 71.4$ years; Erie, $T = 6.05$ years; Ontario, $T = 17.6$ years.

20.

(a) Assuming no air resistance, we have $dv/dt = -9.8$. Therefore, $v(t) = -9.8t + v_0 = -9.8t + 20$ and its position at time t is given by $s(t) = -4.9t^2 + 20t + 30$. When the ball reaches its max height, the velocity will be zero. We see that $v(t) = 0$ implies $t = 20/9.8 \cong 2.04$ seconds. When $t = 2.04$, we see that $s(2.04) \cong 50.4$ meters.

(b) Solving $s(t) = -4.9t^2 + 20t + 30 = 0$, we see that $t = 5.248$ seconds.

(c)

21.

(a) We have $mdv/dt = -v/30 - mg$. Given the conditions from problem 20, we see that the solution is given by $v(t) = -44.1 + 64.1e^{-t/4.5}$. The ball will reach its max height when $v(t) = 0$. This occurs at $t = 1.683$ seconds. The height of the ball is given by $s(t) = -318.45 - 44.1t - 288.45e^{-t/4.5}$. When $t = 1.683$, we have $s(1.683) = 45.78$ meters, the maximum height.

(b) The ball will hit the ground when $s(t) = 0$. This occurs when $t = 5.128$ seconds.

(c)

22.

(a) The equation for the upward motion is $mdv/dt = -\mu v^2 - mg$ where $\mu = 1/1325$. Using the data from exercise 20, and the fact that this equation is separable, we see its solution is given by $v(t) = 44.133 \tan(.425 - .222t)$. Setting $v(t) = 0$, we see the ball will reach its max height at $t = 1.916$ seconds. Integrating $v(t)$, we see the position at time t is given by $s(t) = 198.75 \ln(\cos(0.222t - 0.425)) + 48.57$. Therefore, the max height is given by $s(1.916) = 48.56$ meters.

(b) The differential equation for the downward motion is $mdv/dt = +\mu v^2 - mg$. The solution of this equation is given by $v(T) = 44.13(1 - e^{t/2.25})/(1 + e^{t/2.25})$. Integrating $v(t)$, we see that the position is given by $s(t) = 99.29 \ln(e^{t/2.25}/(1 + e^{t/2.25})^2) + 186.2$. Setting $s(t) = 0$, we see that the ball will spend $t = 3.276$ seconds going downward before hitting the ground. Combining this time with the amount of time the ball spends going upward, 1.916 seconds, we conclude that the ball will hit the ground 5.192 seconds after being thrown upward.

(c)

23.

(a) Measure the positive direction of motion downward. Then the equation of motion is given by

$$m\frac{dv}{dt} = \begin{cases} -0.75v + mg & 0 < t < 10 \\ -12v + mg & t > 10. \end{cases}$$

For the first 10 seconds, the equation becomes $dv/dt = -v/7.5 + 32$ which has solution $v(t) = 240(1 - e^{-t/7.5})$. Therefore, $v(10) = 176.7$ feet per second.

(b) Integrating the velocity function from part (a), we see that the height of the skydiver at time t $(0 < t < 10)$ is given by $s(t) = 240t + 1800e^{-t/7.5} - 1800$. Therefore, $s(10) = 1074.5$ feet.

(c) After the parachute opens, the equation for v is given by $dv/dt = -32v/15 + 32$ (as discussed in part (a)). We will reset t to zero. The solution of this differential equation is given by $v(t) = 15 + 161.7e^{-32t/15}$. As $t \to \infty$, $v(t) \to 15$. Therefore, the limiting velocity is $v_l = 15$ feet/second.

(d) Integrating the velocity function from part (c), we see that the height of the sky diver after falling t seconds with his parachute open is given by $s(t) = 15t - 75.8e^{-32t/15} + 1150.3$. To find how long the skydiver is in the air after the parachute opens, we find T such that $s(T) = 0$. Solving this equation, we have $T = 256.6$ seconds.

(e)

44

24.

(a) The equation of motion is given by $dv/dx = -\mu v$.

(b) The speed of the sled satisfies $\ln(v/v_0) = -\mu x$. Therefore, μ must satisfy $\ln(15/150) = -2000\mu$. Therefore, $\mu = \ln(10)/2000$ f^{-1}.

(c) The solution of $dv/dt = -\mu v^2$ can be expressed as $1/v - 1/v_0 = \mu t$. Using the fact that 1 mi/hour ≈ 1.467 feet/second, the elapsed time is $t \approx 35.56$ seconds.

25.

(a) Measure the positive direction of motion upward. The equation of motion is given by $m dv/dt = -kv - mg$. The solution of this equation is given by $v(t) = -mg/k + (v_0 + mg/k)e^{-kt/m}$. Solving $v(t) = 0$, we see that the mass will reach its max height $t_m = (m/k)\ln[(mg + kv_0)/mg]$ seconds after being projected upward. Integrating the velocity equation, we see that the position of the mass at this time will be given by the position equation

$$s(t) = -mgt/k + \left[\left(\frac{m}{k}\right)^2 g + \frac{mv_0}{k}\right](1 - e^{-kt/m}).$$

Therefore, the max height reached is

$$x_m = s(t_m) = \frac{mv_0}{k} - g\left(\frac{m}{k}\right)^2 \ln\left[\frac{mg + kv_0}{mg}\right].$$

,

(b) These formulas for t_m and x_m come from the fact that for $\delta << 1$, $\ln(1+\delta) = \delta - \frac{1}{2}\delta^2 + \frac{1}{3}\delta^3 - \frac{1}{4}\delta^4 + \ldots$. This formula is just Taylor's formula.

(c) Consider the result for t_m in part (b). Multiplying the equation by $\frac{g}{v_0}$, we have

$$\frac{t_m g}{v_0} = \left[1 - \frac{1}{2}\frac{kv_0}{mg} + \frac{1}{3}\left(\frac{kv_0}{mg}\right)^2 - \cdots\right].$$

The units on the left, must match the units on the right. Since the units for $t_m g/v_0 = (s \cdot m/s^2)/(m/s)$, the units cancel. As a result, we can conclude that kv_0/mg is dimensionless.

26.

(a) The equation of motion is given by $m dv/dt = -kv - mg$. The solution of this equation is given by $v(t) = -mg/k + (v_0 + mg/k)e^{-kt/m}$.

(b) Applying L'Hospital's rule, as $k \to 0$, we have

$$\lim_{k \to 0} -mg/k + (v_0 + mg/k)e^{-kt/m} = v_0 - gt.$$

(c)

$$\lim_{m \to 0} -mg/k + (v_0 + mg/k)e^{-kt/m} = 0.$$

27.

(a) The equation of motion is given by

$$m\frac{dv}{dt} = -6\pi\mu av + \rho'\frac{4}{3}\pi a^3 g - \rho\frac{4}{3}\pi a^3 g.$$

We can rewrite this equation as

$$v' + \frac{6\pi\mu a}{m}v = \frac{4}{3}\frac{\pi a^3 g}{m}(\rho' - \rho).$$

Multiplying by the integrating factor $e^{6\pi\mu at/m}$, we have

$$(e^{6\pi\mu at/m}v)' = \frac{4}{3}\frac{\pi a^3 g}{m}(\rho' - \rho)e^{6\pi\mu at/m}.$$

Integrating this equation, we have

$$v = e^{-6\pi\mu at/m}\left[\frac{2a^2 g(\rho' - \rho)}{9\mu}e^{6\pi\mu at/m} + C\right]$$

$$= \frac{2a^2 g(\rho' - \rho)}{9\mu} + Ce^{-6\pi\mu at/m}.$$

Therefore, we conclude that the limiting velocity is $v_L = (2a^2 g(\rho' - \rho))/9\mu$.

(b) By the equation above, we see that the force exerted on the droplet of oil is given by

$$Ee = -6\pi\mu av + \rho'\frac{4}{3}\pi a^3 g - \rho\frac{4}{3}\pi a^3 g.$$

If $v = 0$, then solving the above equation for e, we have

$$e = \frac{4\pi a^3 g(\rho' - \rho)}{3E}.$$

28.

(a) The equation is given by $mdv/dt = -kv - mg$. The solution of this equation is $v(t) = -(mg/k)(1 - e^{-kt/m})$. Integrating, we see that the position function is given by $x(t) = -(mg/k)t + (m/k)^2 g(1 - e^{-kt/m}) + 30$. First, by setting $x(t) = 0$, we see that the ball will hit the ground $t = 3.63$ seconds after it is dropped. Then $v(3.63) = 11.58$ m/second will be the speed when the mass hits the ground.

(b) In terms of displacement, we have $mvdv/dx = -kv + mg$. This equation comes from applying the chain rule: $dv/dt = dv/dx \cdot dx/dt = vdv/dx$. The solution of this differential equation is given by

$$x(v) = -\frac{mv}{k} - \frac{m^2 g}{k^2} \ln \left| \frac{mg - kv}{mg} \right|.$$

Plugging in the given values for k, m, g, we have $x(v) = -1.25v - 15.31 \ln |0.0816v - 1|$. If $v = 10$, then $x(10) = 13.45$ meters.

(c) Using the equation for $x(v)$ above, we set $x(v) = 30$, $v = 10$, $m = 0.25$, $g = 9.8$. Then solving for k, we have $k = 0.24$.

29.

(a) The equation of motion is given by $mdv/dt = -GMm/(R+x)^2$. By the chain rule,

$$m\frac{dv}{dx} \cdot \frac{dx}{dt} = -G\frac{Mm}{(R+x)^2}.$$

Therefore,

$$mv\frac{dv}{dx} = -G\frac{Mm}{(R+x)^2}.$$

This equation is separable with solution $v^2 = 2GM(R+x)^{-1} + 2gR - 2GM/R$. Here we have used the initial condition $v_0 = \sqrt{2gR}$. From physics, we know that $g = GM/R^2$. Using this substitution, we conclude that $v(x) = \sqrt{2g}[R/\sqrt{R+x}]$.

(b) By part (a), we know that $dx/dt = v(x) = \sqrt{2g}[R/\sqrt{R+x}]$. We want to solve this differential equation with the initial condition $x(0) = 0$. This equation is separable with solution $x(t) = [\frac{3}{2}(\sqrt{2g}Rt + \frac{2}{3}R^{3/2}]^{2/3} - R$. We want to find the time T such that $x(T) = 240,000$. Solving this equation, we conclude that $T \approx 50.6$ hours.

30.

(a) $dv/dt = 0$ implies v is constant, but clearly by the initial condition $v = u \cos A$. $dw/dt = -g$ implies $w = -gt + C$, but also by the initial condition $w = -gt + u \sin A$.

(b) The equation $dx/dt = v = u \cos A$ along with the initial condition implies $x(t) = u \cos At$. The equation $dy/dt = w = -gt + u \sin A$ along with the initial condition implies $y(t) = -gt^2/2 + u \sin At + h$.

(c) Below we have plotted the trajectory of the ball in the cases $\pi/4$, $\pi/3$ and $\pi/6$ respectively.

(d) First, let T be the time it takes for the ball to travel L feet horizontally. Using the equation for x, we know that $x(T) = u \cos AT = L$ implies $T = L/u \cos A$. Then, when the ball reaches this wall, we need the height of the ball to be at least H feet. That is, we need $y(T) \geq H$. Now $y(t) = -16t^2 + u \sin At + 3$ implies we need $y(T) = -16L^2/(u^2 \cos^2(A)) + L \tan A + 3 \geq H$.

(e) If $L = 350$ and $H = 10$, then our inequality becomes

$$-\frac{1,960,000}{(u^2 \cos^2(A))} + 350 \tan A + 3 \geq 10.$$

Now if $u = 110$, then our inequality becomes

$$-\frac{162}{\cos^2(A)} + 350\tan(A) \geq 7.$$

Solving this inequality, we conclude that 0.63 rad $\leq A \leq$ 0.96 rad.

(f) We rewrite the inequality in part (c) as

$$\cos^2(A)(350\tan A - 7) \geq \frac{1,960,000}{u^2}.$$

In order to determine the minimum value necessary, we will maximize the function on the left-hand side. Letting $f(A) = \cos^2(A)(350\tan A - 7)$, we see that $f'(A) = 350\cos(2A) + 7\sin(2A)$. Therefore, $f'(A) = 0$ implies $\tan(2A) = -50$. For $0 < A < \pi/2$, we see that this occurs at $A = 0.7954$ radians. Substituting this value for A into the inequality above, we conclude that

$$u^2 \geq 11426.24.$$

Therefore, the minimum velocity necessary is 106.89 mph and the optimal angle necessary is 0.7954 radians.

31.

(a) The initial conditions are $v(0) = u\cos(A)$ and $w(0) = u\sin(A)$. Therefore, the solutions of the two equations are $v(t) = u\cos(A)e^{-rt}$ and $w(t) = -g/r + (u\sin(A) + g/r)e^{-rt}$.

(b) Now $x(t) = \int v(t) = \frac{u}{r}\cos(A)(1 - e^{-rt})$, and

$$y(t) = \int w(t) = -\frac{gt}{r} + \frac{(g + ur\sin(A) + hr^2)}{r^2} - \left(\frac{u}{r}\sin(A) + \frac{g}{r^2}\right)e^{-rt}.$$

(c) Below we show trajectories for the cases $A = \pi/4, \pi/3$ and $\pi/6$, respectively.

49

(d) Let T be the time it takes the ball to go 350 feet horizontally. Then from above, we see that $e^{-T/5} = (u\cos(A) - 70)/u\cos(A)$. At the same time, the height of the ball is given by $y(T) = -160T + 267 + 125u\sin(A) - (800 + 5u\sin(A))[(u\cos(A) - 70)/u\cos(A)]$. Therefore, u and A must satisfy the inequality

$$800\ln\left[\frac{u\cos(A) - 70}{u\cos(A)}\right] + 267 + 125u\sin(A) - (800 + 5u\sin(A))\left[\frac{u\cos(A) - 70}{u\cos(A)}\right] \geq 10.$$

32.

(a) Solving equation (i), we have $y'(x) = [(k^2 - y)/y]^{1/2}$. The positive answer is chosen since y is an increasing function of x.

(b) $y = k^2\sin^2 t \implies dy/dt = 2k^2\sin t\cos t$. Substituting this into the equation in part (a), we have

$$\frac{2k^2\sin t\cos t\,dt}{dx} = \frac{\cos t}{\sin t}.$$

Therefore, $2k^2\sin^2 t\,dt = dx$.

(c) Letting $\theta = 2t$, we have $k^2\sin^2(\theta/2)d\theta = dx$. Integrating both sides, we have $x(\theta) = k^2(\theta - \sin\theta)/2$. Further, using the fact that $y = k^2\sin^2 t$, we conclude that $y = k^2\sin^2(\theta/2) = k^2(1 - \cos(\theta))/2$.

50

(d) From part (c), we see that $y/x = (1 - \cos\theta)/(\theta - \sin\theta)$. If $x = 1$ and $y = 2$, the solution of the equation is $\theta \approx 1.401$. Substituting that value of θ into either of the equations in part (c), we conclude that $k \approx 2.193$.

Section 2.4

1. Rewriting the equation as

$$y' + \frac{\ln t}{t - 3}y = \frac{2t}{t - 3}$$

and using Theorem 2.4.1, we conclude that a solution is guaranteed to exist in the interval $0 < t < 3$.

2. Rewriting the equation as

$$y' + \frac{1}{t(t - 4)}y = 0$$

and using Theorem 2.4.1, we conclude that a solution is guaranteed to exist in the interval $0 < t < 4$.

3. By Theorem 2.4.1, we conclude that a solution is guaranteed to exist in the interval $\pi/2 < t < 3\pi/2$.

4. Rewriting the equation as

$$y' + \frac{2t}{4 - t^2}y = \frac{3t^2}{4 - t^2}$$

and using Theorem 2.4.1, we conclude that a solution is guaranteed to exist in the interval $-\infty < t < -2$.

5. Rewriting the equation as

$$y' + \frac{2t}{4 - t^2}y = \frac{3t^2}{4 - t^2}$$

and using Theorem 2.4.1, we conclude that a solution is guaranteed to exist in the interval $-2 < t < 2$.

6. Rewriting the equation as

$$y' + \frac{1}{\ln t}y = \frac{\cot t}{\ln t}$$

and using Theorem 2.4.1, we conclude that a solution is guaranteed to exist in the interval $1 < t < \pi$.

7. Using the fact that

$$f = \frac{t - y}{2t + 5y} \implies f_y = \frac{3t - 10y}{(2t + 5y)^2},$$

we see that the hypothesis of Theorem 2.4.2 are satisfied as long as $2t + 5y \neq 0$.

8. Using the fact that

$$f = (1 - t^2 - y^2)^{1/2} \implies f_y = -\frac{y}{(1 - t^2 - y^2)^{1/2}},$$

we see that the hypothesis of Theorem 2.4.2 are satisfied as long as $t^2 + y^2 < 1$.

9. Using the fact that

$$f = \frac{\ln|ty|}{1 - t^2 + y^2} \implies f_y = \frac{1 - t^2 + y^2 - 2y^2 \ln|ty|}{y(1 - t^2 + y^2)^2},$$

we see that the hypothesis of Theorem 2.4.2 are satisfied as long as $y, t \neq 0$ and $1 - t^2 + y^2 \neq 0$.

10. Using the fact that

$$f = (t^2 + y^2)^{3/2} \implies f_y = 3y(t^2 + y^2)^{1/2},$$

we see that the hypothesis of Theorem 2.4.2 are satisfied for all $t \in \mathbb{R}$.

11. Using the fact that

$$f = \frac{1 + t^2}{3y - y^2} \implies f_y = -\frac{(1 + t^2)(3 - 2y)}{(3y - y^2)^2},$$

we see that the hypothesis of Theorem 2.4.2 are satisfied as long as $y \neq 0, 3$.

12. Using the fact that

$$f = \frac{(\cot t)y}{1 + y} \implies f_y = \frac{1}{(1 + y)^2},$$

we see that the hypothesis of Theorem 2.4.2 are satisfied as long as $y \neq -1, t \neq n\pi$ for $n = 0, 1, 2 \dots$.

13. The equation is separable, $y\,dy = -4t\,dt$. Integrating both sides, we conclude that $y^2/2 = -2t^2 + y_0^2/2$ for $y_0 \neq 0$. The solution is defined for $y_0^2 - 4t^2 \geq 0$.

14. The equation is separable and can be written as $dy/y^2 = 2t\,dt$. Integrating both sides, we arrive at the solution $y = y_0/(1 - y_0 t^2)$. For $y_0 > 0$, solutions exist as long as $t^2 < 1/y_0$. For $y_0 \leq 0$, solutions exist for all t.

15. The equation is separable and can be written as $dy/y^3 = -dt$. Integrating both sides, we arrive at the solution $y = y_0/(\sqrt{2ty_0^2 + 1})$. Solutions exist as long as $2y_0^2 t + 1 > 0$.

16. The equation is separable and can be written as $y\,dy = t^2 dt/(1 + t^3)$. Integrating both sides, we arrive at the solution $y = \pm(\frac{2}{3}\ln|1 + t^3| + y_0^2)^{1/2}$. The sign of the solution depends on the sign of the initial data y_0. Solutions exist as long as $\frac{2}{3}\ln|1 + t^3| + y_0^2 \geq 0$; that is, as long as $y_0^2 \geq -\frac{2}{3}\ln|1 + t^3|$. We can rewrite this inequality as $|1 + t^3| \geq e^{-3y_0^2/2}$. In order for the solution to exist, we need $t > -1$ (since the term $t^2/(1 + t^3)$ has a singularity at $t = -1$). Therefore, we can conclude that our solution will exist for $[e^{-3y_0^2/2} - 1]^{1/3} < t < \infty$.

17.

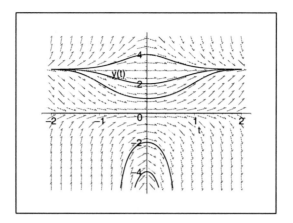

If $y_0 > 0$, then $y \to 3$. If $y_0 = 0$, then $y = 0$. If $y_0 < 0$, then $y \to -\infty$.

18.

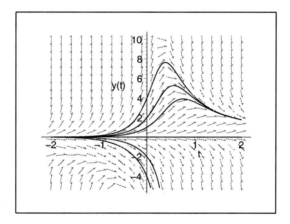

If $y_0 \geq 0$, then $y \to 0$. If $y_0 < 0$, then $y \to -\infty$.

19.

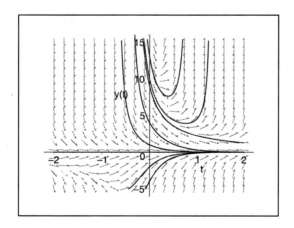

If $y_0 > 9$, then $y \to \infty$. If $y_0 < 9$, then $y \to 0$.

20.

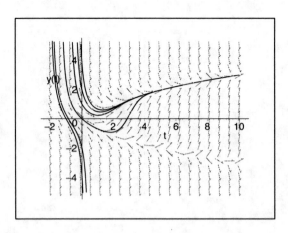

If $y_0 < y_c \approx -0.019$, then $y \to -\infty$. Otherwise, y is asymptotic to $\sqrt{t-1}$.

21.

(a) We know that the family of solutions given by equation (19) are solutions of this initial-value problem. We want to determine if one of these passes through the point $(1, 1)$. That is, we want to find $t_0 > 0$ such that if $y = [\frac{2}{3}(t - t_0)]^{3/2}$, then $(t, y) = (1, 1)$. That is, we need to find $t_0 > 0$ such that $1 = \frac{2}{3}(1 - t_0)$. But, the solution of this equation is $t_0 = -1/2$.

(b) From the analysis in part (a), we find a solution passing through $(2, 1)$ by setting $t_0 = 1/2$.

(c) Since we need $y_0 = \pm[\frac{2}{3}(2 - t_0)]^{3/2}$, we must have $|y_0| \leq [\frac{4}{3}]^{3/2}$.

22.

(a) First, it is clear that $y_1(2) = -1 = y_2(2)$. Further,

$$y_1' = -1 = \frac{-t + [(t-2)^2]^{1/2}}{2} = \frac{-t + (t^2 + 4(1-t))^{1/2}}{2}$$

and

$$y_2' = -\frac{t}{2} = \frac{-t + (t^2 - t^2)^{1/2}}{2}.$$

The function y_1 is a solution for $t \geq 2$. The function y_2 is a solution for all t.

(b) Theorem 2.4.2 requires that f and $\partial f / \partial y$ be continuous in a rectangle about the point $(t_0, y_0) = (2, -1)$. Since f is not continuous if $t < 2$ and $y < -1$, the hypothesis of Theorem 2.4.2 are not satisfied.

(c) If $y = ct + c^2$, then

$$y' = c = \frac{-t + [(t + 2c)^2]^{1/2}}{2} = \frac{-t + (t^2 + 4ct + 4c^2)^{1/2}}{2}.$$

Therefore, y satisfies the equation for $t \geq -2c$.

23.

54

(a) $\phi(t) = e^{2t} \implies \phi' = 2e^{2t}$. Therefore, $\phi' - 2\phi = 0$. Since $(c\phi)' = c\phi'$, we see that $(c\phi)' - 2c\phi = 0$. Therefore, $c\phi$ is also a solution.

(b) $\phi(t) = 1/t \implies \phi' = -1/t^2$. Therefore, $\phi' + \phi^2 = 0$. If $y = c/t$, then $y' = -c/t^2$. Therefore, $y' + y^2 = -c/t^2 + c^2/t^2 = 0$ if and only if $c^2 - c = 0$; that is, if $c = 0$ or $c = 1$.

24. If $y = \phi$ satisfies $\phi' + p(t)\phi = 0$, then $y = c\phi$ satisfies $y' + p(t)y = c\phi' + cp(t)\phi = c(\phi' + p(t)\phi) = 0$.

25. Let $y = y_1 + y_2$, then

$$y' + p(t)y = y_1' + y_2' + p(t)(y_1 + y_2) = y_1' + p(t)y_1 + y_2' + p(t)y_2 = 0.$$

26.

(a)

$$y = \frac{1}{\mu(t)}\left[\int_{t_0}^t \mu(s)g(s)\,ds + c\right] = \frac{1}{\mu(t)}\int_{t_0}^t \mu(s)g(s)\,ds + \frac{c}{\mu(t)}.$$

Therefore, $y_1 = 1/\mu(t)$ and $y_2 = \frac{1}{\mu(t)}\int_{t_0}^t \mu(s)g(s)\,ds$.

(b) For $y_1 = 1/\mu(t) = e^{-\int p(t)\,dt}$, we have

$$y_1' + p(t)y_1 = -p(t)e^{-\int p(t)\,dt} + p(t)e^{-\int p(t)\,dt} = 0.$$

(c) For

$$y_2 = \frac{1}{\mu(t)}\int_{t_0}^t \mu(s)g(s)\,ds = e^{-\int p(t)\,dt}\int_{t_0}^t e^{\int p(s)\,ds}g(s)\,ds,$$

we have

$$y_2' + p(t)y_2 = -p(t)e^{-\int p(t)\,dt}\int_{t_0}^t e^{\int p(s)\,ds}g(s)\,ds + e^{-\int p(t)\,dt}e^{\int p(t)\,dt}g(t)$$

$$+ p(t)e^{-\int p(t)\,dt}\int_{t_0}^t e^{\int p(s)\,ds}g(s)\,ds = g(t).$$

27.

(a) If $n = 0$, then $y(t) = ce^{-\int p(t)\,dt}$. If $n = 1$, then $y(t) = ce^{-\int (p(t)-q(t))\,dt}$.

(b) For $n \neq 0, 1$, let $v = y^{1-n}$. Then

$$v' = (1-n)y^{-n}y' = (1-n)y^{-n}[-p(t)y + q(t)y^n]$$
$$= (1-n)[-p(t)y^{1-n} + q(t)] = (1-n)[-p(t)v + q(t)].$$

That is, $v' + (1-n)p(t)v = (1-n)q(t)$.

28. First, rewrite as

$$y' + \frac{2}{t}y = \frac{1}{t^2}y^3.$$

Here, $n = 3$. Therefore, let $v = y^{1-3} = y^{-2}$. Making this substitution, we see that v satisfies the equation

$$v' - \frac{4}{t}v = -\frac{2}{t^2}.$$

This equation is linear with integrating factor t^{-4}. Therefore, we have

$$\left(\frac{1}{t^4}v' - \frac{4}{t^5}v\right) = -\frac{2}{t^6},$$

which can be written as $(t^{-4}v)' = -2/t^6$. The solution of this equation is given by $v = (2 + ct^5)/5t$. Then, using the fact that $y^2 = 1/v$, we conclude that $y = \pm\sqrt{5t/(2 + ct^5)}$.

29. First, rewrite as

$$y' - ry = -ky^2.$$

Here, $n = 2$. Therefore, let $v = y^{1-2} = y^{-1}$. Making this substitution, we see that v satisfies the equation

$$v' + rv = k.$$

This equation is linear with integrating factor e^{rt}. Therefore, we have

$$\left(e^{rt}v' + re^{rt}v\right) = ke^{rt},$$

which can be written as $(e^{rt}v)' = ke^{rt}$. The solution of this equation is given by $v = (k + cre^{-rt})/r$. Then, using the fact that $y = 1/v$, we conclude that $y = r/(k + cre^{-rt})$.

30. Here $n = 3$. Therefore, v satisfies

$$v' + 2\epsilon v = 2\sigma.$$

This equation is linear with integrating factor $e^{2\epsilon t}$. Its solution is given by $v = (\sigma + c\epsilon e^{-2\epsilon t})/\epsilon$. Then, using the fact that $y^2 = 1/v$, we see that $y = \pm\sqrt{\epsilon}/\sqrt{\sigma + c\epsilon e^{-2\epsilon t}}$.

31. Here $n = 3$. Therefore, v satisfies

$$v' + 2(\Gamma \cos t + T)v = 2.$$

This equation is linear with integrating factor $e^{2(\Gamma \sin t + Tt)}$. Therefore,

$$\left(e^{2(\Gamma \sin t + Tt)}v\right)' = 2e^{2(\Gamma \sin t + Tt)}$$

which implies

$$v = 2e^{-2(\Gamma \sin t + Tt)}\int e^{2(\Gamma \sin t + Tt)}\,dt + ce^{-2(\Gamma \sin t + Tt)}.$$

Then $v = y^{-2}$ implies $y = \pm\sqrt{1/v}$.

32. The solution of the initial value problem $y' + 2y = 1$ is $y = 1/2 + ce^{-2t}$. For $y(0) = 0$, we see that $c = -1/2$. Therefore, $y(t) = \frac{1}{2}(1 - e^{-2t})$ for $0 \le t \le 1$. Then $y(1) = \frac{1}{2}(1 - e^{-2})$.

56

Next, the solution of $y' + 2y = 0$ is given by $y = ce^{-2t}$. The initial condition $y(1) = \frac{1}{2}(1 - e^{-2})$ implies $ce^{-2} = \frac{1}{2}(1 - e^{-2})$. Therefore, $c = \frac{1}{2}(e^2 - 1)$ and we conclude that $y(t) = \frac{1}{2}(e^2 - 1)e^{-2t}$ for $t > 1$.

33. The solution of $y' + 2y = 0$ with $y(0) = 1$ is given by $y(t) = e^{-2t}$ for $0 \leq t \leq 1$. Then $y(1) = e^{-2}$. Then, for $t > 1$, the solution of the equation $y' + y = 0$ is $y = ce^{-t}$. Since we want $y(1) = e^{-2}$, we need $ce^{-1} = e^{-2}$. Therefore, $c = e^{-1}$. Therefore, $y(t) = e^{-1}e^{-t} = e^{-1-t}$ for $t > 1$.

34.

(a) Multiplying the equation by $e^{\int_{t_0}^{t} p(s)\, ds}$, we have

$$\left(e^{\int_{t_0}^{t} p(s)\, ds} y \right)' = e^{\int_{t_0}^{t} p(s)\, ds} g(t).$$

Integrating we have

$$e^{\int_{t_0}^{t} p(s)\, ds} y(t) = y_0 + \int_{t_0}^{t} e^{\int_{t_0}^{s} p(r)\, dr} g(s)\, ds,$$

which implies

$$y(t) = y_0 e^{-\int_{t_0}^{t} p(s)\, ds} + \int_{t_0}^{t} e^{-\int_{s}^{t} p(r)\, dr} g(s)\, ds.$$

(b) Assume $p(t) \geq p_0 > 0$ for all $t \geq t_0$ and $|g(t)| \leq M$ for all $t \geq t_0$. Therefore,

$$\int_{t_0}^{t} p(s)\, ds \geq \int_{t_0}^{t} p_0\, ds = p_0(t - t_0)$$

which implies

$$e^{-\int_{t_0}^{t} p(s)\, ds} \leq e^{-\int_{t_0}^{t} p_0\, ds} = e^{-p_0(t - t_0)} \leq 1 \text{ for } t \geq t_0.$$

Also,

$$\int_{t_0}^{t} e^{-\int_{s}^{t} p(r)\, dr} g(s)\, ds \leq \int_{t_0}^{t} e^{-\int_{s}^{t} p(r)\, dr} |g(s)|\, ds$$

$$\leq \int_{t_0}^{t} e^{-p_0(t-s)} M\, ds$$

$$\leq M \left. \frac{e^{-p_0(t-s)}}{p_0} \right|_{t_0}^{t}$$

$$= M \left[\frac{1}{p_0} - \frac{e^{-p_0(t-t_0)}}{p_0} \right]$$

$$\leq \frac{M}{p_0}$$

57

(c) Let $p(t) = 2t + 1 \geq 1$ for all $t \geq 0$ and let $g(t) = e^{-t^2}$. Therefore, $|g(t)| \leq 1$ for all $t \geq 0$. By the answer to part (a),

$$y(t) = e^{-\int_0^t (2s+1)\, ds} + \int_0^t e^{-\int_s^t (2r+1)\, dr} e^{-s^2}\, ds$$

$$= e^{-(t^2+t)} + e^{-t^2-t} \int_0^t e^s\, ds$$

$$= e^{-t^2}.$$

We see that y satisfies the property that y is bounded for all time $t \geq 0$.

Section 2.5

1.

The only equilibrium point is $y^* = 0$. Since $f'(0) = a > 0$, the equilibrium point is unstable.

2.

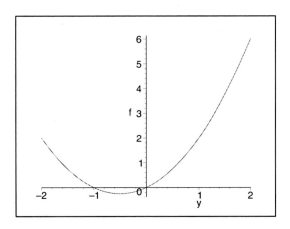

The equilibrium points are $y^* = 0, -a/b$. $y^* = 0$ is unstable and $y^* = -a/b$ is asymptotically stable since $f'(-a/b) < 0$.

3.

The equilibrium points are $y^* = 0, 1, 2$. Since $f'(0), f'(2) > 0$, those equilibrium point are unstable. Since $f'(1) < 0$, $y^* = 1$ is asymptotically stable.

4.

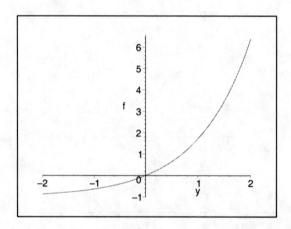

The only equilibrium point is $y^* = 0$. Since $f'(0) > 0$, the equilibrium point is unstable.

5.

60

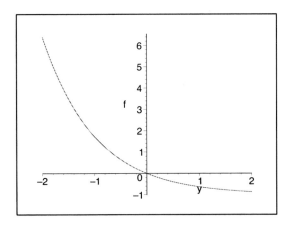

The only equilibrium point is $y^* = 0$. Since $f'(0) < 0$, the equilibrium point is asymptotically stable.

6.

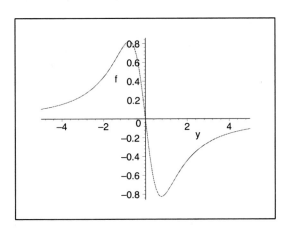

The only equilibrium point is $y^* = 0$. Since $f'(0) < 0$, the equilibrium point is asymptotically stable.

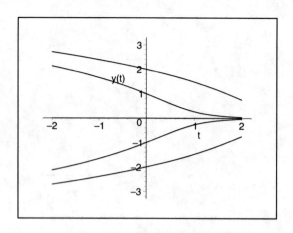

7.

(a) The function $f(y) = k(1 - y)^2 = 0 \implies y = 1$. Therefore, $y^* = 1$ is the only critical point.

(b)

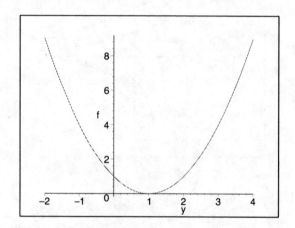

(c) This is a separable equation with solution $y(t) = [y_0 + (1 - y_0)kt]/[1 + (1 - y_0)kt]$. If $y_0 < 1$, then $y \to 1$ as $t \to \infty$. If $y_0 > 1$, then the denominator will go to zero at some finite time $T = 1/(y_0 - 1)$. Therefore, the solution will go towards at infinity.

8.

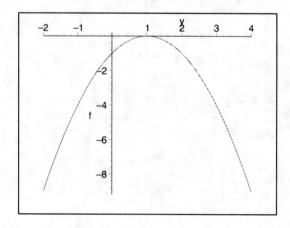

The only equilibrium point is $y^* = 0$. Since $f'(0) < 0$. The equilibrium point is semistable.

9.

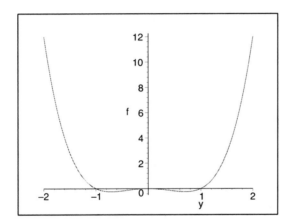

The equilibrium points are $y^* = 0, 1, -1$. Since $f'(-1) < 0$, $y = -1$ is asymptotically stable. Since $f'(1) > 0$, $y = 1$ is unstable. The equilibrium point $y = 0$ is semistable.

10.

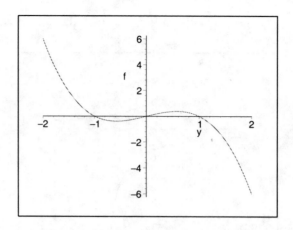

The equilibrium points are $y^* = 0, 1, -1$. Since $f'(-1), f'(1) < 0$, the equilibrium points $y = 1, -1$ are asymptotically stable. Since $f'(0) > 0$, the equilibrium point $y = 0$ is unstable.

11.

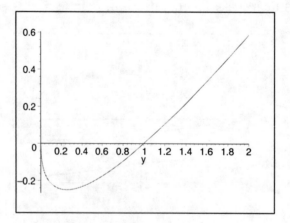

The equilibrium points are $y^* = 0, b^2/a^2$. Since $f'(0) < 0$, the equilibrium point $y = 0$ is asymptotically stable. Since $f'(b^2/a^2) > 0$, the equilibrium point $y = b^2/a^2$ is unstable.

12.

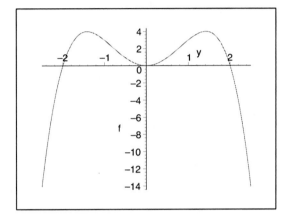

The equilibrium points are $y^* = 0, 2, -2$. The equilibrium point $y = 0$ is semistable. Since $f'(-2) > 0$, the equilibrium point $y = -2$ is unstable. Since $f'(2) < 0$, the equilibrium point $y = 2$ is asymptotically stable.

13.

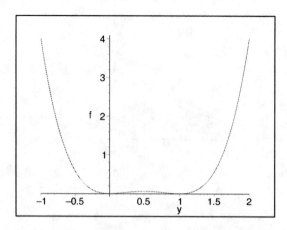

The equilibrium points are $y^* = 0, 1$. They are both semistable.

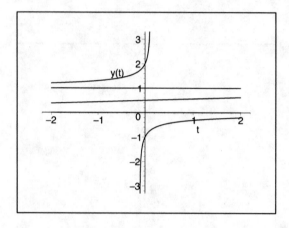

14.

(a) The equation is separable. Using partial fractions, it can be written as

$$\left(\frac{1}{y} + \frac{1/K}{1 - y/K} \right) dy = rdt.$$

Integrating both sides and using the initial condition $y_0 = K/3$, we know the solution y satisfies

$$\ln \left| \frac{y}{1 - y/K} \right| = rt + \ln \left| \frac{K}{2} \right|.$$

To find the time τ such that $y = 2y_0 = 2K/3$, we substitute $y = 2K/3$ and $t = \tau$ into the equation above. Using the properties of logarithm functions, we conclude that $\tau = (\ln 4)/r$. If $r = 0.025$, then $\tau \approx 55.452$ years.

(b) Using the analysis from part (a), we know the general solution satisfies

$$\ln \left| \frac{y}{1 - y/K} \right| = rt + c.$$

The initial condition $y_0 = \alpha K$ implies $c = \ln |\alpha K/(1 - \alpha)|$. Therefore,

$$\ln \left| \frac{y}{1 - y/K} \right| = rt + \ln \left| \frac{\alpha K}{1 - \alpha} \right|.$$

66

In order to find the time T at which $y(T) = \beta K$, we use the equation above. We conclude that

$$T = (1/r) \ln |\beta(1-\alpha)/\alpha(1-\beta)|.$$

15.

(a) Below we sketch the graph of f for $r = 1 = K$.

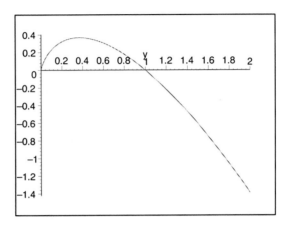

The critical points occur at $y^* = 0, K$. Since $f'(0) > 0$, $y^* = 0$ is unstable. Since $f'(K) < 0$, $y^* = K$ is asymptotically stable.

(b) We calculate y''. Using the chain rule, we see that

$$y'' = ry' \left[\ln\left(\frac{K}{y}\right) - 1 \right].$$

We see that $y'' = 0$ when $y' = 0$ (meaning $y = 0, K$) or when $\ln(K/y) - 1 = 0$, meaning $y = K/e$. Looking at the sign of y'' in the intervals $(0, K/e)$ and $(K/e, K)$, we conclude that y is concave up in the interval $(0, K/e)$ and concave down in the interval $(K/e, K)$.

16.

(a) Using the substitution $u = \ln(y/K)$ and differentiating both sides with respect to t, we conclude that $u' = y'/y$. Substitution into the Gompertz equation yields $u' = -ru$. The solution of this equation is $u = u_0 e^{-rt}$. Therefore,

$$\frac{y}{K} = \exp[\ln(y_0/K)e^{-rt}].$$

(b) For $K = 80.5 \times 10^6$, $y_0/K = 0.25$ and $r = 0.71$, we conclude that $y(2) = 57.58 \times 10^6$.

(c) Solving the equation in part (a) for t, we see that

$$t = -\frac{1}{r} \ln \left[\frac{\ln(y/K)}{\ln(y_0/K)} \right].$$

Plugging in the given values, we conclude that $\tau = 2.21$ years.

17.

(a)

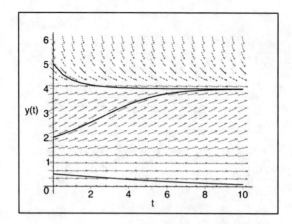

(b) Consider $f(y) = -0.25(1-y)[1-(y/4)]y$. We need to differentiate $f(y)$ with respect to y. We see that

$$f'(y) = -0.25\left(\frac{3}{4}y^2 - \frac{5}{2}y + 1\right).$$

Therefore, $f'(y) = 0$ implies $3y^2 - 10y + 4 = 0$ or $y = \dfrac{5 \pm \sqrt{13}}{3}$.

(c) Since this is a separable equation, we can integrate the equation as follows:

$$\int \frac{dy}{(1-y)(1-(y/4))y} = \int -0.25\, dt.$$

Using partial fractions, we can rewrite the left-hand side as

$$\frac{1}{(1-y)(1-(y/4))y} = \frac{4/3}{1-y} + \frac{-1/12}{1-(y/4)} + \frac{1}{y}.$$

Therefore,

$$\int \frac{dy}{(1-y)(1-(y/4))y} = -\frac{4}{3}\ln|1-y| + \frac{1}{3}\ln|1-(y/4)| + \ln|y|.$$

If $y(0) = 2$, then $y(t) \to 4$ as $t \to \infty$ and moreover, $1 < y(t) < 4$ for all t. Therefore, for $1 < y_0 < 4$,

$$-\frac{4}{3}\ln|1-y| + \frac{1}{3}\ln|1-(y/4)| + \ln|y| = -\frac{4}{3}\ln(y-1) + \frac{1}{3}\ln(1-(y/4)) + \ln(y)$$

$$= \ln\left(\frac{y(1-(y/4))^{1/3}}{(y-1)^{4/3}}\right).$$

We conclude that

$$\frac{y(1-(y/4))^{1/3}}{(y-1)^{4/3}} = Ce^{-0.25t}.$$

68

If $y(0) = 2$, then $C = 2^{2/3}$. Then if $t = 5$, we conclude that $y \approx 3.625$. Similarly, for $y_0 > 4$, we conclude that

$$\frac{y((y/4) - 1)^{1/3}}{(y - 1)^{4/3}} = Ce^{-0.25t}.$$

(d) Consider the equation

$$\frac{y(1 - (y/4))^{1/3}}{(y - 1)^{4/3}} = Ce^{-0.25t}$$

found in part (c). If $y_0 = 2$, then $C = 2^{2/3}$. Letting $y(t) = 3.95$ and solving for t, we see that $t \approx 7.97$. Similarly, using the equation found in part (c) for $y_0 > 4$, we see that if $y_0 = 6$, then $y \leq 4.05$ for $t < 7.97$. For all initial data $2 < y_0 < 6$, the conclusion also holds.

18.

(a) The surface area of the cone is given by

$$S = \pi a \sqrt{h^2 + a^2} + \pi a^2 = \pi a^2 (\sqrt{(h/a)^2 + 1} + 1)$$
$$= \frac{\pi a^2 h}{3} \cdot \frac{3}{h} \left(\sqrt{(h/a)^2 + 1} + 1 \right)$$
$$= c\pi \left(\frac{\pi a^2 h}{3} \right)^{2/3} \cdot \left(\frac{3a}{\pi h} \right)^{2/3}$$
$$= c\pi \left(\frac{3a}{\pi h} \right)^{2/3} V^{2/3}.$$

Therefore, if the rate of evaporation is proportional to the surface area, then rate out $= \alpha \pi (3a/\pi h)^{2/3} V^{2/3}$. Therefore,

$$\frac{dV}{dt} = \text{rate in} - \text{rate out}$$
$$= k - \alpha \pi \left(\frac{3a}{\pi h} \right)^{2/3} \left(\frac{\pi}{3} a^2 h \right)^{2/3}$$
$$= k - \alpha \pi \left(\frac{3a}{\pi h} \right)^{2/3} V^{2/3}.$$

(b) The equilibrium volume can be found by setting $dV/dt = 0$. We see that the equilibrium volume is

$$V = \left(\frac{k}{\alpha \pi} \right)^{3/2} \left(\frac{\pi h}{3a} \right).$$

To find the equilibrium height, we use the fact that the height and radius of the conical pond maintain a constant ratio. Therefore, if h_e, a_e represent the equilibrium values for the h and a, we must have $h_e/a_e = h/a$. Further, we notice that the equilibrium volume can be written as

$$V = \frac{\pi}{3} \left(\frac{k}{\alpha \pi} \right) \left(\frac{k}{\alpha \pi} \right)^{1/2} \left(\frac{h}{a} \right) = \frac{\pi}{3} a_e^2 h_e^2,$$

where $h_e = (k/\alpha\pi)^{1/2}(h/a)$ and $a_e = (k\alpha\pi)^{1/2}$. Since $f'(V) = -\frac{2}{3}\alpha\pi(3a/\pi h)^{2/3}V^{-1/3} < 0$, the equilibrium is asymptotically stable.

(c) In order to guarantee that the pond does not overflow, we need the rate of water in to be less than or equal to the rate of water out. Therefore, we need $k - \alpha\pi a^2 \le 0$.

19.

(a) The rate of increase of the volume is given by

$$\frac{dV}{dt} = k - \alpha a\sqrt{2gh}.$$

Since the cross-section is constant, $dV/dt = Adh/dt$. Therefore,

$$\frac{dh}{dt} = (k - \alpha a\sqrt{2gh})/A.$$

(b) Setting $dh/dt = 0$, we conclude that the equilibrium height of water is $h_e = \frac{1}{2g}\left(\frac{k}{\alpha a}\right)^2$. Since $f'(h_e) < 0$, the equilibrium height is stable.

20.

(a) The equilibrium points are $y^* = 0, 1$. since $f'(0) = \alpha > 0$, the equilibrium solution $y^* = 0$ is unstable. Since $f'(1) = -\alpha < 0$, the equilibrium solution $y^* = 1$ is asymptotically stable.

(b) The equation is separable. The solution is given by

$$y(t) = \frac{y_0 e^{\alpha t}}{1 - y_0 + y_0 e^{\alpha t}}.$$

We see that $\lim_{t\to\infty} y(t) = 1$.

21.

(a) The solution of the separable equation is $y(t) = y_0 e^{-\beta t}$.

(b) Using the result from part (a), we see that $dx/dt = -\alpha x y_0 e^{-\beta t}$. This equation is separable with solution $x(t) = x_0 exp[-\alpha y_0(1 - e^{-\beta t})/\beta]$.

(c) As $t \to \infty$, $y \to 0$ and $x \to x_0 \exp(-\alpha y_0/\beta)$.

22.

(a) Letting $' = d/dt$, we have

$$z' = \frac{nx' - xn'}{n^2}$$

$$= \frac{-\beta nx - \mu nx + \nu\beta x^2 + \mu nx}{n^2}$$

$$= -\beta\frac{x}{n} + \nu\beta\left(\frac{x}{n}\right)^2$$

$$= -\beta z + \nu\beta z^2 = -\beta z(1 - \nu z).$$

70

(b) First, we rewrite the equation as

$$z' + \beta z = \beta \nu z^2.$$

This is a Bernoulli equation with $n = 2$. Let $w = z^{1-n} = z^{-1}$. Then, our equation can be written as

$$w' - \beta w = -\beta \nu.$$

This is a linear equation with solution $w = \nu + ce^{\beta t}$. Then, using the fact that $z = 1/w$, we see that $z = 1/(\nu + ce^{\beta t})$. Finally, the initial condition $z(0) = 1$ implies $c = 1 - \nu$. Therefore, $z(t) = 1/(\nu + (1 - \nu)e^{\beta t})$.

(c) Evaluating $z(20)$ for $\nu = \beta = 1/8$, we conclude that $z(20) = 0.0927$.

23.

(a) The critical points occur when $a - y^2 = 0$. If $a < 0$, there are no critical points. If $a = 0$, then $y^* = 0$ is the only critical point. If $a > 0$, then $y^* = \pm\sqrt{a}$ are the two critical points.

(b) We note that $f'(y) = -2y$. Therefore, $f'(\sqrt{a}) < 0$ which implies that $\sqrt{a}$ is asymptotically stable; $f'(-\sqrt{a}) > 0$ which implies $-\sqrt{a}$ is unstable; the behavior of f' around $y^* = 0$ implies that $y^* = 0$ is semistable.

(c) Below, we graph solutions in the case $a = 1$, $a = 0$ and $a = -1$ respectively.

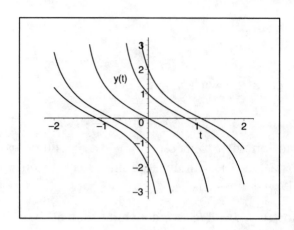

24.

(a) First, for $a < 0$, the only critical point is $y^* = 0$. Second, for $a = 0$, the only critical point is $y^* = 0$. Third, for $a > 0$, the critical points are at $y^* = 0, \pm\sqrt{a}$. Here, $f'(y) = a - 3y^2$. If $a < 0$, then $f'(y) < 0$ for all y, and, therefore, $y^* = 0$ will be asymptotically stable. If $a = 0$, then $f'(0) = 0$. From the behavior on either side of $y^* = 0$, we see that $y^* = 0$ will be asymptotically stable. If $a > 0$, then $f'(0) = a > 0$ which implies that $y^* = 0$ is unstable for $a > 0$. Further, $f'(\pm\sqrt{a}) = -2a < 0$. Therefore, $y^* = \pm\sqrt{a}$ are asymptotically stable for $a > 0$.

(b) Below we sketch solution curves for $a = 1, 0, -1$, respectively.

(c)

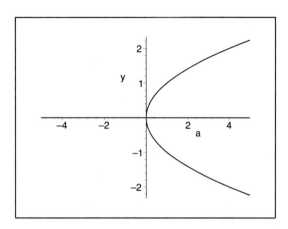

25.

(a) For $a < 0$, the critical points are $y^* = 0, a$. Since $f'(y) = a - 2y$, $f'(0) = a < 0$ and $f'(a) = -a > 0$. Therefore, $y^* = 0$ is asymptotically stable and $y^* = a$ is unstable for $a < 0$. For $a = 0$, the only critical point is $y^* = 0$. which is semistable since $f(y) = -y^2$ is concave down. For $a > 0$, the critical points are $y^* = 0, a$. Since $f'(0) = a > 0$ and $f'(a) = -a < 0$, the critical point $y^* = 0$ is unstable while the critical point $y^* = a$ is asymptotically stable for $a > 0$.

(b) Below we sketch solution curves for $a = 1, 0, -1$, respectively.

(c)

74

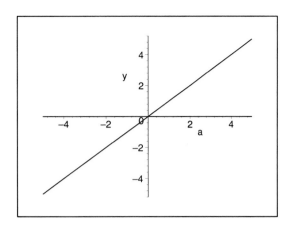

26.

(a) Since the critical points are $x^* = p, q$, we will look at their stability. Since $f'(x) = -\alpha q - \alpha p + 2\alpha x^2$, we see that $f'(p) = \alpha(2p^2 - q - p)$ and $f'(q) = \alpha(2q^2 - q - p)$. Now if $p > q$, then $-p < -q$, and, therefore, $f'(q) = \alpha(2q^2 - q - p) < \alpha(2q^2 - 2q) = 2\alpha q(q-1) < 0$ since $0 < q < 1$. Therefore, if $p > q$, $f'(q) < 0$, and, therefore, $x^* = q$ is asymptotically stable. Similarly, if $p < q$, then $x^* = p$ is asymptotically stable, and therefore, we can conclude that $x(t) \to \min\{p, q\}$ as $t \to \infty$.

The equation is separable. It can be solved by using partial fractions as follows. We can rewrite the equation as

$$\left(\frac{1/(q-p)}{p-x} + \frac{1/(p-q)}{q-x} \right) dx = \alpha dt,$$

which implies

$$\ln \left| \frac{p-x}{q-x} \right| = (p-q)\alpha t + C.$$

The initial condition $x_0 = 0$ implies $C = \ln|p/q|$, and, therefore,

$$\ln \left| \frac{q(p-x)}{p(q-x)} \right| = (p-q)\alpha t.$$

Applying the exponential function and simplifying, we conclude that

$$x(t) = \frac{pq(e^{(p-q)\alpha t} - 1)}{pe^{(p-q)\alpha t} - q}.$$

(b) In this case, $x^* = p$ is the only critical point. Since $f(x) = \alpha(p-x)^2$ is concave up, we conclude that $x^* = p$ is semistable. Further, if $x_0 = 0$, we can conclude that $x \to p$ as $t \to \infty$.

This equation is separable. Its solution is given by

$$x(t) = \frac{p^2 \alpha t}{p\alpha t + 1}.$$

Section 2.6

1. Here $M(x,y) = 2x + 3$ and $N(x,y) = 2y - 2$. Since $M_y = N_x = 0$, the equation is exact. Since $\psi_x = M = 2x + 3$, to solve for ψ, we integrate M with respect to x. We conclude that $\psi = x^2 + 3x + h(y)$. Then $\psi_y = h'(y) = N = 2y - 2$ implies $h(y) = y^2 - 2y$. Therefore, $\psi(x,y) = x^2 + 3x + y^2 - 2y = c$.

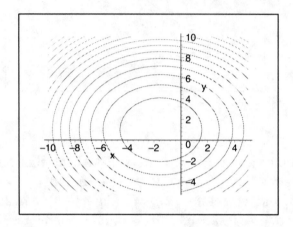

2. Here $M(x,y) = 2x + 4y$ and $N(x,y) = 2x - 2y$. Since $M_y \neq N_x$, the equation is not exact.

3. Here $M(x,y) = 3x^2 - 2xy + 2$ and $N(x,y) = 6y^2 - x^2 + 3$. Since $M_y = -2x = N_x$, the equation is exact. Since $\psi_x = M = 3x^2 - 2xy + 2$, to solve for ψ, we integrate M with respect to x. We conclude that $\psi = x^3 - x^2 y + 2x + h(y)$. Then $\psi_y = -x^2 + h'(y) = N = 6y^2 - x^2 + 3$ implies $h'(y) = 6y^2 + 3$. Therefore, $h(y) = 2y^3 + 3y$ and $\psi(x,y) = x^3 - x^2 y + 2x + 2y^3 + 3y = c$.

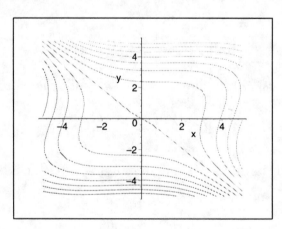

4. Here $M(x,y) = 2xy^2 + 2y$ and $N(x,y) = 2x^2 y + 2x$. Since $M_y = 4xy + 2 = N_x$, the equation is exact. Since $\psi_x = M = 2xy^2 + 2y$, to solve for ψ, we integrate M with respect to x. We conclude that $\psi = x^2 y^2 + 2xy + h(y)$. Then $\psi_y = 2x^2 y + 2x + h'(y) = N = 2x^2 y + 2x$ implies $h'(y) = 0$. Therefore, $h(y) = C$ and $\psi(x,y) = x^2 y^2 + 2xy = c$.

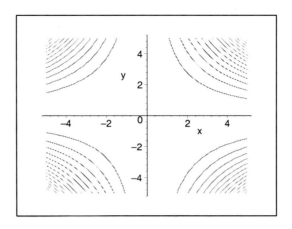

5. Here $M(x,y) = ax + by$ and $N(x,y) = bx + cy$. Since $M_y = b = N_x$, the equation is exact. Since $\psi_x = M = ax + by$, to solve for ψ, we integrate M with respect to x. We conclude that $\psi = ax^2/2 + bxy + h(y)$. Then $\psi_y = bx + h'(y) = N = bx + cy$ implies $h'(y) = cy$. Therefore, $h(y) = cy^2/2$ and $\psi(x,y) = ax^2/2 + bxy + cy^2/2 = c$.

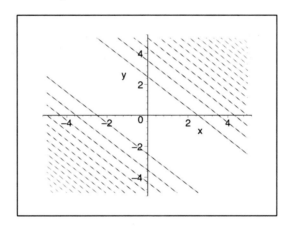

6. Here $M = ax - by$ and $N = bx - cy$. Since $M_y = -b$ and $N_x = b$, the equation is not exact.

7. Here $M(x,y) = e^x \sin y - 2y \sin x$ and $N(x,y) = e^x \cos y + 2 \cos x$. Since $M_y = e^x \cos y - \sin x = N_x$, the equation is exact. Since $\psi_x = M = e^x \sin y - 2y \sin x$, to solve for ψ, we integrate M with respect to x. We conclude that $\psi = e^x \sin y + 2y \cos x + h(y)$. Then $\psi_y = e^x \cos y + 2 \cos x + h'(y) = N = e^x \cos y + 2 \cos x$ implies $h'(y) = 0$. Therefore, $h(y) = C$ and $\psi(x,y) = e^x \sin y + 2y \cos x = c$.

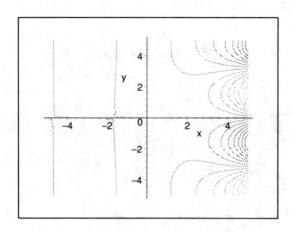

8. Here $M = e^x \sin y + 3y$ and $N = -3x + e^x \sin y$. Therefore, $M_y = e^x \cos y + 3$ and $N_x = -3 + e^x \sin y$. Since $M_y \neq N_x$, therefore, the equation is not exact.

9. Here $M(x, y) = ye^{xy} \cos(2x) - 2e^{xy} \sin(2x) + 2x$ and $N(x, y) = xe^{xy} \cos(2x) - 3$. Since $M_y = e^{xy} \cos(2x) + xye^{xy} \cos(2x) - 2xe^{xy} \sin(2x) = N_x$, the equation is exact. Since $\psi_x = M = ye^{xy} \cos(2x) - 2e^{xy} \sin(2x) + 2x$, to solve for ψ, we integrate M with respect to x. We conclude that $\psi = e^{xy} \cos(2x) + x^2 + h(y)$. Then $\psi_y = xe^{xy} \cos(2x) + h'(y) = N = xe^{xy} \cos(2x) - 3$ implies $h'(y) = -3$. Therefore, $h(y) = -3y$ and $\psi(x, y) = e^{xy} \cos(2x) + x^2 - 3y = c$.

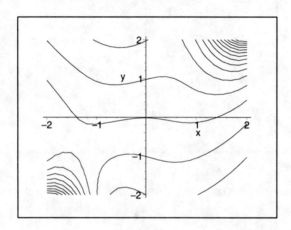

10. Here $M(x, y) = y/x + 6x$ and $N(x, y) = \ln(x) - 2$. Since $M_y = 1/x = N_x$, the equation is exact. Since $\psi_x = M = y/x + 6x$, to solve for ψ, we integrate M with respect to x. We conclude that $\psi = y \ln(x) + 3x^2 + h(y)$. Then $\psi_y = \ln(x) + h'(y) = N = \ln(x) - 2$ implies $h'(y) = -2$. Therefore, $h(y) = -2y$ and $\psi(x, y) = y \ln(x) + 3x^2 - 2y = c$.

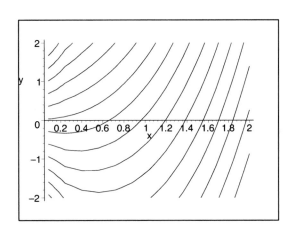

11. Here $M(x, y) = x \ln(y) + xy$ and $N(x, y) = y \ln(x) + xy$. Since $M_y = x/y + x$ and $N_x = y/x + y$, we conclude that the equation is not exact.

12. Here $M(x, y) = x/(x^2+y^2)^{3/2}$ and $N(x, y) = y/(x^2+y^2)^{3/2}$. Since $M_y = N_x$, the equation is exact. Since $\psi_x = M = x/(x^2 + y^2)^{3/2}$, to solve for ψ, we integrate M with respect to x. We conclude that $\psi = -1/(x^2 + y^2)^{1/2} + h(y)$. Then $\psi_y = y/(x^2 + y^2)^{3/2} + h'(y) = N = y/(x^2 + y^2)^{3/2}$ implies $h'(y) = 0$. Therefore, $h(y) = 0$ and $\psi(x, y) = -1/(x^2 + y^2)^{1/2} = c$ which implies that $\psi(x, y) = (x^2 + y^2) = c$.

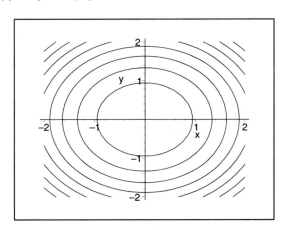

13. Here $M(x, y) = 2x - y$ and $N(x, y) = 2y - x$. Therefore, $M_y = N_x = -1$ which implies that the equation is exact. Integrating M with respect to x, we conclude that $\psi = x^2 - xy + h(y)$. Then $\psi_y = -x + h'(y) = N = 2y - x$ implies $h'(y) = 2y$. Therefore, $h(y) = y^2$ and we conclude that $\psi = x^2 - xy + y^2 = C$. The initial condition $y(1) = 3$ implies $c = 7$. Therefore, $x^2 - xy + y^2 = 7$. Solving for y, we conclude that $y = \dfrac{1}{2}\left[x + \sqrt{28 - 3x^2}\right]$. Therefore, the solution is valid for $3x^2 \leq 28$.

14. Here $M(x, y) = 9x^2 + y - 1$ and $N(x, y) = -4y + x$. Therefore, $M_y = N_x = 1$ which implies that the equation is exact. Integrating M with respect to x, we conclude that $\psi = 3x^3 + xy - x + h(y)$. Then $\psi_y = x + h'(y) = N = -4y + x$ implies $h'(y) = -4y$. Therefore, $h(y) = -2y^2$ and we conclude that $\psi = 3x^3 + xy - x - 2y^2 = C$. The initial condition $y(1) = 0$ implies $c = 2$. Therefore, $3x^3 + xy - x - 2y^2 = 2$. Solving for y, we conclude that $y = [x - (24x^3 + x^2 - 8x - 16)^{1/2}]/4$. The solution is valid for $x > 0.9846$.

79

15. Here $M(x, y) = xy^2 + bx^2y$ and $N(x, y) = x^3 + x^2y$. Therefore, $M_y = 2xy + bx^2$ and $N_x = 3x^2 + 2xy$. In order for the equation to be exact, we need $b = c$. Taking this value for b, we integrating M with respect to x. We conclude that $\psi = x^2y^2/2 + x^3y + h(y)$. Then $\psi_y = x^2y + x^3 + h'(y) = N = x^3 + x^2y$ implies $h'(y) = 0$. Therefore, $h(y) = C$ and $\psi(x, y) = x^2y^2/2 + x^3y = C$. That is, the solution is given implicitly as $x^2y^2/2 + x^3y = c$.

16. Here $M(x, y) = ye^{2xy} + x$ and $N(x, y) = bxe^{2xy}$. Then $M_y = e^{2xy} + 2xye^{2xy}$ and $N_x = be^{2xy} + 2bxye^{2xy}$. The equation will be exact as long as $b = 1$. Integrating M with respect to x, we conclude that $\psi = e^{2xy}/2 + x^2/2 + h(y)$. Then $\psi_y = xe^{2xy} + h'(y) = N = xe^{2xy}$ implies $h'(y) = 0$. Therefore, $h(y) = 0$ and we conclude that the solution is given implicitly by the equation $e^{2xy} + x^2 = C$.

17. We notice that $\psi(x, y) = f(x) + g(y)$. Therefore, $\psi_x = f'(x)$ and $\psi_y = g'(y)$. That is,

$$\psi_x = M(x, y_0) \qquad \psi_y = N(x_0, y).$$

Furthermore, $\psi_{xy} = M_y$ and $\psi_{yx} = N_x$. Based on the hypothesis, $\psi_{xy} = \psi_{yx}$ and $M_y = N_x$.

18. We notice that $(M(x))_y = 0 = (N(y))_x$. Therefore, the equation is exact.

19. Here $M(x, y) = x^2y^3$ and $N(x, y) = x + xy^2$. Therefore, $M_y = 3x^2y^2$ and $N_x = 1 + y^2$. We see that the equation is not exact. Now, multiplying the equation by $\mu(x, y) = 1/xy^3$, the equation becomes

$$x\,dx + (1 + y^2)/y^3\,dy = 0.$$

Now we see that for this equation $M = x$ and $N = (1 + y^2)/y^3$. Therefore, $M_y = 0 = N_x$. Integrating M with respect to x, we see that $\psi = x^2/2 + h(y)$. Further, $\psi_y = h'(y) = N = (1 + y^2)/y^3 = 1/y^3 + 1/y$. Therefore, $h(y) = -1/2y^2 + \ln(y)$ and we conclude that the solution of the equation is given implicitly by $x^2 - 1/y^2 + 2\ln(y) = C$.

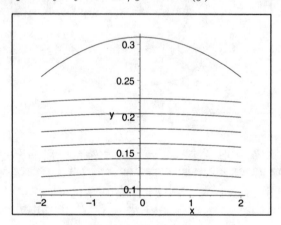

20. Multiplying the equation by $\mu(x, y) = ye^x$, the equation becomes

$$(e^x \sin y - 2y \sin x)dx + (e^x \cos y + 2\cos x)dy = 0.$$

Now we see that for this equation $M = e^x \sin y - 2y \sin x$ and $N = e^x \cos y + 2\cos x$. Therefore, $M_y = e^x \cos y - 2\sin x = N_x$. Integrating M with respect to x, we see that $\psi = e^x \sin y + 2y \cos x + h(y)$. Further, $\psi_y = e^x \cos y + 2\cos x + h'(y) = N = e^x \cos y + 2\cos x$. Therefore, $h(y) = 0$ and we conclude that the solution of the equation is given implicitly by $e^x \sin y + 2y \cos x = C$.

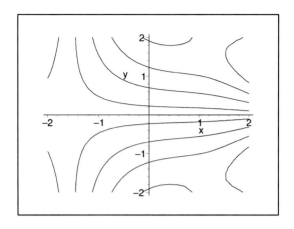

21. Multiplying the equation by $\mu(x, y) = y$, the equation becomes

$$y^2 dx + (2xy - y^2 e^y) dy = 0.$$

Now we see that for this equation $M = y^2$ and $N = 2xy - y^2 e^y$. Therefore, $M_y = 2y = N_x$. Integrating M with respect to x, we see that $\psi = xy^2 + h(y)$. Further, $\psi_y = 2xy + h'(y) = N = 2xy - y^2 e^y$. Therefore, $h'(y) = -y^2 e^y$ which implies that $h(y) = -e^y(y^2 - 2y + 2)$, and we conclude that the solution of the equation is given implicitly by $xy^2 - e^y(y^2 - 2y + 2) = C$.

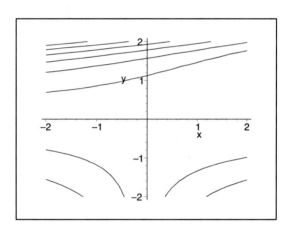

22. Multiplying the equation by $\mu(x, y) = xe^x$, the equation becomes

$$(x^2 + 2x)e^x \sin y\, dx + x^2 e^x \cos y\, dy = 0.$$

Now we see that for this equation $M_y = (x^2 + 2x)e^x \cos y = N_x$. Integrating M with respect to x, we see that $\psi = x^2 e^x \sin y + h(y)$. Further, $\psi_y = x^2 e^x \cos y + h'(y) = N = x^2 e^x \cos y$. Therefore, $h'(y) = 0$ which implies that the solution of the equation is given implicitly by $x^2 e^x \sin y = C$.

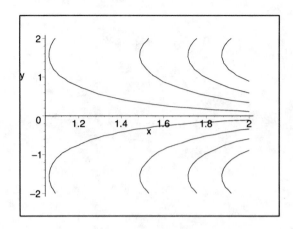

23. Suppose μ is an integrating factor which will make the equation exact. Then multiplying the equation by μ, we have

$$\mu M\,dx + \mu N\,dy = 0.$$

Then we need $(\mu M)_y = (\mu N)_x$. That is, we need $\mu_y M + \mu M_y = \mu_x N + \mu N_x$. Then we rewrite the equation as $\mu(N_x - M_y) = \mu_y M - \mu_x N$. Suppose μ does not depend on x. Then $\mu_x = 0$. Therefore, $\mu(N_x - M_y) = \mu_y M$. Using the assumption that $(N_x - M_y)/M = Q(y)$, we can find an integrating factor μ by choosing μ which satisfies $\mu_y/\mu = Q$. We conclude that $\mu(y) = \exp\int Q(y)\,dy$ is an integrating factor of the differential equation.

24. Suppose μ is an integrating factor which will make the equation exact. Then multiplying the equation by μ, we have

$$\mu M\,dx + \mu N\,dy = 0.$$

Then we need $(\mu M)_y = (\mu N)_x$. That is, we need $\mu_y M + \mu M_y = \mu_x N + \mu N_x$. Then we rewrite the equation as $\mu(N_x - M_y) = \mu_y M - \mu_x N$. By the given assumption, we need μ to satisfy $\mu R(xM - yN) = \mu_y M - \mu_x N$. This equation is satisfied if $\mu_y = (\mu x)R$ and $\mu_x = (\mu y)R$. Consider $\mu = \mu(xy)$. Then $\mu_x = \mu' y$ and $\mu_y = \mu' x$ where $' = d/dz$ for $z = xy$. Therefore, we need to choose μ to satisfy $\mu' = \mu R$. This equation is separable with solution $\mu = \exp(\int R(z)\,dz)$.

25. Since $(M_y - N_x)/N = 3$ is a function of x only, we know that $\mu = e^{3x}$ is an integrating factor for this equation. Multiplying the equation by μ, we have

$$e^{3x}(3x^2 y + 2xy + y^3)dx + e^{3x}(x^2 + y^2)dy = 0.$$

Then $M_y = e^{3x}(3x^2 + 2x + 3y^2) = N_x$. Therefore, this new equation is exact. Integrating M with respect to x, we conclude that $\psi = (x^2 y + y^3/3)e^{3x} + h(y)$. Then $\psi_y = (x^2 + y^2)e^{3x} + h'(y) = N = e^{3x}(x^2 + y^2)$. Therefore, $h'(y) = 0$ and we conclude that the solution is given implicitly by $(3x^2 y + y^3)e^{3x} = c$.

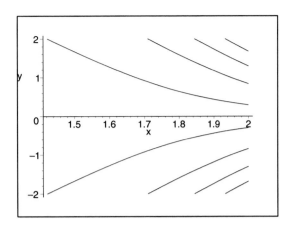

26. Since $(M_y - N_x)/N = -1$ is a function of x only, we know that $\mu = e^{-x}$ is an integrating factor for this equation. Multiplying the equation by μ, we have

$$(e^{-x} - e^x - ye^{-x})dx + e^{-x}dy = 0.$$

Then $M_y = -e^{-x} = N_x$. Therefore, this new equation is exact. Integrating M with respect to x, we conclude that $\psi = -e^{-x} - e^x + ye^{-x} + h(y)$. Then $\psi_y = e^{-x} + h'(y) = N = e^{-x}$. Therefore, $h'(y) = 0$ and we conclude that the solution is given implicitly by $-e^{-x} - e^x + ye^{-x} = c$.

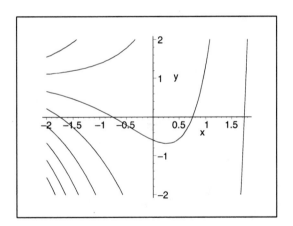

27. Since $(N_x - M_y)/M = 1/y$ is a function of y only, we know that $\mu(y) = e^{\int 1/y \, dy} = y$ is an integrating factor for this equation. Multiplying the equation by μ, we have

$$ydx + (x - y\sin y)dy = 0.$$

Then for this equation, $M_y = 1 = N_x$. Therefore, this new equation is exact. Integrating M with respect to x, we conclude that $\psi = xy + h(y)$. Then $\psi_y = x + h'(y) = N = x - y\sin y$. Therefore, $h'(y) = -y\sin y$ which implies that $h(y) = -\sin y + y\cos y$, and we conclude that the solution is given implicitly by $xy - \sin y + y\cos y = C$.

83

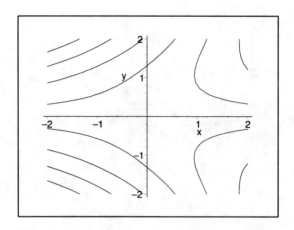

28. Since $(N_x - M_y)/M = (2y - 1)/y$ is a function of y only, we know that $\mu(y) = e^{\int 2 - 1/y\, dy} = e^{2y}/y$ is an integrating factor for this equation. Multiplying the equation by μ, we have

$$e^{2y}dx + (2xe^{2y} - 1/y)dy = 0.$$

Then for this equation, $M_y = N_x$. Therefore, this new equation is exact. Integrating M with respect to x, we conclude that $\psi = xe^{2y} + h(y)$. Then $\psi_y = 2xe^{2y} + h'(y) = N = 2xe^{2y} - 1/y$. Therefore, $h'(y) = -1/y$ which implies that $h(y) = -ln(y)$, and we conclude that the solution is given implicitly by $xe^{2y} - \ln(y) = C$.

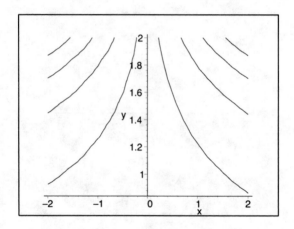

29. Since $(N_x - M_y)/M = \cot(y)$ is a function of y only, we know that $\mu(y) = e^{\int \cot(y)\, dy} = \sin(y)$ is an integrating factor for this equation. Multiplying the equation by μ, we have

$$e^x \sin y dx + (e^x \cos y + 2y)dy = 0.$$

Then for this equation, $M_y = N_x$. Therefore, this new equation is exact. Integrating M with respect to x, we conclude that $\psi = e^x \sin y + h(y)$. Then $\psi_y = e^x \cos y + h'(y) = N = e^x \cos y + 2y$. Therefore, $h'(y) = 2y$ which implies that $h(y) = y^2$, and we conclude that the solution is given implicitly by $e^x siny + y^2 = C$.

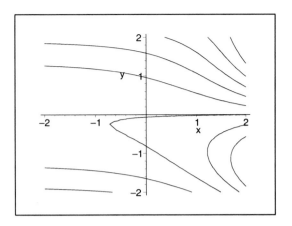

30. Since $(N_x - M_y)/M = 2/y$ is a function of y only, we know that $\mu(y) = e^{\int 2/y\, dy} = y^2$ is an integrating factor for this equation. Multiplying the equation by μ, we have

$$(4x^3 + 3y)dx + (3x + 4y^3)dy = 0.$$

Then for this equation, $M_y = N_x$. Therefore, this new equation is exact. Integrating M with respect to x, we conclude that $\psi = x^4 + 3xy + h(y)$. Then $\psi_y = 3x + h'(y) = N = 3x + 4y^3$. Therefore, $h'(y) = 4y^3$ which implies that $h(y) = y^4$, and we conclude that the solution is given implicitly by $x^4 + 3xy + y^4 = C$.

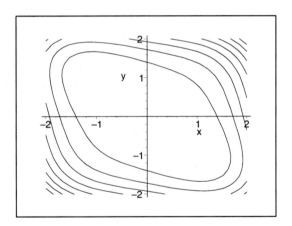

31. Since $(N_x - M_y)/(xM - yN) = 1/xy$ is a function of xy only, we know that $\mu(xy) = e^{\int 1/xy\, dy} = xy$ is an integrating factor for this equation. Multiplying the equation by μ, we have

$$(3x^2y + 6x)dx + (x^3 + 3y^2)dy = 0.$$

Then for this equation, $M_y = N_x$. Therefore, this new equation is exact. Integrating M with respect to x, we conclude that $\psi = x^3y + 3x^2 + h(y)$. Then $\psi_y = x^3 + h'(y) = N = x^3 + 3y^2$. Therefore, $h'(y) = 3y^2$ which implies that $h(y) = y^3$, and we conclude that the solution is given implicitly by $x^3y + 3x^2 + y^3 = C$.

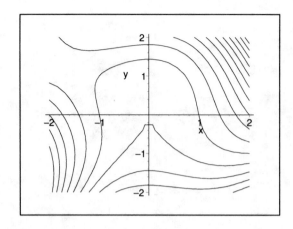

32. Using the integrating factor $\mu = [xy(2x+y)]^{-1}$, this equation can be rewritten as

$$\left[\frac{2}{x} + \frac{2}{2x+y}\right] dx + \left[\frac{1}{y} + \frac{1}{2x+y}\right] dy = 0.$$

Integrating M with respect to x, we see that $\psi = 2\ln|x| + \ln|2x+y| + h(y)$. Then $\psi_y = (2x+y)^{-1} + h'(y) = N = (2x+y)^{-1} + 1/y$. Therefore, $h'(y) = 1/y$ which implies that $h(y) = \ln|y|$. Therefore, $\psi = 2\ln|x| + \ln|2x+y| + \ln|y| = C$. Applying the exponential function, we conclude that the solution is given implicitly be $2x^3y + x^2y^2 = C$.

Section 2.7

1. The Euler formula is $y_{n+1} = y_n + h(3 + t_n - y_n)$ in which $t_n = t_0 + nh$. Since $t_0 = 0$, we have $y_{n+1} = y_n(1-h) + 3h + nh^2$.

(a) For $h = 0.05$, the Euler approximations for y_n at $n = 2, 4, 6, 8$ are given by

$$1.1975, \quad 1.38549, \quad 1.56491, \quad 1.73658$$

(b) For $h = 0.025$, the Euler approximations for y_n at $n = 4, 8, 12, 16$ are given by

$$1.19631, \quad 1.38335, \quad 1.56200, \quad 1.73308$$

2. The Euler formula is $y_{n+1} = y_n + h(5t_n - 3\sqrt{y_n})$ in which $t_n = t_0 + nh$. Since $t_0 = 0$, we have $y_{n+1} = y_n + 5nh^2 - 3h\sqrt{y_n}$ with $y_0 = 2$.

(a) For $h = 0.05$, the Euler approximations for y_n at $n = 2, 4, 6, 8$ are given by

$$1.59980, \quad 1.29288, \quad 1.07242, \quad 0.930175.$$

(b) For $h = 0.025$, the Euler approximations for y_n at $n = 4, 8, 12, 16$ are given by

$$1.61124, \quad 1.31361, \quad 1.10012, \quad 0.962552$$

3. The Euler formula is $y_{n+1} = y_n + h(2y_n - 3t_n)$ in which $t_n = t_0 + nh$. Since $t_0 = 0$, we have $y_{n+1} = y_n(1 + 2h) - 3nh^2$.

(a) For $h = 0.05$, the Euler approximations for y_n at $n = 2, 4, 6, 8$ are given by

$$1.2025, \quad 1.41603, \quad 1.64289, \quad 1.88590$$

(b) For $h = 0.025$, the Euler approximations for y_n at $n = 4, 8, 12, 16$ are given by

$$1.20388, \quad 1.41936, \quad 1.64896, \quad 1.89572$$

4. The Euler formula is $y_{n+1} = y_n + h(2t_n + e^{-t_n y_n})$ in which $t_n = t_0 + nh$. Since $t_0 = 0$, we have $y_{n+1} = y_n + 2nh^2 + he - nhy_n$.

(a) For $h = 0.05$, the Euler approximations for y_n at $n = 2, 4, 6, 8$ are given by

$$1.10244, \quad 1.21426, \quad 1.33484, \quad 1.46399$$

(b) For $h = 0.025$, the Euler approximations for y_n at $n = 4, 8, 12, 16$ are given by

$$1.10365, \quad 1.21656, \quad 1.33817, \quad 1.46832$$

5. The Euler formula is $y_{n+1} = y_n + h(y_n^2 + 2t_n y_n)/(3 + t_n^2)$ in which $t_n = t_0 + nh$. Since $t_0 = 0$, we have $y_{n+1} = y_n + h(y_n^2 + 2nhy_n)/(3 + n^2h^2)$.

(a) For $h = 0.05$, the Euler approximations for y_n at $n = 2, 4, 6, 8$ are given by

$$0.509239, \quad 0.522187, \quad 0.539023, \quad 0.559936$$

(b) For $h = 0.025$, the Euler approximations for y_n at $n = 4, 8, 12, 16$ are given by

$$0.509701, \quad 0.523155, \quad 0.540550, \quad 0.562089$$

6. The Euler formula is $y_{n+1} = y_n + h(t_n^2 - y_n^2)\sin(y_n)$ in which $t_n = t_0 + nh$. Since $t_0 = 0$, we have $y_{n+1} = y_n + h(n^2h^2 - y_n^2)\sin(y_n)$.

(a) For $h = 0.05$, the Euler approximations for y_n at $n = 2, 4, 6, 8$ are given by

$$-0.920498, \quad -0.857538, \quad -0.808030, \quad -0.770038$$

(b) For $h = 0.025$, the Euler approximations for y_n at $n = 4, 8, 12, 16$ are given by

$$-0.922575, \quad -0.860923, \quad -0.812300, \quad -0.774965$$

7. The Euler formula is $y_{n+1} = y_n + h(0.5 - t_n + 2y_n)$ in which $t_n = t_0 + nh$. Since $t_0 = 0$, we have $y_{n+1} = y_n + h(0.5 - nh + 2y_n)$.

(a) For $h = 0.025$, the Euler approximations for y_n at $n = 20, 40, 60, 80$ are given by

$$2.90330, \quad 7.53999, \quad 19.4292, \quad 50.5614$$

(b) For $h = 0.0125$, the Euler approximations for y_n at $n = 40, 80, 120, 160$ are given by

$$2.93506, \quad 7.70957, \quad 20.1081, \quad 52.9779$$

8. The Euler formula is $y_{n+1} = y_n + h(5t_n - 3\sqrt{y_n})$ in which $t_n = t_0 + nh$. Since $t_0 = 0$, we have $y_{n+1} = y_n + h(5nh - 3\sqrt{y_n})$.

(a) For $h = 0.025$, the Euler approximations for y_n at $n = 20, 40, 60, 80$ are given by

$$0.891830, \quad 1.25225, \quad 2.37818, \quad 4.07257$$

(b) For $h = 0.0125$, the Euler approximations for y_n at $n = 40, 80, 120, 160$ are given by

$$0.908902, \quad 1.26872, \quad 2.39336, \quad 4.08799$$

9. The Euler formula is $y_{n+1} = y_n + h\sqrt{t_n + y_n}$ in which $t_n = t_0 + nh$. Since $t_0 = 0$, we have $y_{n+1} = y_n + h\sqrt{nh + y_n}$.

(a) For $h = 0.025$, the Euler approximations for y_n at $n = 20, 40, 60, 80$ are given by

$$3.95713, \quad 5.09853, \quad 6.41548, \quad 7.90174$$

(b) For $h = 0.0125$, the Euler approximations for y_n at $n = 40, 80, 120, 160$ are given by

$$3.95965, \quad 5.10371, \quad 6.42343, \quad 7.91255$$

10. The Euler formula is $y_{n+1} = y_n + h(2t_n + e^{-t_n y_n})$ in which $t_n = t_0 + nh$. Since $t_0 = 0$, we have $y_{n+1} = y_n + h(2nh + e^{-nhy_n})$.

(a) For $h = 0.025$, the Euler approximations for y_n at $n = 20, 40, 60, 80$ are given by

$$1.60729, \quad 2.46830, \quad 3.72167, \quad 5.45963$$

(b) For $h = 0.0125$, the Euler approximations for y_n at $n = 40, 80, 120, 160$ are given by

$$1.60996, \quad 2.47460, \quad 3.73356, \quad 5.47774$$

11. The Euler formula is $y_{n+1} = y_n + h(4 - t_n y_n)/(1 + y_n^2)$ in which $t_n = t_0 + nh$. Since $t_0 = 0$, we have $y_{n+1} = y_n + h(4 - nhy_n)/(1 + y_n^2)$.

(a) For $h = 0.025$, the Euler approximations for y_n at $n = 20, 40, 60, 80$ are given by

$$-1.45865, \quad -0.217545, \quad 1.05715, \quad 1.41487$$

(b) For $h = 0.0125$, the Euler approximations for y_n at $n = 40, 80, 120, 160$ are given by

$$-1.45322, \quad -0.180813, \quad 1.05903, \quad 1.41244$$

12. The Euler formula is $y_{n+1} = y_n + h(y_n^2 + 2t_n y_n)/(3 + t_n^2)$ in which $t_n = t_0 + nh$. Since $t_0 = 0$, we have $y_{n+1} = y_n + h(y_n^2 + 2nhy_n)/(3 + n^2h^2)$.

(a) For $h = 0.025$, the Euler approximations for y_n at $n = 20, 40, 60, 80$ are given by

$$0.587987, \quad 0.791589, \quad 1.14743, \quad 1.70973$$

(b) For $h = 0.0125$, the Euler approximations for y_n at $n = 40, 80, 120, 160$ are given by

$$0.589440, \quad 0.795758, \quad 1.15693, \quad 1.72955$$

13. The Euler formula is
$$y_{n+1} = y_n + h(1 - t_n + 4y_n)$$
in which $t_n = t_0 + nh$. Since $t_0 = 0$, we can write

$$y_{n+1} = y_n + h - nh^2 + 4hy_n$$

with $y_0 = 1$. With $h = 0.01$, a total of 200 iterations is necessary to reach $\bar{t} = 2$. With $h = 0.001$, a total of 2000 iterations is necessary.

14. We will use the first three terms in equation (12),

$$\phi(t_{n+1}) = \phi(t_n) + f[t_n, \phi(t_n)]h + \phi''(t_n)\frac{h^2}{2}.$$

Letting $h = 0.1$, then the approximation is given by

$$y_{n+1} = y_n + h(1 - t_n + 4y_n) + \frac{h^2}{2}(3 - 4t_n + 16y_n).$$

Therefore,

$$y_1 = 1 + 0.1(1 - 0 + 4(1)) + \frac{0.1^2}{2}(3 - 4(0) + 16(1)) = 1 + 0.5 + .095 = 1.595.$$

Then, repeating this argument for y_2, we conclude that $y_2 = 2.4636$.

Solving this linear equation, we conclude that the exact solution is given by

$$y(t) = -\frac{3}{16} + \frac{1}{4}t + \frac{19}{16}e^{4t}.$$

Therefore, $y(0.1) = 1.609$ and $y(0.2) = 2.505$.

15. We know that $e_{n+1} = \frac{1}{2}\phi''(\bar{t}_n)h^2$ where $t_n < \bar{t}_n < t_{n+1}$. Here

$$\phi'(t) = 2\phi(t) - 1.$$

Therefore,

$$\phi''(t) = 2\phi'(t) = 2(2\phi(t) - 1) = 4\phi(t) - 2.$$

Therefore,
$$e_{n+1} = (2\phi(\bar{t}_n) - 1)h^2.$$

Therefore,
$$|e_{n+1}| \le |2M + 1|h^2$$

where $M = \max_{0 \le t \le 1} |\phi(t)|$.

The exact solution of this linear equation is $y(t) = 1/2 + 1/2e^{2t}$. Then, using the fact that the local truncation error is given by $e_{n+1} = \frac{1}{2}\phi(\bar{t}_n)h^2$ and $\phi(t) = 1/2 + 1/2e^{2t}$, we can conclude that
$$e_{n+1} = e^{2\bar{t}_n}h^2.$$

Therefore, $|e_1| \le e^{0.2}(0.1)^2 \approx 0.012$. Similarly, $|e_4| \le e^{0.8}(0.1)^2 \approx 0.022$.

16. We know that $e_{n+1} = \frac{1}{2}\phi''(\bar{t}_n)h^2$ where $t_n < \bar{t}_n < t_{n+1}$. Here
$$\phi'(t) = \frac{1}{2} - t + 2\phi(t).$$

Therefore,
$$\phi''(t) = -1 + 2\phi'(t) = -1 + 2\left(\frac{1}{2} - t + 2\phi(t)\right) = -2t + 4\phi(t).$$

Therefore,
$$e_{n+1} = (-t_n + 2\phi(\bar{t}_n))h^2.$$

Therefore,
$$|e_{n+1}| \le |2M + 1|h^2$$

where $M = \max_{0 \le t \le 1} |\phi(t)|$.

The exact solution of this linear equation is $y(t) = \frac{1}{2}t + e^{2t}$. Then, using the fact that the local truncation error is given by $e_{n+1} = \frac{1}{2}\phi(\bar{t}_n)h^2$ and $\phi(t) = \frac{1}{2}t + e^{2t}$, we can conclude that
$$e_{n+1} = 2e^{2\bar{t}_n}h^2.$$

Therefore, $|e_1| \le 2e^{0.2}(0.1)^2 \approx 0.024$. Similarly, $|e_4| \le 2e^{0.8}(0.1)^2 \approx 0.045$.

17. We know that $e_{n+1} = \frac{1}{2}\phi''(\bar{t}_n)h^2$ where $t_n < \bar{t}_n < t_{n+1}$. Here
$$\phi'(t) = t^2 + (\phi(t))^2.$$

Therefore,
$$\phi''(t) = 2t + 2\phi(t)\phi'(t) = 2t + 2t^2\phi(t) + 2(\phi(t))^3$$

Therefore,
$$e_{n+1} = (\bar{t}_n + \bar{t}_n^2\phi(\bar{t}_n) + (\phi(\bar{t}_n))^3)h^2.$$

Therefore,
$$|e_{n+1}| \le |t_{n+1} + t_{n+1}^2 M_{n+1} + M_{n+1}^3|h^2$$

where $M_{n+1} = \max_{t_n \le t \le t_{n_1}} |\phi(t)|$.

18. We know that $e_{n+1} = \frac{1}{2}\phi''(\bar{t}_n)h^2$ where $t_n < \bar{t}_n < t_{n+1}$. Here

$$\phi'(t) = 5t - 3\sqrt{\phi(t)}.$$

Therefore,

$$\phi''(t) = 5 - \frac{3}{2}(\phi(t))^{-1/2}\phi'(t) = 5 - \frac{3}{2}\frac{5t - 3\sqrt{\phi(t)}}{\sqrt{\phi(t)}}.$$

Therefore,

$$e_{n+1} = \frac{1}{2}\left(5 - \frac{3}{2}\frac{(5\bar{t}_n - 3\sqrt{\phi(\bar{t}_n)})}{\sqrt{\phi(\bar{t}_n)}}\right)h^2$$

$$= \frac{1}{4}\left(19 - 15\frac{\bar{t}_n}{\sqrt{\phi(\bar{t}_n)}}\right)h^2.$$

19. We know that $e_{n+1} = \frac{1}{2}\phi''(\bar{t}_n)h^2$ where $t_n < \bar{t}_n < t_{n+1}$. Here

$$\phi'(t) = \sqrt{t + \phi(t)}.$$

Therefore,

$$\phi''(t) = \frac{1 + \phi'(t)}{2\sqrt{t + \phi(t)}} = \frac{1}{2\sqrt{t + \phi(t)}} + \frac{1}{2}.$$

Therefore,

$$e_{n+1} = \frac{1}{4}\left[1 + \frac{1}{\sqrt{\bar{t}_n + \phi(\bar{t}_n)}}\right]h^2.$$

20. We know that $e_{n+1} = \frac{1}{2}\phi''(\bar{t}_n)h^2$ where $t_n < \bar{t}_n < t_{n+1}$. Here

$$\phi'(t) = 2t + e^{-t\phi(t)}.$$

Therefore,

$$\phi''(t) = 2 + (-\phi(t) - t\phi'(t))e^{-t\phi(t)} = 2 + (-\phi(t) - t(2t + e^{-t\phi(t)}))e^{-t\phi(t)}.$$

Therefore,

$$e_{n+1} = \frac{1}{2}\left[2 + (-\phi(\bar{t}_n) - 2\bar{t}_n^2 - \bar{t}_n e^{-\bar{t}_n\phi(\bar{t}_n)})e^{-\bar{t}_n\phi(\bar{t}_n)}\right]h^2.$$

21.

(a) The solution is given by $\phi(t) = \frac{1}{5\pi}\sin(5\pi t) + 1$.

91

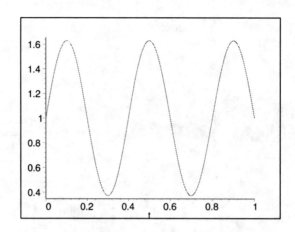

(b) Approximate values at $t = 0.2, 0.4, 0.6$ are given by $1.2, 1.0, 1.2$, respectively.

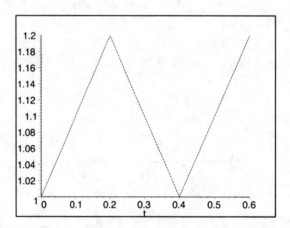

(c) Approximate values at $t = 0.2, 0.4, 0.6$ are given by $1.1, 1.0, 1.1$, respectively.

(d) Since $\phi''(t) = -5\pi \sin(5\pi t)$, the local truncation error for the Euler method is given by

$$e_{n+1} = -\frac{5\pi h^2}{2} \sin(5\pi \bar{t}_n).$$

In order to guarantee that $|e_{n+1}| < 0.05$, we need

$$\frac{5\pi h^2}{2} < 0.05.$$

Solving this inequality, we conclude that we would need $h < 1/\sqrt{50\pi} approx 0.08$.

22.

1. The Euler formula is $y_{n+1} = y_n + h(1 - t_n + 4y_n)$. The approximate values for the solution at $t = 0.1, 0.2, 0.3, 0.4$ are given by

$$1.55, \quad 2.34, \quad 3.46, \quad 5.07.$$

92

2. The Euler formula is $y_{n+1} = y_n + h(3 + t_n - y_n)$. The approximate values for the solution at $t = 0.1, 0.2, 0.3, 0.4$ are given by

$$1.20, \quad 1.39, \quad 1.57, \quad 1.74.$$

3. The Euler formula is $y_{n+1} = y_n + h(2y_n - 3t_n)$. The approximate values for the solution at $t = 0.1, 0.2, 0.3, 0.4$ are given by

$$1.20, \quad 1.42, \quad 1.65, \quad 1.90.$$

23.

(a)

$$1000 \cdot \begin{vmatrix} 6.0 & 18 \\ 2.0 & 6.0 \end{vmatrix} = 1000 \cdot 0 = 0.$$

(b)

$$1000 \cdot \begin{vmatrix} 6.01 & 18.0 \\ 2.00 & 6.00 \end{vmatrix} = 1000(0.06) = 60.$$

(c)

$$1000 \cdot \begin{vmatrix} 6.010 & 18.04 \\ 2.004 & 6.000 \end{vmatrix} = 1000(-0.09216) = -92.16.$$

24. Rounding to three digits, $a(b-c) \approx 0.224$. Similarly, rounding to three digits, $ab \approx 0.702$ and $ac \approx 0.477$. Therefore, $ab - ac \approx 0.225$.

25.

(a) The maximum errors occur at $t = 2$. For $h = 0.001, 0.01, 0.025, 0.05$, they are given by

$$56.0393, \quad 510.8722, \quad 1107.4123, \quad 1794.5339.$$

(b)

93

(c) Yes.

(d) Using a curve-fitting routine, the slope of the least squares line is $\approx .909$.

Section 2.8

1.

(a) The improved Euler formula is

$$y_{n+1} = y_n + h\left(3 + \frac{1}{2}t_n + \frac{1}{2}t_{n+1} - y_n\right) - \frac{h^2}{2}(3 + t_n - y_n).$$

Since $t_n = t_0 + nh$ and $t_0 = 0$, this formula can be simplified to

$$y_{n+1} = y_n + h(3 - y_n) + \frac{h^2}{2}(y_n - 2 + 2n) - \frac{nh^3}{2}$$

with $y_0 = 1$. With $h = 0.05$, the approximate values of the solution at $t = 0.1, 0.2, 0.3, 0.4$ are

$$1.19512, \quad 1.38120, \quad 1.55909, \quad 1.72956$$

(b) Using $h = 0.025$, the approximate values of the solution at $t = 0.1, 0.2, 0.3, 0.4$ are

$$1.19515, \quad 1.38125, \quad 1.55916, \quad 1.72965.$$

(c) Using $h = 0.0125$, the approximate values of the solution at $t = 0.1, 0.2, 0.3, 0.4$ are

$$1.19516, \quad 1.38126, \quad 1.55918, \quad 1.72967.$$

(d) Using the Runge-Kutta method with $h = 0.1$, the approximate values of the solution at $t = 0.1, 0.2, 0.3, 0.4$ are

$$1.19516, \quad 1.38127, \quad 1.55918, \quad 1.72968$$

(e) Using the Runge-Kutta method with $h = 0.05$, the approximate values of the solution at $t = 0.1, 0.2, 0.3, 0.4$ are

$$1.19516, \quad 1.38127, \quad 1.55918, \quad 1.72968$$

2.

(a) The improved Euler formula is

$$y_{n+1} = y_n + \frac{h}{2}\left(5t_n - 3\sqrt{y_n}\right)) + \frac{h}{2}(5t_{n+1} - 3\sqrt{K_n})$$

where $K_n = y_n + h(5t_n - 3\sqrt{y_n})$. Since $t_n = t_0 + nh$ and $t_0 = 0$, this formula can be simplified to

$$y_{n+1} = y_n + \frac{h}{2}(5nh - 3\sqrt{y_n}) + \frac{h}{2}\left[5(n+1)h - 3\sqrt{K_n}\right]$$

with $y_0 = 2$. With $h = 0.05$, the approximate values of the solution at $t = 0.1, 0.2, 0.3, 0.4$ are

$$1.62283, \quad 1.33460, \quad 1.12820, \quad 0.995445$$

94

(b) Using $h = 0.025$, the approximate values of the solution at $t = 0.1, 0.2, 0.3, 0.4$ are

$$1.62243, \quad 1.33386, \quad 1.12718, \quad 0.994215$$

(c) Using $h = 0.0125$, the approximate values of the solution at $t = 0.1, 0.2, 0.3, 0.4$ are

$$1.62234, \quad 1.33368, \quad 1.12693, \quad 0.993921$$

(d) Using the Runge-Kutta method with $h = 0.1$, the approximate values of the solution at $t = 0.1, 0.2, 0.3, 0.4$ are

$$1.62231, \quad 1.33362, \quad 1.12686, \quad 0.993839$$

(e) Using the Runge-Kutta method with $h = 0.05$, the approximate values of the solution at $t = 0.1, 0.2, 0.3, 0.4$ are

$$1.62230, \quad 1.33362, \quad 1.12685, \quad 0.993826$$

3.

(a) The improved Euler formula is

$$y_{n+1} = y_n + \frac{h}{2}\left(4y_n - 3t_n - 3t_{n+1}\right) + h^2(2y_n - 3t_n).$$

Since $t_n = t_0 + nh$ and $t_0 = 0$, this formula can be simplified to

$$y_{n+1} = y_n + 2hy_n + \frac{h^2}{2}(4y_n - 3 - 6n) - 3nh^3$$

with $y_0 = 1$. With $h = 0.05$, the approximate values of the solution at $t = 0.1, 0.2, 0.3, 0.4$ are

$$1.20526, \quad 1.42273, \quad 1.65511, \quad 1.90570$$

(b) Using $h = 0.025$, the approximate values of the solution at $t = 0.1, 0.2, 0.3, 0.4$ are

$$1.20533, \quad 1.42290, \quad 1.65542, \quad 1.90621$$

(c) Using $h = 0.0125$, the approximate values of the solution at $t = 0.1, 0.2, 0.3, 0.4$ are

$$1.20534, \quad 1.42294, \quad 1.65550, \quad 1.90634$$

(d) Using the Runge-Kutta method with $h = 0.1$, the approximate values of the solution at $t = 0.1, 0.2, 0.3, 0.4$ are

$$1.20535, \quad 1.42295, \quad 1.65553, \quad 1.90638$$

(e) Using the Runge-Kutta method with $h = 0.05$, the approximate values of the solution at $t = 0.1, 0.2, 0.3, 0.4$ are

$$1.20535, \quad 1.42296, \quad 1.65553, \quad 1.90638$$

4.

(a) The improved Euler formula is

$$y_{n+1} = y_n + \frac{h}{2}\left(2t_n + e^{-t_n y_n} + 2(t_n + h) + e^{-(t_n+h)K_n}\right)$$

where $K_n = y_n + h(2t_n + e^{-t_n y_n})$. Since $t_n = t_0 + nh$ and $t_0 = 0$, this formula can be simplified to

$$y_{n+1} = y_n + \frac{h}{2}\left(2nh + e^{-nh y_n} + 2h(n+1) + e^{-h(n+1)K_n}\right)$$

with $y_0 = 1$. With $h = 0.05$, the approximate values of the solution at $t = 0.1, 0.2, 0.3, 0.4$ are

$$1.10483, \quad 1.21882, \quad 1.34146, \quad 1.47263$$

(b) Using $h = 0.025$, the approximate values of the solution at $t = 0.1, 0.2, 0.3, 0.4$ are

$$1.10484, \quad 1.21884, \quad 1.34147, \quad 1.47262$$

(c) Using $h = 0.0125$, the approximate values of the solution at $t = 0.1, 0.2, 0.3, 0.4$ are

$$1.10484, \quad 1.21884, \quad 1.34147, \quad 1.47262$$

(d) Using the Runge-Kutta method with $h = 0.1$, the approximate values of the solution at $t = 0.1, 0.2, 0.3, 0.4$ are

$$1.10484, \quad 1.21884, \quad 1.34147, \quad 1.47262$$

(e) Using the Runge-Kutta method with $h = 0.05$, the approximate values of the solution at $t = 0.1, 0.2, 0.3, 0.4$ are

$$1.10484, \quad 1.21884, \quad 1.34147, \quad 1.47262$$

5.

(a) The improved Euler formula is

$$y_{n+1} = y_n + \frac{h}{2}\left(\frac{y_n^2 + 2t_n y_n}{3 + t_n^2}\right) + \frac{h}{2}\left(\frac{K_n^2 + 2t_{n+1}K_n}{3 + t_{n+1}^2}\right)$$

where $K_n = y_n + h(y_n^2 + 2t_n y_n)/(3+t_n^2)$. Since $t_n = t_0 + nh$ and $t_0 = 0$, this formula can be simplified to

$$y_{n+1} = y_n + \frac{h}{2}\left(\frac{y_n^2 + 2nhy_n}{3+n^2h^2}\right) + frach2 \left(\frac{K_n^2 + 2(n+1)hK_n}{3+(n+1)^2h^2}\right)$$

with $y_0 = 0.5$. With $h = 0.05$, the approximate values of the solution at $t = 0.1, 0.2, 0.3, 0.4$ are

$$0.510164, \quad 0.524126, \quad 0.542083, \quad 0.564251$$

(b) Using $h = 0.025$, the approximate values of the solution at $t = 0.1, 0.2, 0.3, 0.4$ are

$$0.510168, \quad 0.524135, \quad 0.542100, \quad 0.564277$$

(c) Using $h = 0.0125$, the approximate values of the solution at $t = 0.1, 0.2, 0.3, 0.4$ are

$$0.510169, \quad 0.524137, \quad 0.542104, \quad 0.564284$$

(d) Using the Runge-Kutta method with $h = 0.1$, the approximate values of the solution at $t = 0.1, 0.2, 0.3, 0.4$ are

$$0.510170, \quad 0.524138, \quad 0.542105, \quad 0.564286$$

(e) Using the Runge-Kutta method with $h = 0.05$, the approximate values of the solution at $t = 0.1, 0.2, 0.3, 0.4$ are

$$0.520169, \quad 0.524138, \quad 0.542105, \quad 0.564286$$

6.

(a) The improved Euler formula is

$$y_{n+1} = y_n + \frac{h}{2}(t_n^2 - y_n^2)\sin y_n + \frac{h}{2}(t_{n+1}^2 - K_n^2)\sin K_n$$

where $K_n = y_n + h(t_n^2 - y_n^2)\sin y_n$. Since $t_n = t_0 + nh$ and $t_0 = 0$, this formula can be simplified to

$$y_{n+1} = y_n + \frac{h}{2}(n^2h^2 - y_n^2)\sin y_n + \frac{h}{2}\left[(n+1)^2h^2 - K_n^2\right]\sin K_n$$

with $y_0 = -1$. With $h = 0.05$, the approximate values of the solution at $t = 0.1, 0.2, 0.3, 0.4$ are

$$-0.924650, \quad -0.864338, \quad -0.816642, \quad -0.780008$$

(b) Using $h = 0.025$, the approximate values of the solution at $t = 0.1, 0.2, 0.3, 0.4$ are

$$-0.924550, \quad -0.864177, \quad -0.816442, \quad -0.779781$$

(c) Using $h = 0.0125$, the approximate values of the solution at $t = 0.1, 0.2, 0.3, 0.4$ are

$$-0.924525, \quad -0.864138, \quad -0.816393, \quad -0.779725$$

(d) Using the Runge-Kutta method with $h = 0.1$, the approximate values of the solution at $t = 0.1, 0.2, 0.3, 0.4$ are

$$-0.924517, \quad -0.864125, \quad -0.816377, \quad -0.779706$$

(e) Using the Runge-Kutta method with $h = 0.05$, the approximate values of the solution at $t = 0.1, 0.2, 0.3, 0.4$ are

$$-0.924517, \quad -0.864125, \quad -0.816377, \quad -0.779706$$

7.

(a) The improved Euler formula is

$$y_{n+1} = y_n + \frac{h}{2}(0.5 - t_n + 2y_n) + \frac{h}{2}(0.5 - t_{n+1} + 2(y_n + h(0.5 - t_n + 2y_n))).$$

Since $t_n = t_0 + nh$ and $t_0 = 0$, this formula can be simplified to

$$y_{n+1} = y_n + h(2y_n + 0.5) + h^2(2y_n - n) - nh^3$$

with $y_0 = 1$. With $h = 0.025$, the approximate values of the solution at $t = 0.5, 1.0, 1.5, 2.0$ are

$$2.96719, \quad 7.88313, \quad 20.8114, \quad 55.5106$$

(b) Using $h = 0.0125$, the approximate values of the solution at $t = 0.5, 1.0, 1.5, 2.0$ are

$$2.96800, \quad 7.88755, \quad 20.8294, \quad 55.5758$$

(c) Using the Runge-Kutta method with $h = 0.1$, the approximate values of the solution at $t = 0.5, 1.0, 1.5, 2.0$ are

$$2.96825, \quad 7.88889, \quad 20.8349, \quad 55.5957$$

(d) Using the Runge-Kutta method with $h = 0.05$, the approximate values of the solution at $t = 0.5, 1.0, 1.5, 2.0$ are

$$2.96828, \quad 7.88904, \quad 20.8355, \quad 55.5980$$

8.

(a) The improved Euler formula is

$$y_{n+1} = y_n + \frac{h}{2}(5t_n - 3\sqrt{y_n}) + \frac{h}{2}(5t_{n+1} - 3\sqrt{K_n})$$

where $K_n = y_n + h(5t_n - 3\sqrt{y_n})$. Since $t_n = t_0 + nh$ and $t_0 = 0$, this formula can be simplified to

$$y_{n+1} = y_n + \frac{h}{2}(5nh - 3\sqrt{y_n}) + \frac{h}{2}\left[5(n+1)h - 3\sqrt{K_n}\right]$$

with $y_0 = 2$. With $h = 0.025$, the approximate values of the solution at $t = 0.5, 1.0, 1.5, 2.0$ are

$$0.926139, \quad 1.28558, \quad 2.40898, \quad 4.10386$$

(b) Using $h = 0.0125$, the approximate values of the solution at $t = 0.5, 1.0, 1.5, 2.0$ are

$$0.925815, \quad 1.28525, \quad 2.40869, \quad 4.10359$$

(c) Using the Runge-Kutta method with $h = 0.1$, the approximate values of the solution at $t = 0.5, 1.0, 1.5, 2.0$ are

$$0.925725, \quad 1.28516, \quad 2.40860, \quad 4.10350$$

(d) Using the Runge-Kutta method with $h = 0.05$, the approximate values of the solution at $t = 0.5, 1.0, 1.5, 2.0$ are

$$0.925711, \quad 1.28515, \quad 2.40860, \quad 4.10350$$

9.

(a) The improved Euler formula is

$$y_{n+1} = y_n + \frac{h}{2}\sqrt{t_n + y_n} + \frac{h}{2}\sqrt{t_{n+1} + K_n}$$

where $K_n = y_n + h\sqrt{t_n + y_n}$. Since $t_n = t_0 + nh$ and $t_0 = 0$, this formula can be simplified to

$$y_{n+1} = y_n + \frac{h}{2}\sqrt{nh + y_n} + \frac{h}{2}\sqrt{(n+1)h + K_n}$$

with $y_0 = 3$. With $h = 0.025$, the approximate values of the solution at $t = 0.5, 1.0, 1.5, 2.0$ are

$$3.96217, \quad 5.10887, \quad 6.43134, \quad 7.92332$$

(b) Using $h = 0.0125$, the approximate values of the solution at $t = 0.5, 1.0, 1.5, 2.0$ are

$$3.96218, \quad 5.10889, \quad 6.43138, \quad 7.92337$$

(c) Using the Runge-Kutta method with $h = 0.1$, the approximate values of the solution at $t = 0.5, 1.0, 1.5, 2.0$ are

$$3.96219, \quad 5.10890, \quad 6.43139, \quad 7.92338$$

(d) Using the Runge-Kutta method with $h = 0.05$, the approximate values of the solution at $t = 0.5, 1.0, 1.5, 2.0$ are

$$3.96219, \quad 5.10890, \quad 6.43139, \quad 7.92338$$

10.

(a) The improved Euler formula is

$$y_{n+1} = y_n + \frac{h}{2}\left(2t_n + e^{-t_n y_n} + 2(t_n + h) + e^{-(t_n + h)K_n}\right)$$

where $K_n = y_n + h(2t_n + e^{-t_n y_n})$. Since $t_n = t_0 + nh$ and $t_0 = 0$, this formula can be simplified to

$$y_{n+1} = y_n + \frac{h}{2}\left(2nh + e^{-nh y_n} + 2h(n + 1) + e^{-h(n+1)K_n}\right)$$

with $y_0 = 1$. With $h = 0.025$, the approximate values of the solution at $t = 0.5, 1.0, 1.5, 2.0$ are

$$1.61263, \quad 2.48097, \quad 3.74556, \quad 5.49595$$

(b) Using $h = 0.0125$, the approximate values of the solution at $t = 0.5, 1.0, 1.5, 2.0$ are

$$1.61263, \quad 2.48092, \quad 3.74550, \quad 5.49589$$

(c) Using the Runge-Kutta method with $h = 0.1$, the approximate values of the solution at $t = 0.5, 1.0, 1.5, 2.0$ are

$$1.61262, \quad 2.48091, \quad 3.74548, \quad 5.49587$$

(d) Using the Runge-Kutta method with $h = 0.05$, the approximate values of the solution at $t = 0.5, 1.0, 1.5, 2.0$ are

$$1.61262, \quad 2.48091, \quad 3.74548, \quad 5.49587$$

11.

(a) The improved Euler formula is

$$y_{n+1} = y_n + \frac{h}{2}\left(\frac{4 - t_n y_n}{1 + y_n^2}\right) + \frac{h}{2}\left(\frac{4 - t_{n+1}K_n}{1 + K_n^2}\right)$$

where $K_n = y_n + h(4 - t_ny_n)/(1 + y_n^2)$. Since $t_n = t_0 + nh$ and $t_0 = 0$, this formula can be simplified to

$$y_{n+1} = y_n + \frac{h}{2}\left(\frac{4 - nhy_n}{1 + y_n^2}\right) + \frac{h}{2}\left(\frac{4 - h(n+1)K_n}{1 + K_n^2}\right)$$

with $y_0 = -2$. With $h = 0.025$, the approximate values of the solution at $t = 0.5, 1.0, 1.5, 2.0$ are

$$-1.44768, \quad -0.144478, \quad 1.06004, \quad 1.40960$$

(b) Using $h = 0.0125$, the approximate values of the solution at $t = 0.5, 1.0, 1.5, 2.0$ are

$$-1.44765, \quad -0.143690, \quad 1.06072, \quad 1.40999$$

(c) Using the Runge-Kutta method with $h = 0.1$, the approximate values of the solution at $t = 0.5, 1.0, 1.5, 2.0$ are

$$-1.44764, \quad -0.143543, \quad 1.06089, \quad 1.41008$$

(d) Using the Runge-Kutta method with $h = 0.05$, the approximate values of the solution at $t = 0.5, 1.0, 1.5, 2.0$ are

$$-1.44764, \quad -0.143427, \quad 1.06095, \quad 1.41011$$

12.

(a) The improved Euler formula is

$$y_{n+1} = y_n + \frac{h}{2}\left(\frac{y_n^2 + 2t_ny_n}{3 + t_n^2}\right) + \frac{h}{2}\left(\frac{K_n^2 + 2t_{n+1}K_n}{3 + t_{n+1}^2}\right)$$

where $K_n = y_n + h(y_n^2 + 2t_ny_n)/(3 + t_n^2)$. Since $t_n = t_0 + nh$ and $t_0 = 0$, this formula can be simplified to

$$y_{n+1} = y_n + \frac{h}{2}\left(\frac{y_n^2 + 2nhy_n}{3 + n^2h^2}\right) + \frac{h}{2}\left(\frac{K_n^2 + 2(n+1)hK_n}{3 + (n+1)^2h^2}\right)$$

with $y_0 = 0.5$. With $h = 0.025$, the approximate values of the solution at $t = 0.5, 1.0, 1.5, 2.0$ are
$$0.590897, \quad 0.799950, \quad 1.16653, \quad 1.74969$$

(b) Using $h = 0.0125$, the approximate values of the solution at $t = 0.5, 1.0, 1.5, 2.0$ are

$$0.590906, \quad 0.799988, \quad 1.16663, \quad 1.74992$$

(c) Using the Runge-Kutta method with $h = 0.1$, the approximate values of the solution at $t = 0.5, 1.0, 1.5, 2.0$ are

$$0.590909, \quad 0.800000, \quad 1.166667, \quad 1.75000$$

(d) Using the Runge-Kutta method with $h = 0.05$, the approximate values of the solution at $t = 0.5, 1.0, 1.5, 2.0$ are

$$0.590909, \quad 0.800000, \quad 1.166667, \quad 1.75000$$

13. The improved Euler method is

$$y_{n+1} = y_n + \frac{h}{2}(1 - t_n + 4y_n) + \frac{h}{2}[1 - (t_n + h) + 4K_n]$$

where $K_n = y_n + h(1 - t_n + 4y_n)$. Since $t_n = nh + t_0$ and $t_0 = 0$, this equation can be simplified to

$$y_{n+1} = y_n + \frac{h}{2}(1 - nh + 4y_n) + \frac{h}{2}[1 - h(n+1) + 4K_n]$$

with $y_0 = 1$.

14. The differential equation is linear. Its exact solution is given by $y(t) = \frac{19}{16}e^{4t} + \frac{1}{4}t - \frac{3}{16}$. The improved Euler method is

$$y_{n+1} = y_n + \frac{h}{2}(1 - t_n + 4y_n) + \frac{h}{2}[1 - (t_n + h) + 4K_n]$$

where $K_n = y_n + h(1 - t_n + 4y_n)$.

15.

(a)

(b) The following are the approximate values of the solution at $t = 0.8, 0.9, 0.95$ using the Runge-Kutta method with $h = 0.01$:

$$5.848616, 14.304785, 50.436365.$$

16.

(a)

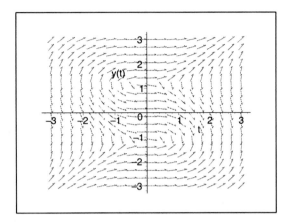

(b) For the integral curve staring at $(0,0)$, the slope becomes infinite near $t_M \approx 1.5$. We note that the exact solution is given implicitly as $y^3 - 4y = t^3$.

(c) Based on the direction field, the solution of the initial value problem with initial condition $y(0) = 0$ should decrease monotonically to the limiting value $y = -2/\sqrt{3}$. Using the Runge-Kutta method, we calculate the approximate value of t_M by looking at the approximate time in the iteration process that the calculated values begin to increase. For $h = 0.1, 0.05, 0.025, 0.01$, the respective times are given by $t_M \approx 1.9, 1.65, 1.55, 1.455$.

(d) These values are not associated with the integral curve starting at $(0,0)$. These values are approximations to nearby integral curves.

(e) Suppose now that $y(0) = 1$. The exact solution is given by $y^3 - 4y = t^3 - 3$. For the integral curve starting at $(0,1)$, the slope becomes infinite near $t_M \approx 2.0$. Using the Runge-Kutta method, we calculate the following approximate values for t_M. For $h = 0.1, 0.05, 0.025, 0.01$, the respective times are given by $t_M \approx 1.85, 1.85, 1.86, 1.835$.

17.

(a) First we notice that

$$\phi'(t_n)h - \frac{f(t_n, y_n)h}{2} = \phi'(t_n)h - \frac{y'_n h}{2}$$
$$= \phi'(t_n)h - \frac{\phi'(t_n)h}{2} = \frac{\phi'(t_n)h}{2}.$$

Using this fact, it follows that $\phi(t_{n+1}) - y_{n+1}$ satisfies the given equation.

(b) First, using the Taylor approximation, we see that

$$f[t_n + h, y_n + hf(t_n, y_n)] - f(t_n, y_n) = f_t(t_n, y_n)h + f_y(t_n, y_n)hf(t_n, y_n)$$
$$+ \frac{1}{2!}(h^2 f_{tt} + 2hk f_{ty} + k^2 f_{yy})\Big|_{x=\xi, y=\eta}$$

103

Next, we see that

$$\phi''(t_n)h = f_t(t_n, \phi(t_n))h + f_y(t_n, \phi(t_n))\phi'(t_n)h$$
$$= f_t(t_n, y_n)h + f_y(t_n, y_n)hf(t_n, y_n).$$

Therefore, we conclude that

$$\frac{1}{2!}\left[\phi''(t_n)h - \{f[t_n + h, y_n + hf(t_n, y_n)] - f(t_n, y_n)\}\right]h$$
$$= h\left(\frac{1}{2!}(h^2 f_{tt} + 2hk f_{ty} + k^2 f_{yy})\Big|_{x=\xi, y=\eta}\right)$$

is proportional to h^3.

(c) If f is linear in t and y, then $f_{tt} = f_{ty} = f_{yy} = 0$. Therefore, the terms from part (b) above are all zero.

18. The exact solution is given by $\phi(t) = -3/16 + t/4 + (19/16)e^{4t}$. Then, by the results of Problem 17(c), the error will be given by

$$e_{n+1} = \frac{\phi'''(\bar{t}_n)h^3}{3!}.$$

Here $\phi' = 1/4 + (19/4)e^{4t}$, $\phi'' = 19e^{4t}$, $\phi''' = 76e^{4t}$. Therefore,

$$e_{n+1} = \frac{38e^{4\bar{t}_n}h^3}{3}.$$

Therefore, on the interval $0 \leq t \leq 2$, we conclude that

$$|e_{n+1}| \leq \frac{38e^8 h^3}{3} = 37,758.8h^3.$$

Then for $h = 0.05$, we conclude that

$$|e_1| \leq \frac{38e^{4(0.05)}(0.05)^3}{3} = 0.00193389.$$

19. The exact solution of the initial value problem is $\phi(t) = \frac{1}{2} + \frac{1}{2}e^{2t}$. Based on the result from problem 17(c), the local truncation error for a linear differential equation is

$$e_{n+1} = \frac{1}{6}\phi'''(\bar{t}_n)h^3.$$

Here $\phi' = e^{2t}$, $\phi'' = 2e^{2t}$, $\phi''' = 4e^{2t}$. Therefore,

$$e_{n+1} = \frac{2}{3}e^{2\bar{t}_n}h^3.$$

Further, on the interval $0 \leq t \leq 1$,

$$|e_{n+1}| \leq \frac{2}{3}e^2 h^3 = 4.92604 h^3.$$

Letting $h = 0.1$,

$$|e_1| \leq \frac{2}{3}e^{2(0.1)}(0.1)^3 = 0.000814269.$$

Using the improved Euler method, with $h = 0.1$, we have $y_1 \approx 1.11000$. The exact value of the solution is $\phi(0.1) = 1.1107014$.

20. The exact solution of the initial value problem is $\phi(t) = \frac{1}{2}t + e^{2t}$. Based on the result from problem 17(c), the local truncation error for a linear differential equation is

$$e_{n+1} = \frac{1}{6}\phi'''(\bar{t}_n)h^3.$$

Here $\phi' = \frac{1}{2} + 2e^{2t}$, $\phi'' = 4e^{2t}$, $\phi''' = 8e^{2t}$. Therefore,

$$e_{n+1} = \frac{4}{3}e^{2\bar{t}_n}h^3.$$

Further, on the interval $0 \leq t \leq 1$,

$$|e_{n+1}| \leq \frac{4}{3}e^2 h^3 = 9.85207 h^3.$$

Letting $h = 0.1$,

$$|e_1| \leq \frac{4}{3}e^{2(0.1)}(0.1)^3 = 0.00162854.$$

21. The Euler formula is

$$y_{n+1} = y_n + h(0.5 - t_n + 2y_n).$$

Since $t_0 = 0$, $y_0 = 1$ and $h = 0.1$, we have

$$y_1 = 1 + 0.1(0.5 - 0 + 2) = 1.25.$$

For $t_0 = 0$, the improved Euler formula is

$$y_{n+1} = y_n + h(2y_n + 0.5) + h^2(2y_n - n) - nh^3.$$

Therefore, for $y_0 = 1$ and $h = 0.1$,

$$y_1 = 1 + 0.1(2 + 0.5) + (0.1)^2(2 - 0) - 0(0.1)^3 = 1.27.$$

Therefore, the estimated error of the Euler method is $e_{n+1}^{\text{ext}} = 1.27 - 1.25 = .02$. If we want the error of the Euler method to be less than 0.0025, we need to multiply the original step size of 0.1 by the factor $\sqrt{0.0025/0.02} \approx 0.35$. Therefore, the required step size is estimated to be $h \approx (0.1)(0.35) = 0.035$.

22. For $t_0 = 0$, the Euler formula is

$$y_{n+1} = y_n + h(5nh - 3\sqrt{y_n}).$$

Therefore, for $y_0 = 2$ and $h = 0.1$, we have

$$y_1 = 2 + 0.1(0 - 3\sqrt{2}) = 1.575736.$$

For $t_0 = 0$, the improved Euler formula is

$$y_{n+1} = y_n + \frac{h}{2}(5nh - 3\sqrt{y_n}) + \frac{h}{2}\left[5(n+1)h - 3\sqrt{K_n}\right]$$

where $K_n = y_n + h(5t_n - 3\sqrt{y_n})$. Therefore, for $y_0 = 2$ and $h = 0.1$,

$$y_1 = 2 + 0.05(0 - 3\sqrt{2}) + 0.05[5(0.1) - 3\sqrt{2 + 0.1(0 - 3\sqrt{2})}] = 1.624575.$$

Therefore, the estimated error of the Euler method is $e_{n+1}^{\text{ext}} = 1.624575 - 1.575736 = 0.048839$. If we want the error of the Euler method to be less than 0.0025, we need to multiply the original step size of 0.1 by the factor $\sqrt{0.0025/0.048839} \approx 0.226$. Therefore, the required step size is estimated to be $h \approx (0.1)(0.226) = 0.0226$.

23. For $t_0 = 0$, the Euler formula is

$$y_{n+1} = y_n + h\sqrt{nh + y_n}.$$

Therefore, for $y_0 = 3$ and $h = 0.1$, we have

$$y_1 = 3 + 0.1\sqrt{0 + 3} = 3.173205.$$

For $t_0 = 0$, the improved Euler formula is

$$y_{n+1} = y_n + \frac{h}{2}\sqrt{nh + y_n} + \frac{h}{2}\sqrt{(n+1)h + K_n}$$

where $K_n = y_n + h\sqrt{t_n + y_n}$. Therefore, for $y_0 = 3$ and $h = 0.1$,

$$y_1 = 3 + 0.05\sqrt{0 + 3} + 0.05\sqrt{0.1 + (3 + 0.1\sqrt{0 + 3})} = 3.177063.$$

Therefore, the estimated error of the Euler method is $e_{n+1}^{\text{ext}} = 3.177063 - 3.173205 = 0.003858$. If we want the error of the Euler method to be less than 0.0025, we need to multiply the original step size of 0.1 by the factor $\sqrt{0.0025/0.003858} \approx 0.805$. Therefore, the required step size is estimated to be $h \approx (0.1)(0.226) = 0.0805$.

24. For $t_0 = 0$, the Euler formula is

$$y_{n+1} = y_n + h(y_n^2 + 2nhy_n)/(3 + n^2h^2).$$

Therefore, for $y_0 = 0.5$ and $h = 0.1$, we have

$$y_1 = 0.5 + 0.1(0.5^2 + 0)/(3 + 0) = 0.508334.$$

106

For $t_0 = 0$, the improved Euler formula is

$$y_{n+1} = y_n + \frac{h}{2}\left(\frac{y_n^2 + 2nhy_n}{3 + n^2 h^2}\right) + \frac{h}{2}\left(\frac{K_n^2 + 2(n+1)hK_n}{3 + (n+1)^2 h^2}\right)$$

where $K_n = y_n + h(y_n^2 + 2t_n y_n)/(3 + t_n^2)$. Therefore, for $y_0 = 0.5$ and $h = 0.1$,

$$y_1 = 0.5 + 0.05\frac{0.5^2 + 0}{3 + 0} + 0.05\frac{(0.5 + 0.1(0.5^2/3))^2 + 2(0.1)(0.5 + 0.1(0.5^2/3))}{3 + 0.1^2} = 0.510148.$$

Therefore, the estimated error of the Euler method is $e_{n+1}^{\text{ext}} = 0.510148 - 0.598334 = 0.0018$. The local truncation error is less than the given tolerance. Therefore, if we allow an error tolerance of 0.0025, we can multiply the original step size of 0.1 by the factor $\sqrt{0.0025/0.0018} \approx$ 1.1785. Therefore, the required step size is estimated to be $h \approx (0.1)(1.1785) = 0.11785$.

Chapter 3
Section 3.1

1.

(a)

$$A = \begin{pmatrix} 2 & 3 \\ -3 & 1 \end{pmatrix} \implies A^{-1} = \frac{1}{11}\begin{pmatrix} 1 & -3 \\ 3 & 2 \end{pmatrix}.$$

Therefore,

$$\mathbf{x} = A^{-1}\mathbf{b} = \frac{1}{11}\begin{pmatrix} 1 & -3 \\ 3 & 2 \end{pmatrix}\begin{pmatrix} 7 \\ -5 \end{pmatrix} = \begin{pmatrix} 2 \\ 1 \end{pmatrix}.$$

That is, $x_1 = 2, x_2 = 1$.

(b) The lines are intersecting, as shown below.

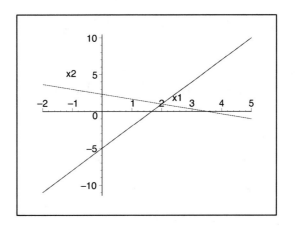

2.

(a)

$$A = \begin{pmatrix} 1 & -2 \\ 2 & 3 \end{pmatrix} \implies A^{-1} = \frac{1}{7}\begin{pmatrix} 3 & 2 \\ -2 & 1 \end{pmatrix}.$$

Therefore,

$$\mathbf{x} = A^{-1}\mathbf{b} = \frac{1}{7}\begin{pmatrix} 3 & 2 \\ -2 & 1 \end{pmatrix}\begin{pmatrix} 10 \\ 6 \end{pmatrix} = \begin{pmatrix} 6 \\ -2 \end{pmatrix}.$$

That is, $x_1 = 6, x_2 = -2$.

(b) Therefore, the lines are intersecting, as shown below.

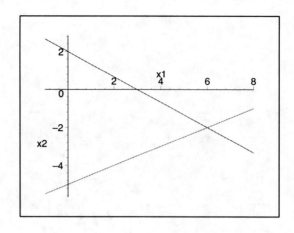

3.

(a)

$$A = \begin{pmatrix} 1 & 3 \\ 2 & -1 \end{pmatrix} \implies A^{-1} = \frac{1}{7} \begin{pmatrix} 1 & 3 \\ 2 & -1 \end{pmatrix}.$$

Therefore,

$$\mathbf{x} = A^{-1}\mathbf{b} = \frac{1}{7} \begin{pmatrix} 1 & 3 \\ 2 & -1 \end{pmatrix} \begin{pmatrix} 0 \\ 0 \end{pmatrix} = \begin{pmatrix} 0 \\ 0 \end{pmatrix}.$$

That is, $x_1 = 0$, $x_2 = 0$.

(b) Therefore, the lines are intersecting, as shown below.

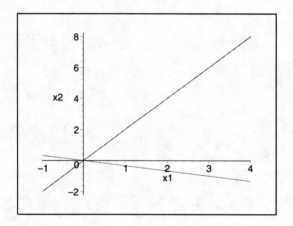

4.

(a)

$$A = \begin{pmatrix} -1 & 2 \\ 2 & -4 \end{pmatrix}$$

$\det A = 0$. Therefore, A is singular which implies either there is no solution or an infinite number of solutions. Multiplying the first equation by 2, we have $-2x_1 + 4x_2 = 8$. Our second equation is $2x_1 - 4x_2 = -6$. Adding the two equations, we have $0 = 2$ which cannot occur. Therefore, we have no solution.

2

(b) Therefore, the lines are parallel, as shown below.

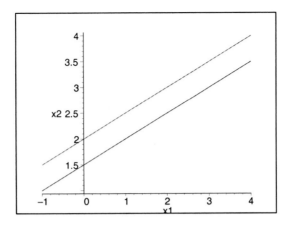

5.

(a)

$$A = \begin{pmatrix} 2 & -3 \\ 1 & 2 \end{pmatrix} \implies A^{-1} = \frac{1}{7}\begin{pmatrix} 2 & 3 \\ -1 & 2 \end{pmatrix}.$$

Therefore,

$$\mathbf{x} = A^{-1}\mathbf{b} = \frac{1}{7}\begin{pmatrix} 2 & 3 \\ -1 & 2 \end{pmatrix}\begin{pmatrix} 4 \\ -5 \end{pmatrix} = \begin{pmatrix} -1 \\ -2 \end{pmatrix}.$$

That is, $x_1 = -1$, $x_2 = -2$.

(b) Therefore, the lines are intersecting, as shown below.

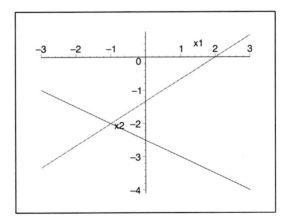

6.

(a)

$$A = \begin{pmatrix} 3 & -2 \\ -6 & 4 \end{pmatrix}$$

$\det(A) = 0$. Therefore, there are no solutions or an infinite number of solutions. Multiplying the first equation by -2, we see the equations represent the same line. Therefore, there are an infinite number of solutions. In particular, any pair (x_1, x_2) such that $x_2 = 3x_1/2$ will satisfy the system of equations.

(b) Therefore, the lines are coincident, as shown below.

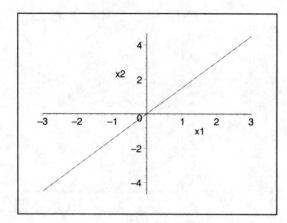

7.

(a)

$$A = \begin{pmatrix} 2 & -3 \\ -4 & 6 \end{pmatrix}$$

$\det(A) = 0$, therefore, there are no solutions or an infinite number of solutions. Multiplying the first equation by -2, we see the two equations are equations for the same line. Therefore, there are an infinite number of solutions. In particular, any numbers (x_1, x_2) such that $x_2 = (2x_1 - 6)/3$.

(b) Therefore, the lines are coincident, as shown below.

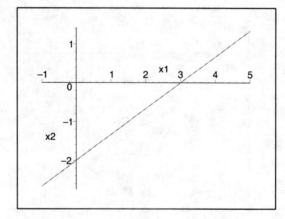

8.

(a)

$$A = \begin{pmatrix} 4 & 1 \\ 4 & -3 \end{pmatrix} \implies A^{-1} = \frac{1}{16} \begin{pmatrix} 3 & 1 \\ 4 & -4 \end{pmatrix}.$$

Therefore,

$$\mathbf{x} = A^{-1}\mathbf{b} = \frac{1}{16} \begin{pmatrix} 3 & 1 \\ 4 & -4 \end{pmatrix} \begin{pmatrix} 0 \\ -12 \end{pmatrix} = \begin{pmatrix} -3/4 \\ 3 \end{pmatrix}.$$

That is, $x_1 = -3/4$, $x_2 = 3$.

(b) Therefore, the lines are intersecting, as shown below.

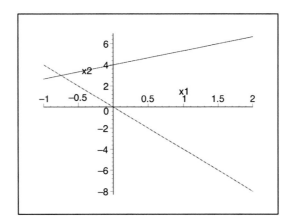

9.

(a)
$$A = \begin{pmatrix} 1 & 4 \\ 4 & 1 \end{pmatrix} \implies A^{-1} = -\frac{1}{15} \begin{pmatrix} 1 & -4 \\ -4 & 1 \end{pmatrix}.$$

Therefore,
$$\mathbf{x} = A^{-1}\mathbf{b} = -\frac{1}{15} \begin{pmatrix} 1 & -4 \\ -4 & 1 \end{pmatrix} \begin{pmatrix} 10 \\ 10 \end{pmatrix} = \begin{pmatrix} 2 \\ 2 \end{pmatrix}.$$

That is, $x_1 = 2$, $x_2 = 2$.

(b) Therefore, the lines are intersecting, as shown below.

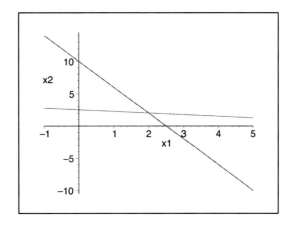

10.

(a)
$$A = \begin{pmatrix} 1 & 1 \\ -1 & 2 \end{pmatrix} \implies A^{-1} = \frac{1}{3} \begin{pmatrix} 2 & -1 \\ 1 & 1 \end{pmatrix}.$$

Therefore,

$$\mathbf{x} = A^{-1}\mathbf{b} = \frac{1}{3}\begin{pmatrix} 2 & -1 \\ 1 & 1 \end{pmatrix}\begin{pmatrix} -1 \\ 4 \end{pmatrix} = \begin{pmatrix} -2 \\ 1 \end{pmatrix}.$$

That is, $x_1 = -2$, $x_2 = 1$.

(b) Therefore, the lines are intersecting, as shown below.

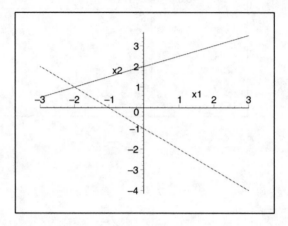

11.

(a)

$$A = \begin{pmatrix} 4 & -3 \\ -2 & 5 \end{pmatrix} \implies A^{-1} = \frac{1}{14}\begin{pmatrix} 5 & 3 \\ 2 & 4 \end{pmatrix}.$$

Therefore,

$$\mathbf{x} = A^{-1}\mathbf{b} = \frac{1}{14}\begin{pmatrix} 5 & 3 \\ 2 & 4 \end{pmatrix}\begin{pmatrix} 0 \\ 0 \end{pmatrix} = \begin{pmatrix} 0 \\ 0 \end{pmatrix}.$$

That is, $x_1 = 0$, $x_2 = 0$.

(b) Therefore, the lines are intersecting, as shown below.

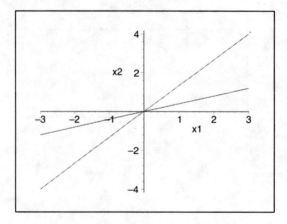

12.

(a)

$$A = \begin{pmatrix} 2 & 5 \\ 4 & 10 \end{pmatrix}$$

$\det(A) = 0$. Therefore, there are no solutions or an infinite many. Multiplying the first equation by 2, we see the two equations are equations of the same line. Therefore, there are an infinite number of solutions. In particular, any pair of the form (x_1, x_2) where $x_2 = -2x_1/5$ will satisfy the system of equations.

(b) Therefore, the lines are coincident, as shown below.

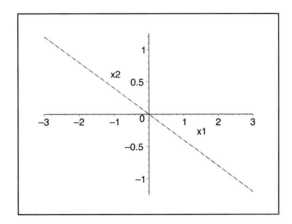

13. $\det(A - \lambda I) = \lambda^2 - \lambda - 2 = 0 \implies \lambda = 2, -1$. First, $\lambda_1 = 2$ implies

$$A - \lambda_1 I = \begin{pmatrix} 1 & -2 \\ 2 & -4 \end{pmatrix}.$$

Therefore,

$$\mathbf{x_1} = \begin{pmatrix} 2 \\ 1 \end{pmatrix}$$

is an eigenvector for λ_1.

Second, $\lambda_2 = -1$ implies

$$A - \lambda_2 I = \begin{pmatrix} 4 & -2 \\ 2 & -1 \end{pmatrix}.$$

Therefore,

$$\mathbf{x_2} = \begin{pmatrix} 1 \\ 2 \end{pmatrix}$$

is an eigenvector for λ_2.

14. $\det(A - \lambda I) = \lambda^2 - 2\lambda + 5 = 0 \implies \lambda = 1 + 2i, 1 - 2i$. First, $\lambda_1 = 1 + 2i$ implies

$$A - \lambda_1 I = \begin{pmatrix} 2 - 2i & -2 \\ 4 & -2 - 2i \end{pmatrix}.$$

Therefore,

$$\mathbf{x_1} = \begin{pmatrix} 1 \\ 1-i \end{pmatrix}$$

is an eigenvector for λ_1.

Second, $\lambda_2 = 1 - 2i$ is the complex conjugate of λ_1 implies that $\mathbf{x_2} \equiv \overline{\mathbf{x_1}}$ is an eigenvector for λ_2. In particular,

$$\mathbf{x_2} = \begin{pmatrix} 1 \\ 1+i \end{pmatrix}$$

is an eigenvector for λ_2.

15. $\det(A - \lambda I) = \lambda^2 - 2\lambda + 1 = 0 \implies \lambda = 1$. Now, $\lambda = 1$ implies

$$A - \lambda_1 I = \begin{pmatrix} 2 & -4 \\ 1 & -2 \end{pmatrix}.$$

Therefore,

$$\mathbf{x_1} = \begin{pmatrix} 2 \\ 1 \end{pmatrix}$$

is an eigenvector for λ.

16. $\det(A - \lambda I) = \lambda^2 + 3\lambda + 2 = 0 \implies \lambda = -1, -2$. First, $\lambda_1 = -1$ implies

$$A - \lambda_1 I = \begin{pmatrix} 2 & -2 \\ 3 & -3 \end{pmatrix}.$$

Therefore,

$$\mathbf{x_1} = \begin{pmatrix} 1 \\ 1 \end{pmatrix}$$

is an eigenvector for λ_1.

Second, $\lambda_2 = -2$ implies

$$A - \lambda_2 I = \begin{pmatrix} 3 & -2 \\ 3 & -2 \end{pmatrix}.$$

Therefore,

$$\mathbf{x_2} = \begin{pmatrix} 2 \\ 3 \end{pmatrix}$$

is an eigenvector for λ_2.

17. $\det(A - \lambda I) = \lambda^2 + 2\lambda + 5 = 0 \implies \lambda = -1 + 2i, -1 - 2i$. First, $\lambda_1 = -1 + 2i$ implies

$$A - \lambda_1 I = \begin{pmatrix} -2i & -4 \\ 1 & -2i \end{pmatrix}.$$

Therefore,

$$\mathbf{x_1} = \begin{pmatrix} 2i \\ 1 \end{pmatrix}$$

is an eigenvector for λ_1.

Second, since $\lambda_2 = -1 - 2i$ is the complex conjugate of λ_1,

$$\mathbf{x_2} \equiv \overline{\mathbf{x_1}} = \begin{pmatrix} -2i \\ 1 \end{pmatrix}$$

is an eigenvector for λ_2.

18. $\det(A - \lambda I) = \lambda^2 - \lambda + 1/4 = 0 \implies \lambda = 1/2$. Now, $\lambda_1 = 1/2$ implies

$$A - \lambda_1 I = \begin{pmatrix} 3/4 & 3/4 \\ -3/4 & -3/4 \end{pmatrix}.$$

Therefore,

$$\mathbf{x_1} = \begin{pmatrix} 1 \\ -1 \end{pmatrix}$$

is an eigenvector for λ.

19. $\det(A - \lambda I) = \lambda^2 + 2\lambda + 1 = 0 \implies \lambda = -1$. Now, $\lambda = -1$ implies

$$A - \lambda_1 I = \begin{pmatrix} -1/2 & 1 \\ -1/4 & 1/2 \end{pmatrix}.$$

Therefore,

$$\mathbf{x_1} = \begin{pmatrix} 1 \\ 1/2 \end{pmatrix}$$

is an eigenvector for λ.

20. $\det(A - \lambda I) = \lambda^2 - 1 = 0 \implies \lambda = 1, -1$. First, $\lambda_1 = 1$ implies

$$A - \lambda_1 I = \begin{pmatrix} 1 & -1 \\ 3 & -3 \end{pmatrix}.$$

Therefore,

$$\mathbf{x_1} = \begin{pmatrix} 1 \\ 1 \end{pmatrix}$$

is an eigenvector for λ_1.

Second, $\lambda_2 = -1$ implies

$$A - \lambda_2 I = \begin{pmatrix} 3 & -1 \\ 3 & -1 \end{pmatrix}.$$

Therefore,

$$\mathbf{x_2} = \begin{pmatrix} 1 \\ 3 \end{pmatrix}$$

is an eigenvector for λ_2.

21. $\det(A - \lambda I) = \lambda^2 + 1 = 0 \implies \lambda = i, -i$. First, $\lambda_1 = i$ implies

$$A - \lambda_1 I = \begin{pmatrix} 2 - i & -5 \\ 1 & -2 - i \end{pmatrix}.$$

9

Therefore,
$$\mathbf{x_1} = \begin{pmatrix} 2 + i \\ 1 \end{pmatrix}$$

is an eigenvector for λ_1.

Second, since λ_2 is the complex conjugate of λ_1,

$$\mathbf{x_2} \equiv \overline{\mathbf{x_1}} = \begin{pmatrix} 2 - i \\ 1 \end{pmatrix}$$

is an eigenvector for λ_2.

22. $\det(A - \lambda I) = \lambda^2 - 7\lambda = 0 \implies \lambda = 0, 7$. First, $\lambda_1 = 0$ implies

$$A - \lambda_1 I = \begin{pmatrix} 6 & 3 \\ 2 & 1 \end{pmatrix}.$$

Therefore,
$$\mathbf{x_1} = \begin{pmatrix} 1 \\ -2 \end{pmatrix}$$

is an eigenvector for λ_1.

Second, $\lambda_2 = 7$ implies

$$A - \lambda_2 I = \begin{pmatrix} -1 & 3 \\ 2 & -6 \end{pmatrix}.$$

Therefore,
$$\mathbf{x_2} = \begin{pmatrix} 3 \\ 1 \end{pmatrix}$$

is an eigenvector for λ_2.

23. $\det(A - \lambda I) = \lambda^2 + \lambda - 6 = 0 \implies \lambda = 2, -3$. First, $\lambda_1 = 2$ implies

$$A - \lambda_1 I = \begin{pmatrix} -1 & 1 \\ 4 & -4 \end{pmatrix}.$$

Therefore,
$$\mathbf{x_1} = \begin{pmatrix} 1 \\ 1 \end{pmatrix}$$

is an eigenvector for λ_1.

Second, $\lambda_2 = -3$ implies

$$A - \lambda_2 I = \begin{pmatrix} 4 & 1 \\ 4 & 1 \end{pmatrix}.$$

Therefore,
$$\mathbf{x_2} = \begin{pmatrix} 1 \\ -4 \end{pmatrix}$$

is an eigenvector for λ_2.

24. $\det(A - \lambda I) = \lambda^2 - \lambda + 5/2 = 0 \implies \lambda = (1 \pm 3i)/2$. First, $\lambda_1 = (1 + 3i)/2$ implies

$$A - \lambda_1 I = \begin{pmatrix} \frac{3}{2} - \frac{3}{2}i & -\frac{5}{2} \\ \frac{9}{5} & -\frac{3}{2} - \frac{3}{2}i \end{pmatrix}.$$

Therefore,

$$\mathbf{x_1} = \begin{pmatrix} 1 \\ \frac{3}{5} - \frac{3}{5}i \end{pmatrix}$$

is an eigenvector for λ_1.

Second, since λ_2 is the complex conjugate of λ_1, then

$$\mathbf{x_2} \equiv \overline{\mathbf{x_1}} = \begin{pmatrix} 1 \\ \frac{3}{5} + \frac{3}{5}i \end{pmatrix}$$

is an eigenvector for λ_2.

25. $\det(A - \lambda I) = \lambda^2 + \lambda + 1/42 = 0 \implies \lambda = -1/2$. Now, $\lambda = 2$ implies

$$A - \lambda_1 I = \begin{pmatrix} -\frac{5}{2} & \frac{5}{2} \\ -\frac{5}{2} & \frac{5}{2} \end{pmatrix}.$$

Therefore,

$$\mathbf{x_1} = \begin{pmatrix} 1 \\ 1 \end{pmatrix}$$

is an eigenvector for λ.

26. $\det(A - \lambda I) = \lambda^2 + 2\lambda + 2 = 0 \implies \lambda = -1 \pm i$. First, $\lambda_1 = -1 + i$ implies

$$A - \lambda_1 I = \begin{pmatrix} 2 - i & -1 \\ 5 & -2 - i \end{pmatrix}.$$

Therefore,

$$\mathbf{x_1} = \begin{pmatrix} 1 \\ 2 - i \end{pmatrix}$$

is an eigenvector for λ_1.

Second, λ_2 is the complex conjugate of λ_1 implies

$$\mathbf{x_2} \equiv \overline{\mathbf{x_1}} = \begin{pmatrix} 1 \\ 2 + i \end{pmatrix}$$

is an eigenvector for λ_2.

27. $\det(A - \lambda I) = \lambda^2 + 2\lambda = 0 \implies \lambda = 0, -2$. First, $\lambda_1 = 0$ implies

$$A - \lambda_1 I = \begin{pmatrix} 1 & \frac{4}{3} \\ -\frac{9}{4} & -3 \end{pmatrix}.$$

Therefore,

$$\mathbf{x_1} = \begin{pmatrix} -\frac{4}{3} \\ 1 \end{pmatrix}$$

11

is an eigenvector for λ_1.

Second, $\lambda_2 = -2$ implies

$$A - \lambda_2 I = \begin{pmatrix} 3 & \frac{4}{3} \\ -\frac{9}{4} & -1 \end{pmatrix}.$$

Therefore,

$$\mathbf{x_2} = \begin{pmatrix} -\frac{4}{9} \\ 1 \end{pmatrix}$$

is an eigenvector for λ_2.

28. $\det(A - \lambda I) = \lambda^2 + 4\lambda + 3 = 0 \implies \lambda = -1, -3$. First, $\lambda_1 = -1$ implies

$$A - \lambda_1 I = \begin{pmatrix} -1 & 1 \\ 1 & -1 \end{pmatrix}.$$

Therefore,

$$\mathbf{x_1} = \begin{pmatrix} 1 \\ 1 \end{pmatrix}$$

is an eigenvector for λ_1.

Second, $\lambda_2 = -3$ implies

$$A - \lambda_2 I = \begin{pmatrix} 1 & 1 \\ 1 & 1 \end{pmatrix}.$$

Therefore,

$$\mathbf{x_2} = \begin{pmatrix} 1 \\ -1 \end{pmatrix}$$

is an eigenvector for λ_2.

29. $\det(A - \lambda I) = \lambda^2 + 9 = 0 \implies \lambda = \pm 3i$. First, $\lambda_1 = 3i$ implies

$$A - \lambda_1 I = \begin{pmatrix} 1 - 3i & 2 \\ -5 & -1 - 3i \end{pmatrix}.$$

Therefore,

$$\mathbf{x_1} = \begin{pmatrix} 1 \\ \frac{-1+3i}{2} \end{pmatrix}$$

is an eigenvector for λ_1.

Second, since λ_2 is the complex conjugate of λ_1,

$$\mathbf{x_2} \equiv \overline{\mathbf{x_1}} = \begin{pmatrix} 1 \\ \frac{-1-3i}{2} \end{pmatrix}$$

is an eigenvector for λ_2.

30. $\det(A - \lambda I) = \lambda^2 + 4\lambda + 4 = 0 \implies \lambda = -2$. Now, $\lambda = -2$ implies

$$A - \lambda I = \begin{pmatrix} 1 & -\frac{1}{2} \\ 2 & -1 \end{pmatrix}.$$

Therefore,
$$\mathbf{x_1} = \begin{pmatrix} 1 \\ 2 \end{pmatrix}$$
is an eigenvector for λ.

31. $\det(A - \lambda I) = \lambda^2 - \dfrac{10}{4}\lambda + 1 = 0 \implies \lambda = 2, 1/2$. First, $\lambda_1 = 2$ implies
$$A - \lambda_1 I = \begin{pmatrix} -\frac{3}{4} & \frac{3}{4} \\ \frac{3}{4} & -\frac{3}{4} \end{pmatrix}.$$

Therefore,
$$\mathbf{x_1} = \begin{pmatrix} 1 \\ 1 \end{pmatrix}$$
is an eigenvector for λ_1.

  Second, $\lambda_2 = 1/2$ implies
$$A - \lambda_2 I = \begin{pmatrix} \frac{3}{4} & \frac{3}{4} \\ \frac{3}{4} & \frac{3}{4} \end{pmatrix}.$$

Therefore,
$$\mathbf{x_2} = \begin{pmatrix} 1 \\ -1 \end{pmatrix}$$
is an eigenvector for λ_2.

32. $\det(A - \lambda I) = \lambda^2 - 3\lambda + \dfrac{9}{4} = 0 \implies \lambda = 3/2$. Now, $\lambda = 3/2$ implies
$$A - \lambda I = \begin{pmatrix} \frac{1}{2} & \frac{1}{2} \\ -\frac{1}{2} & -\frac{1}{2} \end{pmatrix}.$$

Therefore,
$$\mathbf{x_1} = \begin{pmatrix} 1 \\ -1 \end{pmatrix}$$
is an eigenvector for λ.

33.

(a) $\det(A - \lambda I) = \lambda^2 + \lambda - 6 - \alpha = 0 \implies \lambda = (-1 \pm \sqrt{25 + 4\alpha})/2$.

(b) If $25 + 4\alpha > 0$, then there will be two real eigenvalues. If $25 + 4\alpha = 0$, then there will be one real eigenvalue. If $25 + 4\alpha < 0$, then there will be two eigenvalues with non-zero imaginary part (in particular, they will be a complex conjugate pair).

34.

(a) $\det(A - \lambda I) = \lambda^2 - 5\lambda + 6 + 4\alpha = 0 \implies \lambda = (5 \pm \sqrt{1 - 16\alpha})/2$.

(b) If $1 - 16\alpha > 0$ then there will be two real eigenvalues. If $1 = 16\alpha$, then there will be one real eigenvalue. If $1 - 16\alpha < 0$, then there will be two eigenvalues with non-zero imaginary parts (in particular, complex conjugates).

35.

(a) $\det(A - \lambda I) = \lambda^2 - (\alpha+1)\lambda + \alpha - 6 = 0 \implies \lambda = (\alpha + 1 \pm \sqrt{\alpha^2 - 2\alpha + 25})/2$.

(b) Now $\alpha^2 - 2\alpha + 25 > 0$ for all α since $(-2)^2 - 4(25) < 0$. Therefore, there will be two real eigenvalues no matter what the value of α is.

36.

(a) $\det(A - \lambda I) = \lambda^2 - 4\lambda + 3 + 2\alpha^2 = 0 \implies \lambda = 2 \pm \sqrt{1 - 2\alpha^2}$.

(b) If $1 - 2\alpha^2 > 0$, then there will be two real eigenvalues. Now $1 - 2\alpha^2 > 0$ if and only if $-1/\sqrt{2} < \alpha < 1/\sqrt{2}$. If $\alpha = \pm 1/\sqrt{2}$, then there will be one real eigenvalue. If $\alpha > 1/\sqrt{2}$ or $\alpha < -1/\sqrt{2}$, then there will be a complex conjugate pair of eigenvalues with non-zero imaginary part.

37. Let

$$A = \begin{pmatrix} a_{11} & a_{12} \\ a_{21} & a_{22} \end{pmatrix}.$$

Let

$$A^{-1} = \begin{pmatrix} b & c \\ d & e \end{pmatrix}.$$

Now $A^{-1}A = I$ implies

$$\begin{pmatrix} b & c \\ d & e \end{pmatrix} \begin{pmatrix} a_{11} & a_{12} \\ a_{21} & a_{22} \end{pmatrix} = \begin{pmatrix} 1 & 0 \\ 0 & 1 \end{pmatrix}.$$

Therefore, we have the following system of equations

$$ba_{11} + ca_{21} = 1$$
$$ba_{12} + ca_{22} = 0$$
$$da_{11} + ea_{21} = 0$$
$$da_{12} + ea_{22} = 1.$$

Multiplying the first equation by a_{12}, the second equation by $-a_{11}$ and adding them implies $c(a_{21}a_{12} - a_{11}a_{22}) = a_{12}$. Therefore,

$$c = \frac{1}{a_{21}a_{12} - a_{11}a_{22}} a_{12} = \frac{1}{\det(A)} a_{12}.$$

Plugging this value for c into the first equation, we are able to solve for b. In particular,

$$ba_{11} - \frac{a_{12}}{\det(A)} a_{21} = 1 \implies ba_{11} = \frac{a_{11}a_{22}}{\det(A)}.$$

Therefore,

$$b = \frac{a_{22}}{\det(A)}.$$

Similarly, multiplying the third equation by a_{12}, the fourth equation by $-a_{11}$ and adding them, we see that $e(a_{11}a_{22} - a_{21}a_{12}) = a_{11}$. Therefore,

$$e = \frac{a_{11}}{\det(A)}.$$

Plugging this value for e into the third equation, we see that

$$d = -\frac{a_{21}}{\det(A)}.$$

38. Suppose λ is an eigenvalue for A. Then λ is a solution of $\det(A - \lambda I) = 0$. In particular, λ will satisfy $(a_{11} - \lambda)(a_{22} - \lambda) - a_{12}a_{21} = 0$. Therefore, $\lambda^2 - (a_{11} + a_{22})\lambda + a_{11}a_{22} - a_{12}a_{21} = 0$. Now, $\lambda = 0$ is an eigenvalue if and only if λ is a solution of this equation, which is true if and only if $a_{11}a_{22} - a_{12}a_{21} = 0$. In other words, $\lambda = 0$ is an eigenvalue if and only if $\det(A) = 0$.

Section 3.2

1. The system is autonomous, nonhomogeneous.

2. The system is nonautonomous, nonhomogeneous.

3. The system is nonautonomous, homogeneous.

4. The system is autonomous, nonhomogeneous.

5. The system is autonomous, homogeneous.

6. The system is nonautonomous, homogeneous.

7. The system is nonautonomous, nonhomogeneous.

8. The system is autonomous, homogeneous.

9.

(a) The system

$$x' = -x + y + 1 = 0$$
$$y' = x + y - 3 = 0$$

can be rewritten in matrix form as

$$\begin{pmatrix} -1 & 1 \\ 1 & 1 \end{pmatrix} \begin{pmatrix} x \\ y \end{pmatrix} = \begin{pmatrix} -1 & 3 \end{pmatrix}.$$

Now

$$A = \begin{pmatrix} -1 & 1 \\ 1 & 1 \end{pmatrix} \implies A^{-1} = \frac{1}{2} \begin{pmatrix} -1 & 1 \\ 1 & 1 \end{pmatrix}.$$

Therefore,

$$\begin{pmatrix} x \\ y \end{pmatrix} = A^{-1} \begin{pmatrix} -1 \\ 3 \end{pmatrix} = \frac{1}{2} \begin{pmatrix} -1 & 1 \\ 1 & 1 \end{pmatrix} \begin{pmatrix} -1 \\ 3 \end{pmatrix} = \begin{pmatrix} 2 \\ 1 \end{pmatrix}.$$

Therefore, the equilibrium solution is

$$\begin{pmatrix} 2 \\ 1 \end{pmatrix}.$$

(b)

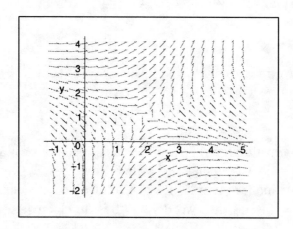

(c) Solutions in the vicinity of the critical point, tend away from the critical point.

10.

(a) The system

$$x' = -x - 4y - 4 = 0$$
$$y' = x - y - 6 = 0$$

can be rewritten in matrix form as

$$\begin{pmatrix} -1 & -4 \\ 1 & -1 \end{pmatrix} \begin{pmatrix} x \\ y \end{pmatrix} = \begin{pmatrix} 4 & 6 \end{pmatrix}.$$

Now

$$A = \begin{pmatrix} -1 & -4 \\ 1 & -1 \end{pmatrix} \implies A^{-1} = \frac{1}{5} \begin{pmatrix} -1 & 4 \\ -1 & -1 \end{pmatrix}.$$

Therefore,

$$\begin{pmatrix} x \\ y \end{pmatrix} = A^{-1} \begin{pmatrix} 4 \\ 6 \end{pmatrix} = \frac{1}{5} \begin{pmatrix} -1 & 4 \\ -1 & -1 \end{pmatrix} \begin{pmatrix} 4 \\ 6 \end{pmatrix} = \begin{pmatrix} 4 \\ -2 \end{pmatrix}.$$

Therefore, the equilibrium solution is

$$\begin{pmatrix} 4 \\ -2 \end{pmatrix}.$$

(b)

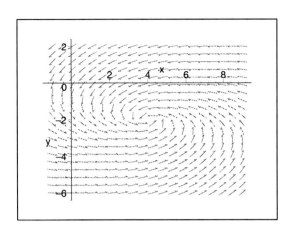

(c) Solutions in the vicinity of the critical point, spiral into the critical point.

11.

(a) The system

$$x' = -0.25x - 0.75y + 8 = 0$$
$$y' = 0.5x + y - 11.5 = 0$$

can be rewritten in matrix form as

$$\begin{pmatrix} -0.25 & -0.75 \\ 0.5 & 1 \end{pmatrix} \begin{pmatrix} x \\ y \end{pmatrix} = \begin{pmatrix} -8 & 11.5 \end{pmatrix}.$$

Now

$$A = \begin{pmatrix} -0.25 & -0.75 \\ 0.5 & 1 \end{pmatrix} \implies A^{-1} = 8 \begin{pmatrix} 1 & 3/4 \\ -1/2 & -1/4 \end{pmatrix}.$$

Therefore,

$$\begin{pmatrix} x \\ y \end{pmatrix} = A^{-1} \begin{pmatrix} -8 \\ 11.5 \end{pmatrix} = 8 \begin{pmatrix} 1 & 3/4 \\ -1/2 & -1/4 \end{pmatrix} \begin{pmatrix} -8 \\ 11.5 \end{pmatrix} = \begin{pmatrix} 4 \\ -2 \end{pmatrix}.$$

Therefore, the equilibrium solution is

$$\begin{pmatrix} 5 \\ 9 \end{pmatrix}.$$

(b)

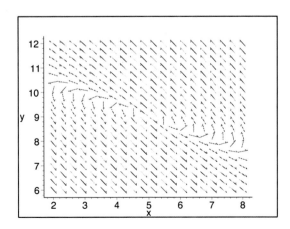

17

(c) Solutions in the vicinity of the critical point, tend away from the critical point.

12.

(a) The system

$$x' = -2x + y - 11 = 0$$
$$y' = -5x + 4y - 35 = 0$$

can be rewritten in matrix form as

$$\begin{pmatrix} -2 & 1 \\ -5 & 4 \end{pmatrix} \begin{pmatrix} x \\ y \end{pmatrix} = \begin{pmatrix} 11 & 35 \end{pmatrix}.$$

Now

$$A = \begin{pmatrix} -2 & 1 \\ -5 & 4 \end{pmatrix} \implies A^{-1} = -\frac{1}{3} \begin{pmatrix} 4 & -1 \\ 5 & -2 \end{pmatrix}.$$

Therefore,

$$\begin{pmatrix} x \\ y \end{pmatrix} = A^{-1} \begin{pmatrix} 11 \\ 35 \end{pmatrix} = -\frac{1}{3} \begin{pmatrix} 4 & -1 \\ 5 & -2 \end{pmatrix} \begin{pmatrix} 11 \\ 35 \end{pmatrix} = \begin{pmatrix} -3 \\ 5 \end{pmatrix}.$$

Therefore, the equilibrium solution is

$$\begin{pmatrix} -3 \\ 5 \end{pmatrix}.$$

(b)

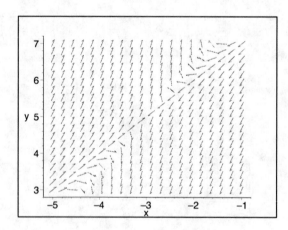

(c) Solutions in the vicinity of the critical point, tend away from the critical point.

13.

18

(a) The system

$$x' = x + y - 3 = 0$$
$$y' = -x + y + 1 = 0$$

can be rewritten in matrix form as

$$\begin{pmatrix} 1 & 1 \\ -1 & 1 \end{pmatrix} \begin{pmatrix} x \\ y \end{pmatrix} = \begin{pmatrix} 3 & -1 \end{pmatrix}.$$

Now

$$A = \begin{pmatrix} 1 & 1 \\ -1 & 1 \end{pmatrix} \implies A^{-1} = \frac{1}{2}\begin{pmatrix} 1 & -1 \\ 1 & 1 \end{pmatrix}.$$

Therefore,

$$\begin{pmatrix} x \\ y \end{pmatrix} = A^{-1}\begin{pmatrix} 3 \\ -1 \end{pmatrix} = \frac{1}{2}\begin{pmatrix} 1 & -1 \\ 1 & 1 \end{pmatrix}\begin{pmatrix} 3 \\ -1 \end{pmatrix} = \begin{pmatrix} 2 \\ 1 \end{pmatrix}.$$

Therefore, the equilibrium solution is

$$\begin{pmatrix} 2 \\ 1 \end{pmatrix}.$$

(b)

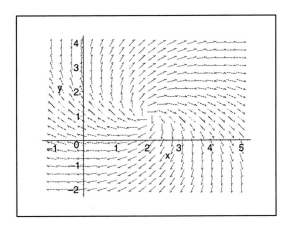

(c) Solutions in the vicinity of the critical point, spiral away from the critical point.

14.

(a) The system

$$x' = -5x + 4y - 35 = 0$$
$$y' = -2x + y - 11 = 0$$

can be rewritten in matrix form as

$$\begin{pmatrix} -5 & 4 \\ -2 & 1 \end{pmatrix} \begin{pmatrix} x \\ y \end{pmatrix} = \begin{pmatrix} 35 & 11 \end{pmatrix}.$$

Now

$$A = \begin{pmatrix} -5 & 4 \\ -2 & 1 \end{pmatrix} \implies A^{-1} = \frac{1}{3} \begin{pmatrix} 1 & -4 \\ 2 & -5 \end{pmatrix}.$$

Therefore,

$$\begin{pmatrix} x \\ y \end{pmatrix} = A^{-1} \begin{pmatrix} 35 \\ 11 \end{pmatrix} = \frac{1}{3} \begin{pmatrix} 1 & -4 \\ 2 & -5 \end{pmatrix} \begin{pmatrix} 35 \\ 11 \end{pmatrix} = \begin{pmatrix} -3 \\ 5 \end{pmatrix}.$$

Therefore, the equilibrium solution is

$$\begin{pmatrix} -3 \\ 5 \end{pmatrix}.$$

(b)

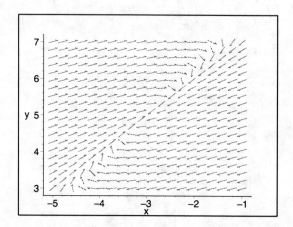

(c) Solutions in the vicinity of the critical point, tend towards the critical point.

15. Let $x_1 = u$ and $x_2 = u'$. Then $x_1' = x_2$ and

$$\begin{aligned} x_2' &= u'' \\ &= -2u - 0.5u \\ &= -2x_1 - 0.5x_2. \end{aligned}$$

Therefore, we obtain the system of equations

$$\begin{aligned} x_1' &= x_2 \\ x_2' &= -2x_1 - 0.5x_2. \end{aligned}$$

16. Let $x_1 = u$ and $x_2 = u'$. Then $x_1' = x_2$ and

$$\begin{aligned} x_2' &= u'' \\ &= -0.5u' - 8u + 6\sin(2t) \\ &= -0.5x_2 - 8x_1 + 6\sin(2t). \end{aligned}$$

Therefore, we obtain the system of equations

$$x_1' = x_2$$
$$x_2' = -8x_1 - 0.5x_2 + 6\sin(2t).$$

17. First divide the equation by t^2. We arrive at the equation

$$u'' = -\frac{1}{t}u' - \left(1 - \frac{1}{4t^2}\right)u.$$

Now let $x_1 = u$ and $x_2 = u'$. Then $x_1' = x_2$ and

$$x_2' = u''$$
$$= -\frac{1}{t}u' - \left(1 - \frac{1}{4t^2}\right)u$$
$$= -\frac{1}{t}x_2 - \left(1 - \frac{1}{4t^2}\right)x_1.$$

Therefore, we obtain the system of equations

$$x_1' = x_2$$
$$x_2' = -\left(1 - \frac{1}{4t^2}\right)x_1 - \frac{1}{t}x_2.$$

18. First, divide the equation by t^2. We arrive at the equation

$$u'' + \frac{3}{t}u' + \frac{5}{t^2}u = 1 + \frac{4}{t^2}.$$

Now let $x_1 = u$ and $x_2 = u'$. Then $x_1' = x_2$ and

$$x_2' = u''$$
$$= -\frac{3}{t}u' - \frac{5}{t^2}u + 1 + \frac{4}{t^2}$$
$$= -\frac{3}{t}x_2 - \frac{5}{t^2}x_1 + 1 + \frac{4}{t^2}.$$

Therefore, we obtain the system of equations

$$x_1' = x_2$$
$$x_2' = -\frac{5}{t^2}x_1 - \frac{3}{t}x_2 + 1 + \frac{4}{t^2}.$$

19. Let $x_1 = u$ and $x_2 = u'$. Then $x_1' = x_2$ and

$$x_2' = u''$$
$$= -0.25u' - 4u + 2\cos 3t$$
$$= -4x_1 - 0.25x_2 + 2\cos 3t.$$

Now $u(0) = 1$ implies $x_1(0) = 1$ and $u'(0) = -2$ implies $x_2(0) = -2$. Therefore, we obtain the system of equations

$$x_1' = x_2$$
$$x_2' = -4x_1 - 0.25x_2 + 2\cos 3t$$

with initial conditions

$$x_1(0) = 1$$
$$x_2(0) = -2.$$

20. First, divide the equation by t. We arrive at the equation

$$u'' + \frac{1}{t}u' + u = 0.$$

Now let $x_1 = u$ and $x_2 = u'$. Then $x_1' = x_2$ and

$$x_2' = -\frac{1}{t}u' - u$$
$$= -\frac{1}{t}x_2 - x_1.$$

Now $u(1) = 1 \implies x_1(1) = u(1) = 1$ and $u'(1) = 0 \implies x_2(1) = u'(1) = 0$. Therefore, we obtain the system of equations

$$x_1' = x_2$$
$$x_2' = -x_1 - \frac{1}{t}x_2$$

with initial conditions

$$x_1(1) = 1$$
$$x_2(1) = 0.$$

21.

(a) Taking a *clockwise* loop around each of the paths, it is easy to see that voltage drops are given by $v_1 - v_2 = 0$ and $v_2 - v_3 = 0$.

(b) Consider the right node. The current in is given by $i_1 + i_2$. The current leaving the node is $-i_3$. Therefore, the current passing through the node is $(i_1 + i_2) - (-i_3)$. Based on Kirchoff's first law, $i_1 + i_2 + i_3 = 0$.

(c) In the capacitor, $Cv_1' = i_1$. In the resistor, $v_2 = Ri_2$. In the inductor, $Li_3' = v_3$.

(d) Based on part (a), $v_3 = v_2 = v_1$. Based on parts (b) and (c),

$$Cv_1' + \frac{1}{R}v_2 + i_3 = 0.$$

It follows that

$$Cv_1' = -\frac{1}{R}v_1 - i_3 \quad \text{and} \quad Li_3' = v_1.$$

22. Let i_1, i_2, i_3 and i_4 be the current through the 1 ohm resistor, 2 ohm resistor, inductor and capacitor, respectively. Assign v_1, v_2, v_3 and v_4 as the respective voltage drops. Based on Kirchoff's second law, the net voltage drops around each loop satisfy

$$v_1 + v_3 + v_4 = 0, v_1 + v_3 + v_2 = 0 \quad \text{and} \quad v_4 - v_2 = 0.$$

Applying Kirchoff's first law to the upper right node, we have

$$i_1 - i_3 = 0.$$

Likewise, in the remaining nodes, we have

$$i_2 + i_4 - i_1 = 0 \quad \text{and} \quad i_3 - i_4 - i_2 = 0.$$

Combining the above equations, we have

$$v_4 - v_2 = 0, v_1 + v_3 + v_4 = 0 \quad \text{and} \quad i_2 + i_4 - i_3 = 0.$$

Using the current-voltage relations, we have

$$v_1 = R_1 i_1, v_2 = R_2 i_2, L i_3' = v_3, C v_4' = i_4.$$

Combining these equations, we have

$$R_1 i_3 + L i_3' + v_4 = 0 \quad \text{and} \quad C v_4' = i_3 - \frac{1}{R_2} v_4.$$

Now set $i_3 = i$ and $v_4 = v$, to obtain the system of equations

$$L i' = -R_1 i - v \quad \text{and} \quad C v' = i - \frac{1}{R_2} v.$$

Finally, using the fact that $R_1 = 1$, $R_2 = 2$, $L = 1$ and $C = 1/2$, we have

$$i' = -i - v \quad \text{and} \quad v' = 2i - v,$$

as claimed.

23. Let i_1, i_2, i_3 and i_4 be the current through the resistors, inductor and capacitor, respectively. Assign v_1, v_2, v_3 and v_4 as the respective voltage drops. Based on Kirchoff's second law, the net voltage drops around each loop satisfy

$$v_1 + v_3 + v_4 = 0, v_1 + v_3 + v_2 = 0 \quad \text{and} \quad v_4 - v_2 = 0.$$

Applying Kirchoff's first law to the upper right node, we have

$$i_3 - (i_2 + i_4) = 0.$$

Likewise, in the remaining nodes, we have

$$i_1 - i_3 = 0 \quad \text{and} \quad i_2 + i_4 - i_1 = 0.$$

Combining the above equations, we have

$$v_4 - v_2 = 0, v_1 + v_3 + v_4 = 0 \quad \text{and} \quad i_2 + i_4 - i_3 = 0.$$

Using the current-voltage relations, we have

$$v_1 = R_1 i_1, v_2 = R_2 i_2, Li_3' = v_3, Cv_4' = i_4.$$

Combining these equations, we have

$$R_1 i_3 + Li_3' + v_4 = 0 \quad \text{and} \quad Cv_4' = i_3 - \frac{1}{R_2} v_4.$$

Now set $i_3 = i$ and $v_4 = v$, to obtain the system of equations

$$Li' = -R_1 i - v \quad \text{and} \quad Cv' = i - \frac{1}{R_2} v.$$

24.

(a) Let $Q_1(t)$ and $Q_2(t)$ be the amount of salt in the respective tanks at time t. The volume of each tank remains constant. Based on conservation of mass, the rate of increase of salt is given by

$$\text{rate of increase} = \text{rate in} - \text{rate out}.$$

For Tank 1, the rate of salt flowing in is $r_{in} = 3q_1 + \frac{Q_2}{100}$. The rate of flow out of Tank 1 is $r_{out} = \frac{Q_1}{15}$. Therefore,

$$\frac{dQ_1}{dt} = 3q_1 + \frac{Q_2}{100} - \frac{Q_1}{15}.$$

Similarly, for Tank 2,

$$\frac{dQ_2}{dt} = q_2 + \frac{Q_1}{30} - \frac{3Q_2}{100}.$$

The initial conditions are $Q_1(0) = Q_1^0$ and $Q_2(0) = Q_2^0$.

(b) The equilibrium values are obtained by solving $-Q_1/15 + Q_2/100 + 3q_1 = 0$ and $Q_1/30 - 3Q_2/100 + q_2 = 0$. Its solution is given by $Q_1^E = 54q_1 + 6q_2$ and $Q_2^E = 60q_1 + 40q_2$.

(c) We would need to be able to solve the system

$$54q_1 + 6q_2 = 60$$
$$60q_1 + 40q_2 = 50.$$

The solution of this system is $q_1 = 7/6$ and $q_2 = -1/2$ which is not a physically realistic solution. Therefore, it is not possible to have those values as an equilibrium state.

(d) We can write

$$q_2 = -9q_1 + \frac{Q_1^E}{6}$$

$$q_2 = -\frac{3}{2} q_1 + \frac{Q_2^E}{40}$$

which are the equations of two lines in the q_1q_2−plane. The intercepts of the first line are $(Q_1^E/54, 0)$ and $(0, Q_1^E/6)$. The intercepts of the second line are $(Q_2^E/60, 0)$ and $(0, Q_2^E/40)$. Therefore, the system will have a unique solution in the first quadrant as long as $Q_1^E/54 \leq Q_2^E/60$ or $Q_2^E/40 \leq Q_1^E/6$. That is, $10/9 \leq Q_2^E/Q_1^E \leq 20/3$.

Section 3.3

1. We look for eigenvalues and eigenvectors of

$$A = \begin{pmatrix} 3 & -2 \\ 2 & -2 \end{pmatrix}.$$

We see that $\det(A - \lambda I) = \lambda^2 - \lambda - 2 = (\lambda - 2)(\lambda + 1)$. Therefore, the eigenvalues are given by $\lambda = 2, -1$.

First, $\lambda_1 = 2$ implies

$$A - \lambda_1 I = \begin{pmatrix} 1 & -2 \\ 2 & -4 \end{pmatrix}.$$

Therefore,

$$\mathbf{v_1} = \begin{pmatrix} 2 \\ 1 \end{pmatrix}$$

is an eigenvector associated with λ_1 and

$$\mathbf{x_1}(t) = e^{2t} \begin{pmatrix} 2 \\ 1 \end{pmatrix}$$

is one solution of the system.

Second, $\lambda_2 = -1$ implies

$$A - \lambda_2 I = \begin{pmatrix} 4 & -2 \\ 2 & -1 \end{pmatrix}.$$

Therefore,

$$\mathbf{v_2} = \begin{pmatrix} 1 \\ 2 \end{pmatrix}$$

is an eigenvector associated with λ_2 and

$$\mathbf{x_2}(t) = e^{-t} \begin{pmatrix} 1 \\ 2 \end{pmatrix}$$

is a second solution of the system.

Therefore, the general solution is given by

$$\mathbf{x}(t) = c_1 e^{2t} \begin{pmatrix} 2 \\ 1 \end{pmatrix} + c_2 e^{-t} \begin{pmatrix} 1 \\ 2 \end{pmatrix}.$$

If the initial condition is a multiple of $\begin{pmatrix} 1 & 2 \end{pmatrix}^t$, then the solution will tend to the origin along the eigenvector $\begin{pmatrix} 1 & 2 \end{pmatrix}^t$. Otherwise, the solution will grow, following the eigenvector $\begin{pmatrix} 2 & 1 \end{pmatrix}^t$.

2. We look for eigenvalues and eigenvectors of

$$A = \begin{pmatrix} 1 & -2 \\ 3 & -4 \end{pmatrix}.$$

We see that $\det(A - \lambda I) = \lambda^2 + 3\lambda + 2 = (\lambda + 2)(\lambda + 1)$. Therefore, the eigenvalues are given by $\lambda = -1, -2$.

First, $\lambda_1 = -1$ implies

$$A - \lambda_1 I = \begin{pmatrix} 2 & -2 \\ 3 & -3 \end{pmatrix}.$$

Therefore,

$$\mathbf{v_1} = \begin{pmatrix} 1 \\ 1 \end{pmatrix}$$

is an eigenvector associated with λ_1 and

$$\mathbf{x_1}(t) = e^{-t} \begin{pmatrix} 1 \\ 1 \end{pmatrix}$$

is one solution of the system.

Second, $\lambda_2 = -2$ implies

$$A - \lambda_2 I = \begin{pmatrix} 3 & -2 \\ 3 & -2 \end{pmatrix}.$$

Therefore,

$$\mathbf{v_2} = \begin{pmatrix} 2 \\ 3 \end{pmatrix}$$

is an eigenvector associated with λ_2 and

$$\mathbf{x_2}(t) = e^{-2t} \begin{pmatrix} 2 \\ 3 \end{pmatrix}$$

is a second solution of the system.

26

Therefore, the general solution is given by

$$\mathbf{x}(t) = c_1 e^{-t} \begin{pmatrix} 1 \\ 1 \end{pmatrix} + c_2 e^{-2t} \begin{pmatrix} 2 \\ 3 \end{pmatrix}.$$

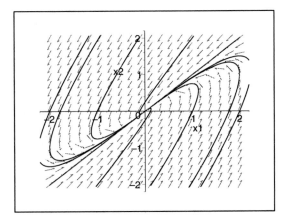

If the initial condition is a multiple of $\begin{pmatrix} 2 & 3 \end{pmatrix}^t$, then the solution will tend to the origin along the eigenvector $\begin{pmatrix} 2 & 3 \end{pmatrix}^t$. Otherwise, the solution will tend to the origin following the eigenvector $\begin{pmatrix} 1 & 1 \end{pmatrix}^t$.

3. We look for eigenvalues and eigenvectors of

$$A = \begin{pmatrix} 2 & -1 \\ 3 & -2 \end{pmatrix}.$$

We see that $\det(A - \lambda I) = \lambda^2 - 1 = (\lambda - 1)(\lambda + 1)$. Therefore, the eigenvalues are given by $\lambda = 1, -1$.

First, $\lambda_1 = 1$ implies

$$A - \lambda_1 I = \begin{pmatrix} 1 & -1 \\ 3 & -3 \end{pmatrix}.$$

Therefore,

$$\mathbf{v_1} = \begin{pmatrix} 1 \\ 1 \end{pmatrix}$$

is an eigenvector associated with λ_1 and

$$\mathbf{x_1}(t) = e^t \begin{pmatrix} 1 \\ 1 \end{pmatrix}$$

is one solution of the system.

Second, $\lambda_2 = -1$ implies

$$A - \lambda_2 I = \begin{pmatrix} 3 & -1 \\ 3 & -1 \end{pmatrix}.$$

Therefore,

$$\mathbf{v_2} = \begin{pmatrix} 1 \\ 3 \end{pmatrix}$$

27

is an eigenvector associated with λ_2 and

$$\mathbf{x_2}(t) = e^{-t} \begin{pmatrix} 1 \\ 3 \end{pmatrix}$$

is a second solution of the system.

Therefore, the general solution is given by

$$\mathbf{x}(t) = c_1 e^t \begin{pmatrix} 1 \\ 1 \end{pmatrix} + c_2 e^{-t} \begin{pmatrix} 1 \\ 3 \end{pmatrix}.$$

If the initial condition is a multiple of $\begin{pmatrix} 1 & 3 \end{pmatrix}^t$, then the solution will tend to the origin along the eigenvector $\begin{pmatrix} 1 & 3 \end{pmatrix}^t$. Otherwise, the solution will grow, following the eigenvector $\begin{pmatrix} 1 & 1 \end{pmatrix}^t$.

4. We look for eigenvalues and eigenvectors of

$$A = \begin{pmatrix} 1 & 1 \\ 4 & -2 \end{pmatrix}.$$

We see that $\det(A - \lambda I) = \lambda^2 + \lambda - 6 = (\lambda + 3)(\lambda - 2)$. Therefore, the eigenvalues are given by $\lambda = 2, -3$.

First, $\lambda_1 = 2$ implies

$$A - \lambda_1 I = \begin{pmatrix} -1 & 1 \\ 4 & -4 \end{pmatrix}.$$

Therefore,

$$\mathbf{v_1} = \begin{pmatrix} 1 \\ 1 \end{pmatrix}$$

is an eigenvector associated with λ_1 and

$$\mathbf{x_1}(t) = e^{2t} \begin{pmatrix} 1 \\ 1 \end{pmatrix}.$$

is one solution of the system.

Second, $\lambda_2 = -3$ implies

$$A - \lambda_2 I = \begin{pmatrix} 4 & 1 \\ 4 & 1 \end{pmatrix}.$$

28

Therefore,

$$\mathbf{v_2} = \begin{pmatrix} 1 \\ -4 \end{pmatrix}$$

is an eigenvector associated with λ_2 and

$$\mathbf{x_2}(t) = e^{-3t} \begin{pmatrix} 1 \\ -4 \end{pmatrix}$$

is a second solution of the system.

Therefore, the general solution is given by

$$\mathbf{x}(t) = c_1 e^{2t} \begin{pmatrix} 1 \\ 1 \end{pmatrix} + c_2 e^{-3t} \begin{pmatrix} 1 \\ -4 \end{pmatrix}.$$

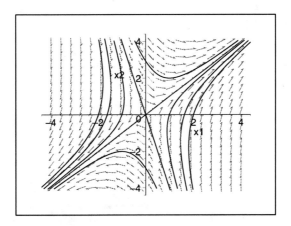

If the initial condition is a multiple of $\begin{pmatrix} 1 & -4 \end{pmatrix}^t$, then the solution will tend to the origin along the eigenvector $\begin{pmatrix} 1 & -4 \end{pmatrix}^t$. Otherwise, the solution will grow, following the eigenvector $\begin{pmatrix} 1 & 1 \end{pmatrix}^t$.

5. We look for eigenvalues and eigenvectors of

$$A = \begin{pmatrix} 4 & -3 \\ 8 & -6 \end{pmatrix}.$$

We see that $\det(A - \lambda I) = \lambda^2 + 2\lambda = \lambda(\lambda + 2)$. Therefore, the eigenvalues are given by $\lambda = 0, -2$.

First, $\lambda_1 = 0$ implies

$$A - \lambda_1 I = \begin{pmatrix} 4 & -3 \\ 8 & -6 \end{pmatrix}.$$

Therefore,

$$\mathbf{v_1} = \begin{pmatrix} 3 \\ 4 \end{pmatrix}$$

is an eigenvector associated with λ_1 and

$$\mathbf{x_1}(t) = \begin{pmatrix} 3 \\ 4 \end{pmatrix}$$

is one solution of the system.

Second, $\lambda_2 = -2$ implies

$$A - \lambda_2 I = \begin{pmatrix} 6 & -3 \\ 8 & -4 \end{pmatrix}.$$

Therefore,

$$\mathbf{v_2} = \begin{pmatrix} 1 \\ 2 \end{pmatrix}$$

is an eigenvector associated with λ_2 and

$$\mathbf{x_2}(t) = e^{-2t} \begin{pmatrix} 1 \\ 2 \end{pmatrix}$$

is a second solution of the system.

Therefore, the general solution is given by

$$\mathbf{x}(t) = c_1 \begin{pmatrix} 3 \\ 4 \end{pmatrix} + c_2 e^{-2t} \begin{pmatrix} 1 \\ 2 \end{pmatrix}.$$

The behavior of the solutions as $t \to \infty$ is similar to the behavior of solutions in Example 5 of the text.

6. We look for eigenvalues and eigenvectors of

$$A = \begin{pmatrix} -2 & 1 \\ 1 & -2 \end{pmatrix}.$$

We see that $\det(A - \lambda I) = \lambda^2 + 4\lambda + 3 = (\lambda + 1)(\lambda + 3)$. Therefore, the eigenvalues are given by $\lambda = -1, -3$.

First, $\lambda_1 = -1$ implies

$$A - \lambda_1 I = \begin{pmatrix} -1 & 1 \\ 1 & -1 \end{pmatrix}.$$

Therefore,

$$\mathbf{v_1} = \begin{pmatrix} 1 \\ 1 \end{pmatrix}$$

30

is an eigenvector associated with λ_1 and

$$\mathbf{x_1}(t) = e^{-t} \begin{pmatrix} 1 \\ 1 \end{pmatrix}$$

is one solution of the system.

Second, $\lambda_2 = -3$ implies

$$A - \lambda_2 I = \begin{pmatrix} 1 & 1 \\ 1 & 1 \end{pmatrix}.$$

Therefore,

$$\mathbf{v_2} = \begin{pmatrix} 1 \\ -1 \end{pmatrix}$$

is an eigenvector associated with λ_2 and

$$\mathbf{x_2}(t) = e^{-3t} \begin{pmatrix} 1 \\ -1 \end{pmatrix}$$

is a second solution of the system.

Therefore, the general solution is given by

$$\mathbf{x}(t) = c_1 e^{-t} \begin{pmatrix} 1 \\ 1 \end{pmatrix} + c_2 e^{-3t} \begin{pmatrix} 1 \\ -1 \end{pmatrix}.$$

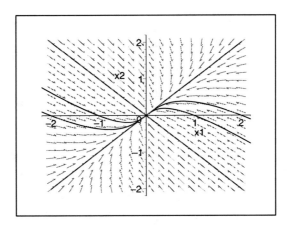

If the initial condition is a multiple of $\begin{pmatrix} 1 & -1 \end{pmatrix}^t$, then the solution will tend to the origin along the eigenvector $\begin{pmatrix} 1 & -1 \end{pmatrix}^t$. Otherwise, the solution will tend to the origin, following the eigenvector $\begin{pmatrix} 1 & 1 \end{pmatrix}^t$.

7. We look for eigenvalues and eigenvectors of

$$A = \begin{pmatrix} 5/4 & 3/4 \\ 3/4 & 5/4 \end{pmatrix}.$$

We see that $\det(A - \lambda I) = \lambda^2 - \dfrac{5}{2}\lambda + 1 = (\lambda - 2)(\lambda - 1/2)$. Therefore, the eigenvalues are given by $\lambda = 2, 1/2$.

First, $\lambda_1 = 2$ implies

$$A - \lambda_1 I = \begin{pmatrix} -3/4 & 3/4 \\ 3/4 & -3/4 \end{pmatrix}.$$

Therefore,

$$\mathbf{v_1} = \begin{pmatrix} 1 \\ 1 \end{pmatrix}$$

is an eigenvector associated with λ_1 and

$$\mathbf{x_1}(t) = e^{2t} \begin{pmatrix} 1 \\ 1 \end{pmatrix}$$

is one solution of the system.

Second, $\lambda_2 = 1/2$ implies

$$A - \lambda_2 I = \begin{pmatrix} 3/4 & 3/4 \\ 3/4 & 3/4 \end{pmatrix}.$$

Therefore,

$$\mathbf{v_2} = \begin{pmatrix} 1 \\ -1 \end{pmatrix}$$

is an eigenvector associated with λ_2 and

$$\mathbf{x_2}(t) = e^{t/2} \begin{pmatrix} 1 \\ -1 \end{pmatrix}$$

is a second solution of the system.

Therefore, the general solution is given by

$$\mathbf{x}(t) = c_1 e^{2t} \begin{pmatrix} 1 \\ 1 \end{pmatrix} + c_2 e^{\frac{t}{2}} \begin{pmatrix} 1 \\ -1 \end{pmatrix}.$$

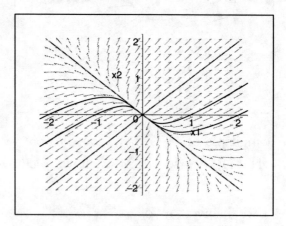

If the initial condition is a multiple of $\begin{pmatrix} 1 & -1 \end{pmatrix}^t$, then the solution will grow following the the the eigenvector $\begin{pmatrix} 1 & -1 \end{pmatrix}^t$. Otherwise, the solution will grow, following the eigenvector $\begin{pmatrix} 1 & 1 \end{pmatrix}^t$.

8. We look for eigenvalues and eigenvectors of

$$A = \begin{pmatrix} -3/4 & -7/4 \\ 1/4 & 5/4 \end{pmatrix}.$$

We see that $\det(A - \lambda I) = \lambda^2 - \dfrac{1}{2}\lambda - \dfrac{1}{2} = (\lambda - 1)(\lambda + 1/2)$. Therefore, the eigenvalues are given by $\lambda = 1, -1/2$.

First, $\lambda_1 = 1$ implies

$$A - \lambda_1 I = \begin{pmatrix} -7/4 & -7/4 \\ 1/4 & 1/4 \end{pmatrix}.$$

Therefore,

$$\mathbf{v_1} = \begin{pmatrix} 1 \\ -1 \end{pmatrix}$$

is an eigenvector associated with λ_1 and

$$\mathbf{x_1}(t) = e^t \begin{pmatrix} 1 \\ -1 \end{pmatrix}$$

is one solution of the system.

Second, $\lambda_2 = -1/2$ implies

$$A - \lambda_2 I = \begin{pmatrix} -1/4 & -7/4 \\ 1/4 & 7/4 \end{pmatrix}.$$

Therefore,

$$\mathbf{v_2} = \begin{pmatrix} -7 \\ 1 \end{pmatrix}$$

is an eigenvector associated with λ_2 and

$$\mathbf{x_2}(t) = e^{-t/2} \begin{pmatrix} -7 \\ 1 \end{pmatrix}$$

is a second solution of the system.

Therefore, the general solution is given by

$$\mathbf{x}(t) = c_1 e^t \begin{pmatrix} 1 \\ -1 \end{pmatrix} + c_2 e^{-\frac{t}{2}} \begin{pmatrix} -7 \\ 1 \end{pmatrix}.$$

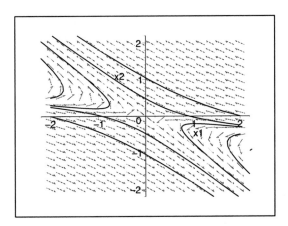

If the initial condition is a multiple of $(-7 \ 1)^t$, then the solution will tend to the origin along the eigenvector $(-7 \ 1)^t$. Otherwise, the solution will grow, following the eigenvector $(1 \ -1)^t$.

9. We look for eigenvalues and eigenvectors of

$$A = \begin{pmatrix} -1/4 & -3/4 \\ 1/2 & 1 \end{pmatrix}.$$

We see that $\det(A - \lambda I) = \lambda^2 - \dfrac{3}{4}\lambda + \dfrac{1}{8} = \left(\lambda - \dfrac{1}{2}\right)\left(\lambda - \dfrac{1}{4}\right)$. Therefore, the eigenvalues are given by $\lambda = 1/2, 1/4$.

First, $\lambda_1 = 1/2$ implies

$$A - \lambda_1 I = \begin{pmatrix} -3/4 & -3/4 \\ 1/2 & 1/2 \end{pmatrix}.$$

Therefore,

$$\mathbf{v_1} = \begin{pmatrix} 1 \\ -1 \end{pmatrix}$$

is an eigenvector associated with λ_1 and

$$\mathbf{x_1}(t) = e^{t/2}\begin{pmatrix} 1 \\ -1 \end{pmatrix}$$

is one solution of the system.

Second, $\lambda_2 = 1/4$ implies

$$A - \lambda_2 I = \begin{pmatrix} -1/2 & -3/4 \\ 1/2 & 3/4 \end{pmatrix}.$$

Therefore,

$$\mathbf{v_2} = \begin{pmatrix} 3 \\ -2 \end{pmatrix}$$

is an eigenvector associated with λ_2 and

$$\mathbf{x_2}(t) = e^{t/4}\begin{pmatrix} 3 \\ -2 \end{pmatrix}$$

is a second solution of the system.

Therefore, the general solution is given by

$$\mathbf{x}(t) = c_1 e^{t/2}\begin{pmatrix} 1 \\ -1 \end{pmatrix} + c_2 e^{t/4}\begin{pmatrix} 3 \\ -2 \end{pmatrix}.$$

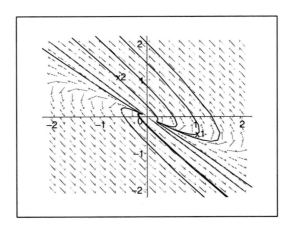

If the initial condition is a multiple of $\begin{pmatrix} 3 & -2 \end{pmatrix}^t$, then the solution will stay on the eigenvector $\begin{pmatrix} 3 & -2 \end{pmatrix}^t$. Otherwise, the solution will grow, following the eigenvector $\begin{pmatrix} 1 & -1 \end{pmatrix}^t$.

10. We look for eigenvalues and eigenvectors of

$$A = \begin{pmatrix} 5 & -1 \\ 3 & 1 \end{pmatrix}.$$

We see that $\det(A - \lambda I) = \lambda^2 - 6\lambda + 8 = (\lambda - 4)(\lambda - 2)$. Therefore, the eigenvalues are given by $\lambda = 4, 2$.

First, $\lambda_1 = 4$ implies

$$A - \lambda_1 I = \begin{pmatrix} 1 & -1 \\ 3 & -3 \end{pmatrix}.$$

Therefore,

$$\mathbf{v_1} = \begin{pmatrix} 1 \\ 1 \end{pmatrix}$$

is an eigenvector associated with λ_1 and

$$\mathbf{x_1}(t) = e^{4t} \begin{pmatrix} 1 \\ 1 \end{pmatrix}$$

is one solution of the system.

Second, $\lambda_2 = 2$ implies

$$A - \lambda_2 I = \begin{pmatrix} 3 & -1 \\ 3 & -1 \end{pmatrix}.$$

Therefore,

$$\mathbf{v_2} = \begin{pmatrix} 1 \\ 3 \end{pmatrix}$$

is an eigenvector associated with λ_2 and

$$\mathbf{x_2}(t) = e^{2t} \begin{pmatrix} 1 \\ 3 \end{pmatrix}$$

is a second solution of the system.

35

Therefore, the general solution is given by

$$\mathbf{x}(t) = c_1 e^{4t} \begin{pmatrix} 1 \\ 1 \end{pmatrix} + c_2 e^{2t} \begin{pmatrix} 1 \\ 3 \end{pmatrix}.$$

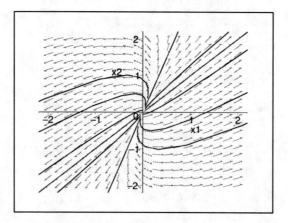

If the initial condition is a multiple of $\begin{pmatrix} 1 & 3 \end{pmatrix}^t$, then the solution will grow, staying on the eigenvector $\begin{pmatrix} 1 & 3 \end{pmatrix}^t$. Otherwise, the solution will grow, following the eigenvector $\begin{pmatrix} 1 & 1 \end{pmatrix}^t$.

11. We look for eigenvalues and eigenvectors of

$$A = \begin{pmatrix} -2 & 1 \\ -5 & 4 \end{pmatrix}.$$

We see that $\det(A - \lambda I) = \lambda^2 - 2\lambda - 3 = (\lambda - 3)(\lambda + 1)$. Therefore, the eigenvalues are given by $\lambda = 3, -1$.

First, $\lambda_1 = 3$ implies

$$A - \lambda_1 I = \begin{pmatrix} -5 & 1 \\ -5 & 1 \end{pmatrix}.$$

Therefore,

$$\mathbf{v_1} = \begin{pmatrix} 1 \\ 5 \end{pmatrix}$$

is an eigenvector associated with λ_1 and

$$\mathbf{x_1}(t) = e^{3t} \begin{pmatrix} 1 \\ 5 \end{pmatrix}$$

is one solution of the system.

Second, $\lambda_2 = -1$ implies

$$A - \lambda_2 I = \begin{pmatrix} -1 & 1 \\ -5 & 5 \end{pmatrix}.$$

Therefore,

$$\mathbf{v_2} = \begin{pmatrix} 1 \\ 1 \end{pmatrix}$$

is an eigenvector associated with λ_2 and

$$\mathbf{x_2}(t) = e^{-t} \begin{pmatrix} 1 \\ 1 \end{pmatrix}$$

is a second solution of the system.

Therefore, the general solution is given by

$$\mathbf{x}(t) = c_1 e^{3t} \begin{pmatrix} 1 \\ 5 \end{pmatrix} + c_2 e^{-t} \begin{pmatrix} 1 \\ 1 \end{pmatrix}.$$

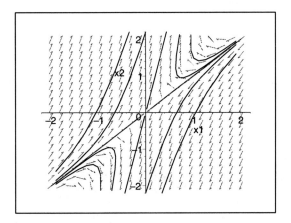

If the initial condition is a multiple of $\begin{pmatrix} 1 & 1 \end{pmatrix}^t$, then the solution will tend to the origin along the eigenvector $\begin{pmatrix} 1 & 1 \end{pmatrix}^t$. Otherwise, the solution will grow, following the eigenvector $\begin{pmatrix} 1 & 5 \end{pmatrix}^t$.

12. We look for eigenvalues and eigenvectors of

$$A = \begin{pmatrix} 3 & 6 \\ -1 & -2 \end{pmatrix}.$$

We see that $\det(A - \lambda I) = \lambda^2 - \lambda = \lambda(\lambda - 1)$. Therefore, the eigenvalues are given by $\lambda = 0, 1$.

First, $\lambda_1 = 0$ implies

$$A - \lambda_1 I = \begin{pmatrix} 3 & 6 \\ -1 & -2 \end{pmatrix}.$$

Therefore,

$$\mathbf{v_1} = \begin{pmatrix} 2 \\ -1 \end{pmatrix}$$

is an eigenvector associated with λ_1 and

$$\mathbf{x_1}(t) = \begin{pmatrix} 2 \\ -1 \end{pmatrix}$$

is one solution of the system.

Second, $\lambda_2 = 1$ implies

$$A - \lambda_2 I = \begin{pmatrix} 2 & 6 \\ -1 & -3 \end{pmatrix}.$$

37

Therefore,

$$\mathbf{v_2} = \begin{pmatrix} 3 \\ 1 \end{pmatrix}$$

is an eigenvector associated with λ_2 and

$$\mathbf{x_2}(t) = e^t \begin{pmatrix} 3 \\ -1 \end{pmatrix}$$

is a second solution of the system.

Therefore, the general solution is given by

$$\mathbf{x}(t) = c_1 \begin{pmatrix} 2 \\ -1 \end{pmatrix} + c_2 e^t \begin{pmatrix} 3 \\ -1 \end{pmatrix}.$$

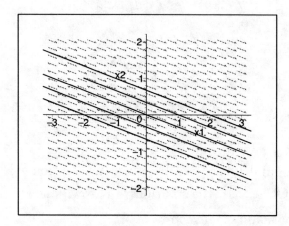

The entire line spanned by the eigenvector $\begin{pmatrix} 2 & -1 \end{pmatrix}^{-t}$ consists of equilibrium points. The direction field changes across the line $x_1 + 2x_2 = 0$.

13.

$$A = \begin{pmatrix} 1 & -2 \\ 3 & -4 \end{pmatrix}$$

implies $\det(A - \lambda I) = \lambda^2 + 3\lambda + 2 = (\lambda + 2)(\lambda + 1)$. Therefore, the eigenvalues are $\lambda = -2$ and $\lambda = -1$.

First, $\lambda_1 = -2$ implies

$$A - \lambda_1 I = \begin{pmatrix} 3 & -2 \\ 3 & -2 \end{pmatrix}.$$

Therefore,

$$\mathbf{v_1} = \begin{pmatrix} 2 \\ 3 \end{pmatrix}$$

is an eigenvector for λ_1. Therefore,

$$\mathbf{x_1}(t) = e^{-2t} \begin{pmatrix} 2 \\ 3 \end{pmatrix}$$

is one solution.

Second, $\lambda_2 = -1$ implies

$$A - \lambda_2 I = \begin{pmatrix} 2 & -2 \\ 3 & -3 \end{pmatrix}.$$

Therefore,

$$\mathbf{v}_2 = \begin{pmatrix} 1 \\ 1 \end{pmatrix}$$

is an eigenvector for λ_2. Therefore,

$$\mathbf{x}_2(t) = e^{-t} \begin{pmatrix} 1 \\ 1 \end{pmatrix}$$

is a second solution.

Therefore, the general solution is

$$\mathbf{x}(t) = c_1 e^{-2t} \begin{pmatrix} 2 \\ 3 \end{pmatrix} + c_2 e^{-t} \begin{pmatrix} 1 \\ 1 \end{pmatrix}.$$

The initial condition $\mathbf{x}(0) = \begin{pmatrix} 3 \\ 1 \end{pmatrix}^t$ implies

$$\begin{pmatrix} 2 & 1 \\ 3 & 1 \end{pmatrix} \begin{pmatrix} c_1 \\ c_2 \end{pmatrix} = \begin{pmatrix} 3 \\ 1 \end{pmatrix}.$$

Therefore,

$$\begin{pmatrix} c_1 \\ c_2 \end{pmatrix} = \begin{pmatrix} -1 & 1 \\ 3 & -2 \end{pmatrix} \begin{pmatrix} 3 \\ 1 \end{pmatrix} = \begin{pmatrix} -2 \\ 7 \end{pmatrix}.$$

Therefore, the solution is

$$\mathbf{x}(t) = -2e^{-2t} \begin{pmatrix} 2 \\ 3 \end{pmatrix} + 7e^{-t} \begin{pmatrix} 1 \\ 1 \end{pmatrix}.$$

Both components tend to zero as $t \to \infty$.

14.

$$A = \begin{pmatrix} 2 & -1 \\ 3 & -2 \end{pmatrix}$$

implies $\det(A - \lambda I) = \lambda^2 - 1 = (\lambda + 1)(\lambda - 1)$. Therefore, the eigenvalues are $\lambda = 1$ and $\lambda = -1$.

First, $\lambda_1 = 1$ implies

$$A - \lambda_1 I = \begin{pmatrix} 1 & 1 \\ 3 & -3 \end{pmatrix}.$$

Therefore,

$$\mathbf{v}_1 = \begin{pmatrix} 1 \\ 1 \end{pmatrix}$$

is an eigenvector for λ_1. Therefore,

$$\mathbf{x}_1(t) = e^t \begin{pmatrix} 1 \\ 1 \end{pmatrix}$$

is one solution.

Second, $\lambda_2 = -1$ implies

$$A - \lambda_2 I = \begin{pmatrix} 3 & -1 \\ 3 & -1 \end{pmatrix}.$$

Therefore,

$$\mathbf{v}_2 = \begin{pmatrix} 1 \\ 3 \end{pmatrix}$$

is an eigenvector for λ_2. Therefore,

$$\mathbf{x}_2(t) = e^{-t} \begin{pmatrix} 1 \\ 3 \end{pmatrix}$$

is a second solution.

Therefore, the general solution is

$$\mathbf{x}(t) = c_1 e^t \begin{pmatrix} 1 \\ 1 \end{pmatrix} + c_2 e^{-t} \begin{pmatrix} 1 \\ 3 \end{pmatrix}.$$

The initial condition $\mathbf{x}(0) = \begin{pmatrix} 2 \\ 5 \end{pmatrix}^t$ implies

$$\begin{pmatrix} 1 & 1 \\ 1 & 3 \end{pmatrix} \begin{pmatrix} c_1 \\ c_2 \end{pmatrix} = \begin{pmatrix} 2 \\ 5 \end{pmatrix}.$$

Therefore,

$$\begin{pmatrix} c_1 \\ c_2 \end{pmatrix} = \frac{1}{2} \begin{pmatrix} 3 & -1 \\ -1 & 1 \end{pmatrix} \begin{pmatrix} 2 \\ 5 \end{pmatrix} = \begin{pmatrix} 1/2 \\ 3/2 \end{pmatrix}.$$

Therefore, the solution is

$$\mathbf{x}(t) = \frac{1}{2} e^t \begin{pmatrix} 1 \\ 1 \end{pmatrix} + \frac{3}{2} e^{-t} \begin{pmatrix} 1 \\ 3 \end{pmatrix}.$$

Both components x_1 and x_2 tend to $+\infty$ as $t \to \infty$.

15.

$$A = \begin{pmatrix} 5 & -1 \\ 3 & 1 \end{pmatrix}$$

implies $\det(A - \lambda I) = \lambda^2 - 6\lambda + 8 = (\lambda - 4)(\lambda - 2)$. Therefore, the eigenvalues are $\lambda = 2$ and $\lambda = 4$.

First, $\lambda_1 = 2$ implies

$$A - \lambda_1 I = \begin{pmatrix} 3 & -1 \\ 3 & -1 \end{pmatrix}.$$

Therefore,

$$\mathbf{v}_1 = \begin{pmatrix} 1 \\ 3 \end{pmatrix}$$

is an eigenvector for λ_1. Therefore,

$$\mathbf{x}_1(t) = e^{2t} \begin{pmatrix} 1 \\ 3 \end{pmatrix}$$

is one solution.

Second, $\lambda_2 = 4$ implies

$$A - \lambda_2 I = \begin{pmatrix} 1 & -1 \\ 3 & -3 \end{pmatrix}.$$

Therefore,

$$\mathbf{v}_2 = \begin{pmatrix} 1 \\ 1 \end{pmatrix}$$

is an eigenvector for λ_2. Therefore,

$$\mathbf{x}_2(t) = e^{4t} \begin{pmatrix} 1 \\ 1 \end{pmatrix}$$

is a second solution.

Therefore, the general solution is

$$\mathbf{x}(t) = c_1 e^{2t} \begin{pmatrix} 1 \\ 3 \end{pmatrix} + c_2 e^{4t} \begin{pmatrix} 1 \\ 1 \end{pmatrix}.$$

41

The initial condition $\mathbf{x}(0) = \begin{pmatrix} 2 \\ -1 \end{pmatrix}^t$ implies

$$\begin{pmatrix} 1 & 3 \\ 1 & 1 \end{pmatrix} \begin{pmatrix} c_1 \\ c_2 \end{pmatrix} = \begin{pmatrix} 2 \\ -1 \end{pmatrix}.$$

Therefore,

$$\begin{pmatrix} c_1 \\ c_2 \end{pmatrix} = \begin{pmatrix} -3/2 \\ 7/2 \end{pmatrix}.$$

Therefore, the solution is

$$\mathbf{x}(t) = -\frac{3}{2} e^{2t} \begin{pmatrix} 1 \\ 3 \end{pmatrix} + \frac{7}{2} e^{4t} \begin{pmatrix} 1 \\ 1 \end{pmatrix}.$$

Both components tend to $+\infty$ as $t \to \infty$.

16.

$$A = \begin{pmatrix} -2 & 1 \\ -5 & 4 \end{pmatrix}$$

implies $\det(A - \lambda I) = \lambda^2 - 2\lambda - 3 = (\lambda + 1)(\lambda - 3)$. Therefore, the eigenvalues are $\lambda = 3$ and $\lambda = -1$.

First, $\lambda_1 = 3$ implies

$$A - \lambda_1 I = \begin{pmatrix} -5 & 1 \\ -5 & 1 \end{pmatrix}.$$

Therefore,

$$\mathbf{v}_1 = \begin{pmatrix} 1 \\ 5 \end{pmatrix}$$

is an eigenvector for λ_1. Therefore,

$$\mathbf{x}_1(t) = e^{3t} \begin{pmatrix} 1 \\ 5 \end{pmatrix}$$

is one solution.

Second, $\lambda_2 = -1$ implies

$$A - \lambda_2 I = \begin{pmatrix} -1 & 1 \\ -5 & 5 \end{pmatrix}.$$

Therefore,

$$\mathbf{v}_2 = \begin{pmatrix} 1 \\ 1 \end{pmatrix}$$

is an eigenvector for λ_2. Therefore,

$$\mathbf{x}_2(t) = e^{-t} \begin{pmatrix} 1 \\ 1 \end{pmatrix}$$

is a second solution.

Therefore, the general solution is

$$\mathbf{x}(t) = c_1 e^{3t} \begin{pmatrix} 1 \\ 5 \end{pmatrix} + c_2 e^{-t} \begin{pmatrix} 1 \\ 1 \end{pmatrix}.$$

The initial condition $\mathbf{x}(0) = \begin{pmatrix} 1 \\ 3 \end{pmatrix}^t$ implies

$$\begin{pmatrix} c_1 \\ c_2 \end{pmatrix} = \begin{pmatrix} 1/2 \\ 1/2 \end{pmatrix}.$$

Therefore, the solution is

$$\mathbf{x}(t) = \frac{1}{2} e^{3t} \begin{pmatrix} 1 \\ 5 \end{pmatrix} + \frac{1}{2} e^{-t} \begin{pmatrix} 1 \\ 1 \end{pmatrix}.$$

Both components tend to $+\infty$ as $t \to \infty$.

17.

(a)

(b)

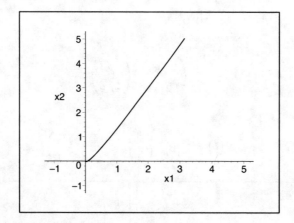

(c) The general solution is given by

$$\mathbf{x}(t) = c_1 e^{-t} \begin{pmatrix} -1 \\ 2 \end{pmatrix} + c_2 e^{-2t} \begin{pmatrix} 1 \\ 2 \end{pmatrix}.$$

The initial condition $\mathbf{x}(0) = \begin{pmatrix} 2 \\ 3 \end{pmatrix}^t$ implies

$$\begin{pmatrix} c_1 \\ c_2 \end{pmatrix} = \begin{pmatrix} -1/4 \\ 7/4 \end{pmatrix}.$$

Therefore, the solution passing through the initial point $(2, 3)$ is given by

$$\mathbf{x}(t) = -\frac{1}{4} e^{-t} \begin{pmatrix} -1 \\ 2 \end{pmatrix} + \frac{7}{4} e^{-2t} \begin{pmatrix} 1 \\ 2 \end{pmatrix}.$$

The component plots are shown below.

18.

(a)

(b)

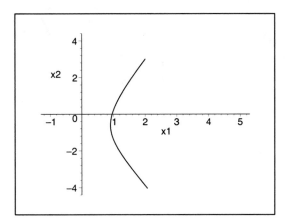

(c) The general solution is given by

$$\mathbf{x}(t) = c_1 e^t \begin{pmatrix} -1 \\ 2 \end{pmatrix} + c_2 e^{-2t} \begin{pmatrix} 1 \\ 2 \end{pmatrix}.$$

The initial condition $\mathbf{x}(0) = \begin{pmatrix} 2 \\ 3 \end{pmatrix}^t$ implies

$$\begin{pmatrix} c_1 \\ c_2 \end{pmatrix} = \begin{pmatrix} -1/4 \\ 7/4 \end{pmatrix}.$$

Therefore, the solution passing through the initial point $(2, 3)$ is given by

$$\mathbf{x}(t) = -\frac{1}{4}e^t \begin{pmatrix} -1 \\ 2 \end{pmatrix} + \frac{7}{4}e^{-2t} \begin{pmatrix} 1 \\ 2 \end{pmatrix}.$$

The component plots are shown below.

19.

(a)

(b)

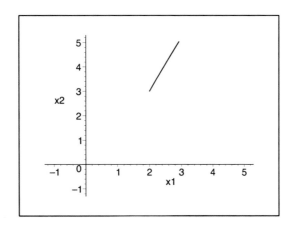

(c) The general solution is given by

$$\mathbf{x}(t) = c_1 e^{-t} \begin{pmatrix} -1 \\ 2 \end{pmatrix} + c_2 e^{2t} \begin{pmatrix} 1 \\ 2 \end{pmatrix}.$$

The initial condition $\mathbf{x}(0) = \begin{pmatrix} 2 \\ 3 \end{pmatrix}^t$ implies

$$\begin{pmatrix} c_1 \\ c_2 \end{pmatrix} = \begin{pmatrix} -1/4 \\ 7/4 \end{pmatrix}.$$

Therefore, the solution passing through the initial point $(2, 3)$ is given by

$$\mathbf{x}(t) = -\frac{1}{4} e^{-t} \begin{pmatrix} -1 \\ 2 \end{pmatrix} + \frac{7}{4} e^{2t} \begin{pmatrix} 1 \\ 2 \end{pmatrix}.$$

The component plots are shown below.

20.

(a)

(b)

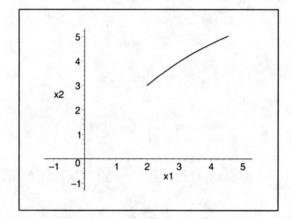

(c) The general solution is given by

$$\mathbf{x}(t) = c_1 e^t \begin{pmatrix} 1 \\ 2 \end{pmatrix} + c_2 e^{2t} \begin{pmatrix} 1 \\ -2 \end{pmatrix}.$$

The initial condition $\mathbf{x}(0) = \begin{pmatrix} 2 \\ 3 \end{pmatrix}^t$ implies

$$\begin{pmatrix} c_1 \\ c_2 \end{pmatrix} = \begin{pmatrix} 7/4 \\ 1/4 \end{pmatrix}.$$

Therefore, the solution passing through the initial point $(2, 3)$ is given by

$$\mathbf{x}(t) = \frac{7}{4} e^t \begin{pmatrix} 1 \\ 2 \end{pmatrix} + \frac{1}{4} e^{2t} \begin{pmatrix} 1 \\ -2 \end{pmatrix}.$$

The component plots are shown below.

21.

(a)

(b)

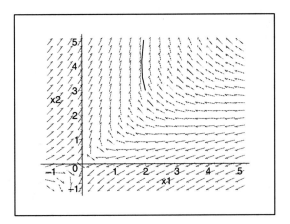

(c) The general solution is given by

$$\mathbf{x}(t) = c_1 e^{0.5t} \begin{pmatrix} 1 \\ 4 \end{pmatrix} + c_2 e^{-0.5t} \begin{pmatrix} 4 \\ 1 \end{pmatrix}.$$

49

The initial condition $\mathbf{x}(0) = \begin{pmatrix} 2 \\ 3 \end{pmatrix}^t$ implies

$$\begin{pmatrix} 1 & 4 \\ 4 & 1 \end{pmatrix} \begin{pmatrix} c_1 \\ c_2 \end{pmatrix} = \begin{pmatrix} 2 \\ 3 \end{pmatrix}.$$

Therefore,

$$\begin{pmatrix} c_1 \\ c_2 \end{pmatrix} = -\frac{1}{15} \begin{pmatrix} 1 & -4 \\ -4 & 1 \end{pmatrix} \begin{pmatrix} 2 \\ 3 \end{pmatrix} = \begin{pmatrix} 2/3 \\ 1/3 \end{pmatrix}.$$

Therefore, the solution passing through the initial point $(2, 3)$ is given by

$$\mathbf{x}(t) = \frac{2}{3} e^{0.5t} \begin{pmatrix} 1 \\ 4 \end{pmatrix} + \frac{1}{3} e^{-0.5t} \begin{pmatrix} 4 \\ 1 \end{pmatrix}.$$

The component plots are shown below.

22.

(a)

(b)

50

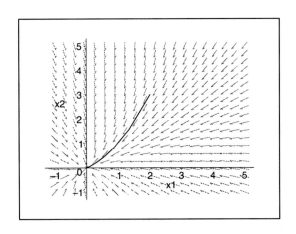

(c) The general solution is given by

$$\mathbf{x}(t) = c_1 e^{-0.5t} \begin{pmatrix} 2 \\ 1 \end{pmatrix} + c_2 e^{-0.8t} \begin{pmatrix} -1 \\ 2 \end{pmatrix}.$$

The initial condition $\mathbf{x}(0) = \begin{pmatrix} 2 \\ 3 \end{pmatrix}^t$ implies

$$\begin{pmatrix} 2 & -1 \\ 1 & 2 \end{pmatrix} \begin{pmatrix} c_1 \\ c_2 \end{pmatrix} = \begin{pmatrix} 2 \\ 3 \end{pmatrix}.$$

Therefore,

$$\begin{pmatrix} c_1 \\ c_2 \end{pmatrix} = \frac{1}{5} \begin{pmatrix} 2 & 1 \\ -1 & 2 \end{pmatrix} \begin{pmatrix} 2 \\ 3 \end{pmatrix} = \begin{pmatrix} 7/5 \\ 4/5 \end{pmatrix}.$$

Therefore, the solution passing through the initial point $(2,3)$ is given by

$$\mathbf{x}(t) = \frac{7}{5} e^{-0.5t} \begin{pmatrix} 2 \\ 1 \end{pmatrix} + \frac{4}{5} e^{-0.8t} \begin{pmatrix} -1 \\ 2 \end{pmatrix}.$$

The component plots are shown below.

23.

(a)

(b)

(c) The general solution is given by

$$\mathbf{x}(t) = c_1 e^{0.3t} \begin{pmatrix} 1 \\ -2 \end{pmatrix} + c_2 e^{0.6t} \begin{pmatrix} 1 \\ 3 \end{pmatrix}.$$

The initial condition $\mathbf{x}(0) = \begin{pmatrix} 2 \\ 3 \end{pmatrix}^t$ implies

$$\begin{pmatrix} 1 & 1 \\ -2 & 3 \end{pmatrix} \begin{pmatrix} c_1 \\ c_2 \end{pmatrix} = \begin{pmatrix} 2 \\ 3 \end{pmatrix}.$$

Therefore,

$$\begin{pmatrix} c_1 \\ c_2 \end{pmatrix} = \frac{1}{5} \begin{pmatrix} 3 & -1 \\ 2 & 1 \end{pmatrix} \begin{pmatrix} 2 \\ 3 \end{pmatrix} = \begin{pmatrix} 3/5 \\ 1 \end{pmatrix}.$$

Therefore, the solution passing through the initial point $(2, 3)$ is given by

$$\mathbf{x}(t) = \frac{7}{5} e^{-0.5t} \begin{pmatrix} 2 \\ 1 \end{pmatrix} + \frac{4}{5} e^{-0.8t} \begin{pmatrix} -1 \\ 2 \end{pmatrix}.$$

The component plots are shown below.

24.

(a)

(b)

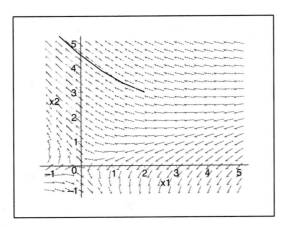

(c) The general solution is given by

$$\mathbf{x}(t) = c_1 e^{1.5t} \begin{pmatrix} -1 \\ 2 \end{pmatrix} + c_2 e^{-t} \begin{pmatrix} 3 \\ 1 \end{pmatrix}.$$

The initial condition $\mathbf{x}(0) = \begin{pmatrix} 2 \\ 3 \end{pmatrix}^t$ implies

$$\begin{pmatrix} -1 & 3 \\ 2 & 1 \end{pmatrix} \begin{pmatrix} c_1 \\ c_2 \end{pmatrix} = \begin{pmatrix} 2 \\ 3 \end{pmatrix}.$$

Therefore,

$$\begin{pmatrix} c_1 \\ c_2 \end{pmatrix} = -\frac{1}{7} \begin{pmatrix} 1 & -3 \\ -2 & -1 \end{pmatrix} \begin{pmatrix} 2 \\ 3 \end{pmatrix} = \begin{pmatrix} 1 \\ 1 \end{pmatrix}.$$

Therefore, the solution passing through the initial point $(2, 3)$ is given by

$$\mathbf{x}(t) = e^{1.5t} \begin{pmatrix} -1 \\ 2 \end{pmatrix} + e^{-t} \begin{pmatrix} 3 \\ 1 \end{pmatrix}.$$

The component plots are shown below.

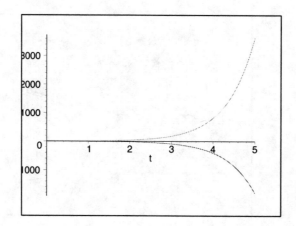

25. The general solution in Example 3 is

$$\mathbf{x}(t) = c_1 e^{-0.25t} \begin{pmatrix} 1 \\ -2 \end{pmatrix} + c_2 e^{-0.05t} \begin{pmatrix} 3 \\ 2 \end{pmatrix}.$$

The initial condition $x_1(0) = 13$, $x_2(0) = -10$ implies that $c_1 = 7$ and $c_2 = 2$. Therefore,

$$x_1(t) = 7e^{-0.25t} + 6e^{-0.05t}$$
$$x_2(t) = -14e^{-0.25t} + 4e^{-0.05t}.$$

The first component x_1 is monotone decreasing. By solving the equation $7e^{-0.25t} + 6e^{-0.05t} = 0.5$, we see that $x_1(T) = 0.5$ when $T = 49.7$. Therefore, $0 \leq x_1(t) \leq 0.5$ for all $t \geq 49.7$. The second component, x_2 is monotone decreasing after reaching its maximum. By solving the equation $-14e^{-0.25t} + 4e^{-0.05t} = 0.5$, we can see that $0 \leq x_2(t) \leq 0.5$ for all $t \geq 41.57$. Therefore, we conclude that the component functions will satisfy the specified bounds for all $t \geq 49.7$.

26.

(a) For $\alpha = 0.5$, the characteristic equation is $2\lambda^2 + 4\lambda + 1 = 0$. Therefore, the eigenvalues are $\lambda = -1 \pm 1/\sqrt{2}$. The eigenvector corresponding to $\lambda_1 = -1 + 1/\sqrt{2}$ is $\mathbf{v_1} = \begin{pmatrix} -\sqrt{2} & 1 \end{pmatrix}^t$. The eigenvector corresponding to $\lambda_2 = -1 - 1/\sqrt{2}$ is $\mathbf{v_2} = \begin{pmatrix} \sqrt{2} & 1 \end{pmatrix}^t$. Therefore, the general solution is

$$\mathbf{x}(t) = c_1 e^{(-1+1/\sqrt{2})t} \begin{pmatrix} -\sqrt{2} \\ 1 \end{pmatrix} + c_2 e^{(-1-1/\sqrt{2})t} \begin{pmatrix} \sqrt{2} \\ 1 \end{pmatrix}.$$

Since both eigenvalues are negative, the equilibrium point is a stable node.

(b) For $\alpha = 2$, the characteristic equation is $\lambda^2 + 2\lambda - 1 = 0$. Therefore, the eigenvalues are $\lambda = -1 \pm \sqrt{2}$. The eigenvector corresponding to $\lambda_1 = -1 + \sqrt{2}$ is $\mathbf{v_1} = \begin{pmatrix} 1 & -\sqrt{2} \end{pmatrix}^t$. The eigenvector corresponding to $\lambda_2 = -1 - \sqrt{2}$ is $\mathbf{v_2} = \begin{pmatrix} 1 & \sqrt{2} \end{pmatrix}^t$. Therefore, the general solution is

$$\mathbf{x}(t) = c_1 e^{(-1+\sqrt{2})t} \begin{pmatrix} 1 \\ -\sqrt{2} \end{pmatrix} + c_2 e^{(-1-\sqrt{2})t} \begin{pmatrix} 1 \\ \sqrt{2} \end{pmatrix}.$$

Since the eigenvalues have opposite sign, the equilibrium point is a saddle point.

(c) For general α, the characteristic equation is $\lambda^2 + 2\lambda + 1 - \alpha = 0$. The eigenvalues are given by $\lambda = -1 \pm \sqrt{\alpha}$. For $0.5 \leq \alpha \leq 2$, both eigenvalues are real and clearly $-1 - \sqrt{\alpha} < 0$. Therefore, we just need to determine whether $-1 + \sqrt{\alpha} = 0$. We see this occurs when $\alpha = 1$. At that value the equilibrium point switches from a saddle point (for $\alpha > 1$) to a stable node (for $\alpha < 1$).

27.

(a) For the given data, the system can be written as

$$\frac{d}{dt}\begin{pmatrix} i \\ v \end{pmatrix} = \begin{pmatrix} -1/2 & -1/2 \\ 3/2 & -5/2 \end{pmatrix}\begin{pmatrix} i \\ v \end{pmatrix}.$$

The characteristic equation for this system is $\lambda^2 + 3\lambda + 2 = 0$. Therefore, the eigenvalues are given by $\lambda = -1, -2$. For $\lambda = -1$, a corresponding eigenvector is given by $\mathbf{v_1} = \begin{pmatrix} 1 & 1 \end{pmatrix}^t$. For $\lambda = -2$, a corresponding eigenvector is given by $\mathbf{v_2} = \begin{pmatrix} 1 & 3 \end{pmatrix}^t$. Therefore, the general solution is

$$\begin{pmatrix} i \\ v \end{pmatrix} = c_1 e^{-t}\begin{pmatrix} 1 \\ 1 \end{pmatrix} + c_2 e^{-2t}\begin{pmatrix} 1 \\ 3 \end{pmatrix}.$$

(b) Since both eigenvalues are real and negative, the equilibrium point $(0,0)$ is a stable node. Therefore, $i(t) \to 0$ and $v(t) \to 0$ as $t \to \infty$.

28.

(a) For the general equation (i), the characteristic equation is given by

$$\lambda^2 + \left(\frac{L + CR_1 R_2}{LCR_2}\right)\lambda + \frac{R_1 + R_2}{LCR_2} = 0.$$

55

The eigenvalues are real and distinct provided that the discriminant is positive. That is,

$$\left(\frac{L + CR_1R_2}{LCR_2}\right)^2 - 4\left(\frac{R_1 + R_2}{LCR_2}\right) > 0,$$

which simplifies to the condition

$$\left(\frac{1}{CR_2} - \frac{R_1}{L}\right)^2 - \frac{4}{LC} > 0.$$

(b) Since the sum of the roots of the characteristic equation is

$$-\frac{L + CR_1R_2}{LCR_2} < 0$$

and the product of the roots is

$$\frac{R_1 + R_2}{LCR_2} > 0,$$

it follows that both roots are negative. Therefore, $i = 0, v = 0$ is a stable node.

29. $\mathbf{u} = \mathbf{x} - \mathbf{x}_c$ implies

$$\begin{aligned}
\mathbf{u}' &= \mathbf{x}' - \mathbf{x}_c' \\
&= A\mathbf{x} + \mathbf{b} \\
&= A(\mathbf{u} + \mathbf{x}_c) + \mathbf{b} \\
&= A\mathbf{u} + A\mathbf{x}_c' + \mathbf{b} \\
&= A\mathbf{u} + A(-A^{-1}\mathbf{b}) + \mathbf{b} \\
&= A\mathbf{u} - \mathbf{b} + \mathbf{b} \\
&= A\mathbf{u}.
\end{aligned}$$

Therefore, $\mathbf{u}' = A\mathbf{u}$ as claimed.

Section 3.4

1.
$$A = \begin{pmatrix} 3 & -2 \\ 4 & -1 \end{pmatrix}$$

implies

$$\det(A - \lambda I) = \lambda^2 - 2\lambda + 5.$$

Therefore, the eigenvalues are given by $\lambda = 1 \pm 2i$. Now, $\lambda_1 = 1 + 2i$ implies

$$A - \lambda_1 I = \begin{pmatrix} 2 - 2i & -2 \\ 4 & -2 - 2i \end{pmatrix}.$$

Therefore,

$$\mathbf{v}_1 = \begin{pmatrix} 1 \\ 1 - i \end{pmatrix}$$

is an eigenvector for λ_1 and

$$\mathbf{x}_1(t) = e^{(1+2i)t}\begin{pmatrix} 1 \\ 1 - i \end{pmatrix}$$

is a solution of our system. Further, we use the fact that the real and imaginary parts of $\mathbf{x}_1(t)$ are linearly independent solutions of our system. Now

$$\text{Re }\mathbf{x}_1(t) = e^t\left[\begin{pmatrix} 1 \\ 1 \end{pmatrix}\cos(2t) - \begin{pmatrix} 0 \\ -1 \end{pmatrix}\sin(2t)\right] = e^t\begin{pmatrix} \cos(2t) \\ \cos(2t) + \sin(2t) \end{pmatrix}$$

$$\text{Im }\mathbf{x}_1(t) = e^t\left[\begin{pmatrix} 1 \\ 1 \end{pmatrix}\sin(2t) + \begin{pmatrix} 0 \\ -1 \end{pmatrix}\cos(2t)\right] = e^t\begin{pmatrix} \sin(2t) \\ \sin(2t) - \cos(2t) \end{pmatrix}.$$

Therefore, the general solution is given by

$$\mathbf{x}(t) = c_1 e^t\begin{pmatrix} \cos(2t) \\ \cos(2t) + \sin(2t) \end{pmatrix} + c_2 e^t\begin{pmatrix} \sin(2t) \\ \sin(2t) - \cos(2t) \end{pmatrix}.$$

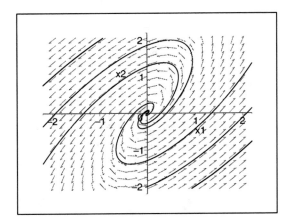

The equilibrium is an unstable spiral point.

2.

$$A = \begin{pmatrix} -1 & -4 \\ 1 & -1 \end{pmatrix}$$

implies

$$\det(A - \lambda I) = \lambda^2 + 2\lambda + 5.$$

Therefore, the eigenvalues are given by $\lambda = -1 \pm 2i$. Now, $\lambda_1 = -1 + 2i$ implies

$$A - \lambda_1 I = \begin{pmatrix} -2i & -4 \\ 1 & -2i \end{pmatrix}.$$

Therefore,

$$\mathbf{v}_1 = \begin{pmatrix} 2i \\ 1 \end{pmatrix}$$

is an eigenvector for λ_1 and

$$\mathbf{x}_1(t) = e^{(-1+2i)t}\begin{pmatrix} 2i \\ 1 \end{pmatrix}$$

57

is a solution of our system. Further, we use the fact that the real and imaginary parts of $\mathbf{x}_1(t)$ are linearly independent solutions of our system. Now

$$\operatorname{Re} \mathbf{x}_1(t) = e^{-t} \left[\begin{pmatrix} 0 \\ 1 \end{pmatrix} \cos(2t) - \begin{pmatrix} 2 \\ 0 \end{pmatrix} \sin(2t) \right] = e^{-t} \begin{pmatrix} -2\sin(2t) \\ \cos(2t) \end{pmatrix}$$

$$\operatorname{Im} \mathbf{x}_1(t) = e^{-t} \left[\begin{pmatrix} 0 \\ 1 \end{pmatrix} \sin(2t) + \begin{pmatrix} 2 \\ 0 \end{pmatrix} \cos(2t) \right] = e^{-t} \begin{pmatrix} 2\cos(2t) \\ \sin(2t) \end{pmatrix}.$$

Therefore, the general solution is given by

$$\mathbf{x}(t) = c_1 e^{-t} \begin{pmatrix} -2\sin(2t) \\ \cos(2t) \end{pmatrix} + c_2 e^{-t} \begin{pmatrix} 2\cos(2t) \\ \sin(2t) \end{pmatrix}.$$

The initial condition, $\mathbf{x}(0) = \begin{pmatrix} 4 & -3 \end{pmatrix}^t$ implies

$$c_1 \begin{pmatrix} 0 \\ 1 \end{pmatrix} + c_2 \begin{pmatrix} 2 \\ 0 \end{pmatrix} = \begin{pmatrix} 4 \\ -3 \end{pmatrix}.$$

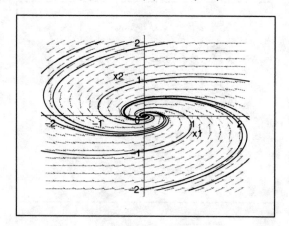

The equilibrium is an asymptotically stable spiral point.

3.

$$A = \begin{pmatrix} 2 & -5 \\ 1 & -2 \end{pmatrix}$$

implies

$$\det(A - \lambda I) = \lambda^2 + 1.$$

Therefore, the eigenvalues are given by $\lambda = \pm i$. Now, $\lambda_1 = i$ implies

$$A - \lambda_1 I = \begin{pmatrix} 2-i & -5 \\ 1 & -2-i \end{pmatrix}.$$

Therefore,

$$\mathbf{v}_1 = \begin{pmatrix} 2+i \\ 1 \end{pmatrix}$$

is an eigenvector for λ_1 and

$$\mathbf{x}_1(t) = e^{it} \begin{pmatrix} 2+i \\ 1 \end{pmatrix}$$

is a solution of our system. Further, we use the fact that the real and imaginary parts of $\mathbf{x}_1(t)$ are linearly independent solutions of our system. Now

$$\operatorname{Re} \mathbf{x}_1(t) = \left[\begin{pmatrix} 2 \\ 1 \end{pmatrix} \cos(t) - \begin{pmatrix} 1 \\ 0 \end{pmatrix} \sin(t)\right] = \begin{pmatrix} 2\cos(t) - \sin(t) \\ \cos(t) \end{pmatrix}$$

$$\operatorname{Im} \mathbf{x}_1(t) = \left[\begin{pmatrix} 2 \\ 1 \end{pmatrix} \sin(t) + \begin{pmatrix} 1 \\ 0 \end{pmatrix} \cos(t)\right] = \begin{pmatrix} 2\sin(t) + \cos(t) \\ \sin(t) \end{pmatrix}.$$

Therefore, the general solution is given by

$$\mathbf{x}(t) = c_1 \begin{pmatrix} 2\cos(t) - \sin(t) \\ \cos(t) \end{pmatrix} + c_2 \begin{pmatrix} 2\sin(t) + \cos(t) \\ \sin(t) \end{pmatrix}.$$

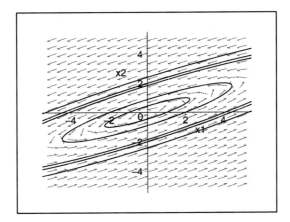

The equilibrium is a stable center.

4.

$$A = \begin{pmatrix} 2 & -5/2 \\ 9/5 & -1 \end{pmatrix}$$

implies

$$\det(A - \lambda I) = \lambda^2 - \lambda + 5/2.$$

Therefore, the eigenvalues are given by $\lambda = (1 \pm 3i)/2$. Now, $\lambda_1 = (1+3i)/2$ has corresponding eigenvector

$$\mathbf{v}_1 = \begin{pmatrix} 5 \\ 3 - 3i \end{pmatrix}$$

is an eigenvector for λ_1 and

$$\mathbf{x}_1(t) = e^{(1+3i)t/2} \begin{pmatrix} 5 \\ 3 - 3i \end{pmatrix}$$

is a solution of our system. Further, we use the fact that the real and imaginary parts of $\mathbf{x}_1(t)$ are linearly independent solutions of our system. Now

$$\operatorname{Re} \mathbf{x}_1(t) = e^{t/2} \begin{pmatrix} 2\cos(3t/2) - \sin(3t/2) \\ \cos(3t/2) \end{pmatrix}$$

$$\operatorname{Im} \mathbf{x}_1(t) = e^{t/2} \begin{pmatrix} \cos(3t/2) + 2\sin(3t/2) \\ \sin(3t/2) \end{pmatrix}.$$

Therefore, the general solution is given by

$$\mathbf{x}(t) = c_1 \begin{pmatrix} 2\cos(3t/2) - \sin(3t/2) \\ \cos(3t/2) \end{pmatrix} + c_2 e^{t/2} \begin{pmatrix} \cos(3t/2) + 2\sin(3t/2) \\ \sin(3t/2) \end{pmatrix}.$$

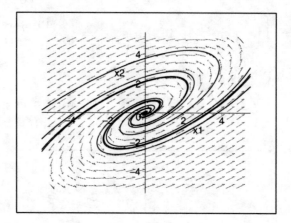

The equilibrium is an unstable spiral point.

5.

$$A = \begin{pmatrix} 1 & -1 \\ 5 & -3 \end{pmatrix}$$

implies

$$\det(A - \lambda I) = \lambda^2 + 2\lambda + 2.$$

Therefore, the eigenvalues are given by $\lambda = -1 \pm i$. Now, $\lambda_1 = -1 - i$ has corresponding eigenvector

$$\mathbf{v}_1 = \begin{pmatrix} 1 \\ 2 + i \end{pmatrix}$$

is an eigenvector for λ_1 and

$$\mathbf{x}_1(t) = e^{(-1-i)t} \begin{pmatrix} 1 \\ 2 + i \end{pmatrix}$$

is a solution of our system. Further, we use the fact that the real and imaginary parts of $\mathbf{x}_1(t)$ are linearly independent solutions of our system. Now

$$\operatorname{Re} \mathbf{x}_1(t) = e^{-t} \begin{pmatrix} \cos(t) \\ 2\cos(t) + \sin(t) \end{pmatrix}$$

$$\operatorname{Im} \mathbf{x}_1(t) = e^{-t} \begin{pmatrix} \sin(t) \\ -\cos(t) + 2\sin(t) \end{pmatrix}.$$

Therefore, the general solution is given by

$$\mathbf{x}(t) = c_1 e^{-t} \begin{pmatrix} \cos(t) \\ 2\cos(t) + \sin(t) \end{pmatrix} + c_2 e^{-t} \begin{pmatrix} \sin(t) \\ -\cos(t) + 2\sin(t) \end{pmatrix}.$$

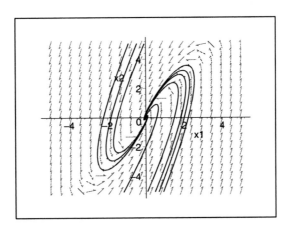

The equilibrium is an asymptotically stable spiral point.

6.

$$A = \begin{pmatrix} 1 & 2 \\ -5 & -1 \end{pmatrix}$$

implies

$$\det(A - \lambda I) = \lambda^2 + 9.$$

Therefore, the eigenvalues are given by $\lambda = \pm 3i$. Now, $\lambda_1 = 3i$ has corresponding eigenvector

$$\mathbf{v}_1 = \begin{pmatrix} -2 \\ 1 - 3i \end{pmatrix}$$

is an eigenvector for λ_1 and

$$\mathbf{x}_1(t) = e^{3it} \begin{pmatrix} -2 \\ 1 - 3i \end{pmatrix}$$

is a solution of our system. Further, we use the fact that the real and imaginary parts of $\mathbf{x}_1(t)$ are linearly independent solutions of our system. Now

$$\text{Re } \mathbf{x}_1(t) = \begin{pmatrix} -2\cos(3t) \\ \cos(3t) + 3\sin(3t) \end{pmatrix}$$

$$\text{Im } \mathbf{x}_1(t) = \begin{pmatrix} -2\sin(3t) \\ -3\cos(3t) + \sin(3t) \end{pmatrix}.$$

Therefore, the general solution is given by

$$\mathbf{x}(t) = c_1 \begin{pmatrix} -2\cos(3t) \\ \cos(3t) + 3\sin(3t) \end{pmatrix} + c_2 \begin{pmatrix} -2\sin(3t) \\ -3\cos(3t) + \sin(3t) \end{pmatrix}.$$

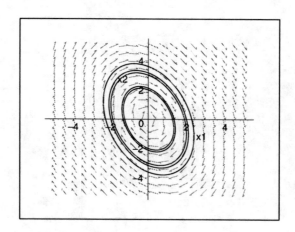

The equilibrium is a stable center.

7.
$$A = \begin{pmatrix} -1 & -4 \\ 1 & -1 \end{pmatrix}$$

implies
$$\det(A - \lambda I) = \lambda^2 + 2\lambda + 5.$$

Therefore, the eigenvalues are given by $\lambda = -1 \pm 2i$. Now, $\lambda_1 = -1 + 2i$ implies
$$A - \lambda_1 I = \begin{pmatrix} -2i & -4 \\ 1 & -2i \end{pmatrix}.$$

Therefore,
$$\mathbf{v}_1 = \begin{pmatrix} 2i \\ 1 \end{pmatrix}$$

is an eigenvector for λ_1 and
$$\mathbf{x}_1(t) = e^{(-1+2i)t} \begin{pmatrix} 2i \\ 1 \end{pmatrix}$$

is a solution of our system. Further, we use the fact that the real and imaginary parts of $\mathbf{x}_1(t)$ are linearly independent solutions of our system. Now

$$\text{Re } \mathbf{x}_1(t) = e^{-t} \left[\begin{pmatrix} 0 \\ 1 \end{pmatrix} \cos(2t) - \begin{pmatrix} 2 \\ 0 \end{pmatrix} \sin(2t) \right] = e^{-t} \begin{pmatrix} -2\sin(2t) \\ \cos(2t) \end{pmatrix}$$

$$\text{Im } \mathbf{x}_1(t) = e^{-t} \left[\begin{pmatrix} 0 \\ 1 \end{pmatrix} \sin(2t) + \begin{pmatrix} 2 \\ 0 \end{pmatrix} \cos(2t) \right] = e^{-t} \begin{pmatrix} 2\cos(2t) \\ \sin(2t) \end{pmatrix}.$$

Therefore, the general solution is given by

$$\mathbf{x}(t) = c_1 e^{-t} \begin{pmatrix} -2\sin(2t) \\ \cos(2t) \end{pmatrix} + c_2 e^{-t} \begin{pmatrix} 2\cos(2t) \\ \sin(2t) \end{pmatrix}.$$

The initial condition, $\mathbf{x}(0) = \begin{pmatrix} 4 & -3 \end{pmatrix}^t$ implies

$$c_1 \begin{pmatrix} 0 \\ 1 \end{pmatrix} + c_2 \begin{pmatrix} 2 \\ 0 \end{pmatrix} = \begin{pmatrix} 4 \\ -3 \end{pmatrix}.$$

Therefore, $c_1 = -3$ and $c_2 = 2$. Therefore, the solution is given by

$$\mathbf{x}(t) = -3e^{-t} \begin{pmatrix} -2\sin(2t) \\ \cos(2t) \end{pmatrix} + 2e^{-t} \begin{pmatrix} 2\cos(2t) \\ \sin(2t) \end{pmatrix}.$$

 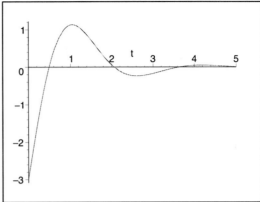

Both components decay to zero as $t \to \infty$.

8.

$$A = \begin{pmatrix} 2 & -5 \\ 1 & -2 \end{pmatrix}$$

implies

$$\det(A - \lambda I) = \lambda^2 + 1.$$

Therefore, the eigenvalues are given by $\lambda = \pm i$. Now, $\lambda_1 = i$ implies

$$A - \lambda_1 I = \begin{pmatrix} 2 - i & -5 \\ 1 & -2 - i \end{pmatrix}.$$

Therefore,

$$\mathbf{v}_1 = \begin{pmatrix} 2 + i \\ 1 \end{pmatrix}$$

is an eigenvector for λ_1 and

$$\mathbf{x}_1(t) = e^{it} \begin{pmatrix} 2 + i \\ 1 \end{pmatrix}$$

is a solution of our system. Further, we use the fact that the real and imaginary parts of $\mathbf{x}_1(t)$ are linearly independent solutions of our system. Now

$$\text{Re } \mathbf{x}_1(t) = \left[\begin{pmatrix} 2 \\ 1 \end{pmatrix} \cos(t) - \begin{pmatrix} 1 \\ 0 \end{pmatrix} \sin(t) \right] = \begin{pmatrix} 2\cos(t) - \sin(t) \\ \cos(t) \end{pmatrix}$$

$$\text{Im } \mathbf{x}_1(t) = \left[\begin{pmatrix} 2 \\ 1 \end{pmatrix} \sin(t) + \begin{pmatrix} 1 \\ 0 \end{pmatrix} \cos(t) \right] = \begin{pmatrix} 2\sin(t) + \cos(t) \\ \sin(t) \end{pmatrix}.$$

Therefore, the general solution is given by

$$\mathbf{x}(t) = c_1 \begin{pmatrix} 2\cos(t) - \sin(t) \\ \cos(t) \end{pmatrix} + c_2 \begin{pmatrix} 2\sin(t) + \cos(t) \\ \sin(t) \end{pmatrix}.$$

The initial condition, $\mathbf{x}(0) = \begin{pmatrix} 3 & 2 \end{pmatrix}^t$ implies

$$c_1 \begin{pmatrix} 2 \\ 1 \end{pmatrix} + c_2 \begin{pmatrix} 1 \\ 0 \end{pmatrix} = \begin{pmatrix} 3 \\ 2 \end{pmatrix}.$$

Therefore, $c_1 = 2$ and $c_2 = -1$. Therefore, the solution is given by

$$\mathbf{x}(t) = 2 \begin{pmatrix} 2\cos(t) - \sin(t) \\ \cos(t) \end{pmatrix} - \begin{pmatrix} 2\sin(t) + \cos(t) \\ \sin(t) \end{pmatrix} = \begin{pmatrix} 3\cos(t) - 4\sin(t) \\ 2\cos(t) - \sin(t) \end{pmatrix}.$$

 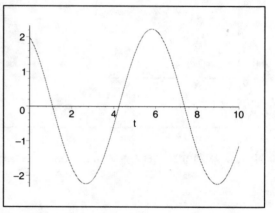

Both components oscillate as $t \to \infty$.

9.

$$A = \begin{pmatrix} 1 & -5 \\ 1 & -3 \end{pmatrix}$$

implies

$$\det(A - \lambda I) = \lambda^2 + 2\lambda + 2.$$

Therefore, the eigenvalues are given by $\lambda = -1 \pm i$. Now, $\lambda_1 = -1 + i$ implies

$$\mathbf{v}_1 = \begin{pmatrix} 2 + i \\ 1 \end{pmatrix}$$

is an eigenvector for λ_1 and

$$\mathbf{x}_1(t) = e^{(-1+i)t} \begin{pmatrix} 2 + i \\ 1 \end{pmatrix}$$

is a solution of our system. Further, we use the fact that the real and imaginary parts of $\mathbf{x}_1(t)$ are linearly independent solutions of our system. Now

$$\text{Re } \mathbf{x}_1(t) = e^{-t} \begin{pmatrix} 2\cos(t) - \sin(t) \\ \cos(t) \end{pmatrix}$$

$$\text{Im } \mathbf{x}_1(t) = e^{-t} \begin{pmatrix} 2\sin(t) + \cos(t) \\ \sin(t) \end{pmatrix}.$$

Therefore, the general solution is given by

$$\mathbf{x}(t) = c_1 e^{-t} \begin{pmatrix} 2\cos(t) - \sin(t) \\ \cos(t) \end{pmatrix} + c_2 e^{-t} \begin{pmatrix} 2\sin(t) + \cos(t) \\ \sin(t) \end{pmatrix}.$$

The initial condition, $\mathbf{x}(0) = \begin{pmatrix} 1 & 1 \end{pmatrix}^t$ implies $c_1 = 1$ and $c_2 = -1$. Therefore, the solution is given by

$$\mathbf{x}(t) = e^{-t} \begin{pmatrix} 2\cos(t) - \sin(t) \\ \cos(t) \end{pmatrix} - e^{-t} \begin{pmatrix} 2\sin(t) + \cos(t) \\ \sin(t) \end{pmatrix} = e^{-t} \begin{pmatrix} \cos(t) - 3\sin(t) \\ \cos(t) - \sin(t) \end{pmatrix}.$$

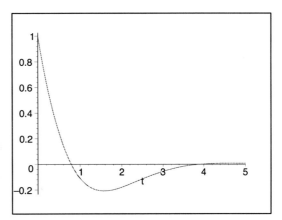

Both components approach zero as $t \to \infty$.

10.

$$A = \begin{pmatrix} -3 & 2 \\ -1 & -1 \end{pmatrix}$$

implies

$$\det(A - \lambda I) = \lambda^2 + 4\lambda + 5.$$

Therefore, the eigenvalues are given by $\lambda = -2 \pm i$. Now, $\lambda_1 = -2 + i$ implies

$$\mathbf{v}_1 = \begin{pmatrix} 1 - i \\ 1 \end{pmatrix}$$

is an eigenvector for λ_1 and

$$\mathbf{x}_1(t) = e^{(-2+i)t} \begin{pmatrix} 1 - i \\ 1 \end{pmatrix}$$

is a solution of our system. Further, we use the fact that the real and imaginary parts of $\mathbf{x}_1(t)$ are linearly independent solutions of our system. Now

$$\operatorname{Re} \mathbf{x}_1(t) = e^{-2t} \begin{pmatrix} \cos(t) + \sin(t) \\ \cos(t) \end{pmatrix}$$

$$\operatorname{Im} \mathbf{x}_1(t) = e^{-2t} \begin{pmatrix} \sin(t) - \cos(t) \\ \sin(t) \end{pmatrix}.$$

Therefore, the general solution is given by

$$\mathbf{x}(t) = c_1 e^{-2t} \begin{pmatrix} \cos(t) + \sin(t) \\ \cos(t) \end{pmatrix} + c_2 e^{-2t} \begin{pmatrix} \sin(t) - \cos(t) \\ \sin(t) \end{pmatrix}.$$

The initial condition, $\mathbf{x}(0) = \begin{pmatrix} 1 & -2 \end{pmatrix}^t$ implies $c_1 = -2$ and $c_2 = -3$. Therefore, the solution is given by

$$\mathbf{x}(t) = -2e^{-2t}\begin{pmatrix} \cos(t) + \sin(t) \\ \cos(t) \end{pmatrix} - 3e^{-2t}\begin{pmatrix} \sin(t) - \cos(t) \\ \sin(t) \end{pmatrix} = e^{-2t}\begin{pmatrix} \cos(t) - 5\sin(t) \\ -2\cos(t) - 3\sin(t) \end{pmatrix}.$$

Both components approach zero as $t \to \infty$.

11.

(a) The eigenvalues are given by $\lambda = -1/4 \pm i$.

(b) Take $\mathbf{x}(0) = \begin{pmatrix} 2 & 2 \end{pmatrix}^t$. Then the trajectory in the $x_1 x_2$-plane is given by

(c)

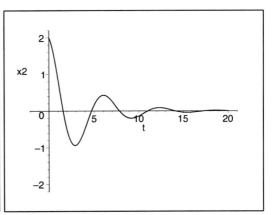

12.

(a) The eigenvalues are given by $\lambda = 1/5 \pm i$.

(b) Take $\mathbf{x}(0) = \begin{pmatrix} 1 & 2 \end{pmatrix}^t$. Then the trajectory in the x_1x_2-plane is given by

(c)

 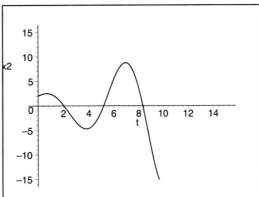

13.

(a) The characteristic equation is $\lambda^2 - 2\alpha\lambda + 1 + \alpha^2 = 0$. Therefore, the eigenvalues are $\lambda = \alpha \pm i$.

(b) For $\alpha > 0$, the equilibrium will be an unstable spiral. For $\alpha < 0$, the equilibrium with be a stable spiral. For $\alpha = 0$, the equilibrium will be a center.

(c) Below we show phase portraits for $\alpha = 1/2$ and $\alpha = -1/2$.

14.

(a) The characteristic equation is $\lambda^2 - \alpha\lambda + 5 = 0$. Therefore, the eigenvalues are $\lambda = (\alpha \pm \sqrt{\alpha^2 - 20})/2$.

(b) The eigenvalues have non-zero imaginary part when $-\sqrt{20} < \alpha < \sqrt{20}$. When $-\sqrt{20} < \alpha < 0$, the equilibrium point is a stable spiral. When $0 < \alpha < \sqrt{20}$, the equilibrium point is an unstable spiral. When $\alpha = 0$, the roots are purely imaginary which means the equilibrium point is a center. When $\alpha^2 > 20$, the roots are real and distinct. In that case, the equilibrium is a stable node if $\alpha < -\sqrt{20}$ and an unstable node if $\sqrt{20} < \alpha$. Therefore, the critical values of α are $\alpha = 0$, $\alpha = -\sqrt{20}$ and $\alpha = \sqrt{20}$.

(c) Below we show phase portraits for $\alpha = -5.5$, $\alpha = -3.5$, $\alpha = -0.5$ and $\alpha = 0.5$.

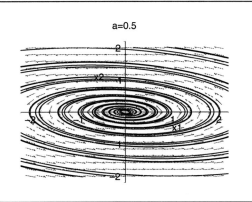

15.

(a) The characteristic equation is $\lambda^2 + 5\alpha - 4 = 0$. Therefore, the eigenvalues are $\lambda = \pm\sqrt{4 - 5\alpha}$.

(b) If $4 - 5\alpha < 0$, then the eigenvalues are purely imaginary. In that case, the equilibrium point is a center. If $4 - 5\alpha > 0$, then the eigenvalues are real and distinct. In that case, the equilibrium point is a center. Therefore, the critical value for α is $\alpha = 4/5$.

(c) Below we show phase portraits for $\alpha = 0$, and $\alpha = 2$.

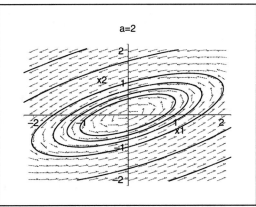

16.

(a) The characteristic equation is $\lambda^2 - \dfrac{5}{2}\lambda + \dfrac{25}{16} - \dfrac{3}{4}\alpha = 0$. Therefore, the eigenvalues are $\lambda = \dfrac{5}{4} \pm \dfrac{1}{2}\sqrt{3\alpha}$.

(b) If $\alpha < 0$, then the eigenvalues have non-zero imaginary part. In this case, the equilibrium point is an unstable spiral. If $\alpha > 0$, then the eigenvalues are real. If in addition $5/4 - \sqrt{3\alpha}/2 > 0$, then the equilibrium point is an unstable node. If $\alpha > 0$, but $5/4 - \sqrt{3\alpha}/2 < 0$, then the equilibrium point is a saddle point. Therefore, the critical values for α are $\alpha = 0$ and $\alpha = 25/12$.

(c) Below we show phase portraits for $\alpha = -0.5$, $\alpha = 0.5$, $\alpha = 1.5$ and $\alpha = 3$.

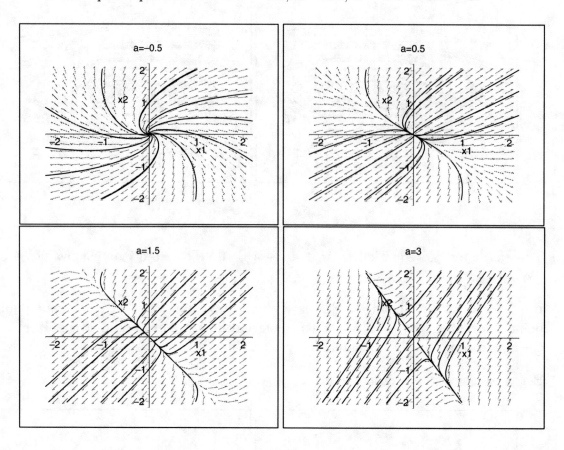

17.

(a) The characteristic equation is $\lambda^2 + 2\lambda + 1 + \alpha = 0$. Therefore, the eigenvalues are $\lambda = -1 \pm \sqrt{-\alpha}$.

(b) If $\alpha \leq 0$, then the eigenvalues are real. If $\alpha < -1$, then the eigenvalues have opposite sign. In this case, the equilibrium point is a saddle. If $-1 < \alpha < 0$, then the eigenvalues are both negative. In this case, the equilibrium is a stable node. If $\alpha > 0$, then the eigenvalues have non-zero imaginary part and negative real part. In this case, the equilibrium point is a stable spiral. Therefore, the critical values for α are $\alpha = 0$ and $\alpha = -1$.

(c) Below we show phase portraits for $\alpha = -1.5$, $\alpha = -0.5$, and $\alpha = 0.5$.

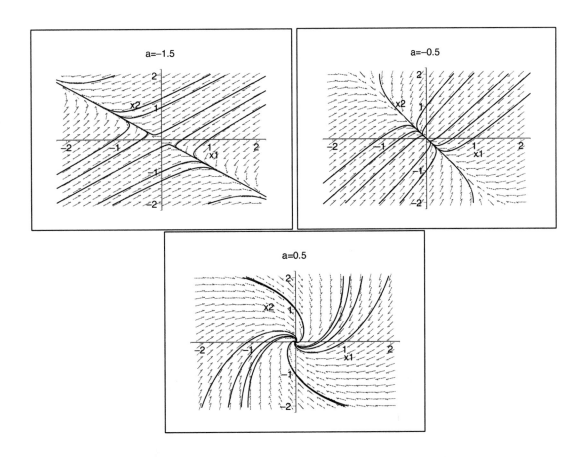

18.

(a) The characteristic equation is $\lambda^2 + \lambda + 6\alpha - 12 = 0$. Therefore, the eigenvalues are $\lambda = (-1 \pm \sqrt{49 - 24\alpha})/2$.

(b) If $49 - 24\alpha < 0$, then the eigenvalues have non-zero imaginary part and negative real part. In this case, the equilibrium point is a stable spiral. If $49 - 24\alpha \geq 0$, then the eigenvalues are real. If, in addition, $-1 + \sqrt{49 - 24\alpha} < 0$, then the eigenvalues are both negative, and the equilibrium point is a stable node. If $-1 + \sqrt{49 - 24\alpha} > 0$, then the eigenvalues have opposite sign, and the equilibrium point is a saddle. Therefore, the critical values for α are $\alpha = 49/24$ and $\alpha = 2$. We conclude that for $\alpha > 49/24$, the equilibrium is a stable spiral. For $49/24 > \alpha > 2$, the equilibrium is a stable node. For $2 > \alpha$, the equilibrium is a saddle point.

(c) Below we show phase portraits for $\alpha = 3$, $\alpha = 97/48$, and $\alpha = 1$.

19.

(a) The characteristic equation is $\lambda^2 + (4 - \alpha)\lambda + 10 - 4\alpha = 0$. Therefore, the eigenvalues are $\lambda = -2 + \dfrac{\alpha}{2} \pm \sqrt{\alpha^2 + 8\alpha - 24}$.

(b) If $\alpha^2 + 8\alpha - 24 < 0$, the eigenvalues have non-zero imaginary part. Otherwise, the eigenvalues are both real. Therefore, two of the critical values for α are $\alpha = -4 + 2\sqrt{10}$ and $-4 - 2\sqrt{10}$. For $-4 - 2\sqrt{10} < \alpha < -4 + 2\sqrt{10}$, $-2 + \alpha/2 < 0$. Therefore, for $-4 - 2\sqrt{10} < \alpha < -4 + 2\sqrt{10}$, the eigenvalues have non-zero imaginary part with negative real part. In this case, the equilibrium is a stable spiral. Now we consider the case when $\alpha^2 + 8\alpha - 24 \geq 0$. In this case, both eigenvalues are real. We need to determine what the sign of these eigenvalues is. We write the eigenvalues as

$$\lambda = \frac{((\alpha - 4) \pm \sqrt{\alpha^2 + 8\alpha - 24})}{2} = \frac{(\alpha - 4) \pm \sqrt{(\alpha - 4)^2 + 16\alpha - 40}}{2}.$$

We notice that if $16\alpha - 40 < 0$, then both eigenvalues have the same sign. Therefore, if $\alpha < -4 - 2\sqrt{10}$, then both eigenvalues are negative, and, therefore, the equilibrium is a stable node. Also if $-4 + 2\sqrt{10} < \alpha < 5/2$, then the eigenvalues are both negative, and the equilibrium is a stable node. On the other hand, if $5/2 < \alpha$, then the eigenvalues have opposite sign and the equilibrium point is a saddle.

(c) Below we show phase portraits for $\alpha = 5$, $\alpha = 2.4$, $\alpha = 1$, $\alpha = -9$ and $\alpha = -12$.

72

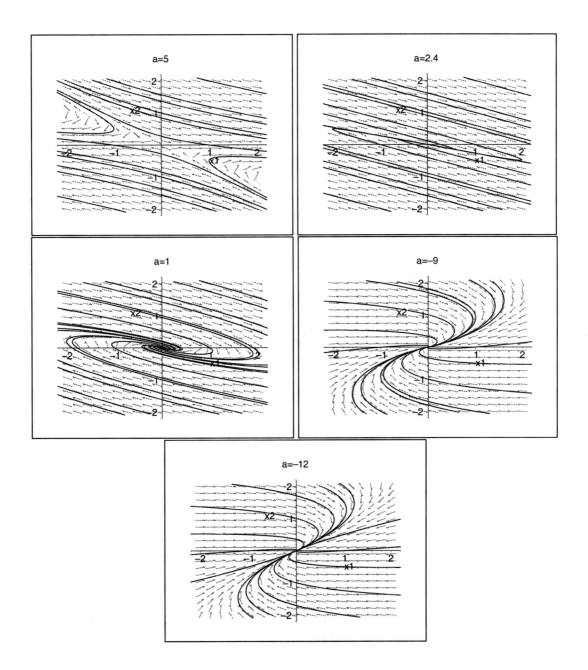

20.

(a) The characteristic equation is $\lambda^2 + 2\lambda - (24 + 8\alpha) = 0$. Therefore, the eigenvalues are $\lambda = -1 \pm \sqrt{25 + 8\alpha}$.

(b) If $\alpha < -25/8$, then the eigenvalues have non-zero imaginary part. The real part will be -1, and, therefore, the equilibrium will be a stable spiral. If $\alpha \geq -25/8$, then the eigenvalues will be real-valued. If $-25/8 \leq \alpha < -3$, the roots are both negative, and, therefore, the equilibrium is a stable node. If $\alpha > -3$, the roots have opposite sign, and, therefore, the equilibrium is a saddle.

(c) Below we show phase portraits for $\alpha = -4$, $\alpha = -3.06$, and $\alpha = -2$.

21.

(a) Based on problems 19-21 of Section 3.2, the system of differential equations is

$$\frac{d}{dt}\begin{pmatrix} i \\ v \end{pmatrix} = \begin{pmatrix} -R_1/L & -1/L \\ 1/C & -1/CR_2 \end{pmatrix}\begin{pmatrix} i \\ v \end{pmatrix}.$$

Plugging in the given values for R_1, R_2, C, L, we arrive at the given system.

(b) The characteristic equation of the system is $\lambda^2 + \lambda + 1/2 = 0$. Therefore, the eigenvalues are $\lambda = (-1 \pm i)/2$. A corresponding eigenvector for $\lambda_1 = (-1+i)/2$ is

$$\mathbf{v_1} = \begin{pmatrix} 1 \\ -4i \end{pmatrix}.$$

Therefore, one solution of this equation is

$$\begin{pmatrix} i \\ v \end{pmatrix} = e^{(-1+i)t/2}\begin{pmatrix} 1 \\ -4i \end{pmatrix}$$

$$= e^{-t/2}\begin{pmatrix} \cos(t/2) \\ 4\sin(t/2) \end{pmatrix} + ie^{-t/2}\begin{pmatrix} \sin(t/2) \\ -4\cos(t/2) \end{pmatrix}.$$

Taking the real and imaginary parts of this solution, we arrive at the general solution,

$$\begin{pmatrix} i \\ v \end{pmatrix} c_1 e^{-t/2}\begin{pmatrix} \cos(t/2) \\ 4\sin(t/2) \end{pmatrix} + c_2 e^{-t/2}\begin{pmatrix} \sin(t/2) \\ -4\cos(t/2) \end{pmatrix}.$$

74

(c) For the given initial conditions, we conclude that $c_1 = 2$ and $c_2 = -3/4$. Therefore,

$$\begin{pmatrix} i \\ v \end{pmatrix} = e^{-t/2} \begin{pmatrix} 2\cos(t/2) - \frac{3}{4}\sin(t/2) \\ 8\sin(t/2) + 3\cos(t/2) \end{pmatrix}.$$

(d) Since the eigenvalues have negative real part, all solutions converge to the origin.

22.

(a) The characteristic equation is $\lambda^2 + \dfrac{1}{RC}\lambda + \dfrac{1}{CL} = 0$. Therefore, the eigenvalues are

$$\lambda = -\frac{1}{2RC} \pm \frac{1}{2RC}\sqrt{1 - \frac{4R^2C}{L}}.$$

The eigenvalues are real and different provided that $1 - \frac{4R^2C}{L} > 0$. Otherwise, they are complex conjugates.

(b) With the specified values, the eigenvalues are $\lambda = -1 \pm i$. An eigenvector corresponding to $\lambda_1 = -1 + i$ is

$$\mathbf{v_1} = \begin{pmatrix} 1 \\ -4i \end{pmatrix}.$$

Therefore, one solution is

$$\begin{pmatrix} i \\ v \end{pmatrix} = e^{(-1+i)t}\begin{pmatrix} 1 \\ -1+i \end{pmatrix}$$

$$= e^{-t}\begin{pmatrix} \cos t \\ -\cos t - \sin t \end{pmatrix} + ie^{-t}\begin{pmatrix} \sin t \\ \cos t - \sin t \end{pmatrix}.$$

Therefore, the general solution is

$$\begin{pmatrix} i \\ v \end{pmatrix} = c_1 e^{-t}\begin{pmatrix} \cos t \\ -\cos t - \sin t \end{pmatrix} + c_2 e^{-t}\begin{pmatrix} \sin t \\ \cos t - \sin t \end{pmatrix}.$$

(c) With the given initial conditions, we see that $c_1 = 2$ and $-c_1 + c_2 = 1$. Therefore, we conclude that $c_1 = 2$ and $c_2 = 3$. Therefore, the solution of the IVP is

$$\begin{pmatrix} i \\ v \end{pmatrix} = e^{-t}\begin{pmatrix} 2\cos t + 3\sin t \\ \cos t - 5\sin t \end{pmatrix}.$$

(d) Since the eigenvalues have negative real part, all solutions converge to the origin.

23.

(a) The characteristic equation is $\lambda^2 - (a_{11} + a_{22})\lambda + a_{11}a_{22} - a_{12}a_{21} = 0$. Therefore, the eigenvalues are

$$\lambda = \frac{(a_{11} + a_{22}) \pm \sqrt{(a_{11} + a_{22})^2 - 4(a_{11}a_{22} - a_{12}a_{21})}}{2}.$$

Therefore, from this equation, we see that in order for the eigenvalues to be purely imaginary, we need $a_{11} + a_{22} = 0$ and $a_{11}a_{22} - a_{12}a_{21} > 0$.

(b) Equation (iii) can be rewritten as

$$-(a_{21}x + a_{22}y)dx + (a_{11}x + a_{12}y)dy = 0.$$

To show that this equation is exact, we need to show that for $M = -(a_{21}x + a_{22}y)$ and $N = (a_{11}x + a_{12}y)$, $M_y = N_x$. We see that $M_y = -a_{22}$ and $N_x = a_{11}$. Then, using the fact that $a_{11} + a_{22} = 0$, we see that $N_x = a_{11} = -a_{22} = M_y$. Therefore, the equation is exact.

(c) Since the equation is exact, we begin by integrating M with respect to x. We conclude that $\psi(x, y) = -a_{21}x^2/2 - a_{22}xy + h(y)$. Then, taking the derivative with respect to y, we have $-a_{22}x + h'(y) = N = a_{11}x + a_{12}y$. Therefore, $h'(y) = a_{12}y$ which implies that $h(y) = a_{12}y^2/2$. Therefore, the solution of this exact equation is given implicitly by $-a_{21}x^2/ - a_{22}xy + a_{12}y^2/2 = k$ or $a_{21}x^2 = 2a_{22}xy - a_{12}y^2 = k$. The discriminant of the quadratic form is

$$(2a_{22})^2 - 4a_{21}(-a_{12}) = 4a_{22}^2 + 4a_{12}a_{21} = 4(-a_{11}a_{22} + a_{12}a_{21}).$$

Since $a_{11}a_{22} - a_{12}a_{21} > 0$, by equation (ii), we can conclude that the discriminant of the quadratic form is always negative, and, therefore, the graph is always an ellipse.

Section 3.5

1.
$$A - \lambda I = \begin{pmatrix} 3 - \lambda & -4 \\ 1 & -1 - \lambda \end{pmatrix}$$

implies $\det(A - \lambda I) = \lambda^2 - 2\lambda + 1 = (\lambda - 1)^2$. Therefore, $\lambda = 1$ is the only eigenvalue. Now $\lambda = 1$ implies

$$A - \lambda I = \begin{pmatrix} 2 & -4 \\ 1 & -2 \end{pmatrix}.$$

Therefore,
$$\mathbf{v}_1 = \begin{pmatrix} 2 \\ 1 \end{pmatrix}$$

is an eigenvector for $\lambda = 1$ and

$$\mathbf{x}_1(t) = e^t \begin{pmatrix} 2 \\ 1 \end{pmatrix}$$

is one solution.

Next, we need to look for a solution $\mathbf{w}$ of

$$(A - I)\mathbf{w} = \begin{pmatrix} 2 \\ 1 \end{pmatrix}.$$

Plugging in for $A - I$, this equation becomes

$$\begin{pmatrix} 2 & -4 \\ 1 & -2 \end{pmatrix} \mathbf{w} = \begin{pmatrix} 2 \\ 1 \end{pmatrix}.$$

We see that
$$\mathbf{w} = \begin{pmatrix} 1 \\ 0 \end{pmatrix}$$
is a solution of this equation. Therefore,
$$\mathbf{x}_2(t) = te^t \begin{pmatrix} 2 \\ 1 \end{pmatrix} + e^t \begin{pmatrix} 1 \\ 0 \end{pmatrix}$$
is also a solution of our original system. Therefore, the general solution of the system is
$$\mathbf{x}(t) = c_1 e^t \begin{pmatrix} 2 \\ 1 \end{pmatrix} + c_2 \left[te^t \begin{pmatrix} 2 \\ 1 \end{pmatrix} + e^t \begin{pmatrix} 1 \\ 0 \end{pmatrix} \right].$$

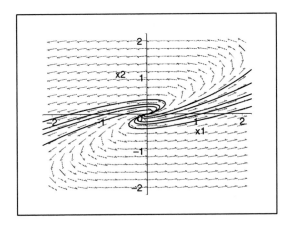

The solution grows as $t \to \infty$.

2.
$$A - \lambda I = \begin{pmatrix} 5/4 - \lambda & 3/4 \\ -3/4 & -1/4 - \lambda \end{pmatrix}$$
implies $\det(A - \lambda I) = \lambda^2 - \lambda + 1/4 = (\lambda - 1/2)^2$. Therefore, $\lambda = 1/2$ is the only eigenvalue. Now $\lambda = 1/2$ implies
$$A - \lambda I = \begin{pmatrix} 3/4 & 3/4 \\ -3/4 & -3/4 \end{pmatrix}.$$
Therefore,
$$\mathbf{v}_1 = \begin{pmatrix} 1 \\ -1 \end{pmatrix}$$
is an eigenvector for $\lambda = 1/2$ and
$$\mathbf{x}_1(t) = e^{t/2} \begin{pmatrix} 1 \\ -1 \end{pmatrix}$$
is one solution.

Next, we need to look for a solution $\mathbf{w}$ of
$$\left(A - \frac{1}{2} I \right) \mathbf{w} = \begin{pmatrix} 1 \\ -1 \end{pmatrix}.$$

Plugging in for $A - 1/2I$, this equation becomes

$$\begin{pmatrix} 3/4 & 3/4 \\ -3/4 & -3/4 \end{pmatrix} \mathbf{w} = \begin{pmatrix} 1 \\ -1 \end{pmatrix}.$$

We see that

$$\mathbf{w} = \begin{pmatrix} 4/3 \\ 0 \end{pmatrix}$$

is a solution of this equation. Therefore,

$$\mathbf{x}_2(t) = te^{t/2} \begin{pmatrix} 1 \\ -1 \end{pmatrix} + e^{t/2} \begin{pmatrix} 4/3 \\ 0 \end{pmatrix}$$

is also a solution of our original system. Therefore, the general solution of the system is

$$\mathbf{x}(t) = c_1 e^{t/2} \begin{pmatrix} 1 \\ -1 \end{pmatrix} + c_2 \left[te^{t/2} \begin{pmatrix} 1 \\ -1 \end{pmatrix} + e^{t/2} \begin{pmatrix} 4/3 \\ 0 \end{pmatrix} \right].$$

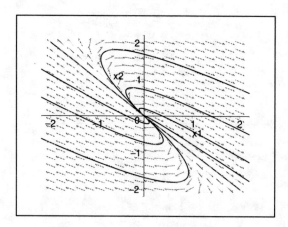

The solution grows as $t \to \infty$.

3.

$$A - \lambda I = \begin{pmatrix} -3/2 - \lambda & 1 \\ -1/4 & -1/2 - \lambda \end{pmatrix}$$

implies $\det(A - \lambda I) = \lambda^2 + 2\lambda + 1 = (\lambda + 1)^2$. Therefore, $\lambda = -1$ is the only eigenvalue. Now $\lambda = -1$ implies

$$A - \lambda I = \begin{pmatrix} -1/2 & 1 \\ -1/4 & 1/2 \end{pmatrix}.$$

Therefore,

$$\mathbf{v}_1 = \begin{pmatrix} 2 \\ 1 \end{pmatrix}$$

is an eigenvector for $\lambda = -1$ and

$$\mathbf{x}_1(t) = e^{-t} \begin{pmatrix} 2 \\ 1 \end{pmatrix}$$

is one solution.

Next, we need to look for a solution $\mathbf{w}$ of

$$(A + I)\,\mathbf{w} = \begin{pmatrix} 2 \\ 1 \end{pmatrix}.$$

Plugging in for $A + I$, this equation becomes

$$\begin{pmatrix} -1/2 & 1 \\ -1/4 & 1/2 \end{pmatrix} \mathbf{w} = \begin{pmatrix} 2 \\ 1 \end{pmatrix}.$$

We see that

$$\mathbf{w} = \begin{pmatrix} 0 \\ 2 \end{pmatrix}$$

is a solution of this equation. Therefore,

$$\mathbf{x}_2(t) = te^{-t} \begin{pmatrix} 2 \\ 1 \end{pmatrix} + e^{-t} \begin{pmatrix} 0 \\ 2 \end{pmatrix}$$

is also a solution of our original system. Therefore, the general solution of the system is

$$\mathbf{x}(t) = c_1 e^{-t} \begin{pmatrix} 2 \\ 1 \end{pmatrix} + c_2 \left[te^{-t} \begin{pmatrix} 2 \\ 1 \end{pmatrix} + e^{-t} \begin{pmatrix} 0 \\ 2 \end{pmatrix} \right].$$

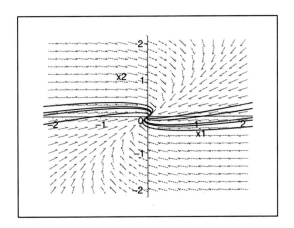

The solutions approach the origin as $t \to \infty$.

4.

$$A - \lambda I = \begin{pmatrix} -3 - \lambda & 5/2 \\ -5/2 & 2 - \lambda \end{pmatrix}$$

implies $\det(A - \lambda I) = \lambda^2 + \lambda + 1/4 = (\lambda + 1/2)^2$. Therefore, $\lambda = -1/2$ is the only eigenvalue. Now $\lambda = -1/2$ implies

$$A - \lambda I = \begin{pmatrix} -5/2 & 5/2 \\ -5/2 & 5/2 \end{pmatrix}.$$

Therefore,

$$\mathbf{v}_1 = \begin{pmatrix} 1 \\ 1 \end{pmatrix}$$

is an eigenvector for $\lambda = -1/2$ and

$$\mathbf{x}_1(t) = e^{-t/2}\begin{pmatrix}1\\1\end{pmatrix}$$

is one solution.

Next, we need to look for a solution $\mathbf{w}$ of

$$\left(A + \frac{1}{2}I\right)\mathbf{w} = \begin{pmatrix}1\\1\end{pmatrix}.$$

Plugging in for $A + \dfrac{1}{2}I$, this equation becomes

$$\begin{pmatrix}-5/2 & 5/2\\-5/2 & 5/2\end{pmatrix}\mathbf{w} = \begin{pmatrix}1\\1\end{pmatrix}.$$

We see that

$$\mathbf{w} = \begin{pmatrix}0\\2/5\end{pmatrix}$$

is a solution of this equation. Therefore,

$$\mathbf{x}_2(t) = te^{-t/2}\begin{pmatrix}1\\1\end{pmatrix} + e^{-t/2}\begin{pmatrix}0\\2/5\end{pmatrix}$$

is also a solution of our original system. Therefore, the general solution of the system is

$$\mathbf{x}(t) = c_1 e^{-t/2}\begin{pmatrix}1\\1\end{pmatrix} + c_2\left[te^{-t/2}\begin{pmatrix}1\\1\end{pmatrix} + e^{-t/2}\begin{pmatrix}0\\2/5\end{pmatrix}\right].$$

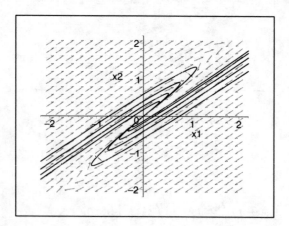

The solutions approach the origin as $t \to \infty$.

5.

$$A - \lambda I = \begin{pmatrix}-1-\lambda & -1/2\\2 & -3-\lambda\end{pmatrix}$$

implies $\det(A - \lambda I) = \lambda^2 + 4\lambda + 4 = (\lambda + 2)^2$. Therefore, $\lambda = -2$ is the only eigenvalue. Now $\lambda = -2$ implies

$$A - \lambda I = \begin{pmatrix} 1 & -1/2 \\ 2 & -1 \end{pmatrix}.$$

Therefore,

$$\mathbf{v}_1 = \begin{pmatrix} 1 \\ 2 \end{pmatrix}$$

is an eigenvector for $\lambda = -2$ and

$$\mathbf{x}_1(t) = e^{-2t} \begin{pmatrix} 1 \\ 2 \end{pmatrix}$$

is one solution.

Next, we need to look for a solution $\mathbf{w}$ of

$$(A + 2I)\,\mathbf{w} = \begin{pmatrix} 1 \\ 2 \end{pmatrix}.$$

Plugging in for $A + 2I$, this equation becomes

$$\begin{pmatrix} 1 & -1/2 \\ 2 & -1 \end{pmatrix} \mathbf{w} = \begin{pmatrix} 1 \\ 2 \end{pmatrix}.$$

We see that

$$\mathbf{w} = \begin{pmatrix} 1 \\ 0 \end{pmatrix}$$

is a solution of this equation. Therefore,

$$\mathbf{x}_2(t) = te^{-2t} \begin{pmatrix} 1 \\ 2 \end{pmatrix} + e^{-2t} \begin{pmatrix} 1 \\ 0 \end{pmatrix}$$

is also a solution of our original system. Therefore, the general solution of the system is

$$\mathbf{x}(t) = c_1 e^{-2t} \begin{pmatrix} 1 \\ 2 \end{pmatrix} + c_2 \left[te^{-2t} \begin{pmatrix} 1 \\ 2 \end{pmatrix} + e^{-2t} \begin{pmatrix} 1 \\ 0 \end{pmatrix} \right].$$

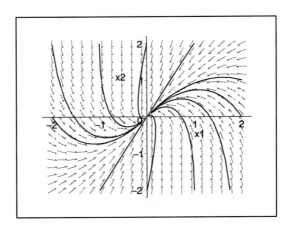

The solutions tend to the origin as $t \to \infty$.

6.

$$A - \lambda I = \begin{pmatrix} 2 - \lambda & 1/2 \\ -1/2 & 1 - \lambda \end{pmatrix}$$

implies $\det(A - \lambda I) = \lambda^2 - 3\lambda + 9/4 = (\lambda - 3/2)^2$. Therefore, $\lambda = 3/2$ is the only eigenvalue. Now $\lambda = 3/2$ implies

$$A - \lambda I = \begin{pmatrix} 1/2 & 1/2 \\ -1/2 & -1/2 \end{pmatrix}.$$

Therefore,

$$\mathbf{v}_1 = \begin{pmatrix} 1 \\ -1 \end{pmatrix}$$

is an eigenvector for $\lambda = 3/2$ and

$$\mathbf{x}_1(t) = e^{3t/2} \begin{pmatrix} 1 \\ -1 \end{pmatrix}$$

is one solution.

Next, we need to look for a solution $\mathbf{w}$ of

$$(A - 3/2I)\,\mathbf{w} = \begin{pmatrix} 1 \\ -1 \end{pmatrix}.$$

Plugging in for $A - 3/2I$, this equation becomes

$$\begin{pmatrix} 1/2 & 1/2 \\ -1/2 & -1/2 \end{pmatrix} \mathbf{w} = \begin{pmatrix} 1 \\ -1 \end{pmatrix}.$$

We see that

$$\mathbf{w} = \begin{pmatrix} 2 \\ 0 \end{pmatrix}$$

is a solution of this equation. Therefore,

$$\mathbf{x}_2(t) = te^{3t/2} \begin{pmatrix} 1 \\ -1 \end{pmatrix} + e^{3t/2} \begin{pmatrix} 2 \\ 0 \end{pmatrix}$$

is also a solution of our original system. Therefore, the general solution of the system is

$$\mathbf{x}(t) = c_1 e^{3t/2} \begin{pmatrix} 1 \\ -1 \end{pmatrix} + c_2 \left[te^{3t/2} \begin{pmatrix} 1 \\ -1 \end{pmatrix} + e^{3t/2} \begin{pmatrix} 2 \\ 0 \end{pmatrix} \right].$$

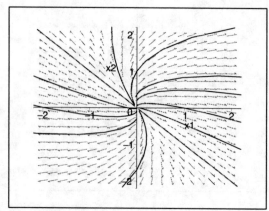

The solutions tend to infinity as $t \to \infty$.

7. The characteristic equation is $(\lambda + 3)^2 = 0$. Therefore, the only eigenvalue is $\lambda = -3$. A corresponding eigenvector is given by

$$\mathbf{v_1} = \begin{pmatrix} 1 \\ 1 \end{pmatrix}$$

and

$$\mathbf{x_1}(t) = e^{-3t} \begin{pmatrix} 1 \\ 1 \end{pmatrix}$$

is one solution of the system. Next, we need to look for a solution $\mathbf{w}$ of

$$(A + 3I)\,\mathbf{w} = \begin{pmatrix} 1 \\ 1 \end{pmatrix}.$$

Plugging in for $A + 3I$, this equation becomes

$$\begin{pmatrix} 4 & -4 \\ 4 & -4 \end{pmatrix} \mathbf{w} = \begin{pmatrix} 1 \\ 1 \end{pmatrix}.$$

We see that

$$\mathbf{w} = \begin{pmatrix} 1/4 \\ 0 \end{pmatrix}$$

is a solution of this equation. Therefore,

$$\mathbf{x_2}(t) = te^{-3t} \begin{pmatrix} 1 \\ 1 \end{pmatrix} + e^{-3t} \begin{pmatrix} 1/4 \\ 0 \end{pmatrix}$$

is also a solution of our original system. Therefore, the general solution of the system is

$$\mathbf{x}(t) = c_1 e^{-3t} \begin{pmatrix} 1 \\ 1 \end{pmatrix} + c_2 \left[te^{-3t} \begin{pmatrix} 1 \\ 1 \end{pmatrix} + e^{-3t} \begin{pmatrix} 1/4 \\ 0 \end{pmatrix} \right].$$

The initial condition implies that $c_1 = 2$ and $c_2 = 4$. Therefore, the solution of the IVP is

$$\mathbf{x}(t) = \begin{pmatrix} 3 + 4t \\ 2 + 4t \end{pmatrix} e^{-3t}.$$

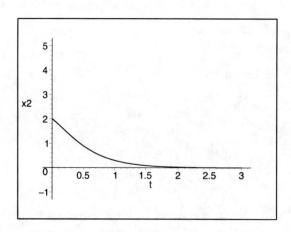

8. The characteristic equation is $(\lambda + 1)^2 = 0$. Therefore, the only eigenvalue is $\lambda = -1$. A corresponding eigenvector is given by

$$\mathbf{v_1} = \begin{pmatrix} 1 \\ 1 \end{pmatrix}$$

and

$$\mathbf{x_1}(t) = e^{-t} \begin{pmatrix} 1 \\ 1 \end{pmatrix}$$

is one solution of the system. Next, we need to look for a solution $\mathbf{w}$ of

$$(A + I)\mathbf{w} = \begin{pmatrix} 1 \\ 1 \end{pmatrix}.$$

Plugging in for $A + I$, this equation becomes

$$\begin{pmatrix} -3/2 & 3/2 \\ -3/2 & 3/2 \end{pmatrix} \mathbf{w} = \begin{pmatrix} 1 \\ 1 \end{pmatrix}.$$

We see that

$$\mathbf{w} = \begin{pmatrix} 0 \\ 2/3 \end{pmatrix}$$

is a solution of this equation. Therefore,

$$\mathbf{x_2}(t) = te^{-t} \begin{pmatrix} 1 \\ 1 \end{pmatrix} + e^{-t} \begin{pmatrix} 0 \\ 2/3 \end{pmatrix}$$

is also a solution of our original system. Therefore, the general solution of the system is

$$\mathbf{x}(t) = c_1 e^{-t} \begin{pmatrix} 1 \\ 1 \end{pmatrix} + c_2 \left[te^{-t} \begin{pmatrix} 1 \\ 1 \end{pmatrix} + e^{-t} \begin{pmatrix} 0 \\ 2/3 \end{pmatrix} \right].$$

The initial condition implies that $c_1 = 3$ and $c_2 = -6$. Therefore, the solution of the IVP is

$$\mathbf{x}(t) = \begin{pmatrix} 3 - 6t \\ -1 - 6t \end{pmatrix} e^{-t}.$$

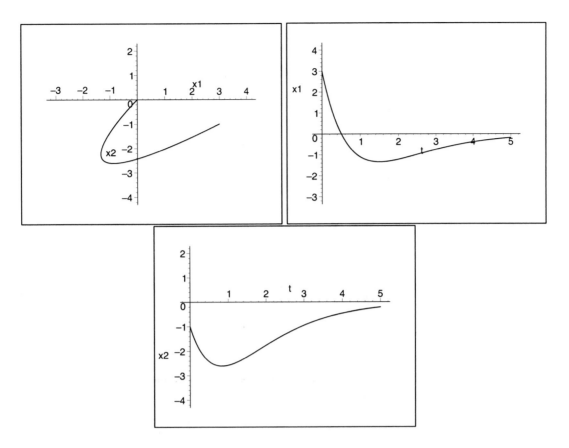

9. The characteristic equation is $(\lambda - 1/2)^2 = 0$. Therefore, the only eigenvalue is $\lambda = 1/2$. A corresponding eigenvector is given by

$$\mathbf{v_1} = \begin{pmatrix} -1 \\ 1 \end{pmatrix}$$

and

$$\mathbf{x_1}(t) = e^{t/2} \begin{pmatrix} -1 \\ 1 \end{pmatrix}$$

is one solution of the system. Next, we need to look for a solution $\mathbf{w}$ of

$$\left(A - \frac{1}{2}I \right) \mathbf{w} = \begin{pmatrix} -1 \\ 1 \end{pmatrix}.$$

Plugging in for $A - \frac{1}{2}I$, this equation becomes

$$\begin{pmatrix} 3/2 & 3/2 \\ -3/2 & -3/2 \end{pmatrix} \mathbf{w} = \begin{pmatrix} -1 \\ 1 \end{pmatrix}.$$

We see that

$$\mathbf{w} = \begin{pmatrix} -2/3 \\ 0 \end{pmatrix}$$

is a solution of this equation. Therefore,

$$\mathbf{x_2}(t) = te^{t/2} \begin{pmatrix} -1 \\ 1 \end{pmatrix} + e^{t/2} \begin{pmatrix} -2/3 \\ 0 \end{pmatrix}$$

85

is also a solution of our original system. Therefore, the general solution of the system is

$$\mathbf{x}(t) = c_1 e^{t/2} \begin{pmatrix} -1 \\ 1 \end{pmatrix} + c_2 \left[t e^{t/2} \begin{pmatrix} -1 \\ 1 \end{pmatrix} + e^{t/2} \begin{pmatrix} -2/3 \\ 0 \end{pmatrix} \right].$$

The initial condition implies that $c_1 = -2$ and $c_2 = -3/2$. Therefore, the solution of the IVP is

$$\mathbf{x}(t) = \begin{pmatrix} 3 + \frac{3}{2}t \\ -2 - \frac{3}{2}t \end{pmatrix} e^{t/2}.$$

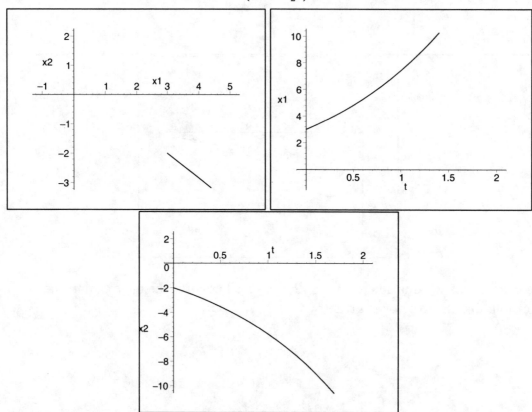

10. From our answer to problem 2, we see that the general solution of this system is given by

$$\mathbf{x}(t) = c_1 e^{t/2} \begin{pmatrix} 1 \\ -1 \end{pmatrix} + c_2 \left[t e^{t/2} \begin{pmatrix} 1 \\ -1 \end{pmatrix} + e^{t/2} \begin{pmatrix} 4/3 \\ 0 \end{pmatrix} \right].$$

Now our initial condition $\mathbf{x}(0) = \begin{pmatrix} 2 \\ 3 \end{pmatrix}^t$ implies

$$\mathbf{x}(0) = c_1 \begin{pmatrix} 1 \\ -1 \end{pmatrix} + c_2 \begin{pmatrix} 4/3 \\ 0 \end{pmatrix} = \begin{pmatrix} 2 \\ 3 \end{pmatrix}.$$

Therefore, $c_1 = -3$ and $c_2 = 15/4$. Therefore,

$$\mathbf{x}(t) = -3 e^{t/2} \begin{pmatrix} 1 \\ -1 \end{pmatrix} + \frac{15}{4} \left[t e^{t/2} \begin{pmatrix} 1 \\ -1 \end{pmatrix} + e^{t/2} \begin{pmatrix} 4/3 \\ 0 \end{pmatrix} \right]$$

$$= \begin{pmatrix} 2 + 15t/4 \\ 3 - 15t/4 \end{pmatrix} e^{t/2}.$$

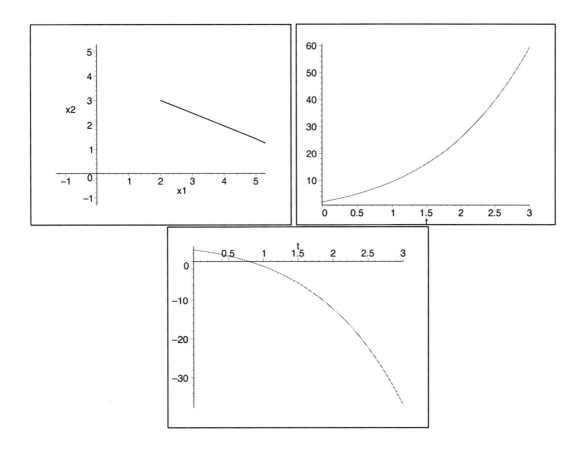

11. From our answer to problem 4, we see that the general solution of this system is given by

$$\mathbf{x}(t) = c_1 e^{-t/2} \begin{pmatrix} 1 \\ 1 \end{pmatrix} + c_2 \left[t e^{-t/2} \begin{pmatrix} 1 \\ 1 \end{pmatrix} + e^{-t/2} \begin{pmatrix} -2/5 \\ 0 \end{pmatrix} \right].$$

Now our initial condition $\mathbf{x}(0) = \begin{pmatrix} 3 \\ 1 \end{pmatrix}^t$ implies

$$\mathbf{x}(0) = c_1 \begin{pmatrix} 1 \\ 1 \end{pmatrix} + c_2 \begin{pmatrix} -2/5 \\ 0 \end{pmatrix} = \begin{pmatrix} 3 \\ 1 \end{pmatrix}.$$

Therefore, $c_1 = 1$ and $c_2 = -5$. Therefore,

$$\mathbf{x}(t) = e^{-t/2} \begin{pmatrix} 1 \\ 1 \end{pmatrix} - 5 \left[t e^{-t/2} \begin{pmatrix} 1 \\ 1 \end{pmatrix} + e^{-t/2} \begin{pmatrix} -2/5 \\ 0 \end{pmatrix} \right]$$

$$= \begin{pmatrix} 3 - 5t \\ 1 - 5t \end{pmatrix} e^{-t/2}$$

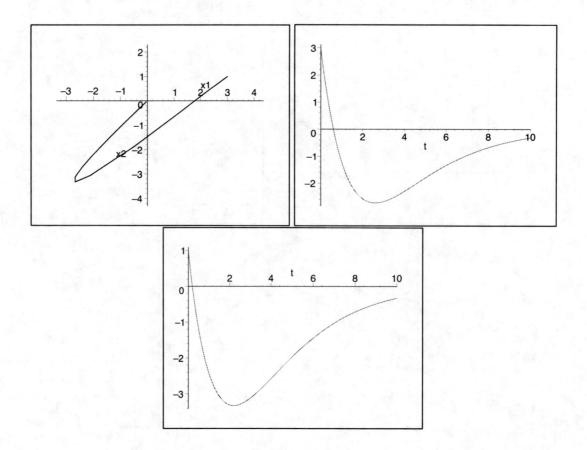

12. From our answer to problem 2, we see that the general solution of this system is given by

$$\mathbf{x}(t) = c_1 e^{3t/2} \begin{pmatrix} 1 \\ -1 \end{pmatrix} + c_2 \left[t e^{3t/2} \begin{pmatrix} 1 \\ -1 \end{pmatrix} + e^{3t/2} \begin{pmatrix} 2 \\ 0 \end{pmatrix} \right].$$

Now our initial condition $\mathbf{x}(0) = \begin{pmatrix} 1 \\ 3 \end{pmatrix}^t$ implies

$$\mathbf{x}(0) = c_1 \begin{pmatrix} 1 \\ -1 \end{pmatrix} + c_2 \begin{pmatrix} 2 \\ 0 \end{pmatrix} = \begin{pmatrix} 1 \\ 3 \end{pmatrix}.$$

Therefore, $c_1 = -3$ and $c_2 = 2$. Therefore,

$$\begin{aligned} \mathbf{x}(t) &= -3 e^{3t/2} \begin{pmatrix} 1 \\ -1 \end{pmatrix} + 2 \left[t e^{3t/2} \begin{pmatrix} 1 \\ -1 \end{pmatrix} + e^{3t/2} \begin{pmatrix} 2 \\ 0 \end{pmatrix} \right] \\ &= \begin{pmatrix} 1 + 2t \\ 3 - 2t \end{pmatrix} e^{3t/2}. \end{aligned}$$

88

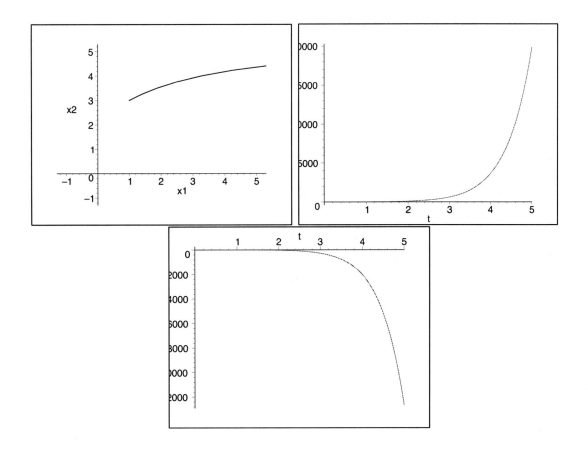

13. The characteristic equation is

$$\lambda^2 + (a+d)\lambda + ad - bc = 0.$$

Therefore, the eigenvalues are

$$\lambda = \frac{a+d}{2} \pm \frac{1}{2}\sqrt{(a+d)^2 - 4(ad-bc)}.$$

In order to guarantee that the solution approaches zero, we need both eigenvalues to be negative. As we can see from the equation above for λ, we need $a + d < 0$ and $ad - bc > 0$.

14.

(a) The eigenvalues of this system are given by

$$\lambda = -\frac{1}{2RC} \pm \frac{\sqrt{L^2 - 4R^2CL}}{2RCL}.$$

The eigenvalues will be real and equal if the discriminant vanishes. The discriminant will vanish when $L = 4R^2C$.

(b) For the given conditions, the system is

$$\frac{d}{dt}\begin{pmatrix} i \\ v \end{pmatrix} = \begin{pmatrix} 0 & 1/4 \\ -1 & -1 \end{pmatrix}\begin{pmatrix} i \\ v \end{pmatrix}.$$

89

The characteristic equation of this system is $\lambda^2 + \lambda + 1/4 = 0$. Therefore, the only eigenvalue is $\lambda = -1/2$. An associated eigenvector is given by

$$\mathbf{v_1} = \left(1 // - 2\right).$$

Therefore, one solution of this system is

$$\mathbf{x_1}(t) = e^{-t/2}\left(1 // - 2\right).$$

Next, we need to look for a solution $\mathbf{w}$ of

$$\left(A + \frac{1}{2}I\right)\mathbf{w} = \begin{pmatrix} 1 \\ -2 \end{pmatrix}.$$

Plugging in for $A + \frac{1}{2}I$, this equation becomes

$$\begin{pmatrix} 1/2 & 1/4 \\ -1 & -1/2 \end{pmatrix}\mathbf{w} = \begin{pmatrix} 1 \\ -2 \end{pmatrix}.$$

We see that

$$\mathbf{w} = \begin{pmatrix} 0 \\ 4 \end{pmatrix}$$

is a solution of this equation. Therefore,

$$\mathbf{x_2}(t) = te^{-t/2}\begin{pmatrix} 1 \\ -2 \end{pmatrix} + e^{-t/2}\begin{pmatrix} 0 \\ 4 \end{pmatrix}$$

is also a solution of our original system. Therefore, the general solution of the system is

$$\mathbf{x}(t) = c_1 e^{-t/2}\begin{pmatrix} 1 \\ -2 \end{pmatrix} + c_2\left[te^{-t/2}\begin{pmatrix} 1 \\ -2 \end{pmatrix} + e^{-t/2}\begin{pmatrix} 0 \\ 4 \end{pmatrix}\right].$$

The initial condition implies that $c_1 = 1$ and $c_2 = 1$. Therefore, the solution of the IVP is

$$\mathbf{x}(t) = \begin{pmatrix} 1+t \\ 2-2t \end{pmatrix} e^{-t/2}.$$

15. The characteristic equation is $\lambda^2 - p\lambda + q = 0$. The eigenvalues are

$$\lambda = \frac{p \pm \sqrt{p^2 - 4q}}{2} = \frac{p \pm \sqrt{\Delta}}{2}.$$

The results can be verified using Table 3.5.1.

16.

(a) If $q > 0$ and $p < 0$, then the roots are either complex conjugates with negative real parts, or both real and negative.

(b) If $q > 0$ and $p = 0$, then the roots are purely imaginary.

90

(c) if $q < 0$, then the roots are real with $\lambda_1 \cdot \lambda_2 > 0$. If $p > 0$, then either the roots are real with $\lambda_1 \cdot \lambda_2 \geq 0$ or the roots are complex conjugates with positive real parts.

Section 3.6

1.

(a) .

$$\frac{dx}{dt} = -x, \frac{dy}{dt} = -2y \implies \frac{dy}{dx} = \frac{2y}{x}$$

$$\implies \int \frac{1}{y} \, dy = 2 \int \frac{1}{x} \, dx$$

$$\implies \ln|y| = 2 \ln|x| + C.$$

Applying the exponential function to this equation, we conclude that $y = Cx^2$ ' for any solution of the system. That is, $H(x, y) = y/x^2$.

(b)

(c)

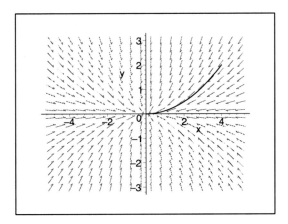

As t increases, the x and y values are decreasing.

91

2.

(a)

$$\frac{dx}{dt} = -x, \frac{dy}{dt} = 2y \implies \frac{dy}{dx} = -\frac{2y}{x}$$

$$\implies \int \frac{1}{y}\, dy = -2 \int \frac{1}{x}\, dx$$

$$\implies \ln|y| = -2\ln|x| + C.$$

Applying the exponential function to this equation, we conclude that $y = C/x^2$ ' for any solution of the system. That is, $H(x, y) = yx^2$.

(b)

(c)

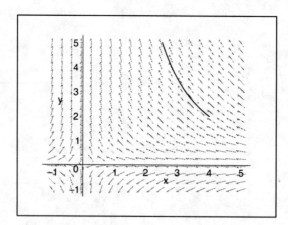

As t increases, the x values are decreasing and the y values are decreasing.

3.

92

(a)

$$\frac{dx}{dt} = -x, \frac{dy}{dt} = 2y \implies \frac{dy}{dx} = -\frac{2y}{x}$$

$$\implies \int \frac{1}{y} \, dy = -2 \int \frac{1}{x} \, dx$$

$$\implies \ln|y| = -2\ln|x| + C.$$

Applying the exponential function to this equation, we conclude that $y = C/x^2$ ' for any solution of the system. That is, $H(x, y) = yx^2$.

(b)

(c)

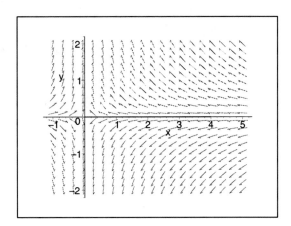

As t increases, the x values decrease and the y values do not change.

4.

93

(a)

$$\frac{dx}{dt} = 2y, \frac{dy}{dt} = 8x \implies \frac{dy}{dx} = \frac{4x}{y}$$

$$\implies \int y \, dy = 4 \int \frac{x}{-} dx$$

$$\implies y^2/2 = 2x^2 + C.$$

We conclude that $y^2 - 4x^2 = C$ for any solution of the system. That is, $H(x, y) = y^2 - 4x^2$.

(b)

(c)

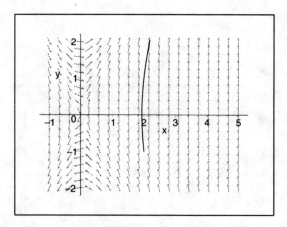

As t increases, the y values are increasing.

5.

(a)

$$\frac{dx}{dt} = 2y, \frac{dy}{dt} = 8x \implies \frac{dy}{dx} = \frac{4x}{y}$$

$$\implies \int y \, dy = \int 4x \, dx$$

$$\implies \frac{y^2}{2} = 2x^2 + C.$$

Therefore, $y^2 - 4x^2 = C$ for any solution of the system. That is, $H(x,y) = y^2 - 4x^2$.

(b)

(c)

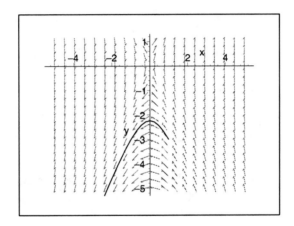

As t increases, the x and y values are decreasing.

6.

(a)

$$\frac{dx}{dt} = 2y, \frac{dy}{dt} = -8x \implies \frac{dy}{dx} = \frac{-4x}{y}$$

$$\implies \int y\, dy = -\int 4x\, dx$$

$$\implies \frac{y^2}{2} = -2x^2 + C.$$

Therefore, $y^2 + 4x^2 = C$ for any solution of the system. That is, $H(x,y) = y^2 + 4x^2$.

(b)

(c)

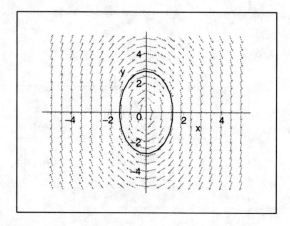

As t increases, the trajectory travels in the clockwise direction.

7.

(a) To find all critical points, we need to solve the following system of equations.

$$x' = 2x - y = 0$$
$$y' = x - 2y = 0.$$

From the second equation, $x = 2y$ implies $4y - y = 0$ which implies the only critical point is $(0,0)$.

(b) From our equations for dx/dt and dy/dt, we see that

$$\frac{dy}{dx} = \frac{x - 2y}{2x - y}$$

$$\implies (2x - y)\, dy = (x - 2y)\, dx$$
$$\implies (2y - x)\, dx + (2x - y)\, dy = 0.$$

96

We notice that this is an exact equation, which can be rewritten as

$$\left(2xy - \frac{x^2}{2} - \frac{y^2}{2}\right)' = 0.$$

Therefore,

$$H(x, y) = 2xy - \frac{x^2}{2} - \frac{y^2}{2}.$$

(c)

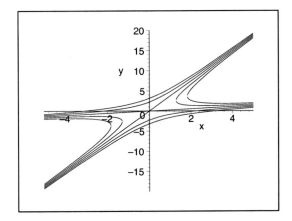

Solutions lying on curves passing through the second quadrant travel to the left. In particular, the x values and y values decrease. Solutions lying on curves solely in the first quadrant will travel to the right along the bottom portion of the curve. Solutions lying on curves solely in the third quadrant will travel to the left along the top portion of the curve. Solutions lying on curves passing through the fourth quadrant will travel to the right with both x and y values increasing.

(d) Solutions lying along the part of the level curve $H(x, y) = 0$ such that $y = (2 + \sqrt{3})x$ will approach the origin. All other solutions will tend away from the critical point as $t \to \infty$.

8.

(a) To find all critical points, we need to solve the following system of equations.

$$x' = -x + y = 0$$
$$y' = x + y = 0.$$

Solving the two equations, we see that the only critical point is $(0, 0)$.

(b) From our equations for dx/dt and dy/dt, we see that

$$\frac{dy}{dx} = \frac{x+y}{-x+y}$$

$$\implies (-x+y)\, dy = (x+y)\, dx$$
$$\implies (x+y)\, dx + (x-y)\, dy = 0.$$

We notice that this is an exact equation, which can be rewritten as

$$\left(\frac{x^2}{2} + xy - \frac{y^2}{2}\right)' = 0.$$

Therefore,

$$H(x, y) = \frac{x^2}{2} + xy - \frac{y^2}{2}.$$

(c)

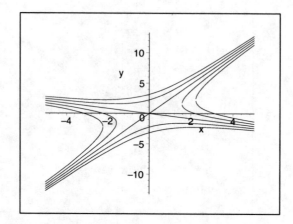

Solutions lying on curves passing through the first and second quadrants travel to the right. In particular, the x values and y values increase. Solutions lying on curves in the first and fourth quadrants will travel up along the curve. In particular, the y values will increase. Solutions lying on curves in the second and third quadrants will travel down along the bottom portion of the curve. In particular, y values will decrease. Solutions lying on curves passing through the third and fourth quadrants will travel to the left.

(d) Solutions lying along the part of the level curve $H(x, y) = 0$ such that $y = (1 - \sqrt{2})x$ will approach the origin. All other solutions will tend away from the critical point as $t \to \infty$.

9.

(a) To find all critical points, we need to solve the following system of equations.

$$x' = 2x - 4y = 0$$
$$y' = 2x - 2y = 0.$$

Solving the two equations, we see that the only critical point is $(0, 0)$.

(b) From our equations for dx/dt and dy/dt, we see that

$$\frac{dy}{dx} = \frac{2x - 2y}{2x - 4y}$$

$$\implies (2x - 4y)\, dy = (2x - 2y)\, dx$$
$$\implies (2x - 2y)\, dx + (4y - 2x)\, dy = 0.$$

98

We notice that this is an exact equation, which can be rewritten as

$$\left(x^2 - 2xy + 2y^2\right)' = 0.$$

Therefore,

$$H(x,y) = x^2 - 2xy + 2y^2.$$

(c)

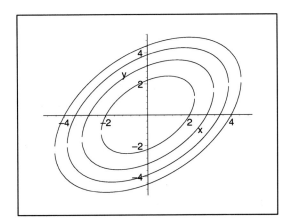

Solutions travel in the counter-clockwise direction along the trajectories shown above.

(d) Solutions orbit the critical point. They do not tend towards or away from the critical point.

10.

(a) To find all critical points, we need to solve the following system of equations.

$$x' = -x + y + x^2 = 0$$
$$y' = y - 2xy = 0.$$

Solving the second equation, we see that we must have $x = 1/2$ or $y = 0$. Plugging these values into the first equation, we see that the critical points are: $(0,0)$, $(1,0)$ and $(1/2, 1/4)$.

(b) From our equations for dx/dt and dy/dt, we see that

$$\frac{dy}{dx} = \frac{y - 2xy}{-x + y + x^2}.$$

Rewriting this equation as $(2xy - y)dx + (x^2 - x + y)dy = 0$, we see that this is an exact equation. For $M = 2xy - y$, we integrate M with respect to x. We conclude that the level curves are given by $H(x,y) = C$ where $H(x,y) = x^2y - xy + h(y)$. Differentiating with respect to y, we have $H_y = x^2 - x + h'(y) = N = x^2 - x + y$. Therefore, $h'(y) = y$ which implies that $h(y) = y^2/2$. Therefore,

$$H(x,y) = x^2y - xy + y^2/2.$$

99

(c)

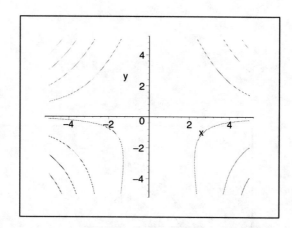

For trajectories in the first quadrant, solutions travel down and to the right. For trajectories in the second quadrant, solutions travel up and to the right. For solutions in the third quadrant, solutions travel to the right and down. For solutions in the fourth quadrant, solutions travel up and to the right.

(d)

The points $(0,0)$ and $(1,0)$ are saddle points. The point $(1/2, 1/4)$ is a center.

11.

(a) To find all critical points, we need to solve the following system of equations.

$$x' = 2x^2 y - 3x^2 - 4y = 0$$
$$y' = -2xy^2 + 6xy = 0.$$

Solving the second equation, we see that we must have $x = 0$, $y = 0$ or $y = 3$. Plugging these values into the first equation, we see that the critical points are: $(0,0)$, $(2,3)$ and $(-2,3)$.

(b) From our equations for dx/dt and dy/dt, we see that

$$\frac{dy}{dx} = \frac{-2xy^2 + 6xy}{2x^2y - 3x^2 - 4y}.$$

Rewriting this equation as $(2xy^2 - 6xy)dx + (2x^2y - 3x^2 - 4y)dy = 0$, we see that this is an exact equation. For $M = 2xy^2 - 6xy$, we integrate M with respect to x. We conclude that the level curves are given by $H(x,y) = C$ where $H(x,y) = x^2y^2 - 3x^2y + h(y)$. Differentiating with respect to y, we have $H_y = 2x^2y - 3x^2 + h'(y) = N = 2x^2y - 3x^2 - 4y$. Therefore, $h'(y) = -4y$ which implies that $h(y) = -2y^2$. Therefore,

$$H(x,y) = x^2y^2 - 3x^2y - 2y^2.$$

(c)

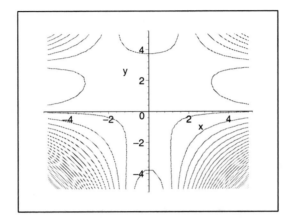

The direction of motion along each trajectory can be seen from the direction field shown below:

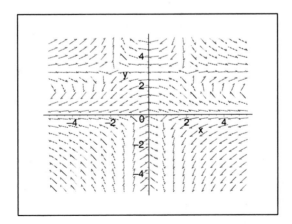

(d) From the direction field above, we see that all critical points are saddle points.

12.

101

(a) To find all critical points, we need to solve the following system of equations.

$$x' = 3x - x^2 = 0$$
$$y' = 2xy - 3y + 2 = 0.$$

The first equation implies $x(3 - x) = 0$. Therefore, $x = 0$ or $x = 3$. Substituting $x = 0$ into the second equation, we see that $y = 2/3$. Substituting $x = 3$ into the second equation, we see that $y = -2/3$. Therefore, the two critical points are $(0, 2/3)$ and $(3, -2/3)$.

(b) From our equations for dx/dt and dy/dt, we see that

$$\frac{dy}{dx} = \frac{2xy - 3y + 2}{3x - x^2}$$

$$\implies (3x - x^2)\, dy = (2xy - 3y + 2)\, dx$$
$$\implies (2xy - 3y + 2)\, dx + (x^2 - 3x)\, dy = 0.$$

We notice that this is an exact equation, which can be rewritten as

$$\left(x^2 y - 3xy + 2x\right)' = 0.$$

Therefore,

$$H(x, y) = x^2 y - 3xy + 2x.$$

(c)

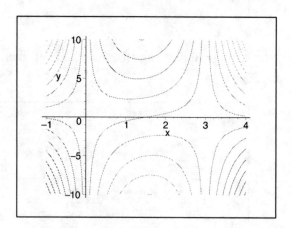

Solutions on trajectories to the right of $x = 3$ travel down. In particular, y will decrease as x decreases. Solutions on trajectories between $x = 0$ and $x = 3$ will travel up. In particular, eventually y will increase as x increases. Solutions on trajectories to the left of $x = 0$ will travel left, approaching the x-axis.

(d) Solutions will tend away from both critical points.

13.

(a) The critical points are given by the solving the equations

$$x(1 - y) = 0$$
$$y(1 + 2x) = 0.$$

The solutions of the first equation are $x = 0$ and $y = 1$. Plugging these values into the second equation, we conclude that the critical points are $(0, 0)$, $(-1/2, 1)$.

(b)

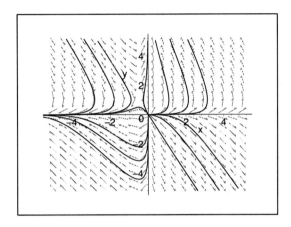

(c) Based on the phase portrait, we conclude that all trajectories starting near the origin diverge. Therefore, $(0, 0)$ is unstable. Near the critical point $(-1/2, 1)$, the trajectories behave like a saddle, and, therefore, $(-1/2, 1)$ is also unstable.

14.

(a) To find the critical points, we need to solve

$$x' = 2 - y = 0$$
$$y' = y - x^2 = 0.$$

By the first equation, we must have $y = 2$. By the second equation, we have $x = \pm 2$. Therefore, the two critical points are $(\sqrt{2}, 2)$ and $(-\sqrt{2}, 2)$.

(b)

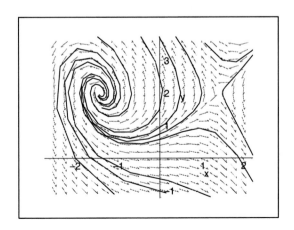

103

(c) The trajectories near the critical point $(\sqrt{2}, 2)$ tend away from the critical point. The trajectories near the critical point $(-\sqrt{2}, 2)$ spiral into the critical point.

15.

(a) The critical points are given by the solving the equations

$$x(1 - x - y) = 0$$
$$y\left(\frac{1}{2} - \frac{1}{4}y - \frac{3}{4}x\right) = 0.$$

Solving these two equations, we see that the critical points are given by $(0, 0)$, $(0, 2)$, $(1/2, 1/2)$ and $(1, 0)$.

(b)

(c) Based on the phase portrait, we conclude that all trajectories starting near the origin diverge. Therefore, $(0, 0)$ is unstable. Near the critical point $(0, 2)$, the trajectories approach the critical point. Similarly, near $(1, 0)$, trajectories approach the critical point. Therefore, both of those critical points are asymptotically stable nodes. Near the critical point $(1/2, 1/2)$, trajectories behave like a saddle.

16.

(a) The critical points are given by the solving the equations

$$-(x - y)(1 - x - y) = 0$$
$$x(2 + y) = 0.$$

The solutions of the second equation are $x = 0$ and $y = -2$. Plugging these values into the first equation, we conclude that the critical points are $(0, 0)$, $(0, 1)$, $(-2, -2)$ and $(3, -2)$.

(b)

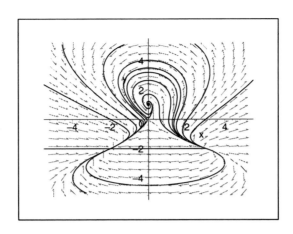

(c) Based on the phase portrait, we conclude that $(0,0)$ is a saddle point, $(0,1)$ is an asymptotically stable spiral point, $(-2,-2)$ is a stable node, and $(3,-2)$ is an unstable node.

17.

(a) The critical points are given by the solving the equations

$$y(2 - x - y) = 0$$
$$-x - y - 2xy = 0.$$

The solutions of the first equation are $y = 0$ and $x + y = 2$. Plugging these values into the second equation, we conclude that the critical points are $(0,0)$, $(1 - \sqrt{2}, 1 + \sqrt{2})$, and $(1 + \sqrt{2}, 1 - \sqrt{2})$.

(b)

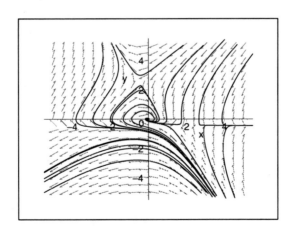

(c) Based on the phase portrait, we conclude that $(0,0)$ is an asymptotically stable spiral point, while $(1 - \sqrt{2}, 1 + \sqrt{2})$ and $(1 + \sqrt{2}, 1 - \sqrt{2})$ are saddle points.

18.

(a) The critical points are given by the solving the equations

$$(2 + x)(y - x) = 0$$
$$y(2 + x - x^2) = 0.$$

The solutions of the second equation are $y = 0$, $x = 2$ and $x = -1$. Plugging these values into the first equation, we conclude that the critical points are $(0, 0)$, $(2, 2)$, $(-1, -1)$ and $(-2, 0)$.

(b)

(c) Based on the phase portrait, we conclude that $(0, 0)$ and $(-2, 0)$ are saddle points and $(2, 2)$ and $(-1, -1)$ are asymptotically stable spiral points.

19.

(a) The critical points are given by the solving the equations

$$- x(1 - 2y) = 0$$
$$y - x^2 - y^2 = 0.$$

The solutions of the first equation are $x = 0$ and $y = 1/2$. Plugging these values into the second equation, we conclude that the critical points are $(0, 0)$, $(0, 1)$, $(1/2, 1/2)$ and $(-1/2, 1/2)$.

(b)

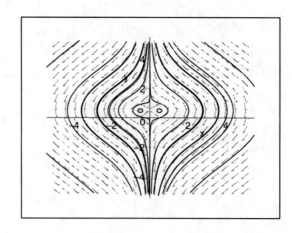

(c) Based on the phase portrait, we conclude that $(0,0)$ and $(0,1)$ are saddle points, while $(1/2, 1/2)$ and $(-1/2, 1/2)$ are centers.

20.

(a) To find the critical points, we need to solve

$$x' = y = 0$$

$$y' = x - \frac{1}{6}x^3 - \frac{1}{5}y = 0.$$

By the first equation, we must have $y = 0$. By the second equation, we have $x - \frac{1}{6}x^3 = 0$ which implies $x = 0$ or $x = \pm\sqrt{6}$. Therefore, the three critical points are $(0,0)$, $(\sqrt{6}, 0)$ and $(-\sqrt{6}, 0)$.

(b)

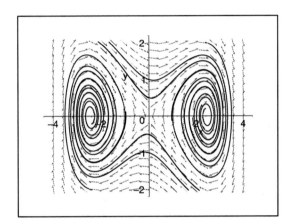

(c) The trajectories near the critical point $(0,0)$ tend away from the critical point. The trajectories near the critical points $(\sqrt{6}, 0)$ and $(-\sqrt{6}, 0)$ spiral into the critical points.

21.

(a)

107

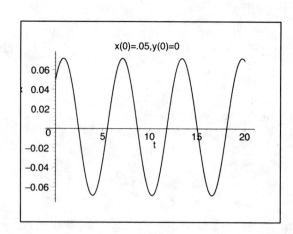

(b) For the first component plot, the period is approximately 6.5 seconds and the amplitude is approximately .46. For the second component plot, the period is approximately 7.5 and the amplitude is approximately 1.06. For the third component plot, the period is approximately 6 seconds and the amplitude is approximately .13. The period decreases as the amplitude decreases.

22.

(a)

$$H(x, y) = x^2 - xy + \frac{1}{2}y^2 - \frac{1}{3}x^3 = c.$$

If $x(0) = 3$ and $y(0) = 0$, then $H(x, y) = 3^2 - 3(0) + 1/2(0)^2 - 1/3(3)^3 = 0$. Therefore, $c = 0$.

Also,

$$x^2 - xy + \frac{1}{2}y^2 - \frac{1}{3}x^3 = 0$$

$$\implies 2x^2 - 2xy + y^2 - \frac{2}{3}x^3 = 0$$

$$\implies x^2 - 2xy + y^2 + x^2 - \frac{2}{3}x^3 = 0$$

$$\implies (x - y)^2 + x^2 - \frac{2}{3}x^3 = 0.$$

(b) Using the fact that $x - y = x'$ and plugging into equation (i) from part (a), we see that

$$(x')^2 + x^2 - \frac{2}{3}x^3 = 0$$

$$\implies (x')^2 = \frac{2}{3}x^3 - x^2$$

$$\implies x' = \pm\sqrt{\frac{2}{3}x^3 - x^2}$$

$$= \pm x\sqrt{\frac{2}{3}x - 1}$$

$$= \pm\frac{1}{3}x\sqrt{6x - 9}.$$

108

Now suppose the negative square root is taken. Then $x' = -\frac{1}{3}x\sqrt{6x-9}$ and $x(0) = 3$ implies $x'(0) = -3$. But $x' = x - y \implies x'(0) = x(0) - y(0) = 3 - 0 = 3$. Therefore, we must take the positive square root. We conclude that

$$x' = \frac{1}{3}x\sqrt{6x-9}.$$

(c) Let $s^2 = 6x - 9$. Then $x' = \frac{1}{3}x\sqrt{s^2} = \frac{1}{3}xs$. Therefore,

$$\frac{dx}{dt} = \frac{xs}{3} \implies \int \frac{dx}{xs} = \int \frac{dt}{3}$$

where $s^2 = 6x - 9$. Now $6x = s^2 + 9$ implies $x = (s^2 + 9)/6$, and, therefore, $dx = s\,ds/3$. Therefore,

$$\int \frac{(s/3)\,ds}{s(s^2+9)/6} = \int \frac{dt}{3}$$

$$\implies \int \frac{2\,ds}{s^2+9} = \int \frac{dt}{3}$$

$$\implies \int \frac{2\,ds}{9((s/3)^2+1)} = \frac{1}{3}t + C.$$

Now, substitute $u = s/3$ which means $du = ds/3$, and, therefore, we have

$$\frac{2}{9}\int \frac{3\,du}{u^2+1} = \frac{1}{3}t + C$$

$$\implies \frac{2}{3}\int \frac{du}{u^2+1} = \frac{1}{3}t + C$$

$$\implies \frac{2}{3}\tan^{-1}(u) = \frac{1}{3}t + C$$

$$\implies \frac{2}{3}\tan^{-1}(s/3) = \frac{1}{3}t + C$$

$$\implies \tan^{-1}(\sqrt{6x-9}/3) = \frac{1}{2}t + C$$

$$\implies \frac{\sqrt{6x-9}}{3} = \tan\left(\frac{1}{2}t + C\right)$$

$$\implies \sqrt{6x-9} = 3\tan\left(\frac{1}{2}t + C\right)$$

$$\implies 6x - 9 = 9\tan^2\left(\frac{1}{2}t + C\right)$$

$$\implies 6x = 9 + 9\tan^2\left(\frac{1}{2}t + C\right)$$

$$\implies x = \frac{3}{2} + \frac{3}{2}\tan^2\left(\frac{1}{2}t + C\right).$$

Now $x(0) = 3$ implies

$$x(0) = \frac{3}{2} + \frac{3}{2}\tan^2(C) = 3$$
$$\implies \tan^2(C) = 1$$
$$\implies \tan(C) = \pm 1$$
$$\implies C = \pm \pi/4.$$

Now, $C = -\pi/4$ is not an admissible solution. This can be seen by looking at the line above, $\sqrt{6x - 9} = 3\tan(t/2 + C)$. Plugging in $t = 0$ and $x(0) = 3$, we must have $3 = 3\tan(C)$, or $\tan(C) = 1$. Therefore, we must have $C = \pi/4$. Therefore, the solution is given by

$$x = \frac{3}{2} + \frac{3}{2}\tan^2\left(\frac{1}{2}t + \frac{\pi}{4}\right).$$

(d) By looking at the equation for x, we see that the solution will have a vertical asymptote when $t/2 + \pi/4 = \pi/2$. That is, there will be a vertical asymptote when $t = \pi/2$. This matches the expectation shown in Figure 3.6.3.

(e) By the same analysis, we see that the solution will have a vertical asymptote when $t/2 + \pi/4 = -\pi/2$. That is, there will be a vertical asymptote when $t = -3\pi/2$.

Section 3.7

1. In vector notation, the initial value problem can be written as

$$\frac{d}{dt}\begin{pmatrix} x \\ y \end{pmatrix} = \begin{pmatrix} x + y + t \\ 4x - 2y \end{pmatrix}$$

with initial condition

$$\mathbf{x}(0) = \begin{pmatrix} 1 \\ 0 \end{pmatrix}.$$

(a) The Euler formula is

$$\begin{pmatrix} x_{n+1} \\ y_{n+1} \end{pmatrix} = \begin{pmatrix} x_n \\ y_n \end{pmatrix} + h\begin{pmatrix} x_n + y_n + t_n \\ 4x_n - 2y_n \end{pmatrix}.$$

With $h = 0.1$, the approximate values at $t_n = 0.2, 0.4, 0.6, 0.8, 1.0$ are given by

$$\begin{pmatrix} x_n \\ y_n \end{pmatrix} = \begin{pmatrix} 1.26 \\ 0.76 \end{pmatrix}, \begin{pmatrix} 1.7714 \\ 1.4824 \end{pmatrix}, \begin{pmatrix} 2.58991 \\ 2.3703 \end{pmatrix}, \begin{pmatrix} 3.82374 \\ 3.60413 \end{pmatrix}, \begin{pmatrix} 5.64246 \\ 5.38885 \end{pmatrix}$$

respectively.

(b) The Runge-Kutta method uses the following intermediate calculations:

$$\mathbf{k}_{n1} = (x_n + y_n + t_n, 4x_n - 2y_n)^T$$

$$\mathbf{k}_{n2} = \left[x_n + \frac{h}{2}k_{n1}^1 + y_n + \frac{h}{2}k_{n1}^2 + t_n + \frac{h}{2}, 4\left(x_n + \frac{h}{2}k_{n1}^1\right) - 2\left(y_n + \frac{h}{2}k_{n1}^2\right)\right]^T$$

$$\mathbf{k}_{n3} = \left[x_n + \frac{h}{2}k_{n2}^1 + y_n + \frac{h}{2}k_{n2}^2 + t_n + \frac{h}{2}, 4\left(x_n + \frac{h}{2}k_{n2}^1\right) - 2\left(y_n + \frac{h}{2}k_{n2}^2\right)\right]^T$$

$$\mathbf{k}_{n4} = [x_n + hk_{n3}^1 + y_n + hk_{n3}^2 + t_n + h, 4(x_n + hk_{n3}^1) - 2(y_n + hk_{n3}^2)]^T.$$

With $h = 0.2$, the approximate values at $t_n = 0.2, 0.4, 0.6, 0.8, 1.0$ are given by

$$\begin{pmatrix} x_n \\ y_n \end{pmatrix} = \begin{pmatrix} 1.32493 \\ 0.758933 \end{pmatrix}, \begin{pmatrix} 1.93679 \\ 1.57919 \end{pmatrix}, \begin{pmatrix} 2.93414 \\ 2.66099 \end{pmatrix}, \begin{pmatrix} 4.48318 \\ 4.22639 \end{pmatrix}, \begin{pmatrix} 6.84236 \\ 6.56452 \end{pmatrix}$$

respectively.

(c) With $h = 0.1$, the approximate values at $t_n = 0.2, 0.4, 0.6, 0.8, 1.0$ are given by

$$\begin{pmatrix} x_n \\ y_n \end{pmatrix} = \begin{pmatrix} 1.32489 \\ 0.759516 \end{pmatrix}, \begin{pmatrix} 1.9369 \\ 1.57999 \end{pmatrix}, \begin{pmatrix} 2.93459 \\ 2.66201 \end{pmatrix}, \begin{pmatrix} 4.48422 \\ 4.22784 \end{pmatrix}, \begin{pmatrix} 6.8444 \\ 6.56684 \end{pmatrix}$$

respectively.

2. In vector notation, the initial value problem can be written as

$$\frac{d}{dt}\begin{pmatrix} x \\ y \end{pmatrix} = \begin{pmatrix} 2x + ty \\ xy \end{pmatrix}$$

with initial condition

$$\mathbf{x}(0) = \begin{pmatrix} 1 \\ 1 \end{pmatrix}.$$

(a) The Euler formula is

$$\begin{pmatrix} x_{n+1} \\ y_{n+1} \end{pmatrix} = \begin{pmatrix} x_n \\ y_n \end{pmatrix} + h\begin{pmatrix} 2x_n + t_n y_n \\ x_n y_n \end{pmatrix}.$$

With $h = 0.1$, the approximate values at $t_n = 0.2, 0.4, 0.6, 0.8, 1.0$ are given by

$$\begin{pmatrix} x_n \\ y_n \end{pmatrix} = \begin{pmatrix} 1.451 \\ 1.232 \end{pmatrix}, \begin{pmatrix} 2.16133 \\ 1.65988 \end{pmatrix}, \begin{pmatrix} 3.29292 \\ 2.55559 \end{pmatrix}, \begin{pmatrix} 5.16361 \\ 4.7916 \end{pmatrix}, \begin{pmatrix} 8.54951 \\ 12.0464 \end{pmatrix}$$

respectively.

(b) The Runge-Kutta method uses the following intermediate calculations:

$$\mathbf{k}_{n1} = (2x_n + t_n y_n, x_n y_n)^T$$

$$\mathbf{k}_{n2} = \left[2x_n + hk_{n1}^1 + \left(t_n + \frac{h}{2}\right)\left(y_n + \frac{h}{2}k_{n1}^2\right), \left(x_n + \frac{h}{2}k_{n1}^1\right)\left(y_n + \frac{h}{2}k_{n1}^2\right)\right]^T$$

$$\mathbf{k}_{n3} = \left[2x_n + hk_{n2}^1 + \left(t_n + \frac{h}{2}\right)\left(y_n + \frac{h}{2}k_{n2}^2\right), \left(x_n + \frac{h}{2}k_{n2}^1\right)\left(y_n + \frac{h}{2}k_{n2}^2\right)\right]^T$$

$$\mathbf{k}_{n4} = \left[2(x_n + hk_{n3}^1) + (t_n + h)\left(y_n + hk_{n3}^2\right), (x_n + hk_{n3}^1)\left(y_n + hk_{n3}^2\right)\right]^T$$

With $h = 0.2$, the approximate values at $t_n = 0.2, 0.4, 0.6, 0.8, 1.0$ are given by

$$\begin{pmatrix} x_n \\ y_n \end{pmatrix} = \begin{pmatrix} 1.51844 \\ 1.28089 \end{pmatrix}, \begin{pmatrix} 2.37684 \\ 1.87711 \end{pmatrix}, \begin{pmatrix} 3.85039 \\ 3.44859 \end{pmatrix}, \begin{pmatrix} 6.6956 \\ 9.50309 \end{pmatrix}, \begin{pmatrix} 15.0987 \\ 64.074 \end{pmatrix}$$

respectively.

(c) With $h = 0.1$, the approximate values at $t_n = 0.2, 0.4, 0.6, 0.8, 1.0$ are given by

$$\begin{pmatrix} x_n \\ y_n \end{pmatrix} = \begin{pmatrix} 1.51855 \\ 1.2809 \end{pmatrix}, \begin{pmatrix} 2.3773 \\ 1.87729 \end{pmatrix}, \begin{pmatrix} 3.85247 \\ 3.45126 \end{pmatrix}, \begin{pmatrix} 6.71282 \\ 9.56846 \end{pmatrix}, \begin{pmatrix} 15.6384 \\ 70.3792 \end{pmatrix}$$

respectively.

3. In vector notation, the initial value problem can be written as

$$\frac{d}{dt} \begin{pmatrix} x \\ y \end{pmatrix} = \begin{pmatrix} -tx - y - 1 \\ x \end{pmatrix}$$

with initial condition

$$\mathbf{x}(0) = \begin{pmatrix} 1 \\ 1 \end{pmatrix}.$$

(a) The Euler formula is

$$\begin{pmatrix} x_{n+1} \\ y_{n+1} \end{pmatrix} = \begin{pmatrix} x_n \\ y_n \end{pmatrix} + h \begin{pmatrix} -t_n x_n - y_n - 1 \\ x_n \end{pmatrix}.$$

With $h = 0.1$, the approximate values at $t_n = 0.2, 0.4, 0.6, 0.8, 1.0$ are given by

$$\begin{pmatrix} x_n \\ y_n \end{pmatrix} = \begin{pmatrix} 0.582 \\ 1.18 \end{pmatrix}, \begin{pmatrix} 0.117969 \\ 1.27344 \end{pmatrix}, \begin{pmatrix} -0.336912 \\ 1.27382 \end{pmatrix}, \begin{pmatrix} -0.730007 \\ 1.18572 \end{pmatrix}, \begin{pmatrix} -1.02134 \\ 1.02371 \end{pmatrix}$$

respectively.

(b) The Runge-Kutta method uses the following intermediate calculations:

$$\mathbf{k}_{n1} = (-t_n x_n - y_n - 1, x_n)^T$$

$$\mathbf{k}_{n2} = \left[-\left(t_n + \frac{h}{2}\right)\left(x_n + \frac{h}{2}k_{n1}^1\right) - \left(y_n + \frac{h}{2}k_{n1}^2\right) - 1, x_n + \frac{h}{2}k_{n1}^1 \right]^T$$

$$\mathbf{k}_{n3} = \left[-\left(t_n + \frac{h}{2}\right)\left(x_n + \frac{h}{2}k_{n2}^1\right) - \left(y_n + \frac{h}{2}k_{n2}^2\right) - 1, x_n + \frac{h}{2}k_{n2}^1 \right]^T$$

$$\mathbf{k}_{n4} = \left[-\left(t_n + h\right)\left(x_n + hk_{n3}^1\right) - \left(y_n + hk_{n3}^2\right) - 1, x_n + hk_{n3}^1 \right]^T$$

With $h = 0.2$, the approximate values at $t_n = 0.2, 0.4, 0.6, 0.8, 1.0$ are given by

$$\begin{pmatrix} x_n \\ y_n \end{pmatrix} = \begin{pmatrix} 0.568451 \\ 1.15775 \end{pmatrix}, \begin{pmatrix} 0.109776 \\ 1.22556 \end{pmatrix}, \begin{pmatrix} -0.32208 \\ 1.20347 \end{pmatrix}, \begin{pmatrix} -0.681296 \\ 1.10162 \end{pmatrix}, \begin{pmatrix} -0.937852 \\ 0.937852 \end{pmatrix}$$

respectively.

(c) With $h = 0.1$, the approximate values at $t_n = 0.2, 0.4, 0.6, 0.8, 1.0$ are given by

$$\begin{pmatrix} x_n \\ y_n \end{pmatrix} = \begin{pmatrix} 0.56845 \\ 1.15775 \end{pmatrix}, \begin{pmatrix} 0.109773 \\ 1.22557 \end{pmatrix}, \begin{pmatrix} -0.322081 \\ 1.20347 \end{pmatrix}, \begin{pmatrix} -0.681291 \\ 1.10161 \end{pmatrix}, \begin{pmatrix} -0.937841 \\ 0.93784 \end{pmatrix}$$

respectively.

4. In vector notation, the initial value problem can be written as

$$\frac{d}{dt}\begin{pmatrix} x \\ y \end{pmatrix} = \begin{pmatrix} x - y + xy \\ 3x - 2y - xy \end{pmatrix}$$

with initial condition

$$\mathbf{x}(0) = \begin{pmatrix} 0 \\ 1 \end{pmatrix}.$$

(a) The Euler formula is

$$\begin{pmatrix} x_{n+1} \\ y_{n+1} \end{pmatrix} = \begin{pmatrix} x_n \\ y_n \end{pmatrix} + h \begin{pmatrix} x_n - y_n + x_n y_n \\ 3x_n - 2y_n - x_n y_n \end{pmatrix}.$$

With $h = 0.1$, the approximate values at $t_n = 0.2, 0.4, 0.6, 0.8, 1.0$ are given by

$$\begin{pmatrix} x_n \\ y_n \end{pmatrix} = \begin{pmatrix} -0.198 \\ 0.618 \end{pmatrix}, \begin{pmatrix} -0.378796 \\ 0.28329 \end{pmatrix}, \begin{pmatrix} -0.51932 \\ -0.0321025 \end{pmatrix}, \begin{pmatrix} -0.594324 \\ -0.326801 \end{pmatrix}, \begin{pmatrix} -0.588278 \\ -0.57545 \end{pmatrix}$$

respectively.

(b) Given

$$f(t, x, y) = x - y + xy$$
$$g(t, x, y) = 3x - 2y - xy,$$

the Runge-Kutta method uses the following intermediate calculations:

$$\mathbf{k}_{n1} = [f(t_n, x_n, y_n), g(t_n, x_n, y_n)]^T$$

$$\mathbf{k}_{n2} = \left[f\left(t_n + \frac{h}{2}, x_n + \frac{h}{2}k_{n1}^1, y_n + \frac{h}{2}k_{n1}^2\right), g\left(t_n + \frac{h}{2}, x_n + \frac{h}{2}k_{n1}^1, y_n + \frac{h}{2}k_{n1}^2\right) \right]^T$$

$$\mathbf{k}_{n3} = \left[f\left(t_n + \frac{h}{2}, x_n + \frac{h}{2}k_{n2}^1, y_n + \frac{h}{2}k_{n2}^2\right), g\left(t_n + \frac{h}{2}, x_n + \frac{h}{2}k_{n2}^1, y_n + \frac{h}{2}k_{n2}^2\right) \right]^T$$

$$\mathbf{k}_{n4} = \left[f\left(t_n + h, x_n + hk_{n3}^1, y_n + hk_{n3}^2\right), g\left(t_n + h, x_n + hk_{n3}^1, y_n + hk_{n3}^2\right) \right]^T$$

With $h = 0.2$, the approximate values at $t_n = 0.2, 0.4, 0.6, 0.8, 1.0$ are given by

$$\begin{pmatrix} x_n \\ y_n \end{pmatrix} = \begin{pmatrix} -0.196904 \\ 0.630936 \end{pmatrix}, \begin{pmatrix} -0.372643 \\ 0.298888 \end{pmatrix}, \begin{pmatrix} -0.501302 \\ -0.0111429 \end{pmatrix}, \begin{pmatrix} -0.561270 \\ -0.288943 \end{pmatrix}, \begin{pmatrix} -0.547053 \\ -0.508303 \end{pmatrix}$$

respectively.

(c) With $h = 0.1$, the approximate values at $t_n = 0.2, 0.4, 0.6, 0.8, 1.0$ are given by

$$\begin{pmatrix} x_n \\ y_n \end{pmatrix} = \begin{pmatrix} -0.196935 \\ 0.630939 \end{pmatrix}, \begin{pmatrix} -0.372687 \\ 0.298866 \end{pmatrix}, \begin{pmatrix} -0.501345 \\ -0.012184 \end{pmatrix}, \begin{pmatrix} -0.561292 \\ -0.28907 \end{pmatrix}, \begin{pmatrix} -0.547031 \\ -0.508427 \end{pmatrix}$$

respectively.

5. In vector notation, the initial value problem can be written as

$$\frac{d}{dt} \begin{pmatrix} x \\ y \end{pmatrix} = \begin{pmatrix} x(1 - 0.5x - 0.5y) \\ y(-0.25 + 0.5x) \end{pmatrix}$$

with initial condition

$$\mathbf{x}(0) = \begin{pmatrix} 4 \\ 1 \end{pmatrix}.$$

(a) The Euler formula is

$$\begin{pmatrix} x_{n+1} \\ y_{n+1} \end{pmatrix} = \begin{pmatrix} x_n \\ y_n \end{pmatrix} + h \begin{pmatrix} x_n(1 - 0.5x_n - 0.5y_n) \\ y_n(-0.25 + 0.5x_n) \end{pmatrix}.$$

With $h = 0.1$, the approximate values at $t_n = 0.2, 0.4, 0.6, 0.8, 1.0$ are given by

$$\begin{pmatrix} x_n \\ y_n \end{pmatrix} = \begin{pmatrix} 2.96225 \\ 1.34538 \end{pmatrix}, \begin{pmatrix} 2.34119 \\ 1.67121 \end{pmatrix}, \begin{pmatrix} 1.90236 \\ 1.97158 \end{pmatrix}, \begin{pmatrix} 1.56602 \\ 2.23895 \end{pmatrix}, \begin{pmatrix} 1.29768 \\ 2.46732 \end{pmatrix}$$

respectively.

(b) Given

$$f(t, x, y) = x(1 - 0.5x - 0.5y)$$
$$g(t, x, y) = y(-0.25 + 0.5x),$$

the Runge-Kutta method uses the following intermediate calculations:

$$\mathbf{k}_{n1} = [f(t_n, x_n, y_n), g(t_n, x_n, y_n)]^T$$

$$\mathbf{k}_{n2} = \left[f\left(t_n + \frac{h}{2}, x_n + \frac{h}{2}k_{n1}^1, y_n + \frac{h}{2}k_{n1}^2\right), g\left(t_n + \frac{h}{2}, x_n + \frac{h}{2}k_{n1}^1, y_n + \frac{h}{2}k_{n1}^2\right) \right]^T$$

$$\mathbf{k}_{n3} = \left[f\left(t_n + \frac{h}{2}, x_n + \frac{h}{2}k_{n2}^1, y_n + \frac{h}{2}k_{n2}^2\right), g\left(t_n + \frac{h}{2}, x_n + \frac{h}{2}k_{n2}^1, y_n + \frac{h}{2}k_{n2}^2\right) \right]^T$$

$$\mathbf{k}_{n4} = \left[f\left(t_n + h, x_n + hk_{n3}^1, y_n + hk_{n3}^2\right), g\left(t_n + h, x_n + hk_{n3}^1, y_n + hk_{n3}^2\right) \right]^T$$

With $h = 0.2$, the approximate values at $t_n = 0.2, 0.4, 0.6, 0.8, 1.0$ are given by

$$\begin{pmatrix} x_n \\ y_n \end{pmatrix} = \begin{pmatrix} 3.06339 \\ 1.34858 \end{pmatrix}, \begin{pmatrix} 2.44497 \\ 1.68638 \end{pmatrix}, \begin{pmatrix} 1.9911 \\ 2.00036 \end{pmatrix}, \begin{pmatrix} 1.63818 \\ 2.27981 \end{pmatrix}, \begin{pmatrix} 1.3555 \\ 2.5175 \end{pmatrix}$$

respectively.

(c) With $h = 0.1$, the approximate values at $t_n = 0.2, 0.4, 0.6, 0.8, 1.0$ are given by

$$\begin{pmatrix} x_n \\ y_n \end{pmatrix} = \begin{pmatrix} 3.06314 \\ 1.34899 \end{pmatrix}, \begin{pmatrix} 2.44465 \\ 1.68699 \end{pmatrix}, \begin{pmatrix} 1.99075 \\ 2.00107 \end{pmatrix}, \begin{pmatrix} 1.63781 \\ 2.28057 \end{pmatrix}, \begin{pmatrix} 1.35514 \\ 2.51827 \end{pmatrix}$$

respectively.

6. In vector notation, the initial value problem can be written as

$$\frac{d}{dt} \begin{pmatrix} x \\ y \end{pmatrix} = \begin{pmatrix} \exp(-x + y) - \cos x \\ \sin(x - 3y) \end{pmatrix}$$

with initial condition

$$\mathbf{x}(0) = \begin{pmatrix} 1 \\ 2 \end{pmatrix}.$$

(a) The Euler formula is

$$\begin{pmatrix} x_{n+1} \\ y_{n+1} \end{pmatrix} = \begin{pmatrix} x_n \\ y_n \end{pmatrix} + h \begin{pmatrix} \exp(-x_n + y_n) - \cos x_n \\ \sin(x_n - 3y_n) \end{pmatrix}.$$

With $h = 0.1$, the approximate values at $t_n = 0.2, 0.4, 0.6, 0.8, 1.0$ are given by

$$\begin{pmatrix} x_n \\ y_n \end{pmatrix} = \begin{pmatrix} 1.42386 \\ 2.18957 \end{pmatrix}, \begin{pmatrix} 1.82234 \\ 2.36791 \end{pmatrix}, \begin{pmatrix} 2.21728 \\ 2.53329 \end{pmatrix}, \begin{pmatrix} 2.61118 \\ 2.68763 \end{pmatrix}, \begin{pmatrix} 2.9955 \\ 2.83354 \end{pmatrix}$$

respectively.

(b) Given

$$f(t, x, y) = \exp(-x + y) - \cos x$$
$$g(t, x, y) = \sin(x - 3y),$$

the Runge-Kutta method uses the following intermediate calculations:

$$\mathbf{k}_{n1} = [f(t_n, x_n, y_n), g(t_n, x_n, y_n)]^T$$

$$\mathbf{k}_{n2} = \left[f\left(t_n + \frac{h}{2}, x_n + \frac{h}{2}k_{n1}^1, y_n + \frac{h}{2}k_{n1}^2\right), g\left(t_n + \frac{h}{2}, x_n + \frac{h}{2}k_{n1}^1, y_n + \frac{h}{2}k_{n1}^2\right) \right]^T$$

$$\mathbf{k}_{n3} = \left[f\left(t_n + \frac{h}{2}, x_n + \frac{h}{2}k_{n2}^1, y_n + \frac{h}{2}k_{n2}^2\right), g\left(t_n + \frac{h}{2}, x_n + \frac{h}{2}k_{n2}^1, y_n + \frac{h}{2}k_{n2}^2\right) \right]^T$$

$$\mathbf{k}_{n4} = \left[f\left(t_n + h, x_n + hk_{n3}^1, y_n + hk_{n3}^2\right), g\left(t_n + h, x_n + hk_{n3}^1, y_n + hk_{n3}^2\right) \right]^T$$

With $h = 0.2$, the approximate values at $t_n = 0.2, 0.4, 0.6, 0.8, 1.0$ are given by

$$\begin{pmatrix} x_n \\ y_n \end{pmatrix} = \begin{pmatrix} 1.41513 \\ 2.18699 \end{pmatrix}, \begin{pmatrix} 1.81208 \\ 2.36233 \end{pmatrix}, \begin{pmatrix} 2.20635 \\ 2.5258 \end{pmatrix}, \begin{pmatrix} 2.59826 \\ 2.6794 \end{pmatrix}, \begin{pmatrix} 2.97806 \\ 2.82487 \end{pmatrix}$$

respectively.

(c) With $h = 0.1$, the approximate values at $t_n = 0.2, 0.4, 0.6, 0.8, 1.0$ are given by

$$\begin{pmatrix} x_n \\ y_n \end{pmatrix} = \begin{pmatrix} 1.41513 \\ 2.18699 \end{pmatrix}, \begin{pmatrix} 1.81209 \\ 2.36233 \end{pmatrix}, \begin{pmatrix} 2.20635 \\ 2.52581 \end{pmatrix}, \begin{pmatrix} 2.59826 \\ 2.67941 \end{pmatrix}, \begin{pmatrix} 2.97806 \\ 2.82488 \end{pmatrix}$$

respectively.

7. Given

$$f(t, x, y) = -x + 4y$$
$$g(t, x, y) = x - y,$$

the Runge-Kutta method uses the following intermediate calculations,

$$\mathbf{k}_{n1} = [f(t_n, x_n, y_n), g(t_n, x_n, y_n)]^T$$

$$\mathbf{k}_{n2} = \left[f\left(t_n + \frac{h}{2}, x_n + \frac{h}{2}k_{n1}^1, y_n + \frac{h}{2}k_{n1}^2\right), g\left(t_n + \frac{h}{2}, x_n + \frac{h}{2}k_{n1}^1, y_n + \frac{h}{2}k_{n1}^2\right) \right]^T$$

$$\mathbf{k}_{n3} = \left[f\left(t_n + \frac{h}{2}, x_n + \frac{h}{2}k_{n2}^1, y_n + \frac{h}{2}k_{n2}^2\right), g\left(t_n + \frac{h}{2}, x_n + \frac{h}{2}k_{n2}^1, y_n + \frac{h}{2}k_{n2}^2\right) \right]^T$$

$$\mathbf{k}_{n4} = \left[f\left(t_n + h, x_n + hk_{n3}^1, y_n + hk_{n3}^2\right), g\left(t_n + h, x_n + hk_{n3}^1, y_n + hk_{n3}^2\right) \right]^T.$$

With $h = 0.2$, the approximate values of the solution at $t = 1$ are given by

$$\begin{pmatrix} x_n \\ y_n \end{pmatrix} = \begin{pmatrix} 1.43421 \\ 0.64202 \end{pmatrix}.$$

With $h = 0.1$, the approximate values of the solution at $t = 1$ are given by

$$\begin{pmatrix} x_n \\ y_n \end{pmatrix} = \begin{pmatrix} 1.43384 \\ 0.64222 \end{pmatrix}.$$

With $h = 0.05$ and $h = 0.025$ the approximate values of the solution (approximating up to the first 6 digits) at $t = 1$ are given by

$$\begin{pmatrix} x_n \\ y_n \end{pmatrix} = \begin{pmatrix} 1.43382 \\ 0.64223 \end{pmatrix}.$$

The exact solution is given by

$$\phi(1) = \frac{e + 3e^{-3}}{2} = 1.43382$$

$$\psi(1) = \frac{e - 3e^{-3}}{4} = 0.64223.$$

8. Let $y = x'$. The second order ODE can be transformed into the first order system

$$x' = y$$
$$y' = t - 3x - t^2 y$$

with initial conditions $x(0) = 1$, $y(0) = 2$. Given

$$f(t, x, y) = y$$
$$g(t, x, y) = t - 3x - t^2 y,$$

the Runge-Kutta method uses the following intermediate calculations,

$$\mathbf{k}_{n1} = [f(t_n, x_n, y_n), g(t_n, x_n, y_n)]^T$$

$$\mathbf{k}_{n2} = \left[f\left(t_n + \frac{h}{2}, x_n + \frac{h}{2}k_{n1}^1, y_n + \frac{h}{2}k_{n1}^2 \right), g\left(t_n + \frac{h}{2}, x_n + \frac{h}{2}k_{n1}^1, y_n + \frac{h}{2}k_{n1}^2 \right) \right]^T$$

$$\mathbf{k}_{n3} = \left[f\left(t_n + \frac{h}{2}, x_n + \frac{h}{2}k_{n2}^1, y_n + \frac{h}{2}k_{n2}^2 \right), g\left(t_n + \frac{h}{2}, x_n + \frac{h}{2}k_{n2}^1, y_n + \frac{h}{2}k_{n2}^2 \right) \right]^T$$

$$\mathbf{k}_{n4} = \left[f\left(t_n + h, x_n + hk_{n3}^1, y_n + hk_{n3}^2 \right), g\left(t_n + h, x_n + hk_{n3}^1, y_n + hk_{n3}^2 \right) \right]^T.$$

With $h = 0.1$, the approximate values at $t_n = 0.5, 1.0$ are given by

$$\begin{pmatrix} x_n \\ y_n \end{pmatrix} = \begin{pmatrix} 1.543 \\ 1.14743 \end{pmatrix}, \begin{pmatrix} 0.07075 \\ -1.3885 \end{pmatrix},$$

respectively.

Chapter 4

Section 4.1

1. The differential equation is linear.

2. The differential equation is nonlinear.

3. The differential equation is linear.

4. The differential equation is linear.

5. The differential equation is nonlinear.

6. We will use the equation $mg - kL = 0$. If the mass is 8 lb, then $mg = 8$. If the mass stretches the spring 6 inches, then $L = 6$. Therefore, $8 - 6k = 0$ which means $k = 4/3$ lb/inch.

7. Again, we use the equation $mg - kL = 0$. Here the mass is 10 kg. The force due to gravity is $g = 9.8$ m/s^2. Therefore, $mg = 98$ Newtons. The mass stretches the spring .7 meters. Therefore, $k = 98/.7 = 140$ N/m.

8. The spring constant is $k = 2/(1/2) = 4$ lb/ft. The mass $m = 2/32 = 1/16$ lb·s^2/ft. The equation of motion is

$$\frac{1}{16}y'' + 4y = 0$$

or

$$y'' + 64y = 0$$

with initial conditions $y(0) = 1/4$ ft, $y'(0) = 0$ ft/sec.

9. The spring constant is $k = .98/.05 = 19.6$ N/m. The mass $m = .1$ kg. The equation of motion is

$$.1y'' + 19.6y = 0$$

or

$$y'' + 196y = 0$$

with initial conditions $y(0) = 0$ m, $y'(0) = .1$ m/sec.

10. The spring constant is $k = 3/(1/4) = 12$ lb/ft. The mass $m = 3/32$ lb·s^2/ft. The equation of motion is

$$\frac{3}{32}y'' + 12y = 0$$

or

$$y'' + 128y = 0$$

with initial conditions $y(0) = -1/12$ ft, $y'(0) = 2$ ft/sec.

11. The inductance $L = 1$ henry. The resistance $R = 0$. The capacitance $C = 0.25 \times 10^{-6}$ farads. Therefore, the equation for charge q is

$$q'' + (4 \times 10^6)q = 0$$

with initial conditions $q(0) = 10^{-6}$ coulombs, $q'(0) = 0$ coulombs/sec.

12. The spring constant is $k = .196/.05 = 3.92$ N/m. The mass $m = .02$ kg. The damping constant is $\gamma = 400$ dyne-sec/cm $= .4$ N-sec/cm. Therefore, the equation of motion is

$$.02y'' + .4y' + 3.92y = 0$$

or

$$y'' + 20y' + 196y = 0$$

with initial conditions $y(0) = .02$ m, $y'(0) = 0$ m/sec.

13. The spring constant is $k = 16/(1/4) = 64$ lb/ft. The mass $m = 1/2$ lb·s^2/ft. The damping coefficient is $\gamma = 2$ lb-sec/ft. Therefore, the equation of motion is

$$\frac{1}{2}y'' + 2y' + 64y = 0$$

or

$$y'' + 4y' + 128y = 0$$

with initial conditions $y(0) = 0$ ft, $y'(0) = 1/4$ ft/sec.

14. The spring constant is $k = 3/.1 = 30$ N/m. The mass $m = 2$ kg. The damping coefficient is $\gamma = 3/5$ N-sec/m. Therefore, the equation of motion is

$$2y'' + \frac{3}{5}y' + 30y = 0$$

or

$$y'' + .3y' + 15y = 0$$

with initial conditions $y(0) = .05$ m, $y'(0) = .1$ m/sec.

15. The inductance $L = 0.2$ henry. The resistance $R = 3 \times 10^2$ ohms. The capacitance $C = 10^{-5}$ farads. Therefore, the equation for charge q is

$$0.2q'' + 300q' + 10^5 q = 0$$

or

$$q'' + 1500q' + 500,000q = 0$$

with initial conditions $q(0) = 10^{-6}$ coulombs, $q'(0) = 0$ coulombs/sec.

16. We know the net force acting on the mass satisfies

$$F_{\text{net}} = mx''(t)$$

where x'' is the acceleration of the mass. Here, we assume the spring is pulling the mass to the left. We will consider that to be the negative direction. The force due to the spring is given by

$$F_s = -k(L + x).$$

The force due to air resistance, F_d, is acting in the direction opposite to the direction of motion of the mass, and is proportional to the velocity of the mass. Therefore,

$$F_d = -\gamma x'(t)$$

2

where $\gamma \geq 0$. In particular, if the mass is moving to the left, then the velocity $x'(t) < 0$ in which case, $F_d \geq 0$, while if the mass is moving to the right, then the velocity $x'(t) > 0$ in which case, $F_d \leq 0$. Finally, we are assuming that an external force $F(t)$ is acting on the mass. Putting together these factors, we have

$$F_{\text{net}} = F_s + F_d + F(t)$$
$$\implies mx''(t) = -k(L+x) - \gamma x'(t) + F(t).$$

From equation (13), we know that $mg = kL$. Since the mass lies on a frictionless, horizontal surface, the force due to gravity is zero. Therefore, $kL = 0$, meaning $L = 0$. In particular, the spring is not stretched by the mass when the spring is in equilibrium. Therefore, we conclude that

$$mx''(t) + \gamma x'(t) + kx(t) = F(t),$$

as in equation (18). The derivation differs here since it is assumed there is no force due to gravity acting on the spring.

17.

(a) We know that the net force is given by $F_{\text{net}} = mx''(t)$. Here, we are assuming there are no damping of external forces present. Also, we are assuming there is no force due to gravity. Therefore,

$$F_{\text{net}} = F_s$$
$$\implies mx'' = -kx - \epsilon x^3$$
$$\implies mx'' + kx + \epsilon x^3 = 0.$$

(b) If we assume that the maximum displacement is small, then $x^3 \approx 0$, in which case the linearized equation is
$$mx'' + kx = 0.$$

18. We know the net force acting on the mass is given by $F_{\text{net}} = mx''$. The force exerted on the mass by spring 1 is given by $F_{s_1} = -k_1 x$. The force exerted on the mass by spring 2 is given by $F_{s_2} = -k_2 x$. Also, we are assuming air resistance is acting on the mass with coefficient γ. The air resistance is proportional to the velocity $x'(t)$ of the mass. Therefore, we have

$$F_{\text{net}} = F_{s_1} + F_{s_2} + F_d$$
$$\implies mx'' = -k_1 x - k_2 x - \gamma x'$$
$$\implies mx'' + \gamma x' + (k_1 + k_2)x = 0.$$

19. Let $u(t)$ be the depth of the block into the water. Let m be the mass of the block and $F_B(t)$ be the upward buoyant force exerted by the fluid. The equation of motion is given by

$$mu''(t) = F_B(t).$$

Since the density of the cubic block is ρ, the mass $m = \rho \cdot l^3$. Since the buoyant force is equal to the weight of the displaced fluid, we see that $F_B = -\rho_0 \cdot u(t) \cdot l^2 \cdot g$. Therefore,

$$\rho \cdot l^3 u'' = -\rho_0 \cdot u(t) \cdot l^2 \cdot g,$$

or

$$\rho l u'' + \rho_0 g u = 0.$$

20.

(a)

(b)

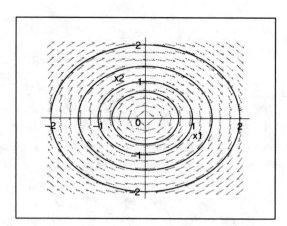

(c) The critical point is stable.

21.

(a)

(b)

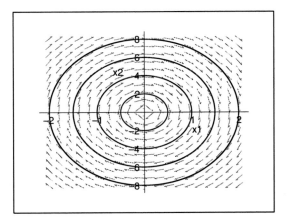

(c) The critical point is stable.

22.

(a)

(b)

(c) The critical point is asymptotically stable.

23.

(a)

(b)

(c) The critical point is asymptotically stable.

24.

(a)

(b)

(c) The critical point is unstable.

25.

(a)

(b)

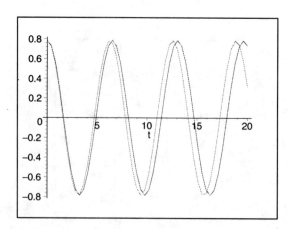

26.

(a) If $y_1 = \cos(rt)$, then $y_1' = -r\sin(rt)$ and $y_1'' = -r^2\cos(rt)$. Therefore, in order for y_1 to be a solution of our equation, we need

$$-mr^2\cos(rt)k\cos(rt) = 0$$

or $r = \pm\sqrt{k/m}$. Similarly, for $r = \pm\sqrt{k/m}$, $y_2 = \sin(rt)$ will be a solution of this equation. If $y = c_1y_1 + c_2y_2$, then

$$y'' = c_1y_1'' + c_2y_2''$$

implies

$$my'' + ky = m(c_1y_1'' + c_2y_2'') + k(c_1y_1 + c_2y_2) = c_1(my_1'' + ky_1) + c_2(my_2'' + ky_2) = 0.$$

Therefore, $y = c_1y_1 + c_2y_2$ will be a solution if y_1 and y_2 are solutions.

(b) By part (a), $y_1(t) = \cos(rt)$ and $y_2(t) = \sin(rt)$ will be solutions as long as $r = \pm\sqrt{k/m}$. Further, any solution of the form $y = c_1\cos(rt) + c_2\sin(rt)$ will be a solution.

(i) In this case, $r = \pm 1$. Note that we can take $r = 1$ as $r = -1$ will just change the coefficients c_1 and c_2 above. Now if $r = 1$, then $y(t) = c_1\cos(t) + c_2\sin(t)$. Then $y(0) = c_1 = 1$ and $y'(t) = -\sin(t) + c_2\cos(t)$ implies $y'(0) = c_2 = -1$. Therefore, our solution is $y(t) = \cos(t) - \sin(t)$.

(ii) In this case, $r = \pm 4$. As above, we will take $r = 4$. Then $y(t) = c_1\cos(4t) + c_2\sin(4t)$. Therefore, $y(0) = c_1 = 1$ and $y'(t) = -4\sin(4t) + 4c_2\cos(4t)$ implies $y'(0) = 4c_2 = -1 \implies c_2 = -1/4$. Therefore, $y(t) = \cos(4t) - \frac{1}{4}\sin(4t)$.

8

(c)

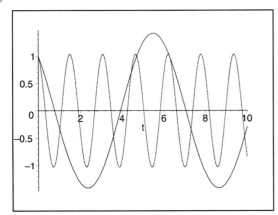

The soft spring has a higher amplitude and a longer period. The stiff spring has a smaller amplitude and a shorter period.

Section 4.2

1. Write the IVP as
$$y'' + \frac{3}{t}y' = 1.$$
Since the function $p(t) = 3/t$ is continuous for all $t > 0$ and $t_0 = 1 > 0$, the IVP is guaranteed to have a unique solution for all $t > 0$.

2. Write the IVP as
$$y'' - \frac{3t}{t-1}y' + \frac{4}{t-1}y = \frac{\sin t}{t-1}.$$
Since the coefficient functions are continuous for all $t < 1$ and $t_0 = -2 < 1$, the IVP is guaranteed to have a unique solution for all $t < 1$.

3. Write the IVP as
$$y'' + \frac{3t}{t(t-4)}y' + \frac{4}{t(t-4)}y = \frac{2}{t(t-4)}.$$
Since the coefficient functions are continuous for all t such that $t \neq 0, 4$ and $t_0 = 3$, the IVP is guaranteed to have a unique solution for all t such that $0 < t < 4$.

4. Since the coefficient function $3\ln(|t|)$ is continuous for all $t \neq 0$ and $t_0 = 2$, the IVP is guaranteed to have a unique solution for all $t > 0$.

5. Write the IVP as
$$y'' + \frac{x}{x-3}y' + \frac{\ln|x|}{x-3}y = 0.$$
Since the coefficient functions are continuous for all x such that $x \neq 0, 3$ and $x_0 = 1$, the IVP is guaranteed to have a unique solution for all x such that $0 < x < 3$.

6. Write the IVP as
$$y'' + \frac{1}{x-2}y' + \tan x y = 0.$$
Since the coefficient functions are continuous for all x such that $x \neq 2, n\pi + \pi/2$ and $x_0 = 3$, the IVP is guaranteed to have a unique solution for all x such that $2 < t < 3\pi/2$.

9

7.

$$W(e^{2t}, e^{-3t/2}) = \begin{vmatrix} e^{2t} & e^{-3t/2} \\ 2e^{2t} & -\frac{3}{2}e^{-3t/2} \end{vmatrix} = -\frac{7}{2}e^{t/2}.$$

8.

$$W(\cos t, \sin t) = \begin{vmatrix} \cos t & \sin t \\ -\sin t & \cos t \end{vmatrix} = 1.$$

9.

$$W(e^{-2t}, te^{-2t}) = \begin{vmatrix} e^{-2t} & te^{-2t} \\ -2e^{-2t} & (1-2t)e^{-2t} \end{vmatrix} = e^{-4t}.$$

10.

$$W(x, xe^x) = \begin{vmatrix} x & xe^x \\ 1 & (1+x)e^x \end{vmatrix} = x^2 e^x.$$

11.

$$W(e^t \sin t, e^t \cos t) = \begin{vmatrix} e^t \sin t & e^t \cos t \\ e^t(\sin t + \cos t) & e^t(\cos t - \sin t) \end{vmatrix} = -e^{2t}.$$

12.

$$W(\cos^2\theta, 1 + \cos 2\theta) = \begin{vmatrix} \cos^2\theta & 1 + \cos 2\theta \\ -2\sin\theta\cos\theta & -2\sin 2\theta \end{vmatrix} = 0.$$

13. If $y_1 = t^2$, then $y_1'' = 2$. Therefore, $t^2 y_1'' - 2y_1 = t^2(2) - 2t^2 = 0$. If $y_2 = t^{-1}$, then $y_2' = 2t^{-3}$. Therefore, $t^2 y_2'' - 2y_2 = t^2(2t^{-3}) - 2t^{-1} = 0$. Since the equation is linear, the function $y_3 = c_1 t^2 + c_2 t^{-1}$ will also be a solution.

14.

(a) Dividing the equation by y, we have

$$y'' + \frac{1}{y}(y')^2 = 0.$$

This equation will not have the form required.

(b) Consider $T[y_1 + y_2]$. We see that $T[y_1 + y_2] = (y_1 + y_2)(y_1'' + y_2'') + (y_1' + y_2')^2 \neq y_1 y_1'' + (y_1')^2 + y_2 y_2'' + (y_2')^2 = T[y_1] + T[y_2]$.

(c) If $y_1 = 1$, then $y_1' = 0 = y_1''$. Therefore, $T[y_1] = 0$. If $y_2 = t^{1/2}$, then $y_2' = t^{-1/2}/2$ and $y_2'' = -t^{-3/2}/4$. Therefore, $T[y_2] = -t^{1/2}(t^{-3/2}/4) + (t^{-1/2}/2)^2 = -t^{-1}/4 + t^{-1}/4 = 0$. But $T[c_1 y_1 + c_2 y_2] = (c_1 + c_2 t^{1/2})(-c_2 t^{-3/2}/4) + (c_2 t^{-1/2}/2)^2 = -c_1 c_2 t^{-3/2}/4 \neq 0$ in general.

15. No. Substituting $y = \sin(t^2)$ into the differential equation, we have

$$-4t^2 \sin(t^2) + 2\cos(t^2) + 2t \cos(t^2)p(t)\sin(t^2)q(t) = 0.$$

For the equation to be valid, we must have $p(t) = -1/t$, which is not continuous, or even defined, at $t = 0$.

16. $W(e^{2t}, g(t)) = e^{2t}g'(t) - 2e^{2t}g(t) = 3e^{4t}$. Dividing both sides by e^{2t}, we find that g must satisfy the ODE $g' - 2g = 3e^{2t}$. Therefore, $g(t) = 3te^{2t} + ce^{2t}$.

17. $W(t, g(t)) = tg'(t) - g(t) = t^2e^t$. Dividing both sides of the equation by t, we have $g' - g/t = te^t$. Therefore, $g(t) = te^t + ct$.

18. $W(f, g) = fg' - f'g$. Also, $W(u, v) = W(2f - g, f + 2g)$. Upon evaluation, $W(u, v) = 5fg' - 5f'g = 5W(f, g)$.

19. $W(f, g) = fg' - f'g = t\cos t - \sin t$ and $W(u, v) = -4fg' + 4f'g$. Therefore, $W(u, v) = -4t\cos t + 4\sin t$.

20. For $y_1 = \cos 2t$, $y_1'' = -4\cos 2t$. Therefore, $y_1'' + 4y_1 = 0$. Similarly, for $y_2 = \sin 2t$, $y_2'' = -4\sin 2t$. Therefore, $y_2'' + 4y_2 = 0$. Since $W(\cos 2t, \sin 2t) = 2$, they form a fundamental set.

21. For $y_1 = e^t$, $y_1' = y_1'' = e^t$. Therefore, $y_1'' - 2y_1' + y_1 = e^t - 2e^t + e^t = 0$. For $y_2 = te^t$, $y_2' = (1 + t)e^t$ and $y_2'' = (2 + t)e^t$. Therefore, $y_2'' - 2y_2' + y_2 = (2 + t)e^t - 2(1 + t)e^t + te^t = 0$. Further, $W(e^t, te^t) = e^{2t}$. Therefore, the solutions form a fundamental set.

22. For $y_1 = x$, $y_1' = 1$ and $y_1'' = 0$. Therefore, $x^2y_1'' - x(x + 2)y_1' + (x + 2)y_1 = -x(x + 2) + (x + 2)x = 0$. For $y_2 = xe^x$, $y_2' = (1 + x)e^x$ and $y_2'' = (2 + x)e^x$. Therefore, $x^2y_2'' - x(x + 2)y_2' + (x + 2)y_2 = x^2(2 + x)e^x - x(x + 2)(1 + x)e^x + (x + 2)xe^x = 0$. Further, $W(x, xe^x) = x^2e^x$. Therefore, the solutions form a fundamental set.

23. For $y_1 = x$, $y_1' = 1$ and $y_1'' = 0$. Therefore, $(1 - x\cot x)y_1'' - xy_1' + y_1 = -x + x = 0$. For $y_2 = \sin x$, $y_2' = \cos x$ and $y_2'' = -\sin x$. Therefore, $(1 - x\cot x)y_2'' - xy_2' + y_2 = -(1 - x\cot x)\sin x - x\cos x + \sin x = 0$. Further, $W(x, \sin x) = x\cos x - \sin x$ which is nonzero for $0 < x < \pi$. Therefore, the functions form a fundamental set of solutions.

24.

(a) For $y_1 = e^{-t}$, $y_1' = -e^{-t}$ and $y_1'' = e^{-t}$. Therefore, $y_1'' - y_1' - 2y_1 = e^{-t} + e^{-t} - 2e^{-t} = 0$. For $y_2 = e^{2t}$, $y_2' = 2e^{2t}$ and $y_2'' = 4e^{2t}$. Therefore, $y_2'' - y_2' - 2y_2 = 4e^{2t} - 2e^{2t} - 2e^{2t} = 0$. Further, $W(e^{-t}, e^{2t}) = 3e^t$. Therefore, the functions form a fundamental set of solutions.

(b) Since the equation is linear and y_3, y_4, y_5 are all linear combinations of solutions, they are also solutions of the differential equation.

(c) $W(y_1, y_3) = -6e^t$, $W(y_2, y_3) = 0$, $W(y_1, y_4) = 6e^t$ and $W(y_4, y_5) = 0$. Therefore, $\{y_1, y_3\}$ and $\{y_1, y_4\}$ form fundamental sets of solutions, but $\{y_2, y_3\}$ and $\{y_4, y_5\}$ do not.

Section 4.3

1.

(a) The characteristic equation is given by $\lambda^2 + 2\lambda - 3 = 0$. Therefore, the two distinct roots are $\lambda = -3, 1$. Therefore, the general solution is given by

$$y(t) = c_1e^t + c_2e^{-3t}.$$

(b) For y above, we see that

$$y' = c_1e^t - 3c_2e^{-3t}.$$

11

Therefore, we can rewrite our solution as the two parameter family

$$\begin{pmatrix} x_1 \\ x_2 \end{pmatrix} = \begin{pmatrix} y \\ y' \end{pmatrix} = \begin{pmatrix} c_1 e^t + c_2 e^{-3t} \\ c_1 e^t - 3c_2 e^{-3t} \end{pmatrix}$$

$$= c_1 \begin{pmatrix} 1 \\ 1 \end{pmatrix} e^t + c_2 \begin{pmatrix} 1 \\ -3 \end{pmatrix} e^{-3t}.$$

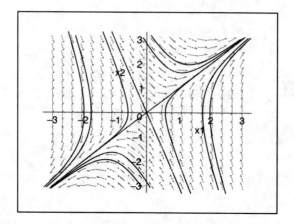

(c) The critical point $(0,0)$ is a saddle point, therefore, unstable.

2.

(a) The characteristic equation is given by $\lambda^2 + 3\lambda + 2 = 0$. Therefore, the two distinct roots are $\lambda = -1, -2$. Therefore, the general solution is given by

$$y(t) = c_1 e^{-t} + c_2 e^{-2t}.$$

(b) For y above, we see that

$$y' = -c_1 e^{-t} - 2c_2 e^{-2t}.$$

Therefore, we can rewrite our solution as the two parameter family

$$\begin{pmatrix} x_1 \\ x_2 \end{pmatrix} = \begin{pmatrix} y \\ y' \end{pmatrix} = \begin{pmatrix} c_1 e^{-t} + c_2 e^{-2t} \\ -c_1 e^{-t} - 2c_2 e^{-2t} \end{pmatrix}$$

$$= c_1 \begin{pmatrix} 1 \\ -1 \end{pmatrix} e^{-t} + c_2 \begin{pmatrix} 1 \\ -2 \end{pmatrix} e^{-2t}.$$

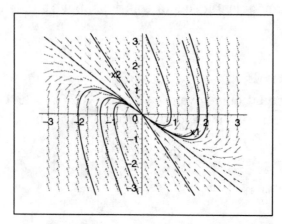

(c) The critical point $(0,0)$ is an asymptotically stable node.

3.

(a) The characteristic equation is given by $6\lambda^2 - \lambda - 1 = 0$. Therefore, the two distinct roots are $\lambda = 1/2, -1/3$. Therefore, the general solution is given by

$$y(t) = c_1 e^{t/2} + c_2 e^{-t/3}.$$

(b) For y above, we see that

$$y' = \frac{c_1}{2} e^{t/2} - \frac{c_2}{3} e^{-t/3}.$$

Therefore, we can rewrite our solution as the two parameter family

$$\begin{pmatrix} x_1 \\ x_2 \end{pmatrix} = \begin{pmatrix} y \\ y' \end{pmatrix} = \begin{pmatrix} c_1 e^{t/2} + c_2 e^{-t/3} \\ \frac{c_1}{2} e^{t/2} - \frac{c_2}{3} e^{-t/3} \end{pmatrix}$$

$$= c_1 \begin{pmatrix} 1 \\ \frac{1}{2} \end{pmatrix} e^{t/2} + c_2 \begin{pmatrix} 1 \\ -\frac{1}{3} \end{pmatrix} e^{-t/3}.$$

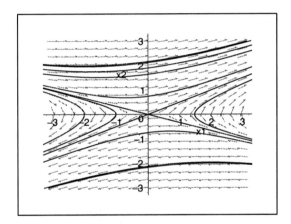

(c) The critical point $(0,0)$ is a saddle point, and, therefore, unstable.

4.

(a) The characteristic equation is given by $2\lambda^2 - 3\lambda + 1 = 0$. Therefore, the two distinct roots are $\lambda = 1/2, 1$. Therefore, the general solution is given by

$$y(t) = c_1 e^{t/2} + c_2 e^t.$$

(b) For y above, we see that

$$y' = \frac{c_1}{2} e^{t/2} + c_2 e^t.$$

Therefore, we can rewrite our solution as the two parameter family

$$\begin{pmatrix} x_1 \\ x_2 \end{pmatrix} = \begin{pmatrix} y \\ y' \end{pmatrix} = \begin{pmatrix} c_1 e^{t/2} + c_2 e^t \\ \frac{c_1}{2} e^{t/2} + c_2 e^t \end{pmatrix}$$

$$= c_1 \begin{pmatrix} 1 \\ \frac{1}{2} \end{pmatrix} e^{t/2} + c_2 \begin{pmatrix} 1 \\ 1 \end{pmatrix} e^t.$$

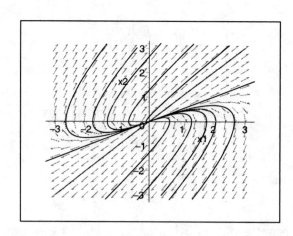

(c) The critical point $(0,0)$ is an unstable node.

5.

(a) The characteristic equation is given by $\lambda^2 + 5\lambda = 0$. Therefore, the two distinct roots are $\lambda = 0, -5$. Therefore, the general solution is given by

$$y(t) = c_1 + c_2 e^{-5t}.$$

(b) For y above, we see that

$$y' = -5c_2 e^{-5t}.$$

Therefore, we can rewrite our solution as the two parameter family

$$\begin{pmatrix} x_1 \\ x_2 \end{pmatrix} = \begin{pmatrix} y \\ y' \end{pmatrix} = \begin{pmatrix} c_1 + c_2 e^{-5t} \\ -5c_2 e^{-5t} \end{pmatrix}$$

$$= c_1 \begin{pmatrix} 1 \\ 0 \end{pmatrix} + c_2 \begin{pmatrix} 1 \\ -5 \end{pmatrix} e^{-5t}.$$

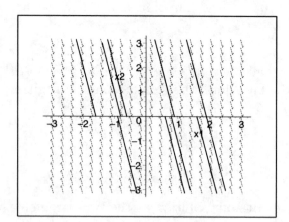

(c) Every point of the form $(x_0, 0)$ is a critical point. They are each nonisolated, stable points.

6.

(a) The characteristic equation is given by $4\lambda^2 - 9 = 0$. Therefore, the two distinct roots are $\lambda = 3/2, -3/2$. Therefore, the general solution is given by

$$y(t) = c_1 e^{3t/2} + c_2 e^{-3t/2}.$$

(b) For y above, we see that

$$y' = \frac{3c_1}{2} e^{3t/2} - \frac{3c_2}{2} e^{-3t/2}.$$

Therefore, we can rewrite our solution as the two parameter family

$$\begin{pmatrix} x_1 \\ x_2 \end{pmatrix} = \begin{pmatrix} y \\ y' \end{pmatrix} = \begin{pmatrix} c_1 e^{3t/2} + c_2 e^{-3t/2} \\ \frac{3c_1}{2} e^{3t/2} - \frac{3c_2}{2} e^{-3t/2} \end{pmatrix}$$

$$= c_1 \begin{pmatrix} 1 \\ \frac{3}{2} \end{pmatrix} e^{3t/2} + c_2 \begin{pmatrix} 1 \\ -\frac{3}{2} \end{pmatrix} e^{-3t/2}.$$

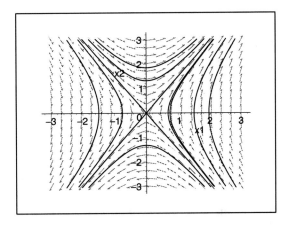

(c) The critical point $(0, 0)$ is a saddle point, and, therefore, unstable.

7.

(a) The characteristic equation is given by $\lambda^2 - 9\lambda + 9 = 0$. Therefore, the two distinct roots are $\lambda = (9 + 3\sqrt{5})/2, (9 - 3\sqrt{5})/2$. Therefore, the general solution is given by

$$y(t) = c_1 e^{(9+3\sqrt{5})t/2} + c_2 e^{(9-3\sqrt{5})t/2}.$$

(b) For y above, we see that

$$y' = \frac{(9 + 3\sqrt{5})c_1}{2} e^{(9+3\sqrt{5})t/2} + \frac{(9 - 3\sqrt{5})c_2}{2} e^{(9-3\sqrt{5})t/2}.$$

Therefore, we can rewrite our solution as the two parameter family

$$\begin{pmatrix} x_1 \\ x_2 \end{pmatrix} = \begin{pmatrix} y \\ y' \end{pmatrix} = \begin{pmatrix} c_1 e^{(9+3\sqrt{5})t/2} + c_2 e^{(9-3\sqrt{5})t/2} \\ \frac{(9+3\sqrt{5})c_1}{2} e^{(9+3\sqrt{5})t/2} + \frac{(9-3\sqrt{5})c_2}{2} e^{(9-3\sqrt{5})t/2} \end{pmatrix}$$

$$= c_1 \begin{pmatrix} 1 \\ \frac{9+3\sqrt{5}}{2} \end{pmatrix} e^{(9+3\sqrt{5})t/2} + c_2 \begin{pmatrix} 1 \\ \frac{9-3\sqrt{5}}{2} \end{pmatrix} e^{(9-3\sqrt{5})t/2}.$$

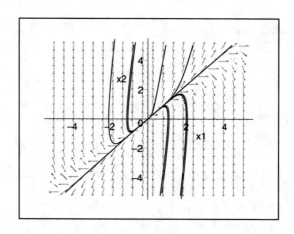

(c) The critical point $(0,0)$ is an unstable node.

8.

(a) The characteristic equation is given by $\lambda^2 - 2\lambda - 2 = 0$. Therefore, the two distinct roots are $\lambda = 1 + \sqrt{3}, 1 - \sqrt{3}$. Therefore, the general solution is given by

$$y(t) = c_1 e^{(1+\sqrt{3})t} + c_2 e^{(1-\sqrt{3})t}.$$

(b) For y above, we see that

$$y' = (1 + \sqrt{3})c_1 e^{(1+\sqrt{3})t} + (1 - \sqrt{3})c_2 e^{(1-\sqrt{3})t}.$$

Therefore, we can rewrite our solution as the two parameter family

$$\begin{pmatrix} x_1 \\ x_2 \end{pmatrix} = \begin{pmatrix} y \\ y' \end{pmatrix} = \begin{pmatrix} c_1 e^{(1+\sqrt{3})t} + c_2 e^{(1-\sqrt{3})t} \\ (1 + \sqrt{3})c_1 e^{(1+\sqrt{3})t} + (1 - \sqrt{3})c_2 e^{(1-\sqrt{3})t} \end{pmatrix}$$

$$= c_1 \begin{pmatrix} 1 \\ 1 + \sqrt{3} \end{pmatrix} e^{(1+\sqrt{3})t} + c_2 \begin{pmatrix} 1 \\ 1 - \sqrt{3} \end{pmatrix} e^{(1-\sqrt{3})t}.$$

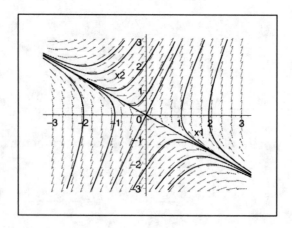

(c) The critical point $(0,0)$ is a saddle point, therefore, unstable.

9.

(a) The characteristic equation is given by $\lambda^2 - 2\lambda + 1 = 0$. Therefore, we have one repeated root $\lambda = 1$. Therefore, the general solution is given by

$$y(t) = c_1 e^t + c_2 t e^t.$$

(b) For y above, we see that

$$y' = c_1 e^t + c_2(1 + t)e^t.$$

Therefore, we can rewrite our solution as the two parameter family

$$\begin{pmatrix} x_1 \\ x_2 \end{pmatrix} = \begin{pmatrix} y \\ y' \end{pmatrix} = \begin{pmatrix} c_1 e^t + c_2 t e^t \\ c_1 e^t + c_2(1 + t)e^t \end{pmatrix}$$

$$= c_1 \begin{pmatrix} 1 \\ 1 \end{pmatrix} e^t + c_2 \begin{pmatrix} t \\ 1 + t \end{pmatrix} e^t.$$

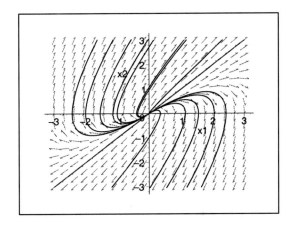

(c) The critical point $(0, 0)$ is an unstable improper node.

10.

(a) The characteristic equation is given by $9\lambda^2 + 6\lambda + 1 = 0$. Therefore, we have one repeated root $\lambda = -1/3$. Therefore, the general solution is given by

$$y(t) = c_1 e^{-t/3} + c_2 t e^{-t/3}.$$

(b) For y above, we see that

$$y' = -\frac{c_1}{3} e^{-t/3} + c_2 \left(1 - \frac{t}{3}\right) e^{-t/3}.$$

Therefore, we can rewrite our solution as the two parameter family

$$\begin{pmatrix} x_1 \\ x_2 \end{pmatrix} = \begin{pmatrix} y \\ y' \end{pmatrix} = \begin{pmatrix} c_1 e^{-t/3} + c_2 t e^{-t/3} \\ -\frac{c_1}{3} e^{-t/3} + c_2 \left(1 - \frac{t}{3}\right) e^{-t/3} \end{pmatrix}$$

$$= c_1 \begin{pmatrix} 1 \\ -\frac{1}{3} \end{pmatrix} e^{-t/3} + c_2 \begin{pmatrix} t \\ 1 - \frac{t}{3} \end{pmatrix} e^{-t/3}.$$

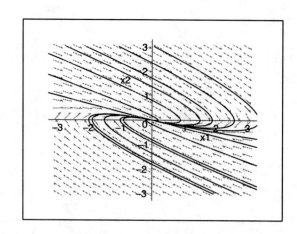

(c) The critical point $(0, 0)$ is an asymptotically stable improper node.

11.

(a) The characteristic equation is given by $4\lambda^2 - 4\lambda + 1 = 0$. Therefore, we have one repeated root $\lambda = 1/2$. Therefore, the general solution is given by

$$y(t) = c_1 e^{t/2} + c_2 t e^{t/2}.$$

(b) For y above, we see that

$$y' = \frac{c_1}{2} e^{t/2} + c_2 \left(1 + \frac{t}{2}\right) e^{t/2}.$$

Therefore, we can rewrite our solution as the two parameter family

$$\begin{pmatrix} x_1 \\ x_2 \end{pmatrix} = \begin{pmatrix} y \\ y' \end{pmatrix} = \begin{pmatrix} c_1 e^{t/2} + c_2 t e^{t/2} \\ \frac{c_1}{2} e^{t/2} + c_2 \left(1 + \frac{t}{2}\right) e^{t/2} \end{pmatrix}$$

$$= c_1 \begin{pmatrix} 1 \\ \frac{1}{2} \end{pmatrix} e^{t/2} + c_2 \begin{pmatrix} t \\ 1 + \frac{t}{2} \end{pmatrix} e^{t/2}.$$

(c) The critical point $(0, 0)$ is an unstable improper node.

12.

(a) The characteristic equation is given by $4\lambda^2 + 12\lambda + 9 = 0$. Therefore, we have one repeated root $\lambda = -3/2$. Therefore, the general solution is given by

$$y(t) = c_1 e^{-3t/2} + c_2 t e^{-3t/2}.$$

(b) For y above, we see that

$$y' = -\frac{3c_1}{2}e^{-3t/2} + c_2\left(1 - \frac{3t}{2}\right)e^{-3t/2}.$$

Therefore, we can rewrite our solution as the two parameter family

$$\begin{pmatrix} x_1 \\ x_2 \end{pmatrix} = \begin{pmatrix} y \\ y' \end{pmatrix} = \begin{pmatrix} c_1 e^{-3t/2} + c_2 t e^{-3t/2} \\ -\frac{3c_1}{2}e^{-3t/2} + c_2\left(1 - \frac{3t}{2}\right)e^{-3t/2} \end{pmatrix}$$

$$= c_1 \begin{pmatrix} 1 \\ -\frac{3}{2} \end{pmatrix} e^{-3t/2} + c_2 \begin{pmatrix} t \\ 1 - \frac{3t}{2} \end{pmatrix} e^{-3t/2}.$$

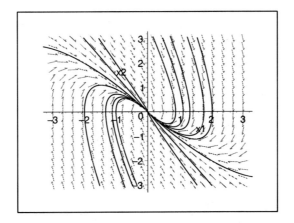

(c) The critical point $(0,0)$ is an asymptotically stable improper node.

13.

(a) The characteristic equation is given by $25\lambda^2 - 20\lambda + 4 = 0$. Therefore, we have one repeated root $\lambda = 2/5$. Therefore, the general solution is given by

$$y(t) = c_1 e^{2t/5} + c_2 t e^{2t/5}.$$

(b) For y above, we see that

$$y' = \frac{2c_1}{5}e^{2t/5} + c_2\left(1 + \frac{2t}{5}\right)e^{2t/5}.$$

Therefore, we can rewrite our solution as the two parameter family

$$\begin{pmatrix} x_1 \\ x_2 \end{pmatrix} = \begin{pmatrix} y \\ y' \end{pmatrix} = \begin{pmatrix} c_1 e^{2t/5} + c_2 t e^{2t/5} \\ \frac{2c_1}{5}e^{2t/5} + c_2\left(1 + \frac{2t}{5}\right)e^{2t/5} \end{pmatrix}$$

$$= c_1 \begin{pmatrix} 1 \\ \frac{2}{5} \end{pmatrix} e^{2t/5} + c_2 \begin{pmatrix} t \\ 1 + \frac{2t}{5} \end{pmatrix} e^{2t/5}.$$

19

(c) The critical point $(0,0)$ is an unstable improper node.

14.

(a) The characteristic equation is given by $\lambda^2 - 6\lambda + 9 = 0$. Therefore, we have one repeated root $\lambda = 3$. Therefore, the general solution is given by

$$y(t) = c_1 e^{3t} + c_2 t e^{3t}.$$

(b) For y above, we see that

$$y' = 3c_1 e^{3t} + c_2 (1 + 3t) e^{3t}.$$

Therefore, we can rewrite our solution as the two parameter family

$$\begin{pmatrix} x_1 \\ x_2 \end{pmatrix} = \begin{pmatrix} y \\ y' \end{pmatrix} = \begin{pmatrix} c_1 e^{3t} + c_2 t e^{3t} \\ 3c_1 e^{3t} + c_2 (1 + 3t) e^{3t} \end{pmatrix}$$
$$= c_1 \begin{pmatrix} 1 \\ 3 \end{pmatrix} e^{3t} + c_2 \begin{pmatrix} t \\ 1 + 3t \end{pmatrix} e^{3t}.$$

(c) The critical point $(0,0)$ is an unstable improper node.

15.

(a) The characteristic equation is given by $\lambda^2+4\lambda+4=0$. Therefore, we have one repeated root $\lambda=-2$. Therefore, the general solution is given by

$$y(t)=c_1e^{-2t}+c_2te^{-2t}.$$

(b) For y above, we see that

$$y'=-2c_1e^{-2t}+c_2e^{-2t}-2c_2te^{-2t}.$$

Therefore, we can rewrite our solution as the two parameter family

$$\begin{pmatrix}x_1\\x_2\end{pmatrix}=\begin{pmatrix}y\\y'\end{pmatrix}=\begin{pmatrix}c_1e^{-2t}+c_2te^{-2t}\\-2c_1e^{-2t}+c_2e^{-2t}-2c_2te^{-2t}\end{pmatrix}$$

$$=c_1\begin{pmatrix}1\\-2\end{pmatrix}e^{-2t}+c_2\left[\begin{pmatrix}0\\1\end{pmatrix}+\begin{pmatrix}1\\-2\end{pmatrix}t\right]e^{-2t}.$$

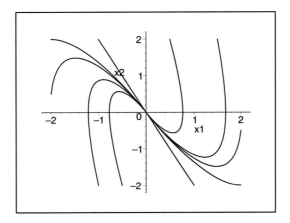

(c) The critical point $(0,0)$ is asymptotically stable.

16.

(a) The characteristic equation is given by $9\lambda^2-24\lambda+16=0$. Therefore, we have one repeated root $\lambda=4/3$. Therefore, the general solution is given by

$$y(t)=c_1e^{4t/3}+c_2te^{4t/3}.$$

(b) For y above, we see that

$$y'=\frac{4}{3}c_1e^{4t/3}+c_2e^{4t/3}+\frac{4}{3}c_2te^{4t/3}.$$

Therefore, we can rewrite our solution as the two parameter family

$$\begin{pmatrix}x_1\\x_2\end{pmatrix}=\begin{pmatrix}y\\y'\end{pmatrix}=\begin{pmatrix}c_1e^{4t/3}+c_2te^{4t/3}\\\frac{4}{3}c_1e^{4t/3}+c_2e^{4t/3}+\frac{4}{3}c_2te^{4t/3}\end{pmatrix}$$

$$=c_1\begin{pmatrix}1\\4/3\end{pmatrix}e^{4t/3}+c_2\left[\begin{pmatrix}0\\1\end{pmatrix}+\begin{pmatrix}1\\4/3\end{pmatrix}t\right]e^{4t/3}.$$

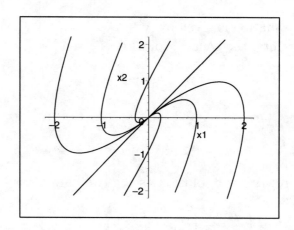

(c) The critical point $(0, 0)$ is unstable.

17. We want to solve the equation

$$\mathbf{x}' = \begin{pmatrix} 0 & 1 \\ -c/a & -b/a \end{pmatrix} \mathbf{x}$$

in the case that $b^2 - 4ac > 0$. First, we look for eigenvalues of A. We have

$$A - \lambda I = \begin{pmatrix} -\lambda & 1 \\ -c/a & -b/a - \lambda \end{pmatrix}.$$

Therefore, $\det(A - \lambda I) = \lambda^2 + \dfrac{b}{a}\lambda + \dfrac{c}{a}$. Therefore, $\det(A - \lambda I) = 0 \implies a\lambda^2 + b\lambda + c = 0$. Therefore,

$$\lambda_{1,2} = \frac{-b \pm \sqrt{b^2 - 4ac}}{2a}.$$

Here we are assuming that $b^2 - 4ac > 0$. Therefore, the two eigenvalues are real and distinct. Now we look for eigenvectors. For λ_i, $i = 1, 2$, we have

$$A - \lambda_i I = \begin{pmatrix} -\lambda_i & 1 \\ -c/a & -b/a - \lambda_i \end{pmatrix}.$$

Since λ_i is an eigenvalue, the matrix $A - \lambda_i I$ is singular. Further, $\mathbf{v}_i = \begin{pmatrix} x_{1i} & x_{2i} \end{pmatrix}^t$ will be an eigenvector as long as $-\lambda_i x_{1i} + x_{2i} = 0$. In particular, we see that

$$\mathbf{v}_i = \begin{pmatrix} 1 \\ \lambda_i \end{pmatrix}$$

satisfies the necessary equation. We conclude that

$$\mathbf{x}_1(t) = e^{\lambda_1 t} \begin{pmatrix} 1 \\ \lambda_1 \end{pmatrix}$$

and

$$\mathbf{x}_2(t) = e^{\lambda_2 t} \begin{pmatrix} 1 \\ \lambda_2 \end{pmatrix}$$

are two linearly independent solutions. Consequently, the general solution of equation (1) is given by the first component of $c_1\mathbf{x}_1 + c_2\mathbf{x}_2$.

18. As in the solution to exercise 17 above, we note that the eigenvalues are given by

$$\lambda_{1,2} = \frac{-b \pm \sqrt{b^2 - 4ac}}{2a}.$$

In this case, we are assuming that $b^2 - 4ac = 0$. Therefore, we have one repeated eigenvalue $\lambda = -b/2a$. As in the solution to exercise 17 above, an associated eigenvector is given by $\mathbf{v} = \begin{pmatrix} 1 & \lambda_1 \end{pmatrix}^t$. Therefore,

$$\mathbf{x}_1(t) = e^{\lambda_1 t}\begin{pmatrix} 1 \\ \lambda_1 \end{pmatrix}$$

is one solution. In order to look for another solution, we need to look for $\mathbf{w}$ satisfying the equation

$$(A - \lambda I)\mathbf{w} = \mathbf{v}$$

for $\lambda = -b/2a$ and $\mathbf{v} = \begin{pmatrix} 1 & \lambda_1 \end{pmatrix}^t$. By substituting these values in for λ and $\mathbf{v}$, our equation reduces to

$$\begin{pmatrix} -b/2a & 1 \\ -c/a & -b/2a \end{pmatrix}\begin{pmatrix} w_1 \\ w_2 \end{pmatrix} = \begin{pmatrix} 1 \\ -b/2a \end{pmatrix}.$$

We see that

$$\mathbf{w} = \begin{pmatrix} 0 \\ 1 \end{pmatrix}$$

is a solution of this equation. Therefore,

$$\mathbf{x}_2(t) = \left[t\begin{pmatrix} 1 \\ \lambda_1 \end{pmatrix} + \begin{pmatrix} 0 \\ 1 \end{pmatrix} \right] e^{\lambda_1 t}$$

is another solution of our original equation. Therefore,

$$\mathbf{x}_1(t) = e^{\lambda_1 t}\begin{pmatrix} 1 \\ \lambda_1 \end{pmatrix}$$

and

$$\mathbf{x}_2(t) = e^{\lambda_2 t}\begin{pmatrix} t \\ 1 + \lambda_1 t \end{pmatrix}$$

form a fundamental set of solutions of Equation (2). Further, the general solution of Equation (1) corresponds to the first component of $c_1\mathbf{x}_1 + c_2\mathbf{x}_2$.

19. The characteristic equation is given by $\lambda^2 + \lambda - 2 = 0$. Therefore, the two distinct roots are $\lambda = -2, 1$. Therefore, the general solution is given by

$$y(t) = c_1 e^{-2t} + c_2 e^t.$$

Therefore,

$$y'(t) = -2c_1 e^{-2t} + c_2 e^t.$$

Now using the initial conditions, we need

$$c_1 + c_2 = 1$$
$$-2c_1 + c_2 = 1.$$

The solution of this system of equations is $c_1 = 0$ and $c_2 = 1$. Therefore, the specific solution is $y(t) = e^t$.

The solution $y \to \infty$ as $t \to \infty$.

20. The characteristic equation is given by $9\lambda^2 - 12\lambda + 4 = 0$. Therefore, there is one repeated root, $\lambda = 2/3$. Therefore, the general solution is given by

$$y(t) = c_1 e^{2t/3} + c_2 t e^{2t/3}.$$

Therefore,

$$y'(t) = \frac{2c_1}{3} e^{2t/3} + c_2 \left(1 + \frac{2t}{3}\right) e^{2t/3}.$$

Now using the initial conditions, we need

$$c_1 = 2$$
$$\frac{2c_1}{3} + c_2 = -1.$$

The solution of this system of equations is $c_1 = 2$ and $c_2 = -7/3$. Therefore, the specific solution is $y(t) = 2e^{2t/3} - \frac{7}{3} t e^{2t/3}$.

24

The solution $y \to -\infty$ as $t \to \infty$.

21. The characteristic equation is given by $\lambda^2 + 4\lambda + 3 = 0$. Therefore, the two distinct roots are $\lambda = -1, -3$. Therefore, the general solution is given by

$$y(t) = c_1 e^{-t} + c_2 e^{-3t}.$$

Therefore,

$$y'(t) = -c_1 e^{-t} - 3c_2 e^{-3t}.$$

Now using the initial conditions, we need

$$c_1 + c_2 = 2$$
$$-c_1 - 3c_2 = -1.$$

The solution of this system of equations is $c_1 = 5/2$ and $c_2 = -1/2$. Therefore, the specific solution is $y(t) = \dfrac{5}{2} e^{-t} - \dfrac{1}{2} e^{-3t}$.

The solution $y \to 0$ as $t \to \infty$.

22. The characteristic equation is given by $6\lambda^2 - 5\lambda + 1 = 0$. Therefore, the two distinct real roots are $\lambda = 1/3, 1/2$. Therefore, the general solution is given by

$$y(t) = c_1 e^{t/3} + c_2 e^{t/2}.$$

Therefore,

$$y'(t) = \frac{c_1}{3} e^{t/3} + \frac{c_2}{2} e^{t/2}.$$

Now using the initial conditions, we need

$$c_1 + c_2 = 4$$
$$\frac{c_1}{3} + \frac{c_2}{2} = 0.$$

The solution of this system of equations is $c_1 = 12$ and $c_2 = -8$. Therefore, the specific solution is $y(t) = 12e^{t/3} - 8e^{t/2}$.

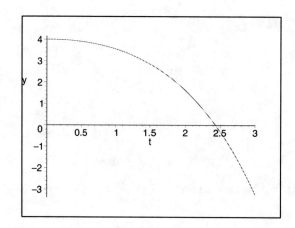

The solution $y \to -\infty$ as $t \to \infty$.

23. The characteristic equation is given by $\lambda^2 - 6\lambda + 9 = 0$. Therefore, there is one repeated root, $\lambda = 3$. Therefore, the general solution is given by

$$y(t) = c_1 e^{3t} + c_2 t e^{3t}.$$

Therefore,

$$y'(t) = 3c_1 e^{3t} + c_2(1 + 3t)e^{3t}.$$

Now using the initial conditions, we need

$$c_1 = 0$$
$$3c_1 + c_2 = 2.$$

The solution of this system of equations is $c_1 = 0$ and $c_2 = 2$. Therefore, the specific solution is $y(t) = 2te^{3t}$.

The solution $y \to \infty$ as $t \to \infty$.

24. The characteristic equation is given by $\lambda^2 + 3\lambda = 0$. Therefore, the two distinct real roots are $\lambda = 0, -3$. Therefore, the general solution is given by

$$y(t) = c_1 + c_2 e^{-3t}.$$

Therefore,

$$y'(t) = -3c_2 e^{-3t}.$$

Now using the initial conditions, we need

$$c_1 + c_2 = -2$$
$$-3c_2 = 3.$$

The solution of this system of equations is $c_1 = -1$ and $c_2 = -1$. Therefore, the specific solution is $y(t) = -1 - e^{-3t}$.

The solution $y \to -1$ as $t \to \infty$.

25. The characteristic equation is given by $\lambda^2 + 4\lambda + 4 = 0$. Therefore, there is one repeated root, $\lambda = -2$. Therefore, the general solution is given by

$$y(t) = c_1 e^{-2t} + c_2 t e^{-2t}.$$

Therefore,

$$y'(t) = -2c_1 e^{-2t} + c_2(1 - 2t)e^{-2t}.$$

Now using the initial conditions, we need

$$c_1 e^2 - c_2 e^2 = 2$$
$$-2c_1 e^2 + 3c_2 e^2 = 1.$$

The solution of this system of equations is $c_1 = 7e^{-2}$ and $c_2 = 5e^{-2}$. Therefore, the specific solution is $y(t) = 7e^{-2(1+t)} + 5te^{-2(1+t)}$.

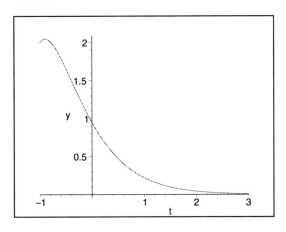

27

The solution $y \to 0$ as $t \to \infty$.

26. The characteristic equation is given by $\lambda^2 + 5\lambda + 3 = 0$. Therefore, the two distinct real roots are $\lambda = (-5 \pm \sqrt{13})/2$. Therefore, the general solution is given by

$$y(t) = c_1 e^{(-5+\sqrt{13})t/2} + c_2 e^{(-5-\sqrt{13})t/2}.$$

Therefore,

$$y'(t) = \frac{(-5 + \sqrt{13})c_1}{2} e^{(-5+\sqrt{13})t/2} + \frac{(-5 - \sqrt{13})c_2}{2} e^{(-5-\sqrt{13})t/2}.$$

Now using the initial conditions, we need

$$c_1 + c_2 = 1$$

$$\frac{(-5 + \sqrt{13})c_1}{2} + \frac{(-5 - \sqrt{13})c_2}{2} = 0.$$

The solution of this system of equations is $c_1 = (1 + 5/\sqrt{13})/2$ and $c_2 = (1 - 5/\sqrt{13})/2$. Therefore, the specific solution is

$$y(t) = \frac{1 + 5/\sqrt{13}}{2} e^{(-5+\sqrt{13})t/2} + \frac{1 - 5/\sqrt{13}}{2} e^{(-5-\sqrt{13})t/2}.$$

The solution $y \to 0$ as $t \to \infty$.

27. The characteristic equation is given by $2\lambda^2 + \lambda - 4 = 0$. Therefore, the two distinct real roots are $\lambda = (-1 \pm \sqrt{33})/4$. Therefore, the general solution is given by

$$y(t) = c_1 e^{(-1+\sqrt{33})t/4} + c_2 e^{(-1-\sqrt{33})t/4}.$$

Therefore,

$$y'(t) = \frac{(-1 + \sqrt{33})c_1}{4} e^{(-1+\sqrt{33})t/4} + \frac{(-1 - \sqrt{33})c_2}{4} e^{(-1-\sqrt{33})t/4}.$$

Now using the initial conditions, we need

$$c_1 + c_2 = 0$$

$$\frac{(-1 + \sqrt{33})c_1}{4} + \frac{(-1 - \sqrt{33})c_2}{4} = 1.$$

The solution of this system of equations is $c_1 = 2/\sqrt{33}$ and $c_2 = -2/\sqrt{33}$. Therefore, the specific solution is

$$y(t) = 2\sqrt{33}e^{(-1+\sqrt{33})t/4} - \frac{2}{\sqrt{33}}e^{(-1-\sqrt{33})t/4}.$$

The solution $y \to \infty$ as $t \to \infty$.

28. The characteristic equation is given by $\lambda^2 + 8\lambda - 9 = 0$. Therefore, the two distinct real roots are $\lambda = -9, 1$. Therefore, the general solution is given by

$$y(t) = c_1 e^{-9t} + c_2 e^t.$$

Therefore,

$$y'(t) = -9c_1 e^{-9t} + c_2 e^t.$$

Now using the initial conditions, we need

$$c_1 e^{-9} + c_2 e = 1$$
$$-9c_1 e^{-9} + c_2 e = 0.$$

The solution of this system of equations is $c_1 = \dfrac{1}{10}e^9$ and $c_2 = \dfrac{9}{10}e^{-1}$. Therefore, the specific solution is

$$y(t) = \frac{1}{10}e^{9-9t} + \frac{9}{10}e^{-1+t}.$$

The solution $y \to \infty$ as $t \to \infty$.

29. The characteristic equation is given by $4\lambda^2 - 1 = 0$. Therefore, the two distinct real roots are $\lambda = 1/2, -1/2$. Therefore, the general solution is given by

$$y(t) = c_1 e^{t/2} + c_2 e^{-t/2}.$$

Therefore,

$$y'(t) = \frac{c_1}{2} e^{t/2} - \frac{c_2}{2} e^{-t/2}.$$

Now using the initial conditions, we need

$$c_1 e^{-1} + c_2 e = 1$$
$$\frac{c_1}{2} e^{-1} - \frac{c_2}{2} e = -1.$$

The solution of this system of equations is $c_1 = -e/2$ and $c_2 = \frac{3}{2} e^{-1}$. Therefore, the specific solution is

$$y(t) = -\frac{1}{2} e^{1+t/2} + \frac{3}{2} e^{-1-t/2}.$$

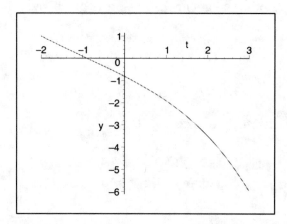

The solution $y \to -\infty$ as $t \to \infty$.

30. We need to find a characteristic equation of degree two with roots $\lambda = 2, -3$. We take $p(\lambda) = (\lambda - 2)(\lambda + 3) = \lambda^2 + \lambda - 6$. Therefore, the differential equation is $y'' + y' - 6y = 0$.

31. We need to find a characteristic equation of degree two with one repeated root $\lambda = -2$. We take $p(\lambda) = (\lambda + 2)^2 = \lambda^2 + 4\lambda + 4$. Therefore, the differential equation is $y'' + 4y' + 4y = 0$.

32. The characteristic equation is $\lambda^2 - (2\alpha - 1)\lambda + \alpha(\alpha - 1) = 0$. Solving this equation, we see that the roots are $\lambda = \alpha, \alpha - 1$. Therefore, the general solution is

$$y(t) = c_1 e^{\alpha t} + c_2 e^{(\alpha-1)t}.$$

In order for the solution to tend to zero, we need $\alpha, \alpha - 1 < 0$. Therefore, the solutions will all tend to zero as long as $\alpha < 0$. All solutions will become unbounded if $\alpha > 0$ and $\alpha - 1 > 0$. This will occur exactly when $\alpha > 1$.

33. The characteristic equation is $\lambda^2 + (3 - \alpha)\lambda - 2(\alpha - 1) = 0$. Solving this equation, we see that the roots are $\lambda = \alpha - 1, -2$. Therefore, the general solution is

$$y(t) = c_1 e^{(\alpha-1)t} + c_2 e^{-2t}.$$

In order for the solution to tend to zero, we need $\alpha - 1 < 0$. Therefore, the solutions will all tend to zero as long as $\alpha < 1$. Due the term $c_2 e^{-2t}$, we can never guarantee that all solutions will become unbounded as $t \to \infty$.

34. Suppose the roots are distinct, λ_1, λ_2. Then the solution is

$$y(t) = c_1 e^{\lambda_1 t} + c_2 e^{\lambda_2 t}.$$

Solving the equation $y(t) = 0$, we see that we must have $c_1 e^{\lambda_1 t} = -c_2 e^{\lambda_2 t}$ which implies $e^{(\lambda_1 - \lambda_2)t} = -c_2/c_1$. First, in order to guarantee any solution of this equation, we would need $c_2/c_1 < 0$. Then, applying the natural logarithm function to the equation, we see that $t = \ln(-c_2/c_1)/(\lambda_1 - \lambda_2)$.

If the roots are not distinct, then the solution is given by

$$y(t) = c_1 e^{\lambda t} + c_2 t e^{\lambda t}.$$

Therefore, $y(t) = 0$ implies $(c_1 + c_2 t)e^{\lambda t} = 0$. Since $e^{\lambda t} \neq 0$, we must have $c_1 + c_2 t = 0$. Therefore, the solution will be zero only when $t = -c_1/c_2$.

35.

(a) If y is a constant solution, then the equation reduces to $cy = d$, which implies $y = d/c$.

(b) If y_e is an equilibrium solution, then $y_e = d/c$. Therefore, $Y = y - d/c$ satisfies

$$aY'' + bY' + cY = ay'' + by' + c(y - d/c) = 0$$

since y is a solution of

$$ay'' + by' + cy = d.$$

36. The roots of the characteristic equation are

$$\lambda = \frac{-b + \sqrt{b^2 - 4ac}}{2a}.$$

Therefore, the roots will be

(a) real, different and negative if (1) $b^2 - 4ac > 0$ and (2) $b > \sqrt{b^2 - 4ac}$.

(b) real with opposite signs if (1) $b^2 - 4ac > 0$ and (2) $b < \sqrt{b^2 - 4ac}$.

(c) real, different and positive if (1) $b^2 - 4ac > 0$ and (2) $-b > \sqrt{b^2 - 4ac}$.

37. Let $y_2(t) = t^2 v(t)$. Since y satisfies the differential equation, we have

$$t^2(t^2 v'' + 4tv' + 2v) - 4t(t^2 v' + 2tv) + 6t^2 v = 0.$$

After collecting terms, we have $t^4 v'' = 0$. Therefore, $v(t) = c_1 + c_2 t$. Thus $y_2(t) = c_1 t^2 + c_2 t^3$. Since we already have the solution $y_1(t) = t^2$, we set $c_1 = 0$ and $c_2 = 1$. Therefore, we get the solution $y_2(t) = t^3$.

38. Let $y_2(t) = tv(t)$. Since y satisfies the differential equation, we have

$$t^2(tv'' + 2v') + 2t(tv' + v) - 2tv = 0.$$

After collecting terms, we have $t^3 v'' + 4t^2 v' = 0$. This equation is linear in v'. Solving this equation for v', we have $v'(t) = ct^{-4}$, and, therefore, $v(t) = c_1 t^{-3} + c_2$. Therefore, $y_2(t) = c_1 t^{-2} + c_2 t$. Since we already have the solution $y_1(t) = t$, we set $c_1 = 1$ and $c_2 = 0$. Therefore, we get the solution $y_2(t) = t^{-2}$.

39. Let $y_2(t) = t^{-1} v(t)$. Since y satisfies the differential equation, we have

$$t^2(t^{-1} v'' - 2t^{-2} v' + 2t^{-3} v) + 3t(t^{-1} v' - t^{-2} v) + t^{-1} v = 0.$$

After collecting terms, we have $tv'' + v' = 0$. This equation is linear in v'. Solving this equation for v', we have $v'(t) = ct^{-1}$, and, therefore, $v(t) = c_1 \ln t + c_2$. Therefore, $y_2(t) = c_1 t^{-1} \ln t + c_2 t^{-1}$. Since we already have the solution $y_1(t) = t^{-1}$, we set $c_1 = 1$ and $c_2 = 0$. Therefore, we get the solution $y_2(t) = t^{-1} \ln t$.

40. Let $y_2(t) = tv(t)$. Substituting y into the differential equation, we conclude that $v'' - v' = 0$. This equation is linear in v'. We conclude that $v'(t) = ce^t$ and, therefore, $v(t) = c_1 e^t + c_2$. Therefore, $y_2(t) = c_1 te^t + c_2 t$. Since we already have the solution $y_1(t) = t$, we set $c_1 = 1$ and $c_2 = 0$. Therefore, we get the solution $y_2(t) = te^t$.

41. Let $y_2(x) = \sin(x^2) v(x)$. Substituting y into the differential equation, we conclude that

$$v'' + [4x^2 \cos(x^2) - \sin(x^2)]v' = 0.$$

This equation is linear in v'. We conclude that $v'(x) = cx/[\sin(x^2)]^2$ and, therefore,

$$v(x) = c_1 \frac{\cos(x^2)}{\sin(x^2)} + c_2.$$

Therefore,

$$y_2(x) = c_1 \cos(x^2) + c_2 \sin(x^2).$$

Since we already have the solution $y_1(x) = \sin(x^2)$, we take the solution

$$y_2(x) = \cos(x^2).$$

42. Let $y_2(x) = e^x v(x)$. Substituting y into the differential equation, we conclude that

$$v'' + \frac{x-2}{x-1} v' = 0.$$

This equation is linear in v'. An integrating factor is $\mu(x) = \dfrac{e^x}{x-1}$. Rewriting the equation as

$$\left[\frac{e^x v'}{x-1}\right]' = 0,$$

we conclude that $v'(x) = c(x-1)e^{-x}$. Therefore, $v(x) = c_1 x e^{-x} + c_2$ and $y_2(x) = c_1 x + c_2 e^x$. Since we already have the solution $y_1(x) = e^x$, we take the solution $y_2(x) = x$.

43. Let $y_2(x) = x^{1/4} e^{2\sqrt{x}} v(x)$. Substituting y into the differential equation, we conclude that

$$2x^{9/4} v'' + (4x^{7/4} + x^{5/4}) v' = 0.$$

This equation is linear in v'. An integrating factor is

$$\mu(x) = \exp\left(\int \left[2x^{-1/2} + \frac{1}{2x}\right] dx\right)$$
$$= \sqrt{x}\exp(4\sqrt{x}).$$

Rewriting the equation as

$$\left[\sqrt{x}\exp(4\sqrt{x}v')\right]' = 0,$$

we conclude that $v'(x) = c\exp(-4\sqrt{x})/\sqrt{x}$. Integrating, we have $v(x) = c_1 \exp(-4\sqrt{x}) + c_2$, and, therefore, $y_2(x) = c_1 x^{1/4} e^{(-2\sqrt{x})} + c_2 x^{1/4} e^{(2\sqrt{x})}$. Since we already have the solution $y_1(x) = x^{1/4} e^{2\sqrt{x}}$, we take the solution $y_2(x) = x^{1/4} e^{-2\sqrt{x}}$.

44. Let $y_2(x) = x^{-1/2}\sin x v(x)$. Substituting y into the differential equation, we conclude that

$$\sin x v'' + 2\cos x v' = 0.$$

This equation is linear in v'. Its solution is given by $v'(x) = c/\sin(x)^2$. Integrating with respect to x, we have $v(x) = c_1 \cot(x) + c_2$. Therefore, $y_2(x) = c_1 x^{-1/2}\cos(x) + c_2 x^{-1/2}\sin(x)$. Since we already have the solution $y_1(x) = x^{-1/2}\sin(x)$, we take the solution $y_2(x) = x^{-1/2}\cos(x)$.

45.

(a) For $y_1 = e^x$, we have $y_1' = y_1'' = e^x$. Therefore,

$$xy_1'' - (x+N)y_1' + Ny_1 = xe^x - (x+N)e^x + Ne^x = 0.$$

(b) Let $y_2 = e^x v$. Then in order for y_2 to satisfy the differential equation we need

$$xv'' + (x - N)v' = 0.$$

Solving this equation, we have $v' = Cx^N e^{-x}$. Therefore,

$$v = c_1 \int x^N e^{-x}\, dx + c_2,$$

which implies

$$y_2 = c_1 e^x \int x^N e^{-x}\, dx + c_2 e^x.$$

Since we already have the solution $y_1(x) = e^x$, we take

$$y_2(x) = ce^x \int x^N e^{-x} \, dx.$$

For $N = 1$, we have

$$y_2(x) = c(-1 - x).$$

For $N = 2$, we have

$$y_2(x) = c(-2 - 2x - x^2).$$

Therefore, letting $c = -1/N!$, for $N = 1$, we have

$$y_2(x) = 1 + x$$

and for $N = 2$, we have

$$y_2(x) = 1 + x + \frac{1}{2}x^2.$$

46. Direct substitution verifies that $y_1(x) = e^{-\delta x^2/2}$ is a solution of this differential equation. To find another solution, we set $y_2(x) = e^{-\delta x^2/2}v(x)$. Then, in order for y_2 to satisfy the differential equation, we need v to satisfy

$$v'' - \delta x v' = 0.$$

Solving this equation for v', we see that the solution is given by

$$v'(x) = c_1 e^{\delta x^2/2}.$$

Integrating, we obtain

$$v(x) = c_1 \int_0^x e^{\delta u^2/2} \, du + c_2.$$

Therefore,

$$y_2(x) = c_1 e^{-\delta x^2/2} \int_0^x e^{\delta u^2/2} \, du + c_2 e^{-\delta x^2/2}.$$

Setting $c_2 = 0$, we arrive at the solution

$$y_2(x) = e^{-\delta x^2/2} \int_0^x e^{\delta u^2/2} \, du.$$

Section 4.4

1. For all parts below, we let $z_1 = a_1 + ib_1$ and $z_2 = a_2 + ib_2$.

(a)

$$z_1 + z_2 = a_1 + ib_1 + a_2 + ib_2$$
$$= (a_1 + a_2) + i(b_1 + b_2).$$

Therefore,

$$\overline{z_1 + z_2} = (a_1 + a_2) - i(b_1 + b_2)$$
$$= (a_1 - ib_1) + (a_2 - ib_2)$$
$$= \overline{z_1} + \overline{z_2}.$$

(b)

$$z_1 z_2 = (a_1 + ib_1)(a_2 + ib_2)$$
$$= (a_1 a_2 - b_1 b_2) + i(a_1 b_2 + a_2 b_1).$$

Therefore,

$$\overline{z_1 z_2} = (a_1 a_2 - b_1 b_2) - i(a_1 b_2 + a_2 b_1)$$
$$= a_1 a_2 - i a_1 b_2 - b_1 b_2 - i a_2 b_1$$
$$= a_1(a_2 - ib_2) - b_1 i(a_2 - ib_2)$$
$$= (a_2 - ib_2)(a_1 - ib_1)$$
$$= \overline{z_1} \cdot \overline{z_2}$$

(c)

$$\frac{z_1}{z_2} = \frac{a_1 + ib_1}{a_2 + ib_2}$$
$$= \frac{a_1 + ib_1}{a_2 + ib_2} \cdot \frac{a_2 - ib_2}{a_2 - ib_2}$$
$$= \frac{(a_1 a_2 + b_1 b_2) + i(a_2 b_1 - a_1 b_2)}{a_2^2 + b_2^2}.$$

Therefore,

$$\overline{\left(\frac{z_1}{z_2}\right)} = \frac{(a_1 a_2 + b_1 b_2) + i(a_1 b_2 - a_2 b_1)}{a_2^2 + b_2^2}$$

Next, we see that

$$\frac{\overline{z_1}}{\overline{z_2}} = \frac{a_1 - ib_1}{a_2 - ib_2}$$
$$= \frac{a_1 - ib_1}{a_2 - ib_2} \cdot \frac{a_2 + ib_2}{a_2 + ib_2}$$
$$= \frac{(a_1 a_2 + b_1 b_2) + i(a_1 b_2 - a_2 b_1)}{a_2^2 + b_2^2}.$$

Therefore, the result holds.

(d) If $z = 0$, then $z = 0 + 0i$ which implies $|z| = 0^2 + 0^2 = 0$. Also, if $|z| = 0$, then $a^2 + b^2 = 0$ which implies $a = 0, b = 0$, and, therefore, $z = 0 + 0i$.

(e) As in the solution to part (b), we know that

$$z_1 z_2 = (a_1 a_2 - b_1 b_2) + i(a_1 b_2 + a_2 b_1).$$

Therefore,

$$
\begin{aligned}
|z_1 z_2| &= (a_1 a_2 - b_1 b_2)^2 + (a_1 b_2 + a_2 b_1)^2 \\
&= (a_1 a_2)^2 - 2a_1 a_2 b_1 b_2 + (b_1 b_2)^2 + (a_1 b_2)^2 + 2a_1 b_2 a_2 b_1 + (a_2 b_1)^2 \\
&= a_1^2 a_2^2 + b_1^2 b_2^2 + a_1^2 b_2^2 + a_2^2 b_1^2.
\end{aligned}
$$

Also,

$$
\begin{aligned}
|z_1||z_2| &= |a_1 + ib_1||a_2 + ib_2| \\
&= (a_1^2 + b_1^2)(a_2^2 + b_2^2) \\
&= a_1^2 a_2^2 + a_1^2 b_2^2 + b_1^2 a_2^2 + b_1^2 b_2^2.
\end{aligned}
$$

Therefore, the desired result holds.

(f) As in the solution to part (c), we know that

$$
\frac{z_1}{z_2} = \frac{(a_1 a_2 + b_1 b_2) + i(a_2 b_1 - a_1 b_2)}{a_2^2 + b_2^2}.
$$

Therefore,

$$
\begin{aligned}
\left|\frac{z_1}{z_2}\right| &= \left|\frac{(a_1 a_2 + b_1 b_2) + i(a_2 b_1 - a_1 b_2)}{a_2^2 + b_2^2}\right| \\
&= \left(\frac{a_1 a_2 + b_1 b_2}{a_2^2 + b_2^2}\right)^2 + \left(\frac{a_2 b_1 - a_1 b_2}{a_2^2 + b_2^2}\right)^2 \\
&= \frac{a_1^2 a_2^2 + 2a_1 a_2 b_1 b_2 + b_1^2 b_2^2 + a_2^2 b_1^2 - 2a_2 b_1 a_1 b_2 + a_1^2 b_2^2}{(a_2^2 + b_2^2)^2} \\
&= \frac{a_1^2 a_2^2 + b_1^2 b_2^2 + a_2^2 b_1^2 + a_1^2 b_2^2}{(a_2^2 + b_2^2)} \\
&= \frac{a_1^2(a_2^2 + b_2^2) + b_1^2(a_2^2 + b_2^2)}{(a_2^2 + b_2^2)^2} \\
&= \frac{a_1^2 + b_1^2}{a_2^2 + b_2^2} \\
&= \frac{|z_1|}{|z_2|}.
\end{aligned}
$$

2. By the law of cosines,

$$
\begin{aligned}
|z_1 + z_2|^2 &= |z_1|^2 + |z_2|^2 - 2|z_1||z_2|\cos\phi \\
\implies |z_1 + z_2|^2 &\leq |z_1|^2 + 2|z_1||z_2| + |z_2|^2 = (|z_1| + |z_2|)^2.
\end{aligned}
$$

Taking the square root of both sides above, we see that

$$
|z_1 + z_2| \leq |z_1| + |z_2|.
$$

We notice that $|z_1 + z_2| = |z_1| + |z_2|$ if and only if $\cos\phi = -1$. Under that condition, the law of cosines leads us to the equation

$$|z_1 + z_2|^2 = |z_1|^2 + |z_2|^2 + 2|z_1||z_2| = (|z_1| + |z_2|)^2 \implies |z_1 + z_2| = |z_1| + |z_2|.$$

The condition that $\cos\phi = -1$ means $\phi = \pi$.

3.

(a)

$$
\begin{aligned}
z_1 z_2 &= (r_1 \cos\theta_1 + ir_1 \sin\theta_1)(r_2 \cos\theta_2 + ir_2 \sin\theta_2) \\
&= r_1 r_2 \cos\theta_1 \cos\theta_2 + ir_1 r_2 \cos\theta_1 \sin\theta_2 + ir_1 r_2 \sin\theta_1 \cos\theta_2 - r_1 r_2 \sin\theta_1 \sin\theta_2 \\
&= r_1 r_2 (\cos\theta_1 \cos\theta_2 - \sin\theta_1 \sin\theta_2) + ir_1 r_2 (\cos\theta_1 \sin\theta_2 + \sin\theta_1 \cos\theta_2) \\
&= |z_1||z_2| \cos(\theta_1 + \theta_2) + i|z_1||z_2| \sin(\theta_1 + \theta_2) \\
&= |z_1||z_2|(\cos(\theta_1 + \theta_2) + i\sin(\theta_1 + \theta_2)) \\
&= |z_1||z_2| e^{i(\theta_1 + \theta_2)}.
\end{aligned}
$$

(b)

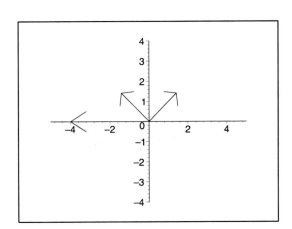

4. $e^{1+2i} = e^1 e^{2i} = e(\cos(2) + i\sin(2)) = e\cos(2) + ie\sin(2)$.

5. $e^{2-3i} = e^2 e^{-3i} = e^2(\cos(-3) + i\sin(-3)) = e^2 \cos(3) - ie^2 \sin(3)$.

6. $e^{i\pi} = (\cos(\pi) + i\sin(\pi)) = -1$.

7. $e^{2-(\pi/2)i} = e^2 e^{-(\pi/2)i} = e^2(\cos(-\pi/2) + i\sin(-\pi/2)) = -ie^2$.

8. $2^{1-i} = e^{(1-i)(\ln 2)} = e^{\ln 2} e^{-i\ln 2} = e^{\ln 2}(\cos(-\ln 2) + i\sin(-\ln 2)) = 2(\cos(\ln 2) - i\sin(\ln 2)) = 2\cos(\ln 2) - 2i\sin(\ln 2)$.

9. $\pi^{-1+2i} = e^{(-1+2i)\ln\pi} = e^{-\ln\pi} e^{2i\ln\pi} = e^{-\ln\pi}(\cos(2\ln\pi) + i\sin(2\ln\pi)) = \dfrac{1}{\pi}\cos(2\ln\pi) + \dfrac{i}{\pi}\sin(2\ln\pi)$.

10.

(a) If $y(0) = y_0$ and $y'(0) = y_1$ are real numbers, then

$$y(0) = c_1 + c_2 = y_0$$

is real, and

$$y'(0) = c_1(\mu + i\nu) + c_2(\mu - i\nu) = y_1$$

is real. We need to show that $c_2 = \overline{c_1}$. Let $c_1 = a_1 + ib_1$ and $c_2 = a_2 + ib_2$. We need to show that $a_1 = a_2$ and $b_2 = -b_1$. We begin by substituting the complex expressions for c_1 and c_2 into the equation for y_0. Upon doing so, we have

$$(a_1 + a_2) + i(b_1 + b_2) = y_0$$

is real. In order for this to be real, we need the imaginary part to be zero. In particular, we need $b_1 + b_2 = 0$, or $b_2 = -b_1$. Then using the equation for y_1, we have

$$(a_1 + ib_1)(\mu + i\nu) + (a_2 + ib_2)(\mu - i\nu) = y_1$$

is real. Multiplying and simplifying, we have

$$(a_1\mu - b_1\nu + a_2\mu + b_2\nu) + i(a_1\nu + b_1\mu - a_2\nu + b_2\mu) = y_1.$$

Since, $b_2 = -b_1$, this equation reduces to

$$(a_1\mu - b_1\nu + a_2\mu - b_1\nu) + i(a_1\nu - a_2\nu) = y_1$$

is real. In order for this to be real, we need the imaginary part to be zero. In particular, we need $a_1 = a_2$. Therefore, $c_2 = \overline{c_1}$ as claimed.

(b) In order to show that $y(t)$ is real, we will show that $\overline{y(t)} = y(t)$. We see that

$$\overline{y(t)} = \overline{ce^{(\mu+i\nu)t}} + \overline{\overline{c}e^{(\mu-i\nu)t}}$$
$$= \overline{c}e^{(\mu-i\nu)t} + ce^{(\mu+i\nu)t}$$
$$= y(t).$$

Therefore, $y(t)$ is real.

(c) If $c = \alpha + i\beta$, then

$$y(t) = (\alpha + i\beta)e^{(\mu+i\nu)t} + (\alpha - i\beta)e^{(\mu-i\nu)t}$$
$$= (\alpha + i\beta)e^{\mu t}(\cos(\nu t) + i\sin(\nu t)) + (\alpha - i\beta)e^{\mu t}(\cos(\nu t) - i\sin(\nu t))$$
$$= e^{\mu t}\left[\alpha\cos(\nu t) + i\alpha\sin(\nu t) + i\beta\cos(\nu t) - \beta\sin(\nu t) + \alpha\cos(\nu t) - i\alpha\sin(\nu t) - i\beta\cos(\nu t) - \beta\sin\right.$$
$$= e^{\mu t}[2\alpha\cos(\nu t) - 2\beta\sin(\nu t)].$$

Therefore, $c_1 = 2\alpha$ and $c_2 = -2\beta$ in Equation 10.

11.

(a) The characteristic equation is $\lambda^2 - 2\lambda + 2 = 0$. Therefore, the roots are $\lambda = 1 \pm i$. Therefore, one complex solution is

$$y(t) = e^{(1+i)t} = e^t(\cos(t) + i\sin(t)).$$

Considering the real and imaginary parts of this function, we arrive at the general solution

$$y(t) = c_1 e^t \cos(t) + c_2 e^t \sin(t).$$

(b) From our solution in part (a), we see that

$$y'(t) = c_1 e^t(\cos(t) - \sin(t)) + c_2 e^t(\sin(t) + \cos(t)).$$

Therefore,

$$\mathbf{x} = \begin{pmatrix} y \\ y' \end{pmatrix} = c_1 e^t \begin{pmatrix} \cos(t) \\ \cos(t) - \sin(t) \end{pmatrix} + c_2 e^t \begin{pmatrix} \sin(t) \\ \sin(t) + \cos(t) \end{pmatrix}.$$

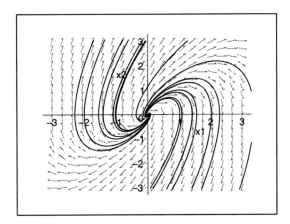

(c) We see that the critical point $(0,0)$ is an unstable spiral point.

12.

(a) The characteristic equation is $\lambda^2 - 2\lambda + 6 = 0$. Therefore, the roots are $\lambda = 1 \pm i\sqrt{5}$. Therefore, one complex solution is

$$y(t) = e^{(1+i\sqrt{5})t} = e^t(\cos(\sqrt{5}t) + i\sin(\sqrt{5}t)).$$

Considering the real and imaginary parts of this function, we arrive at the general solution

$$y(t) = c_1 e^t \cos(\sqrt{5}t) + c_2 e^t \sin(\sqrt{5}t).$$

(b) From our solution in part (a), we see that

$$y'(t) = c_1 e^t(\cos(\sqrt{5}t) - \sqrt{5}\sin(\sqrt{5}t)) + c_2 e^t(\sin(\sqrt{5}t) + \sqrt{5}\cos(\sqrt{5}t)).$$

Therefore,

$$\mathbf{x} = \begin{pmatrix} y \\ y' \end{pmatrix} = c_1 e^t \begin{pmatrix} \cos(\sqrt{5}t) \\ \cos(\sqrt{5}t) - \sqrt{5}\sin(\sqrt{5}t) \end{pmatrix} + c_2 e^t \begin{pmatrix} \sin(\sqrt{5}t) \\ \sin(\sqrt{5}t) + \sqrt{5}\cos(\sqrt{5}t) \end{pmatrix}.$$

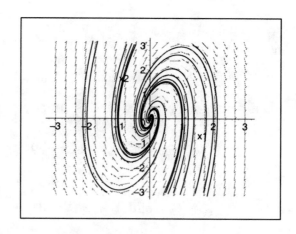

(c) We see that the critical point $(0, 0)$ is an unstable spiral point.

13.

(a) The characteristic equation is $\lambda^2 + 2\lambda - 8 = 0$. Therefore, the roots are $\lambda = -4, 2$. Therefore, the general solution is

$$y(t) = c_1 e^{-4t} + c_2 e^{2t}$$

(b) From our solution in part (a), we see that

$$y'(t) = -4c_1 e^{-4t} + 2c_2 e^{2t}.$$

Therefore,

$$\mathbf{x} = \begin{pmatrix} y \\ y' \end{pmatrix} = c_1 e^{-4t} \begin{pmatrix} 1 \\ -4 \end{pmatrix} + c_2 e^{2t} \begin{pmatrix} 1 \\ 2 \end{pmatrix}.$$

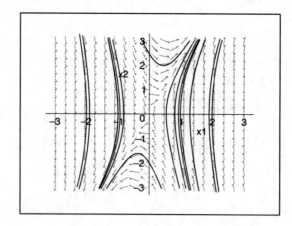

(c) We see that the critical point $(0, 0)$ is an unstable saddle point.

14.

(a) The characteristic equation is $\lambda^2 + 2\lambda + 2 = 0$. Therefore, the roots are $\lambda = -1 \pm i$. Therefore, one complex solution is

$$y(t) = e^{(-1+i)t} = e^{-t}(\cos(t) + i\sin(t)).$$

Considering the real and imaginary parts of this function, we arrive at the general solution

$$y(t) = c_1 e^{-t} \cos(t) + c_2 e^{-t} \sin(t).$$

(b) From our solution in part (a), we see that

$$y'(t) = c_1 e^{-t}(-\cos(t) - \sin(t)) + c_2 e^{-t}(-\sin(t) + \cos(t)).$$

Therefore,

$$\mathbf{x} = \begin{pmatrix} y \\ y' \end{pmatrix} = c_1 e^{-t} \begin{pmatrix} \cos(t) \\ \cos(t) - \sin(t) \end{pmatrix} + c_2 e^{-t} \begin{pmatrix} \sin(t) \\ -\sin(t) + \cos(t) \end{pmatrix}.$$

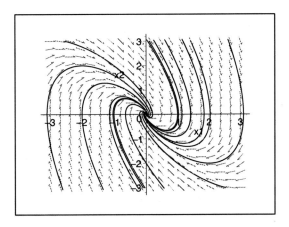

(c) We see that the critical point $(0,0)$ is an asymptotically stable spiral point.

15.

(a) The characteristic equation is $\lambda^2 + 6\lambda + 13 = 0$. Therefore, the roots are $\lambda = -3 \pm 2i$. Therefore, one complex solution is

$$y(t) = e^{(-3+2i)t} = e^{-3t}(\cos(2t) + i\sin(2t)).$$

Considering the real and imaginary parts of this function, we arrive at the general solution

$$y(t) = c_1 e^{-3t} \cos(2t) + c_2 e^{-3t} \sin(2t).$$

(b) From our solution in part (a), we see that

$$y'(t) = c_1 e^{-3t}(-3\cos(2t) - 2\sin(2t)) + c_2 e^{-3t}(-3\sin(2t) + 2\cos(2t)).$$

Therefore,

$$\mathbf{x} = \begin{pmatrix} y \\ y' \end{pmatrix} = c_1 e^{-3t} \begin{pmatrix} \cos(2t) \\ -3\cos(2t) - 2\sin(2t) \end{pmatrix} + c_2 e^{-3t} \begin{pmatrix} \sin(2t) \\ -3\sin(2t) + 2\cos(2t) \end{pmatrix}.$$

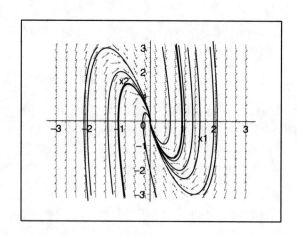

(c) We see that the critical point $(0,0)$ is an asymptotically stable spiral point.

16.

(a) The characteristic equation is $\lambda^2 + 2\lambda + 1.25 = 0$. Therefore, the roots are $\lambda = -1 \pm \dfrac{1}{2}i$. Therefore, one complex solution is

$$y(t) = e^{(-1+i/2)t} = e^{-t}(\cos(t/2) + i\sin(t/2)).$$

Considering the real and imaginary parts of this function, we arrive at the general solution

$$y(t) = c_1 e^{-t}\cos(t/2) + c_2 e^{-t}\sin(t/2).$$

(b) From our solution in part (a), we see that

$$y'(t) = c_1 e^{-t}\left(-\cos(t/2) - \frac{1}{2}\sin(t/2)\right) + c_2 e^{-t}\left(-\sin(t/2) + \frac{1}{2}\cos(t/2)\right).$$

Therefore,

$$\mathbf{x} = \begin{pmatrix} y \\ y' \end{pmatrix} = c_1 e^{-t}\begin{pmatrix} \cos(t/2) \\ -\cos(t/2) - \frac{1}{2}\sin(t/2) \end{pmatrix} + c_2 e^{-t}\begin{pmatrix} \sin(t/2) \\ -\sin(t/2) + \frac{1}{2}\cos(t/2) \end{pmatrix}.$$

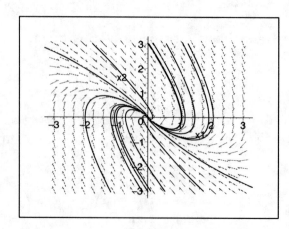

42

(c) We see that the critical point $(0,0)$ is an asymptotically stable spiral point.

17.

(a) The characteristic equation is $4\lambda^2 + 9 = 0$. Therefore, the roots are $\lambda = \pm\dfrac{3}{2}i$. Therefore, one complex solution is

$$y(t) = e^{3it/2} = \cos(3t/2) + i\sin(3t/2).$$

Considering the real and imaginary parts of this function, we arrive at the general solution

$$y(t) = c_1\cos(3t/2) + c_2\sin(3t/2).$$

(b) From our solution in part (a), we see that

$$y'(t) = -\frac{3c_1}{2}\sin(3t/2) + \frac{3c_2}{2}\cos(3t/2).$$

Therefore,

$$\mathbf{x} = \begin{pmatrix} y \\ y' \end{pmatrix} = c_1 \begin{pmatrix} \cos(3t/2) \\ -\frac{3}{2}\sin(3t/2) \end{pmatrix} + c_2 \begin{pmatrix} \sin(3t/2) \\ \frac{3}{2}\cos(3t/2) \end{pmatrix}.$$

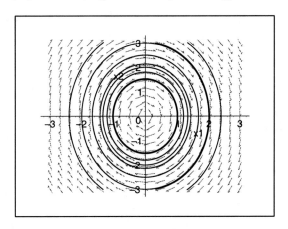

(c) We see that the critical point $(0,0)$ is a stable center.

18.

(a) The characteristic equation is $9\lambda^2 + 9\lambda - 4 = 0$. Therefore, the roots are $\lambda = -4/3, 1/3$. Therefore, the general solution is

$$y(t) = c_1 e^{-4t/3} + c_2 e^{t/3}.$$

(b) From our solution in part (a), we see that

$$y'(t) = -\frac{4c_1}{3}e^{-4t/3} + \frac{c_2}{3}e^{t/3}.$$

Therefore,

$$\mathbf{x} = \begin{pmatrix} y \\ y' \end{pmatrix} = c_1 e^{-4t/3} \begin{pmatrix} 1 \\ -4/3 \end{pmatrix} + c_2 e^{t/3} \begin{pmatrix} 1 \\ 1/3 \end{pmatrix}.$$

43

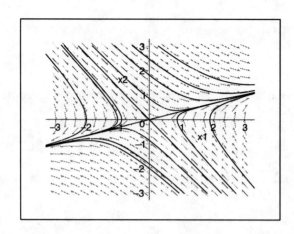

(c) We see that the critical point $(0,0)$ is an unstable saddle point.

19.

(a) The characteristic equation is $\lambda^2 + \lambda + 1.25 = 0$. Therefore, the roots are $\lambda = -\dfrac{1}{2} \pm i$. Therefore, one complex solution is

$$y(t) = e^{(-\frac{1}{2}+i)t} = e^{-t/2}(\cos(t) + i\sin(t)).$$

Considering the real and imaginary parts of this function, we arrive at the general solution

$$y(t) = c_1 e^{-t/2}\cos(t) + c_2 e^{-t/2}\sin(t).$$

(b) From our solution in part (a), we see that

$$y'(t) = c_1 e^{-t/2}\left(-\frac{1}{2}\cos(t) - \sin(t)\right) + c_2 e^{-t/2}\left(-\frac{1}{2}\sin(t) + \cos(t)\right).$$

Therefore,

$$\mathbf{x} = \begin{pmatrix} y \\ y' \end{pmatrix} = c_1 e^{-t/2}\begin{pmatrix} \cos(t) \\ -\frac{1}{2}\cos(t) - \sin(t) \end{pmatrix} + c_2 e^{-t/2}\begin{pmatrix} \sin(t) \\ -\frac{1}{2}\sin(t) + \cos(t) \end{pmatrix}.$$

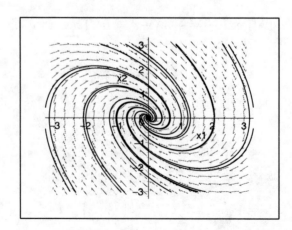

(c) We see that the critical point $(0, 0)$ is an asymptotically stable spiral point.

20.

(a) The characteristic equation is $\lambda^2 + 4\lambda + 6.25 = 0$. Therefore, the roots are $\lambda = -2 \pm \dfrac{3}{2}i$. Therefore, one complex solution is

$$y(t) = e^{\left(-2 + \frac{3}{2}i\right)t} = e^{-2t}(\cos(3t/2) + i\sin(3t/2)).$$

Considering the real and imaginary parts of this function, we arrive at the general solution

$$y(t) = c_1 e^{-2t}\cos(3t/2) + c_2 e^{-2t}\sin(3t/2).$$

(b) From our solution in part (a), we see that

$$y'(t) = c_1 e^{-2t}\left(-2\cos(3t/2) - \frac{3}{2}\sin(3t/2)\right) + c_2 e^{-2t}\left(-2\sin(3t/2) + \frac{3}{2}\cos(3t/2)\right).$$

Therefore,

$$\mathbf{x} = \begin{pmatrix} y \\ y' \end{pmatrix} = c_1 e^{-2t}\begin{pmatrix} \cos(3t/2) \\ -2\cos(3t/2) - \frac{3}{2}\sin(3t/2) \end{pmatrix} + c_2 e^{-2t}\begin{pmatrix} \sin(3t/2) \\ -2\sin(3t/2) + \frac{3}{2}\cos(3t/2) \end{pmatrix}.$$

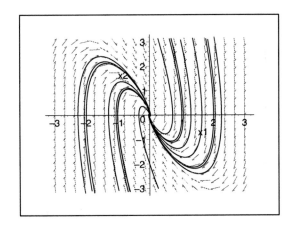

(c) We see that the critical point $(0, 0)$ is an asymptotically stable spiral point.

21. The characteristic equation is $\lambda^2 + 4 = 0$, which has roots $\lambda = \pm 2i$. Therefore, the general solution is $y(t) = c_1\cos(2t) + c_2\sin(2t)$. The derivative of y is $y'(t) = -2c_1\sin(2t) + 2c_2\cos(2t)$. Using the initial conditions, we have $c_1 = 0$ and $2c_2 = 1$. Therefore, the solution is

$$y(t) = \frac{1}{2}\sin(2t).$$

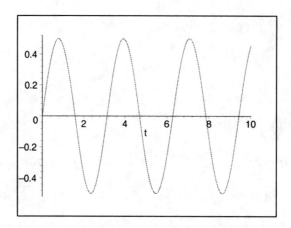

The solution will continue to oscillate with the same amplitude as $t \to \infty$.

22. The characteristic equation is $\lambda^2 + 4\lambda + 5 = 0$, which has roots $\lambda = -2 \pm i$. Therefore, the general solution is $y(t) = c_1 e^{-2t} \cos(t) + c_2 e^{-2t} \sin(t)$. The derivative of y is $y'(t) = c_1 e^{-2t}(-2\cos(t) - \sin(t)) + c_2 e^{-2t}(-2\sin(t) + \cos(t))$. Using the initial conditions, we have $c_1 = 1$ and $-2c_1 + c_2 = 0$. Therefore, $c_1 = 1$ and $c_2 = 2$, and we conclude that the solution is

$$y(t) = e^{-2t} \cos(t) + 2e^{-2t} \sin(t).$$

The solution $y \to 0$ as $t \to \infty$.

23. The characteristic equation is $\lambda^2 - 2\lambda + 5 = 0$, which has roots $\lambda = 1 \pm 2i$. Therefore, the general solution is $y(t) = c_1 e^t \cos(2t) + c_2 e^t \sin(2t)$. The derivative of y is $y'(t) = c_1 e^t(\cos(2t) - 2\sin(2t)) + c_2 e^t(\sin(2t) + 2\cos(2t))$. Using the initial conditions, we have $c_1 = 0$ and $c_2 = -e^{-\pi/2}$. Therefore,

$$y(t) = -e^{t-\pi/2} \sin(2t).$$

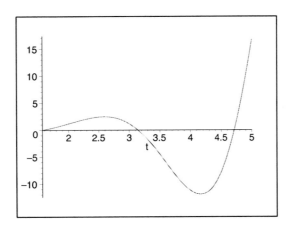

The solution $y \to \infty$ as $t \to \infty$.

24. The characteristic equation is $\lambda^2 + 1 = 0$, which has roots $\lambda = \pm i$. Therefore, the general solution is $y(t) = c_1 \cos(t) + c_2 \sin(t)$. The derivative of y is $y'(t) = -c_1 \sin(t) + c_2 \cos(t)$. Using the initial conditions, we have

$$\frac{1}{2}c_1 + \frac{\sqrt{3}}{2}c_2 = 2$$

$$-\frac{\sqrt{3}}{2}c_1 + \frac{1}{2}c_2 = -4.$$

The solution of this system is $c_1 = 1 + 2\sqrt{3}$ and $c_2 = \sqrt{3} - 2$, and we conclude that the solution is

$$y(t) = (1 + 2\sqrt{3})\cos(t) + (\sqrt{3} - 2)\sin(t).$$

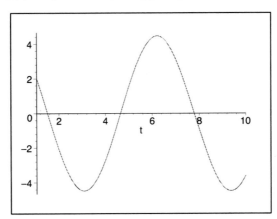

The solution continues to oscillate as $t \to \infty$.

25. The characteristic equation is $\lambda^2 + \lambda + 1.25 = 0$, which has roots $\lambda = -\frac{1}{2} \pm i$. Therefore, the general solution is $y(t) = c_1 e^{-t/2} \cos(t) + c_2 e^{-t/2} \sin(t)$. The derivative of y is $y'(t) = c_1 e^{-t/2} \left(-frac12 \cos(t) - \sin(t) \right) + c_2 e^{-t/2} \left(-\frac{1}{2}\sin(t) + \cos(t) \right)$. Using the initial conditions, we have $c_1 = 3$ and $-c_1/2 + c_2 = 1$. Therefore, $c_1 = 3$ and $c_2 = 5/2$, and we conclude that the solution is

$$y(t) = 3e^{-t/2}\cos(t) + \frac{5}{2}e^{-t/2}\sin(t).$$

47

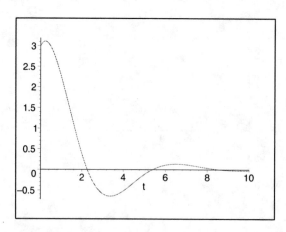

The solution y oscillates as decays to zero as $t \to \infty$.

26. The characteristic equation is $\lambda^2 + 2\lambda + 2 = 0$, which has roots $\lambda = -1 \pm i$. Therefore, the general solution is $y(t) = c_1 e^{-t}\cos(t) + c_2 e^{-t}\sin(t)$. The derivative of y is $y'(t) = c_1 e^{-t}(-\cos(t) - \sin(t)) + c_2 e^{-t}(-\sin(t) + \cos(t))$. Using the initial conditions, we have

$$\frac{\sqrt{2}}{2}c_1 e^{-\pi/4} + \frac{\sqrt{2}}{2}c_2 e^{-\pi/4} = 2$$
$$-\sqrt{2}c_1 e^{-\pi/4} = -2.$$

Therefore, $c_1 = \sqrt{2}e^{\pi/4}$ and $c_2 = \sqrt{2}e^{\pi/4}$, and we conclude that the solution is

$$y(t) = \sqrt{2}e^{-(t-\pi/4)}\cos(t) + \sqrt{2}e^{-(t-\pi/4)}\sin(t).$$

The solution y oscillates as it decays to zero as $t \to \infty$.

27. Direct calculation gives the result. On the other hand, using the fact that $W(fg, fh) = f^2 W(g, h)$, we see that

$$W(e^{\mu t}\cos(\nu t), e^{\mu t}\sin(\nu t)) = e^{2\mu t}W(\cos(\nu t), \sin(\nu t))$$
$$= e^{2\mu t}[\cos(\nu t)(\sin(\nu t))' - (\cos(\nu t))'\sin(\nu t)]$$
$$= \nu e^{2\mu t}.$$

28. As in the solution to 4.3.17, we see that the eigenvalues are given by

$$\lambda_{1,2} = \frac{-b \pm \sqrt{b^2 - 4ac}}{2a}.$$

48

Here, we are assuming $b^2 - 4ac < 0$. Therefore, our eigenvalues are complex conjugates. As in the solution to 4.3.17, an eigenvalue for λ_i is given by

$$\mathbf{v}_i = \begin{pmatrix} 1 \\ \lambda_i \end{pmatrix}.$$

Therefore,

$$\mathbf{x}_1(t) = e^{\lambda_1 t} \begin{pmatrix} 1 \\ \lambda_1 \end{pmatrix}$$

and

$$\mathbf{x}_2(t) = e^{\lambda_2 t} \begin{pmatrix} 1 \\ \lambda_2 \end{pmatrix}$$

form a fundamental set of complex-valued solutions. Therefore, the general solution of Equation (1) is given by the real and imaginary parts of the first component of $\mathbf{x}_1(t)$.

29.

(a) Clearly, y_1 and y_2 are solutions. Further, $W(\cos t, \sin t) = \cos^2 t + \sin^2 t = 1$. Therefore, the form a fundamental set of solutions.

(b) $y' = ie^{it}$ and $y'' = -e^{it}$. Therefore,

$$y'' + y = -e^{it} + e^{it} = 0.$$

Since $y_1 = \cos(t)$ and $y_2 = \sin(t)$ are a fundamental set of solutions. Any other solution must be a linear combination of those two. Since $y = e^{it}$ is also a solution, we must have $e^{it} = c_1 \cos t + c_2 \sin t$ for some constants c_1, c_2.

(c) Setting $t = 0$, we have $1 = c_1 \cos(0) + c_2 \sin(0) = c_1$.

(d) Further, differentiating equation (i), we have

$$ie^{it} = -c_1 \sin t + c_2 \cos t.$$

Then setting $t = 0$, we have

$$i = -c_1 \sin(0) + c_2 \cos(0) = c_2.$$

We conclude that $c_1 = 1$ and $c_2 = i$. Therefore,

$$e^{it} = \cos(t) + i \sin(t).$$

30. From Euler's formula, $e^{it} = \cos t + i \sin t$, it follows that $e^{-it} = \cos t - i \sin t$. Adding these equations, we have $e^{it} + e^{-it} = 2 \cos t$. Subtracting these two equations, we have $e^{it} - e^{-it} = 2i \sin t$.

31. Let $\lambda_1 = \mu_1 + i\nu_1$ and $\lambda_2 = \mu_2 + i\nu_2$. Therefore,

$$e^{(\lambda_1 + \lambda_2)t} = e^{(\mu_1 + \mu_2)t + i(\nu_1 + \nu_2)t}$$
$$= e^{(\mu_1 + \mu_2)t}[\cos((\nu_1 + \nu_2)t) + i\sin((\nu_1 + \nu_2)t)]$$
$$= e^{(\mu_1 + \mu_2)t}[(\cos(\nu_1 t) + i\sin(\nu_1 t))(\cos(\nu_2 t) + i\sin(\nu_2 t))]$$
$$= e^{\mu_1 t}(\cos(\nu_1 t) + i\sin(\nu_1 t)) \cdot e^{\mu_2 t}(\cos(\nu_1 t) + i\sin(\nu_1 t)).$$

Therefore, $e^{(\lambda_1 + \lambda_2)t} = e^{\lambda_1 t}e^{\lambda_2 t}$.

32.

$$\frac{d}{dt}e^{\lambda t} = \frac{d}{dt}(e^{\mu t}(\cos(\nu t) + i\sin(\nu t)))$$
$$= \mu e^{\mu t}(\cos(\nu t) + i\sin(\nu t)) + e^{\mu t}(-\nu\sin(\nu t) + i\nu\cos(\nu t))$$
$$= \mu e^{\mu t}(\cos(\nu t) + i\sin(\nu t)) + i\nu e^{\mu t}(\cos(\nu t) + i\sin(\nu t))$$
$$= (\mu + i\nu)e^{\mu t}(\cos(\nu t) + i\sin(\nu t))$$
$$= \lambda e^{\lambda t}.$$

33. If $\phi(t) = u(t) + iv(t)$ is a solution, then

$$(u + iv)'' + p(t)(u + iv)' + q(t)(u + iv) = 0,$$

which implies

$$(u'' + iv'') + p(t)(u' + iv') + q(t)(u + iv) = 0.$$

From this equation, we conclude that

$$(u'' + p(t)u' + q(t)u) + i(v'' + p(t)v' + q(t)v) = 0.$$

From this equation, we conclude that both the real and imaginary parts are zero, and, therefore, u and v are both solutions of equation (ii).

34. Let $z = \ln x$. Then

$$\frac{dy}{dx} = \frac{1}{x}\frac{dy}{dz}$$

and

$$\frac{d^2y}{dx^2} = \frac{1}{x^2}\frac{d^2y}{dz^2} - \frac{1}{x^2}\frac{dy}{dz}.$$

Therefore,

$$x^2\left[\frac{1}{x^2}\frac{d^2y}{dz^2} - \frac{1}{x^2}\frac{dy}{dz}\right] + x\left[\frac{1}{x}\frac{dy}{dz}\right] + y = 0,$$

which reduces to

$$y'' + y = 0.$$

The associated characteristic equation is given by $\lambda^2 + 1 = 0$. Therefore, the roots are $\lambda = \pm i$. Consequently, the general solution is given by

$$y(z) = c_1 e^{iz} + c_2 e^{-iz},$$

or
$$y(z) = c_1 \cos(z) + c_2 \sin(z).$$

Rewriting the equation back in terms of x, we have

$$y(x) = c_1 \cos(\ln x) + c_2 \sin(\ln x).$$

35. Letting $z = \ln x$ and using the equations for dy/dx, d^2/dx^2 in terms of dy/dz and d^2y/dz^2 (as described in the solution to exercise 34 above), our equation reduces to

$$y'' + 3y' + 2y = 0$$

where $' = d/dz$. The associated characteristic equation is given by $\lambda^2 + 3\lambda + 2 = 0$. The roots of this equation are $\lambda = -1, -2$. Therefore, the solution is given by

$$\begin{aligned} y(z) &= c_1 e^{-2z} + c_2 e^{-z} \\ &= c_1 e^{-2\ln x} + c_2 e^{-\ln x} \\ &= c_1 x^{-2} + c_2 x^{-1}. \end{aligned}$$

36. As in the solution to exercise 34, we make the substitution $z = \ln x$. As a result, our equation can be rewritten as

$$y'' + 2y' + 1.25y = 0$$

where $' = d/dz$. The associated characteristic equation is given by $\lambda^2 + 2\lambda + 5/4 = 0$. The roots are $-1 \pm \dfrac{1}{2}i$. Therefore,

$$y(z) = c_1 e^{(-1+\frac{1}{2}i)z} + c_2 e^{(-1-\frac{1}{2}i)z},$$

or

$$y(z) = e^{-z}[c_1 \cos(z/2) + c_2 \sin(z/2)].$$

Rewriting the equation back in terms of x, we have

$$y(x) = x^{-1}\left[c_1 \cos\left(\frac{1}{2}\ln x\right) + c_2 \sin\left(\frac{1}{2}\ln x\right)\right].$$

37. Let $z = \ln x$. Then our equation can be rewritten as

$$y'' - 5y' - 6y = 0$$

where $' = d/dz$. The associated characteristic equation is $\lambda^2 - 5\lambda - 6 = 0$ which has roots $\lambda = 6, -1$. Therefore, the solution is

$$y(z) = c_1 e^{6z} + c_2 e^{-z}.$$

Rewriting in terms of x, the general solution is

$$y(x) = c_1 x^6 + c_2 x^{-1}.$$

38. Let $z = \ln x$. Then our equation can be rewritten as

$$y'' - y' - 2y = 0$$

where $' = d/dz$. The associated characteristic equation is $\lambda^2 - \lambda - 2 = 0$ which has roots $\lambda = 2, -1$. Therefore, the solution is

$$y(z) = c_1 e^{2z} + c_2 e^{-z}.$$

Rewriting in terms of x, the general solution is

$$y(x) = c_1 x^2 + c_2 x^{-1}.$$

39. Letting $z = \ln x$, our equation can be rewritten as

$$y'' - 4y' + 4y = 0$$

where $' = d/dz$. The associated characteristic equation is $\lambda^2 - 4\lambda + 4 = 0$ which has root $\lambda = 2$. Therefore, the solution is

$$y(z) = c_1 e^{2z} + c_2 z e^{2z}.$$

Rewriting in terms of x, the general solution is

$$y(x) = c_1 x^2 + c_2 (\ln x) x^2.$$

40. Letting $z = \ln x$, our equation can be rewritten as

$$y'' + y' + 4y = 0$$

where $' = d/dz$. The associated characteristic equation is $\lambda^2 + \lambda + 4 = 0$ which has roots $\lambda = (-1 \pm \sqrt{15}i)/2$. Therefore, the solution is

$$y(z) = e^{-z/2}[c_1 \cos(\sqrt{15}z/2) + c_2 \sin(\sqrt{15}z/2)].$$

Rewriting in terms of x, the general solution is

$$y(x) = x^{-1/2}\left[c_1 \cos\left(\frac{\sqrt{15}}{2}\ln x\right) + c_2 \sin\left(\frac{\sqrt{15}}{2}\ln x\right)\right].$$

41. Letting $z = \ln x$, our equation can be rewritten as

$$2y'' - 6y' + 6y = 0$$

where $' = d/dz$. The associated characteristic equation is $\lambda^2 - 3\lambda + 3 = 0$ which has roots $\lambda = (3 \pm \sqrt{3}i)/2$. Therefore, the solution is

$$y(z) = e^{3z/2}[c_1 \cos(\sqrt{3}z/2) + c_2 \sin(\sqrt{3}z/2)].$$

Rewriting in terms of x, the general solution is

$$y(x) = x^{3/2}\left[c_1 \cos\left(\frac{\sqrt{3}}{\ln}x\right) + c_2 \sin\left(\frac{\sqrt{3}}{2}\ln x\right)\right].$$

42. Letting $z = \ln x$, our equation can be rewritten as

$$2y'' - y' - 3y = 0$$

where $' = d/dz$. The associated characteristic equation is $2\lambda^2 - \lambda - 3 = 0$ which has roots $\lambda = 3/2, -1$. Therefore, the solution is

$$y(z) = c_1 e^{3z/2} + c_2 e^{-z}.$$

Rewriting in terms of x, the general solution is

$$y(x) = c_1 x^{3/2} + c_2 x^{-1}.$$

The initial condition $y(1) = 1$ implies

$$y(1) = c_1 + c_2 = 1.$$

Using the fact that

$$y'(x) = \frac{3}{2}c_1 x^{1/2} - c_2 x^{-2},$$

and the initial condition $y'(1) = 1$, we have

$$y'(1) = \frac{3}{2}c_1 - c_2 = 1.$$

Combining these two equations, we have $\frac{5}{2}c_1 = 2$ which implies $c_1 = 4/5$, and, therefore, $c_2 = 1/5$. We conclude that the solution of the IVP is

$$y(x) = \frac{4}{5}x^{3/2} + \frac{1}{5}x^{-1}.$$

As $x \to 0$, the solution $y \to +\infty$.

43. Letting $z = \ln x$, our equation can be rewritten as

$$4y'' + 4y' + 17y = 0$$

where $' = d/dz$. The associated characteristic equation is $4\lambda^2 + 4\lambda + 17 = 0$ which has roots $\lambda = (-1 \pm 4i)/2$. Therefore, the solution is

$$y(z) = e^{-z/2}[c_1 \cos(2z) + c_2 \sin(2z)].$$

Rewriting in terms of x, the general solution is

$$y(x) = x^{-1/2}[c_1 \cos(2\ln x) + c_2 \sin(2\ln x)].$$

The initial condition $y(1) = 2$ implies

$$y(1) = c_1 = 2.$$

Using the fact that

$$y'(x) = -\frac{1}{2}x^{-3/2}[c_1 \cos(2\ln x) + c_2 \sin(2\ln x)] + x^{-1/2}\left[-c_1 \sin(2\ln x) \cdot \frac{2}{x} + c_2 \cos(2\ln x) \cdot \frac{2}{x}\right]$$

and the initial condition $y'(1) = -3$, we have

$$y'(1) = -\frac{1}{2}[c_1 \cos(0) + c_2 \sin(0)] + [-c_1 \sin(0) \cdot 2 + c_2 \cos(0) \cdot 2] = -\frac{1}{2}c_1 + 2c_2 = -1 + 2c_2 = -3.$$

Therefore, $c_2 = -1$. We conclude that the solution of the IVP is

$$y(x) = x^{-1/2}[2\cos(2\ln x) - \sin(2\ln x)]$$

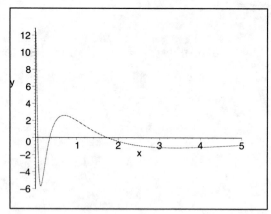

As $x \to 0$, the solution $y \to +\infty$.

44. Letting $z = \ln(-x)$, our equation can be rewritten as

$$y'' - 4y' + 4y = 0$$

where $' = d/dz$. The associated characteristic equation is $\lambda^2 - 4\lambda + 4 = 0$ which has root $\lambda = 2$. Therefore, the solution is

$$y(z) = c_1 e^{2z} + c_2 z e^{2z}.$$

54

Rewriting in terms of x, the general solution is

$$y(x) = c_1 x^2 + c_2 (\ln(-x)) x^2.$$

The initial condition $y(-1) = 2$ implies

$$y(-1) = c_1 + c_2 \ln(1) = c_1 = 2.$$

Using the fact that

$$y'(x) = 2c_1 x + c_2 x + 2c_2 x \ln(-x)$$

and the initial condition $y'(-1) = 3$, we have

$$y'(-1) = -2c_1 - c_2 - 2c_2 \ln(1) = -2(2) - c_2 = -4 - c_2 = 3.$$

Therefore, $c_2 = -7$. We conclude that the solution of the IVP is

$$y(x) = 2x^2 - 7x^2 \ln(-x).$$

As $x \to 0^-$, the solution $y \to 0$.

45. Letting $z = \ln x$, our equation can be rewritten as

$$y'' + 2y' + 5y = 0$$

where $' = d/dz$. The associated characteristic equation is $\lambda^2 + 2\lambda + 5 = 0$ which has roots $\lambda = -1 \pm 2i$. Therefore, the solution is

$$y(z) = e^{-z}[c_1 \cos(2z) + c_2 \sin(2z)].$$

Rewriting in terms of x, the general solution is

$$y(x) = x^{-1}[c_1 \cos(2 \ln x) + c_2 \sin(2 \ln x)].$$

The initial condition $y(1) = 1$ implies

$$y(1) = c_1 \cos(0) + c_2 \sin(0) = c_1 = 1.$$

Using the fact that

$$y'(x) = -x^{-2}[c_1 \cos(2 \ln x) + c_2 \sin(2 \ln x)] + x^{-1} \left[-c_1 \sin(2 \ln x) \cdot \frac{2}{x} + c_2 \cos(2 \ln x) \cdot \frac{2}{x} \right]$$

and the initial condition $y'(1) = -1$, we have

$$y'(1) = -[c_1 \cos(0) + c_2 \sin(0)] + [-c_1 \sin(0) \cdot 2 + c_2 \cos(0) \cdot 2] = -1 + 2c_2 = -1.$$

Therefore, $c_2 = 0$. We conclude that the solution of the IVP is

$$y(x) = x^{-1} \cos(2 \ln x).$$

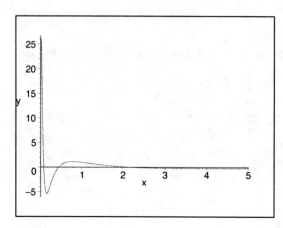

As $x \to 0$, the solution $y \to \infty$.

46. Consider

$$ax^2 y'' + bxy' + cy = 0$$

for $x < 0$. Let $\hat{x} = -x$. then

$$\frac{dy}{dx} = \frac{dy}{d\hat{x}} \frac{d\hat{x}}{dx} = -\frac{dy}{d\hat{x}}.$$

Further,

$$\frac{d^2 y}{dx^2} = \frac{d}{dx} \left[-\frac{dy}{d\hat{x}} \right] = -\frac{d^2 y}{d\hat{x}^2} \frac{d\hat{x}}{dx} = -\frac{d^2 y}{d\hat{x}^2}(-1) = \frac{d^2 y}{d\hat{x}^2}.$$

Therefore, for $\hat{x} > 0$,

$$ax^2 y'' + bxy' + cy = a(-\hat{x})^2 \frac{d^2 y}{d\hat{x}^2} + b(-\hat{x}) \left(-\frac{dy}{d\hat{x}} \right) + cy$$

$$= a\hat{x}^2 y'' + b\hat{x}y' + cy = 0 \qquad \text{where } ' = \frac{d}{d\hat{x}}.$$

Section 4.5

1. $R \cos \delta = 3$ and $R \sin \delta = 4$ implies $R = \sqrt{25} = 5$ and $\delta = \arctan(4/3) \approx 0.9273$. Therefore,

$$y = 5 \cos(2t - 0.9273).$$

56

2. $R\cos\delta = -1$ and $R\sin\delta = \sqrt{3}$ implies $R = \sqrt{4} = 2$ and $\delta = \pi + \arctan(-\sqrt{3}) = 2\pi/3$. Therefore,

$$y = 2\cos(t - 2\pi/3).$$

3. $R\cos\delta = 4$ and $R\sin\delta = -2$ implies $R = \sqrt{20} = 2\sqrt{5}$ and $\delta = \arctan(-1/2) \approx -0.4636$. Therefore,

$$y = 2\sqrt{5}\cos(3t + 0.46362).$$

4. $R\cos\delta = -2$ and $R\sin\delta = -3$ implies $R = \sqrt{13}$ and $\delta = \pi + \arctan(3/2) \approx 4.1244$. Therefore,

$$y = \sqrt{13}\cos(\pi t - 4.1244).$$

5.

(a) The spring constant is $k = 2/(1/2) = 4$ lb/ft. The mass $m = 2/32 = 1/16$ lb-s^2/ft. Therefore, the equation of motion is

$$\frac{1}{16}y'' + 4y = 0$$

which can be simplified to

$$y'' + 64y = 0.$$

The initial conditions are $y(0) = 1/4$ ft, $y'(0) = 0$ ft/sec. The general solution of the differential equation is $y(t) = A\cos(8t) + B\sin(8t)$. The initial condition implies $A = 1/4$ and $B = 0$. Therefore, the solution is $y(t) = \dfrac{1}{4}\cos(8t)$.

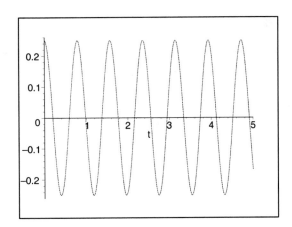

The frequency is $\omega = 8$ rad/sec. The period is $T = \pi/4$ seconds. The amplitude is $R = 1/4$ feet.

(b)

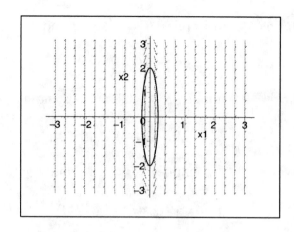

6.

(a) The spring constant is $k = .98/.05 = 19.6$ N/m. The mass $m = 0.1$ kg. Therefore, the equation of motion is

$$0.1y'' + 19.6y = 0$$

which can be simplified to

$$y'' + 196y = 0.$$

The initial conditions are $y(0) = 0$ m, $y'(0) = 10$ cm/sec. The general solution of the differential equation is $y(t) = A\cos(14t) + B\sin(14t)$. The initial condition implies $A = 0$ and $B = 5/7$. Therefore, the solution is $y(t) = \dfrac{5}{7}\sin(14t)$ cm. The period is $T = 2\pi/14$ seconds. Therefore, the mass will first return to its equilibrium position in half that time; that is, in $\pi/14$ seconds.

(b)

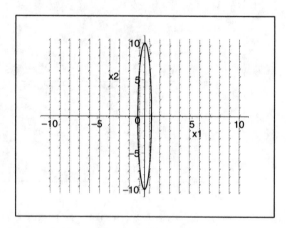

7. The spring constant is $k = 3/(1/4) = 12$ lb/ft. The mass is $3/32$ lb-s^2/ft. Therefore, the equation of motion is

$$\frac{3}{32}y'' + 12y = 0,$$

which can be simplified to

$$y'' + 128y = 0.$$

58

The initial conditions are $y(0) = -1/12$ ft, $y'(0) = 2$ ft/sec. The general solution is $y(t) = A\cos(8\sqrt{2}t) + B\sin(8\sqrt{2}t)$. Considering the initial conditions, we arrive at the solution

$$y(t) = -\frac{1}{12}\cos(8\sqrt{2}t) + \frac{1}{4\sqrt{2}}\sin(8\sqrt{2}t).$$

The frequency $\omega_0 = 8\sqrt{2}$ rad/sec. The period is $T = \pi/(4\sqrt{2})$ seconds. The amplitude is $R = \sqrt{(1/12)^2 + (1/4\sqrt{2})^2} = \sqrt{11}/12$ ft. The phase is $\delta = \pi - \arctan(3/\sqrt{2})$.

8. The inductance $L = 1$ henry. The resistance $R = 0$. The capacitance $C = 0.25 \times 10^{-6}$ farads. Therefore, the equation for charge q is

$$q'' + (4 \times 10^6)q = 0.$$

The initial conditions are $q(0) = 10^{-6}$ coulombs, $q'(0) = 0$ coulombs/sec. The general solution of this equation is $q(t) = A\cos(2000t) + B\sin(2000t)$. The initial conditions imply the specific solution is $q(t) = 10^{-6}\cos(2000t)$ coulombs.

9.

(a) The spring constant is $k = .196/.05 = 3.92$ N/m. The mass $m = .02$ kg. The damping constant is $\gamma = 400$ dyne-sec/cm $= .4$ N-sec/m. Therefore, the equation of motion is

$$.02y'' + .4y' + 3.92y = 0$$

or

$$y'' + 20y' + 196y = 0,$$

with initial conditions $y(0) = .02$ m, $y'(0) = 0$ m/sec. The solution of this equation is $y(t) = e^{-10t}(A\cos(4\sqrt{6}t) + B\sin(4\sqrt{6}t))$. The initial conditions imply

$$y(t) = e^{-10t}\left[2\cos(4\sqrt{6}t) + (5/\sqrt{6})\sin(4\sqrt{6}t)\right] \text{ cm}$$

The quasi frequency is $\nu = 4\sqrt{6}$ rad/sec. The quasi period is $T_d = \pi/2\sqrt{6}$. For undamped motion, the equation would be

$$y'' + 196y = 0,$$

which has general solution $y(t) = A\cos(14t) + B\sin(14t)$. Therefore, the period for the undamped motion is $T = \pi/7$. The ratio of the quasi period to the period of the undamped motion is $T_d/T = 7/2\sqrt{6} \approx 1.4289$. The solution y will satisfy $|y(\tau)| < 0.05$ for all $t > .4045$ seconds.

(b)

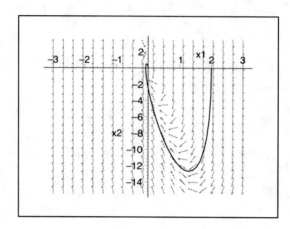

10. The spring constant is $k = 16/(1/4) = 64$ lb/ft. The mass is $m = 16/32 = 1/2$ lb-s^2/ft. The damping coefficient is $\gamma = 2$ lb-sec/ft. Therefore, the equation of motion is

$$\frac{1}{2}y'' + 2y' + 64y = 0,$$

which can be rewritten as

$$y'' + 4y' + 128y = 0.$$

The initial conditions are $y(0) = 0$ ft, $y'(0) = 1/4$ ft/sec. The general solution is $y(t) = e^{-2t}(A\cos(2\sqrt{31}t) + B\sin(2\sqrt{31}t))$. The initial conditions imply the specific solution is

$$y(t) = \frac{1}{8\sqrt{31}}e^{-2t}\sin(2\sqrt{31}t).$$

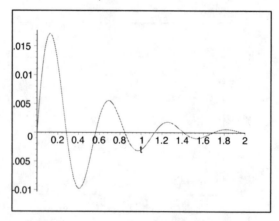

The quasi period is $T_d = \pi/\sqrt{31}$ seconds. Therefore, the solution will first reach its equilibrium position in half that time; that is, in $t = \pi/2\sqrt{31}$ seconds. Based on the graph, and the solution of $y(t) = -0.01$ inches, we see that $|y(t)| < 0.01$ for $t \geq \tau = 1.5927$.

11.

(a) The spring constant is $k = 3/.1 = 30$ N/m. The mass is $m = 2$ kg. The damping coefficient is $\gamma = 3/5$ N-sec/m. Therefore, the equation of motion is

$$2y'' + \frac{3}{5}y' + 30y = 0$$

which can be rewritten as

$$y'' + 0.3y' + 15y = 0.$$

The initial conditions are $y(0) = 0.05$ m and $y'(0) = 0.01$ m/sec. The general solution is $y(t) = e^{-0.15t}(A\cos(\mu t) + B\sin(\mu t))$ where $\mu \approx 3.87008$. Based on the initial conditions, we have $A = 0.05$ and $B = 0.00452$. Therefore, the specific solution is

$$y(t) = e^{-0.15t}(0.05\cos(\mu t) + 0.00452\sin(\mu t)).$$

The quasi frequency is $\nu = \mu = 3.87008$ radians/second. The undamped equation is

$$y'' + 15y = 0,$$

which has general solution $y(t) = A\cos(\sqrt{15}t) + B\sin(\sqrt{15}t)$. Therefore, the frequency for undamped motion is $\omega_0 = \sqrt{15}$. Therefore, the ratio of ν to ω_0 is $\nu/\omega_0 = 3.87008/\sqrt{15} \approx 0.99925$.

(b)

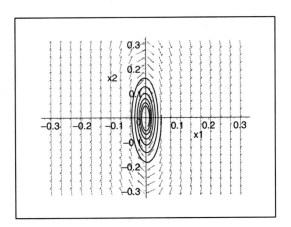

12.

(a) The inductance $L = 0.2$ henry. The resistance $R = 3 \times 10^2$ ohms. The capacitance $C = 10^{-5}$ farads. Therefore, the equation for charge q is

$$0.2q'' + 300q' + 10^5 q = 0$$

which can be rewritten as

$$q'' + 1500q' + 500,000q = 0.$$

The initial conditions are $q(0) = 10^{-6}$ coulombs, $q'(0) = 0$ coulombs/sec. The general solution of the differential equation is $q(t) = Ae^{-500t} + Be^{-1000t}$. The initial conditions imply $A + B = 10^{-6}$ and $-500A - 1000B = 0$. Therefore, $A = 2 \times 10^{-6}$ and $B = -10^{-6}$ and the specific solution is

$$y(t) = 10^{-6}(2e^{-500t} - e^{-1000t}).$$

(b)

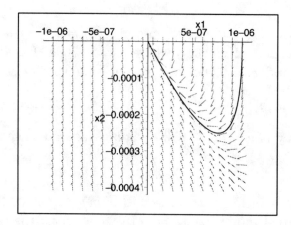

13. The frequency of the undamped motion is $\omega_0 = 1$. Therefore, the period of the undamped motion is $T = 2\pi$. The quasi frequency of the damped motion is $\mu = \dfrac{1}{2}\sqrt{4 - \gamma^2}$. Therefore, the period of the undamped motion is $T_d = 4\pi/2\sqrt{4 - \gamma^2}$. We want to find γ such that $T_d = 1.5T$. That is, we want γ to satisfy $4\pi/2\sqrt{4 - \gamma^2} = 3\pi$. Solving this equation, we have $\gamma = 2\sqrt{5}/3$.

14. The spring constant is $k = mg/L$. The equation of motion for an undamped system is

$$my'' + \frac{mg}{L}y = 0$$

which simplifies to

$$y'' + \frac{g}{L}y = 0.$$

Therefore, the frequency is $\omega_0 = \sqrt{g/L}$, and, therefore, the period is $T = 2\pi/\sqrt{g/L} = 2\pi\sqrt{L/g}$.

15. Suppose y is a solution of the differential equation

$$my'' + \gamma y' + ky = 0$$

and satisfies the initial conditions $y(t_0) = y_0$, $y'(t_0) = y_1$. Then suppose v is a solution of the same differential equation, but satisfies the initial conditions $v(t_0) = v_0$ and $v'(t_0) = 0$. Let $w = y - v$. Then w satisfies the same differential equation, since the equation is linear, and, further, $w(t_0) = 0$, $w'(t_0) = y_1$. Therefore, $y = v + w$ where v, w satisfy the specified conditions.

16. Using trig identities, we note that $r\sin(\omega_0 t - \theta) = r\sin(\omega_0 t)\cos\theta - r\cos(\omega_0 t)\sin\theta$. In order to write $A\cos(\omega_0 t) + B\sin(\omega_0 t)$ in the form $r\sin(\omega_0 t - \theta)$, we need $A = -r\sin\theta$ and

$B = r \cos \theta$. Therefore, we must have $A^2 + B^2 = r^2$ and $\tan \theta = -A/B$. From the text, we know that in order to write $A \cos(\omega_0 t) + B \sin(\omega_0 t) = R \cos(\omega_0 t - \delta)$, we need $R = \sqrt{A^2 + B^2}$ and $\tan \delta = B/A$. Therefore, in order for $R \cos(\omega_0 t - \delta) = r \sin(\omega_0 t - \theta)$, we need $r = R$ and $\tan \theta = -1/\tan \delta$.

17. The spring constant is $k = 8/(1.5/12) = 64$ lb/ft. The mass is $8/32 = 1/4$ lb-s^2/ft. Therefore, the equation of motion is

$$\frac{1}{4}y'' + \gamma y' + 64y = 0.$$

The motion will experience critical damping when $\gamma = 2\sqrt{km} = 8$ lb-s/ft.

18. The inductance $L = 0.2$ henry. The capacitance $C = 0.8 \times 10^{-6}$ farads. Therefore, the equation for charge q is

$$0.2q'' + Rq' + \frac{10^7}{8}q = 0$$

The motion will experience critical damping when $R = 2\sqrt{0.2 \times 10^7/8} = 10^3$ ohms.

19. If the system is critically damped or overdamped, then $\gamma \geq 2\sqrt{km}$. If $\gamma = 2\sqrt{km}$ (critically damped), then the solution is given by

$$y(t) = (A + Bt)e^{-\gamma t/2m}.$$

In this case, if $y = 0$, then we must have $A + Bt = 0$, that is, $t = -A/B$ (assuming $B \neq 0$). If $B = 0$, the solution is never zero (unless $A = 0$). If $\gamma > 2\sqrt{km}$ (overdamped), then the solution is given by

$$y(t) = Ae^{\lambda_1 t} + Be^{\lambda_2 t}$$

where λ_1, λ_2 are given by equation (19) in the text. Assume for the moment, that $A, B \neq 0$. Then $y = 0$ implies $Ae^{\lambda_1 t} = -Be^{\lambda_2 t}$ which implies $e^{(\lambda_1 - \lambda_2)t} = -B/A$. There is only one solution to this equation. If $A = 0$ or $B = 0$, then there are no solutions to the equation $y = 0$ (unless they are both zero, in which case, the solution is identically zero).

20. If the system is critically damped, then the general solution is

$$y(t) = (A + Bt)e^{-\gamma t/2m}.$$

The initial conditions imply $A = y_0$ and $B = v_0 + (\gamma y_0/2m)$. If $v_0 = 0$, then $B = \gamma y_0/2m$. In this case, the specific solution is

$$y(t) = (y_0 + (\gamma y_0/2m)t)e^{-\gamma t/2m}.$$

As $t \to \infty$, $y \to 0$. In order for $y = 0$, we would need $y_0 + (\gamma y_0/2m)t = 0$, but there are no positive times t satisfying this equation. If $v_0 \neq 0$, the specific solution is

$$y(t) = (y_0 + (v_0 + (\gamma y_0/2m))t)e^{-\gamma t/2m}.$$

In order to find a positive time such that $y(t) = 0$, we need

$$y_0 + (v_0 + (\gamma y_0/2m))t = 0.$$

The solution of this equation is

$$t = -\frac{y_0}{v_0 + (\gamma y_0/2m)}.$$

In order for there to be a positive time t satisfying this equation, we need $v_0 + (\gamma y_0/2m) < 0$. That is, we need $v_0 < -\gamma y_0/2m$.

21.

(a) Let $y = Re^{-\gamma t/2m}\cos(\nu t - \delta)$. Then y attains a maximum when $\nu t_k - \delta = 2k\pi$. Therefore, the time between successive maxima is $T_d = t_{k+1} - t_k = \frac{1}{\nu}(2(k+1)\pi - 2k\pi) = 2\pi/nu$.

(b)

$$\begin{aligned}
\frac{y(t_k)}{y(t_{k+1})} &= \frac{e^{-\gamma t_k/2m}}{e^{-\gamma t_{k+1}/2m}} \\
&= e^{(\gamma t_{k+1} - \gamma t_k)/2m} \\
&= e^{\gamma T_d/2m}.
\end{aligned}$$

(c) From parts (a) and (b), we have $\Delta = \ln[y(t_k)/y(t_{k+1})] = \gamma T_d/2m = 2\pi\gamma/2m\nu = \pi\gamma/m\nu$.

22. The spring constant is $k = 16/(1/4) = 64$ lb/ft. The mass $m = 1/2$ lb-s^2/ft. The damping coefficient is $\gamma = 2$ lb-sec/ft. The quasi frequency is $\nu = 2\sqrt{31}$ rad/sec. Therefore,

$$\Delta = \frac{2\pi}{(1/2)2\sqrt{31}} = \frac{2\pi}{\sqrt{31}} \approx 1.1285.$$

23. For problem 17, the mass is $m = 1/4$ lb-s^2/ft. We suppose that $\Delta = 3$ and $T_d = 0.3$ seconds. From problem 21, we know that $\Delta = \gamma T_d/2m$. Therefore, we have

$$3 = \frac{0.3\gamma}{2(1/4)} = 0.6\gamma.$$

Therefore, the damping coefficient is $\gamma = 5$ lb-sec/ft.

24. The general solution of this equation is

$$y(t) = A\cos(\sqrt{2k/3}\,t) + B\sin(\sqrt{2k/3}\,t).$$

The period of this system is $T = 2\pi/\sqrt{2k/3}$ sec. Therefore, if the period is $T = \pi$ seconds, then k must satisfy $2\pi/\sqrt{2k/3} = \pi$. The solution of this equation is $k = 6$. The initial conditions $y(0) = 2$ and $y'(0) = v$ imply $A = 2$ and $B\sqrt{2k/3} = v$. Since $k = 6$, we know that $2B = v$. The amplitude of the system is $\sqrt{A^2 + B^2} = \sqrt{2^2 + (v/2)^2}$. Since the amplitude is assumed to be 3, we must have $\sqrt{2^2 + (v/2)^2} = 3$ which implies $v = 2\sqrt{5}$.

25.

(a) The solution of the IVP is $y(t) = e^{-t/8}\left(2\cos\left(\frac{3}{8}\sqrt{7}t\right) + 0.252\sin\left(\frac{3}{8}\sqrt{7}t\right)\right)$.

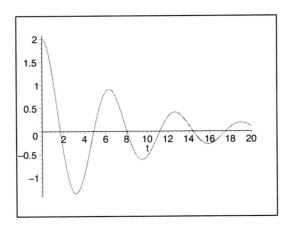

Using the graph above and numerical evidence, we conclude that $\tau \approx 41.715$ seconds.

(b) For $\gamma = 0.5$, $\tau \approx 20.402$; for $\gamma = 1.0$, $\tau \approx 9.168$; for $\gamma = 1.5$, $\tau \approx 7.184$.

(c)

(d) For $\gamma = 1.7$, $\tau \approx 6.767$; for $\gamma = 1.8$, $\tau \approx 5.473$; for $\gamma = 1.9$, $\tau \approx 6.460$. τ steadily decreases to about $\tau_{min} \approx 4.873$, corresponding to the critical value $\gamma_0 \approx 1.73$.

(e) We have
$$y(t) = \frac{4e^{-\gamma t/2}}{\sqrt{4 - \gamma^2}}\cos(\nu t - \delta)$$

where $\nu = \frac{1}{2}\sqrt{4 - \gamma^2}$, and $\delta = \tan^{-1}(\gamma/\sqrt{4 - \gamma^2})$. Therefore, $|y(t)| \leq \dfrac{4e^{-\gamma t/2}}{\sqrt{4 - \gamma^2}}$. To find the critical τ, we need to solve $4e^{-\gamma\tau/2}/\sqrt{4 - \gamma^2} = 0.01$ Solving this equation, we see that $\tau = (2/\gamma)\ln(400/\sqrt{4 - \gamma^2})$.

26.

(a) The characteristic equation is $m\lambda^2 + \gamma\lambda + k = 0$. Since $\gamma^2 < 4km$, the roots are

$$\lambda_{1,2} = -\frac{\gamma}{2m} \pm i\frac{\sqrt{4mk - \gamma^2}}{2m}.$$

Therefore, the general solution is

$$y(t) = e^{-\gamma t/2m}\left[A\cos\frac{\sqrt{4mk - \gamma^2}}{2m}t + B\sin\frac{\sqrt{4mk - \gamma^2}}{2m}t\right].$$

our initial conditions imply $A = y_0$ and

$$B = \frac{(2mv_0 - \gamma y_0)}{\sqrt{4mk - \gamma^2}}.$$

(b) We can write $y(t) = Re^{-\gamma t/2m}\cos(\nu t - \delta)$, in which

$$R = \sqrt{y_0^2 + \frac{(2mv_0 - \gamma y_0)^2}{4mk - \gamma^2}}$$

and

$$\delta = \arctan\left[\frac{(2mv_0 - \gamma y_0)}{y_0\sqrt{4mk - \gamma^2}}\right].$$

(c) We see that

$$R = \sqrt{y_0^2 + \frac{(2mv_0 - \gamma y_0)^2}{4mk - \gamma^2}} = 2\sqrt{\frac{m(ky_0^2 + \gamma y_0 v_0 + mv_0^2)}{4mk - \gamma^2}}.$$

We see that R increases without bound as $\gamma \to 2\sqrt{mk}$.

27. Referring to problem 19, the equation of motion is given by

$$\rho l u'' + \rho_0 g u = 0.$$

The general solution of this differential equation is

$$u(t) = A\cos\left(\sqrt{\rho_0 g/\rho l}\,t\right) + B\sin\left(\sqrt{\rho_0 g/\rho l}\,t\right).$$

Therefore, the frequency is $\omega_0 = \sqrt{\rho_0 g/\rho l}$ rad/sec., and the period is $T = 2\pi\sqrt{\rho l/\rho_0 g}$ seconds.

28.

(a) The general solution is $y(t) = A\cos(\sqrt{2}t) + B\sin(\sqrt{2}t)$. The initial conditions imply $A = 0$ and $B = \sqrt{2}$. Therefore, the specific solution is $y(t) = \sqrt{2}\sin(\sqrt{2}t)$.

(b)

(c)

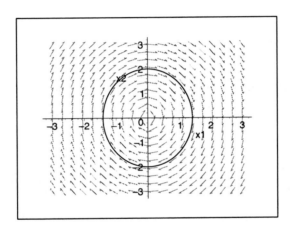

28.

(a) The roots of the characteristic equation are $\lambda = -1/8 \pm i\sqrt{127}/8$. Therefore, considering the initial conditions, the solution of this equation is

$$y = \frac{16}{\sqrt{127}} e^{-t/8} \sin(\sqrt{127}t/8).$$

(b)

(c)

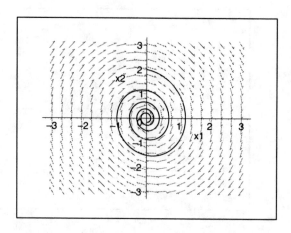

30.

(a) The kinetic energy is given by $K.E. = \frac{1}{2}mv^2$, where v is velocity. Initially, $v = b$, therefore, the kinetic energy initially is $\frac{1}{2}mb^2$. The potential energy is given by $P.E. = \frac{1}{2}ky^2$ where y is the distance the spring is stretched. Therefore, $P.E. = ka^2/2$.

(b) The general solution is

$$y(t) = A\cos(\sqrt{k/m}t) + B\sin(\sqrt{k/m}t).$$

The initial conditions imply that $A = a$ and $B = b\sqrt{m/k}$. Therefore, the specific solution is

$$y(t) = a\cos(\sqrt{k/m}t) + b\sqrt{m/k}\sin(\sqrt{k/m}t).$$

(c) By the equation for y, we see that

$$\frac{ky^2}{2} = \frac{k}{2}(A^2\cos^2(\sqrt{k/m}t) + 2AB\cos(\sqrt{k/m}t)\sin(\sqrt{k/m}t) + B^2\sin^2(\sqrt{k/m}t).$$

Further,

$$y'(t) = -A\sqrt{k/m}\sin(\sqrt{k/m}t) + B\sqrt{k/m}\cos(\sqrt{k/m}t)$$

implies

$$\frac{m(y')^2}{2} = \frac{m}{2}(A^2(k/m)\sin^2(\sqrt{k/m}t) - 2AB(k/m)\sin(\sqrt{k/m}t)\cos(\sqrt{k/m}t) + B^2(k/m)\cos^2(\sqrt{k/m}t)$$

Therefore, the total energy satisfies

$$\frac{ky^2}{2} + \frac{m(y')^2}{2} = \frac{k}{2}(A^2 + B^2) = \frac{k}{2}(a^2 + b^2(m/k)) = ka^2/2 + mb^2/2.$$

Therefore, energy is conserved.

31.

(a) For $m = k = 1$, $\gamma = 0$ and $\epsilon = 0$, the equation of motion is $x'' + x = 0$. The general solution of this equation is $x(t) = A \cos t + B \sin t$. Considering the initial conditions, we arrive at the specific solution $x(t) = \sin t$. We see that the amplitude is $R = 1$ and the period is $T = 2\pi$ seconds.

(b)

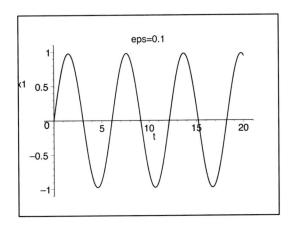

We estimate that the amplitude is $A = 0.98$ and the period is $T = 6.07$ seconds.

(c)

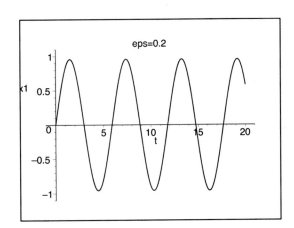

We estimate that the amplitude is $A = 0.96$ and the period is $T = 5.90$ seconds.

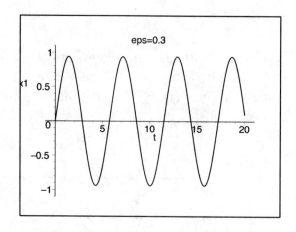

We estimate that the amplitude is $A = 0.94$ and the period is $T = 5.74$ seconds.

(d) The amplitude and the period both seem to decrease as ϵ increases.

(e)

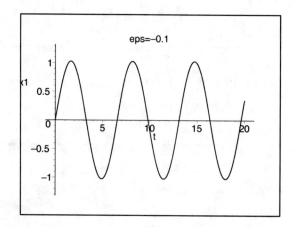

We estimate that the amplitude is $A = 1.03$ and the period is $T = 6.55$ seconds.

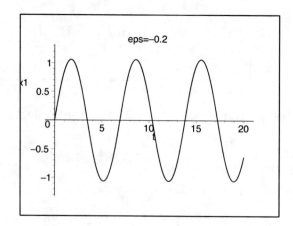

We estimate that the amplitude is $A = 1.06$ and the period is $T = 6.90$ seconds.

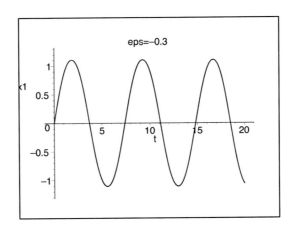

We estimate that the amplitude is $A = 1.11$ and the period is $T = 7.41$ seconds. The amplitude and period seem to increase as the value of ϵ decreases.

Section 4.6

1. The characteristic equation for the homogeneous problem is $\lambda^2 - 2\lambda - 3 = 0$, which has roots $\lambda = 3, -1$. Therefore, the solution of the homogeneous problem is $y_h(t) = c_1 e^{3t} + c_2 e^{-t}$. To find a solution of the inhomogeneous problem, we look for a solution of the form $y_p(t) = Ae^{2t}$. Substituting a function of this form into the differential equation, we have

$$4Ae^{2t} - 4Ae^{2t} - 3Ae^{2t} = 3e^{2t}.$$

Therefore, we need $-3A = 3$, or $A = -1$. Therefore, the solution of the inhomogeneous problem is

$$y(t) = c_1 e^{3t} + c_2 e^{-t} - e^{2t}.$$

2. The characteristic equation for the homogeneous problem is $\lambda^2 + 2\lambda + 5 = 0$, which has roots $\lambda = -1 \pm 2i$. Therefore, the solution of the homogeneous problem is $y_h(t) = e^{-t}(c_1 \cos(2t) + c_2 \sin(2t))$. To find a solution of the inhomogeneous problem, we look for a solution of the form $y_p(t) = A \cos(2t) + B \sin(2t)$. Substituting a function of this form into the differential equation, and equating like terms, we have $B - 4A = 3$ and $A + 4B = 0$. The solution of these equations is $A = -12/17$ and $B = 3/17$. Therefore, the solution of the inhomogeneous problem is

$$y(t) = e^{-t}(c_1 \cos(2t) + c_2 \sin(2t)) - \frac{12}{17} \cos(2t) + \frac{3}{17} \sin(2t).$$

3. The characteristic equation for the homogeneous problem is $\lambda^2 - 2\lambda - 3 = 0$, which has roots $\lambda = 3, -1$. Therefore, the solution of the homogeneous problem is $y_h(t) = c_1 e^{3t} + c_2 e^{-t}$. Since $y(t) = e^{-t}$ is a solution of the homogeneous problem, to find a solution of the inhomogeneous problem, we look for a solution of the form $y_p(t) = Ate^{-t} + Bt^2 e^{-t}$. Substituting a function of this form into the differential equation, and equating like terms, we have $-4A + 2B = 0$ and $-8B = -3$. The solution of these equations is $A = 3/16$ and $B = 3/8$. Therefore, the solution of the inhomogeneous problem is

$$y(t) = e^{-t}(c_1 e^{3t} + c_2 e^{-t}) + \frac{3}{16} te^{-t} + \frac{3}{8} t^2 e^{-t}.$$

71

4. The characteristic equation for the homogeneous problem is $\lambda^2 + 2\lambda = 0$, which has roots $\lambda = 0, -2$. Therefore, the solution of the homogeneous problem is $y_h(t) = c_1 + c_2 e^{-2t}$. To find a solution of the inhomogeneous problem, we look for a solution of the form $y_p(t) = At + B\cos(2t) + C\sin(2t)$. Substituting a function of this form into the differential equation, and equating like terms, we have $2A = 3$, $-4B + 4C = 0$ and $-4B - 4C = 4$. The solution of these equations is $A = 3/2$, $B = -1/2$ and $C = -1/2$. Therefore, the solution of the inhomogeneous problem is

$$y(t) = c_1 + c_2 e^{-2t} + \frac{3}{2}t - \frac{1}{2}\cos(2t) - \frac{1}{2}\sin(2t).$$

5. The characteristic equation for the homogeneous problem is $\lambda^2 + 9 = 0$, which has roots $\lambda = \pm 3i$. Therefore, the solution of the homogeneous problem is $y_h(t) = c_1 \cos(3t) + c_2 \sin(3t)$. To find a solution of the inhomogeneous problem, we look for a solution of the form $y_p(t) = Ae^{3t} + Bte^{3t} + Ct^2 e^{3t} + D$. Substituting a function of this form into the differential equation, and equating like terms, we have $18A + 6B + 2C = 0$, $18B + 12C = 0$, $18C = 1$ and $9D = 6$. The solution of these equations is $A = 1/162$, $B = -1/27$, $C = 1/18$ and $D = 2/3$. Therefore, the solution of the inhomogeneous problem is

$$y(t) = c_1 \cos(3t) + c_2 \sin(3t) + \frac{1}{162}e^{3t} - \frac{1}{27}te^{3t} + \frac{1}{18}t^2 e^{3t} + \frac{2}{3}.$$

6. The characteristic equation for the homogeneous problem is $\lambda^2 + 2\lambda + 1 = 0$, which has the repeated root $\lambda = -1$. Therefore, the solution of the homogeneous problem is $y_h(t) = c_1 e^{-t} + c_2 e^{-t}$. To find a solution of the inhomogeneous problem, we look for a solution of the form $y_p(t) = At^2 e^{-t}$. Substituting a function of this form into the differential equation, and equating like terms, we have $2A = 2$. Therefore, $A = 1$ and the solution of the inhomogeneous problem is

$$y(t) = c_1 e^{-t} + c_2 e^{-t} + t^2 e^{-t}.$$

7. The characteristic equation for the homogeneous problem is $2\lambda^2 + 3\lambda + 1 = 0$, which has roots $\lambda = -1, -1/2$. Therefore, the solution of the homogeneous problem is $y_h(t) = c_1 e^{-t} + c_2 e^{-t/2}$. To find a solution of the inhomogeneous problem, we will first look for a solution of the form $Y_1(t) = A + Bt + Ct^2$ to account for the inhomogeneous term t^2. Substituting a function of this form into the differential equation, and equating like terms, we have $A + 3B + 4C = 0$, $B + 6C = 0$ and $C = 1$. The solution of these equations is $A = 14$, $B = -6$, $C = 1$. Therefore, $Y_1 = 14 - 6t + t^2$. Next, we look for a solution of the inhomogeneous problem of the form $Y_2(t) = D\cos t + E\sin t$. Substituting this function into the ODE and equating like terms, we find that $D = -9/10$ and $E = -3/10$. Therefore, the solution of the inhomogeneous problem is

$$y(t) = c_1 e^{-t} + c_2 e^{-t/2} + 14 - 6t + t^2 - \frac{9}{10}\cos t - \frac{3}{10}\sin t.$$

8. The characteristic equation for the homogeneous problem is $\lambda^2 + 1 = 0$, which has roots $\lambda = \pm i$. Therefore, the solution of the homogeneous problem is $y_h(t) = c_1 \cos(t) + c_2 \sin(t)$.

To find a solution of the inhomogeneous problem, we look for a solution of the form $y_p(t) = A\cos(2t) + B\sin(2t) + Ct\cos(2t) + Dt\sin(2t)$. Substituting a function of this form into the differential equation, and equating like terms, we have $4D - 3A = 0$, $-3B - 4C = 3$, $-3C = 1$, and $-3D = 0$. The solution of these equations is $A = 0$, $B = -5/9$, $C = -1/3$ and $D = 0$. Therefore, the solution of the inhomogeneous problem is

$$y(t) = c_1\cos(t) + c_2\sin(t) - \frac{5}{9}\sin(2t) - \frac{1}{3}t\cos(2t).$$

9. The characteristic equation for the homogeneous problem is $\lambda^2 + \omega_0^2 = 0$, which has roots $\lambda = \pm\omega_0 i$. Therefore, the solution of the homogeneous problem is $y_h(t) = c_1\cos(\omega_0 t) + c_2\sin(\omega_0 t)$. To find a solution of the inhomogeneous problem, we look for a solution of the form $y_p(t) = A\cos(\omega t) + B\sin(\omega t)$. Substituting a function of this form into the differential equation, and equating like terms, we have $(\omega_0^2 - \omega^2)A = 1$ and $(\omega_0^2 - \omega^2)B = 0$. The solution of these equations is $A = 1/(\omega_0^2 - \omega^2)$ and $B = 0$. Therefore, the solution of the inhomogeneous problem is

$$y(t) = c_1\cos(\omega_0 t) + c_2\sin(\omega_0 t) + \frac{1}{\omega_0^2 - \omega^2}\cos(\omega t).$$

10. From problem 9, the solution of the homogeneous problem is $y_h(t) = c_1\cos(\omega_0 t) + c_2\sin(\omega_0 t)$. Since $\cos(\omega_0 t)$ is a solution of the homogeneous problem, we look for a solution of the inhomogeneous problem of the form $y_p(t) = At\cos(\omega_0 t) + Bt\sin(\omega_0 t)$. Substituting a function of this form into the differential equation, and equating like terms, we have $A = 0$ and $B = \dfrac{1}{2\omega_0}$. Therefore, the solution of the inhomogeneous problem is

$$y(t) = c_1\cos(\omega_0 t) + c_2\sin(\omega_0 t) + \frac{1}{2\omega_0}t\sin(\omega_0 t).$$

11. The characteristic equation for the homogeneous problem is $\lambda^2 + \lambda + 4 = 0$, which has roots $\lambda = (-1 \pm i\sqrt{15})/2$. Therefore, the solution of the homogeneous problem is $y_h(t) = e^{-t/2}(c_1\cos(\sqrt{15}t/2) + c_2\sin(\sqrt{15}t/2))$. To find a solution of the inhomogeneous problem, we look for a solution of the form $y_p(t) = Ae^t + Be^{-t}$. Substituting a function of this form into the differential equation, and equating like terms, we have $6A = 1$ and $4B = -1$. The solution of these equations is $A = 1/6$ and $B = -1/4$. Therefore, the solution of the inhomogeneous problem is

$$y(t) = e^{-t/2}(c_1\cos(\sqrt{15}t/2) + c_2\sin(\sqrt{15}t/2)) + \frac{1}{6}e^t - \frac{1}{4}e^{-t}.$$

12. The characteristic equation for the homogeneous problem is $\lambda^2 - \lambda - 2 = 0$, which has roots $\lambda = -1, 2$. Therefore, the solution of the homogeneous problem is $y_h(t) = c_1 e^{-t} + c_2 e^{2t}$. To find a solution of the inhomogeneous problem, we look for a solution of the form $y_p(t) = Ate^{2t} + Be^{-2t}$. Substituting a function of this form into the differential equation, and equating like terms, we have $A = 1/6$ and $B = 1/8$. Therefore, the solution of the inhomogeneous problem is

$$y(t) = c_1 e^{-t} + c_2 e^{2t} + \frac{1}{6}te^{2t} + \frac{1}{8}e^{-2t}.$$

13. The characteristic equation for the homogeneous problem is $\lambda^2 + \lambda - 2 = 0$, which has roots $\lambda = 1, -2$. Therefore, the solution of the homogeneous problem is $y_h(t) = c_1 e^t + c_2 e^{-2t}$. To find a solution of the inhomogeneous problem, we look for a solution of the form $y_p(t) = At + B$. Substituting a function of this form into the differential equation, and equating like terms, we have $-2A = 2$ and $A - 2B = 0$. The solution of these equations is $A = -1$ and $B = -1/2$. Therefore, the solution of the inhomogeneous problem is

$$y(t) = c_1 e^t + c_2 e^{-2t} - t - \frac{1}{2}.$$

The initial conditions imply

$$c_1 + c_2 - \frac{1}{2} = 0$$
$$c_1 - 2c_2 - 1 = 1.$$

Therefore, $c_1 = 1$ and $c_2 = -1/2$ which implies the particular solution of the IVP is

$$y(t) = e^t - \frac{1}{2}e^{-2t} - t - \frac{1}{2}.$$

14. The characteristic equation for the homogeneous problem is $\lambda^2 + 4 = 0$, which has roots $\lambda = \pm 2i$. Therefore, the solution of the homogeneous problem is $y_h(t) = c_1 \cos(2t) + c_2 \sin(2t)$. First, to find a solution of the inhomogeneous problem, we look for a solution of the form $Y_1(t) = A + Bt + Ct^2$ to correspond with the term t^2. Substituting a function of this form into the differential equation, and equating like terms, we have $A = -1/8$, $B = 0$ and $C = 1/4$. Next, considering the inhomogeneous term $3e^t$, we look for a solution of the form $Y_2(t) = De^t$. Substituting this function into the equation and equating like terms, we have $D = 3/5$. Therefore, the solution of the inhomogeneous problem is

$$y(t) = c_1 \cos(2t) + c_2 \sin(2t) - \frac{1}{8} + \frac{1}{4}t^2 + \frac{3}{5}e^t.$$

The initial conditions imply

$$\frac{19}{40} + c_1 = 0$$
$$\frac{3}{5} + 2c_2 = 2.$$

Therefore, $c_1 = -19/40$ and $c_2 = 7/10$ which implies the particular solution of the IVP is

$$y(t) = -\frac{19}{40} \cos(2t) + \frac{7}{10} \sin(2t) - \frac{1}{8} + \frac{1}{4}t^2 + \frac{3}{5}e^t.$$

15. The characteristic equation for the homogeneous problem is $\lambda^2 - 2\lambda + 1 = 0$, which has the repeated root $\lambda = 1$. Therefore, the solution of the homogeneous problem is $y_h(t) = c_1 e^t + c_2 t e^t$. First, to find a solution of the inhomogeneous problem, we look for a solution of the form $Y_1(t) = At^2 e^t + Bt^3 e^t$ to correspond with the term $t e^t$. Substituting a function

74

of this form into the differential equation, and equating like terms, we have $Y_1(t) = t^3 e^t / 6$. Next, considering the inhomogeneous term 4, we look for a solution of the form $Y_2(t) = C$. Substituting this function into the equation and equating like terms, we have $Y_2 = 4$. Therefore, the solution of the inhomogeneous problem is

$$y(t) = c_1 e^t + c_2 t e^t + \frac{1}{6} t^3 e^t + 4.$$

The initial conditions imply

$$c_1 + 4 = 0$$
$$c_1 + c_2 = 1.$$

Therefore, $c_1 = -3$ and $c_2 = 4$ which implies the particular solution of the IVP is

$$y(t) = -3 e^t + 4 t e^t + \frac{1}{6} t^3 e^t + 4.$$

16. The characteristic equation for the homogeneous problem is $\lambda^2 - 2\lambda - 3 = 0$, which has roots $\lambda = 3, -1$. Therefore, the solution of the homogeneous problem is $y_h(t) = c_1 e^{3t} + c_2 e^{-t}$. To find a solution of the inhomogeneous problem, we look for a solution of the form $y_p(t) = Ae^{2t} + Bte^{2t}$. Substituting a function of this form into the differential equation, and equating like terms, we have $y_p(t) = 2e^{2t} + 3te^{2t}$. Therefore, the solution of the inhomogeneous problem is

$$y(t) = c_1 e^{3t} + c_2 e^{-t} + 2e^{2t} + 3te^{2t}.$$

The initial conditions imply

$$c_1 + c_2 + 2 = 1$$
$$3c_1 - c_2 + 4 + 3 = 0.$$

Therefore, $c_1 = -2$ and $c_2 = 1$ which implies the particular solution of the IVP is

$$y(t) = -2e^{3t} + e^{-t} + 2e^{2t} + 3te^{2t}.$$

17. The characteristic equation for the homogeneous problem is $\lambda^2 + 4 = 0$, which has roots $\lambda = \pm 2i$. Therefore, the solution of the homogeneous problem is $y_h(t) = c_1 \cos(2t) + c_2 \sin(2t)$. To find a solution of the inhomogeneous problem, we look for a solution of the form $y_p(t) = At \cos(2t) + Bt \sin(2t)$. Substituting a function of this form into the differential equation, and equating like terms, we have $y_p(t) = -\frac{3}{4} t \cos(2t)$. Therefore, the solution of the inhomogeneous problem is

$$y(t) = c_1 \cos(2t) + c_2 \sin(2t) - \frac{3}{4} t \cos(2t).$$

The initial conditions imply

$$c_1 = 2$$
$$2c_2 - \frac{3}{4} = -1.$$

Therefore, $c_1 = 2$ and $c_2 = -1/8$ which implies the particular solution of the IVP is

$$y(t) = 2\cos(2t) - \frac{1}{8}\sin(2t) - \frac{3}{4}t\cos(2t).$$

18. The characteristic equation for the homogeneous problem is $\lambda^2 + 2\lambda + 5 = 0$, which has roots $\lambda = -1 \pm 2i$. Therefore, the solution of the homogeneous problem is $y_h(t) = e^{-t}(c_1\cos(2t) + c_2\sin(2t))$. To find a solution of the inhomogeneous problem, we look for a solution of the form $y_p(t) = Ate^{-t}\cos(2t) + Bte^{-t}\sin(2t)$. Substituting a function of this form into the differential equation, and equating like terms, we have $y_p(t) = te^{-t}\sin(2t)$. Therefore, the solution of the inhomogeneous problem is

$$y(t) = e^{-t}(c_1\cos(2t) + c_2\sin(2t)) + te^{-t}\sin(2t).$$

The initial conditions imply

$$c_1 = 1$$
$$-c_1 + 2c_2 = 0.$$

Therefore, $c_1 = 1$ and $c_2 = 1/2$ which implies the particular solution of the IVP is

$$y(t) = e^{-t}\left(\cos(2t) + \frac{1}{2}\sin(2t)\right) + te^{-t}\sin(2t).$$

19.

(a) The characteristic equation for the homogeneous problem is $\lambda^2 + 3\lambda = 0$, which has roots $\lambda = 0, -3$. Therefore, the solution of the homogeneous problem is $y_h(t) = c_1 + c_2 e^{-3t}$. Therefore, to find a solution of the inhomogeneous problem, we look for a solution of the form $Y(t) = t(A_0 t^4 + A_1 t^3 + A_2 t^2 + A_3 t + A_4) + t(B_0 t^2 + B_1 t + B_2)e^{-3t} + D\sin 3t + E\cos 3t$.

(b) Substituting a function of this form into the differential equation, and equating like terms, we have $A_0 = 2/15$, $A_1 = -2/9$, $A_2 = 8/27$, $A_3 = -8/27$, $A_4 = 16/81$, $B_0 = -1/9$, $B_1 = -1/9$, $B_2 = -2/27$, $D = -1/18$, $E = -1/18$ Therefore, the solution of the inhomogeneous problem is

$$y(t) = c_1 + c_2 e^{-3t} + t((2/15)t^4 - (2/9)t^3 + (8/27)t^2 - (8/27)t + (16/81)) + t((-1/9)t^2 - (1/9)t - 2/27)e^{-}$$

20.

(a) The characteristic equation for the homogeneous problem is $\lambda^2 + 1 = 0$, which has roots $\lambda = \pm i$. Therefore, the solution of the homogeneous problem is $y_h(t) = c_1\cos(t) + c_2\sin(t)$. Therefore, to find a solution of the inhomogeneous problem, we look for a solution of the form $Y(t) = A_0 t + A_1 + t(B_0 t + B_1)\sin t + t(D_0 t + D_1)\cos t$.

(b) Substituting a function of this form into the differential equation, and equating like terms, we have $A_0 = 1$, $A_1 = 0$, $B_0 = 0$, $B_1 = 1/4$, $D_0 = -1/4$, $D_1 = 0$. Therefore, the solution of the inhomogeneous problem is

$$y(t) = c_1 \cos(t) + c_2 \sin(t) + t + \frac{1}{4}t \sin t - \frac{1}{4}t^2 \cos t$$

21.

(a) The characteristic equation for the homogeneous problem is $\lambda^2 - 5\lambda + 6 = 0$, which has roots $\lambda = 2, 3$. Therefore, the solution of the homogeneous problem is $y_h(t) = c_1 e^{2t} + c_2 e^{3t}$. Therefore, to find a solution of the inhomogeneous problem, we look for a solution of the form $Y(t) = e^t(A \cos 2t + B \sin 2t) + (D_0 t + D_1)e^{2t} \sin t + (E_0 t + E_1)e^{2t} \cos t$.

(b) Substituting a function of this form into the differential equation, and equating like terms, we have $A = -1/20$, $B = -3/20$, $D_0 = -3/2$, $D_1 = -5$, $E_0 = 3/2$, $E_1 = 1/2$. Therefore, the solution of the inhomogeneous problem is

$$y(t) = c_1 e^{2t} + c_2 e^{3t} + e^t((-1/20) \cos 2t - (3/20) \sin 2t) + ((-3/2)t - 5)e^{2t} \sin t + ((3/2)t + (1/2))e^{2t} \cos t.$$

22.

(a) The characteristic equation for the homogeneous problem is $\lambda^2 + 2\lambda + 2 = 0$, which has roots $\lambda = -1 \pm i$. Therefore, the solution of the homogeneous problem is $y_h(t) = e^{-t}(c_1 \cos(t) + c_2 \sin(t))$. Therefore, to find a solution of the inhomogeneous problem, we look for a solution of the form $Y(t) = Ae^{-t} + t(B_0 t^2 + B_1 t + B_2)e^{-t} \cos t + t(D_0 t^2 + D_1 t + D_2)e^{-t} \sin t$.

(b) Substituting a function of this form into the differential equation, and equating like terms, we have $A = 3$, $B_0 = -2/3$, $B_1 = 0$, $B_2 = 1$, $D_0 = 0$, $D_1 = 1$, $D_2 = 1$. Therefore, the solution of the inhomogeneous problem is

$$y(t) = e^{-t}(c_1 \cos(t) + c_2 \sin(t)) + 3e^{-t} + t((-2/3)t^2 + 1)e^{-t} \cos t + t(t + 1)e^{-t} \sin t.$$

23.

(a) The characteristic equation for the homogeneous problem is $\lambda^2 - 4\lambda + 4 = 0$, which has the repeated root $\lambda = 2$. Therefore, the solution of the homogeneous problem is $y_h(t) = c_1 e^{2t} + c_2 t e^{2t}$. Therefore, to find a solution of the inhomogeneous problem, we look for a solution of the form $Y(t) = A_0 t^2 + A_1 t + A_2 + t^2(B_0 t + B_1)e^{2t} + (D_0 t + D_1) \sin 2t + (E_0 t + E_1) \cos 2t$.

(b) Substituting a function of this form into the differential equation, and equating like terms, we have $A_0 = 1/2$, $A_1 = 1$, $A_2 = 3/4$, $B_0 = 2/3$, $B_1 = 0$, $D_0 = 0$, $D_1 = -1/16$, $E_0 = 1/8$, $E_1 = 1/16$. Therefore, the solution of the inhomogeneous problem is

$$y(t) = c_1 e^{2t} + c_2 t e^{2t} + (1/2)t^2 + t + 3/4 + (2/3)t^3 e^{2t} - (1/16) \sin 2t + ((1/8)t + (1/16)) \cos 2t$$

24.

(a) The characteristic equation for the homogeneous problem is $\lambda^2 + 4 = 0$, which has roots $\lambda = \pm 2i$. Therefore, the solution of the homogeneous problem is $y_h(t) = c_1 \cos(2t) + c_2 \sin(2t)$. Therefore, to find a solution of the inhomogeneous problem, we look for a solution of the form $Y(t) = t(A_0 t^2 + A_1 t + A_2)\sin 2t + t(B_0 t^2 + B_1 t + B_2)\cos 2t$.

(b) Substituting a function of this form into the differential equation, and equating like terms, we have $A_0 = 0$, $A_1 = 13/16$, $A_2 = 7/4$, $B_0 = -1/12$, $B_1 = 0$, $B_2 = 13/32$. Therefore, the solution of the inhomogeneous problem is

$$y(t) = c_1 \cos(2t) + c_2 \sin(2t) + t((13/16)t + (7/4))\sin 2t + t((-1/12)t^2 + (13/32))\cos 2t$$

25.

(a) The characteristic equation for the homogeneous problem is $\lambda^2 + 3\lambda + 2 = 0$, which has roots $\lambda = -1, -2$. Therefore, the solution of the homogeneous problem is $y_h(t) = c_1 e^{-t} + c_2 e^{-2t}$. Therefore, to find a solution of the inhomogeneous problem, we look for a solution of the form $Y(t) = (A_0 t^2 + A_1 t + A_2)e^t \sin 2t + (B_0 t^2 + B_1 t + B_2)e^t \cos 2t + e^{-t}(D \cos t + E \sin t) + F e^t$.

(b) Substituting a function of this form into the differential equation, and equating like terms, we have $A_0 = 1/52$, $A_1 = 10/169$, $A_2 = -1233/35152$, $B_0 = -5/52$, $B_1 = 73/676$, $B_2 = -4105/35152$, $D = -3/2$, $E = 3/2$, $F = 2/3$ Therefore, the solution of the inhomogeneous problem is

$$y(t) = c_1 e^{-t} + c_2 e^{-2t} + (A_0 t^2 + A_1 t + A_2)e^t \sin 2t + (B_0 t^2 + B_1 t + B_2)e^t \cos 2t + e^{-t}(D \cos t + E \sin t) + F e^t$$

with coefficients given above.

26.

(a) The characteristic equation for the homogeneous problem is $\lambda^2 + 2\lambda + 5 = 0$, which has roots $\lambda = -1 \pm 2i$. Therefore, the solution of the homogeneous problem is $y_h(t) = e^{-t}(c_1 \cos(2t) + c_2 \sin(2t))$. Therefore, to find a solution of the inhomogeneous problem, we look for a solution of the form $Y(t) = t(A_0 t + A_1)e^{-t}\cos 2t + t(B_0 t + B_1)\sin 2t + (D_0 t + D_1)e^{-2t}\cos t + (E_0 t + E_1)e^{-2t}\sin t$.

(b) Substituting a function of this form into the differential equation, and equating like terms, we have $A_0 = 0$, $A_1 = 3/16$, $B_0 = 3/8$, $B_1 = 0$, $D_0 = -2/5$, $D_1 = -7/25$, $E_0 = 1/5$, $E_1 = 1/25$. Therefore, the solution of the inhomogeneous problem is

$$y(t) = e^{-t}(c_1 \cos(2t) + c_2 \sin(2t)) + t(A_0 t + A_1)e^{-t}\cos 2t + t(B_0 t + B_1)\sin 2t + (D_0 t + D_1)e^{-2t}\cos t + (E_0 t +$$

with coefficients given above.

27.

(a) For $Y = ve^{-t}$, we have $Y' = v'e^{-t} - ve^{-t}$ and $Y'' = v''e^{-t} - 2v'e^{-t} + ve^{-t}$. Then,

$$Y'' - 3Y' - 4Y = 2e^{-t}$$
$$\implies v''e^{-t} - 2v'e^{-t} + ve^{-t} - 3v'e^{-t} + 3ve^{-t} - 4ve^{-t} = 2e^{-t}.$$

Simplifying this equation, we have

$$v'' - 5v' = 2.$$

(b) We see that $(v')' - 5(v') = 2$. Therefore, letting $w = v'$, we see that w must satisfy $w' - 5w = 2$. This equation is linear with integrating factor $\mu(t) = e^{-5t}$. Therefore, we have $[e^{-5t}w]' = 2e^{-5t}$ which implies that $w = -2/5 + ce^{5t}$.

(c) Integrating w, we see that $v = (-2/5)t + \frac{c_1}{5}e^{5t} + c_2$. Then using the fact that $Y = ve^{-t}$, we conclude that

$$Y(t) = -\frac{2}{5}te^{-t} + \frac{c_1}{t}e^{4t} + c_2 e^{-t}.$$

28. Letting $z = \ln x$, our equation can be rewritten as

$$y'' - 4y' + 4y = z$$

where $' = d/dz$. First, we will solve the homogeneous equation. The associated characteristic equation is $\lambda^2 - 4\lambda + 4 = 0$ which has root $\lambda = 2$. Therefore, the solution of the homogeneous equation is

$$y(z) = c_1 e^{2z} + c_2 z e^{2z}.$$

Now we solve the inhomogeneous equation by the method of undetermined coefficients. We look for a solution of the form $y = Az + B$. If $y = Az + B$, then $y' = A$ and $y'' = 0$. Plugging this into the inhomogeneous equation, we have

$$0 - 4A + 4(Az + B) = z \implies 4Az + 4B - 4A = z \implies 4A = 1, 4B - 4A = 0.$$

Therefore, $A = 1/4$ and $B = 1/4$. Therefore,

$$y(z) = c_1 e^{2z} + c_2 z e^{2z} + \frac{1}{4}z + \frac{1}{4}.$$

Rewriting the equation in terms of x, we have

$$y(x) = c_1 x^2 + c_2 x^2 \ln x + \frac{1}{4}\ln x + \frac{1}{4}.$$

29. Letting $z = \ln x$, our equation can be rewritten as

$$y'' + 6y' + 5y = e^z$$

where $' = d/dz$. First, we will solve the homogeneous equation. The associated characteristic equation is $\lambda^2 + 6\lambda + 5 = 0$ which has roots $\lambda = -5, -1$. Therefore, the solution of the homogeneous equation is

$$y(z) = c_1 e^{-5z} + c_2 e^{-z}.$$

Now we solve the inhomogeneous equation by the method of undetermined coefficients. We look for a solution of the form $y = Ae^z$. If $y = Ae^z$, then $y' = Ae^z$ and $y'' = Ae^z$. Plugging this into the inhomogeneous equation, we have

$$Ae^z + 6Ae^z + 5Ae^z = e^z \implies 12A = 1 \implies A = 1/12.$$

Therefore,

$$y(z) = c_1 e^{-5z} + c_2 e^{-z} + \frac{1}{12} e^z.$$

Rewriting the equation in terms of x, we have

$$y(x) = c_1 x^{-5} + c_2 x^{-1} + \frac{1}{12} x.$$

30. Letting $z = \ln x$, our equation can be rewritten as

$$y'' - 3y' + 2y = 3e^{2z} + 2z$$

where $' = d/dz$. First, we will solve the homogeneous equation. The associated characteristic equation is $\lambda^2 - 3\lambda + 2 = 0$ which has roots $\lambda = 1, 2$. Therefore, the solution of the homogeneous equation is

$$y(z) = c_1 e^z + c_2 e^{2z}.$$

Now we solve the inhomogeneous equation by the method of undetermined coefficients. Since e^{2z} is a solution of the homogeneous equation, we look for a solution of the form $y = Aze^{2z} + Bz + C$. If $y = Aze^{2z} + Bz + C$, then $y' = Ae^{2z} + 2Aze^{2z} + B$ and $y'' = 4Ae^{2z} + 4Aze^{2z}$. Plugging this into the inhomogeneous equation, we have

$$4Ae^{2z} + 4Aze^{2z} - 3[Ae^{2z} + 2Aze^{2z} + B] + 2[Aze^{2z} + Bz + C] = 3e^{2z} + 2z.$$

Equating like coefficients, we have

$$A = 3, 2B = 2, -3B + 2C = 0.$$

Therefore, $A = 3$, $B = 1$ and $C = 3/2$. Therefore,

$$y(z) = c_1 e^z + c_2 e^{2z} + \frac{3}{z} e^{2z} + z + \frac{3}{2}.$$

Rewriting the equation in terms of x, we have

$$y(x) = c_1 x + c_2 x^2 + 3x^2 \ln x + \ln x + \frac{3}{2}.$$

31. Letting $z = \ln x$, our equation can be rewritten as

$$y'' + 4y = \sin z$$

where $' = d/dz$. First, we will solve the homogeneous equation. The associated characteristic equation is $\lambda^2 + 4 = 0$ which has roots $\lambda = \pm 2i$. Therefore, the solution of the homogeneous equation is

$$y(z) = c_1 \cos(2z) + c_2 \sin(2z).$$

Now we solve the inhomogeneous equation by the method of undetermined coefficients. We look for a solution of the form $y = A \sin z + B \cos z$. If $y = A \sin z + B \cos z$, then $y = A \cos z - B \sin z$ and $y'' = -A \sin z - B \cos z$. Plugging this into the inhomogeneous equation, we have

$$y'' + 4y = -A \sin z - B \cos z + 4(A \sin z + B \cos z) = 3A \sin z + 3B \cos z = \sin z.$$

Equating like coefficients, we have

$$3A = 1, 3B = 0.$$

Therefore, $A = 1/3$ and $B = 0$. Therefore,

$$y(z) = c_1 \cos(2z) + c_2 \sin(2z) + \frac{1}{3} \sin z.$$

Rewriting the equation in terms of x, we have

$$y(x) = c_1 \cos(2 \ln x) + c_2 \sin(2 \ln x) + \frac{1}{3} \sin(\ln x).$$

32. The solution of the homogeneous problem is $y_h(t) = c_1 \cos(\lambda t) + c_2 \sin(\lambda t)$. Since the differential operator does not contain a first derivative term, to solve the inhomogeneous problem, we look for a solution of the form $Y(t) = \sum_{m=1}^{N} C_m \sin(m\pi t)$. Substitution of Y into the ODE leads to the equation

$$-\sum_{m=1}^{N} m^2\pi^2 C_m \sin(m\pi t) + \lambda^2 \sum_{m=1}^{N} C_m \sin(m\pi t) = \sum_{m=1}^{N} a_m \sin(m\pi t).$$

Equating coefficients of the individual terms, we see that

$$C_m = \frac{a_m}{\lambda^2 - m^2\pi^2}.$$

Therefore, the solution of the inhomogeneous problem is

$$y(t) = c_1 \cos(\lambda t) + c_2 \sin(\lambda t) + \sum_{m=1}^{N} C_m \sin(m\pi t)$$

where the coefficients C_m are given above.

33. The solution of the homogeneous problem is $y_h(t) = c_1 \cos t + c_2 \sin t$. To solve the inhomogeneous IVP (starting at $t = 0$), we begin by looking for a solution of the inhomogeneous equation of the form $Y = A + Bt$. Substituting a function of this form into the ODE leads to the equation $A + Bt = t$. Therefore, $A = 0$ and $B = 1$, and the solution of the inhomogeneous problem is $y(t) = c_1 \cos t + c_2 \sin t + t$. Now, we consider the initial conditions. The initial conditions imply

$$c_1 = 0$$
$$c_2 + 1 = 1.$$

Therefore, the solution of the inhomogeneous problem for $0 \leq t \leq \pi$ is given by $y_1(t) = t$. Now we need to solve the inhomogeneous problem starting at $t = \pi$. We look for a solution of the inhomogeneous problem of the form $Y(t) = Ce^{-t}$. Substituting a function of this form into the ODE leads to the equation $Ae^{-t} + Ae^{-t} = \pi e^{\pi - t}$. Therefore, $2A = \pi e^{\pi}$, or $A = \pi e^{\pi}/2$. Therefore, a solution of the inhomogeneous problem (starting at $t = \pi$) is given by $y(t) = d_1 \cos t + d_2 \sin t + \dfrac{\pi}{2} e^{\pi - t}$. Using the solution of the IVP starting at $t = 0$, $y_1(t) = t$, we see that at time $t = \pi$, $y_1(\pi) = \pi$ and $y_1'(\pi) = 1$. Using these as our new initial conditions for $t = \pi$, we see that d_1, d_2 must satisfy

$$-d_1 + \frac{\pi}{2} = \pi$$

$$-d_2 - \frac{\pi}{2} = 1.$$

The solution of these equations is $d_1 = -\pi/2$, $d_2 = -1 - \pi/2$. Therefore, we conclude that the solution of the inhomogeneous IVP is

$$y(t) = \begin{cases} t & 0 \leq t \leq \pi \\ -\frac{\pi}{2}\cos t - \left(1 + \frac{\pi}{2}\right)\sin t + \frac{\pi}{2}e^{\pi - t} & t > \pi. \end{cases}$$

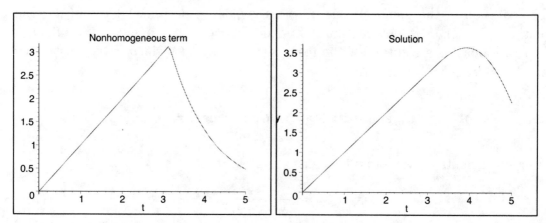

34.   The solution of the homogeneous problem is $y_h(t) = e^{-t}(c_1 \cos(2t) + c_2 \sin(2t))$. To solve the inhomogeneous IVP (starting at $t = 0$), we begin by looking for a solution of the inhomogeneous equation of the form $Y = A$. Substituting a function of this form into the ODE leads to the equation $5A = 1$. Therefore, $A = 1/5$, and the solution of the inhomogeneous problem is $y(t) = e^{-t}(c_1 \cos(2t) + c_2 \sin(2t)) + 1/5$. Now, we consider the initial conditions. The initial conditions imply

$$c_1 + \frac{1}{5} = 0$$

$$-c_1 + 2c_2 = 0.$$

Therefore, $c_1 = -1/5$ and $c_2 = -1/10$. Therefore, the solution of the inhomogeneous problem for $0 \leq t \leq \pi/2$ is given by

$$y_1(t) = e^{-t}\left(-\frac{1}{5}\cos(2t) - \frac{1}{10}\sin(2t)\right) + \frac{1}{5}.$$

Now for $t > \pi/2$, the inhomogeneous term is zero. Therefore, the solution of the problem starting at $t = \pi/2$ will be of the form $y_2(t) = e^{-t}(d_1 cos(2t) + d_2 \sin(2t))$. Using the function y_1, we know that at time $t = \pi/2$, the solution $y_1(\pi/2) = (1 + e^{-\pi/2})/5$ and $y'(\pi/2) = 0$. Using these values as our new initial conditions for $t = \pi/2$, we see that d_1, d_2 must satisfy

$$-e^{-\pi/2}d_1 = (1 + e^{-\pi/2})/5$$
$$e^{-\pi/2}d_1 - 2e^{-\pi/2}d_2 = 0.$$

The solution of these equations is $d_1 = -(e^{\pi/2} + 1)/5$, $d_2 = -(e^{\pi/2} + 1)/10$. Therefore, we conclude that the solution of the inhomogeneous IVP is

$$y(t) = \begin{cases} e^{-t}\left(-\frac{1}{5}\cos(2t) - \frac{1}{10}\sin(2t)\right) + \frac{1}{5} & 0 \le t \le \pi/2 \\ e^{-t}\left(-\frac{e^{\pi/2}+1}{5}\cos(2t) - \frac{e^{\pi/2}+1}{10}\sin(2t)\right) & t > \pi/2. \end{cases}$$

Section 4.7

1. Consider the trig identities $\cos(\alpha \pm \beta) = \cos\alpha\cos\beta \mp \sin\alpha\sin\beta$. Subtracting these two identities we obtain $\cos(\alpha - \beta) - \cos(\alpha + \beta) = 2\sin\alpha\sin\beta$. Here, our expression is $\cos(9t) - \cos(7t)$. Therefore, we let $\alpha - \beta = 9t$ and $\alpha + \beta = 7t$. Solving this system of equations, we have $\alpha = 8t$ and $\beta = -t$. Therefore, we can write $\cos(9t) - \cos(7t) = 2\sin(8t)\sin(-t) = -2\sin(8t)\sin(t)$.

2. Consider the trig identities $\sin(\alpha \pm \beta) = \sin\alpha\cos\beta \pm \cos\alpha\sin\beta$. Subtracting these two identities we obtain $\sin(\alpha + \beta) - \sin(\alpha - \beta) = 2\cos\alpha\sin\beta$. Here, our expression is $\sin(7t) - \sin(6t)$. Therefore, we let $\alpha + \beta = 7t$ and $\alpha - \beta = 6t$. Solving this system of equations, we have $\alpha = 13t/2$ and $\beta = t/2$. Therefore, we can write $\sin(7t) - \sin(6t) = 2\sin(t/2)\cos(13t/2)$.

3. Consider the trig identities $\cos(\alpha \pm \beta) = \cos\alpha\cos\beta \mp \sin\alpha\sin\beta$. Adding these two identities we obtain $\cos(\alpha - \beta) + \cos(\alpha + \beta) = 2\cos\alpha\cos\beta$. Here, our expression is $\cos(\pi t) - \cos(2\pi t)$. Therefore, we let $\alpha - \beta = \pi t$ and $\alpha + \beta = 2\pi t$. Solving this system of equations, we have $\alpha = 3\pi t/2$ and $\beta = \pi t/2$. Therefore, we can write $\cos(\pi t) + \cos(2\pi t) = 2\cos(3\pi t/2)\cos(\pi t/2)$.

4. Consider the trig identities $\sin(\alpha \pm \beta) = \sin\alpha\cos\beta \pm \cos\alpha\sin\beta$. Adding these two identities we obtain $\sin(\alpha - \beta) + \sin(\alpha + \beta) = 2\sin\alpha\cos\beta$. Here, our expression is $\sin(3t) + \sin(4t)$. Therefore, we let $\alpha + \beta = 4t$ and $\alpha - \beta = 3t$. Solving this system of equations, we have $\alpha = 7t/2$ and $\beta = t/2$. Therefore, we can write $\sin(3t) + \sin(4t) = 2\sin(7t/2)\cos(t/2)$.

5. The spring constant is $k = 4/(1.5/12) = 32$ lb/ft. The mass is $m = 4/32 = 1/8$ lb-s^2/ft. Assuming no damping, but an external force, $F(t) = 2\cos(3t)$, the equation describing the motion is

$$\frac{1}{8}y'' + 32y = 2\cos(3t)$$

which can be rewritten as

$$y'' + 256y = 16\cos(3t).$$

The initial conditions are $y(0) = 2/12 = 1/6$ ft. and $y'(0) = 0$.

6. The spring constant is $k = 5(9.8)/.1 = 490$ N/m. The mass is $m = 5$ kg. The damping force is $\gamma = 2/.04 = 50$ N-sec/m. The external force is $F(t) = 10\sin(t/2)$. Therefore, the equation describing the motion is

$$5y'' + 50y' + 490y = 10\sin(t/2)$$

which can be rewritten as

$$y'' + 10y' + 98y = 2\sin(t/2).$$

The initial conditions are $y(0) = 0$ m and $y'(0) = 0.03$ m/s.

7.

(a) The solution of the homogeneous problem is $y_h(t) = c_1\cos(16t) + c_2\sin(16t)$. To find a solution of the inhomogeneous problem, we look for a solution of the form $Y(t) = A\cos(3t)$ (since there is no first derivative term, we may exclude the $\sin(3t)$ function). Looking for a solution of this form, we arrive at the equation

$$-9A\cos(3t) + 256A\cos(3t) = 16\cos(3t).$$

Therefore, we need A to satisfy $247A = 16$ or $A = 16/247$. Therefore, the solution of the inhomogeneous problem is

$$y(t) = c_1\cos(16t) + c_2\sin(16t) + \frac{16}{247}\cos(3t).$$

The initial conditions are $y(0) = 1/6$ and $y'(0) = 0$. Therefore, c_1, c_2 must satisfy

$$c_1 + \frac{16}{247} = \frac{1}{6}$$
$$16c_2 = 0.$$

Therefore, we have $c_1 = 151/1482$ and $c_2 = 0$. Therefore, the solution is

$$y(t) = \frac{151}{1482}\cos(16t) + \frac{16}{247}\cos(3t).$$

(b)

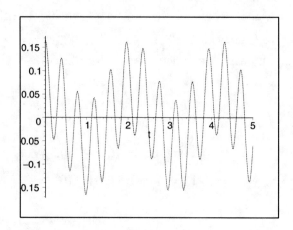

84

(c) If we replace the given external force with a force $Ae^{i\omega t}$, the inhomogeneous problem to solve becomes
$$y'' + 256y = Ae^{i\omega t}.$$

The frequency for the homogeneous problem is $\omega_0 = 16$. Therefore, using equation (5), we know that a particular solution of this inhomogeneous problem will be given by $Y(t) = G(i\omega)Ae^{i\omega t}$ where $G(i\omega) = \dfrac{1}{256 - \omega^2}$ is the frequency response. The gain $|G(i\omega)| = |1/(256 - \omega^2)|$ and the angle

$$\phi(\omega) = \arccos\left\{ \frac{256 - \omega^2}{\sqrt{(256 - \omega^2)^2}} \right\}$$
$$= \arccos\left\{ \frac{256 - \omega^2}{|256 - \omega^2|} \right\}$$

implies

$$\phi(\omega) = \begin{cases} \arccos(1) = 0 & 16 > \omega \\ \arccos(0) = \pi/2 & 16 = \omega \\ \arccos(-1) = \pi & 16 < \omega. \end{cases}$$

Resonance will occur at $\omega_{\max} = \omega_0 = 16$.

8.

(a) The solution of the homogeneous problem is $y_h(t) = e^{-5t}(c_1 \cos(\sqrt{73}t) + c_2 \sin(\sqrt{73}t))$. To find a solution of the inhomogeneous problem, we look for a solution of the form $Y(t) = A\cos(t/2) + B\sin(t/2)$. Substituting a function of this form into the ODE and equating like coefficients, we find a particular solution

$$y_p(t) = \frac{1}{153281}[-160\cos(t/2) + 3128\sin(t/2)].$$

Therefore, the solution of the inhomogeneous problem is

$$y(t) = e^{-5t}(c_1 \cos(\sqrt{73}t) + c_2 \sin(\sqrt{73}t)) + \frac{1}{153281}[-160\cos(t/2) + 3128\sin(t/2)].$$

The initial conditions are $y(0) = 0$ and $y'(0) = 0.03$. Using these initial conditions, we conclude that $c_1 = 160/153281$ and $c_2 = 383443\sqrt{73}/1118951300$. Therefore, the solution of the IVP is

$$y(t) = \frac{1}{153281}e^{-5t}\left[160\cos(\sqrt{73}t) + \frac{383443\sqrt{73}}{7300}\sin(\sqrt{73}t)\right]$$
$$+ \frac{1}{153281}[-160\cos(t/2) + 3128\sin(t/2)].$$

(b) The terms with the factor e^{-5t} are the transient terms. The terms coming from y_p are the steady-state parts of the solution.

(c)

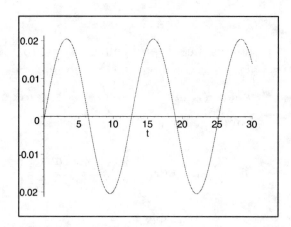

(d) If we replace the given external force with a force $Ae^{i\omega t}$, the inhomogeneous problem to solve becomes

$$y'' + 10y' + 98y = Ae^{i\omega t}.$$

Considering equation (3), we see that $\delta = 5$ and $\omega_0 = 7\sqrt{2}$. Therefore, using equation (5), we know that a particular solution of this inhomogeneous problem will be given by $Y(t) = G(i\omega)Ae^{i\omega t}$ where

$$G(i\omega) = \frac{98 - \omega^2 - 10\omega i}{(98 - \omega^2)^2 + 100\omega^2}$$

is the frequency response. The gain

$$|G(i\omega)| = \frac{1}{\sqrt{(98 - \omega^2)^2 + 100\omega^2}}$$

and the angle

$$\phi(\omega) = \arccos\left\{\frac{98 - \omega^2}{\sqrt{(98 - \omega^2)^2 + 100\omega^2}}\right\}.$$

The gain will be a maximum at $\omega_{\max} = \sqrt{\omega_0^2 - 2\delta^2} = \sqrt{48} = 4\sqrt{3}$.

86

9. The spring constant is $k = 12$ lb/ft and the mass is $6/32$ lb-s^2/ft. The forcing term is $4\cos(7t)$. Therefore, the equation of motion is

$$\frac{6}{32}y'' + 12y = 4\cos(7t)$$

which can be simplified to

$$y'' + 64y = \frac{64}{3}\cos(7t).$$

The solution of the homogeneous problem is $y_h(t) = c_1\cos(8t) + c_2\sin(8t)$. Then we look for a particular solution of the form $y_p(t) = A\cos(7t)$. (We do not need to include a term of the form $\sin(7t)$ since there are no first derivative terms in the equation.) Substituting y_p into the ODE, we conclude that $A = 64/45$. Therefore, the general solution of this differential equation is $y(t) = c_1\cos(8t) + c_2\sin(8t) + \dfrac{64}{45}\cos(7t)$. The initial conditions $y(0) = 0$ ft and $y'(0) = 0$ ft/sec imply that $c_1 = -64/45$ and $c_2 = 0$. Therefore, the solution of this IVP is $y(t) = -\dfrac{64}{45}\cos(8t) + \dfrac{64}{45}\cos(7t)$.

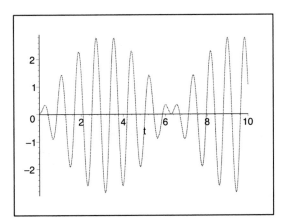

10. The spring constant is $k = 8/(1/2) = 16$ lb/ft and the mass is $8/32 = 1/4$ lb-s^2/ft. The forcing term is $8\sin(8t)$. Therefore, the equation of motion is

$$\frac{1}{4}y'' + 16y = 8\sin(8t)$$

which can be simplified to

$$y'' + 64y = 32\sin(8t).$$

The solution of the homogeneous problem is $y_h(t) = c_1\cos(8t) + c_2\sin(8t)$. Then we look for a particular solution of the form $y_p(t) = At\sin(8t) + Bt\cos(8t)$. Substituting y_p into the ODE, we conclude that $A = 0$ and $B = -2$. Therefore, the general solution of this differential equation is $y(t) = c_1\cos(8t) + c_2\sin(8t) - 2t\cos(8t)$. The initial conditions $y(0) = 1/4$ ft and $y'(0) = 0$ ft/sec imply that $c_1 = 1/4$ and $c_2 = 1/4$. Therefore, the solution of this IVP is $y(t) = \dfrac{1}{4}\cos(8t) + \dfrac{1}{4}\sin(8t) - 2t\cos(8t)$. Solving for y', we see that $y'(t) = (-2 + 16t)\sin(8t)$. We see that $y'(t) = 0$ when $t = 1/8$ or $\sin(8t) = 0$. Therefore, the first four times the velocity is zero are $t = 1/8, \pi/8, \pi/4, 3\pi/8$.

11. The spring constant is $k = 8/(1/2) = 16$ lb/ft and the mass is $8/32 = 1/4$ lb-s^2/ft. The damping constant is $\gamma = 0.25$ lb-sec/ft. The external force is $4\cos(2t)$ lbs. Therefore, the equation of motion is

$$\frac{1}{4}y'' + \frac{1}{4}y' + 16y = 4\cos(2t)$$

which can be simplified to

$$y'' + y' + 64y = 16\cos(2t).$$

(a) The roots of the characteristic equation are $\lambda = (-1 \pm \sqrt{255}i)/2$. Therefore, the solution of the homogeneous equation will be transient. To find the steady-state solution, we look for a particular solution of the form $y_p(t) = A\cos(2t) + B\sin(2t)$. Substituting y_p into the ODE, we conclude that $A = 240/901$ and $B = 8/901$. Therefore, the steady-state response is

$$y(t) = \frac{240}{901}\cos(2t) + \frac{8}{901}\sin(2t).$$

(b) With a forcing term of the form $F_0 \cos(\omega t)$, the steady-state response can be written as $Y(t) = R\cos(\omega t - \delta)$ where the amplitude

$$R = \frac{F_0}{\sqrt{m^2(\omega_0^2 - \omega^2)^2 + \gamma^2\omega^2}}$$

$$= \frac{F_0}{\sqrt{(k - m\omega^2)^2 + \gamma^2\omega^2}}.$$

The amplitude will be maximized when the denominator is minimized. This will occur when $k = m\omega^2$; that is, when $m = k/\omega^2 = 16/4 = 4$ slugs.

12. The spring constant is $k = 3$ N/m. The mass is 2 kg. The damping equals y'. The external force is $3\cos(3t) - 2\sin(3t)$ N. Therefore, the equation of motion is

$$2y'' + y' + 3y = 3\cos(3t) - 2\sin(3t).$$

Since the system is damped, the steady state response is equal to the particular solution. We look for a solution of the form $y_p(t) = A\cos(3t) + B\sin(3t)$. Substituting an equation of this form into the ODE, we conclude that $A = -1/6$ and $B = 1/6$. Therefore, the steady-state response is

$$y(t) = \frac{1}{6}(\sin(3t) - \cos(3t)).$$

13. The amplitude of the steady-state response is given by

$$R = \frac{1}{\sqrt{(\omega_0^2 - \omega^2)^2 + 4\delta^2\omega^2}}$$

$$= \frac{m}{\sqrt{m^2(\omega_0^2 - \omega^2)^2 + 4m^2\delta^2\omega^2}}$$

$$= \frac{m}{\sqrt{m^2(\omega_0^2 - \omega^2)^2 + \gamma^2\omega^2}}.$$

The amplitude will be a maximum when the denominator is a minimum. Consider the function $f(z) = m^2(\omega_0^2 - z)^2 + \gamma^2 z$. This function has a minimum when $z = \omega_0^2 - \gamma^2/2m^2$. Therefore, the amplitude reaches a maximum at $\omega_{max}^2 = \omega_0^2 - \gamma^2/2m^2$. Since $\omega_0^2 = k/m$, we can conclude that

$$\omega_{max}^2 = \omega_0^2 \left(1 - \frac{\gamma^2}{2km} \right).$$

Substituting $\omega^2 = \omega_{max}^2$ into the expression for the amplitude, we have

$$
\begin{aligned}
R &= \frac{m}{\sqrt{\gamma^4/4m^2 + gamma^2(\omega_0^2 - \gamma^2/2m^2)}} \\
&= \frac{m}{\sqrt{\omega_0^2\gamma^2 - \gamma^4/4m^2}} \\
&= \frac{m}{\gamma\omega_0\sqrt{1 - \gamma^2/4mk}}.
\end{aligned}
$$

14. First, for $0 \leq t \leq \pi$, the solution of the homogeneous problem is $y_h(t) = c_1 \cos(t) + c_2 \sin(t)$. Then we look for a particular solution of the form $y_p(t) = Bt$. We see that a particular solution is given by $y_p(t) = At$. Therefore, the general solution for $0 \leq t \leq \pi$ is $y(t) = c_1 \cos(t) + c_2 \sin(t) + At$. The initial condition $y(0) = 0$ implies $c_1 = 0$. Then $y'(t) = -c_1 \sin(t) + c_2 \cos(t) + A$. Therefore, $y'(0) = c_2 + A = 0$ implies $c_2 = -A$. Therefore, the solution of this IVP is $y(t) = -A \sin(t) + At$ for $0 \leq t \leq \pi$.

Then at time $t = \pi$, $y(\pi) = A\pi$ and $y'(\pi) = 2A$. In this time interval, the forcing term is $F(t) = A(2\pi - t)$. Therefore, we look for a particular solution of the form $y_p(t) = C + Dt$. Substituting a function of this form into the ODE, we see that $C = 2\pi A$ and $D = -A$. Therefore, the general solution for $\pi < t \leq 2\pi$ is $y(t) = c_1 \cos(t) + c_2 \sin(t) + 2\pi A - At$. Now considering the initial conditions $y(\pi) = A\pi$ and $y'(\pi) = 2A$, we need

$$- c_1 + \pi A = A\pi$$
$$- c_2 - A = 2A.$$

Therefore, $c_1 = 0$ and $c_2 = -3A$. Therefore, the solution of the IVP for $\pi < t \leq 2\pi$ is $y(t) = -3A \sin(t) + 2\pi A - At$.

Then at time $t = 2\pi$, $y(2\pi) = 0$ and $y'(2\pi) = -4A$. In this time interval, the forcing term is $F(t) = 0$. Therefore, the general solution is given by the solution of the homogeneous problem, $y(t) = c_1 \cos(t) + c_2 \sin(t)$. Considering our initial conditions, we need $c_1 = 0$ and $c_2 = -4A$. Therefore, the solution of the IVP for $2\pi < t$ is given by $y(t) = -4A \sin(t)$.

To summarize, we conclude that the solution is given by

$$
y(t) = \begin{cases}
-A \sin(t) + At & 0 \leq t \leq \pi \\
-3A \sin(t) + 2\pi A - At & \pi < t \leq 2\pi \\
-4A \sin(t) & 2\pi < t.
\end{cases}
$$

15. The inductance is $L = 1$ henry. The resistance $R = 5 \times 10^3$ ohms. The capacitance $C = 0.25 \times 10^{-6}$ farads. The forcing term is due to the 12−volt battery. Therefore, the equation for charge q is

$$q'' + 5000q' + (4 \times 10^6)q = 12.$$

The initial conditions are $q(0) = 0$, $q'(0) = 0$. The solution of the homogeneous problem is $q_h(t) = c_1 e^{-1000t} + c_2 e^{-4000t}$. Looking for a solution of the inhomogeneous problem, we find a particular solution of the form $q_p(t) = 3 \times 10^{-6}$. Therefore, the general solution is given by $q(t) = c_1 e^{-1000t} + c_2 e^{-4000t} + 3 \times 10^{-6}$. Considering our initial conditions, we conclude that $c_1 = -4 \times 10^{-6}$ and $c_2 = 10^{-6}$. Therefore, the solution of the IVP is $q(t) = 10^{-6}(-4e^{-1000t} + e^{-4000t} + 3)$. At $t = 0.001, 0.01$, we have $q(0.001) \cong 1.5468 \times 10^{-6}$ and $q(0.01) \cong 2.9998 \times 10^{-6}$, respectively. From our function q, we see that $q(t) \to 3 \times 10^{-6}$ as $t \to \infty$.

16.

(a) As the system is damped, the homogeneous solution will be the transient solution, and the steady-state solution will be given by a particular solution of the inhomogeneous problem. We look for a particular solution of the form $A\cos(\omega t) + B\sin(\omega t)$. Substituting this function into the ODE, we see that A, B must satisfy

$$(2 - \omega^2)A + \frac{1}{4}\omega B = 2$$

$$-\frac{1}{4}\omega A + (2 - \omega^2)B = 0.$$

The solution of this system is $A = 32(2 - \omega^2)/[16\omega^4 - 63\omega^2 + 64]$ and $B = 8\omega/[16\omega^4 - 63\omega^2 + 64]$. Therefore, the steady-state part of the solution is

$$Y(t) = [32(2 - \omega^2)\cos(\omega t) + 8\omega\sin(\omega t)]/[16\omega^4 - 63\omega^2 + 64].$$

(b) Here $\omega_0^2 = 2$ and $\delta = 1/8$. Therefore,

$$|G(i\omega)| = \frac{1}{\sqrt{(2 - \omega^2)^2 + 4\omega^2/64}}$$

$$= \frac{4}{\sqrt{16\omega^4 - 63\omega^2 + 64}}.$$

Also,

$$\phi(\omega) = \cos^{-1}\left[4(2 - \omega^2)/\sqrt{16\omega^4 - 63\omega^2 + 64}\right].$$

(c)

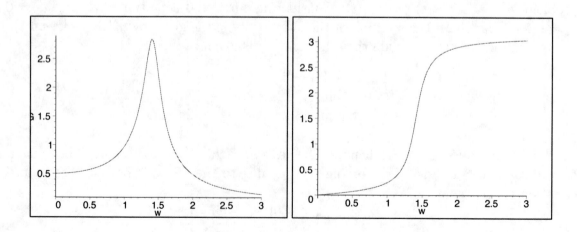

(d) The amplitude is maximum when $\omega_{max}^2 = \omega_0^2 - 2\delta^2$. Since $\omega_0^2 = 2$ and $\delta = 1/8$, we see that $\omega_{max}^2 = 2 - 1/32 = 63/32$. Therefore, $\omega_{max} = \sqrt{63/32} = 3\sqrt{14}/8 \approx 1.4031$. At ω_{max},

$$|G(i\omega_{max})| = \frac{4}{\sqrt{16\omega_{max}^4 - 63\omega_{max}^2 + 64}} = \frac{32}{\sqrt{127}} \approx 2.8395.$$

17.

(a) The solution of the homogeneous problem is $y_h(t) = c_1\cos(t) + c_2\sin(t)$. To find a particular solution of the inhomogeneous problem, we look for a solution of the form $y_p(t) = A\cos(\omega t)$. Substituting a function of this form into the ODE, we conclude that $A = 3/(1-\omega^2)$. Therefore, the general solution of this ODE is given by $y(t) = c_1\cos(t) + c_2\sin(t) + \dfrac{3}{1-\omega^2}\cos(\omega t)$. Applying the initial conditions, we see that $c_1 + 3/(1-\omega^2) = 0$ and $c_2 = 0$. Therefore, the solution of this IVP is

$$y(t) = \frac{3(\cos(\omega t) - \cos(t))}{1 - \omega^2}.$$

(b)

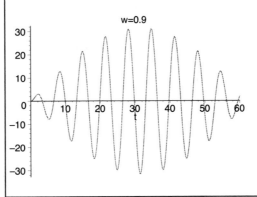

18.

91

(a) The solution of the homogeneous problem is $y_h(t) = c_1 \cos(t) + c_2 \sin(t)$. To find a particular solution of the inhomogeneous problem, we look for a solution of the form $y_p(t) = A\cos(\omega t)$. Substituting a function of this form into the ODE, we conclude that $A = 3/(1-\omega^2)$. Therefore, the general solution of this ODE is given by $y(t) = c_1 \cos(t) +$ $c_2 \sin(t) + \dfrac{3}{1-\omega^2} \cos(\omega t)$. Applying the initial conditions, we see that $c_1 + 3/(1-\omega^2) = 1$ and $c_2 = 1$. Therefore, the solution of this IVP is

$$y(t) = \frac{1}{1-\omega^2}\left[3\cos(\omega t) - (\omega^2 + 2)\cos(t)\right] + \sin(t).$$

(b)

19.

20.

(a)

(b)

21.

(a)

(b)

22.

(a)

94

(b)

23.

(a)

(b)

24.

(a)

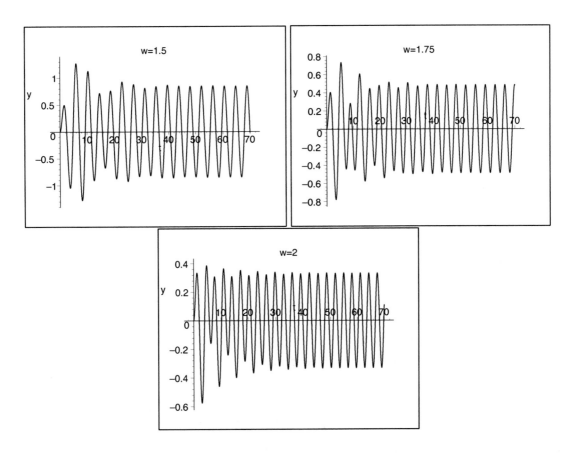

We estimate that the amplitudes of the steady responses are $G_H(0.5) \approx 1.1129$, $G_H(0.75) \approx 1.3856$, $G_H(1) \approx 1.8632$, $G_H(1.25) \approx 2.5001$, $G_H(1.5) \approx 0.8495$, $G_H(1.75) \approx 0.4887$, $G_H(2) \approx 0.3351$.

(b)

(c) The amplitude for a similar system with a linear spring is given by

$$R = \frac{5}{\sqrt{25 - 49\omega^2 + 25\omega^4}}.$$

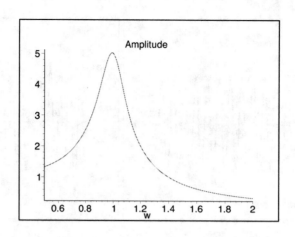

Section 4.8

1.

(a)

$$\mathbf{Xu} = \begin{pmatrix} x_{11}(t) & x_{12}(t) \\ x_{21}(t) & x_{22}(t) \end{pmatrix} \begin{pmatrix} u_1(t) \\ u_2(t) \end{pmatrix} = \begin{pmatrix} x_{11}(t)u_1(t) + x_{12}(t)u_2(t) \\ x_{21}(t)u_1(t) + x_{22}(t)u_2(t) \end{pmatrix}.$$

Therefore,

$$(\mathbf{Xu})' = \begin{pmatrix} x_{11}'u_1 + x_{11}u_1' + x_{12}'u_2 + x_{12}u_2' \\ x_{21}'u_1 + x_{21}u_1' + x_{22}'u_2 + x_{22}u_2' \end{pmatrix}.$$

Now

$$\mathbf{X}' = \begin{pmatrix} x_{11}' & x_{12}' \\ x_{21}' & x_{22}' \end{pmatrix} \qquad \mathbf{u}' = \begin{pmatrix} u_1' \\ u_2' \end{pmatrix}.$$

Therefore,

$$\begin{aligned} \mathbf{X}'\mathbf{u} + \mathbf{X}\mathbf{u}' &= \begin{pmatrix} x_{11}' & x_{12}' \\ x_{21}' & x_{22}' \end{pmatrix} \begin{pmatrix} u_1 \\ u_2 \end{pmatrix} + \begin{pmatrix} x_{11} & x_{12} \\ x_{21} & x_{22} \end{pmatrix} \begin{pmatrix} u_1' \\ u_2' \end{pmatrix} \\ &= \begin{pmatrix} x_{11}'u_1 + x_{12}'u_2 + x_{11}u_1 + x_{12}u_2' \\ x_{21}'u_1 + x_{22}'u_2 + x_{21}u_1' + x_{22}u_2' \end{pmatrix} \\ &= \mathbf{X}'\mathbf{u} + \mathbf{X}\mathbf{u}'. \end{aligned}$$

(b) Now

$$\mathbf{x}_p(t) = \mathbf{X}(t) \int \mathbf{X}^{-1}(t)\mathbf{g}(t), \, dt.$$

Let

$$\mathbf{u}(t) = \int \mathbf{X}(t)\mathbf{g}(t) \, dt.$$

Therefore, $\mathbf{x}_p(t) = \mathbf{X}(t)\mathbf{u}(t)$, which implies

$$\begin{aligned} \mathbf{x}_p' &= (\mathbf{Xu})' \\ &= \mathbf{X}'\mathbf{u} + \mathbf{X}\mathbf{u}' \\ &= \mathbf{P}(t)\mathbf{Xu} + \mathbf{X}(\mathbf{X}^{-1}\mathbf{g}) \\ &= \mathbf{P}(t)\mathbf{x}_p + \mathbf{g}. \end{aligned}$$

2. Let $\mathbf{X}(t)$ be the fundamental matrix,

$$\mathbf{X}(t) = \begin{pmatrix} e^{-2t} & e^{-4t} \\ e^{-2t} & -e^{-4t} \end{pmatrix}.$$

Then

$$\mathbf{X}^{-1}(t) = \frac{1}{2}\begin{pmatrix} e^{2t} & e^{2t} \\ e^{4t} & -e^{4t} \end{pmatrix}.$$

Therefore,

$$\mathbf{x}_p(t) = \mathbf{X}(t)\int \mathbf{X}^{-1}(t)g(t)\,dt$$
$$= \begin{pmatrix} e^{-2t} & e^{-4t} \\ e^{-2t} & -e^{-4t} \end{pmatrix}\int \frac{1}{2}\begin{pmatrix} e^{2t} & e^{2t} \\ e^{4t} & -e^{4t} \end{pmatrix}\begin{pmatrix} 1 \\ 4t \end{pmatrix}\,dt$$
$$= \frac{1}{2}\begin{pmatrix} e^{-2t} & e^{-4t} \\ e^{-2t} & -e^{-4t} \end{pmatrix}\int \begin{pmatrix} e^{2t} + 4te^{2t} \\ e^{4t} - 4te^{4t} \end{pmatrix}\,dt.$$

Integrating by parts, we have

$$\int \begin{pmatrix} e^{2t} + 4te^{2t} \\ e^{4t} - 4te^{4t} \end{pmatrix}\,dt = \begin{pmatrix} -\frac{1}{2}e^{2t} + 2te^{2t} \\ \frac{1}{2}e^{4t} - te^{4t} \end{pmatrix}.$$

Therefore,

$$\mathbf{x}_p(t) = \frac{1}{2}\begin{pmatrix} e^{-2t} & e^{-4t} \\ e^{-2t} & -e^{-4t} \end{pmatrix}\begin{pmatrix} -\frac{1}{2}e^{2t} + 2te^{2t} \\ \frac{1}{2}e^{4t} - te^{4t} \end{pmatrix}$$
$$= \begin{pmatrix} t/2 \\ (3t-1)/2 \end{pmatrix}.$$

Therefore,

$$\mathbf{x}_p(t) = \begin{pmatrix} t/2 \\ (3t-1)/2 \end{pmatrix}.$$

3. Let $\mathbf{X}(t)$ be the fundamental matrix,

$$\mathbf{X}(t) = \begin{pmatrix} e^{-t} & -e^{t} \\ 0 & e^{t} \end{pmatrix}.$$

Then

$$\mathbf{X}^{-1}(t) = \begin{pmatrix} e^{t} & e^{t} \\ 0 & e^{-t} \end{pmatrix}.$$

Therefore,

$$\mathbf{X}^{-1}(t)\mathbf{g}(t) = \begin{pmatrix} e^{t} & e^{t} \\ 0 & e^{-t} \end{pmatrix}\begin{pmatrix} e^{-t} \\ t \end{pmatrix}$$
$$= \begin{pmatrix} 1 + te^{t} \\ te^{-t} \end{pmatrix}.$$

Therefore, integrating by parts, we have

$$\int \mathbf{X}^{-1}(t)\mathbf{g}(t)\,dt = \int \begin{pmatrix} 1 + te^t \\ te^{-t} \end{pmatrix} dt$$

$$= \begin{pmatrix} t + te^t - e^t \\ -te^{-t} - e^{-t} \end{pmatrix}.$$

Therefore,

$$\mathbf{x}_p(t) = \mathbf{X}(t) \int \mathbf{X}^{-1}(t)g(t)\,dt$$

$$= \begin{pmatrix} e^{-t} & -e^t \\ 0 & e^t \end{pmatrix} \begin{pmatrix} t + te^t - e^t \\ -te^{-t} - e^{-t} \end{pmatrix}$$

$$= \begin{pmatrix} te^{-t} + 2t \\ -t - 1 \end{pmatrix}.$$

Therefore,

$$\mathbf{x}_p(t) = \begin{pmatrix} te^{-t} + 2t \\ -t - 1 \end{pmatrix}.$$

4. Let $\mathbf{X}(t)$ be the fundamental matrix,

$$\mathbf{X}(t) = \begin{pmatrix} 0 & -e^{-t} \\ e^{-t} & te^{-t} \end{pmatrix}.$$

Then

$$\mathbf{X}^{-1}(t) = \begin{pmatrix} te^t & e^t \\ -e^t & 0 \end{pmatrix}.$$

Therefore,

$$\mathbf{X}^{-1}(t)\mathbf{g}(t) = \begin{pmatrix} te^t & e^t \\ -e^t & 0 \end{pmatrix} \begin{pmatrix} -1 \\ t \end{pmatrix}$$

$$= \begin{pmatrix} 0 \\ e^t \end{pmatrix}.$$

Therefore,

$$\int \mathbf{X}^{-1}(t)\mathbf{g}(t)\,dt = \int \begin{pmatrix} 0 \\ e^t \end{pmatrix} dt$$

$$= \begin{pmatrix} 0 \\ e^t \end{pmatrix}.$$

Therefore,

$$\mathbf{x}_p(t) = \mathbf{X}(t) \int \mathbf{X}^{-1}(t)g(t)\,dt$$

$$= \begin{pmatrix} 0 & -e^{-t} \\ e^{-t} & te^{-t} \end{pmatrix} \begin{pmatrix} 0 \\ e^t \end{pmatrix}$$

$$= \begin{pmatrix} -1 \\ t \end{pmatrix}.$$

Therefore,

$$\mathbf{x}_p(t) = \begin{pmatrix} -1 \\ t \end{pmatrix}.$$

5. Let $\mathbf{X}(t)$ be the fundamental matrix,

$$\mathbf{X}(t) = \begin{pmatrix} \cos t & \sin t \\ -\sin t & \cos t \end{pmatrix}.$$

Then

$$\mathbf{X}^{-1}(t) = \begin{pmatrix} \cos t & -\sin t \\ \sin t & \cos t \end{pmatrix}.$$

Therefore,

$$\mathbf{X}^{-1}(t)\mathbf{g}(t) = \begin{pmatrix} \cos t & -\sin t \\ \sin t & \cos t \end{pmatrix} \begin{pmatrix} \cos t \\ -\sin t \end{pmatrix}$$

$$= \begin{pmatrix} 1 \\ 0 \end{pmatrix}.$$

Therefore,

$$\int \mathbf{X}^{-1}(t)\mathbf{g}(t)\,dt = \int \begin{pmatrix} 1 \\ 0 \end{pmatrix} dt$$

$$= \begin{pmatrix} t \\ 0 \end{pmatrix}.$$

Therefore,

$$\mathbf{x}_p(t) = \mathbf{X}(t) \int \mathbf{X}^{-1}(t)g(t)\,dt$$

$$= \begin{pmatrix} \cos t & \sin t \\ -\sin t & \cos t \end{pmatrix} \begin{pmatrix} t \\ 0 \end{pmatrix}$$

$$= \begin{pmatrix} t\cos t \\ -t\sin t \end{pmatrix}.$$

Therefore,

$$\mathbf{x}_p(t) = \begin{pmatrix} t\cos t \\ -t\sin t \end{pmatrix}.$$

6. The general solution of the equation in problem 2 is

$$\mathbf{x}(t) = c_1 \begin{pmatrix} e^{-2t} \\ e^{-2t} \end{pmatrix} + c_2 \begin{pmatrix} e^{-4t} \\ -e^{-4t} \end{pmatrix} + \begin{pmatrix} t/2 \\ (3t-1)/2 \end{pmatrix}.$$

The initial condition $\mathbf{x}(0) = \begin{pmatrix} 2 & -1 \end{pmatrix}^t$ implies

$$\mathbf{x}(0) = c_1 \begin{pmatrix} 1 \\ 1 \end{pmatrix} + c_2 \begin{pmatrix} 1 \\ -1 \end{pmatrix} + \begin{pmatrix} 0 \\ -1/2 \end{pmatrix} = \begin{pmatrix} 2 \\ -1 \end{pmatrix}.$$

101

Therefore, $c_1 + c_2 = 2$ and $c_1 - c_2 - \dfrac{1}{2} = -1$. Solving this system of equations, we have $c_1 = 3/4$ and $c_2 = 5/4$. Therefore, the solution of the IVP is

$$x(t) = \frac{3}{4}\begin{pmatrix} e^{-2t} \\ e^{-2t} \end{pmatrix} + \frac{5}{4}\begin{pmatrix} e^{-4t} \\ -e^{-4t} \end{pmatrix} + \begin{pmatrix} t/2 \\ (3t-1)/2 \end{pmatrix}.$$

7. The general solution of the equation in problem 3 is

$$x(t) = c_1\begin{pmatrix} e^{-t} \\ 0 \end{pmatrix} + c_2\begin{pmatrix} -e^{t} \\ e^{t} \end{pmatrix} + \begin{pmatrix} te^{-t} + 2t \\ -t - 1 \end{pmatrix}.$$

The initial condition $x(0) = \begin{pmatrix} -1 & 1 \end{pmatrix}^t$ implies

$$x(0) = c_1\begin{pmatrix} 1 \\ 0 \end{pmatrix} + c_2\begin{pmatrix} -1 \\ 1 \end{pmatrix} + \begin{pmatrix} 0 \\ -1 \end{pmatrix} = \begin{pmatrix} -1 \\ 1 \end{pmatrix}.$$

Therefore, $c_1 - c_2 = -1$ and $c_2 - 1 = 1$. Solving this system of equations, we have $c_1 = 1$ and $c_2 = 2$. Therefore, the solution of the IVP is

$$x(t) = \begin{pmatrix} e^{-t} \\ 0 \end{pmatrix} + 2\begin{pmatrix} -e^{t} \\ e^{t} \end{pmatrix} + \begin{pmatrix} te^{-t} + 2t \\ -t - 1 \end{pmatrix}.$$

8. The general solution of the equation in problem 4 is

$$x(t) = c_1\begin{pmatrix} 0 \\ e^{-t} \end{pmatrix} + c_2\begin{pmatrix} -e^{-t} \\ te^{-t} \end{pmatrix} + \begin{pmatrix} -1 \\ t \end{pmatrix}.$$

The initial condition $x(0) = \begin{pmatrix} 1 & 0 \end{pmatrix}^t$ implies

$$x(0) = c_1\begin{pmatrix} 0 \\ 1 \end{pmatrix} + c_2\begin{pmatrix} -1 \\ 0 \end{pmatrix} + \begin{pmatrix} -1 \\ 0 \end{pmatrix} = \begin{pmatrix} 1 \\ 0 \end{pmatrix}.$$

Therefore, $-c_2 - 1 = 1$ and $c_1 = 0$. Solving this system of equations, we have $c_1 = 0$ and $c_2 = -2$. Therefore, the solution of the IVP is

$$x(t) = -2\begin{pmatrix} -e^{-t} \\ te^{-t} \end{pmatrix} + \begin{pmatrix} -1 \\ t \end{pmatrix}.$$

9. The general solution of the equation in problem 5 is

$$x(t) = c_1\begin{pmatrix} \cos t \\ -\sin t \end{pmatrix} + c_2\begin{pmatrix} \sin t \\ \cos t \end{pmatrix} + \begin{pmatrix} t\cos t \\ -t\sin t \end{pmatrix}.$$

The initial condition $x(0) = \begin{pmatrix} 1 & 1 \end{pmatrix}^t$ implies

$$x(0) = c_1\begin{pmatrix} 1 \\ 0 \end{pmatrix} + c_2\begin{pmatrix} 0 \\ 1 \end{pmatrix} + \begin{pmatrix} 0 \\ 0 \end{pmatrix} = \begin{pmatrix} 1 \\ 1 \end{pmatrix}.$$

Therefore, $c_1 = 1$ and $c_2 = 1$. Therefore, the solution of the IVP is

$$\mathbf{x}(t) = \begin{pmatrix} \cos t \\ -\sin t \end{pmatrix} + \begin{pmatrix} \sin t \\ \cos t \end{pmatrix} + \begin{pmatrix} t \cos t \\ -t \sin t \end{pmatrix}.$$

10. The solution of the homogeneous equation is $y_h(t) = c_1 e^{2t} + c_2 e^{3t}$. The functions $y_1(t) = e^{2t}$ and $y_2(t) = e^{3t}$ form a fundamental set of solutions. The Wronskian of these functions is $W(y_1, y_2) = e^{5t}$. Using the method of variation of parameters, a particular solution is given by $Y(t) = u_1(t)y_1(t) + u_2(t)y_2(t)$ where

$$u_1(t) = -\int \frac{e^{3t}(2e^t)}{W(t)} \, dt = 2e^{-t}$$

$$u_2(t) = \int \frac{e^{2t}(2e^t)}{W(t)} \, dt = -e^{-2t}.$$

Therefore, the particular solution is $Y(t) = 2e^t - e^t = e^t$.

11. The solution of the homogeneous equation is $y_h(t) = c_1 e^{2t} + c_2 e^{-t}$. The functions $y_1(t) = e^{2t}$ and $y_2(t) = e^{-t}$ form a fundamental set of solutions. The Wronskian of these functions is $W(y_1, y_2) = -3e^t$. Using the method of variation of parameters, a particular solution is given by $Y(t) = u_1(t)y_1(t) + u_2(t)y_2(t)$ where

$$u_1(t) = -\int \frac{e^{-t}(2e^{-t})}{W(t)} \, dt = -\frac{2}{9} e^{-3t}$$

$$u_2(t) = \int \frac{e^{2t}(2e^{-t})}{W(t)} \, dt = -\frac{2t}{3}.$$

Therefore, a particular solution is $Y(t) = -\frac{2}{9} e^{-t} - \frac{2}{3} t e^{-t}$. As $-2e^{-t}/9$ is a solution of the homogeneous equation, we can omit it and conclude that $y_p(t) = -\frac{2}{3} t e^{-t}$ is a particular solution of the inhomogeneous problem.

12. The solution of the homogeneous equation is $y_h(t) = c_1 e^{-t} + c_2 t e^{-t}$. The functions $y_1(t) = e^{-t}$ and $y_2(t) = t e^{-t}$ form a fundamental set of solutions. The Wronskian of these functions is $W(y_1, y_2) = e^{-2t}$. Using the method of variation of parameters, a particular solution is given by $Y(t) = u_1(t)y_1(t) + u_2(t)y_2(t)$ where

$$u_1(t) = -\int \frac{te^{-t}(3e^{-t})}{W(t)} \, dt = -\frac{3}{2} t^2$$

$$u_2(t) = \int \frac{e^{-t}(3e^{-t})}{W(t)} \, dt = 3t.$$

Therefore, a particular solution is $Y(t) = -\frac{3}{2} t^2 e^{-t} + 3t^2 e^{-t} = \frac{3}{2} t^2 e^{-t}$.

13. The solution of the homogeneous equation is $y_h(t) = c_1 e^{t/2} + c_2 t e^{t/2}$. The functions $y_1(t) = e^{t/2}$ and $y_2(t) = t e^{t/2}$ form a fundamental set of solutions. The Wronskian of these

functions is $W(y_1, y_2) = e^t$. Using the method of variation of parameters, a particular solution is given by $Y(t) = u_1(t)y_1(t) + u_2(t)y_2(t)$ where

$$u_1(t) = -\int \frac{te^{t/2}(4e^{t/2})}{W(t)} \, dt = -2t^2$$

$$u_2(t) = \int \frac{e^{t/2}(4e^{t/2})}{W(t)} \, dt = 4t.$$

Therefore, a particular solution is $Y(t) = -2t^2 e^{t/2} + 4t^2 e^{t/2} = 2t^2 e^{t/2}$.

14. The solution of the homogeneous equation is $y_h(t) = c_1 \cos(t) + c_2 \sin(t)$. The functions $y_1(t) = \cos(t)$ and $y_2(t) = \sin(t)$ form a fundamental set of solutions. The Wronskian of these functions is $W(y_1, y_2) = 1$. Using the method of variation of parameters, a particular solution is given by $Y(t) = u_1(t)y_1(t) + u_2(t)y_2(t)$ where

$$u_1(t) = -\int \frac{\sin(t)\tan(t)}{W(t)} \, dt = \sin(t) - \ln(\sec(t) + \tan(t))$$

$$u_2(t) = \int \frac{\cos(t)\tan(t)}{W(t)} \, dt = -\cos(t).$$

Therefore, a particular solution is $Y(t) = (\sin(t) - \ln(\sec(t) + \tan(t)))\cos(t) - \cos(t)\sin(t) = -\cos(t)\ln(\sec(t) + \tan(t))$. Therefore, the general solution is $y(t) = c_1 \cos(t) + c_2 \sin(t) - \cos(t)\ln(\sec(t) + \tan(t))$.

15. The solution of the homogeneous equation is $y_h(t) = c_1 \cos(3t) + c_2 \sin(3t)$. The functions $y_1(t) = \cos(3t)$ and $y_2(t) = \sin(3t)$ form a fundamental set of solutions. The Wronskian of these functions is $W(y_1, y_2) = 3$. Using the method of variation of parameters, a particular solution is given by $Y(t) = u_1(t)y_1(t) + u_2(t)y_2(t)$ where

$$u_1(t) = -\int \frac{\sin(3t)(9\sec^2(3t))}{W(t)} \, dt = -\csc(3t)$$

$$u_2(t) = \int \frac{\cos(3t)(9\sec^2(3t))}{W(t)} \, dt = \ln(\sec(3t) + \tan(3t)).$$

Therefore, a particular solution is $Y(t) = -1 + \sin(3t)\ln(\sec(3t) + \tan(3t))$. Therefore, the general solution is $y(t) = c_1 \cos(3t) + c_2 \sin(3t) + \sin(3t)\ln(\sec(3t) + \tan(3t)) - 1$.

16. The solution of the homogeneous equation is $y_h(t) = c_1 e^{-2t} + c_2 t e^{-2t}$. The functions $y_1(t) = e^{-2t}$ and $y_2(t) = t e^{-2t}$ form a fundamental set of solutions. The Wronskian of these functions is $W(y_1, y_2) = e^{-4t}$. Using the method of variation of parameters, a particular solution is given by $Y(t) = u_1(t)y_1(t) + u_2(t)y_2(t)$ where

$$u_1(t) = -\int \frac{te^{-2t}(t^{-2}e^{-2t})}{W(t)} \, dt = -\ln(t)$$

$$u_2(t) = \int \frac{e^{-2t}(t^{-2}e^{-2t})}{W(t)} \, dt = -\frac{1}{t}.$$

Therefore, a particular solution is $Y(t) = -e^{-2t}\ln(t) - e^{-2t}$. Therefore, the general solution is $y(t) = c_1 e^{-2t} + c_2 t e^{-2t} - e^{-2t}\ln(t)$.

17. The solution of the homogeneous equation is $y_h(t) = c_1 \cos(2t) + c_2 \sin(2t)$. The functions $y_1(t) = \cos(2t)$ and $y_2(t) = \sin(2t)$ form a fundamental set of solutions. The Wronskian of these functions is $W(y_1, y_2) = 2$. Using the method of variation of parameters, a particular solution is given by $Y(t) = u_1(t)y_1(t) + u_2(t)y_2(t)$ where

$$u_1(t) = -\int \frac{\sin(2t)(3\csc(2t))}{W(t)}\, dt = -\frac{3}{2}t$$

$$u_2(t) = \int \frac{\cos(2t)(3\csc(2t))}{W(t)}\, dt = \frac{3}{4}\ln|\sin(2t)|.$$

Therefore, a particular solution is $Y(t) = -\frac{3}{2}t\cos(2t) + \frac{3}{4}(\sin(2t))\ln|\sin(2t)|$. Therefore, the general solution is $y(t) = c_1 \cos(2t) + c_2 \sin(2t) - \frac{3}{2}t\cos(2t) + \frac{3}{4}\sin(2t)\ln|\sin(2t)|$.

18. The solution of the homogeneous equation is $y_h(t) = c_1 \cos(t/2) + c_2 \sin(t/2)$. The functions $y_1(t) = \cos(t/2)$ and $y_2(t) = \sin(t/2)$ form a fundamental set of solutions. The Wronskian of these functions is $W(y_1, y_2) = 1/2$. Writing the ODE in standard form, we see that $g(t) = \sec(t/2)/2$. Using the method of variation of parameters, a particular solution is given by $Y(t) = u_1(t)y_1(t) + u_2(t)y_2(t)$ where

$$u_1(t) = -\int \frac{\cos(t/2)(\sec(t/2))}{2W(t)}\, dt = 2\ln[\cos(t/2)]$$

$$u_2(t) = \int \frac{\sin(t/2)(\sec(t/2))}{2W(t)}\, dt = t.$$

Therefore, a particular solution is $Y(t) = 2\cos(t/2)\ln[\cos(t/2)] + t\sin(t/2)$. Therefore, the general solution is $y(t) = c_1 \cos(t/2) + c_2 \sin(t/2) + 2\cos(t/2)\ln[\cos(t/2)] + t\sin(t/2)$.

19. The solution of the homogeneous equation is $y_h(t) = c_1 e^t + c_2 t e^t$. The functions $y_1(t) = e^t$ and $y_2(t) = te^t$ form a fundamental set of solutions. The Wronskian of these functions is $W(y_1, y_2) = e^{2t}$. Using the method of variation of parameters, a particular solution is given by $Y(t) = u_1(t)y_1(t) + u_2(t)y_2(t)$ where

$$u_1(t) = -\int \frac{te^t(e^t)}{W(t)(1+t^2)}\, dt = -\frac{1}{2}\ln(1+t^2)$$

$$u_2(t) = \int \frac{e^t(e^t)}{W(t)(1+t^2)}\, dt = \arctan(t).$$

Therefore, a particular solution is $Y(t) = -\frac{1}{2}e^t \ln(1+t^2) + te^t \arctan(t)$. Therefore, the general solution is $y(t) = c_1 e^t + c_2 t e^t - \frac{1}{2}e^t \ln(1+t^2) + te^t \arctan(t)$.

20. The solution of the homogeneous equation is $y_h(t) = c_1 e^{3t} + c_2 e^{2t}$. The functions $y_1(t) = e^{3t}$ and $y_2(t) = e^{2t}$ form a fundamental set of solutions. The Wronskian of these functions is $W(y_1, y_2) = -e^{5t}$. Using the method of variation of parameters, a particular

solution is given by $Y(t) = u_1(t)y_1(t) + u_2(t)y_2(t)$ where

$$u_1(t) = -\int \frac{e^{2s}(g(s))}{W(s)} \, ds = \int e^{-3s} g(s) \, ds$$

$$u_2(t) = \int \frac{e^{3s}(g(s))}{W(s)} \, ds = -\int e^{-2s} g(s) \, ds.$$

Therefore, a particular solution is $Y(t) = \int e^{3(t-s)} g(s) \, ds - \int e^{2(t-s)} g(s) \, ds$. Therefore, the general solution is $y(t) = c_1 e^{3t} + c_2 e^{2t} + \int e^{3(t-s)} g(s) \, ds - \int e^{2(t-s)} g(s) \, ds$.

21. The solution of the homogeneous equation is $y_h(t) = c_1 \cos(2t) + c_2 \sin(2t)$. The functions $y_1(t) = \cos(2t)$ and $y_2(t) = \sin(2t)$ form a fundamental set of solutions. The Wronskian of these functions is $W(y_1, y_2) = 2$. Using the method of variation of parameters, a particular solution is given by $Y(t) = u_1(t)y_1(t) + u_2(t)y_2(t)$ where

$$u_1(t) = -\int \frac{\sin(2s)(g(s))}{W(s)} \, ds = -\frac{1}{2} \int \sin(2s) g(s) \, ds$$

$$u_2(t) = \int \frac{\cos(2s)(g(s))}{W(s)} \, ds = \frac{1}{2} \int \cos(2s) g(s) \, ds.$$

Therefore, a particular solution is

$$Y(t) = -\frac{1}{2} \int \cos(2t) \sin(2s) g(s) \, ds + \frac{1}{2} \int \sin(2t) \cos(2s) g(s) \, ds.$$

Using the fact that $\sin(2t) \cos(2s) - \cos(2t) \sin(2s) = \sin(2t - 2s)$, we see that the general solution is

$$y(t) = c_1 \cos(2t) + c_2 \sin(2t) + \frac{1}{2} \int \sin(2t - 2s) g(s) \, ds.$$

22. By direct substitution, it can be verified that $y_1(t) = t$ and $y_2(t) = te^t$ are solutions of the homogeneous equations. The Wronskian of these functions is $W(y_1, y_2) = t^2 e^t$. Rewriting the equation in standard form, we have

$$y'' - \frac{t(t+2)}{t^2} y' + \frac{t+2}{t^2} y = 2t.$$

Therefore, $g(t) = 2t$. Using the method of variation of parameters, a particular solution is given by $Y(t) = u_1(t)y_1(t) + u_2(t)y_2(t)$ where

$$u_1(t) = -\int \frac{te^t(2t)}{W(t)} \, dt = -2t$$

$$u_2(t) = \int \frac{t(2t)}{W(t)} \, dt = -2e^{-t}.$$

Therefore, a particular solution is $Y(t) = -2t^2 - 2t^2 = -4t^2$.

23. By direct substitution, it can be verified that $y_1(t) = 1 + t$ and $y_2(t) = e^t$ are solutions of the homogeneous equations. The Wronskian of these functions is $W(y_1, y_2) = te^t$. Rewriting the equation in standard form, we have

$$y'' - \frac{1+t}{t}y' + \frac{1}{t}y = te^{2t}.$$

Therefore, $g(t) = te^{2t}$. Using the method of variation of parameters, a particular solution is given by $Y(t) = u_1(t)y_1(t) + u_2(t)y_2(t)$ where

$$u_1(t) = -\int \frac{e^t(te^{2t})}{W(t)}\, dt = -\frac{1}{2}e^{2t}$$

$$u_2(t) = \int \frac{(1+t)(te^{2t})}{W(t)}\, dt = te^t.$$

Therefore, a particular solution is $Y(t) = -\frac{1}{2}(1+t)e^{2t} + te^{2t} = \frac{1}{2}(t-1)e^{2t}$.

24. By direct substitution, it can be verified that $y_1(t) = e^t$ and $y_2(t) = t$ are solutions of the homogeneous equations. The Wronskian of these functions is $W(y_1, y_2) = (1-t)e^t$. Rewriting the equation in standard form, we have

$$y'' - \frac{t}{1-t}y' - \frac{1}{1-t}y = 2(1-t)e^{-t}.$$

Therefore, $g(t) = 2(1-t)e^{-t}$. Using the method of variation of parameters, a particular solution is given by $Y(t) = u_1(t)y_1(t) + u_2(t)y_2(t)$ where

$$u_1(t) = -\int \frac{t(2(1-t)e^{-t})}{W(t)}\, dt = te^{-2t} + \frac{1}{2}e^{-2t}$$

$$u_2(t) = \int \frac{2(1-t)}{W(t)}\, dt = -2e^{-t}.$$

Therefore, a particular solution is $Y(t) = te^{-t} + \frac{1}{2}e^{-t} - 2te^{-t} = \left(\frac{1}{2} - t\right)e^{-t}$.

25. By direct substitution, it can be verified that $y_1(x) = x^{-1/2}\sin x$ and $y_2(x) = x^{-1/2}\cos x$ are solutions of the homogeneous equations. The Wronskian of these functions is $W(y_1, y_2) = -1/x$. Rewriting the equation in standard form, we have

$$y'' + \frac{1}{x}y' + \frac{x^2 - 0.25}{x^2}y = 3\frac{\sin x}{x^{1/2}}.$$

Therefore, $g(x) = 3\sin(x)x^{-1/2}$. Using the method of variation of parameters, a particular solution is given by $Y(x) = u_1(x)y_1(x) + u_2(x)y_2(x)$ where

$$u_1(x) = -\int \frac{x^{-1/2}\cos(x)(3\sin(x)x^{-1/2})}{W(x)}\, dx = -\frac{3}{2}\cos^2(x)$$

$$u_2(x) = \int \frac{x^{-1/2}\sin(x)(3\sin(x)x^{-1/2})}{W(x)}\, dx = \frac{3}{2}\cos(x)\sin(x) - \frac{3x}{2}.$$

Therefore, a particular solution is

$$Y(x) = -\frac{3}{2}\cos^2(x)\sin(x)x^{-1/2} + \left(\frac{3}{2}\cos(x)\sin(x) - \frac{3x}{2}\right)\cos(x)x^{-1/2}$$

$$= -\frac{3}{2}x^{1/2}\cos(x).$$

26. By direct substitution, it can be verified that $y_1(x) = e^x$ and $y_2(x) = x$ are solutions of the homogeneous equations. The Wronskian of these functions is $W(y_1, y_2) = (1 - x)e^x$. Rewriting the equation in standard form, we have

$$y'' + \frac{x}{1-x}y' - \frac{1}{1-x}y = \frac{g(x)}{1-x}.$$

Therefore, the inhomogeneous term is $g(x)/(1-x)$. Using the method of variation of parameters, a particular solution is given by $Y(x) = u_1(x)y_1(x) + u_2(x)y_2(x)$ where

$$u_1(x) = -\int \frac{sg(s)}{(1-s)W(s)}\,ds$$

$$u_2(x) = \int \frac{e^s g(s)}{(1-s)W(s)}\,ds.$$

Therefore, a particular solution is

$$Y(x) = -e^x \int \frac{sg(s)}{(1-s)W(s)}\,ds + x \int \frac{e^s g(s)}{(1-s)W(s)}\,ds$$

$$= \int \frac{(xe^s - se^x)g(s)}{(1-s)^2 e^s}\,ds.$$

27. By direct substitution, it can be verified that $y_1(x) = x^{-1/2}\sin x$ and $y_2(x) = x^{-1/2}\cos x$ are solutions of the homogeneous equations. The Wronskian of these functions is $W(y_1, y_2) = -1/x$. Rewriting the equation in standard form, we have

$$y'' + \frac{1}{x}y' + \frac{x^2 - 0.25}{x^2}y = \frac{g(x)}{x^2}.$$

Therefore, the inhomogeneous term is $g(x)/(x^2)$. Using the method of variation of parameters, a particular solution is given by $Y(x) = u_1(x)y_1(x) + u_2(x)y_2(x)$ where

$$u_1(x) = -\int \frac{s^{-1/2}\cos(s)g(s)}{s^2 W(s)}\,ds$$

$$u_2(x) = \int \frac{s^{-1/2}\sin(s)g(s)}{s^2 W(s)}\,ds.$$

Therefore, a particular solution is

$$Y(x) = -x^{-1/2}\sin(x)\int \frac{s^{-1/2}\cos(s)g(s)}{s^2 W(s)}\,ds + x^{-1/2}\cos(x)\int \frac{s^{-1/2}\sin(s)g(s)}{s^2 W(s)}\,ds$$

$$= \frac{1}{\sqrt{x}}\int \frac{(\sin(x)\cos(s) - \cos(x)\sin(s))g(s)}{s\sqrt{s}}\,ds$$

$$= \frac{1}{\sqrt{x}}\int \frac{\sin(x-s)g(s)}{s\sqrt{s}}\,ds.$$

28. First, we rewrite the equation as

$$y'' - \frac{2}{t^2}y = \frac{3t^2 - 1}{t^2}.$$

Then, we look for a solution of the homogeneous equation

$$y'' - \frac{2}{t^2}y = 0.$$

We look for a solution of the form t^α. Upon substituting a function of this form into the ODE, we have

$$\alpha(\alpha - 1)t^{\alpha-2} - \frac{2}{t^2}t^\alpha = 0,$$

which implies $\alpha(\alpha - 1) - 2 = 0$. This equation can be rewritten as $\alpha^2 - \alpha - 2 = 0$. The solutions of this equation are $\alpha = 2, -1$. Therefore, $y_1(t) = t^2$ and $y_2(t) = t^{-1}$ are solutions of the homogeneous equation.

Now to solve the inhomogeneous equation, we look for a solution of the form $y(t) = v(t)t^2$, using the fact that $y_1(t) = t^2$ is a solution of the homogeneous problem. Substituting a function of this form into the ODE, we see that v must satisfy

$$t^2 v'' + 4t v' = \frac{3t^2 - 1}{t^2}.$$

Dividing this equation by t^2, we have

$$v'' + \frac{4}{t}v' = \frac{3t^2 - 1}{t^4}.$$

This equation is linear in $w = v'$ with integrating factor $\mu(t) = e^{\int 4/t\, dt} = t^4$. Therefore,

$$t^4 v'' + 4t^3 v' = 3t^2 - 1 \implies (t^4 v')' = 3t^2 - 1 \implies t^4 v' = t^3 - t + c \implies v' = t^{-1} - t^{-3} + ct^{-4}.$$

Then integrating v', we conclude that $v = \ln(t) + \frac{1}{2}t^{-2} + c_1 t^{-3} + c_2$. Therefore, the general solution is given by

$$y(t) = v y_1 = t^2 \ln(t) + \frac{1}{2} + c_1 t^{-1} + c_2 t^2.$$

29. First, we rewrite the equation as

$$y'' - \frac{3}{x}y' + \frac{4}{x^2}y = \ln x.$$

Then, we look for a solution of the homogeneous equation

$$y'' - \frac{3}{x}y' + \frac{4}{x^2}y = 0.$$

We look for a solution of the form x^α. Upon substituting a function of this form into the ODE, we have

$$\alpha(\alpha - 1)x^{\alpha-2} - 3\alpha x^{\alpha-2} + 4x^{\alpha-2} = 0,$$

109

which implies $\alpha(\alpha-1)-3\alpha+4=0$. This equation can be rewritten as $\alpha^2-4\alpha+4=0$. The solution of this equation is $\alpha=2$. Therefore, $y_1(x)=x^2$ is a solution of the homogeneous equation.

Now to solve the inhomogeneous equation, we look for a solution of the form $y(x)=v(x)x^2$, using the fact that $y_1(x)=x^2$ is a solution of the homogeneous problem. Substituting a function of this form into the ODE, we see that v must satisfy

$$x^2 v'' + x v' = \ln x.$$

Dividing this equation by x^2, we have

$$v'' + \frac{1}{x}v' = \frac{\ln x}{x^2}.$$

This equation is linear in $w=v'$ with integrating factor $\mu(x)=e^{\int 1/x\,dx}=x$. Therefore,

$$x v'' + v' = \frac{\ln x}{x} \implies (xv')' = \frac{\ln x}{x} \implies xv' = \frac{1}{2}(\ln(x))^2 + c \implies v' = \frac{\ln(x)^2}{2x} + \frac{c}{x}.$$

Then integrating v', we conclude that $v = \frac{1}{6}(\ln(x))^3 + c_1\ln(x) + c_2$. Therefore, the general solution is given by

$$y(t) = vy_1 = \frac{1}{6}x^2(\ln(x))^3 + c_1 x^2 \ln(x) + c_2 x^2.$$

30. First, we rewrite the equation as

$$y'' - \frac{2}{t}y' + \frac{2}{t^2}y = 4.$$

Then, we look for a solution of the homogeneous equation

$$y'' - \frac{2}{t}y' + \frac{2}{t^2}y = 0.$$

We look for a solution of the form t^α. Upon substituting a function of this form into the ODE, we have

$$\alpha(\alpha-1)t^{\alpha-2} - 2\alpha t^{\alpha-2} + 2t^{\alpha-2} = 0,$$

which implies $\alpha(\alpha-1)-2\alpha+2=0$. This equation can be rewritten as $\alpha^2-3\alpha+2=0$. The solutions of this equation are $\alpha=2,1$. Therefore, $y_1(t)=t$ is a solution of the homogeneous equation.

Now to solve the inhomogeneous equation, we look for a solution of the form $y(t)=v(t)t$, using the fact that $y_1(t)=t$ is a solution of the homogeneous problem. Substituting a function of this form into the ODE, we see that v must satisfy

$$tv'' = 4.$$

Dividing this equation by t and integrating, we see that $v'=4\ln(t)+c_1$. Integrating again, we conclude that $v=4t\ln(t)-4t+c_1 t+c_2$. Therefore, the general solution is given by

$$y(t) = vy_1 = 4t^2\ln(t) - 4t^2 + c_1 t^2 + c_2 t.$$

31. First, we rewrite the equation as

$$y'' + \frac{7}{t}y' + \frac{5}{t^2}y = \frac{1}{t}.$$

Then, we look for a solution of the homogeneous equation

$$y'' + \frac{7}{t}y' + \frac{5}{t^2}y = 0.$$

We look for a solution of the form t^α. Upon substituting a function of this form into the ODE, we have

$$\alpha(\alpha - 1)t^{\alpha-2} + 7\alpha t^{\alpha-2} + 5t^{\alpha-2} = 0,$$

which implies $\alpha(\alpha - 1) + 7\alpha + 5 = 0$. This equation can be rewritten as $\alpha^2 + 6\alpha + 5 = 0$. The solutions of this equation are $\alpha = -5, -1$. In particular, $y_1(t) = t^{-1}$ is a solution of the homogeneous equation.

Now to solve the inhomogeneous equation, we look for a solution of the form $y(t) = v(t)t^{-1}$, using the fact that $y_1(t) = t^{-1}$ is a solution of the homogeneous problem. Substituting a function of this form into the ODE, we see that v must satisfy

$$v'' + \frac{5}{t}v' = 1.$$

This equation is linear with integrating factor $\mu(t) = \int e^{5/t}\, dt = t^5$. Therefore, the equation can be written as $t^5 v'' + 5t^4 v' = t^5$ which implies $(t^5 v')' = t^5$. Then integrating, we have $t^5 v' = t^6/6 + c_1$. Therefore, $v' = t/6 + c_1 t^{-5}$. Now integrating this equation, we conclude that $v = t^2/12 + c_1 t^{-4} + c_2$. Therefore, the general solution is given by

$$y(t) = vy_1 = \frac{t}{12} + c_1 t^{-5} + c_2 t^{-1}.$$

32. Suppose u is a solution of $L[u] = 0$ with initial conditions $u(t_0) = y_0$ and $u'(t_0) = y_1$. Suppose y is a solution of $L[y] = g(t)$ with initial conditions $y(t_0) = y_0$, $y'(t_0) = y_1$. Let $v(t) = y - u(t)$. We will show that v satisfies $L[v] = g(t)$ with initial conditions $v(t_0) = 0$, $v'(t_0) = 0$. First, $v(t_0) = y(t_0) - u(t_0) = y_0 - y_0 = 0$. Second, $v'(t_0) = y'(t_0) - u'(t_0) = y_1 - y_1 = 0$. Next, we see that

$$L[v(t)] = L[y-u(t)] = (y-u)''+p(t)(y-u)'+q(t)(y-u) = [y''+p(t)y'+q(t)y]-[u''+p(t)u'+q(t)u] = g(t)-0$$

Therefore, v is a solution of the indicated IVP.

33. We rewrite (27) as

$$Y(t) = -y_1(t) \int_{t_0}^{t} \frac{y_2(\tau)g(\tau)}{W(y_1, y_2)(\tau)}\, d\tau + y_2(t) \int_{t_0}^{t} \frac{y_1(\tau)g(\tau)}{W(y_1, y_2)(\tau)}\, d\tau.$$

Bringing the terms $y_1(t)$ and $y_2(t)$ inside the integrals and using the fact that $W(y_1, y_2)(\tau) = y_1(\tau)y_2'(\tau) - y_1'(\tau)y_2(\tau)$, the desired result holds. Then, using the product rule and the

111

Fundamental Theorem of Calculus, we see that

$$
\begin{aligned}
Y'(t) &= \frac{y_1(t)y_2(t) - y_1(t)y_2(t)}{y_1(t)y_2'(t) - y_1'(t)y_2(t)} g(t) + \int_{t_0}^t \frac{y_1(\tau)y_2'(t) - y_1'(t)y_2(\tau)}{y_1(\tau)y_2'(\tau) - y_1'(\tau)y_2(\tau)} g(\tau)\, d\tau \\
&= \int_{t_0}^t \frac{y_1(\tau)y_2'(t) - y_1'(t)y_2(\tau)}{y_1(\tau)y_2'(\tau) - y_1'(\tau)y_2(\tau)} g(\tau)\, d\tau.
\end{aligned}
$$

Differentiating again, we see that

$$
\begin{aligned}
Y''(t) &= \frac{y_1(t)y_2'(t) - y_1'(t)y_2(t)}{y_1(t)y_2'(t) - y_1'(t)y_2(t)} g(t) + \int_{t_0}^t \frac{y_1(\tau)y_2''(t) - y_1''(t)y_2(\tau)}{y_1(\tau)y_2'(\tau) - y_1'(\tau)y_2(\tau)} g(\tau)\, d\tau \\
&= g(t) + \int_{t_0}^t \frac{y_1(\tau)y_2''(t) - y_1''(t)y_2(\tau)}{y_1(\tau)y_2'(\tau) - y_1'(\tau)y_2(\tau)} g(\tau)\, d\tau.
\end{aligned}
$$

Now substituting Y back into our equation, we see that

$$
\begin{aligned}
L[Y] &= Y'' + p(t)Y' + q(t)Y \\
&= g(t) + \int_{t_0}^t \frac{y_1(\tau)(y_2''(t) - p(t)y_2'(t) + q(t)y_2(t)) - y_2(\tau)(y_1''(t) + p(t)y_1'(t) + q(t)y_1(t))}{y_1(\tau)y_2'(\tau) - y_1'(\tau)y_2(\tau)} g(\tau)\, d\tau \\
&= g(t).
\end{aligned}
$$

Further, $Y(t_0) = 0$ and $Y'(t_0) = 0$ since the upper and lower limits of integration are both t_0.

34.

(a) The general solution of the homogeneous problem is $y_h(t) = c_1 \cos(t) + c_2 \sin(t)$. First, we note that $W(y_1, y_2) = 1$. Therefore, using Problem 33, the solution of the IVP is given by

$$
\begin{aligned}
Y(t) &= \int_{t_0}^t \cos(\tau)\sin(t) - \cos(t)\sin(\tau) g(\tau)\, d\tau \\
&= \int_{t_0}^t \sin(t - \tau)g(\tau)\, d\tau.
\end{aligned}
$$

(b) Then by problem 32, the solution of the IVP with non-zero initial data is given by $y = u(t) + v(t)$ where u is a solution of (38) and v is a solution of (39). In part (a), we found a solution to equation (39). Therefore, we just need to find a solution of the homogeneous problem which satisfies the specified initial conditions. Given that the general solution of the homogeneous problem is $y_h(t) = c_1 \cos(t) + c_2 \sin(t)$, we just need the constants c_1 and c_2 to satisfy $c_1 = y_0$ and $c_2 = y_1$. Therefore, the solution of this IVP is

$$
y(t) = y_0 \cos(t) + y_1 \sin(t) + \int_0^t \sin(t - \tau)g(\tau)\, d\tau.
$$

35. The given linear operator $L[y] = (D - a)(D - b)y$ can be written as $L[y] = y'' - (a + b)y' + aby$. Therefore, $p(t) = -(a + b)$ and $q(t) = ab$. To solve the given inhomogeneous problem, we first need to solve the associated homogeneous problem. In particular, we need to look for the general solution of

$$y'' - (a + b)y' + aby = 0.$$

The roots of the associated equation are $\lambda = a, b$. Therefore, the general solution of the homogeneous problem is $y(t) = c_1 e^{at} + c_2 e^{bt}$. Letting $y_1(t) = e^{at}$ and $y_2(t) = e^{bt}$, we now use the result in problem 33. In particular, a solution of the indicated problem will be given by

$$Y(t) = \int_{t_0}^{t} \frac{e^{a\tau} e^{bt} - e^{at} e^{b\tau}}{be^{a\tau} e^{b\tau} - ae^{a\tau} e^{b\tau}} g(\tau)\, d\tau$$

$$= \int_{t_0}^{t} \frac{1}{b - a} [e^{b(t-\tau)} - e^{a(t-\tau)}] g(\tau)\, d\tau.$$

36. The given linear operator $L[y] = [D^2 - 2\lambda D + (\lambda^2 + \mu^2)]y$ can be written as $L[y] = y'' - 2\lambda y' + (\lambda^2 + \mu^2)y$. Therefore, $p(t) = -2\lambda$ and $q(t) = \lambda^2 + \mu^2$. To solve the given inhomogeneous problem, we first need to solve the associated homogeneous problem. In particular, we need to look for the general solution of

$$y'' - 2\lambda y' + (\lambda^2 + \mu^2)y = 0.$$

The roots of the associated equation are $\lambda = \lambda \pm i\mu$. Therefore, the general solution of the homogeneous problem is $y(t) = e^{\lambda t}(c_1 \cos(\mu t) + c_2 \sin(\mu t))$. Letting $y_1(t) = e^{\lambda t} \cos(\mu t)$ and $y_2(t) = e^{\lambda t} \sin(\mu t)$, we now use the result in problem 33. In particular, a solution of the indicated problem will be given by

$$Y(t) = \int_{t_0}^{t} \frac{e^{\lambda \tau} \cos(\mu \tau) e^{\lambda t} \sin(\mu t) - e^{\lambda t} \cos(\mu t) e^{\lambda \tau} \sin(\mu \tau)}{\mu e^{2\lambda \tau}} g(\tau)\, d\tau$$

$$= \frac{1}{\mu} \int_{t_0}^{t} e^{\lambda(t-\tau)} \sin(\mu(t - \tau)) g(\tau)\, d\tau.$$

37. The given linear operator $L[y] = (D - a)^2 y$ can be written as $L[y] = y'' - 2ay' + a^2 y$. Therefore, $p(t) = -2a$ and $q(t) = a^2$. To solve the given inhomogeneous problem, we first need to solve the associated homogeneous problem. In particular, we need to look for the general solution of

$$y'' - 2ay' + a^2 y = 0.$$

The repeated root of the associated equation is $\lambda = a$. Therefore, the general solution of the homogeneous problem is $y(t) = c_1 e^{at} + c_2 t e^{at}$. Letting $y_1(t) = e^{at}$ and $y_2(t) = t e^{at}$, we now use the result in problem 33. In particular, a solution of the indicated problem will be given by

$$Y(t) = \int_{t_0}^{t} e^{a(t-\tau)}(t - \tau)g(\tau)\, d\tau.$$

113

38. Consider the associated characteristic equation $\lambda^2 + b\lambda + c = 0$. If there are two distinct, real roots, then the solution of this IVP will be given by the result in problem (35). If there is one repeated root, then the solution will be given by the result in problem (37). If the roots have non-zero imaginary part, then the solution will be given by the result in problem (36). In any of these three cases, we see that the solution is of the form

$$\int_{t_0}^{t} K(t - s)g(s)\, ds.$$

39. Let $y(t) = v(t)y_1(t)$ where y_1 is a solution of the homogeneous equation. Substituting a function of this form into the ODE, we have

$$v''y_1 + 2v'y_1' + vy_1'' + p(t)[v'y_1 + vy_1'] + q(t)vy_1 = g(t).$$

By assumption, $y_1'' + p(t)y_1 + q(t)y_1 = 0$. Therefore, v must be a solution of the ODE

$$v''y_1 + [2y_1' + p(t)y_1]v' = g(t).$$

40. Write the equation as $y'' - t^{-1}(1+t)y' + t^{-1}y = te^{2t}$. Using the fact that $y_1(t) = 1+t$ is a solution of the homogeneous equation, we consider a function of the form $y(t) = (1+t)v(t)$. From the method in problem 39, we know that y will be a solution of the given ODE as long as v satisfies

$$(1 + t)v'' + \left[2 - t^{-1}(1 + t)^2\right]v' = te^{2t}.$$

We can simplify this equation to

$$v'' - \frac{1 + t^2}{t(t + 1)}v' = \frac{t}{t + 1}e^{2t}.$$

Using the fact that this equation is linear in v' and has an integrating factor $\mu(t) = t^{-1}(1 + t)^2 e^{-t}$, we find that $v' = (t^2 e^{2t} + c_1 t e^t)/(1+t)^2$. Integrating this equation, we conclude that $v(t) = e^{2t}/2 - e^{2t}/(t + 1) + c_1 e^t/(t + 1) + c_2$. Therefore, the solution of the original ODE is

$$y(t) = \frac{1}{2}(t - 1)e^{2t} + c_1 e^t + c_2(t + 1).$$

41. Write the equation as $y'' + t(1 - t)^{-1}y' - (1 - t)^{-1}y = 2(1 - t)e^{-t}$. Using the fact that $y_1(t) = e^t$ is a solution of the homogeneous equation, we consider a function of the form $y(t) = e^t v(t)$. From the method in problem 39, we know that y will be a solution of the given ODE as long as v satisfies

$$e^t v'' + \left[2e^t + t(1 - t)^{-1}e^t\right]v' = 2(1 - t)e^{-t}.$$

We can simplify this equation to

$$v'' + \frac{2 - t}{1 - t}v' = 2(1 - t)e^{-2t}.$$

Using the fact that this equation is linear in v' and has an integrating factor $\mu(t) = e^t/(t-1)$, we find that $v' = (t - 1)(2e^{-2t} + c_1 e^{-t})$. Integrating this equation, we conclude that $v(t) = (1/2 - t)e^{-2t} - c_1 t e^{-t} + c_2$. Therefore, the solution of the original ODE is

$$y(t) = \left(\frac{1}{2} - t\right)e^{-t} - c_1 t + c_2 e^t.$$

114

Chapter 5
Section 5.1

1.

The function is piecewise continuous.

2.

The function is neither continuous nor piecewise continuous.

3.

The function is continuous.

4.

The function is piecewise continuous.

5. The function is of exponential order. We can take $K = 3$, $a = 5$ and $M = 0$.

6. Since
$$f(t) = -e^{-3t} \implies |f(t)| = e^{-3t},$$
the function is of exponential order. We can take $K = 1$, $a = -3$ and $M = 0$.

7. Since
$$|f(t)| = |e^{2t} \sin(3t)| \leq e^{2t},$$
the function is of exponential order. We can take $K = 1$, $a = 2$ and $M = 0$.

8. Since
$$t^{10} = e^{\ln t^{10}} = e^{10 \ln t} \leq e^{10t}$$
for all $t \geq 0$, the function is of exponential order. We can take $K = 1$, $a = 10$ and $M = 0$.

9.
$$\cosh(t^2) = \frac{e^{t^2} + e^{-t^2}}{2}$$
is not of exponential order since for all $a > 0$, there exists an M such that $e^{t^2} \geq e^{at}$ for all $t \geq M$.

10. Since
$$|\sin(t^4)| \leq 1 = e^{0 \cdot t}$$
is of exponential order. We can take $K = 1$, $a = 0$ and $M = 0$.

11. Since
$$\frac{1}{1+t} \leq 1$$
for all $t \geq 0$, the function is of exponential order. We can take $K = 1$, $a = 0$ and $M = 0$.

12. e^{t^3} is not of exponential order since for all $a > 0$, there exists an $M > 0$ such that $e^{t^3} \geq e^{at}$ for all $t \geq M$.

13.

(a) For $f(t) = t$, the Laplace transform is

$$F(s) = \int_0^\infty e^{-st} t \, dt = \left. \frac{-e^{-st} t - e^{-st}}{s^2} \right|_{t=0}^\infty = \frac{1}{s^2}.$$

(b) For $f(t) = t^2$, the Laplace transform is

$$F(s) = \int_0^\infty e^{-st} t^2 \, dt = \left. -\frac{e^{-st} s^2 t^2 + 2e^{-st} st + 2e^{-st}}{s^3} \right|_{t=0}^\infty = \frac{2}{s^3}.$$

(c) For $f(t) = t^n$, the Laplace transform is

$$F(s) = \int_0^\infty e^{-st} t^n \, dt = \frac{n!}{s^{n+1}}.$$

14. For

$$f(t) = \begin{cases} t & 0 \le t \le 1 \\ 1 & 1 < t \end{cases}$$

the Laplace transform

$$F(s) = \int_0^\infty e^{-st} f(t) \, dt = \int_0^1 e^{-st} t \, dt + \int_1^\infty e^{-st} \, dt.$$

We integrate the first integral on the right-hand side by parts. In particular, we have

$$\int_0^t e^{-st} t \, dt = \left. \frac{te^{-st}}{-s} \right|_0^1 - \int_0^1 \frac{e^{-st}}{-s} \, dt$$

$$= \frac{e^{-s}}{-s} + \frac{1}{s} \int_0^1 e^{-st} \, dt$$

$$= \frac{e^{-s}}{-s} + \frac{1}{s} \cdot \left. \frac{e^{-st}}{-s} \right|_0^1$$

$$= \frac{e^{-s}}{-s} - \frac{e^{-s}}{s^2} + \frac{1}{s^2}.$$

The second integral above can be evaluated as follows:

$$\int_1^\infty e^{-st}, dt = \left. \frac{e^{-st}}{-s} \right|_1^\infty = \frac{e^{-s}}{s}.$$

Combining these two integrals, we have

$$F(s) = \frac{e^{-s}}{-s} - \frac{e^{-s}}{s^2} + \frac{1}{s^2} + \frac{e^{-s}}{s} = \frac{1}{s^2} - \frac{e^{-s}}{s^2}.$$

15. For

$$f(t) = \begin{cases} 0 & 0 \le t \le 1 \\ 1 & 1 < t \le 2 \\ 0 & 2 < t, \end{cases}$$

3

the Laplace transform is given by

$$F(s) = \int_0^1 e^{-st} \cdot 0 \, dt + \int_1^2 e^{-st} \, dt + \int_2^\infty e^{-st} \cdot 0 \, dt$$

$$= \int_1^2 e^{-st} \, dt = \frac{e^{-st}}{-s} \Big|_1^2$$

$$= \frac{e^{-2s}}{-s} - \frac{e^{-s}}{-s}$$

$$= \frac{e^{-s} - e^{-2s}}{s}.$$

16. For

$$f(t) = \begin{cases} 0 & 0 \le t \le 1 \\ e^{-t} & 1 < t \end{cases}$$

the Laplace transform is given by

$$F(s) = \int_1^\infty e^{-st} e^{-t} \, dt = \int_1^\infty e^{-(s+1)t} \, dt$$

$$= \frac{e^{-(s+1)t}}{-(s+1)} \Big|_1^\infty = 0 - \frac{e^{-(s+1)}}{-(s+1)}$$

$$= \frac{e^{-(s+1)}}{s+1}.$$

17. For

$$f(t) = \begin{cases} t^2 & 0 \le t \le 1 \\ 3 - t & 1 < t \le 2 \\ 1 & 2 < t \end{cases}$$

the Laplace transform is given by

$$F(s) = \int_0^1 e^{-st} t^2 \, dt + \int_1^2 e^{-st} (3 - t) \, dt + \int_2^\infty e^{-st} \, dt.$$

For the first integral on the right-hand side, we proceed as follows:

$$\int_0^1 e^{-st} t^2 \, dt = \frac{t^2 e^{-st}}{-s} \Big|_0^1 - \int_0^1 \frac{e^{-st}}{-s} \cdot 2t \, dt$$

$$= \frac{e^{-s}}{-s} + \frac{2}{s} \int_0^1 t e^{-st} \, dt.$$

By the solution to exercise 14, we know that

$$\int_0^1 t e^{-st} \, dt = \frac{e^{-s}}{-s} - \frac{e^{-s}}{s^2} + \frac{1}{s^2}.$$

Therefore,

$$\int_0^1 e^{-st}t^2\,dt = \frac{e^{-s}}{-s} + \frac{2}{s}\left[\frac{e^{-s}}{-s} - \frac{e^{-s}}{s^2} + \frac{1}{s^2}\right]$$

$$= \frac{e^{-s}}{-s} - \frac{2e^{-s}}{s^2} - \frac{2e^{-s}}{s^3} + \frac{2}{s^3}.$$

For the second integral on the right-hand side, we begin by separating it into two pieces.

$$\int_1^2 e^{-st}(3-t)\,dt = 3\int_1^2 e^{-st}\,dt - \int_1^2 te^{-st}\,dt \equiv A + B.$$

Now, for A, we have

$$3\int_1^2 e^{-st}\,dt = \left.\frac{3e^{-st}}{-s}\right|_1^2$$

$$= \frac{3e^{-s}}{s} - \frac{3e^{-2s}}{s}.$$

For term B, we have

$$-\int_1^2 te^{-st}\,dt = -\left[\left.\frac{te^{-st}}{-s}\right|_1^2 - \int_1^2 \frac{e^{-st}}{-s}\,dt\right]$$

$$= -\left[\frac{2e^{-2s}}{-s} - \frac{e^{-s}}{-s} + \frac{1}{s}\int_1^2 e^{-st}\,dt\right]$$

$$= -\left[\frac{2e^{-2s}}{-s} + \frac{e^{-s}}{s} = \frac{1}{s}\cdot\left.\frac{e^{-st}}{-s}\right|_1^2\right]$$

$$= -\left[\frac{2e^{-2s}}{-s} + \frac{e^{-s}}{s} - \frac{e^{-2s}}{s^2} + \frac{e^{-s}}{s^2}\right].$$

Now combining the results for A and B, we have

$$\int_1^2 e^{-st}(3-t)\,dt = \frac{3e^{-s}}{s} - \frac{3e^{-2s}}{s} - \left[\frac{2e^{-2s}}{-s} + \frac{e^{-s}}{s} - \frac{e^{-2s}}{s^2} + \frac{e^{-s}}{s^2}\right].$$

Finally, for the third integral above, we have

$$\int_2^\infty e^{-st}\,dt = \left.\frac{e^{-st}}{-s}\right|_2^\infty = 0 - \frac{e^{-2s}}{-s} = \frac{e^{-2s}}{s}.$$

Combining these integrals, we have

$$F(s) = \frac{e^{-s}}{-s} - \frac{2e^{-s}}{s^2} - \frac{2e^{-s}}{s^3} + \frac{2}{s^3} + \frac{3e^{-s}}{s} - \frac{3e^{-2s}}{s} + \frac{2e^{-2s}}{s} - \frac{e^{-s}}{s} + \frac{e^{-2s}}{s^2} - \frac{e^{-s}}{s^2} + \frac{e^{-2s}}{s}.$$

Simplifying, we conclude that the Laplace transform is given by

$$F(s) = \frac{e^{-s}}{s} - \frac{3e^{-s}}{s^2} + \frac{e^{-2s}}{s^2} - \frac{2e^{-s}}{s^3} + \frac{2}{s^3}.$$

18.

$$\int_0^a \cosh(bt)e^{-st}\,dt = \frac{1}{2}\int_0^a e^{(b-s)t}\,dt + \frac{1}{2}\int_0^a e^{-(b+s)t}\,dt$$

$$= \frac{1}{2}\left[\frac{1 - e^{(b-s)a}}{s-b}\right] + \frac{1}{2}\left[\frac{1 - e^{-(b+s)a}}{s+b}\right].$$

Taking the limit as $a \to \infty$, we conclude that

$$\int_0^\infty \cosh(bt)e^{-st}\,dt = \frac{1}{2}\left[\frac{1}{s-b}\right] + \frac{1}{2}\left[\frac{1}{s+b}\right] = \frac{s}{s^2 - b^2}.$$

This function is valid as long as $s > |b|$.

19.

$$\int_0^a \sinh(bt)e^{-st}\,dt = \frac{1}{2}\int_0^a e^{(b-s)t}\,dt - \frac{1}{2}\int_0^a e^{-(b+s)t}\,dt$$

$$= \frac{1}{2}\left[\frac{1 - e^{(b-s)a}}{s-b}\right] - \frac{1}{2}\left[\frac{1 - e^{-(b+s)a}}{s+b}\right].$$

Taking the limit as $a \to \infty$, we conclude that

$$\int_0^\infty \cosh(bt)e^{-st}\,dt = \frac{1}{2}\left[\frac{1}{s-b}\right] - \frac{1}{2}\left[\frac{1}{s+b}\right] = \frac{b}{s^2 - b^2}.$$

This function is valid as long as $s > |b|$.

20. We note that $e^{at}\cosh(bt) = (e^{(a+b)t} + e^{(a-b)t})/2$. Thefore,

$$\int_0^A e^{at}\sinh(bt)e^{-st}\,dt = \frac{1}{2}\left[\frac{1 - e^{(a+b-s)A}}{s-a+b}\right] + \frac{1}{2}\left[\frac{1 - e^{-(b-a+s)A}}{s+b-a}\right].$$

Taking the limit as $A \to \infty$,

$$\int_0^\infty e^{at}\sinh(bt)e^{-st}\,dt = \frac{1}{2}\left[\frac{1}{s-a+b}\right] + \frac{1}{2}\left[\frac{1}{s+b-a}\right] = \frac{s-a}{(s-a)^2 - b^2}.$$

This limit exists as long as $s - a > |b|$.

21. We note that $e^{at}\sinh(bt) = (e^{(a+b)t} - e^{(a-b)t})/2$. Therefore,

$$\int_0^A e^{at}\sinh(bt)e^{-st}\,dt = \frac{1}{2}\left[\frac{1 - e^{(a+b-s)A}}{s-a+b}\right] - \frac{1}{2}\left[\frac{1 - e^{-(b-a+s)A}}{s+b-a}\right].$$

Taking the limit as $A \to \infty$,

$$\int_0^\infty e^{at}\sinh(bt)e^{-st}\,dt = \frac{1}{2}\left[\frac{1}{s-a+b}\right] - \frac{1}{2}\left[\frac{1}{s+b-a}\right] = \frac{b}{(s-a)^2 - b^2}.$$

This limit exists as long as $s - a > |b|$.

22. Since the Laplace transform is a linear operator, we have

$$\mathcal{L}[\cos(bt)] = \frac{1}{2}\mathcal{L}[e^{ibt}] + \frac{1}{2}\mathcal{L}[e^{-ibt}].$$

Now

$$\int_0^\infty e^{\pm ibt} e^{-st}\, dt = \frac{1}{s \mp ib}.$$

Therefore,

$$\mathcal{L}[\cos(bt)] = \frac{1}{2}\left[\frac{1}{s-ib} + \frac{1}{s+ib}\right] = \frac{s}{s^2+b^2}.$$

23. Since the Laplace transform is a linear operator, we have

$$\mathcal{L}[e^{at}\sin(bt)] = \frac{1}{2i}\mathcal{L}[e^{(a+ib)t}] - \frac{1}{2}\mathcal{L}[e^{(a-ib)t}].$$

Now

$$\int_0^\infty e^{(a\pm ib)t} e^{-st}\, dt = \frac{1}{s-a\mp ib}.$$

Therefore,

$$\mathcal{L}[e^{at}\cos(bt)] = \frac{1}{2i}\left[\frac{1}{s-a-ib} - \frac{1}{s-a+ib}\right] = \frac{b}{(s-a)^2+b^2}.$$

24. Since the Laplace transform is a linear operator, we have

$$\mathcal{L}[e^{at}\cos(bt)] = \frac{1}{2}\mathcal{L}[e^{(a+ib)t}] + \frac{1}{2}\mathcal{L}[e^{(a-ib)t}].$$

Now

$$\int_0^\infty e^{(a\pm ib)t} e^{-st}\, dt = \frac{1}{s-a\mp ib}.$$

Therefore,

$$\mathcal{L}[e^{at}\cos(bt)] = \frac{1}{2}\left[\frac{1}{s-a-ib} + \frac{1}{s-a+ib}\right] = \frac{s-a}{(s-a)^2+b^2}.$$

25. Integrating by parts, we have

$$\int_0^A te^{at} \cdot e^{-st}\, dt = -\left.\frac{te^{(a-s)t}}{s-a}\right|_0^A + \int_0^A \frac{1}{s-a}e^{(a-s)t}\, dt$$

$$= \frac{1-e^{A(a-s)}+A(a-s)e^{A(a-s)}}{(s-a)^2}.$$

Taking the limit as $A \to \infty$,

$$\int_0^\infty e^{at} \cdot e^{-st}\, dt = \frac{1}{(s-a)^2}.$$

26. We will use the fact that $\sin(at) = (e^{iat} - e^{-iat})/2i$. Therefore,

$$\int_0^A t\sin(at)e^{-st}\,dt = \frac{1}{2i}\left[\int_0^A te^{(ia-s)t}\,dt - \int_0^A te^{-(ia+s)t}\,dt\right]$$

$$= \frac{1}{2i}\left[-\frac{te^{(ia-s)t}}{s-ia}\bigg|_0^A + \int_0^A \frac{1}{s-ia}e^{(ia-s)t}\,dt\right]$$

$$-\frac{1}{2i}\left[-\frac{te^{(-ia-s)t}}{s+ia}\bigg|_0^A + \int_0^A \frac{1}{s+ia}e^{(-ia-s)t}\,dt\right]$$

$$= \frac{1}{2i}\left[-\frac{Ae^{(ia-s)A}}{s-ia} + \frac{Ae^{(-ia-s)A}}{s+ia} + \frac{1}{(s-ia)^2} - \frac{1}{(s+ia)^2}\right].$$

Taking the limit as $A \to \infty$, we have

$$\int_0^\infty t\sin(at)e^{-st}\,dt = \frac{1}{2i}\left[\frac{(s+ia)^2 - (s-ia)^2}{(s-ia)^2(s+ia)^2}\right]$$

$$= \frac{2as}{(s^2+a^2)^2}.$$

27. We will use the fact that $t\cosh(at) = (te^{at} + te^{-at})/2$. We note that for any value of c,

$$\int_0^A te^{ct}\cdot e^{-st}\,dt = -\frac{te^{(c-s)t}}{s-c}\bigg|_0^A + \int_0^A \frac{1}{s-c}e^{(c-s)t}\,dt$$

$$= \frac{1 - e^{A(c-s)} + A(c-s)e^{A(c-s)}}{(s-c)^2}.$$

Taking the limit as $A \to \infty$, we have

$$\int_0^\infty te^{ct}\cdot e^{-st}\,dt = \frac{1}{(s-c)^2}.$$

Therefore,

$$\int_0^\infty t\cosh(at)\cdot e^{-st}\,dt = \frac{1}{2}\left[\frac{1}{(s-a)^2} + \frac{1}{(s+a)^2}\right] = \frac{s^2+a^2}{(s-a)^2(s+a)^2}.$$

28. Integrating by parts, we see that

$$\int_0^A t^n e^{at}\cdot e^{-st}\,dt = -\frac{t^n e^{(a-s)t}}{s-a}\bigg|_0^A + \int_0^A \frac{n}{s-a}t^{n-1}e^{(a-s)t}\,dt$$

$$= -\frac{A^n e^{-(s-a)A}}{s-a} + \int_0^A \frac{n}{s-a}t^{n-1}e^{(a-s)t}\,dt.$$

Continuing to integrate by parts, we conclude that

$$\int_0^A t^n e^{at}\cdot e^{-st}\,dt = -\frac{A^n e^{(a-s)A}}{s-a} - \frac{nA^{n-1}e^{(a-s)A}}{(s-a)^2} - \frac{n!Ae^{(a-s)A}}{(n-2)!(s-a)^3} - \cdots - \frac{n!(e^{(a-s)A}-1)}{(s-a)^{n+1}}.$$

We notice that as $A \to \infty$, all of the terms above except the last one go to zero. We conclude that

$$\int_0^\infty t^n e^{at} \cdot e^{-st} \, dt = \frac{n!}{(s-a)^{n+1}}.$$

29. We will use the fact that $\sin(at) = (e^{iat} - e^{-iat})/2i$. Therefore,

$$\int_0^A t^2 \sin(at) e^{-st} \, dt = \frac{1}{2i} \left[\int_0^A t^2 e^{(ia-s)t} \, dt - \int_0^A t^2 e^{-(ia+s)t} \, dt \right].$$

Now for any value of c, we note that

$$\int_0^A t^2 e^{(c-s)t} \, dt = \frac{e^{(c-s)t}(c-s)^2 t^2 - 2e^{(c-s)t}(c-s)t + 2e^{(c-s)t}}{(c-s)^3} \bigg|_0^A$$

$$= \frac{e^{(c-s)A}(c-s)^2 A^2 - 2e^{(c-s)A}(c-s)A + 2e^{(c-s)A} - 2}{(c-s)^3}.$$

Taking the limit as $A \to \infty$, we have

$$\int_0^\infty t^2 e^{(c-s)t} \, dt = -\frac{2}{(c-s)^3}.$$

Therefore,

$$\int_0^\infty t^2 \sin(at) e^{-st} \, dt = \frac{1}{2i} \left[-\frac{2}{(ia-s)^3} + \frac{2}{(ia+s)^3} \right] = \frac{2a(3s^2 - a^2)}{(s^2 + a^2)^3}.$$

30. We will use the fact that $t^2 \sinh(at) = (t^2 e^{at} - t^2 e^{-at})/2$. Using the fact that for any c,

$$\int_0^\infty t^2 e^{(c-s)t} \, dt = -\frac{2}{(c-s)^3}.$$

we see that

$$\int_0^\infty \sinh(at) e^{-st} \, dt = \frac{1}{2} \left[\frac{2}{(s-a)^3} - \frac{2}{(s+a)^3} \right] = \frac{2a(3s^2 + a^2)}{(s^2 - a^2)^3}.$$

31. First, we note that

$$\int_0^A (t^2 + 1)^{-1} \, dt = \tan^{-1} t \big|_0^A = \tan^{-1}(A) - \tan^{-1}(0) = \tan^{-1}(A).$$

Therefore, as $A \to \infty$, we have

$$\int_0^\infty (t^2 + 1)^{-1} \, dt = \lim_{A \to \infty} \tan^{-1}(A) = \frac{\pi}{2}.$$

Therefore, the integral converges.

9

32. Integrating by parts, we see that

$$\int_0^A te^{-t}\,dt = 1 - e^{-A} - Ae^{-A}.$$

As $A \to \infty$, $\int_0^\infty te^{-t}\,dt = 1$. Therefore, the integral converges.

33. Based on a series expansion, we see that for $t > 0$,

$$e^t > 1 + t + \frac{t^2}{2} > \frac{t^2}{2}.$$

Therefore, $t^{-2}e^t > 1/2$. Therefore,

$$\int_1^A t^{-2}e^t\,dt > \int_1^A \frac{1}{2}\,dt = \frac{A-1}{2}.$$

As $A \to \infty$, we see that $(A-1)/2 \to \infty$. Therefore, the integral diverges.

34. We see that

$$\left| \int_0^\infty e^{-t}\cos(t)\,dt \right| \leq \int_0^\infty |e^{-t}\cos(t)|\,dt \leq \int_0^\infty e^{-t}\,dt = 1.$$

Therefore, the integral converges.

35.

(a)

$$\begin{aligned}
|F(s)| &= \left| \int_0^{M+1} e^{-st}f(t)\,dt + \int_{M+1}^\infty e^{-st}f(t)\,dt \right| \\
&\leq \max_{0\leq t\leq M+1} |f(t)| \int_0^{M+1} e^{-st}\,dt + K\int_{M+1}^\infty e^{-st}e^{at}\,dt \\
&= \max_{0\leq t\leq M+1} |f(t)| \left.\frac{e^{-st}}{-s}\right|_0^{M+1} + K\left.\frac{e^{-(s-a)t}}{-(s-a)}\right|_{M+1}^\infty \\
&= \max_{0\leq t\leq M+1} |f(t)| \frac{1 - e^{-(M+1)t}}{s} + \frac{K}{s-a}e^{-(s-a)(M+1)}.
\end{aligned}$$

(b) Since $e^{-(s-a)(M+1)} \to 0$ as $s \to \infty$ and $(s-a)/s \to 1$ as $s \to \infty$, there exists a constant K_1 such that $Ke^{-(s-a)(M+1)} < K_1(s-a)/s$ for s sufficiently large.

(c) Using the results from parts (a) and (b) and the fact that $(1 - e^{-(M+1)t})/s < 1/s$, we conclude that

$$|F(s)| \leq \max_{0\leq t\leq M+1} |f(t)|\frac{1}{s} + \frac{K_1}{s}$$

for s sufficiently large.

36.

(a) Integrating by parts,

$$\int_0^A e^{-x} x^p \, dx = -e^{-x} x^p \Big|_0^A + p \int_0^A e^{-x} x^{p-1} \, dx$$

$$= -A^p e^{-A} + p \int_0^A e^{-x} x^{p-1} \, dx.$$

Taking the limit as $A \to \infty$, we see that

$$\Gamma(p+1) = \int_0^\infty e^{-x} x^p \, dx = p \int_0^\infty e^{-x} x^{p-1} \, dx = p\Gamma(p).$$

(b) We see that

$$\Gamma(1) = \int_0^\infty e^{-x} \, dx = 1.$$

(c) Let $p = n$. Using the result in part (a), we see that

$$\Gamma(n+1) = n\Gamma(n) = n(n-1)\Gamma(n-1) = \ldots = n(n-1)(n-2) \cdot 2 \cdot \Gamma(1).$$

Since $\Gamma(1) = 1$, we conclude that $\Gamma(n+1) = n!$.

(d) Using the result in part (a), we see that

$$\Gamma(p+n) = (p+n-1)\Gamma(p+n-1) = (p+n-1)(p+n-2)\Gamma(p+n-2) = (p+n-1)(p+n-2)\cdots(p+1)p\Gamma(p).$$

Therefore,

$$\frac{\Gamma(p+n)}{\Gamma(p)} = p(p+1)(p+2)\cdots(p+n-1).$$

Given that $\Gamma(1/2) = \sqrt{\pi}$, it follows that

$$\Gamma\left(\frac{3}{2}\right) = \frac{1}{2}\Gamma\left(\frac{1}{2}\right) = \frac{\sqrt{\pi}}{2}$$

and

$$\Gamma\left(\frac{11}{2}\right) = \frac{9}{2} \cdot \frac{7}{2} \cdot \frac{5}{2} \cdot \frac{3}{2}\Gamma\left(\frac{3}{2}\right) = \frac{945\sqrt{\pi}}{32}.$$

37.

(a) By definition,

$$\mathcal{L}(t^p) = \int_0^\infty e^{-st} t^p \, dt.$$

Now letting $x = st$, we see that $dx = s \, dt$. Therefore,

$$\mathcal{L}(t^p) = \int_0^\infty e^{-st} t^p \, dt = \int_0^\infty e^{-st} \left(\frac{x}{s}\right)^p \frac{dx}{s} = \frac{1}{s^{p+1}} \int_0^\infty e^{-st} x^p \, dx = \frac{\Gamma(p+1)}{s^{p+1}}.$$

11

(b) Using part (a) and the fact that for n a positive integer, $\Gamma(n+1) = n!$, we conclude that $\mathcal{L}(t^n) = n!/s^{n+1}$.

(c) By definition,

$$\mathcal{L}(t^{-1/2}) = \int_0^\infty e^{-st} t^{-1/2} \, dt.$$

Making the change of variables $x = \sqrt{st}$, we see that $x^2 = st$ implies $2x \, dx = s \, dt$. Therefore,

$$\int_0^\infty e^{-st} t^{-1/2} \, dt = \int_0^\infty e^{-x^2} \left(\frac{x^2}{s}\right)^{-1/2} \frac{2x \, dx}{s} = \frac{2}{\sqrt{s}} \int_0^\infty e^{-x^2} \, dx.$$

(d) Using the result from part (a), we see that

$$\mathcal{L}(t^{1/2}) = \frac{\Gamma(3/2)}{s^{3/2}}.$$

Then, using the result from problem 36, we know that $\Gamma(3/2) = \sqrt{\pi}/2$. Therefore,

$$\mathcal{L}(t^{1/2}) = \frac{\sqrt{\pi}}{2s^{3/2}}.$$

Section 5.2

1. By example 7 in Section 5.1, for $f(t) = \sin(4t)$, $F(s) = \dfrac{4}{s^2 + 16}$. Therefore, by Theorem 5.2.1,

$$\mathcal{L}(e^{-2t} \sin(4t)) = F(s+2) = \frac{4}{(s+2)^2 + 16}.$$

2. First, we compute the Laplace transform of $f(t) = \cos(2t)$. For $f(t) = \cos(2t)$,

$$F(s) = \int_0^\infty e^{-st} \cos(2t) \, dt$$

$$= \frac{1}{2} \int_0^\infty e^{-st} (e^{2it} + e^{-2it}) \, dt$$

$$= \frac{1}{2} \int_0^\infty (e^{-(s-2i)t} + e^{-(s+2i)t}) \, dt$$

$$= \lim_{b \to \infty} \frac{1}{2} \left[\frac{e^{-(s-2i)t}}{-(s-2i)} + \frac{e^{-(s+2i)t}}{-(s+2i)} \Big|_0^b \right]$$

$$= \frac{1}{2} \left[\frac{1}{s-2i} + \frac{1}{s+2i} \right]$$

$$= \frac{2}{s^2 + 4}.$$

Therefore, by Theorem 5.2.1,

$$\mathcal{L}(e^{3t} \cos(2t)) = F(s-3) = \frac{s-3}{(s-3)^2 + 4}.$$

3. By Corollary 5.2.5, $\mathcal{L}(t^n) = \dfrac{n!}{s^{n+1}}$. Combining this fact with linearity, we have

$$\mathcal{L}(t^3 - 4t^2 + 5) = \mathcal{L}(t^3) - 4\mathcal{L}(t^2) + 5\mathcal{L}(1)$$
$$= \frac{3!}{s^4} - 4 \cdot \frac{2!}{s^3} + 5 \cdot \frac{0!}{s^1}$$
$$= \frac{6}{s^4} - \frac{8}{s^3} + \frac{5}{s}.$$

4. By Theorem 5.2.4, $\mathcal{L}(t\cos(t)) = (-1)F'(s)$ where $F(s) = \mathcal{L}(\cos t)$.

$$\mathcal{L}(\cos t) = \frac{s}{s^2 + 1} \implies F'(s) = \frac{d}{ds}\left[\frac{s}{s^2 + 1}\right] = \frac{1 - s^2}{(s^2 + 1)^2}.$$

Therefore,

$$\mathcal{L}(t\cos t) = \frac{s^2 - 1}{(s^2 + 1)^2}.$$

5. By Theorem 5.2.1, $\mathcal{L}(e^{-4t}(t^2 + 1)^2) = F(s + 4)$ where $F(s) = \mathcal{L}((t^2 + 1)^2)$. Using linearity and Corollary 5.2.5, we have

$$\mathcal{L}((t^2 + 1)^2) = \mathcal{L}(t^4 + 2t^2 + 1)$$
$$= \frac{4!}{s^5} + \frac{4}{s^3} + \frac{1}{s}.$$

Therefore,

$$\mathcal{L}(e^{-4t}(t^2 + 1)^2) = \frac{4!}{(s + 4)^5} + \frac{4}{(s + 4)^3} + \frac{1}{s + 4}.$$

6. By Theorem 5.2.1, $\mathcal{L}(t^5 e^{2t}) = F(s - 2)$ where $F(s) = \mathcal{L}(t^5)$. By Corollary 5.2.5,

$$\mathcal{L}(t^5) = \frac{5!}{s^6}.$$

Therefore,

$$\mathcal{L}(t^5 e^{2t}) = \frac{5!}{(s - 2)^6}.$$

7. By Theorem 5.2.4, $\mathcal{L}(t^2 \sin(bt)) = F''(s)$ where $F(s) = \mathcal{L}(\sin(bt))$. By example 7 of Section 5.1,

$$\mathcal{L}(\sin(bt)) = \frac{b}{s^2 + b^2}.$$

Therefore,

$$F(s) = \frac{b}{s^2 + b^2} \implies F'(s) = \frac{-2sb}{(s^2 + b^2)^2}$$
$$\implies F''(s) = \frac{6bs^2 - 2b^3}{(s^2 + b^2)^3}.$$

Therefore,
$$\mathcal{L}(t^2 \sin(bt)) = \frac{6bs^2 - 2b^3}{(s^2 + b^2)^3}.$$

8. By Theorem 5.2.1, $\mathcal{L}(e^{at}t^n) = F(s-a)$ where $F(s) = \mathcal{L}(t^n) = \frac{n!}{s^{n+1}}$. Therefore,
$$\mathcal{L}(e^{at}t^n) = \frac{n!}{(s-a)^{n+1}}.$$

9. By Theorem 5.2.4, $\mathcal{L}(te^{at} \sin(bt)) = (-1)F'(s)$ where $F(s) = \mathcal{L}(e^{at} \sin(bt))$. Further, $\mathcal{L}(e^{at} \sin(bt)) = G(s-a)$ where $G(s) = \mathcal{L}(\sin(bt)) = \frac{b}{s^2 + b^2}$. Therefore,
$$\mathcal{L}(e^{at} \sin(bt)) = \frac{b}{(s-a)^2 + b^2}$$

and
$$\frac{d}{ds}\left[\frac{b}{(s-a)^2 + b^2}\right] = \frac{-2b(s-a)}{((s-a)^2 + b^2)^2}.$$
Therefore,
$$\mathcal{L}(te^{at} \sin(bt)) = \frac{2b(s-a)}{((s-a)^2 + b^2)^2}.$$

10. By Theorem 5.2.4, $\mathcal{L}(te^{at} \cos(bt)) = (-1)F'(s)$ where $F(s) = \mathcal{L}(e^{at} \cos(bt))$. Further, $\mathcal{L}(e^{at} \cos(bt)) = G(s-a)$ where $G(s) = \mathcal{L}(\cos(bt))$. Now
$$\mathcal{L}(\cos(bt)) = \frac{s}{s^2 + b^2} \implies \mathcal{L}(e^{at} \cos(bt)) = \frac{s-a}{(s-a)^2 + b^2}$$

and
$$\frac{d}{ds}\left[\frac{s-a}{(s-a)^2 + b^2}\right] = \frac{b^2 - (s-a)^2}{((s-a)^2 + b^2)^2}.$$
Therefore,
$$\mathcal{L}(te^{at} \cos(bt)) = \frac{(s-a)^2 - b^2}{((s-a)^2 + b^2)^2}.$$

11.

(a) Let $g_1(t) = \int_0^t f(t_1)\, dt_1$. Therefore, $g_1'(t) = f(t)$ which implies $\mathcal{L}(g_1'(t)) = \mathcal{L}(f(t)) = F(s)$. But also, by Theorem 5.2.2, $\mathcal{L}(g_1'(t)) = s\mathcal{L}(g_1(t)) - g_1(0) = s\mathcal{L}(g_1(t))$. Therefore,
$$s\mathcal{L}(g_1(t)) = \mathcal{L}(g_1'(t)) = F(s) \implies \mathcal{L}(g_1(t)) = \frac{F(s)}{s}.$$

That is,
$$\mathcal{L}\left(\int_0^t f(t_1)\, dt_1\right) = \frac{1}{s}F(s).$$

14

(b) Let
$$g_n(t) = \int_0^t \int_0^{t_n} \cdots \int_0^{t_2} f(t_1)\, dt_1 \cdots dt_n.$$

Therefore,
$$g_n'(t) = \int_0^t \int_0^{t_{n-1}} \cdots \int_0^{t_2} f(t_1)\, dt_1 \cdots dt_{n-1}.$$

Continuing, we conclude that $g_n^{(n)}(t) = f(t)$. Therefore, $\mathcal{L}(g_n^{(n)}(t)) = \mathcal{L}(f(t)) = F(s)$. But also, by Corollary 5.2.3,

$$\mathcal{L}(g_n^{(n)}(t)) = s^n \mathcal{L}(g_n(t)) - s^{n-1} g_n(0) - \ldots - s g_n^{(n-2)}(0) - g_n^{(n-1)}(0) = s^n \mathcal{L}(g_n(t)).$$

Therefore,
$$\mathcal{L}(g_n(t)) = \frac{1}{s^n} \mathcal{L}(g_n^{(n)}(t)) = \frac{1}{s^n} F(s).$$

12.

$$
\begin{aligned}
y'' + 2y' - 2y = 0 \implies & \mathcal{L}(y'') + 2\mathcal{L}(y') - 2\mathcal{L}(y) = 0 \\
\implies & [s^2 \mathcal{L}(y) - sy(0) - y'(0)] + 2[s\mathcal{L}(y) - y(0)] - 2\mathcal{L}(y) = 0 \\
\implies & [s^2 + 2s - 2]\mathcal{L}(y) - s - 1 - 2 = 0 \\
\implies & [s^2 + 2s - 2]\mathcal{L}(y) - s - 3 = 0 \\
\implies & \mathcal{L}(y) = \frac{s+3}{s^2 + 2s - 2}.
\end{aligned}
$$

13.

$$
\begin{aligned}
9y'' + 12y' + 4y = 0 \implies & 9\mathcal{L}(y'') + 12\mathcal{L}(y') + 4\mathcal{L}(y) = 0 \\
\implies & 9[s^2 \mathcal{L}(y) - sy(0) - y'(0)] + 12[s\mathcal{L}(y) - y(0)] + 4\mathcal{L}(y) = 0 \\
\implies & 9[s^2 \mathcal{L}(y) - 2s + 1] + 12[s\mathcal{L}(y) - 2] + 4\mathcal{L}(y) = 0 \\
\implies & [9s^2 + 12s + 4]\mathcal{L}(y) - 18s + 9 - 24 = 0 \\
\implies & \mathcal{L}(y) = \frac{18s + 15}{9s^2 + 12s + 4}.
\end{aligned}
$$

14.

$$
\begin{aligned}
y'' + 4y' + 3y = 0 \implies & \mathcal{L}(y'') + 4\mathcal{L}(y') + 3\mathcal{L}(y) = 0 \\
\implies & [s^2 \mathcal{L}(y) - sy(0) - y'(0)] + 4[s\mathcal{L}(y) - y(0)] + 3\mathcal{L}(y) = 0 \\
\implies & [s^2 \mathcal{L}(y) - 2s + 1] + 4[s\mathcal{L}(y) - 2] + 3\mathcal{L}(y) = 0 \\
\implies & [s^2 + 4s + 3]\mathcal{L}(y) - 2s + 1 - 8 = 0 \\
\implies & \mathcal{L}(y) = \frac{2s + 7}{s^2 + 4s + 3}.
\end{aligned}
$$

15.

$$6y'' + 5y' + y = 0 \implies 6\mathcal{L}(y'') + 5\mathcal{L}(y') + \mathcal{L}(y) = 0$$
$$\implies 6[s^2\mathcal{L}(y) - sy(0) - y'(0)] + 5[s\mathcal{L}(y) - y(0)] + \mathcal{L}(y) = 0$$
$$\implies 6[s^2\mathcal{L}(y) - 4s] + 5[s\mathcal{L}(y) - 4] + \mathcal{L}(y) = 0$$
$$\implies [6s^2 + 5s + 1]\mathcal{L}(y) - 24s - 20 = 0$$
$$\implies \mathcal{L}(y) = \frac{24s + 20}{6s^2 + 5s + 1}.$$

16.

$$y'' - 2y' + 2y = t^2 e^t + 4 \implies \mathcal{L}(y'') - 2\mathcal{L}(y') + 2\mathcal{L}(y) = \mathcal{L}(t^2 e^t + 4)$$
$$\implies [s^2\mathcal{L}(y) - sy(0) - y'(0)] - 2[s\mathcal{L}(y) - y(0)] + 2\mathcal{L}(y) = \mathcal{L}(t^2 e^t + 4).$$

Now $\mathcal{L}(t^2 e^t + 4) = \mathcal{L}(t^2 e^t + 4\mathcal{L}(1)$. First, $\mathcal{L}(t^2 e^t) = F(s-1)$ where $F(s) = \mathcal{L}(t^2) = 2/s^3$. Therefore, $\mathcal{L}(t^2 e^t) = 2/(s-1)^3$. Further, $4\mathcal{L}(1) = \frac{4}{s}$. Therefore,

$$\mathcal{L}(t^2 e^t + 4) = \frac{2}{(s-1)^3} + \frac{4}{s}.$$

Therefore,

$$[s^2\mathcal{L}(y) - sy(0) - y'(0)] - 2[s\mathcal{L}(y) - y(0)] + 2\mathcal{L}(y) = \frac{2}{(s-1)^3} + \frac{4}{s}$$
$$\implies [s^2\mathcal{L}(y) - s - 1] - 2[s\mathcal{L}(y) - 1] + 2\mathcal{L}(y) = \frac{2}{(s-1)^3} + \frac{4}{s}$$
$$\implies [s^2 - 2s + 2]\mathcal{L}(y) - s - 1 + 2 = \frac{2}{(s-1)^3} + \frac{4}{s}$$
$$\implies \mathcal{L}(y) = \frac{1}{(s^2 - 2s + 2)} \cdot \left[\frac{2}{(s-1)^3} + \frac{4}{s} + s - 1\right].$$

17.

$$y'' - 2y' - 3y = t^2 + 4 \implies \mathcal{L}(y'') - 2\mathcal{L}(y') - 3\mathcal{L}(y) = \mathcal{L}(t^2 + 4)$$
$$\implies [s^2\mathcal{L}(y) - sy(0) - y'(0)] - 2[s\mathcal{L}(y) - y(0)] - 3\mathcal{L}(y) = \frac{2!}{s^3} + \frac{4}{s}$$
$$\implies [s^2\mathcal{L}(y) - s] - 2[s\mathcal{L}(y) - 1] - 3\mathcal{L}(y) = \frac{2}{s^3} + \frac{4}{s}$$
$$\implies [s^2 - 2s - 3]\mathcal{L}(y) - s + 2 = \frac{2}{s^3} + \frac{4}{s}$$
$$\implies \mathcal{L}(y) = \frac{1}{s^2 - 2s - 3}\left[\frac{2}{s^3} + \frac{4}{s} + s - 2\right].$$

18.

$$y'' + 4y = 3e^{-2t}\sin(2t) \implies \mathcal{L}(y'') + 4\mathcal{L}(y) = 3\mathcal{L}(e^{-2t}\sin(2t))$$
$$\implies [s^2\mathcal{L}(y) - sy(0) - y'(0)] + 4\mathcal{L}(y) = 3F(s+2)$$

16

where $F(s) = \mathcal{L}(\sin(2t)) = 2/(s^2 + 4)$. Therefore,

$$[s^2\mathcal{L}(y) - sy(0) - y'(0)] + 4\mathcal{L}(y) = \frac{6}{(s+2)^2 + 4}$$

$$\implies [s^2\mathcal{L}(y) - 2s + 1] + 4\mathcal{L}(y) = \frac{6}{(s+2)^2 + 4}$$

$$\implies [s^2 + 4]\mathcal{L}(y) - 2s + 1 = \frac{6}{(s+2)^2 + 4}$$

$$\implies \mathcal{L}(y) = \frac{1}{s^2 + 4}\left[\frac{6}{(s+2)^2 + 4} + 2s - 1\right].$$

19.

$$y'' + 2y' + 5y = t\cos(2t) \implies \mathcal{L}(y'') + 2\mathcal{L}(y') + 5\mathcal{L}(y) = \mathcal{L}(t\cos(2t))$$
$$\implies [s^2\mathcal{L}(y) - sy(0) - y'(0)] + 2[s\mathcal{L}(y) - y(0)] + 5\mathcal{L}(y) = (-1)F'(s)$$

where $F(s) = \mathcal{L}(\cos(2t)) = s/(s^2 + 4)$. Therefore, $F'(s) = (4 - s^2)/(s^2 + 4)^2$ implies

$$\mathcal{L}(t\cos(2t)) = \frac{s^2 - 4}{(s^2 + 4)^2}.$$

Therefore,

$$[s^2\mathcal{L}(y) - s] + 2[s\mathcal{L}(y) - 1] + 5\mathcal{L}(y) = \frac{s^2 - 4}{(s^2 + 4)^2}$$

$$\implies [s^2 + 2s + 5]\mathcal{L}(y) - s - 2 = \frac{s^2 - 4}{(s^2 + 4)^2}$$

$$\implies \mathcal{L}(y) = \frac{1}{s^2 + 2s + 5}\left[\frac{s^2 - 4}{(s^2 + 4)^2} + s + 2\right].$$

20.

$$y''' + y'' + y' + y = 0 \implies \mathcal{L}(y''') + \mathcal{L}(y'') + \mathcal{L}(y') + \mathcal{L}(y) = 0$$
$$\implies [s^3\mathcal{L}(y) - s^2y(0) - sy'(0) - y''(0)] + [s^2\mathcal{L}(y) - sy(0) - y'(0)] + [s\mathcal{L}(y) - y(0)] + \mathcal{L}(y) = 0$$
$$\implies [s^3\mathcal{L}(y) - s^2 + 2] + [s^2\mathcal{L}(y) - s] + [s\mathcal{L}(y) - 1] + \mathcal{L}(y) = 0$$
$$\implies [s^3 + s^2 + s + 1]\mathcal{L}(y) - s^2 + 2 - s - 1 = 0$$
$$\implies \mathcal{L}(y) = \frac{s^2 + s - 1}{s^3 + s^2 + s + 1}.$$

21.

$$y'''' - y = te^{-t} \implies \mathcal{L}(y'''') - \mathcal{L}(y) = \mathcal{L}(te^{-t})$$
$$\implies [s^4\mathcal{L}(y) - s^3y(0) - s^2y'(0) - sy''(0) - y'''(0)] - \mathcal{L}(y) = F(s+1)$$

17

where $F(s) = \mathcal{L}(t) = 1/s^2$. Therefore, $\mathcal{L}(te^{-t}) = 1/(s+1)^2$ which implies

$$[s^4 \mathcal{L}(y) - 1] - \mathcal{L}(y) = \frac{1}{(s+1)^2} \implies [s^4 - 1]\mathcal{L}(y) - 1 = \frac{1}{(s+1)^2}$$

$$\implies \mathcal{L}(y) = \frac{1}{s^4 - 1}\left[\frac{1}{(s+1)^2} + 1\right].$$

22.

$$Lq'' + Rq' + \frac{1}{c}q = e(t) \implies L\mathcal{L}(q'') + R\mathcal{L}(q') + \frac{1}{c}\mathcal{L}(q) = \mathcal{L}(e)$$

$$\implies L[s^2\mathcal{L}(q) - sq(0) - q'(0)] + R[s\mathcal{L}(q) - q(0)] + \frac{1}{c}\mathcal{L}(q) = E(s)$$

$$\implies \left[Ls^2 + Rs + \frac{1}{c}\right]\mathcal{L}(q) - Lsq(0) - Lq'(0) - Rq(0) = E(s)$$

which implies

$$Q(s) = \frac{Lsq(0) + Lq'(0) + Rq(0) + E(s)}{Ls^2 + Rs + \frac{1}{c}}$$

$$= \frac{(Ls + R)q(0) + Li(0) + E(s)}{Ls^2 + Rs + \frac{1}{c}}.$$

Then from eqn. (9), $I(s) = sQ(s) - q(0)$ implies

$$I(s) = s\left[\frac{(Ls + R)q(0) + Li(0) + E(s)}{Ls^2 + Rs + \frac{1}{c}}\right] - q(0)$$

$$= \frac{Ls^2q(0) + Rsq(0) + Lsi(0) + sE(s) - Ls^2q(0) - Rsq(0) - \frac{1}{c}q(0)}{Ls^2 + Rs + \frac{1}{c}}$$

$$= \frac{Lsi(0) + sE(s) - \frac{1}{c}q(0)}{Ls^2 + Rs + \frac{1}{c}}.$$

23.

$$y'' + 4y = f(t) \implies \mathcal{L}(y'') + 4\mathcal{L}(y) = \mathcal{L}(f(t))$$

$$\implies [s^2\mathcal{L}(y) - sy(0) - y'(0)] + 4\mathcal{L}(y) = \mathcal{L}(f(t)).$$

Now

$$\mathcal{L}(f(t)) = \int_0^\infty f(t)e^{-st}\, dt$$

$$= \int_0^\pi e^{-st}\, dt$$

$$= \frac{1}{s} - \frac{e^{-\pi s}}{s}.$$

Therefore,

$$[s^2 \mathcal{L}(y) - s] + 4\mathcal{L}(y) = \frac{1}{s} - \frac{e^{-\pi s}}{s}$$

$$\implies [s^2 + 4]\mathcal{L}(y) - s = \frac{1 - e^{-\pi s}}{s}$$

$$\implies \mathcal{L}(y) = \frac{s^2 + 1 - e^{-\pi s}}{s(s^2 + 4)}.$$

24.

$$y'' + y = f(t) \implies \mathcal{L}(y'') + \mathcal{L}(y) = \mathcal{L}(f(t))$$
$$\implies [s^2\mathcal{L}(y) - sy(0) - y'(0)] + \mathcal{L}(y) = \mathcal{L}(f(t)).$$

Now

$$\mathcal{L}(f(t)) = \int_0^\infty f(t)e^{-st}\, dt$$

$$= \int_0^1 te^{-st}\, dt$$

$$= \frac{1}{s^2} - \frac{e^{-s}}{s} - \frac{e^{-s}}{s^2}.$$

Therefore,

$$s^2\mathcal{L}(y) + \mathcal{L}(y) = \frac{1}{s^2} - \frac{e^{-s}}{s} - \frac{e^{-s}}{s^2}$$

$$\implies [s^2 + 1]\mathcal{L}(y) = \frac{1 - se^{-s} - e^{-s}}{s^2}$$

$$\implies \mathcal{L}(y) = \frac{1 - se^{-s} - e^{-s}}{s^2(s^2 + 1)}.$$

25.

$$y'' + 4y = f(t) \implies \mathcal{L}(y'') + 4\mathcal{L}(y) = \mathcal{L}(f(t))$$
$$\implies [s^2\mathcal{L}(y) - sy(0) - y'(0)] + 4\mathcal{L}(y) = \mathcal{L}(f(t)).$$

Now

$$\mathcal{L}(f(t)) = \int_0^\infty f(t)e^{-st}\, dt$$

$$= \int_0^1 te^{-st}\, dt + \int_1^\infty e^{-st}\, dt$$

$$= \frac{1 - e^{-s}}{s^2}.$$

Therefore,

$$s^2 \mathcal{L}(y) + 4\mathcal{L}(y) = \frac{1 - e^{-s}}{s^2}$$

$$\implies [s^2 + 4]\mathcal{L}(y) = \frac{1 - e^{-s}}{s^2}$$

$$\implies \mathcal{L}(y) = \frac{1 - e^{-s}}{s^2(s^2 + 4)}.$$

26. The equation describing the amount of salt in the water at time t is given by

$$\frac{dy}{dt} = \text{ rate in } - \text{ rate out.}$$

Here, the rate in $= \dfrac{1}{2}$ lb/gal $\cdot 2$ gal/min $= 1$ lb/min. The rate out $= y(t)/100$ lb/gal $\cdot 2$ gal/min $= y/50$ lb/min. Therefore, for the first 10 minutes, the equation describing the amount of salt in the mixture is given by

$$\frac{dy}{dt} + \frac{y}{50} = 1 \qquad 0 \le t \le 10.$$

Then after 10 minutes, the rate of salt in becomes 0 gal/min. The rate of salt out remains the same. Therefore,

$$\frac{dy}{dt} + \frac{y}{50} = 0 \qquad 10 < t < \infty.$$

Combining these two equations, we have

$$\frac{dy}{dt} + \frac{y}{50} = \begin{cases} 1 & 0 \le t \le 10 \\ 0 & 10 < t < \infty. \end{cases}$$

Since the tank is assumed to initially contain fresh water, our initial condition is $y(0) = 0$. Applying the Laplace transform, we have

$$y' + \frac{1}{50}y = f(t) \implies \mathcal{L}(y') + \frac{1}{50}\mathcal{L}(y) = \mathcal{L}(f(t))$$

$$\implies [s\mathcal{L}(y) - y(0)] + \frac{1}{50}\mathcal{L}(y) = \mathcal{L}(f(t)).$$

Now

$$\mathcal{L}(f(t)) = \int_0^\infty f(t)e^{-st}\, dt$$

$$= \int_0^{10} e^{-st}\, dt$$

$$= \frac{1 - e^{-10s}}{s}.$$

Therefore,

$$sL(y) + \frac{1}{50}L(y) = \frac{1 - e^{-10s}}{s}$$

$$\implies [s + 1/50]L(y) = \frac{1 - e^{-10s}}{s}$$

$$\implies L(y) = \frac{1 - e^{-10s}}{s(s + 1/50)}.$$

27. The equation of motion is given by

$$my'' + \gamma y' + ky = \begin{cases} F_0 t & 0 \le t \le T \\ 0 & T < t < \infty. \end{cases}$$

Since the mass is initially in the equilibrium state, the initial conditions are $y(0) = 0$ and $y'(0) = 0$. Applying the Laplace transform, we have

$$my'' + \gamma y' + ky = f(t) \implies mL(y'') + \gamma L(y') + kL(y) = L(f(t))$$
$$\implies m[s^2 L(y) - sy(0) - y'(0)] + \gamma[sL(y) - y(0)] + kL(y) = L(f(t)).$$

Now

$$L(f(t)) = \int_0^\infty f(t)e^{-st}\,dt$$

$$= \int_0^T F_0 t e^{-st}\,dt$$

$$= F_0 \frac{1 - sTe^{-sT} - e^{-sT}}{s^2}.$$

Therefore,

$$ms^2 L(y) + \gamma sL(y) + kL(y) = F_0 \frac{1 - sTe^{-sT} - e^{-sT}}{s^2}$$

$$\implies [ms^2 + \gamma s + k]L(y) = F_0 \frac{1 - sTe^{-sT} - e^{-sT}}{s^2}$$

$$\implies L(y) = F_0 \frac{1 - sTe^{-sT} - e^{-sT}}{s^2(ms^2 + \gamma s + k)}.$$

28.

(a)

$$\mathcal{L}(\sin(t)) = \int_0^\infty \sin(t)e^{-st}\,dt$$

$$= \int_0^\infty \sum_{n=0}^\infty \frac{(-1)^n t^{2n+1}}{(2n+1)!} e^{-st}\,dt$$

$$= \sum_{n=0}^\infty \frac{(-1)^n}{(2n+1)!} \int_0^\infty t^{2n+1} e^{-st}\,dt$$

$$= \sum_{n=0}^\infty \frac{(-1)^n}{(2n+1)!} \left[\frac{(2n+1)!}{s^{2n+2}} \right]$$

$$= \sum_{n=0}^\infty \frac{(-1)^n}{s^{2n+2}}$$

$$= \frac{1}{s^2} \sum_{n=0}^\infty \left(\frac{-1}{s^2} \right)^n$$

$$= \frac{1}{s^2} \left[\frac{1}{1 + 1/s^2} \right]$$

$$= \frac{1}{s^2 + 1},$$

where we used the fact that the sum of the geometric series $\sum_{n=0}^\infty r^n = 1/(1-r)$.

(b) Since the Taylor series for $\sin t = \sum_{n=0}^\infty \frac{(-1)^n t^{2n+1}}{(2n+1)!}$, we see that the Taylor series for $f(t)$ is given by

$$\sum_{n=0}^\infty \frac{(-1)^n t^{2n}}{(2n+1)!}.$$

Then

$$\mathcal{L}(f(t)) = \int_0^\infty f(t)e^{-st}\,dt$$

$$= \int_0^\infty \sum_{n=0}^\infty \frac{(-1)^n t^{2n}}{(2n+1)!} e^{-st}\,dt$$

$$= \sum_{n=0}^\infty \frac{(-1)^n}{(2n+1)!} \int_0^\infty t^{2n} e^{-st}\,dt$$

$$= \sum_{n=0}^\infty \frac{(-1)^n}{(2n+1)!} \left[\frac{(2n)!}{s^{2n+1}} \right]$$

$$= \sum_{n=0}^\infty \frac{(-1)^n}{(2n+1)s^{2n+1}}$$

$$= \arctan(1/s),$$

22

where we have used the fact that the Taylor series for $\arctan(t) = \sum_{n=0}^{\infty} \frac{(-1)^n t^{2n+1}}{(2n+1)}$.

(c)

$$\mathcal{L}(J_0(t)) = \int_0^{\infty} J_0(t) e^{-st}\, dt$$

$$= \int_0^{\infty} \sum_{n=0}^{\infty} \frac{(-1)^n t^{2n}}{2^{2n}(n!)^2} e^{-st}\, dt$$

$$= \sum_{n=0}^{\infty} \frac{(-1)^n}{2^{2n}(n!)^2} \int_0^{\infty} t^{2n} e^{-st}\, dt$$

$$= \sum_{n=0}^{\infty} \frac{(-1)^n}{2^{2n}(n!)^2} \left[\frac{(2n)!}{s^{2n+1}}\right]$$

$$= (s^2 + 1)^{-1/2}.$$

$$\mathcal{L}(J_0(\sqrt{t})) = \int_0^{\infty} J_0(\sqrt{t}) e^{-st}\, dt$$

$$= \int_0^{\infty} \sum_{n=0}^{\infty} \frac{(-1)^n t^n}{2^{2n}(n!)^2} e^{-st}\, dt$$

$$= \sum_{n=0}^{\infty} \frac{(-1)^n}{2^{2n}(n!)^2} \int_0^{\infty} t^n e^{-st}\, dt$$

$$= \sum_{n=0}^{\infty} \frac{(-1)^n}{2^{2n}(n!)^2} \left[\frac{n!}{s^{n+1}}\right]$$

$$= \sum_{n=0}^{\infty} \frac{(-1)^n}{2^{2n}(n!)s^{n+1}}$$

$$= s^{-1} e^{-1/4s}.$$

29.

(a) Taking the Laplace transform of the Airy equation, we have

$$\mathcal{L}(y'') - \mathcal{L}(ty) = 0.$$

Using the differentiation property of the transform, we have

$$\mathcal{L}(y'') + \frac{d}{ds}\mathcal{L}(y) = 0.$$

Therefore,

$$[s^2 Y(s) - sy(0) - y'(0)] + \frac{d}{ds}Y(s) = 0.$$

23

Using the initial conditions, we have

$$[s^2 Y(s) - s] + \frac{d}{ds} Y(s) = 0,$$

which we can rewrite as

$$Y'(s) + s^2 Y(s) = s.$$

(b) Taking the Laplace transform of the Legendre equation, we have

$$\mathcal{L}(y'') - \mathcal{L}(t^2 y'') - 2\mathcal{L}(ty') + \alpha(\alpha + 1)\mathcal{L}(y) = 0.$$

Using the differentiation property of the transform, we have

$$\mathcal{L}(y'') - \frac{d^2}{ds^2} \mathcal{L}(y'') + 2\frac{d}{ds} \mathcal{L}(y') + \alpha(\alpha + 1)\mathcal{L}(y) = 0.$$

Therefore,

$$[s^2 Y(s) - sy(0) - y'(0)] - \frac{d^2}{ds^2}[s^2 Y(s) - sy(0) - y'(0)] + +2\frac{d}{ds}[sY(s) - y(0)] + \alpha(\alpha+1)Y(s) = 0.$$

Using the initial conditions, we have

$$s^2 Y(s) - 1 - \frac{d^2}{ds^2}[s^2 Y(s) - 1] + 2\frac{d}{ds}[sY(s)] + \alpha(\alpha + 1)Y(s) = 0.$$

After carrying out the differentiation, the equation simplifies to

$$s^2 Y'' + 2sY' - [s^2 + \alpha(\alpha + 1)]Y = -1.$$

Section 5.3

1. We need to find constants a, b such that

$$a(s - 3) + b(s + 2) = s - 18,$$

which implies

$$(a + b)s - 3a + 2b = s - 18.$$

Equating like coefficients, we need $a + b = 1$ and $-3a + 2b = -18$. Solving the first equation, we have $a = 1 - b$. Plugging this equation into the second equation, we have $-3(1 - b) + 2b = -18$. Therefore, $b = -3$ and $a = 4$.

2. We need a_1, a_2 to satisfy the following equation,

$$a_1(s + 2) + a_2 = 3s + 4.$$

Equating like coefficients, we need $a_1 = 3$ and $2a_1 + a_2 = 4$. Therefore, $a_1 = 3$ and $a_2 = -2$.

3. We need a, b, c to satisfy the following equation,

$$a(s^2 + 4) + (bs + c \cdot 2)(s + 4) = -3s^2 + 32 - 14s.$$

24

We can rewrite this equation as

$$(a+b)s^2 + (4b+2c)s + 4a + 8c = -3s^2 + 32 - 14s.$$

Equating like coefficients, we need $a+b = -3$, $4b+2c = -14$ and $4a+8c = 32$. Solving this system of equations, we have $a = 2$, $b = -5$ and $c = 3$.

4. We need a_1, a_2, a_3 to satisfy the equation

$$a_1 s^2 + a_2 s + a_3 = s^2 + 2s + 2.$$

Equating like coefficients, we have $a_1 = 1$, $a_2 = 2$ and $a_3 = 2$.

5. We need a, b, c to satisfy the following equation

$$a(s^2 - 2s + 5) + (b(s-1) + 2c)(s+1) = 3s^2 - 8s + 5.$$

This equation can be rewritten as

$$(a+b)s^2 + (-2a + 2c)s + (5a - b + 2c) = 3s^2 - 8s + 5.$$

Equating like coefficients, we have $a+b = 3$, $-2a + 2c = -8$ and $5a - b + 2c = 5$. Solving this system of equations, we have $a = 2$, $b = 1$ and $c = -2$.

6. We need a_1, a_2, b and c to satisfy the following equation,

$$a_1 s(s+3)(s+1) + a_2(s+3)(s+1) + b(s+1)s^2 + c(s+3)s^2 = -2s^3 - 8s^2 + 8s + 6.$$

We can rewrite this equation as

$$(a_1 + b + c)s^3 + (4a_1 + a_2 + b + 3c)s^2 + (3a_1 + 4a_2)s + 3a_2 = -2s^3 - 8s^2 + 8s + 6.$$

Equating like coefficients, we have $3a_2 = 6$ which implies $a_2 = 2$. Then $3a_1 + 4a_2 = 8$ which implies $a_1 = 0$. Using this values for a_1 and a_2 and plugging back into the equations $a_1 + b + c = -2$ and $4a_1 + a_2 + b + 3c = -8$, we conclude that $c = -4$ and $b = 2$. Therefore, we have $a_1 = 0$, $a_2 = 2$, $b = 2$ and $c = -4$.

7. We need $a_1, b_1, a_2, b_2, a_3, b_3$ to satisfy the following equation

$$(a_1 s + b_1)(s^2 + 1)^2 + (a_2 s + b_2)(s^2 + 1) + (a_3 s + b_3) = 8s^3 - 15s - s^5.$$

We can rewrite this equation as

$$a_1 s^5 + b_1 s^4 + (2a_1 + a_2)s^3 + (2b_1 + b_2)s^2 + (a_1 + a_2 + a_3)s + (b_1 + b_2 + b_3) = -s^5 + 8s^3 - 15s.$$

Equating like coefficients, we have $a_1 = -1$ and $b_1 = 0$. Then, $2a_1 + a_2 = 8$ implies $a_2 = 10$, while $2b_1 + b_2 = 0$ implies $b_2 = 0$. Finally, $a_1 + a_2 + a_3 = -15$ implies $a_3 = -24$ and $b_1 + b_2 + b_3 = 0$ implies $b_3 = 0$. We conclude that $a_1 = -1$, $b_1 = 0$, $a_2 = 10$, $b_2 = 0$, $a_3 = -24$ and $b_3 = 0$.

8. We need a_1, b_1, a_2, b_2 to satisfy the following equation,

$$(a_1(s+1) + 2b_1)((s+1)^2 + 4) + a_2(s+1) + 2b_2 = s^3 + 3s^2 + 3s + 1.$$

25

We can rewrite this equation as

$$a_1 s^3 + (3a_1 + 2b_1)s^2 + (7a_1 + 4b_1 + a_2)s + (5a_1 + 10b_1 + a_2 + 2b_2) = s^3 + 3s^2 + 3s + 1.$$

Equating like coefficients, we have $a_1 = 1$. Then $3a_1 + 2b_1 = 3$ implies $b_1 = 0$. Then $7a_1 + 4b_1 + a_2 = 3$ implies $a_2 = -4$. Then $5a_1 + 10b_1 + a_2 + 2b_2 = 1$ implies $b_2 = 0$. To summarize, we have $a_1 = 1$, $b_1 = 0$, $a_2 = -4$ and $b_2 = 0$.

9. Writing the function as

$$\frac{3}{s^2 + 4} = \frac{3}{2}\frac{2}{s^2 + 4},$$

we see that $\mathcal{L}^{-1}(Y(s)) = \dfrac{3}{2}\sin(2t)$.

10. Writing the function as

$$\frac{4}{(s-1)^3} = 2\frac{2}{(s-1)^3},$$

we see that $\mathcal{L}^{-1}(Y(s)) = 2t^2 e^t$.

11. Using partial fractions, we write

$$\frac{2}{s^2 + 3s - 4} = \frac{2}{5}\left[\frac{1}{s-1} - \frac{1}{s+4}\right].$$

Therefore, we see that $\mathcal{L}^{-1}(Y(s)) = \dfrac{2}{5}(e^t - e^{-4t})$.

12. Using partial fractions, we write

$$\frac{3s}{s^2 - s - 6} = \frac{9}{5}\frac{1}{s-3} + \frac{6}{5}\frac{1}{s+2}.$$

Therefore, we see that $\mathcal{L}^{-1}(Y(s)) = \dfrac{9}{5}e^{3t} + \dfrac{6}{5}e^{-2t}$.

13. Completing the square in the denominator, we have

$$\frac{2s + 2}{s^2 + 2s + 5} = \frac{2(s + 1)}{(s+1)^2 + 4}.$$

Therefore, we see that $\mathcal{L}^{-1}(Y(s)) = 2e^{-t}\cos(2t)$.

14. We rewrite

$$\frac{2s - 3}{s^2 - 4} = 2\frac{s}{s^2 - 4} - \frac{3}{2}\frac{2}{s^2 - 4}.$$

Therefore, we see that $\mathcal{L}^{-1}(Y(s)) = 2\cosh(2t) - \dfrac{3}{2}\sinh(2t)$.

15. Completing the square in the denominator, we have

$$\frac{2s + 1}{s^2 - 2s + 2} = \frac{2s + 1}{(s-1)^2 + 1} = \frac{2(s - 1)}{(s-1)^2 + 1} + \frac{3}{(s-1)^2 + 1}.$$

Therefore, we see that $\mathcal{L}^{-1}(Y(s)) = 2e^t\cos(t) + 3e^t\sin(t)$.

16. Using partial fractions, we write

$$\frac{8s^2 - 4s + 12}{s(s^2 + 4)} = 3\frac{1}{s} + 5\frac{s}{s^2 + 4} - 2\frac{2}{s^2 + 4}.$$

Therefore, we see that $\mathcal{L}^{-1}(Y(s)) = 3 + 5\cos(2t) - 2\sin(2t)$.

17. Completing the square in the denominator, we have

$$\frac{1 - 2s}{s^2 + 4s + 5} = \frac{5 - 2(s + 2)}{(s + 2)^2 + 1}.$$

Therefore, we see that $\mathcal{L}^{-1}(Y(s)) = 5e^{-2t}\sin(t) - 2e^{-2t}\cos(t)$.

18. Completing the square in the denominator, we have

$$\frac{2s - 3}{s^2 + 2s + 10} = \frac{2(s + 1) - 5}{(s + 1)^2 + 9}.$$

Therefore, we see that $\mathcal{L}^{-1}(Y(s)) = 2e^{-t}\cos(3t) - \frac{5}{3}e^{-t}\sin(3t)$.

19. First we write

$$\frac{3s + 2}{(s - 2)(s + 2)(s + 1)} = \frac{a}{s - 2} + \frac{b}{s + 2} + \frac{c}{s + 1}.$$

Then, a, b, c must satisfy the following equation

$$a(s + 2)(s + 1) + b(s - 2)(s + 1) + c(s - 2)(s + 2) = 3s + 2.$$

We can rewrite this equation as

$$(a + b + c)s^2 + (3a - b)s + (2a - 2b - 4c) = 3s + 2.$$

Equating like coefficients, we have $a + b + c = 0$, $3a - b = 3$ and $2a - 2b - 4c = 2$. Solving this system of equations, we have $a = 2/3$, $b = -1$ and $c = 1/3$. Therefore, we can write

$$\frac{3s + 2}{(s - 2)(s + 2)(s + 1)} = \frac{2/3}{s - 2} - \frac{1}{s + 2} + \frac{1/3}{s + 1},$$

which implies

$$F(s) = \frac{2}{s - 2} - \frac{3}{s + 2} + \frac{1}{s + 1}.$$

By the linearity of $\mathcal{L}^{-1}$, we have

$$\mathcal{L}^{-1}(F(s)) = 2\mathcal{L}^{-1}\left(\frac{1}{s - 2}\right) - 3\mathcal{L}^{-1}\left(\frac{1}{s + 2}\right) + \mathcal{L}^{-1}\left(\frac{1}{s + 1}\right)$$
$$= 2e^{2t} - 3e^{-2t} + e^{-t}.$$

20. First we write
$$\frac{s^3 - 2s^2 + 8}{s^2(s-2)(s+2)} = \frac{a}{s} + \frac{b}{s^2} + \frac{c}{s-2} + \frac{d}{s+2}.$$

Then, a, b, c, d must satisfy the following equation

$$as(s-2)(s+2) + b(s-2)(s+2) + cs^2(s+2) + ds^2(s-2) = s^3 - 2s^2 + 8.$$

We can rewrite this equation as

$$(a+c+d)s^3 + (b+2c-2d)s^2 + (-4a)s - 4b = s^3 - 2s^2 + 8.$$

Equating like coefficients, we have $a + c + d = 1$, $b + 2c - 2d = -2$, $-4a = 0$ and $-4b = 8$. Solving this system of equations, we have $a = 0$, $b = -2$, $c = 1/2$ and $d = 1/2$. Therefore, we can write

$$\frac{s^3 - 2s^2 + 8}{s^2(s-2)(s+2)} = \frac{-2}{s^2} + \frac{1/2}{s-2} + \frac{1/2}{s+2},$$

which implies

$$F(s) = \frac{-4}{s^2} + \frac{1}{s-2} + \frac{1}{s+2}.$$

By the linearity of $\mathcal{L}^{-1}$, we have

$$\mathcal{L}^{-1}(F(s)) = 4\mathcal{L}^{-1}\left(\frac{1}{s^2}\right) + \mathcal{L}^{-1}\left(\frac{1}{s-2}\right) + \mathcal{L}^{-1}\left(\frac{1}{s+2}\right)$$
$$= 4t + e^{2t} + e^{-2t}.$$

21. First we write
$$\frac{s^2 - 3s + 11}{(s^2 - 4s + 8)(s+3)} = \frac{as+b}{s^2 - 4s + 8} + \frac{c}{s+3}$$

Then, a, b, c must satisfy the following equation

$$(as+b)(s+3) + c(s^2 - 4s + 8) = s^2 - 3s + 11.$$

We can rewrite this equation as

$$(a+c)s^2 + (3a+b-4c)s + (3b+8c) = s^2 - 3s + 11.$$

Equating like coefficients, we have $a + c = 1$, $3a + b - 4c = -3$, and $3b + 8c = 11$. Solving this system of equations, we have $a = 0$, $b = 1$ and $c = 1$. Therefore, we can write

$$\frac{s^2 - 3s + 11}{(s^2 - 4s + 8)(s+3)} = \frac{1}{s^2 - 4s + 8} + \frac{1}{s+3}$$

which implies

$$F(s) = \frac{4}{s^2 - 4s + 8} + \frac{4}{s+3}.$$

By the linearity of $\mathcal{L}^{-1}$, we have

$$\mathcal{L}^{-1}(F(s)) = 4\mathcal{L}^{-1}\left(\frac{1}{s^2 - 4s + 8}\right) + 4\mathcal{L}^{-1}\left(\frac{1}{s+3}\right)$$

$$= 4\mathcal{L}^{-1}\left(\frac{1}{(s-2)^2 + 2^2}\right) + 4\mathcal{L}^{-1}\left(\frac{1}{s+3}\right)$$

$$= 2\mathcal{L}^{-1}\left(\frac{2}{(s-2)^2 + 2^2}\right) + 4\mathcal{L}^{-1}\left(\frac{1}{s+3}\right)$$

$$= 2e^{2t}\sin(2t) + 4e^{-3t}.$$

22. First we write

$$\frac{s^3 + 3s^2 + 4s + 3}{(s^2 + 1)(s^2 + 4)} = \frac{as + b}{s^2 + 1} + \frac{cs + d}{s^2 + 4}.$$

Then, a, b, c, d must satisfy the equation

$$(as + b)(s^2 + 4) + (cs + d)(s^2 + 1) = s^3 + 3s^2 + 4s + 3.$$

This equation can be rewritten as

$$(a + c)s^3 + (b + d)s^2 + (4a + c)s + (4b + d) = s^3 + 3s^2 + 4s + 3.$$

Equating like coefficients, we have $a + c = 1$, $b + d = 3$, $4a + c = 4$ and $4b + d = 3$. Solving this system of equations, we have $a = 1$, $b = 0$, $c = 0$, and $d = 3$. Therefore,

$$F(s) = 2\left[\frac{s}{s^2 + 1} + \frac{3}{s^2 + 4}\right] = 2\left[\frac{s}{s^2 + 1}\right] + 3\left[\frac{2}{s^2 + 2^2}\right].$$

By the linearity of $\mathcal{L}^{-1}$, we have

$$\mathcal{L}^{-1}(F(s)) = 2\mathcal{L}^{-1}\left(\frac{s}{s^2 + 1}\right) + 3\mathcal{L}^{-1}\left(\frac{2}{s^2 + 2^2}\right)$$

$$= 2\cos(t) + 3\sin(2t).$$

23. First we write

$$\frac{s^3 - 2s^2 - 6s - 6}{(s^2 + 2s + 2)s^2} = \frac{as + b}{s^2 + 2s + 2} + \frac{c}{s} + \frac{d}{s^2}.$$

Then, a, b, c, d must satisfy the equation

$$(as + b)s^2 + cs(s^2 + 2s + 2) + d(s^2 + 2s + 2) = s^3 - 2s^2 - 6s - 6.$$

This equation can be rewritten as

$$(a + c)s^3 + (b + 2c + d)s^2 + (2c + 2d)s + 2d = s^3 - 2s^2 - 6s - 6.$$

Equating like coefficients, we have $a + c = 1$, $b + 2c + d = -2$, $2c + 2d = -6$ and $2d = -6$. Solving this system of equations, we have $a = 1$, $b = 1$, $c = 0$, and $d = -3$. Therefore,

$$F(s) = \frac{s + 1}{s^2 + 2s + 2} - \frac{3}{s^2} = \frac{s + 1}{(s + 1)^2 + 1} - \frac{3}{s^2}$$

29

By the linearity of $\mathcal{L}^{-1}$, we have

$$\mathcal{L}^{-1}(F(s)) = \mathcal{L}^{-1}\left(\frac{s+1}{(s+1)^2+1}\right) - 3\mathcal{L}^{-1}\left(\frac{1}{s^2}\right)$$
$$= e^{-t}\cos(t) - 3t.$$

24. First, we write

$$\frac{s^2+3}{(s^2+2s+2)^2} = \frac{as+b}{s^2+2s+2} + \frac{cs+d}{(s^2+2s+2)^2}.$$

Then, a, b, c, d must satisfy the following equation,

$$(as+b)(s^2+2s+3) + cs+d = s^2+3.$$

We can rewrite this equation as

$$as^3 + (2a+b)s^2 + (2a+2b+c)s + (2b+d) = s^2+3.$$

Equating like coefficients, we have $a = 0$, $2a+b = 1$, $2a+2b+c = 0$ and $2b+d = 3$. Solving this system of equations, we have $a = 0$, $b = 1$, $c = -2$ and $d = 1$. Therefore,

$$F(s) = \frac{1}{s^2+2s+2} + \frac{-2s+1}{(s^2+2s+2)^2}$$
$$= \frac{1}{(s+1)^2+1} + \frac{-2s-2}{(s^2+2s+2)^2} + \frac{3}{(s^2+2s+2)^2}.$$

By the linearity of $\mathcal{L}^{-1}$, we have

$$\mathcal{L}^{-1}(F(s)) = \mathcal{L}^{-1}\left(\frac{1}{(s+1)^2+1}\right) + \mathcal{L}^{-1}\left(\frac{-2s-2}{(s^2+2s+2)^2}\right) + 3\mathcal{L}^{-1}\left(\frac{1}{(s^2+2s+2)^2}\right).$$

First,

$$\mathcal{L}^{-1}\left(\frac{1}{(s+1)^2+1}\right) = e^{-t}\sin(t).$$

Next, using the fact that

$$\frac{-2s-2}{(s^2+2s+2)^2} = \frac{d}{ds}\left(\frac{1}{s^2+2s+2}\right),$$

we see that

$$\mathcal{L}^{-1}\left(\frac{-2s-2}{(s^2+2s+2)^2}\right) = -te^{-t}\sin(t).$$

Finally, using the rule for the product of Laplace transforms, we have

$$\mathcal{L}^{-1}\left(\frac{1}{(s^2+2s+2)^2}\right) = \int_0^t e^{-(t-\tau)}\sin(t-\tau)e^{-\tau}\sin(\tau)\,d\tau$$

$$= e^{-t}\int_0^t \sin(t-\tau)\sin(\tau)\,d\tau$$

$$= \frac{1}{2}e^{-t}\int_0^t [\cos(t-2\tau)-\cos(t)]\,d\tau$$

$$= \frac{1}{2}e^{-t}\left[\frac{\sin(t-2\tau)}{-2}-\tau\cos(t)\right]\Big|_0^t$$

$$= \frac{1}{2}e^{-t}\left[\frac{\sin(-t)}{-2}-t\cos(t)-\frac{\sin(t)}{-2}\right]$$

$$= \frac{1}{2}e^{-t}[\sin(t)-t\cos(t)]$$

$$= \frac{1}{2}e^{-t}\sin(t)-\frac{1}{2}te^{-t}\cos(t).$$

Combining these results, we conclude that

$$\mathcal{L}^{-1}(F(s)) = e^{-t}\sin(t)-te^{-t}\sin(t)+\frac{3}{2}e^{-t}\sin(t)-\frac{3}{2}te^{-t}\cos(t)$$

$$= \frac{5}{2}e^{-t}\sin(t)-te^{-t}\sin(t)-\frac{3}{2}te^{-t}\cos(t).$$

25. Using a computer algebra system, we see that

$$\mathcal{L}^{-1}(Y(s)) = \left(1-\frac{7}{3}t\right)e^{-t}\cos 3t - \left(\frac{8}{9}+\frac{4}{3}t\right)e^{-t}\sin 3t.$$

26. Using a computer algebra system, we see that

$$\mathcal{L}^{-1}(Y(s)) = -\left(\frac{1016}{3125}+\frac{549}{625}t+\frac{118}{125}t^2+\frac{9}{25}t^3+\frac{1}{20}t^4\right)+\frac{1016}{3125}e^{2t}\cos t+\frac{713}{3125}e^{2t}\sin t.$$

27. Using a computer algebra system, we see that

$$\mathcal{L}^{-1}(Y(s)) = \left(\frac{1}{250}+\frac{13}{25}t-\frac{1}{40}t^2\right)e^t\cos t + \left(-\frac{261}{500}+\frac{3}{200}t+\frac{1}{5}t^2\right)e^t\sin t$$

$$-\left(\frac{1}{250}+\frac{1}{100}t\right)e^{-2t}.$$

28. Using a computer algebra system, we see that

$$\mathcal{L}^{-1}(Y(s)) = -\left(\frac{59}{4000}-\frac{21}{400}t-\frac{3}{160}t^2\right)e^{3t}+\left(\frac{59}{4000}+\frac{1}{80}t\right)e^{-t}\cos 2t$$

$$-\left(\frac{3}{1000}+\frac{7}{320}t\right)e^{-t}\sin 2t.$$

Section 5.4

1.

$$\mathcal{L}(y'') - \mathcal{L}(y') - 6\mathcal{L}(y) = 0$$
$$\implies [s^2\mathcal{L}(y) - sy(0) - y'(0)] - [s\mathcal{L}(y) - y(0)] - 6\mathcal{L}(y) = 0$$
$$\implies [s^2\mathcal{L}(y) - s + 1] - [s\mathcal{L}(y) - 1] - 6\mathcal{L}(y) = 0$$
$$\implies [s^2 - s - 6]\mathcal{L}(y) = s - 2$$
$$\implies \mathcal{L}(y) = \frac{s-2}{s^2 - s - 6}.$$

Using partial fractions, we write

$$\mathcal{L}(y) = \frac{1}{5}\frac{1}{s-3} + \frac{4}{5}\frac{1}{s+2},$$

which implies

$$y(t) = \frac{1}{5}e^{3t} + \frac{4}{5}e^{-2t}.$$

2.

$$\mathcal{L}(y'') + 3\mathcal{L}(y') + 2\mathcal{L}(y) = \mathcal{L}(t)$$
$$\implies [s^2\mathcal{L}(y) - sy(0) - y'(0)] + 3[s\mathcal{L}(y) - y(0)] + 2\mathcal{L}(y) = \frac{1}{s^2}$$
$$\implies [s^2\mathcal{L}(y) - s] + 3[s\mathcal{L}(y) - 1] + 2\mathcal{L}(y) = \frac{1}{s^2}$$
$$\implies [s^2 + 3s + 2]\mathcal{L}(y) = s + 3 + \frac{1}{s^2}$$
$$\implies \mathcal{L}(y) = \frac{s+3}{s^2 + 3s + 2} + \frac{1}{s^2(s^2 + 3s + 2)} = \frac{s^3 + 3s^2 + 1}{s^2(s+2)(s+1)}.$$

Using partial fractions, we write

$$\mathcal{L}(y) = \frac{s^3 + 3s^2 + 1}{s^2(s+2)(s+1)} = \frac{a}{s} + \frac{b}{s^2} + \frac{c}{s+2} + \frac{d}{s+1}.$$

Then a, b, c, d must satisfy the equation

$$as(s+2)(s+1) + b(s+2)(s+1) + cs^2(s+1) + ds^2(s+2) = s^3 + 3s^2 + 1.$$

This equation can be rewritten as

$$(a + c + d)s^3 + (3a + b + c + 2d)s^2 + (2a + 3b)s + 2b = s^3 + 3s^2 + 1.$$

Equating like coefficients, we have $a + c + d = 1$, $3a + b + c + 2d = 3$, $2a + 3b = 0$ and $2b = 1$. Solving this system of equations, we have $a = -3/4$, $b = 1/2$, $c = -5/4$ and $d = 3$. Therefore,

$$\mathcal{L}(y) = \frac{-3/4}{s} + \frac{1/2}{s^2} + \frac{-5/4}{s+2} + \frac{3}{s+1},$$

which implies

$$y = -\frac{3}{4}\mathcal{L}^{-1}\left(\frac{1}{s}\right) + \frac{1}{2}\mathcal{L}^{-1}\left(\frac{1}{s^2}\right) - \frac{5}{4}\mathcal{L}^{-1}\left(\frac{1}{s+2}\right) + 3\mathcal{L}^{-1}\left(\frac{1}{s+1}\right)$$
$$= -\frac{3}{4} + \frac{1}{2}t - \frac{5}{4}e^{-2t} + 3e^{-t}.$$

3.

$$\mathcal{L}(y'') - 2\mathcal{L}(y') + 2\mathcal{L}(y) = 0$$
$$\implies [s^2\mathcal{L}(y) - sy(0) - y'(0)] - 2[s\mathcal{L}(y) - y(0)] + 2\mathcal{L}(y) = 0$$
$$\implies [s^2\mathcal{L}(y) - 1] - 2s\mathcal{L}(y) + 2\mathcal{L}(y) = 0$$
$$\implies [s^2 - 2s + 2]\mathcal{L}(y) = 1$$
$$\implies \mathcal{L}(y) = \frac{1}{s^2 - 2s + 2}.$$

Completing the square in the denominator, we have

$$\mathcal{L}(y) = \frac{1}{(s-1)^2 + 1},$$

which implies

$$y(t) = e^t \sin(t).$$

4.

$$\mathcal{L}(y'') - 4\mathcal{L}(y') + 4\mathcal{L}(y) = 0$$
$$\implies [s^2\mathcal{L}(y) - sy(0) - y'(0)] - 4[s\mathcal{L}(y) - y(0)] + 4\mathcal{L}(y) = 0$$
$$\implies [s^2\mathcal{L}(y) - s - 1] - 4[s\mathcal{L}(y) - 1] + 4\mathcal{L}(y) = 0$$
$$\implies [s^2 - 4s + 4]\mathcal{L}(y) = s - 3$$
$$\implies \mathcal{L}(y) = \frac{s-3}{s^2 - 4s + 4}.$$

Therefore, we have

$$\mathcal{L}(y) = \frac{s-3}{(s-2)^2} = \frac{s-2}{(s-2)^2} - \frac{1}{(s-2)^2} = \frac{1}{s-2} - \frac{1}{(s-2)^2}$$

which implies

$$y(t) = e^{2t} - te^{2t}.$$

5.

$$\mathcal{L}(y'') - 2\mathcal{L}(y') + 4\mathcal{L}(y) = 0$$
$$\implies [s^2\mathcal{L}(y) - sy(0) - y'(0)] - 2[s\mathcal{L}(y) - y(0)] + 4\mathcal{L}(y) = 0$$
$$\implies [s^2\mathcal{L}(y) - 2s] - 2[s\mathcal{L}(y) - 2] + 4\mathcal{L}(y) = 0$$
$$\implies [s^2 - 2s + 4]\mathcal{L}(y) = 2s - 4$$
$$\implies \mathcal{L}(y) = \frac{2s-4}{s^2 - 2s + 4}.$$

33

Completing the square in the denominator, we have

$$\mathcal{L}(y) = \frac{2s - 4}{(s - 1)^2 + 3} = \frac{2(s - 1)}{(s - 1)^2 + 3} - \frac{2}{(s - 1)^2 + 3}$$

which implies

$$y(t) = 2e^t \cos(\sqrt{3}t) - \frac{2}{\sqrt{3}}e^t \sin(\sqrt{3}t).$$

6.

$$\mathcal{L}(y'') + 2\mathcal{L}(y') + 5\mathcal{L}(y) = \mathcal{L}(e^{-t}\sin(2t))$$

$$\implies [s^2\mathcal{L}(y) - sy(0) - y'(0)] + 2[s\mathcal{L}(y) - y(0)] + 5\mathcal{L}(y) = \frac{2}{(s + 1)^2 + 4}$$

$$\implies [s^2\mathcal{L}(y) - 2s + 1] + 2[s\mathcal{L}(y) - 2] + 5\mathcal{L}(y) = \frac{2}{(s + 1)^2 + 4}$$

$$\implies [s^2 + 2s + 5]\mathcal{L}(y) - 2s + 1 - 4 = \frac{2}{(s + 1)^2 + 4}$$

$$\implies \mathcal{L}(y) = \frac{2s + 3}{s^2 + 2s + 5} + \frac{2}{(s^2 + 2s + 5)[(s + 1)^2 + 4]}.$$

We rewrite $\mathcal{L}(y)$ as

$$\mathcal{L}(y) = \frac{2(s + 1)}{(s + 1)^2 + 4} + \frac{1}{(s + 1)^2 + 4} + \frac{2}{((s + 1)^2 + 4)^2}$$

which implies

$$y = 2\mathcal{L}^{-1}\left(\frac{s + 1}{(s + 1)^2 + 4}\right) + \mathcal{L}^{-1}\left(\frac{1}{(s + 1)^2 + 4}\right) + 2\mathcal{L}^{-1}\left(\frac{1}{((s + 1)^2 + 4)^2}\right)$$

$$= 2e^{-t}\cos(2t) + \frac{1}{2}e^{-t}\sin(2t) + \frac{1}{2}\mathcal{L}^{-1}\left(\frac{2}{(s + 1)^2 + 4} \cdot \frac{2}{(s + 1)^2 + 4}\right).$$

By the rule for the product of Laplace transforms, we have

$$\frac{1}{2}\mathcal{L}^{-1}\left(\frac{2}{(s + 1)^2 + 4} \cdot \frac{2}{(s + 1)^2 + 4}\right) = \frac{1}{2}\int_0^t e^{-(t-\tau)}\sin(2(t - \tau))e^{-\tau}\sin(2\tau)\,d\tau$$

$$= \frac{1}{2}e^{-t}\int_0^t \sin(2t - 2\tau)\sin(2\tau)\,d\tau$$

$$= -\frac{1}{4}e^{-t}\int_0^t [\cos(2t) - \cos(2t - 4\tau)]\,d\tau$$

$$= -\frac{1}{4}e^{-t}\left[\tau\cos(2t) + \frac{\sin(2t - 4\tau)}{4}\right]\Big|_0^t$$

$$= -\frac{1}{4}e^{-t}\left[t\cos(2t) + \frac{\sin(-2t)}{4} - \frac{\sin(2t)}{4}\right]$$

$$= -\frac{1}{4}e^{-t}\left[t\cos(2t) - \frac{\sin(2t)}{2}\right]$$

$$= \frac{1}{8}e^{-t}\sin(2t) - \frac{1}{4}te^{-t}\cos(2t).$$

34

Therefore, we conclude that

$$y = 2e^{-t}\cos(2t) + \frac{5}{8}e^{-t}\sin(2t) - \frac{1}{4}te^{-t}\cos(2t).$$

7.

$$\mathcal{L}(y'') + \omega^2\mathcal{L}(y) = \mathcal{L}(\cos(2t))$$
$$\implies [s^2\mathcal{L}(y) - sy(0) - y'(0)] + \omega^2\mathcal{L}(y) = \frac{s}{s^2+4}$$
$$\implies [s^2\mathcal{L}(y) - s] + \omega^2\mathcal{L}(y) = \frac{s}{s^2+4}$$
$$\implies [s^2 + \omega^2]\mathcal{L}(y) = s + \frac{s}{s^2+4}$$
$$\implies \mathcal{L}(y) = \frac{s}{s^2+\omega^2} + \frac{s}{(s^2+\omega^2)(s^2+4)}.$$

Using partial fractions on the second term, we rewrite $\mathcal{L}(y)$ as

$$\mathcal{L}(y) = \frac{s}{s^2+\omega^2} + \frac{1}{4-\omega^2}\left[\frac{s}{s^2+\omega^2} - \frac{s}{s^2+4}\right]$$

which implies

$$y = \cos(\omega t) + \frac{1}{4-\omega^2}\left[\cos(\omega t) - \cos(2t)\right]$$
$$= \frac{1}{4-\omega^2}[(5-\omega^2)\cos(\omega t) - \cos(2t)].$$

8.

$$\mathcal{L}(y'') - 2\mathcal{L}(y') + 2\mathcal{L}(y) = \mathcal{L}(\cos(t))$$
$$\implies [s^2\mathcal{L}(y) - sy(0) - y'(0)] - 2[s\mathcal{L}(y) - y(0)] + 2\mathcal{L}(y) = \frac{s}{s^2+1}$$
$$\implies [s^2\mathcal{L}(y) - s] - 2[s\mathcal{L}(y) - 1] + 2\mathcal{L}(y) = \frac{s}{s^2+1}$$
$$\implies [s^2 - 2s + 2]\mathcal{L}(y) = s + 2 + \frac{s}{s^2+1}$$
$$\implies \mathcal{L}(y) = \frac{s+2}{s^2-2s+2} + \frac{s}{(s^2-2s+2)(s^2+1)}.$$

Using partial fractions on the second term, we have

$$\frac{s}{(s^2-2s+2)(s^2+1)} = \frac{1}{5}\left[\frac{s-2}{s^2+1} - \frac{s-4}{s^2-2s+2}\right].$$

Therefore, we can write

$$\mathcal{L}(y) = \frac{1}{5}\frac{s}{s^2+1} - \frac{2}{5}\frac{1}{s^2+1} + \frac{2}{5}\frac{2s-3}{s^2-2s+2}.$$

Completing the square in the denominator for the last term, we have

$$\frac{2s-3}{s^2-2s+2} = \frac{2(s-1)-1}{(s-1)^2+1}.$$

Therefore,

$$\mathcal{L}(y) = \frac{1}{5}\frac{s}{s^2+1} - \frac{2}{5}\frac{1}{s^2+1} + \frac{2}{5}\frac{2(s-1)-1}{(s-1)^2+1},$$

which implies

$$y = \frac{1}{5}\cos(t) - \frac{2}{5}\sin(t) + \frac{4}{5}e^t\cos(t) - \frac{2}{5}e^t\sin(t).$$

9.

$$\mathcal{L}(y'') - 2\mathcal{L}(y') + 2\mathcal{L}(y) = \mathcal{L}(e^{-t})$$

$$\implies [s^2\mathcal{L}(y) - sy(0) - y'(0)] - 2[s\mathcal{L}(y) - y(0)] + 2\mathcal{L}(y) = \frac{1}{s+1}$$

$$\implies [s^2\mathcal{L}(y) - 1] - 2[s\mathcal{L}(y)] + 2\mathcal{L}(y) = \frac{1}{s+1}$$

$$\implies [s^2 - 2s + 2]\mathcal{L}(y) = 1 + \frac{1}{s+1}$$

$$\implies \mathcal{L}(y) = \frac{1}{s^2-2s+2} + \frac{1}{(s^2-2s+2)(s+1)}.$$

Using partial fractions on the second term, we have

$$\frac{1}{(s^2-2s+2)(s+1)} = \frac{1}{5}\frac{1}{s+1} + \frac{1}{5}\frac{3-s}{s^2-2s+2}.$$

Therefore, we can write

$$\mathcal{L}(y) = \frac{1}{5}\frac{1}{s+1} + \frac{1}{5}\frac{8-s}{s^2-2s+2}.$$

Completing the square in the denominator for the last term, we have

$$\frac{8-s}{s^2-2s+2} = -\frac{(s-1)-7}{(s-1)^2+1}.$$

Therefore,

$$\mathcal{L}(y) = \frac{1}{5}\frac{1}{s+1} - \frac{1}{5}\frac{(s-1)-7}{(s-1)^2+1},$$

which implies

$$y = \frac{1}{5}e^{-t} - \frac{1}{5}e^t\cos(t) + \frac{7}{5}e^t\sin(t).$$

36

10.

$$\mathcal{L}(y'') + 2\mathcal{L}(y') + \mathcal{L}(y) = 4\mathcal{L}(e^{-t})$$

$$\implies [s^2\mathcal{L}(y) - sy(0) - y'(0)] + 2[s\mathcal{L}(y) - y(0)] + \mathcal{L}(y) = \frac{4}{s+1}$$

$$\implies [s^2\mathcal{L}(y) - 2s + 1] + 2[s\mathcal{L}(y) - 2] + \mathcal{L}(y) = \frac{4}{s+1}$$

$$\implies [s^2 + 2s + 1]\mathcal{L}(y) = 2s + 3 + \frac{4}{s+1}$$

$$\implies \mathcal{L}(y) = \frac{2s+3}{s^2+2s+1} + \frac{4}{(s^2+2s+1)(s+1)}$$

$$\implies \mathcal{L}(y) = \frac{2(s+1)+1}{(s+1)^2} + \frac{4}{(s+1)^3}$$

$$\implies \mathcal{L}(y) = \frac{2}{s+1} + \frac{1}{(s+1)^2} + \frac{4}{(s+1)^3}.$$

Therefore,

$$y = 2e^{-t} + te^{-t} + 2t^2 e^{-t}.$$

11.

$$\mathcal{L}(y^{(4)}) - 4\mathcal{L}(y''') + 6\mathcal{L}(y'') - 4\mathcal{L}(y') + \mathcal{L}(y) = 0$$

implies

$$[s^4\mathcal{L}(y) - s^3 y(0) - s^2 y'(0) - s y''(0) - y'''(0)] - 4[s^3\mathcal{L}(y) - s^2 y(0) - s y'(0) - y''(0)]$$
$$+ 6[s^2\mathcal{L}(y) - s y(0) - y'(0)] - 4[s\mathcal{L}(y) - y(0)] + \mathcal{L}(y) = 0.$$

Plugging in the initial conditions, we have

$$[s^4\mathcal{L}(y) - s^2 - 1] - 4[s^3\mathcal{L}(y) - s] + 6[s^2\mathcal{L}(y) - 1] - 4[s\mathcal{L}(y)] + \mathcal{L}(y) = 0$$
$$\implies [s^4 - 4s^3 + 6s^2 - 4s + 1]\mathcal{L}(y) - s^2 - 1 + 4s - 6 = 0$$
$$\implies \mathcal{L}(y) = \frac{s^2 - 4s + 7}{s^4 - 4s^3 + 6s^2 - 4s + 1} = \frac{s^2 - 4s + 7}{(s-1)^4}.$$

Using partial fractions, we write

$$\mathcal{L}(y) = \frac{s^2 - 4s + 7}{(s-1)^4} = \frac{a}{s-1} + \frac{b}{(s-1)^2} + \frac{c}{(s-1)^3} + \frac{d}{(s-1)^4}.$$

Then a, b, c, d must satisfy the equation

$$a(s-1)^3 + b(s-1)^2 + c(s-1) + d = s^2 - 4s + 7.$$

This equation can be rewritten as

$$as^3 + (-3a+b)s^2 + (3a-2b+c)s + (-a+b-c+d) = s^2 - 4s + 7.$$

Equating like coefficients, we have $a = 0$, $-3a + b = 1$, $3a - 2b + c = -4$ and $-a + b - c + d = 7$. Solving this system of equations, we have $a = 0$, $b = 1$, $c = -2$ and $d = 4$. Therefore,

$$\mathcal{L}(y) = \frac{1}{(s-1)^2} - \frac{2}{(s-1)^3} + \frac{4}{(s-1)^4}$$

which implies

$$y = \mathcal{L}^{-1}\left(\frac{1}{(s-1)^2}\right) - \mathcal{L}^{-1}\left(\frac{2}{(s-1)^3}\right) + \frac{2}{3}\mathcal{L}^{-1}\left(\frac{6}{(s-1)^4}\right)$$

$$= te^t - t^2 e^t + \frac{2}{3}t^3 e^t.$$

12.
$$\mathcal{L}(y^{(4)}) - \mathcal{L}(y) = 0$$

implies

$$[s^4 \mathcal{L}(y) - s^3 y(0) - s^2 y'(0) - s y''(0) - y'''(0)] - \mathcal{L}(y) = 0.$$

Plugging in the initial conditions, we have

$$[s^4 \mathcal{L}(y) - s^3 - s] - \mathcal{L}(y) = 0$$
$$\implies [s^4 - 1]\mathcal{L}(y) = s^3 + s$$
$$\implies \mathcal{L}(y) = \frac{s^3 + s}{s^4 - 1} = \frac{s^3 + s}{(s^2 - 1)(s^2 + 1)} = \frac{s}{s^2 - 1}.$$

Using partial fractions, we write

$$\mathcal{L}(y) = \frac{s}{s^2 - 1} = \frac{a}{s-1} + \frac{b}{s+1}.$$

Then a, b must satisfy the equation

$$a(s+1) + b(s-1) = s$$

This equation can be rewritten as

$$(a + b)s + (a - b) = s.$$

Equating like coefficients, we have $a + b = 1$, $a - b = 0$. Solving this system of equations, we have $a = 1/2$ and $b = 1/2$. Therefore,

$$\mathcal{L}(y) = \frac{1/2}{s-1} + \frac{1/2}{s+1}$$

which implies

$$y = \frac{1}{2}\mathcal{L}^{-1}\left(\frac{1}{s-1}\right) + \frac{1}{2}\mathcal{L}^{-1}\left(\frac{1}{s+1}\right)$$

$$= \frac{1}{2}e^t + \frac{1}{2}e^{-t}.$$

13.

$$\mathcal{L}(y^{(4)}) - 4\mathcal{L}(y) = 0$$

implies

$$[s^4\mathcal{L}(y) - s^3 y(0) - s^2 y'(0) - s y''(0) - y'''(0)] - 4\mathcal{L}(y) = 0$$
$$\implies [s^4\mathcal{L}(y) - s^3 + 2s] - 4\mathcal{L}(y) = 0$$
$$\implies (s^4 - 4)\mathcal{L}(y) = s^3 - 2s$$
$$\implies \mathcal{L}(y) = \frac{s^3 - 2s}{s^4 - 4} = \frac{s}{s^2 + 2}.$$

Therefore,

$$y = \mathcal{L}^{-1}\left(\frac{s}{s^2 + 2}\right)$$
$$= \cos(\sqrt{2}t).$$

14. This system can be written as

$$y_1' = -3y_1 + 4y_2$$
$$y_2' = -2y_1 + 3y_2$$

with the initial conditions $y_1(0) = 1$ and $y_2(0) = 0$. Taking the Laplace transform of each equation, we have

$$\mathcal{L}(y_1') = -3\mathcal{L}(y_1) + 4\mathcal{L}(y_2)$$
$$\mathcal{L}(y_2') = -2\mathcal{L}(y_1) + 3\mathcal{L}(y_2)$$

which implies

$$s\mathcal{L}(y_1) - y_1(0) = -3\mathcal{L}(y_1) + 4\mathcal{L}(y_2)$$
$$s\mathcal{L}(y_2) - y_2(0) = -2\mathcal{L}(y_1) + 3\mathcal{L}(y_2).$$

Letting $Y_1 = \mathcal{L}(y_1)$, $Y_2 = \mathcal{L}(y_2)$ and plugging in the initial conditions, we have

$$sY_1 - 1 = -3Y_1 + 4Y_2$$
$$sY_2 - 0 = -2Y_1 + 3Y_2.$$

These equations can be written in matrix form as

$$\begin{pmatrix} s+3 & -4 \\ 2 & s-3 \end{pmatrix} \begin{pmatrix} Y_1 \\ Y_2 \end{pmatrix} = \begin{pmatrix} 1 \\ 0 \end{pmatrix}.$$

The solution of this system is given by

$$\begin{pmatrix} Y_1 \\ Y_2 \end{pmatrix} = \frac{1}{(s+3)(s-3)+8} \begin{pmatrix} s-3 & 4 \\ -2 & s+3 \end{pmatrix} \begin{pmatrix} 1 \\ 0 \end{pmatrix}$$
$$= \frac{1}{s^2-1} \begin{pmatrix} s-3 \\ -2 \end{pmatrix}.$$

Now

$$Y_1(s) = \frac{s-3}{s^2-1} \implies y_1(t) = \frac{2}{s+1} - \frac{1}{s-1}$$

$$\implies y_1(t) = \mathcal{L}^{-1}\left(\frac{2}{s+1}\right) - \mathcal{L}^{-1}\left(\frac{1}{s-1}\right) = 2e^{-t} - e^t.$$

and

$$Y_2(s) = \frac{-2}{s^2-1} = \frac{1}{s+1} - \frac{1}{s-1}$$

$$\implies y_2(t) = \mathcal{L}^{-1}\left(\frac{1}{s+1}\right) - \mathcal{L}^{-1}\left(\frac{1}{s-1}\right) = e^{-t} - e^t.$$

15. This system can be written as

$$y_1' = 5y_1 - 2y_2$$
$$y_2' = 6y_1 - 2y_2$$

with the initial conditions $y_1(0) = 1$ and $y_2(0) = 0$. Taking the Laplace transform of each equation, we have

$$\mathcal{L}(y_1') = 5\mathcal{L}(y_1) - 2\mathcal{L}(y_2)$$
$$\mathcal{L}(y_2') = 6\mathcal{L}(y_1) - 2\mathcal{L}(y_2)$$

which implies

$$s\mathcal{L}(y_1) - y_1(0) = 5\mathcal{L}(y_1) - 2\mathcal{L}(y_2)$$
$$s\mathcal{L}(y_2) - y_2(0) = 6\mathcal{L}(y_1) - 2\mathcal{L}(y_2).$$

Letting $Y_1 = \mathcal{L}(y_1)$, $Y_2 = \mathcal{L}(y_2)$ and plugging in the initial conditions, we have

$$sY_1 - 1 = 5Y_1 - 2Y_2$$
$$sY_2 - 0 = 6Y_1 - 2Y_2.$$

These equations can be written in matrix form as

$$\begin{pmatrix} s-5 & 2 \\ -6 & s+2 \end{pmatrix} \begin{pmatrix} Y_1 \\ Y_2 \end{pmatrix} = \begin{pmatrix} 1 \\ 0 \end{pmatrix}.$$

The solution of this system is given by

$$\begin{pmatrix} Y_1 \\ Y_2 \end{pmatrix} = \frac{1}{(s-5)(s+2)+12} \begin{pmatrix} s+2 & -2 \\ 6 & s-5 \end{pmatrix} \begin{pmatrix} 1 \\ 0 \end{pmatrix}$$

$$= \frac{1}{s^2-3s+2} \begin{pmatrix} s+2 \\ 6 \end{pmatrix}$$

$$= \frac{1}{(s-2)(s-1)} \begin{pmatrix} s+2 \\ 6 \end{pmatrix}.$$

We first look at $Y_1(s)$. Using partial fractions, we write

$$Y_1(s) = \frac{s+2}{(s-2)(s-1)} = \frac{a}{s-2} + \frac{b}{s-1}.$$

Then a, b must satisfy the equation $a(s-1) + b(s-2) = s+2$. We can simplify this equation, rewriting it as $(a+b)s + (-a-2b) = s+2$. Equating like coefficients, we have $a+b = 1$ and $-a-2b = 2$. The solution of this system is $a = 4$ and $b = -3$. Therefore,

$$Y_1(s) = \frac{4}{s-2} - \frac{3}{s-1} \implies y_1(t) = 4\mathcal{L}^{-1}\left(\frac{1}{s-2}\right) - 3\mathcal{L}^{-1}\left(\frac{1}{s-1}\right) = 4e^{2t} - 3e^t.$$

Next, we look at $Y_2(s)$. Using partial fractions, we write

$$Y_2(s) = \frac{6}{(s-2)(s-1)} = \frac{6}{s-2} - \frac{6}{s-1}$$

$$\implies y_2(t) = 6\mathcal{L}^{-1}\left(\frac{1}{s-2}\right) - 6\mathcal{L}^{-1}\left(\frac{1}{s-1}\right) = 6e^{2t} - 6e^t.$$

16. This system can be written as

$$y_1' = 4y_1 - 4y_2$$
$$y_2' = 5y_1 - 4y_2$$

with the initial conditions $y_1(0) = 1$ and $y_2(0) = 0$. Taking the Laplace transform of each equation, we have

$$\mathcal{L}(y_1') = 4\mathcal{L}(y_1) - 4\mathcal{L}(y_2)$$
$$\mathcal{L}(y_2') = 5\mathcal{L}(y_1) - 4\mathcal{L}(y_2)$$

which implies

$$s\mathcal{L}(y_1) - y_1(0) = 4\mathcal{L}(y_1) - 4\mathcal{L}(y_2)$$
$$s\mathcal{L}(y_2) - y_2(0) = 5\mathcal{L}(y_1) - 4\mathcal{L}(y_2).$$

Letting $Y_1 = \mathcal{L}(y_1)$, $Y_2 = \mathcal{L}(y_2)$ and plugging in the initial conditions, we have

$$sY_1 - 1 = 4Y_1 - 4Y_2$$
$$sY_2 - 0 = 5Y_1 - 4Y_2.$$

These equations can be written in matrix form as

$$\begin{pmatrix} s-4 & 4 \\ -5 & s+4 \end{pmatrix} \begin{pmatrix} Y_1 \\ Y_2 \end{pmatrix} = \begin{pmatrix} 1 \\ 0 \end{pmatrix}.$$

The solution of this system is given by

$$\begin{pmatrix} Y_1 \\ Y_2 \end{pmatrix} = \frac{1}{(s-4)(s+4)+20} \begin{pmatrix} s+4 & -4 \\ 5 & s-4 \end{pmatrix} \begin{pmatrix} 1 \\ 0 \end{pmatrix}$$

$$= \frac{1}{s^2+4} \begin{pmatrix} s+4 \\ 5 \end{pmatrix}.$$

41

Now

$$Y_1(s) = \frac{s+4}{s^2+4}$$

$$\implies y_1(t) = \mathcal{L}^{-1}\left(\frac{s}{s^2+4}\right) + 2\mathcal{L}^{-1}\left(\frac{2}{s^2+2^2}\right) = \cos(2t) + 2\sin(2t).$$

Next,

$$Y_2(s) = \frac{5}{s^2+4}$$

$$\implies y_2(t) = \frac{5}{2}\mathcal{L}^{-1}\left(\frac{2}{s^2+2^2}\right) = \frac{5}{2}\sin(2t).$$

17. This system can be written as

$$y_1' = y_2$$
$$y_2' = -y_1$$

with the initial conditions $y_1(0) = 1$ and $y_2(0) = 1$. Taking the Laplace transform of each equation, we have

$$\mathcal{L}(y_1') = \mathcal{L}(y_2)$$
$$\mathcal{L}(y_2') = -\mathcal{L}(y_1)$$

which implies

$$s\mathcal{L}(y_1) - y_1(0) = \mathcal{L}(y_2)$$
$$s\mathcal{L}(y_2) - y_2(0) = -\mathcal{L}(y_1).$$

Letting $Y_1 = \mathcal{L}(y_1)$, $Y_2 = \mathcal{L}(y_2)$ and plugging in the initial conditions, we have

$$sY_1 - 1 = Y_2$$
$$sY_2 - 1 = -Y_1.$$

These equations can be written in matrix form as

$$\begin{pmatrix} s & -1 \\ 1 & s \end{pmatrix} \begin{pmatrix} Y_1 \\ Y_2 \end{pmatrix} = \begin{pmatrix} 1 \\ 1 \end{pmatrix}.$$

The solution of this system is given by

$$\begin{pmatrix} Y_1 \\ Y_2 \end{pmatrix} = \frac{1}{s^2+1}\begin{pmatrix} s & 1 \\ -1 & s \end{pmatrix}\begin{pmatrix} 1 \\ 1 \end{pmatrix}$$

$$= \frac{1}{s^2+1}\begin{pmatrix} s+1 \\ s-1 \end{pmatrix}.$$

Now

$$Y_1(s) = \frac{s+1}{s^2+1}$$

$$\implies y_1(t) = \mathcal{L}^{-1}\left(\frac{s}{s^2+1}\right) + \mathcal{L}^{-1}\left(\frac{1}{s^2+1}\right) = \cos(t) + \sin(t).$$

Next,

$$Y_2(s) = \frac{s-1}{s^2+1}$$

$$\implies y_2(t) = \mathcal{L}^{-1}\left(\frac{s}{s^2+1}\right) - \mathcal{L}^{-1}\left(\frac{1}{s^2+1}\right) = \cos(t) - \sin(t).$$

18. This system can be written as

$$y_1' = -4y_1 - y_2$$
$$y_2' = y_1 - 2y_2$$

with the initial conditions $y_1(0) = 1$ and $y_2(0) = 0$. Taking the Laplace transform of each equation, we have

$$\mathcal{L}(y_1') = -4\mathcal{L}(y_1) - \mathcal{L}(y_2)$$
$$\mathcal{L}(y_2') = \mathcal{L}(y_1) - 2\mathcal{L}(y_2)$$

which implies

$$s\mathcal{L}(y_1) - y_1(0) = -4\mathcal{L}(y_1) - \mathcal{L}(y_2)$$
$$s\mathcal{L}(y_2) - y_2(0) = \mathcal{L}(y_1) - 2\mathcal{L}(y_2).$$

Letting $Y_1 = \mathcal{L}(y_1)$, $Y_2 = \mathcal{L}(y_2)$ and plugging in the initial conditions, we have

$$sY_1 - 1 = -4Y_1 - Y_2$$
$$sY_2 - 0 = Y_1 - 2Y_2.$$

These equations can be written in matrix form as

$$\begin{pmatrix} s+4 & 1 \\ -1 & s+2 \end{pmatrix} \begin{pmatrix} Y_1 \\ Y_2 \end{pmatrix} = \begin{pmatrix} 1 \\ 0 \end{pmatrix}.$$

The solution of this system is given by

$$\begin{pmatrix} Y_1 \\ Y_2 \end{pmatrix} = \frac{1}{(s+4)(s+2)+1} \begin{pmatrix} s+2 & -1 \\ 1 & s+4 \end{pmatrix} \begin{pmatrix} 1 \\ 0 \end{pmatrix}$$

$$= \frac{1}{s^2+6s+9} \begin{pmatrix} s+2 \\ 1 \end{pmatrix}$$

$$= \frac{1}{(s+3)^2} \begin{pmatrix} s+2 \\ 1 \end{pmatrix}.$$

Now

$$Y_1(s) = \frac{s+2}{(s+3)^2} = \frac{s+3-1}{(s+3)^2} = \frac{1}{s+3} - \frac{1}{(s+3)^2}$$

$$\implies y_1(t) = \mathcal{L}^{-1}\left(\frac{1}{s+3}\right) - \mathcal{L}^{-1}\left(\frac{1}{(s+3)^2}\right) = e^{-3t} - te^{-3t}.$$

Next,

$$Y_2(s) = \frac{1}{(s+3)^2}$$

$$\implies y_2(t) = \mathcal{L}^{-1}\left(\frac{1}{(s+3)^2}\right) = te^{-3t}.$$

19. This system can be written as

$$y_1' = 5y_1 - 7y_2$$
$$y_2' = 7y_1 - 9y_2$$

with the initial conditions $y_1(0) = 0$ and $y_2(0) = 1$. Taking the Laplace transform of each equation, we have

$$\mathcal{L}(y_1') = 5\mathcal{L}(y_1) - 7\mathcal{L}(y_2)$$
$$\mathcal{L}(y_2') = 7\mathcal{L}(y_1) - 9\mathcal{L}(y_2)$$

which implies

$$s\mathcal{L}(y_1) - y_1(0) = 5\mathcal{L}(y_1) - 7\mathcal{L}(y_2)$$
$$s\mathcal{L}(y_2) - y_2(0) = 7\mathcal{L}(y_1) - 9\mathcal{L}(y_2).$$

Letting $Y_1 = \mathcal{L}(y_1)$, $Y_2 = \mathcal{L}(y_2)$ and plugging in the initial conditions, we have

$$sY_1 - 0 = 5Y_1 - 7Y_2$$
$$sY_2 - 1 = 7Y_1 - 9Y_2.$$

These equations can be written in matrix form as

$$\begin{pmatrix} s-5 & 7 \\ -7 & s+9 \end{pmatrix} \begin{pmatrix} Y_1 \\ Y_2 \end{pmatrix} = \begin{pmatrix} 0 \\ 1 \end{pmatrix}.$$

The solution of this system is given by

$$\begin{pmatrix} Y_1 \\ Y_2 \end{pmatrix} = \frac{1}{(s-5)(s+9)+49} \begin{pmatrix} s+9 & -7 \\ 7 & s-5 \end{pmatrix} \begin{pmatrix} 0 \\ 1 \end{pmatrix}$$

$$= \frac{1}{s^2+4s+4} \begin{pmatrix} -7 \\ s-5 \end{pmatrix}$$

$$= \frac{1}{(s+2)^2} \begin{pmatrix} -7 \\ s-5 \end{pmatrix}.$$

Now

$$Y_1(s) = \frac{-7}{(s+2)^2}$$

$$\implies y_1(t) = -7\mathcal{L}^{-1}\left(\frac{1}{(s+2)^2}\right) = -7te^{-2t}.$$

Next,

$$Y_2(s) = \frac{s-5}{(s+2)^2} = \frac{s+2-7}{(s+2)^2} = \frac{1}{s+2} - \frac{7}{(s+2)^2}$$

$$\implies y_2(t) = \mathcal{L}^{-1}\left(\frac{1}{s+2}\right) - 7\mathcal{L}^{-1}\left(\frac{1}{(s+2)^2}\right) = e^{-2t} - 7te^{-2t}.$$

20. Applying the Laplace transform, we have

$$\mathcal{L}(y_1') = -4\mathcal{L}(y_1) - \mathcal{L}(y_2) + \mathcal{L}(e^t)$$
$$\mathcal{L}(y_2') = \mathcal{L}(y_1) - 2\mathcal{L}(y_2) + \mathcal{L}(\sin(2t))$$

Letting $Y_1 = \mathcal{L}(y_1)$ and $Y_2 = \mathcal{L}(y_2)$ and using the property of the transform of a derivative, we have

$$sY_1 - y_1(0) = -4Y_1 - Y_2 + \frac{1}{s-1}$$

$$sY_2 - y_2(0) = Y_1 - 2Y_2 + \frac{2}{s^2+4}.$$

Plugging in the initial conditions, we have

$$sY_1 - 1 = -4Y_1 - Y_2 + \frac{1}{s-1}$$

$$sY_2 - 2 = Y_1 - 2Y_2 + \frac{2}{s^2+4}.$$

This system can be written in matrix form as

$$\begin{pmatrix} s+4 & 1 \\ -1 & s+2 \end{pmatrix}\begin{pmatrix} Y_1 \\ Y_2 \end{pmatrix} = \begin{pmatrix} \frac{s}{s-1} \\ \frac{2s^2+10}{s^2+4} \end{pmatrix}.$$

The solution of this system is given by

$$\begin{pmatrix} Y_1 \\ Y_2 \end{pmatrix} = \frac{1}{(s+3)^2}\begin{pmatrix} \frac{(s+2)s}{s-1} - \frac{2s^2+10}{s^2+4} \\ \frac{s}{s-1} + \frac{(s+4)(2s^2+10)}{s^2+4} \end{pmatrix}.$$

Taking the inverse Laplace transform, we have

$$y_1(t) = \mathcal{L}^{-1}\left(\frac{(s+2)s}{(s-1)(s+3)^2} - \frac{2s^2+10}{(s+3)^2(s^2+4)}\right)$$

$$= \left(-\frac{151t}{52} + \frac{2005}{2704}\right)e^{-3t} + \frac{3}{16}e^t + \frac{12}{169}\cos(2t) - \frac{5}{169}\sin(2t)$$

45

and

$$y_2(t) = \mathcal{L}^{-1}\left(\frac{s}{(s+3)^2(s-1)} + \frac{(s+4)(2s^2+10)}{(s+3)^2(s^2+4)}\right)$$
$$= \left(\frac{151t}{52} + \frac{5847}{2704}\right)e^{-3t} + \frac{1}{16}e^t - \frac{38}{169}\cos(2t) + \frac{44}{169}\sin(2t).$$

21. Applying the Laplace transform, we have

$$\mathcal{L}(y_1') = 5\mathcal{L}(y_1) - \mathcal{L}(y_2) + \mathcal{L}(2e^{-t})$$
$$\mathcal{L}(y_2') = \mathcal{L}(y_1) + 3\mathcal{L}(y_2) + \mathcal{L}(e^t)$$

Letting $Y_1 = \mathcal{L}(y_1)$ and $Y_2 = \mathcal{L}(y_2)$ and using the property of the transform of a derivative, we have

$$sY_1 - y_1(0) = 5Y_1 - Y_2 + \frac{2}{s+1}$$
$$sY_2 - y_2(0) = Y_1 + 3Y_2 + \frac{1}{s-1}.$$

Plugging in the initial conditions, we have

$$sY_1 + 3 = 5Y_1 - Y_2 + \frac{2}{s+1}$$
$$sY_2 - 2 = Y_1 + 3Y_2 + \frac{1}{s-1}.$$

This system can be written in matrix form as

$$\begin{pmatrix} s-5 & 1 \\ -1 & s-3 \end{pmatrix}\begin{pmatrix} Y_1 \\ Y_2 \end{pmatrix} = \begin{pmatrix} (-3s-1)/(s+1) \\ (2s-1)/(s-1) \end{pmatrix}.$$

The solution of this system is given by

$$\begin{pmatrix} Y_1 \\ Y_2 \end{pmatrix} = \frac{1}{(s-4)^2}\begin{pmatrix} \frac{(s-3)(-3s-1)}{(s+1)} - \frac{(2s-1)}{(s-1)} \\ \frac{-3s-1}{s+1} + \frac{(s-5)(2s-1)}{s-1} \end{pmatrix}.$$

Taking the inverse Laplace transform, we have

$$y_1(t) = \mathcal{L}^{-1}\left(\frac{1}{(s-4)^2} \cdot \left[\frac{(s-3)(-3s-1)}{s+1} - \frac{2s-1}{s-1}\right]\right)$$
$$= \left(-\frac{74t}{15} - \frac{578}{225}\right)e^{4t} - \frac{8}{25}e^{-t} - \frac{1}{9}e^t$$

and

$$y_2(t) = \mathcal{L}^{-1}\left(\frac{-3s-1}{(s-4)^2(s+1)} + \frac{(s-5)(2s-1)}{(s-4)^2(s-1)}\right)$$
$$= \left(-\frac{74t}{15} + \frac{532}{225}\right)e^{4t} + \frac{2}{25}e^{-t} - \frac{4}{9}e^t.$$

22. Applying the Laplace transform, we have

$$\mathcal{L}(y_1') = -\mathcal{L}(y_1) - 5\mathcal{L}(y_2) + \mathcal{L}(3)$$
$$\mathcal{L}(y_2') = \mathcal{L}(y_1) + 3\mathcal{L}(y_2) + \mathcal{L}(5\cos(t))$$

Letting $Y_1 = \mathcal{L}(y_1)$ and $Y_2 = \mathcal{L}(y_2)$ and using the property of the transform of a derivative, we have

$$sY_1 - y_1(0) = -Y_1 - 5Y_2 + \frac{3}{s}$$
$$sY_2 - y_2(0) = Y_1 + 3Y_2 + \frac{5s}{s^2 + 1}.$$

Plugging in the initial conditions, we have

$$sY_1 - 1 = -Y_1 - 5Y_2 + \frac{3}{s}$$
$$sY_2 + 1 = Y_1 + 3Y_2 + \frac{5s}{s^2 + 1}.$$

This system can be written in matrix form as

$$\begin{pmatrix} s+1 & 5 \\ -1 & s-3 \end{pmatrix} \begin{pmatrix} Y_1 \\ Y_2 \end{pmatrix} = \begin{pmatrix} \frac{s+3}{s} \\ \frac{-s^2+5s-1}{s^2+1} \end{pmatrix}.$$

The solution of this system is given by

$$\begin{pmatrix} Y_1 \\ Y_2 \end{pmatrix} = \frac{1}{s^2 - 2s + 2} \begin{pmatrix} \frac{(s-3)(s+3)}{s} + \frac{5(s^2-5s+1)}{(s^2+1)} \\ \frac{s+3}{s} + \frac{(s+1)(-s^2+5s-1)}{s^2+1} \end{pmatrix}.$$

Taking the inverse Laplace transform, we have

$$y_1(t) = \mathcal{L}^{-1}\left(\frac{1}{s^2 - 2s + 2} \cdot \left[\frac{(s-3)(s+3)}{s} + \frac{5(s^2 - 5s + 1)}{s^2 + 1} \right] \right)$$
$$= -\frac{9}{2} + \frac{11}{2}e^t \cos(t) - \frac{7}{2}e^t \sin(t) + 5(2\cos(t) - 4\sin(t))e^{t/2}\sinh(t/2).$$

and

$$y_2(t) = \mathcal{L}^{-1}\left(\frac{1}{s^2 - 2s + 2} \left[\frac{s+3}{s} + \frac{(s+1)(-s^2 + 5s - 1)}{s^2 + 1} \right] \right)$$
$$= -\frac{3}{2}e^t \cos(t) + \frac{15}{2}e^t \sin(t) + \frac{3}{2} - \cos(t) - 3\sin(t).$$

23. Applying the Laplace transform, we have

$$\mathcal{L}(y_1') = -2\mathcal{L}(y_1) + \mathcal{L}(y_2)$$
$$\mathcal{L}(y_2') = \mathcal{L}(y_1) - 2\mathcal{L}(y_2) + \mathcal{L}(\sin(t))$$

Letting $Y_1 = \mathcal{L}(y_1)$ and $Y_2 = \mathcal{L}(y_2)$ and using the property of the transform of a derivative, we have

$$sY_1 - y_1(0) = -2Y_1 + Y_2$$
$$sY_2 - y_2(0) = Y_1 - 2Y_2 + \frac{1}{s^2 + 1}.$$

Plugging in the initial conditions, we have

$$sY_1 = -2Y_1 + Y_2$$
$$sY_2 = Y_1 - 2Y_2 + \frac{1}{s^2 + 1}.$$

This system can be written in matrix form as

$$\begin{pmatrix} s+2 & -1 \\ -1 & s+2 \end{pmatrix} \begin{pmatrix} Y_1 \\ Y_2 \end{pmatrix} = \begin{pmatrix} 0 \\ \frac{1}{s^2+1} \end{pmatrix}.$$

The solution of this system is given by

$$\begin{pmatrix} Y_1 \\ Y_2 \end{pmatrix} = \frac{1}{s^2 + 4s + 3} \begin{pmatrix} \frac{1}{s^2+1} \\ \frac{s+2}{s^2+1} \end{pmatrix}.$$

Taking the inverse Laplace transform, we have

$$y_1(t) = \mathcal{L}^{-1} \left(\frac{1}{(s^2 + 4s + 3)(s^2 + 1)} \right)$$
$$= \frac{1}{4}e^{-t} - \frac{1}{20}e^{-3t} - \frac{1}{5}\cos(t) + \frac{1}{10}\sin(t)$$

and

$$y_2(t) = \mathcal{L}^{-1} \left(\frac{s+2}{(s^2 + 4s + 3)(s^2 + 1)} \right)$$
$$= \frac{1}{20}e^{-3t} - \frac{3}{10}\cos(t) + \frac{2}{5}\sin(t) + \frac{1}{4}e^{-t}.$$

24. Applying the Laplace transform, we have

$$\mathcal{L}(y_1') = \mathcal{L}(y_2) - \mathcal{L}(y_3)$$
$$\mathcal{L}(y_2') = \mathcal{L}(y_1) + \mathcal{L}(y_3) - \mathcal{L}(e^{-t})$$
$$\mathcal{L}(y_3') = \mathcal{L}(y_1) + \mathcal{L}(y_2) + \mathcal{L}(e^t).$$

Letting $Y_1 = \mathcal{L}(y_1)$, $Y_2 = \mathcal{L}(y_2)$, $Y_3 = \mathcal{L}(y_3)$ and using the property of the transform of a derivative, we have

$$sY_1 - y_1(0) = Y_2 - Y_3$$
$$sY_2 - y_2(0) = Y_1 + Y_3 - \frac{1}{s+1}$$
$$sY_3 - y_3(0) = Y_1 + Y_2 + \frac{1}{s-1}.$$

Plugging in the initial conditions, we have

$$sY_1 - 1 = Y_2 - Y_3$$

$$sY_2 - 2 = Y_1 + Y_3 - \frac{1}{s+1}$$

$$sY_3 - 3 = Y_1 + Y_2 + \frac{1}{s-1}.$$

This system can be written in matrix form as

$$\begin{pmatrix} s & -1 & 1 \\ -1 & s & -1 \\ -1 & -1 & s \end{pmatrix} \begin{pmatrix} Y_1 \\ Y_2 \\ Y_3 \end{pmatrix} = \begin{pmatrix} 1 \\ \frac{2s+1}{s+1} \\ \frac{3s-2}{s-1} \end{pmatrix}.$$

The solution of this system is given by

$$\begin{pmatrix} Y_1 \\ Y_2 \\ Y_3 \end{pmatrix} = \begin{pmatrix} \frac{1}{s} + \frac{2s+1}{(s+1)^2 s} - \frac{3s-2}{(s+1)s(s-1)} \\ \frac{1}{(s-1)s} + \frac{(s^2+1)(2s+1)}{s(s^2-1)(s+1)} + \frac{3s-2}{(s+1)s(s-1)} \\ \frac{1}{(s-1)s} + \frac{2s+1}{(s-1)s(s+1)} + \frac{3s-2}{s(s-1)} \end{pmatrix}.$$

Taking the inverse Laplace transform, we have

$$y_1(t) = \mathcal{L}^{-1}\left(\frac{1}{s} + \frac{2s+1}{(s+1)^2 s} - \frac{3s-2}{(s+1)s(s-1)} \right)$$

$$= (-1+t)e^{-t} + 2\cosh(t) - 3\sinh(t)$$

and

$$y_2(t) = \mathcal{L}^{-1}\left(\frac{1}{(s-1)s} + \frac{(s^2+1)(2s+1)}{s(s^2-1)(s+1)} + \frac{3s-2}{(s+1)s(s-1)} \right)$$

$$= 1 + (t+3)\sinh(t) + (1-t)\cosh(t) + 2e^{t/2}\sinh(t/2)$$

and

$$y_3(t) = \mathcal{L}^{-1}\left(\frac{1}{(s-1)s} + \frac{2s+1}{(s-1)s(s+1)} + \frac{3s-2}{s(s-1)} \right)$$

$$= 2e^{t/2}\sinh(t/2) + 1 + \cosh(t) + 2\sinh(t) + e^t.$$

25. First taking the Laplace transform of each equation, we have

$$\mathcal{L}(x'') - \mathcal{L}(y'') + \mathcal{L}(x) - 4\mathcal{L}(y) = 0$$
$$\mathcal{L}(x') + \mathcal{L}(y') = \mathcal{L}(\cos(t)).$$

Letting $X = \mathcal{L}(x)$ and $Y = \mathcal{L}(y)$, we have

$$[s^2 X - sx(0) - x'(0)] - [s^2 Y - sy(0) - y'(0)] + X - 4Y = 0$$

$$[sX - x(0)] + [sY - y(0)] = \frac{s}{s^2+1}.$$

Applying our initial conditions, we have

$$[s^2X - 1] - [s^2Y - 2] + X - 4Y = 0$$
$$sX + sY = \frac{s}{s^2 + 1}.$$

Therefore,

$$(s^2 + 1)X - (s^2 + 4)Y = -1$$
$$sX + sY = \frac{s}{s^2 + 1}.$$

Multiplying the first equation by s, the second equation by $s^2 + 1$ and subtracting the first equation from the second, we have

$$[2s^2 + 5]Y = 2 \implies Y = \frac{2}{2s^2 + 5} = \frac{1}{s^2 + 5/2}.$$

Therefore,

$$y(t) = \sqrt{\frac{2}{5}} \sin\left(\sqrt{\frac{5}{2}}t\right)$$
$$= \frac{\sqrt{10}}{5} \sin\left(\frac{\sqrt{10}}{2}t\right).$$

Then using the fact that $sX + sY = s/(s^2 + 1)$, we have

$$X = \frac{3/2}{(s^2 + 5/2)(s^2 + 1)}$$
$$= \frac{1}{s^2 + 1} - \frac{2}{2s^2 + 5}$$
$$= \frac{1}{s^2 + 1} - \frac{1}{s^2 + 5/2}.$$

Therefore,

$$x(t) = \sin(t) - \sqrt{\frac{2}{5}} \sin\left(\sqrt{\frac{5}{2}}t\right)$$
$$= \sin(t) - \frac{\sqrt{10}}{5} \sin\left(\frac{\sqrt{10}}{2}t\right).$$

26. First, taking the Laplace transform of each equation, we have

$$\mathcal{L}(m_1') = -k_1\mathcal{L}(m_1)$$
$$\mathcal{L}(m_2') = k_1\mathcal{L}(m_1) - k_2\mathcal{L}(m_2)$$
$$\mathcal{L}(m_3') = k_2\mathcal{L}(m_2).$$

Letting $M_1 = \mathcal{L}(m_1)$, $M_2 = \mathcal{L}(m_2)$, $M_3 = \mathcal{L}(m_3)$ and using the property involving the Laplace transform of a derivative, we have

$$sM_1 - m_1(0) = -k_1 M_1$$
$$sM_2 - m_2(0) = k_1 M_1 - k_2 M_2$$
$$sM_3 - m_3(0) = k_2 M_2.$$

Plugging in the initial conditions, we have

$$sM_1 - m_0 = -k_1 M_1$$
$$sM_2 = k_1 M_1 - k_2 M_2$$
$$sM_3 = k_2 M_2.$$

The first equation implies

$$(s + k_1)M_1 = m_0 \implies M_1 = \frac{m_0}{s + k_1}$$
$$\implies m_1(t) = m_0 \mathcal{L}^{-1}\left(\frac{1}{s + k_1}\right) = m_0 e^{-k_1 t}.$$

The second equation implies

$$(s + k_2)M_2 = \frac{k_1 m_0}{s + k_1} \implies M_2 = \frac{k_1 m_0}{(s + k_2)(s + k_1)}.$$

Using partial fractions, we have

$$M_2 = \frac{a}{s + k_2} + \frac{b}{s + k_1}.$$

Then a, b must satisfy the equation $a(s + k_1) + b(s + k_2) = k_1 m_0$. This equation can be rewritten as $(a + b)s + (ak_1 + bk_2) = k_1 m_0$. Equating like coefficients, we have $a + b = 0$, $ak_1 + bk_2 = k_1 m_0$. Solving this system of equations, we have $a = k_1 m_0/(k_1 - k_2)$ and $b = k_1 m_0/(k_2 - k_1)$. Therefore,

$$M_2 = \frac{k_1 m_0/(k_1 - k_2)}{s + k_2} + \frac{k_1 m_0/(k_2 - k_1)}{s + k_1}$$

which implies

$$m_2(t) = \frac{k_1 m_0}{k_1 - k_2}\mathcal{L}^{-1}\left(\frac{1}{s + k_2}\right) + \frac{k_1 m_0}{k_2 - k_1}\mathcal{L}^{-1}\left(\frac{1}{s + k_1}\right)$$
$$= \frac{k_1 m_0}{k_1 - k_2}e^{-k_2 t} + \frac{k_1 m_0}{k_2 - k_1}e^{-k_1 t}.$$

Finally, the equation for M_3 becomes

$$sM_3 = k_2 M_2 = \frac{k_2 k_1 m_0}{(s + k_2)(s + k_1)},$$

which implies
$$M_3 = \frac{k_2 k_1 m_0}{s(s + k_2)(s + k_1)}.$$

Using partial fractions, we rewrite this equation as
$$\frac{k_2 k_1 m_0}{s(s + k_2)(s + k_1)} = \frac{a}{s} + \frac{b}{s + k_2} + \frac{c}{s + k_1}.$$

Then a, b, c must satisfy the equation $a(s + k_2)(s + k_1) + bs(s + k_1) + cs(s + k_2) = k_2 k_1 m_0$. This equation can be rewritten as $(a + b + c)s^2 + (ak_1 + ak_2 + bk_1 + ck_2)s + ak_1 k_2 = k_1 k_2 m_0$. Equating like coefficients, we have $a + b + c = 0$, $ak_1 + ak_2 + bk_1 + ck_2 = 0$ and $ak_1 k_2 = k_1 k_2 m_0$. The solution of this system is given by $a = m_0$, $b = -m_0 k_1/(k_1 - k_2)$ and $c = m_0 k_2/(k_1 - k_2)$. Therefore,

$$M_3 = \frac{m_0}{s} + \frac{-m_0 k_1/(k_1 - k_2)}{s + k_2} + \frac{m_0 k_2/(k_1 - k_2)}{s + k_1}$$

$$\implies m_3(t) = m_0 \mathcal{L}^{-1} \frac{1}{s} - \frac{m_0 k_1}{k_1 - k_2} \mathcal{L}^{-1} \left(\frac{1}{s + k_2} \right) + \frac{m_0 k_2}{k_1 - k_2} \mathcal{L}^{-1} \left(\frac{1}{s + k_1} \right)$$

$$\implies m_3(t) = m_0 - \frac{m_0 k_1}{k_1 - k_2} e^{-k_2 t} + \frac{m_0 k_2}{k_1 - k_2} e^{-k_1 t}.$$

Section 5.5

1.

2.

3.

4.

5.

6.

7. Using the Heaviside function, we can write

$$f(t) = (t - 2)^2 u_2(t).$$

The Laplace transform has the property that

$$\mathcal{L}(u_c(t)f(t - c)) = e^{-cs}\mathcal{L}(f(t)).$$

Therefore,

$$\mathcal{L}(u_2(t)(t - 2)^2) = e^{-2s}\mathcal{L}(t^2) = \frac{2e^{-2s}}{s^3}.$$

8. Using the Heaviside function, we can write

$$f(t) = u_1(t)[(t - 1)^2 + 1].$$

The Laplace transform has the property that

$$\mathcal{L}(u_c(t)f(t - c)) = e^{-cs}\mathcal{L}(f(t)).$$

Therefore,

$$\mathcal{L}(u_1(t)[(t - 1)^2 + 1]) = e^{-s}\mathcal{L}(t^2 + 1) = e^{-s}\left[\frac{2}{s^3} + \frac{1}{s}\right].$$

9. Using the Heaviside function, we can write
$$f(t) = (t - \pi)[u_\pi(t) - u_{2\pi}(t)].$$

The Laplace transform has the property that
$$\mathcal{L}(u_c(t)f(t - c)) = e^{-cs}\mathcal{L}(f(t)).$$

Therefore,
$$\mathcal{L}((t - \pi)[u_\pi(t) - u_{2\pi}(t)]) = \mathcal{L}(u_\pi(t)(t - \pi)) - \mathcal{L}(u_{2\pi}(t)(t - 2\pi)) + \pi\mathcal{L}(u_{2\pi}(t))$$
$$= e^{-\pi s}\mathcal{L}(t) - e^{-2\pi s}\mathcal{L}(t) + \pi e^{-2\pi s}\mathcal{L}(1)$$
$$= \frac{e^{-\pi s}}{s^2} - \frac{e^{-2\pi s}}{s^2} + \frac{\pi e^{-2\pi s}}{s}.$$

10. The Laplace transform has the property that
$$\mathcal{L}(u_c(t)f(t - c)) = e^{-cs}\mathcal{L}(f(t)).$$

Therefore,
$$\mathcal{L}(u_1 + 2u_3 - 6u_4) = \mathcal{L}(u_1) + 2\mathcal{L}(u_3) - 6\mathcal{L}(u_4)$$
$$= [e^{-s} + 2e^{-3s} - 6e^{-4s}]\mathcal{L}(1)$$
$$= \frac{e^{-s} + 2e^{-3s} - 6e^{-4s}}{s}.$$

11. The Laplace transform has the property that
$$\mathcal{L}(u_c(t)f(t - c)) = e^{-cs}\mathcal{L}(f(t)).$$

Therefore,
$$\mathcal{L}((t - 3)u_2(t) - (t - 2)u_3(t)]) = \mathcal{L}((t - 2)u_2) - \mathcal{L}(u_2) - \mathcal{L}((t - 3)u_3) - \mathcal{L}(u_3)$$
$$= e^{-2s}\mathcal{L}(t) - e^{-2s}\mathcal{L}(1) - e^{-3s}\mathcal{L}(t) - e^{-3s}\mathcal{L}(1)$$
$$= e^{-2s}\left[\frac{1}{s^2} - \frac{1}{s}\right] - e^{-3s}\left[\frac{1}{s^2} + \frac{1}{s}\right].$$

12. The Laplace transform has the property that
$$\mathcal{L}(u_c(t)f(t - c)) = e^{-cs}\mathcal{L}(f(t)).$$

Therefore,
$$\mathcal{L}(t - (t - 1)u_1(t)) = \mathcal{L}(t) - \mathcal{L}((t - 1)u_1(t))$$
$$= \frac{1}{s^2}[1 - e^{-s}]$$

55

13. Using the fact that

$$\mathcal{L}^{-1}(e^{-cs}G(s)) = u_c(t)g(t-c),$$

we see that

$$\mathcal{L}^{-1}\left(\frac{3!e^{-s}}{(s-2)^4}\right) = u_1(t)g(t-1)$$

where

$$g(t) = \mathcal{L}^{-1}\left(\frac{3!}{(s-2)^4}\right) = t^3 e^{2t}.$$

Therefore,

$$\mathcal{L}^{-1}(F(s)) = u_1(t)(t-1)^3 e^{2(t-1)}.$$

14. Using the fact that

$$\mathcal{L}^{-1}(e^{-cs}G(s)) = u_c(t)g(t-c),$$

we see that

$$\mathcal{L}^{-1}\left(\frac{e^{-2s}}{s^2+s-2}\right) = u_2(t)g(t-2)$$

where

$$g(t) = \mathcal{L}^{-1}\left(\frac{1}{s^2+s-2}\right).$$

Using partial fractions, we write

$$\frac{1}{s^2+s-2} = \frac{1}{3}\left[\frac{1}{s-1} - \frac{1}{s+2}\right].$$

Therefore,

$$g(t) = \frac{1}{3}[e^t - e^{-2t}].$$

Therefore,

$$\mathcal{L}^{-1}(F(s)) = \frac{1}{3}u_2(t)[e^{(t-2)} - e^{-2(t-2)}].$$

15. Using the fact that

$$\mathcal{L}^{-1}(e^{-cs}G(s)) = u_c(t)g(t-c),$$

we see that

$$\mathcal{L}^{-1}\left(\frac{e^{-2s}2(s-1)}{s^2-2s+2}\right) = u_2(t)g(t-2)$$

where

$$g(t) = \mathcal{L}^{-1}\left(\frac{2(s-1)}{s^2-2s+2}\right).$$

Completing the square in the denominator, we write

$$\frac{2(s-1)}{s^2 - 2s + 2} = \frac{2(s-1)}{(s-1)^2 + 1}.$$

Therefore,

$$g(t) = 2e^t \cos(t).$$

Therefore,

$$\mathcal{L}^{-1}(F(s)) = 2u_2(t)e^{t-2}\cos(t-2).$$

16. Using the fact that

$$\mathcal{L}^{-1}(e^{-cs}G(s)) = u_c(t)g(t-c),$$

we see that

$$\mathcal{L}^{-1}\left(\frac{2e^{-2s}}{s^2 - 4}\right) = u_2(t)g(t-2)$$

where

$$g(t) = \mathcal{L}^{-1}\left(\frac{2}{s^2 - 4}\right).$$

Using partial fractions, we write

$$\frac{2}{s^2 - 4} = \frac{1}{2}\left[\frac{1}{s-2} - \frac{1}{s+2}\right].$$

Therefore,

$$g(t) = \frac{1}{2}[e^{2t} - e^{-2t}].$$

Therefore,

$$\mathcal{L}^{-1}(F(s)) = \frac{1}{2}u_2(t)[e^{2(t-2)} - e^{-2(t-2)}].$$

17. Using the fact that

$$\mathcal{L}^{-1}(e^{-cs}G(s)) = u_c(t)g(t-c),$$

we see that

$$\mathcal{L}^{-1}\left(\frac{(s-2)e^{-s}}{s^2 - 4s + 3}\right) = u_1(t)g(t-1)$$

where

$$g(t) = \mathcal{L}^{-1}\left(\frac{s-2}{s^2 - 4s + 3}\right).$$

Using partial fractions, we write

$$\frac{s-2}{s^2 - 4s + 3} = \frac{1}{2}\left[\frac{1}{s-1} + \frac{1}{s-3}\right].$$

Therefore,

$$g(t) = \frac{1}{2}[e^t + e^{3t}].$$

Therefore,

$$\mathcal{L}^{-1}(F(s)) = \frac{1}{2}[e^{t-1} + e^{3(t-1)}]u_1(t).$$

18. Using the fact that

$$\mathcal{L}^{-1}(e^{-cs}G(s)) = u_c(t)g(t-c),$$

we see that

$$\mathcal{L}^{-1}\left(\frac{e^{-s} + e^{-2s} - e^{-3s} - e^{-4s}}{s}\right) = u_1(t)g(t-1) + u_2(t)g(t-2) - u_3(t)g(t-3) - u_4(t)g(t-4),$$

where

$$g(t) = \mathcal{L}^{-1}\left(\frac{1}{s}\right) = 1.$$

Therefore,

$$\mathcal{L}^{-1}(F(s)) = u_1(t) + u_2(t) - u_3(t) - u_4(t).$$

19. By definition of the Laplace transform,

$$\mathcal{L}(f(t)) = \int_0^\infty f(t)e^{-st}\, dt$$

$$= \int_0^1 e^{-st}\, dt$$

$$= \frac{1 - e^{-s}}{s}.$$

20. Using the definition of the Laplace transform, we have

$$\mathcal{L}(f(t)) = \int_0^\infty f(t)e^{-st}\, dt$$

$$= \int_0^1 e^{-st}\, dt + \int_2^3 e^{-st}\, dt$$

$$= \frac{1 - e^{-s}}{s} + \frac{e^{-2s} - e^{-3s}}{s}$$

$$= \frac{1 - e^{-s} + e^{-2s} - e^{-3s}}{s}.$$

21. Using the definition of the Laplace transform, we have

$$\mathcal{L}(f(t)) = \int_0^\infty f(t)e^{-st}\,dt$$

$$= \int_0^1 e^{-st}\,dt + \int_2^3 e^{-st}\,dt + \int_4^5 e^{-st}\,dt + \ldots + \int_{2n}^{2n+1} e^{-st}\,dt$$

$$= \frac{1}{s}\left[1 - e^{-s} + e^{-2s} - e^{-3s} + \ldots + e^{-2ns} - e^{-(2n+1)s}\right]$$

$$= \frac{1 - e^{-(2n+2)s}}{s(1 + e^{-s})}$$

22. Using Theorem 5.5.3, we know that

$$\mathcal{L}(f(t)) = \frac{F_T(s)}{1 - e^{-sT}}$$

where T is the period and

$$F_T(s) = \int_0^T e^{-st} f(t)\,dt.$$

Here, $T = 2$. Therefore,

$$F_T(s) = \int_0^2 e^{-st} f(t)\,dt = \int_0^1 e^{-st}\,dt - \int_1^2 e^{-st}\,dt$$

$$= \frac{1 - e^{-s}}{s} - \frac{e^{-s} - e^{-2s}}{s}$$

$$= \frac{1 - 2e^{-s} + e^{-2s}}{s}.$$

Therefore,

$$\mathcal{L}(f(t)) = \frac{1 - 2e^{-s} + e^{-2s}}{s(1 - e^{-2s})}$$

$$= \frac{(1 - e^{-s})^2}{s(1 - e^{-s})(1 + e^{-s})}$$

$$= \frac{1 - e^{-s}}{s(1 + e^{-s})}.$$

23. Using Theorem 5.5.3, we know that

$$\mathcal{L}(f(t)) = \frac{F_T(s)}{1 - e^{-sT}}$$

where T is the period and

$$F_T(s) = \int_0^T e^{-st} f(t)\,dt.$$

Here, $T = 1$. Therefore,

$$F_T(s) = \int_0^1 e^{-st} f(t) \, dt = \int_0^1 e^{-st} t \, dt$$
$$= \frac{1 - se^{-s} - e^{-s}}{s^2}.$$

Therefore,

$$\mathcal{L}(f(t)) = \frac{1 - se^{-s} - e^{-s}}{s^2(1 - e^{-s})}.$$

24. Using Theorem 5.5.3, we know that

$$\mathcal{L}(f(t)) = \frac{F_T(s)}{1 - e^{-sT}}$$

where T is the period and

$$F_T(s) = \int_0^T e^{-st} f(t) \, dt.$$

Here, $T = \pi$. Therefore,

$$F_T(s) = \int_0^\pi e^{-st} f(t) \, dt = \int_0^\pi e^{-st} \sin(t) \, dt$$
$$= \frac{1 + e^{-\pi s}}{s^2 + 1}.$$

Therefore,

$$\mathcal{L}(f(t)) = \frac{1 + e^{-\pi s}}{(s^2 + 1)(1 - e^{-\pi s})}.$$

25.

(a)

$$\mathcal{L}(f(t)) = \mathcal{L}(1) - \mathcal{L}(u_1(t))$$
$$= \frac{1}{s} - \frac{e^{-s}}{s}.$$

(b)

Let $F(s) = \mathcal{L}[1 - u_1(t)]$. Then

$$\mathcal{L}\left[\int_0^t [1 - u_1(\xi)]\,d\xi\right] = \frac{F(s)}{s} = \frac{1 - e^{-s}}{s^2}.$$

(c)

Let $G(s) = \mathcal{L}(g(t))$. Then

$$\begin{aligned}
\mathcal{L}(h(t)) &= G(s) - e^{-s}G(s) \\
&= \frac{1 - e^{-s}}{s^2} - e^{-s}\frac{1 - e^{-s}}{s^2} \\
&= \frac{(1 - e^{-s})^2}{s^2}.
\end{aligned}$$

26.

(a)

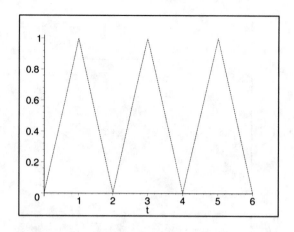

(b) The given function is periodic. Therefore, by Theorem 5.5.3,

$$\mathcal{L}(p(t)) = \frac{F_T(s)}{1 - e^{-sT}}$$

where T is the period and

$$F_T(s) = \int_0^T e^{-st} f(t)\, dt.$$

Here the period is $T = 2$ and by problem 25(c),

$$F_2(s) = \frac{(1 - e^{-s})^2}{s^2}.$$

Therefore,

$$\mathcal{L}(p(t)) = \frac{(1 - e^{-s})^2}{s^2(1 - e^{-2s})}$$

$$= \frac{(1 - e^{-s})^2}{s^2(1 - e^{-s})(1 + e^{-s})}$$

$$= \frac{1 - e^{-s}}{s^2(1 + e^{-s})}.$$

(c) By Theorem 5.2.2,

$$\mathcal{L}(p'(t)) = s\mathcal{L}(p(t)) - p(0).$$

Further, $p'(t) = f(t)$ where f is the periodic square wave in problem 22. Therefore,

$$\mathcal{L}(p'(t)) = \mathcal{L}(f(t)) = \frac{1 - e^{-s}}{s(1 + e^{-s})}.$$

We also note that $p(0) = 0$. Therefore,

$$\mathcal{L}(p(t)) = \frac{1}{s}\mathcal{L}(p'(t)) = \frac{1 - e^{-s}}{s^2(1 + e^{-s})}.$$

62

Section 5.6

1. Let $Y(s) = \mathcal{L}(y)$. Applying the Laplace transform to the equation, we have

$$\mathcal{L}(y'') + \mathcal{L}(y) = \mathcal{L}(f(t))$$
$$\implies [s^2 Y(s) - sy(0) - y'(0)] + Y(s) = \mathcal{L}(f(t)).$$

Applying the initial conditions, we have

$$s^2 Y(s) - 1 + Y(s) = \mathcal{L}(f(t)).$$

The forcing function $f(t)$ can be written as $f(t) = u_0(t) - u_{\pi/2}(t)$. Therefore,

$$\mathcal{L}(f(t)) = \mathcal{L}(u_0(t)) - \mathcal{L}(u_{\pi/2}(t)) = \frac{1 - e^{-\pi s/2}}{s}.$$

Therefore, the equation for Y becomes

$$[s^2 + 1]Y(s) = 1 + \frac{1 - e^{-\pi s/2}}{s}$$
$$\implies Y(s) = \frac{1}{s^2 + 1} + \frac{1 - e^{-\pi s/2}}{s(s^2 + 1)}.$$

Using partial fractions, we write the second term as

$$(1 - e^{-\pi s/2}) \left[\frac{1}{s} - \frac{s}{s^2 + 1} \right].$$

Therefore,

$$Y(s) = \frac{1}{s^2 + 1} + \frac{1}{s} - \frac{s}{s^2 + 1} - \frac{e^{-\pi s/2}}{s} + \frac{se^{-\pi s/2}}{s^2 + 1}.$$

Then, using the fact that

$$\mathcal{L}^{-1}(e^{-cs}G(s)) = u_c(t)g(t - c),$$

we conclude that

$$y(t) = \sin(t) + 1 - \cos(t) - u_{\pi/2}(t) + u_{\pi/2}(t)\cos(t - \pi/2)$$
$$= \sin(t) + 1 - \cos(t) - u_{\pi/2}(t)[1 - \sin(t)].$$

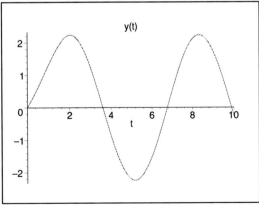

63

2. Let $Y(s) = \mathcal{L}(y)$. Applying the Laplace transform to the equation, we have

$$\mathcal{L}(y'') + 2\mathcal{L}(y') + 2\mathcal{L}(y) = \mathcal{L}(h(t))$$
$$\implies [s^2 Y(s) - sy(0) - y'(0)] + 2[sY(s) - y(0)] + 2Y(s) = \mathcal{L}(h(t)).$$

Applying the initial conditions, we have

$$s^2 Y(s) - 1 + 2sY(s) + 2Y(s) = \mathcal{L}(h(t)).$$

The forcing function $h(t)$ can be written as $h(t) = u_\pi(t) - u_{2\pi}(t)$. Therefore,

$$\mathcal{L}(h(t)) = \mathcal{L}(u_\pi(t)) - \mathcal{L}(u_{2\pi}(t)) = \frac{e^{-\pi s} - e^{-2\pi s}}{s}.$$

Therefore, the equation for Y becomes

$$[s^2 + 2s + 2]Y(s) = 1 + \frac{e^{-\pi s} - e^{-2\pi s}}{s}$$
$$\implies Y(s) = \frac{1}{s^2 + 2s + 2} + \frac{e^{-\pi s} - e^{-2\pi s}}{s(s^2 + 2s + 2)}.$$

Using partial fractions, we write the second term as

$$\frac{1}{2}(e^{-\pi s} - e^{-2\pi s})\left[\frac{1}{s} - \frac{2 + s}{s^2 + 2s + 2}\right].$$

Therefore,

$$Y(s) = \frac{1}{s^2 + 2s + 2} + \frac{1}{2}(e^{-\pi s} - e^{-2\pi s})\left[\frac{1}{s} - \frac{2 + s}{s^2 + 2s + 2}\right]$$
$$= \frac{1}{2}\left[\frac{e^{-\pi s} - e^{-2\pi s}}{s}\right] + \left[\frac{1 - e^{-\pi s} + e^{-2\pi s}}{s^2 + 2s + 2}\right] - \frac{1}{2}\left[\frac{(e^{-\pi s} - e^{-2\pi s})s}{s^2 + 2s + 2}\right].$$

Completing the square in the denominator for the last terms, we conclude that

$$Y(s) = \frac{1}{2}\left[\frac{e^{-\pi s} - e^{-2\pi s}}{s}\right] + \frac{1}{2}\left[\frac{2 - e^{-\pi s} + e^{-2\pi s}}{(s+1)^2 + 1}\right] - \frac{1}{2}\left[\frac{(e^{-\pi s} - e^{-2\pi s})(s + 1)}{(s+1)^2 + 1}\right].$$

Then, using the fact that

$$\mathcal{L}^{-1}(e^{-cs}G(s)) = u_c(t)g(t - c),$$

we conclude that

$$y(t) = \frac{1}{2}[2e^{-t}\sin(t) - u_\pi(t)e^{-(t-\pi)}\sin(t - \pi) + u_{2\pi}(t)e^{-(t-2\pi)}\sin(t - 2\pi)]$$
$$- \frac{1}{2}[u_\pi(t)e^{-t}\cos(t - \pi) - u_{2\pi}(t)\cos(t - 2\pi)] + \frac{1}{2}[u_\pi(t) - u_{2\pi}(t)] +$$
$$= e^{-t}\sin t + \frac{1}{2}u_\pi(t)[1 + e^{-(t-\pi)}\sin t + e^{-(t-\pi)}\cos t]$$
$$- \frac{1}{2}u_{2\pi}(t)[1 - e^{-(t-2\pi)}\sin t - e^{-(t-2\pi)}\cos t]$$

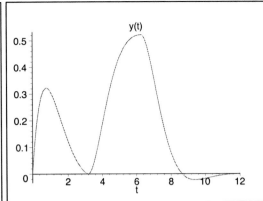

The solution starts out as free oscillation, due to the initial conditions. The amplitude increases while the forcing is present. After the forcing is removed, the solution decays due to the damping term.

3. Let $Y(s) = \mathcal{L}(y)$. Applying the Laplace transform to the equation, we have

$$\mathcal{L}(y'') + 4\mathcal{L}(y) = \mathcal{L}(\sin(t) - u_{2\pi}(t)\sin(t - 2\pi))$$
$$\implies [s^2 Y(s) - sy(0) - y'(0)] + 4Y(s) = \frac{1 - e^{-2\pi s}}{s^2 + 1}.$$

Applying the initial conditions, we have

$$s^2 Y(s) + 4Y(s) = \frac{1 - e^{-2\pi s}}{s^2 + 1}.$$

Therefore, the equation for Y becomes

$$[s^2 + 4]Y(s) = \frac{1 - e^{-2\pi s}}{s^2 + 1}$$
$$\implies Y(s) = \frac{1 - e^{-2\pi s}}{(s^2 + 4)(s^2 + 1)}.$$

Using partial fractions, we can write

$$Y(s) = \frac{1}{3}(1 - e^{-2\pi s})\left[\frac{1}{s^2 + 1} - \frac{1}{s^2 + 4}\right].$$

Therefore, we conclude that

$$y(t) = \frac{1}{3}\left[\sin(t) - \frac{1}{2}\sin(2t) - u_{2\pi}(t)\sin(t - 2\pi) + \frac{1}{2}u_{2\pi}(t)\sin(2(t - 2\pi))\right]$$
$$= \frac{1}{3}\left[\sin(t) - \frac{1}{2}\sin(2t)\right] - \frac{1}{3}u_{2\pi}(t)\left[\sin(t) - \frac{1}{2}\sin(2t)\right]$$
$$= \frac{1}{6}(1 - u_{2\pi}(t))[2\sin(t) - \sin(2t)].$$

65

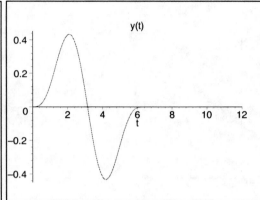

4. Let $Y(s) = \mathcal{L}(y)$. Applying the Laplace transform to the equation, we have

$$\mathcal{L}(y'') + 4\mathcal{L}(y) = \mathcal{L}(\sin(t) + u_\pi(t)\sin(t - \pi))$$

$$\implies [s^2 Y(s) - sy(0) - y'(0)] + 4Y(s) = \frac{1 + e^{-\pi s}}{s^2 + 1}.$$

Applying the initial conditions, we have

$$s^2 Y(s) + 4Y(s) = \frac{1 + e^{-\pi s}}{s^2 + 1}.$$

Therefore, the equation for Y becomes

$$[s^2 + 4]Y(s) = \frac{1 + e^{-\pi s}}{s^2 + 1}$$

$$\implies Y(s) = \frac{1 + e^{-\pi s}}{(s^2 + 4)(s^2 + 1)}.$$

Using partial fractions, we can write

$$Y(s) = \frac{1}{3}(1 + e^{-\pi s})\left[\frac{1}{s^2 + 1} - \frac{1}{s^2 + 4}\right].$$

Therefore, we conclude that

$$y(t) = \frac{1}{3}\left[\sin(t) + u_\pi(t)\sin(t - \pi) - \frac{1}{2}\sin(2t) - \frac{1}{2}u_\pi(t)\sin(2(t - \pi))\right]$$

$$= \frac{1}{6}[2\sin(t) - \sin(2t)] - \frac{1}{6}u_\pi(t)[2\sin(t) + \sin(2t)].$$

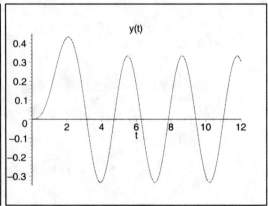

66

Since there is no damping term, the solution follows the forcing function, after which the solution is a steady oscillation about $y = 0$.

5. First, we write the inhomogeneous term as $f(t) = u_0(t) - u_{10}(t)$. Then, let $Y(s) = \mathcal{L}(y)$. Applying the Laplace transform to the equation, we have

$$\mathcal{L}(y'') + 3\mathcal{L}(y') + 2\mathcal{L}(y) = \mathcal{L}(u_0(t) - u_{10}(t))$$

$$\implies [s^2 Y(s) - sy(0) - y'(0)] + 3[sY(s) - y(0)] + 2Y(s) = \frac{1 - e^{-10s}}{s}.$$

Applying the initial conditions, we have

$$s^2 Y(s) + 3sY(s) + 2Y(s) = \frac{1 - e^{-10s}}{s}.$$

Therefore, the equation for Y becomes

$$[s^2 + 3s + 2]Y(s) = \frac{1 - e^{-10s}}{s}$$

$$\implies Y(s) = \frac{1 - e^{-10s}}{s(s^2 + 3s + 2)}.$$

Using partial fractions, we can write

$$Y(s) = \frac{1}{2}(1 - e^{-10s}) \left[\frac{1}{s} + \frac{1}{s+2} - \frac{2}{s+1} \right].$$

Therefore, we conclude that

$$y(t) = \frac{1}{2}[1 + e^{-2t} - 2e^{-t}] - \frac{1}{2}u_{10}(t)[1 + e^{-2(t-10)} - 2e^{-(t-10)}].$$

The solution increases to a temporary steady value of $y = 1/2$. After the forcing ceases, the response decays exponentially to $y = 0$.

6. Let $Y(s) = \mathcal{L}(y)$. Applying the Laplace transform to the equation, we have

$$\mathcal{L}(y'') + 3\mathcal{L}(y') + 2\mathcal{L}(y) = \mathcal{L}(u_2(t))$$

$$\implies [s^2 Y(s) - sy(0) - y'(0)] + 3[sY(s) - y(0)] + 2Y(s) = \frac{e^{-2s}}{s}.$$

Applying the initial conditions, we have

$$s^2 Y(s) - 1 + 3sY(s) + 2Y(s) = \frac{e^{-2s}}{s}.$$

Therefore, the equation for Y becomes

$$[s^2 + 3s + 2]Y(s) = 1 + \frac{e^{-2s}}{s}$$

$$\implies Y(s) = \frac{1}{s^2 + 3s + 2} + \frac{e^{-2s}}{s(s^2 + 3s + 2)}.$$

Using partial fractions, we can write

$$Y(s) = \frac{1}{s+1} - \frac{1}{s+2} + \frac{1}{2}(e^{-2s})\left[\frac{1}{s} + \frac{1}{s+2} - \frac{2}{s+1}\right].$$

Therefore, we conclude that

$$y(t) = e^{-t} - e^{-2t} + \frac{1}{2}u_2(t)[1 + e^{-2(t-2)} - 2e^{-(t-2)}].$$

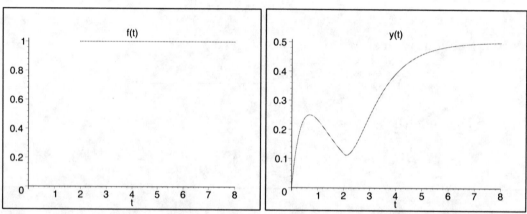

Due to the initial conditions, the response has a transient overshoot, followed by an exponential convergence to a steady value of $y_s = 1/2$.

7. Let $Y(s) = \mathcal{L}(y)$. Applying the Laplace transform to the equation, we have

$$\mathcal{L}(y'') + \mathcal{L}(y) = \mathcal{L}(u_{3\pi}(t))$$

$$\implies [s^2 Y(s) - sy(0) - y'(0)] + Y(s) = \frac{e^{-3\pi s}}{s}.$$

Applying the initial conditions, we have

$$s^2 Y(s) - s + Y(s) = \frac{e^{-3\pi s}}{s}.$$

Therefore, the equation for Y becomes

$$[s^2 + 1]Y(s) = s + \frac{e^{-3\pi s}}{s}$$

$$\implies Y(s) = \frac{s}{s^2 + 1} + \frac{e^{-3\pi s}}{s(s^2 + 1)}.$$

68

Using partial fractions, we can write

$$Y(s) = \frac{s}{s^2+1} + e^{-3\pi s}\left[\frac{1}{s} - \frac{s}{s^2+1}\right].$$

Therefore, we conclude that

$$y(t) = \cos(t) + u_{3\pi}(t)[1 - \cos(t-3\pi)]$$
$$= \cos(t) + u_{3\pi}(t)[1 + \cos(t)].$$

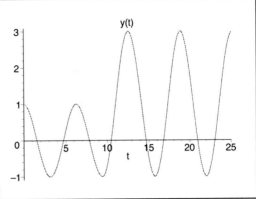

Due to the initial conditions, the solution temporarily oscillated about $y = 0$. After the forcing is applied, the response is a steady oscillation about $y_m = 1$.

8. Let $Y(s) = \mathcal{L}(y)$. Applying the Laplace transform to the equation, we have

$$\mathcal{L}(y'') + \mathcal{L}(y') + \frac{5}{4}\mathcal{L}(y) = \mathcal{L}(t - u_{\pi/2}(t)(t - \pi/2))$$

$$\implies [s^2 Y(s) - sy(0) - y'(0)] + [sY(s) - y(0)] + \frac{5}{4}Y(s) = \frac{1 - e^{-\pi s/2}}{s^2}.$$

Applying the initial conditions, we have

$$s^2 Y(s) + sY(s) + \frac{5}{4}Y(s) = \frac{1 - e^{-\pi s/2}}{s^2}.$$

Therefore, the equation for Y becomes

$$[s^2 + s + 5/4]Y(s) = \frac{1 - e^{-\pi s/2}}{s^2}$$

$$\implies Y(s) = \frac{1 - e^{-\pi s/2}}{s^2(s^2 + s + 5/4)}.$$

Using partial fractions, we can write

$$Y(s) = (1 - e^{-\pi s/2})\left[\frac{64s - 16}{25(4s^2 + 4s + 5)} + \frac{4}{5s^2} - \frac{16}{25s}\right].$$

69

Now, completing the square in the denominator of the first term on the right-hand side above, we have

$$\frac{64s - 16}{25(4s^2 + 4s + 5)} = \frac{64s - 16}{100((s + 1/2)^2 + 1)} = \frac{16}{25}\left[\frac{s + 1/2}{(s + 1/2)^2 + 1} - \frac{3/4}{(s + 1/2)^2 + 1}\right].$$

Therefore,

$$Y(s) = (1 - e^{-\pi s/2}) \cdot \frac{16}{25}\left[\frac{s + 1/2}{(s + 1/2)^2 + 1} - \frac{3/4}{(s + 1/2)^2 + 1} + \frac{5}{4s^2} - \frac{1}{s}\right].$$

Therefore, we conclude that

$$y(t) = \frac{16}{25}\left[e^{-t/2}\cos(t) - \frac{3}{4}e^{-t/2}\sin(t) + \frac{5}{4}t - 1\right]$$
$$- \frac{16}{25}u_{\pi/2}(t)\left[e^{-(t-\pi/2)/2}\cos(t - \pi/2) - \frac{3}{4}e^{-(t-\pi/2)/2}\sin(t - \pi/2) + \frac{5}{4}(t - \pi/2) - 1\right]$$

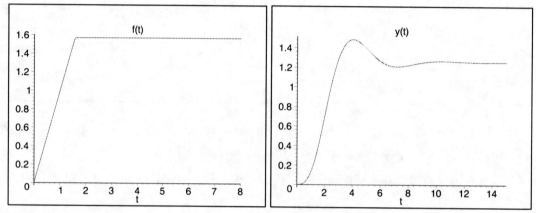

The forcing term causes the initial oscillation, but then the solution approaches the steady state $y_s(t) = \pi/2$.

9. First, we can write the inhomogeneous term as $g(t) = \frac{t}{2} - \frac{1}{2}u_6(t)[t - 6]$. Then, let $Y(s) = \mathcal{L}(y)$. Applying the Laplace transform to the equation, we have

$$\mathcal{L}(y'') + \mathcal{L}(y) = \frac{1}{2}\mathcal{L}(t - u_6(t)(t - 6))$$
$$\implies [s^2 Y(s) - sy(0) - y'(0)] + Y(s) = \frac{1}{2}\left[\frac{1 - e^{-6s}}{s^2}\right].$$

Applying the initial conditions, we have

$$s^2 Y(s) - 1 + Y(s) = \frac{1}{2}\frac{1 - e^{-6s}}{s^2}.$$

Therefore, the equation for Y becomes

$$[s^2 + 1]Y(s) = 1 + \frac{1}{2}\frac{1 - e^{-6s}}{s^2}$$
$$\implies Y(s) = \frac{1}{s^2 + 1} + \frac{1}{2}\frac{1 - e^{-6s}}{s^2(s^2 + 1)}.$$

70

Using partial fractions, we can write

$$Y(s) = \frac{1}{s^2+1} + \frac{1}{2}(1 - e^{-6s})\left[\frac{1}{s^2} - \frac{1}{s^2+1}\right].$$

Therefore, we conclude that

$$y(t) = \sin(t) + \frac{1}{2}\left[t - \sin(t) - u_6(t)(t-6) + u_6(t)\sin(t-6)\right]$$
$$= \frac{1}{2}[t + \sin(t)] - \frac{1}{2}[(t-6) - \sin(t-6)]u_6(t).$$

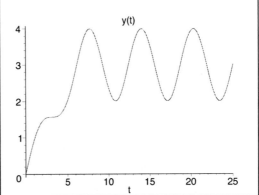

The solution increases in response to the ramp input, and, thereafter oscillated about a mean value of $y_m = 3$.

10. First, we can write the inhomogeneous term as $g(t) = \sin(t)[1 - u_\pi(t)] = \sin(t) + u_\pi(t)\sin(t-\pi)$. Then, let $Y(s) = \mathcal{L}(y)$. Applying the Laplace transform to the equation, we have

$$\mathcal{L}(y'') + \mathcal{L}(y') + \frac{5}{4}\mathcal{L}(y) = \mathcal{L}(\sin(t) + u_\pi(t)\sin(t-\pi))$$
$$\implies [s^2Y(s) - sy(0) - y'(0)] + [sY(s) - y(0)] + \frac{5}{4}Y(s) = \frac{1 + e^{-\pi s}}{s^2+1}.$$

Applying the initial conditions, we have

$$s^2Y(s) + sY(s) + \frac{5}{4}Y(s) = \frac{1 + e^{-\pi s}}{s^2+1}.$$

Therefore, the equation for Y becomes

$$[s^2 + s + 5/4]Y(s) = \frac{1 + e^{-\pi s}}{s^2+1}$$
$$\implies Y(s) = \frac{1 + e^{-\pi s}}{(s^2 + s + 5/4)(s^2+1)}.$$

Using partial fractions, we can write

$$Y(s) = (1 + e^{-\pi s}) \left[\frac{64s + 48}{17(4s^2 + 4s + 5)} + \frac{-16s + 4}{17(s^2 + 1)} \right]$$

$$= \frac{1}{17}(1 + e^{-\pi s}) \left[\frac{16(s + 1/2) + 4}{(s + 1/2)^2 + 1} - \frac{16s}{s^2 + 1} + \frac{4}{s^2 + 1} \right].$$

Therefore, we conclude that

$$y(t) = \frac{1}{17} \left[16e^{-t/2} \cos(t) + 4e^{-t/2} \sin(t) - 16 \cos(t) + 4 \sin(t) \right]$$

$$+ \frac{1}{17} u_\pi(t) \left[16e^{-(t-\pi)/2} \cos(t - \pi) + 4e^{-(t-\pi)/2} \sin(t - \pi) - 16 \cos(t - \pi) + 4 \sin(t - \pi) \right]$$

$$= \frac{1}{17} [16 \cos(t)(e^{-t/2} - 1) + 4 \sin(t)(e^{-t/2} + 1)]$$

$$+ \frac{1}{17} u_\pi(t)[16 \cos(t)(1 - e^{-(t-\pi)/2}) - 4 \sin(t)(1 + e^{-(t-\pi)/2})].$$

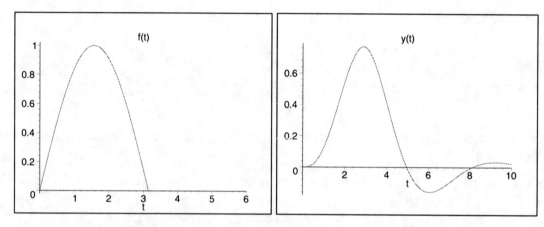

The initial oscillation is a response to the forcing term, but after the forcing term is taken away, the solution decays to zero due to the damping term.

11. Let $Y(s) = \mathcal{L}(y)$. Applying the Laplace transform to the equation, we have

$$\mathcal{L}(y'') + 4\mathcal{L}(y) = \mathcal{L}(u_\pi(t) - u_{3\pi}(t))$$

$$\implies [s^2 Y(s) - sy(0) - y'(0)] + 4Y(s) = \frac{e^{-\pi s} - e^{-3\pi s}}{s}.$$

Applying the initial conditions, we have

$$s^2 Y(s) + 4Y(s) = \frac{e^{-\pi s} - e^{-3\pi s}}{s}.$$

Therefore, the equation for Y becomes

$$[s^2 + 4]Y(s) = \frac{e^{-\pi s} - e^{-3\pi s}}{s}$$

$$\implies Y(s) = \frac{e^{-\pi s} - e^{-3\pi s}}{s(s^2 + 4)}.$$

72

Using partial fractions, we can write

$$Y(s) = (e^{-\pi s} - e^{-3\pi s})\frac{1}{4}\left[\frac{1}{s} - \frac{s}{s^2 + 4}\right]$$

Therefore, we conclude that

$$y(t) = \frac{1}{4}[1 - \cos(2(t - \pi))]u_\pi(t) - \frac{1}{4}[1 - \cos(2(t - 3\pi))]u_{3\pi}(t).$$

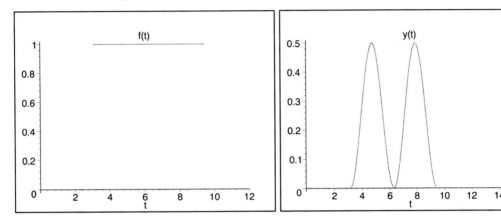

Since there is no damping term, the solution responds immediately to the forcing term. There is a temporary oscillation about $y = 1/4$.

12. Let $Y(s) = \mathcal{L}(y)$. Applying the Laplace transform to the equation, we have

$$\mathcal{L}(y'''') - \mathcal{L}(y) = \mathcal{L}(u_1(t) - u_2(t))$$

$$\implies [s^4 Y(s) - s^3 y(0) - s^2 y'(0) - s y''(0) - y'''(0)] - Y(s) = \frac{e^{-s} - e^{-2s}}{s}.$$

Applying the initial conditions, we have

$$s^4 Y(s) - Y(s) = \frac{e^{-s} - e^{-2s}}{s}.$$

Therefore, the equation for Y becomes

$$[s^4 - 1]Y(s) = \frac{e^{-s} - e^{-2s}}{s}$$

$$\implies Y(s) = \frac{e^{-s} - e^{-2s}}{s(s^4 - 1)}.$$

Using partial fractions, we can write

$$Y(s) = (e^{-s} - e^{-2s})\frac{1}{4}\left[\frac{-4}{s} + \frac{1}{s+1} + \frac{1}{s-1} + \frac{2s}{s^2 + 1}\right]$$

Therefore, we conclude that

$$y(t) = -[u_1(t) - u_2(t)] + \frac{1}{4}[e^{-(t-1)} + e^{(t-1)} + 2\cos(t - 1)]u_1(t)$$

$$- \frac{1}{4}[e^{-(t-2)} + e^{(t-2)} + 2\cos(t - 2)]u_2(t).$$

73

The solution increases exponentially.

13. Let $Y(s) = \mathcal{L}(y)$. Applying the Laplace transform to the equation, we have

$$\mathcal{L}(y'''') + 5\mathcal{L}(y'') + 4\mathcal{L}(y) = \mathcal{L}(1 - u_\pi(t))$$
$$\implies [s^4 Y(s) - s^3 y(0) - s^2 y'(0) - sy''(0) - y'''(0)] + 5[s^2 Y(s) - sy(0) - y'(0)]$$
$$+ 4Y(s) = \frac{1 - e^{-\pi s}}{s}.$$

Applying the initial conditions, we have

$$s^4 Y(s) + 5s^2 Y(s) + 4Y(s) = \frac{1 - e^{-\pi s}}{s}.$$

Therefore, the equation for Y becomes

$$[s^4 + 5s^2 + 4]Y(s) = \frac{1 - e^{-\pi s}}{s}$$
$$\implies Y(s) = \frac{1 - e^{-\pi s}}{s(s^4 + 5s^2 + 4)}.$$

Using partial fractions, we can write

$$Y(s) = (1 - e^{-\pi s})\frac{1}{12}\left[\frac{3}{s} + \frac{s}{s^2 + 4} - \frac{4s}{s^2 + 1}\right].$$

Therefore, we conclude that

$$y(t) = \frac{1}{12}[3 + \cos(2t) - 4\cos(t)] - u_\pi(t)\frac{1}{12}[3 + \cos(2(t - \pi)) - 4\cos(t - \pi)]$$
$$= \frac{1}{12}[3 + \cos(2t) - 4\cos(t)] - u_\pi(t)\frac{1}{12}[3 + \cos(2t) + 4\cos(t)].$$

74

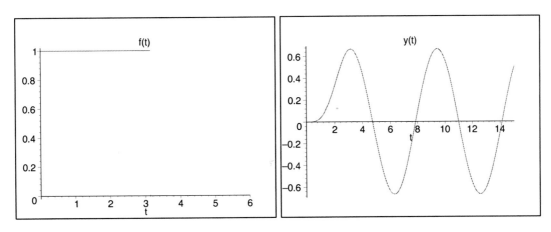

After an initial transient, the solution oscillates about $y_m = 0$.

14. The specified function is defined by

$$f(t) = \begin{cases} 0 & 0 \le t < t_0 \\ \frac{h}{k}(t - t_0) & t_0 \le t < t_0 + k \\ h & t_0 + k \le t. \end{cases}$$

Using u_c, this function can be written as

$$f(t) = \frac{h}{k}(t - t_0)u_{t_0}(t) - \frac{h}{k}(t - t_0 - k)u_{t_0+k}(t).$$

15. The specified function is defined by

$$f(t) = \begin{cases} 0 & 0 \le t < t_0 \\ \frac{h}{k}(t - t_0) & t_0 \le t < t_0 + k \\ -\frac{h}{k}(t - t_0 - 2k) & t_0 + k \le t < t_0 + 2kk \\ h & t_0 + 2k \le t. \end{cases}$$

Using u_c, this function can be written as

$$f(t) = \frac{h}{k}(t - t_0)u_{t_0}(t) - \frac{2h}{k}(t - t_0 - k)u_{t_0+k}(t) + \frac{h}{k}(t - t_0 - 2k)u_{t_0+2k}(t).$$

16.

(a)

(b) Let $U(s) = \mathcal{L}(u)$. Applying the Laplace transform to the equation, we have

$$\mathcal{L}(u'') + \frac{1}{4}\mathcal{L}(u') + \mathcal{L}(y) = k\mathcal{L}(g(t))$$

$$\implies [s^2 U(s) - su(0) - u'(0)] + \frac{1}{4}[sU(s) - u(0)] + U(s) = k\frac{e^{-3s/2} - e^{-5s/2}}{s}.$$

Applying the initial conditions, we have

$$s^2 U(s) + \frac{1}{4}sU(s) + U(s) = k\frac{e^{-3s/2} - e^{-5s/2}}{s}.$$

Therefore, the equation for U becomes

$$[s^2 + 1/4s + 1]U(s) = k\frac{e^{-3s/2} - e^{-5s/2}}{s}$$

$$\implies U(s) = k\frac{e^{-3s/2} - e^{-5s/2}}{s(s^2 + 1/4s + 1)}.$$

Using partial fractions, we can write

$$U(s) = k(e^{-3s/2} - e^{-5s/2})\left[\frac{1}{s} - \frac{4s+1}{4s^2 + s + 4}\right]$$

$$= k(e^{-3s/2} - e^{-5s/2})\left[\frac{1}{s} - \frac{4s+1}{4s^2 + s + 4}\right]$$

$$= k(e^{-3s/2} - e^{-5s/2})\left[\frac{1}{s} - \frac{(s+1/8) + 1/8}{(s+1/8)^2 + 63/64}\right].$$

Therefore, we conclude that

$$u(t) = ku_{\frac{3}{2}}(t)\left[1 - e^{-(t-3/2)/8}\cos\left(\frac{\sqrt{63}(t-3/2)}{8}\right) - \frac{e^{-(t-3/2)/8}}{\sqrt{63}}\sin\left(\frac{\sqrt{63}(t-3/2)}{8}\right)\right]$$

$$- ku_{\frac{5}{2}}(t)\left[1 - e^{-(t-5/2)/8}\cos\left(\frac{\sqrt{63}(t-5/2)}{8}\right) - \frac{e^{-(t-5/2)/8}}{\sqrt{63}}\sin\left(\frac{\sqrt{63}(t-5/2)}{8}\right)\right]$$

(c)

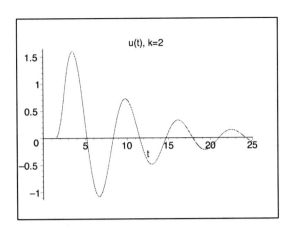

(d) Due to the damping term, the solution will decay to zero. The maximum will occur shortly after the forcing ceases. By plotting the various solutions, it appears that the solution will reach a value of $u = 2$ as long as $k > 2.51$.

(e) Based on the graph and numerical calculation, $|u(t)| < 0.1$ for $t > 25.6773$.

17. In this problem, we consider the IVP

$$y'' + 4y = \frac{1}{k}[(t-5)u_5(t) - (t-5-k)u_{5+k}(t)].$$

(a)

The problem is identical to example 1 when $k = 5$.

(b) Let $Y(s) = \mathcal{L}(y)$. Applying the Laplace transform to the equation, we have

$$\mathcal{L}(y'') + 4\mathcal{L}(y) = \mathcal{L}(f(t))$$

$$\implies [s^2 Y(s) - sy(0) - y'(0)] + 4Y(s) = \frac{e^{-5s}}{ks^2} - \frac{e^{-(5+k)s}}{ks^2}.$$

Applying the initial conditions, we have

$$s^2 Y(s) + 4Y(s) = \frac{e^{-5s}}{ks^2} - \frac{e^{-(5+k)s}}{ks^2}.$$

Therefore, the equation for Y becomes

$$Y(s) = \frac{e^{-5s}}{ks^2(s^2+4)} - \frac{e^{-(5+k)s}}{ks^2(s^2+4)}.$$

Using partial fractions, we can write

$$Y(s) = (e^{-5s} - e^{-(5+k)s})\frac{1}{4k}\left[\frac{1}{s^2} - \frac{1}{s^2+4}\right].$$

Therefore, we conclude that

$$y(t) = \frac{u_5(t)}{4k}\left[(t-5) - \frac{1}{2}\sin(2(t-5))\right]$$
$$- \frac{u_{5+k}(t)}{4k}\left[(t-5-k) + \frac{1}{2}\sin(2(t-5-k))\right].$$

(c) We note that for $t > 5+k$, the solution is given by

$$y(t) = \frac{1}{4} - \frac{1}{8k}\sin(2t-10) + \frac{1}{8k}\sin(2t-10-2k)$$
$$= \frac{1}{4} - \frac{\sin(k)}{4k}\cos(2t-10-k).$$

Therefore, for $t > 5+k$, the solution oscillates about $y_m = 1/4$ with an amplitude of $A = |\sin(k)|/4k$.

18.

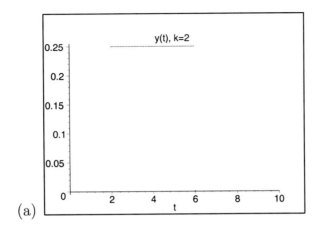

(a)

(b) We can write $f_k(t) = \dfrac{1}{2k}[u_{4-k}(t) - u_{4+k}(t)]$. Let $Y(s) = \mathcal{L}(y)$. Applying the Laplace transform to the equation, we have

$$\mathcal{L}(y'') + \frac{1}{3}\mathcal{L}(y') + 4\mathcal{L}(y) = \frac{1}{2k}\mathcal{L}(u_{4-k}(t) - u_{4+k}(t))$$

$$\implies [s^2 Y(s) - sy(0) - y'(0)] + \frac{1}{3}[sY(s) - y(0)] + 4Y(s) = \frac{e^{-(4-k)s}}{2ks} - \frac{e^{-(4+k)s}}{2ks}.$$

Applying the initial conditions, we have

$$s^2 Y(s) + \frac{1}{3}sY(s) + 4Y(s) = \frac{e^{-(4-k)s}}{2ks} - \frac{e^{-(4+k)s}}{2ks}.$$

Therefore, the equation for Y becomes

$$Y(s) = \frac{3e^{-(4-k)s}}{2ks(3s^2 + s + 12)} - \frac{3e^{-(4+k)s}}{2ks(3s^2 + s + 12)}.$$

Using partial fractions, we can write

$$\frac{1}{s(3s^2 + s + 12)} = \frac{1}{12}\left[\frac{1}{s} - \frac{1}{3s^2 + s + 12}\right]$$

$$= \frac{1}{12}\left[\frac{1}{s} - \frac{1}{6}\frac{1 + 6(s + 1/6)}{(s + 1/6)^2 + 143/36}\right].$$

Let

$$H(s) = \frac{1}{8k}\left[\frac{1}{s} - \frac{1/6}{(s + 1/6)^2 + 143/36} - \frac{s + 1/6}{(s + 1/6)^2 + 143/36}\right].$$

Therefore, letting

$$h(t) = \mathcal{L}^{-1}(H(s)) = \frac{1}{8k} - \frac{e^{-t/6}}{8k}\left[\frac{1}{\sqrt{143}}\sin\left(\frac{\sqrt{143}t}{6}\right) + \cos\left(\frac{\sqrt{143}t}{6}\right)\right],$$

we conclude that

$$y(t) = h(t - 4 + k)u_{4-k}(t) - h(t - 4 - k)u_{4+k}(t).$$

79

(c)

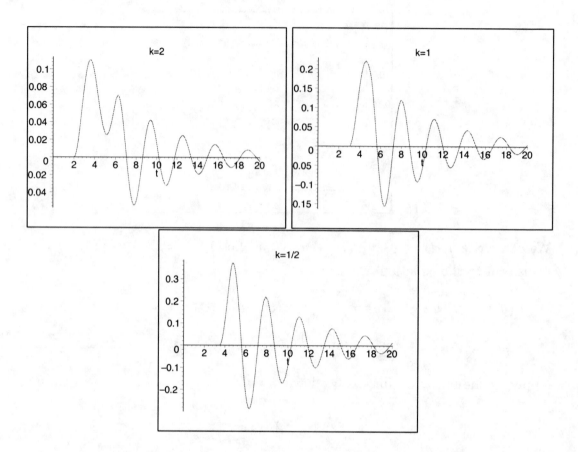

As k decreases, the solution remains null for a longer period of time. In addition, the amplitude increases.

19.

(a) Let $Y(s) = \mathcal{L}(y)$. Applying the Laplace transform to the equation, we have

$$[s^2 Y(s) - sy(0) - y'(0)] + Y(s) = \mathcal{L}(1 + 2\sum_{k=1}^{n}(-1)^k u_{k\pi}(t))$$
$$= \frac{1 + 2\sum_{k=1}^{n}(-1)^k e^{-k\pi s}}{s}.$$

Applying the initial conditions, we have

$$[s^2 + 1]Y(s) = \frac{1 + 2\sum_{k=1}^{n}(-1)^k e^{-k\pi s}}{s}.$$

Therefore,

$$Y(s) = \frac{1 + 2\sum_{k=1}^{n}(-1)^k e^{-k\pi s}}{s(s^2 + 1)}.$$

Using partial fractions, we see that

$$Y(s) = \left(1 + 2\sum_{k=1}^{n}(-1)^k e^{-k\pi s}\right)\left[\frac{1}{s} - \frac{s}{s^2+1}\right].$$

Therefore, we conclude that

$$y(t) = 1 + 2\sum_{k=1}^{n}(-1)^k u_{k\pi}(t) - \cos(t) - 2\sum_{k=1}^{n}(-1)^k u_{k\pi}(t)\cos(t - k\pi)$$

$$= 1 - \cos(t) + 2\sum_{k=1}^{n}(-1)^k[1 - \cos(t - k\pi)]u_{k\pi}(t).$$

(b)

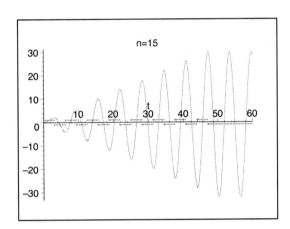

(c) As $n \to \infty$, the amplitude of the solution increases. Below is a graph for the case $n = 30$.

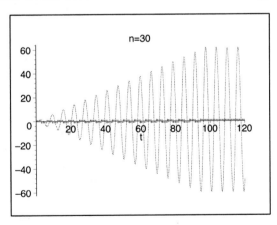

20.

(a) Let $Y(s) = \mathcal{L}(y)$. Applying the Laplace transform to the equation, we have

$$[s^2 Y(s) - sy(0) - y'(0)] + 0.1[sY(s) - y(0)] + Y(s) = \mathcal{L}(1 + 2\sum_{k=1}^{n}(-1)^k u_{k\pi}(t))$$

$$= \frac{1 + 2\sum_{k=1}^{n}(-1)^k e^{-k\pi s}}{s}.$$

Applying the initial conditions, we have

$$[s^2 + 0.1s + 1]Y(s) = \frac{1 + 2\sum_{k=1}^{n}(-1)^k e^{-k\pi s}}{s}.$$

Therefore,

$$Y(s) = \frac{1 + 2\sum_{k=1}^{n}(-1)^k e^{-k\pi s}}{s(s^2 + 0.1s + 1)}.$$

Using partial fractions, we see that

$$Y(s) = \left(1 + 2\sum_{k=1}^{n}(-1)^k e^{-k\pi s}\right)\left[\frac{1}{s} - \frac{s + 0.1}{s^2 + 0.1s + 1}\right]$$

$$= \left(1 + 2\sum_{k=1}^{n}(-1)^k e^{-k\pi s}\right)\left[\frac{1}{s} - \frac{s + 0.05}{(s + 0.05)^2 + .9975} - \frac{0.05}{(s + 0.05)^2 + .9975}\right]$$

Therefore, we conclude that

$$y(t) = h(t) + 2\sum_{k=1}^{n}(-1)^k u_{k\pi}(t)h(t - k\pi)$$

where

$$h(t) = 1 - e^{-0.05t}\cos(\sqrt{.9975}t) - \frac{0.05}{\sqrt{.9975}}e^{-0.05t}\sin(\sqrt{.9975}t).$$

(b)

 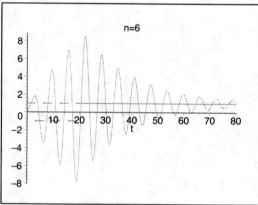

(c) Due to the damping term, for n odd, the solution will approach $y = -1$ while for n even, the solution will approach $y = 1$. The frequency is $\sqrt{.9975} \approx 1$.

21.

(a)

(b) Let $Y(s) = \mathcal{L}(y)$. Applying the Laplace transform to the equation, we have

$$[s^2 Y(s) - sy(0) - y'(0)] + Y(s) = \frac{1 + \sum_{k=1}^{n}(-1)^k e^{-k\pi s}}{s}.$$

Applying the initial conditions, we have

$$[s^2 + 1]Y(s) = \frac{1 + \sum_{k=1}^{n}(-1)^k e^{-k\pi s}}{s}.$$

Therefore,

$$Y(s) = \frac{1 + \sum_{k=1}^{n}(-1)^k e^{-k\pi s}}{s(s^2 + 1)}.$$

Using partial fractions, we see that

$$Y(s) = \left(1 + \sum_{k=1}^{n}(-1)^k e^{-k\pi s}\right)\left[\frac{1}{s} - \frac{s}{s^2 + 1}\right].$$

Therefore, we conclude that

$$y(t) = 1 + \sum_{k=1}^{n}(-1)^k u_{k\pi}(t) - \cos(t) - \sum_{k=1}^{n}(-1)^k u_{k\pi}(t)\cos(t - k\pi)$$

$$= 1 - \cos(t) + \sum_{k=1}^{n}(-1)^k[1 - \cos(t - k\pi)]u_{k\pi}(t).$$

(c)

There is no damping there. Each interval of forcing adds energy to the system. Therefore, the amplitude will continue to increase until increase. For $n = 15$, $g(t) = 0$ when $t > 15\pi$. Therefore, the oscillation will eventually become steady.

(d) If n is even, the forcing term $g(t) = 1$ for large values of t, while for n odd, the forcing term $g(t) = 0$ for large values of t. Therefore, for n even, the solution will eventually oscillate about the line $y = 1$, while for n odd, the solution will eventually oscillate about the line $y = 0$. As n increases, the amount of forcing increases, causing the amplitude of the oscillations to increase until they reach a steady state after $t > n\pi$.

22.

(a)

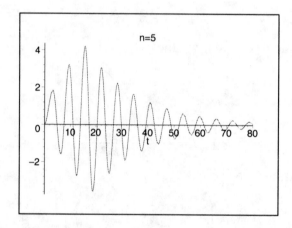

(b) For n odd, the solution approaches $y = 0$. For n even, the solution approaches $y = 1$. The solution is a sum of damped sinusoids, each of frequency $\omega \approx 1$.

(c) In problem 20, the solution approaches $y = 1$ if n is even and $y = -1$ if n is odd.

23.

(a) Let $Y(s) = \mathcal{L}(y)$. Applying the Laplace transform to the equation, we have

$$[s^2 Y(s) - sy(0) - y'(0)] + Y(s) = \frac{1 + 2\sum_{k=1}^{n}(-1)^k e^{(-11k/4)s}}{s}.$$

Applying the initial conditions, we have

$$[s^2 + 1]Y(s) = \frac{1 + 2\sum_{k=1}^{n}(-1)^k e^{(-11k/4)s}}{s}.$$

Therefore,

$$Y(s) = \frac{1 + 2\sum_{k=1}^{n}(-1)^k e^{(-11k/4)s}}{s(s^2 + 1)}.$$

Using partial fractions, we see that

$$Y(s) = \left(1 + 2\sum_{k=1}^{n}(-1)^k e^{(-11k/4)s}\right)\left[\frac{1}{s} - \frac{s}{s^2+1}\right].$$

Therefore, we conclude that

$$y(t) = 1 + 2\sum_{k=1}^{n}(-1)^k u_{11k/4}(t) - \cos(t) - 2\sum_{k=1}^{n}(-1)^k u_{11k/4}(t)\cos(t - 11k/4)$$

$$= 1 - \cos(t) + 2\sum_{k=1}^{n}(-1)^k[1 - \cos(t - 11k/4)]u_{11k/4}(t).$$

(b)

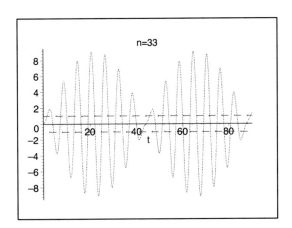

(c) Based on the plot, the "slow period" appears to be approximately 88 and the "fast period" appears to be about 6.

(d) The natural frequency of the system is $\omega_0 = 1$. The forcing function is initially periodic with period $T = 11/2$. Therefore, the corresponding forcing frequency is $\omega = 1.1424$. Therefore, the "slow frequency" is

$$\omega_s = \frac{|\omega - \omega_0|}{2} = 0.0712$$

and the "fast frequency' is

$$\omega_f = \frac{|\omega + \omega_0|}{2} = 1.0712.$$

Based on these values, the slow period is predicted as 88.247 and the fast period is predicted to be 5.8656.

Section 5.7

1. Let $Y(s) = \mathcal{L}(y)$ and take the Laplace transform of the ODE. We arrive at

$$[s^2 Y(s) - sy(0) - y'(0)] + 2[sY(s) - y(0)] + 2Y(s) = e^{-\pi s}.$$

Applying the initial conditions, we have

$$[s^2Y(s) - s] + 2[sY(s) - 1] + 2Y(s) = e^{-\pi s},$$

which can be rewritten as

$$[s^2 + 2s + 2]Y(s) - s - 2 = e^{-\pi s}.$$

Therefore,

$$\begin{aligned}
Y(s) &= \frac{s + 2 + e^{-\pi s}}{s^2 + 2s + 2} \\
&= \frac{s + 2 + e^{-\pi s}}{(s + 1)^2 + 1} \\
&= \frac{s + 1}{(s + 1)^2 + 1} + \frac{1}{(s + 1)^2 + 1} + \frac{e^{-\pi s}}{(s + 1)^2 + 1}.
\end{aligned}$$

Therefore,

$$y(t) = e^{-t}\cos(t) + e^{-t}\sin(t) + u_\pi(t)e^{-(t-\pi)}\sin(t - \pi).$$

2. Let $Y(s) = \mathcal{L}(y)$ and take the Laplace transform of the ODE. We arrive at

$$[s^2Y(s) - sy(0) - y'(0)] + 4Y(s) = e^{-\pi s} - e^{-2\pi s}.$$

Applying the initial conditions, we have

$$[s^2 + 4]Y(s) = e^{-\pi s} - e^{-2\pi s}.$$

Therefore,

$$Y(s) = \frac{e^{-\pi s} - e^{-2\pi s}}{s^2 + 4}$$

Therefore,

$$\begin{aligned}
y(t) &= \frac{1}{2}\sin(2t - 2\pi)u_\pi(t) - \frac{1}{2}\sin(2t - 4\pi)u_{2\pi}(t) \\
&= \frac{1}{2}\sin(2t)[u_\pi(t) - u_{2\pi}(t)].
\end{aligned}$$

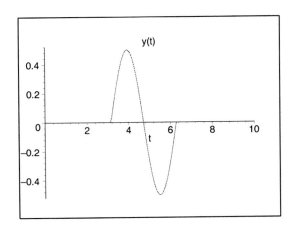

3. Let $Y(s) = \mathcal{L}(y)$ and take the Laplace transform of the ODE. We arrive at

$$[s^2 Y(s) - sy(0) - y'(0)] + 3[sY(s) - y(0)] + 2Y(s) = e^{-5s} + \frac{e^{-10s}}{s}.$$

Applying the initial conditions, we have

$$[s^2 Y(s) - 1/2] + 3sY(s) + 2Y(s) = e^{-5s} + \frac{e^{-10s}}{s},$$

which can be rewritten as

$$[s^2 + 3s + 2]Y(s) - 1/2 = e^{-5s} + \frac{e^{-10s}}{s}.$$

Therefore,

$$\begin{aligned}
Y(s) &= \frac{1/2 + e^{-5s}}{s^2 + 3s + 2} + \frac{e^{-10s}}{s(s^2 + 3s + 2)} \\
&= 1/2 + e^{-5s}\left[\frac{1}{s+1} - \frac{1}{s+2}\right] + e^{-10s}\left[\frac{1}{2(s+2)} - \frac{1}{s+1} + \frac{1}{2s}\right]
\end{aligned}$$

Therefore,

$$y(t) = \frac{1}{2}[e^{-t} - e^{-2t}] + u_5(t)[e^{-(t-5)} - e^{-2(t-5)}] + u_{10}(t)\left[\frac{1}{2}e^{-2(t-10)} - e^{-(t-10)} + \frac{1}{2}\right].$$

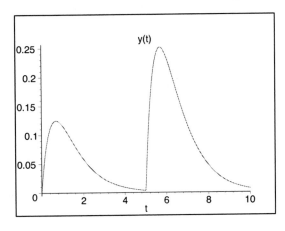

87

4. Let $Y(s) = \mathcal{L}(y)$ and take the Laplace transform of the ODE. We arrive at

$$[s^2 Y(s) - sy(0) - y'(0)] - Y(s) = -20e^{-3s}.$$

Applying the initial conditions, we have

$$[s^2 Y(s) - s] - Y(s) = -20e^{-3s},$$

which can be rewritten as

$$[s^2 - 1]Y(s) - s = -20e^{-3s}.$$

Therefore,

$$Y(s) = \frac{s}{s^2 - 1} - \frac{20e^{-3s}}{s^2 - 1}.$$

Therefore,

$$y(t) = \cosh(t) - 20\sinh(t - 3)u_3(t).$$

5. Let $Y(s) = \mathcal{L}(y)$ and take the Laplace transform of the ODE. We arrive at

$$[s^2 Y(s) - sy(0) - y'(0)] + 2[sY(s) - y(0)] + 3Y(s) = \frac{1}{s^2 + 1} + \frac{e^{-3\pi s}}{.}$$

Applying the initial conditions, we have

$$[s^2 + 2s + 3]Y(s) = \frac{1}{s^2 + 1} + \frac{e^{-3\pi s}}{.}$$

Therefore,

$$Y(s) = \frac{1}{(s^2 + 1)(s^2 + 2s + 3)} + \frac{e^{-3\pi s}}{s^2 + 2s + 3}$$

$$= \frac{1}{4}\left[\frac{-s + 1}{s^2 + 1} + \frac{s + 1}{(s + 1)^2 + 2}\right] + \frac{e^{-3\pi s}}{(s + 1)^2 + 2}.$$

Therefore,

$$y(t) = \frac{1}{4}\left[-\cos(t) + \sin(t) + e^{-t}\cos(\sqrt{2}t)\right] + \frac{1}{\sqrt{2}}u_{3\pi}(t)e^{-(t-3\pi)}\sin(\sqrt{2}(t - 3\pi)).$$

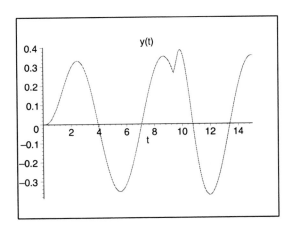

6. Let $Y(s) = \mathcal{L}(y)$ and take the Laplace transform of the ODE. We arrive at

$$[s^2Y(s) - sy(0) - y'(0)] + 4Y(s) = e^{-4\pi s}.$$

Applying the initial conditions, we have

$$[s^2Y(s) - s/2] + 4Y(s) = e^{-4\pi s},$$

which can be rewritten as

$$[s^2 + 4]Y(s) - s/2 = e^{-4\pi s}.$$

Therefore,

$$Y(s) = \frac{s}{2(s^2 + 4)} + \frac{e^{-4\pi s}}{s^2 + 4}.$$

Therefore,

$$y(t) = \frac{1}{2}\cos(2t) + \frac{1}{2}\sin(2t - 8\pi)u_{4\pi}(t)$$

$$= \frac{1}{2}\cos(2t) + \frac{1}{2}\sin(2t)u_{4\pi}(t).$$

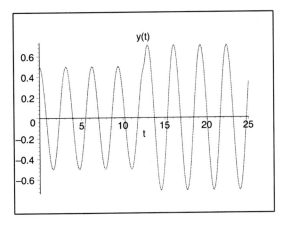

7. Let $Y(s) = \mathcal{L}(y)$ and take the Laplace transform of the ODE. We arrive at

$$[s^2Y(s) - sy(0) - y'(0)] + Y(s) = e^{-2\pi s}.$$

Applying the initial conditions, we have

$$[s^2Y(s) - 1] + Y(s) = e^{-2\pi s},$$

which can be rewritten as

$$[s^2 + 1]Y(s) - 1 = e^{-2\pi s}.$$

Therefore,

$$Y(s) = \frac{1}{s^2 + 1} + \frac{e^{-2\pi s}}{s^2 + 1}.$$

Therefore,

$$y(t) = \sin(t) + u_{2\pi}(t)\sin(t - 2\pi)$$
$$= \sin(t)[1 + u_{2\pi}(t)].$$

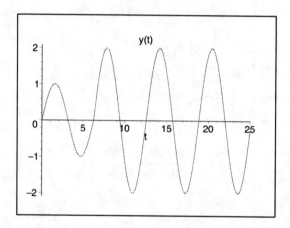

8. Let $Y(s) = \mathcal{L}(y)$ and take the Laplace transform of the ODE. We arrive at

$$[s^2Y(s) - sy(0) - y'(0)] + 4Y(s) = 2e^{-(\pi/4)s}.$$

Applying the initial conditions, we have

$$[s^2 + 4]Y(s) = 2e^{-(\pi/4)s}.$$

Therefore,

$$Y(s) = \frac{2e^{-(\pi/4)s}}{s^2 + 4}.$$

Therefore,

$$y(t) = u_{\pi/4}(t)\sin(2(t - \pi/4)).$$

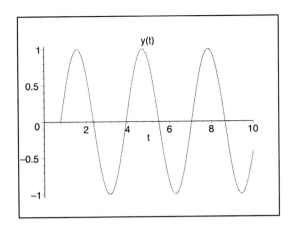

9. Let $Y(s) = \mathcal{L}(y)$ and take the Laplace transform of the ODE. We arrive at

$$[s^2Y(s) - sy(0) - y'(0)] + Y(s) = \frac{e^{-(\pi/2)s}}{s} + 3e^{-(3\pi/2)s} - \frac{e^{-2\pi s}}{s}.$$

Applying the initial conditions, we have

$$[s^2 + 1]Y(s) = \frac{e^{-(\pi/2)s}}{s} + 3e^{-(3\pi/2)s} - \frac{e^{-2\pi s}}{s}.$$

Therefore,

$$Y(s) = \frac{e^{-(\pi/2)s}}{s(s^2 + 1)} + \frac{3e^{-(3\pi/2)s}}{s^2 + 1} - \frac{e^{-2\pi s}}{s(s^2 + 1)}$$
$$= e^{-(\pi/2)s}\left[\frac{1}{s} - \frac{s}{s^2 + 1}\right] + \frac{3e^{-(3\pi/2)s}}{s^2 + 1} - e^{-2\pi s}\left[\frac{1}{s} - \frac{s}{s^2 + 1}\right].$$

Therefore,

$$y(t) = u_{\pi/2}(t)[1 - \cos(t - \pi/2)] + 3u_{3\pi/2}(t)\sin(t - 3\pi/2) - u_{2\pi}(t)[1 - \cos(t - 2\pi)].$$

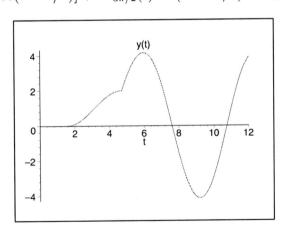

10. Let $Y(s) = \mathcal{L}(y)$ and take the Laplace transform of the ODE. We arrive at

$$2[s^2Y(s) - sy(0) - y'(0)] + [sY(s) - y(0)] + 4Y(s) = \frac{1}{2}e^{-(\pi/6)s}.$$

91

Applying the initial conditions, we have

$$[2s^2 + s + 4]Y(s) = \frac{1}{2}e^{-(\pi/6)s}.$$

Therefore,

$$Y(s) = \frac{e^{-(\pi/6)s}}{2(2s^2 + s + 4)}$$

$$= e^{-(\pi/6)s}\left[\frac{1/4}{(s+1/4)^2 + (31/16)}\right].$$

Therefore,

$$y(t) = u_{\pi/6}(t)\frac{1}{\sqrt{31}}e^{-(t-\pi/6)/4}\sin\left(\frac{\sqrt{31}}{4}(t-\pi/6)\right).$$

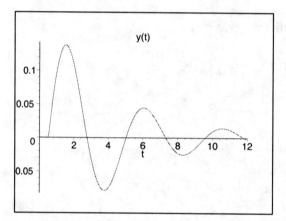

11. Let $Y(s) = \mathcal{L}(y)$ and take the Laplace transform of the ODE. We arrive at

$$[s^2 Y(s) - sy(0) - y'(0)] + 2[sY(s) - y(0)] + 2Y(s) = \frac{s}{s^2+1} + e^{-(\pi/2)s}.$$

Applying the initial conditions, we have

$$[s^2 + 2s + 2]Y(s) = \frac{s}{s^2+1} + e^{-(\pi/2)s}.$$

Therefore,

$$Y(s) = \frac{s}{(s^2+1)(s^2+2s+2)} + \frac{e^{-(\pi/2)s}}{s^2+2s+2}$$

$$= \frac{1}{5}\left[\frac{s}{s^2+1} + \frac{2}{s^2+1} - \frac{s+1}{(s+1)^2+1} - \frac{3}{(s+1)^2+1}\right] + \frac{e^{-(\pi/2)s}}{(s+1)^2+1}$$

Therefore,

$$y(t) = \frac{1}{5}\left[\cos(t) + 2\sin(t) - e^{-t}\cos(t) - 3e^{-t}\sin(t)\right] + u_{\pi/2}(t)e^{-(t-\pi/2)}\sin(t-\pi/2).$$

92

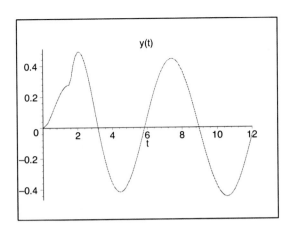

12. Let $Y(s) = \mathcal{L}(y)$ and take the Laplace transform of the ODE. We arrive at

$$[s^4 Y(s) - s^3 y(0) - s^2 y'(0) - sy''(0) - y'''(0)] - Y(s) = e^{-s}.$$

Applying the initial conditions, we have

$$[s^4 - 1]Y(s) = e^{-s}.$$

Therefore,

$$Y(s) = \frac{e^{-s}}{s^4 - 1}$$
$$= e^{-s} \frac{1}{2}\left[\frac{1}{s^2 - 1} - \frac{1}{s^2 + 1}\right].$$

Therefore,

$$y(t) = \frac{1}{2}[\sinh(t - 1) - \sin(t - 1)]u_1(t).$$

13.

(a) Our equation will be of the form

$$2y'' + y' + 2y = \delta(t - 5) + k\delta(t - t_0).$$

93

We need to find k and t_0 so that the system will rest again after exactly one cycle. Applying the Laplace transform and using the initial conditions, this equation becomes

$$[2s^2 + s + 2]Y(s) = e^{-5s} + ke^{-t_0 s}.$$

Therefore,

$$Y(s) = \frac{e^{-5s} + ke^{-t_0 s}}{2s^2 + s + 2}$$

$$= (e^{-5s} + ke^{-t_0 s})\frac{1}{2}\left[\frac{1}{(s + 1/4)^2 + (15/16)}\right].$$

Therefore,

$$y(t) = u_5(t)\frac{2}{\sqrt{15}}e^{-(t-5)/4}\sin\left(\frac{\sqrt{15}}{4}(t - 5)\right) + ku_{t_0}(t)\frac{2}{\sqrt{15}}e^{-(t-t_0)/4}\sin\left(\frac{\sqrt{15}}{4}(t - t_0)\right).$$

In order for this system to return to equilibrium after one cycle, we need the magnitude of these sinusoidal waves to be the same. Therefore, we need $e^{5/4} = -ke^{t_0/4}$, and we need the phase to differ by 2π. That is, we need

$$\frac{\sqrt{15}}{4}(t - 5) = \frac{\sqrt{15}}{4}(t - t_0) + 2\pi.$$

Solving this equation, we see that $t_0 = 5 + \dfrac{8\pi}{\sqrt{15}}$. Solving the equation above for k, we see that $k = -e^{-2\pi/\sqrt{15}}$.

(b) From the analysis above, we see that the solution will be zero after time t_0.

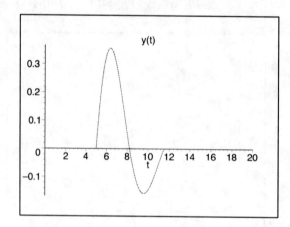

14.

(a) Applying the Laplace transform and using the initial conditions, the ODE becomes

$$s^2 Y(s) + \frac{1}{2}sY(s) + Y(s) = e^{-s}.$$

94

Solving for the transform, we have

$$Y(s) = \frac{e^{-s}}{s^2 + s/2 + 1}$$

$$= \frac{e^{-s}}{(s + 1/4)^2 + (15/16)}.$$

Therefore,

$$y(t) = u_1(t) \frac{4}{\sqrt{15}} e^{-(t-1)/4} \sin\left(\frac{\sqrt{15}}{4}(t-1)\right).$$

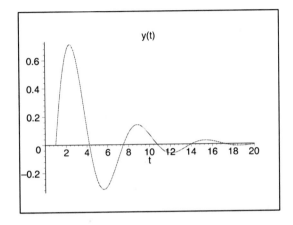

(b) As shown on the graph, the maximum is attained at some $t_1 > 2$. Using the equation for $y(t)$ and calculating $y'(t)$, we see that $t_1 \approx 2.3613$. The maximum value is $y(2.3613) \approx 0.71153$.

(c) Setting $\gamma = 1/4$, the transform of the solution is

$$Y(s) = \frac{e^{-s}}{s^2 + s/4 + 1}.$$

Using the same analysis as in part (a), we conclude that

$$y(t) = \frac{8}{3\sqrt{7}} e^{-(t-1)/8} \sin\left(\frac{3\sqrt{7}}{8}(t-1)\right) u_1(t).$$

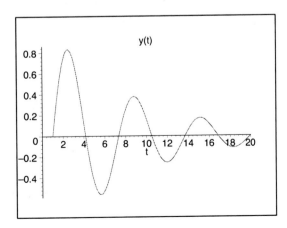

95

By setting $y'(t) = 0$, we see that the maximum is attained at $t_1 \approx 2.4569$ with $y(t_1) \approx 0.8335$.

(d) Suppose that $0 < \gamma < 1$. Then the transform of the solution is

$$Y(s) = \frac{e^{-s}}{s^2 + \gamma s + 1},$$

which implies

$$y(t) = \frac{2}{\sqrt{4 - \gamma^2}} e^{-\gamma(t-1)/2} \sin\left(\sqrt{1 - \gamma^2/4} \cdot (t - 1)\right) u_1(t).$$

Setting $y'(t) = 0$, we obtain

$$\tan\left[\sqrt{1 - \gamma^2/4} \cdot (t - 1)\right] = \frac{1}{\gamma}\sqrt{4 - \gamma^2}.$$

As $\gamma \to 0$, we obtain the formal equation $\tan(t - 1) = \infty$. Hence $t_1 \to 1 + \pi/2$. Setting $t_1 = 1 + \pi/2$ in $y(t)$ and letting $\gamma \to 0$, we find that $y_1 \to 1$. These conclusions agree with the case $\gamma = 0$, for which it is straightforward to show that $y(t) = \sin(t - 1)u_1(t)$.

15.

(a) The solution of this IVP is

$$y(t) = \frac{4k}{\sqrt{15}} e^{-(t-1)/4} \sin\left(\frac{\sqrt{15}}{4}(t - 1)\right) u_1(t).$$

This function is a multiple of the answer in problem 14(a). Therefore, the peak value occurs at $t_1 \approx 2.3613$. The maximum value is calculated as $y(2.3613) \approx 0.71153k$. We find that the appropriate value of k is $k_1 = 2/0.71153 \approx 2.8108$.

(b) Based on problem 14(c), the solution is

$$y(t) = \frac{8k}{3\sqrt{7}} e^{-(t-1)/8} \sin\left(\frac{3\sqrt{7}}{8}(t - 1)\right) u_1(t).$$

Since this function is a multiple of the solution in problem 14(c), we have $t_1 \approx 2.4569$ with $y(t_1) \approx 0.8335k$. The solution reaches a maximum value of $y = 2$ for $k_1 = 2/0.8335 \approx 2.3995$.

(c) Similar to problem 14(d), for $0 < \gamma < 1$, the solution is given by

$$y(t) = \frac{2k}{\sqrt{4 - \gamma^2}} e^{-\gamma(t-1)/2} \sin\left(\sqrt{1 - \gamma^2/4}(t - 1)\right) u_1(t).$$

As $\gamma \to 0$, we see that $y'(t) = 0$ at $t_1 \to 1 + \pi/2$. Setting $t_1 = 1 + \pi/2$ in $y(t)$ and letting $\gamma \to 0$, we find that $y_1 \to k$. Requiring that the peak value remains at $y = 2$, the limiting value of k is $k_1 = 2$. These conclusions agree with the case $\gamma = 0$ for which it is straightforward to show that $y(t) = k\sin(t - 1)u_1(t)$.

16.

(a) Applying the Laplace transform and using the initial conditions, this equation becomes

$$[s^2 + 1]Y(s) = \frac{1}{2k} \left[\frac{e^{-(4-k)s}}{s} - \frac{e^{-(4+k)s}}{s} \right].$$

Therefore,

$$Y(s) = \frac{1}{2k} \left[\frac{e^{-(4-k)s} - e^{-(4+k)s}}{s(s^2+1)} \right]$$

$$= \frac{1}{2k}(e^{-(4-k)s} - e^{-(4+k)s}) \left[\frac{1}{s} - \frac{s}{s^2+1} \right].$$

Therefore,

$$y(t) = \phi(t,k) = \frac{1}{2k} \left[u_{4-k}(t)[1 - \cos(t-4+k)] - u_{4+k}(t)[1 - \cos(t-4-k)] \right].$$

(b) For any $t < 4$, $\phi(t,k) = 0$ as long as $4 - k > t$. If $t \geq 4$, then for $4 + k < t$,

$$\phi(t,k) = -\frac{1}{2k}[\cos(t-4+k) - \cos(t-4-k)].$$

It follows that

$$\lim_{k \to 0} \phi(t,k) = \lim_{k \to 0} -\frac{\cos(t-4+k) - \cos(t-4-k)}{2k} = \sin(t-4).$$

Therefore,

$$\lim_{k \to 0} \phi(t,k) = \sin(t-4)u_4(t).$$

(c) Replacing the inhomogeneous term with $\delta(t-4)$, our equation becomes

$$y'' + y = \delta(t-4).$$

Applying the Laplace transform and using our initial conditions, we have

$$[s^2 + 1]Y(s) = e^{-4s}.$$

Therefore,

$$Y(s) = \frac{e^{-4s}}{s^2 + 1}.$$

It follows that the solution is $\phi_0(t) = \sin(t-4)u_4(t)$.

(d)

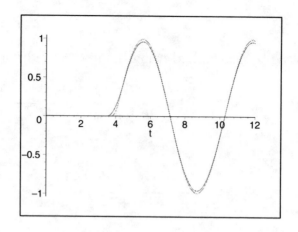

17.

(b) Applying the Laplace transform to the equation and using the initial conditions, we have

$$[s^2 + 1]Y(s) = \sum_{k=1}^{20} e^{-k\pi s}.$$

Therefore,

$$Y(s) = \frac{1}{s^2 + 1} \sum_{k=1}^{20} e^{-k\pi s},$$

which implies

$$y(t) = \sum_{k=1}^{20} \sin(t - k\pi) u_{k\pi}(t).$$

(c) After the sequence of impulses ends, the oscillator returns to equilibrium.

18.

(b) Applying the Laplace transform to the equation and using the initial conditions, we have

$$[s^2 + 1]Y(s) = \sum_{k=1}^{20} (-1)^{k+1} e^{-k\pi s}.$$

Therefore,

$$Y(s) = \frac{1}{s^2+1} \sum_{k=1}^{20} (-1)^{k+1} e^{-k\pi s},$$

which implies

$$y(t) = \sum_{k=1}^{20} (-1)^{k+1} \sin(t - k\pi) u_{k\pi}(t).$$

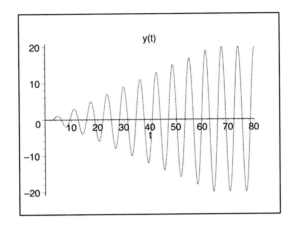

(c) After the sequence of impulses ends, the oscillator continues to oscillate at a constant amplitude.

19.

(b) Applying the Laplace transform to the equation and using the initial conditions, we have

$$[s^2 + 1]Y(s) = \sum_{k=1}^{20} e^{-(k\pi/2)s}.$$

Therefore,

$$Y(s) = \frac{1}{s^2+1} \sum_{k=1}^{20} e^{-(k\pi/2)s},$$

which implies

$$y(t) = \sum_{k=1}^{20} \sin(t - k\pi/2) u_{k\pi/2}(t).$$

(c) After the sequence of impulses ends, the oscillator returns to equilibrium.

20.

(b) Applying the Laplace transform to the equation and using the initial conditions, we have

$$[s^2 + 1]Y(s) = \sum_{k=1}^{20}(-1)^{k+1}e^{-(k\pi/2)s}.$$

Therefore,

$$Y(s) = \frac{1}{s^2 + 1}\sum_{k=1}^{20}(-1)^{k+1}e^{-(k\pi/2)s},$$

which implies

$$y(t) = \sum_{k=1}^{20}(-1)^{k+1}\sin(t - k\pi/2)u_{k\pi/2}(t).$$

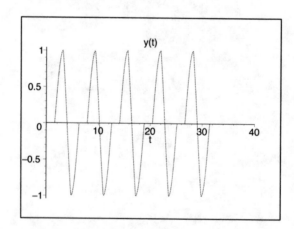

(c) After the sequence of impulses ends, the oscillator returns to equilibrium.

21.

100

(b) Applying the Laplace transform to the equation and using the initial conditions, we have

$$[s^2 + 1]Y(s) = \sum_{k=1}^{20} e^{-(2k-1)\pi s}.$$

Therefore,

$$Y(s) = \frac{1}{s^2+1} \sum_{k=1}^{20} e^{-(2k-1)\pi s},$$

which implies

$$y(t) = \sum_{k=1}^{20} \sin(t - (2k-1)\pi) u_{(2k-1)\pi}(t).$$

(c) After the sequence of impulses ends, the oscillator continues to oscillate at a constant amplitude.

22.

(b) Applying the Laplace transform to the equation and using the initial conditions, we have

$$[s^2 + 1]Y(s) = \sum_{k=1}^{20} (-1)^{k+1} e^{-(11k/4)s}.$$

Therefore,

$$Y(s) = \frac{1}{s^2+1} \sum_{k=1}^{20} (-1)^{k+1} e^{-(11k/4)s},$$

which implies

$$y(t) = \sum_{k=1}^{20} (-1)^{k+1} \sin(t - 11k/4) u_{11k/4}(t).$$

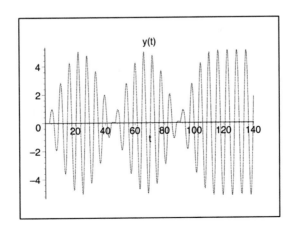

(c) After the sequence of impulses ends, the oscillator continues to oscillate at a constant amplitude.

23.

(b) Applying the Laplace transform to the equation and using the initial conditions, we have

$$[s^2 + 0.1s + 1]Y(s) = \sum_{k=1}^{20}(-1)^{k+1}e^{-k\pi s}.$$

Therefore,

$$Y(s) = \frac{1}{s^2 + 0.1s + 1}\sum_{k=1}^{20}(-1)^{k+1}e^{-k\pi s}$$

$$= \frac{1}{(s+1/20)^2 + (399/400)}\sum_{k=1}^{20}(-1)^{k+1}e^{-k\pi s},$$

which implies

$$y(t) = \sum_{k=1}^{20}(-1)^{k+1}\frac{20}{\sqrt{399}}e^{-(t-k\pi)/20}\sin\left(\frac{\sqrt{399}}{20}(t - k\pi)\right)u_{k\pi}(t).$$

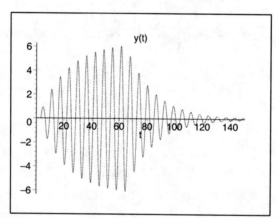

(c) After the sequence of impulses ends, the solution oscillates to zero due to the damping term.

24.

(b) Applying the Laplace transform to the equation and using the initial conditions, we have

$$[s^2 + 0.1s + 1]Y(s) = \sum_{k=1}^{15}e^{-(2k-1)\pi s}.$$

Therefore,

$$Y(s) = \frac{1}{s^2 + 0.1s + 1}\sum_{k=1}^{15}e^{-(2k-1)\pi s}$$

$$= \frac{1}{(s+1/20)^2 + (399/400)}\sum_{k=1}^{15}e^{-(2k-1)\pi s},$$

102

which implies

$$y(t) = \sum_{k=1}^{15} \frac{20}{\sqrt{399}} e^{-(t-(2k-1)\pi)/20} \sin\left(\frac{\sqrt{399}}{20}(t - (2k-1)\pi)\right) u_{(2k-1)\pi}(t).$$

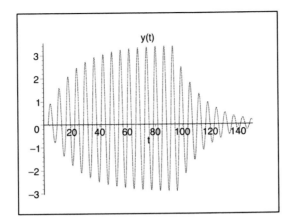

(c) After the sequence of impulses ends, the solution oscillates to zero due to the damping term.

25.

(a) A fundamental set of solutions to the homogeneous problem is $y_1(t) = e^{-t}\cos(t)$ and $y_2(t) = e^{-t}\sin(t)$. Based on problem 22 from Section 4.8, a particular solution is given by

$$y_p(t) = \int_0^t \frac{y_1(s)y_2(t) - y_1(t)y_2(s)}{W(y_1, y_2)(s)} f(s)\, ds.$$

Using the functions y_1, y_2 above, we see that

$$y_p(t) = \int_0^t \frac{e^{-(s-t)}[\cos(s)\sin(t) - \sin(s)\cos(t)]}{e^{-2s}} f(s)\, ds$$
$$= \int_0^t e^{-(t-s)} \sin(t - s) f(s)\, ds.$$

Given the fact that the initial conditions are zero, the solution of this IVP is

$$y_p(t) = \int_0^t e^{-(t-s)} \sin(t - s) f(s)\, ds.$$

(b) If $f(t) = \delta(t - \pi)$, then for $t < \pi$, $y(t) = 0$. On the other hand, for $t > \pi$,

$$y(t) = \int_0^t e^{-(t-s)} \sin(t - s)\delta(s - \pi)\, ds = e^{-(t-\pi)} \sin(t - \pi).$$

Therefore, $y(t) = u_\pi(t)e^{-(t-\pi)} \sin(t - \pi)$.

(c) Applying the Laplace transform to the ODE with $f(t) = \delta(t - \pi)$, we conclude that the Laplace transform satisfies

$$Y(s) = \frac{e^{-\pi s}}{s^2 + 2s + 2}$$

$$= \frac{e^{-\pi s}}{(s+1)^2 + 1}.$$

Therefore,

$$y(t) = u_\pi(t) \sin(t - \pi).$$

Section 5.8

1.

(a)

$$f * (g_1 + g_2)(t) = \int_0^t f(t - \tau)[g_1 + g_2](\tau), d\tau$$

$$= \int_0^t f(t - \tau)g_1(\tau) \, d\tau + \int_0^t f(t - \tau)g_2(\tau) \, d\tau. \quad = f * g_1(t) + f * g_2(t).$$

(b)

$$f * (g * h)(t) = \int_0^t f(t - \tau)[g * h(\tau)] \, d\tau$$

$$= \int_0^t f(t - \tau) \left[\int_0^\tau g(\tau - \eta)h(\eta) \, d\eta \right] d\tau$$

$$= \int_0^t \int_0^\tau f(t - \tau)g(\tau - \eta)h(\eta) \, d\eta \, d\tau.$$

Interchanging the order of integration, we have

$$\int_0^t \int_0^\tau f(t - \tau)g(\tau - \eta)h(\eta) \, d\eta \, d\tau = \int_0^t \int_\eta^t f(t - \tau)g(\tau - \eta)h(\eta) \, d\tau \, d\eta$$

$$= \int_0^t \left[\int_\eta^t f(t - \tau)g(\tau - \eta) \, d\tau \right] h(\eta) \, d\eta.$$

Now letting $\tau - \eta = u$, we have

$$\int_\eta^t f(t - \tau)g(\tau - \eta) \, d\tau = \int_0^{t-\eta} f(t - \eta - u)g(u) \, du$$

$$= f * g(t - \eta).$$

Therefore,

$$\int_0^t f(t - \tau)[g * h(\tau)] \, d\tau = \int_0^t [f * g(t - \tau)]h(\tau) \, d\tau.$$

104

2. For $f(t) = \sin(t)$,

$$f * f(t) = \int_0^t \sin(t - \tau) \sin(\tau) \, d\tau$$

$$= \frac{1}{2} \int_0^t [\cos(t - 2\tau) - \cos(t)] \, d\tau$$

$$= \frac{1}{2}[\sin(t) - t\cos(t)].$$

Clearly, $f * f$ is not necessarily nonnegative.

3. $\mathcal{L}[t^2] = 2/s^3$ and $\mathcal{L}[\cos(2t)] = s/(s^2 + 4)$. Therefore,

$$\mathcal{L}\left[\int_0^t (t - \tau)^2 \cos(2\tau) \, d\tau\right] = \left(\frac{2}{s^3}\right) \cdot \left(\frac{s}{s^2 + 4}\right)$$

$$= \frac{2}{s^2(s^2 + 4)}.$$

4. $\mathcal{L}[e^{-t}] = 1/(s + 1)$ and $\mathcal{L}[\sin(t)] = 1/(s^2 + 1)$. Therefore,

$$\mathcal{L}\left[\int_0^t e^{-(t-\tau)} \sin(\tau) \, d\tau\right] = \left(\frac{1}{s + 1}\right) \cdot \left(\frac{1}{s^2 + 1}\right)$$

$$= \frac{1}{(s + 1)(s^2 + 1)}.$$

5. $\mathcal{L}[t] = 1/s^2$ and $\mathcal{L}[e^t] = 1/(s - 1)$. Therefore,

$$\mathcal{L}\left[\int_0^t (t - \tau)e^\tau \, d\tau\right] = \left(\frac{1}{s^2}\right) \cdot \left(\frac{1}{s - 1}\right)$$

$$= \frac{1}{s^2(s - 1)}.$$

6. $\mathcal{L}[\sin(t)] = 1/(s^2 + 1)$ and $\mathcal{L}[\cos(t)] = s/(s^2 + 1)$. Therefore,

$$\mathcal{L}\left[\int_0^t \sin(t - \tau) \cos(\tau) \, d\tau\right] = \left(\frac{1}{s^2 + 1}\right) \cdot \left(\frac{s}{s^2 + 1}\right)$$

$$= \frac{s}{(s^2 + 1)^2}.$$

7. $\mathcal{L}^{-1}[1/s^4] = t^3/6$ and $\mathcal{L}^{-1}[1/(s^2 + 1)] = \sin(t)$. Therefore,

$$\mathcal{L}^{-1}\left[\frac{1}{s^4(s^2 + 1)}\right] = \int_0^t \frac{1}{6}(t - \tau)^3 \sin(\tau) \, d\tau.$$

8. $\mathcal{L}^{-1}[1/(s + 1)] = e^{-t}$ and $\mathcal{L}^{-1}[2/(s^2 + 4)] = \cos(2t)$. Therefore,

$$\mathcal{L}^{-1}\left[\frac{s}{(s + 1)(s^2 + 4)}\right] = \int_0^t e^{-(t-\tau)} \cos(2\tau) \, d\tau.$$

9. $\mathcal{L}^{-1}[1/(s+1)^2] = te^{-t}$ and $\mathcal{L}^{-1}[1/(s^2+4)] = \dfrac{1}{2}\sin(2t)$. Therefore,

$$\mathcal{L}^{-1}\left[\frac{1}{(s+1)^2(s^2+4)}\right] = \int_0^t \frac{1}{2}(t-\tau)e^{-(t-\tau)}\sin(2\tau)\,d\tau.$$

10. Let $g(t) = \mathcal{L}^{-1}[G(s)]$. Since $\mathcal{L}^{-1}[1/(s^2+1)] = \sin(t)$,

$$\mathcal{L}^{-1}\left[\frac{G(s)}{s^2+1}\right] = \int_0^t g(t-\tau)\sin(\tau)\,d\tau.$$

11.

$$F(s) = \frac{1}{(s^2+1)^2} = \frac{1}{s^2+1} \cdot \frac{1}{s^2+1} = G(s) \cdot G(s)$$

$$\implies \mathcal{L}^{-1}(F(s)) = \int_0^t g(t-\tau)g(\tau)\,d\tau$$

where $g(t) = \mathcal{L}^{-1}(G(s)) = \sin(t)$. Therefore,

$$\mathcal{L}^{-1}(F(s)) = \int_0^t \sin(t-\tau)\sin(\tau)\,d\tau.$$

12.

$$F(s) = \frac{s}{(s^2+1)^2} = G(s) \cdot H(s)$$

where $G(s) = s/(s^2+1)$ and $H(s) = 1/(s^2+1)$. Therefore,

$$\mathcal{L}^{-1}(F(s)) = \int_0^t g(t-\tau)h(\tau)\,d\tau$$

where $g(t) = \mathcal{L}^{-1}(G(s)) = \cos(t)$ and $h(t) = \mathcal{L}^{-1}(H(s)) = \sin(t)$. Therefore,

$$\mathcal{L}^{-1}(F(s)) = \int_0^t \cos(t-\tau)\sin(\tau)\,d\tau.$$

13.

(a)

$$f * g = \int_0^t (t-\tau)^m \tau^n\,d\tau.$$

Introducing a new variable u such that $\tau = t - tu$, we have $d\tau = -t\,du$. Therefore,

$$\int_0^t (t-\tau)^m \tau^n\,d\tau = -\int_1^0 (tu)^m(t-tu)^n t\,du$$

$$= t^{m+n+1}\int_0^1 u^m(1-u)^n\,du.$$

(b) From part (a),

$$\mathcal{L}[f * g] = \mathcal{L}\left[t^{m+n+1}\int_0^1 u^m(1-u)^n\, du\right]$$

$$= \frac{(m+n+1)!}{s^{m+n+2}}\int_0^1 u^m(1-u)^n\, du.$$

By the Convolution Theorem, we know that $\mathcal{L}[f * g] = \mathcal{L}[f]\cdot\mathcal{L}[g]$. Therefore, calculating $\mathcal{L}[f]\cdot\mathcal{L}[g]$, we have

$$\mathcal{L}[f]\cdot\mathcal{L}[g] = \frac{m!}{s^{m+1}}\cdot\frac{n!}{s^{n+1}}.$$

Equating $\mathcal{L}[f * g]$ and $\mathcal{L}[f]\cdot\mathcal{L}[g]$, we have

$$\frac{(m+n+1)!}{s^{m+n+2}}\int_0^1 u^m(1-u)^n\, du = \frac{m!}{s^{m+1}}\cdot\frac{n!}{s^{n+1}}.$$

Therefore, we conclude that

$$\int_0^1 u^m(1-u)^n\, du = \frac{m!\,n!}{(m+n+1)!}.$$

(c) For the case when m and n are positive numbers, but not necessarily integers, we have

$$\mathcal{L}[f * g] = \mathcal{L}\left[t^{m+n+1}\int_0^1 u^m(1-u)^n\, du\right]$$

$$= \frac{\Gamma(m+n+2)}{s^{m+n+2}}\int_0^1 u^m(1-u)^n\, du.$$

Further,

$$\mathcal{L}[f]\cdot\mathcal{L}[g] = \frac{\Gamma(m+1)}{s^{m+1}}\cdot\frac{\Gamma(n+1)}{s^{n+1}}.$$

Equating $\mathcal{L}[f * g]$ and $\mathcal{L}[f]\cdot\mathcal{L}[g]$, we have

$$\frac{\Gamma(m+n+2)}{s^{m+n+2}}\int_0^1 u^m(1-u)^n\, du = \frac{\Gamma(m+1)}{s^{m+1}}\cdot\frac{\Gamma(n)}{s^{n+1}}.$$

Therefore, we conclude that

$$\int_0^1 u^m(1-u)^n\, du = \frac{\Gamma(m+1)\Gamma(n+1)}{\Gamma(m+n+2)}.$$

14. Applying the Laplace transform to the equation, we have

$$[s^2 Y(s) - s y(0) - y'(0)] + \omega^2 Y(s) = \mathcal{L}[g(t)].$$

Letting $G(s) = \mathcal{L}[g(t)]$ and applying the initial conditions, we have

$$[s^2 + \omega^2]Y(s) = 1 + G(s).$$

Therefore,

$$Y(s) = \frac{1}{s^2 + \omega^2} + \frac{G(s)}{s^2 + \omega^2}.$$

Now

$$\mathcal{L}^{-1}\left[\frac{G(s)}{s^2 + \omega^2}\right] = \frac{1}{\omega}\int \sin(\omega(t - \tau))g(\tau)\, d\tau.$$

Therefore,

$$y(t) = \frac{1}{\omega}\sin(\omega t) + \frac{1}{\omega}\int \sin(\omega(t - \tau))g(\tau)\, d\tau.$$

15. Applying the Laplace transform to the equation, we have

$$[s^2 Y(s) - sy(0) - y'(0)] + 2[sY(s) - y(0)] + 2Y(s) = \frac{\alpha}{s^2 + \alpha^2}.$$

Applying the initial conditions, we have

$$[s^2 + 2s + 2]Y(s) = \frac{\alpha}{s^2 + \alpha^2}.$$

Therefore,

$$Y(s) = \frac{1}{s^2 + 2s + 2} \cdot \frac{\alpha}{s^2 + \alpha^2}$$
$$= \frac{1}{(s + 1)^2 + 1} \cdot \frac{\alpha}{s^2 + \alpha^2}.$$

Therefore,

$$y(t) = \int_0^t e^{-(t-\tau)}\sin(t - \tau)\sin(\alpha\tau)\, d\tau.$$

16. Applying the Laplace transform to the equation, we have

$$4[s^2 Y(s) - sy(0) - y'(0)] + 4[sY(s) - y(0)] + 17Y(s) = G(s)$$

where $G(s) = \mathcal{L}[g(t)]$. Applying the initial conditions, we have

$$[4s^2 + 4s + 17]Y(s) = G(s).$$

Therefore,

$$Y(s) = \frac{G(s)}{4s^2 + 4s + 17}$$
$$= \frac{1}{4}\frac{G(s)}{(s + 1/2)^2 + 4}.$$

Therefore,

$$y(t) = \frac{1}{8}\int_0^t e^{-(t-\tau)/2}\sin(2(t - \tau))g(\tau)\, d\tau.$$

108

17. Applying the Laplace transform to the equation, we have

$$[s^2Y(s) - sy(0) - y'(0)] + [sY(s) - y(0)] + \frac{5}{4}Y(s) = \frac{1}{s} - \frac{e^{-\pi s}}{s}.$$

Applying the initial conditions, we have

$$[s^2 + s + 5/4]Y(s) = s + \frac{1}{s} - \frac{e^{-\pi s}}{s}.$$

Therefore,

$$Y(s) = \frac{s}{s^2 + s + 5/4} + \frac{1}{s^2 + s + 5/4} \cdot \frac{1 - e^{-\pi s}}{s}$$

$$= \frac{s + 1/2}{(s + 1/2)^2 + 1} - \frac{1/2}{(s + 1/2)^2 + 1} + \frac{1}{(s + 1/2)^2 + 1} \cdot \frac{1 - e^{-\pi s}}{s}$$

Therefore,

$$y(t) = e^{-t/2}\cos(t) - \frac{1}{2}e^{-t/2}\sin(t) + \int_0^t e^{-(t-\tau)/2}\sin(t - \tau)(1 - u_\pi(\tau))\,d\tau.$$

18. Applying the Laplace transform to the equation, we have

$$[s^2Y(s) - sy(0) - y'(0)] + 4[sY(s) - y(0)] + 4Y(s) = G(s)$$

where $G(s) = \mathcal{L}[g(t)]$. Applying the initial conditions, we have

$$[s^2 + 4s + 4]Y(s) = 2s + 5 + G(s).$$

Therefore,

$$Y(s) = \frac{2s + 5}{(s + 2)^2} + \frac{G(s)}{(s + 2)^2}.$$

Using partial fractions, we write

$$\frac{2s + 5}{(s + 2)^2} = \frac{2}{s + 2} + \frac{1}{(s + 2)^2},$$

we can conclude that

$$y(t) = 2e^{-2t} + te^{-2t} + \int_0^t (t - \tau)e^{-2(t-\tau)}g(\tau)\,d\tau.$$

19. Applying the Laplace transform to the equation, we have

$$[s^2Y(s) - sy(0) - y'(0)] + 3[sY(s) - y(0)] + 2Y(s) = \frac{s}{s^2 + \alpha^2}$$

Applying the initial conditions, we have

$$[s^2 + 3s + 2]Y(s) = s + 3 + \frac{s}{s^2 + \alpha^2}.$$

Therefore,

$$Y(s) = \frac{s + 3}{s^2 + 3s + 2} + \frac{s}{(s^2 + 3s + 2)(s^2 + \alpha^2)}.$$

Using partial fractions, we write

$$\frac{s + 3}{s^2 + 3s + 2} = \frac{2}{s + 1} - \frac{1}{s + 2},$$

and

$$\frac{1}{s^2 + 3s + 2} = \frac{1}{s + 1} - \frac{1}{s + 2}.$$

Therefore, we can conclude that

$$y(t) = 2e^{-t} - e^{-2t} + \int_0^t (e^{-(t-\tau)} - e^{-2(t-\tau)}) \cos(\alpha\tau)\, d\tau.$$

20. Applying the Laplace transform to the equation and applying the initial conditions, we have

$$[s^4 - 1]Y(s) = G(s).$$

Therefore,

$$Y(s) = \frac{G(s)}{s^4 - 1}.$$

Using partial fractions, we can write

$$\frac{1}{s^4 - 1} = \frac{1}{2}\left[\frac{1}{s^2 - 1} - \frac{1}{s^2 + 1}\right].$$

Therefore, we can conclude that

$$y(t) = \frac{1}{2}\int_0^t [\sinh(t - \tau) - \sin(t - \tau)]g(\tau)\, d\tau.$$

21. Applying the Laplace transform to the equation, we have

$$[s^4 Y(s) - s^3 y(0) - s^2 y'(0) - s y''(0) - y'''(0)] + 5[s^2 Y(s) - s y(0) - y'(0)] + 4Y(s) = G(s).$$

Applying the initial conditions, we have

$$[s^4 + 5s^2 + 4]Y(s) = s^3 + 5s + G(s).$$

110

Therefore,

$$Y(s) = \frac{s^3 + 5s}{s^4 + 5s^2 + 4} + \frac{G(s)}{s^4 + 5s^2 + 4}.$$

Using partial fractions, we write

$$\frac{s^3 + 5s}{s^4 + 5s^2 + 4} = \frac{4s}{3(s^2 + 1)} - \frac{s}{3(s^2 + 4)}$$

and

$$\frac{1}{s^4 + 5s^2 + 4} = \frac{1}{3(s^2 + 1)} - \frac{1}{3(s^2 + 4)}.$$

Therefore, we can conclude that

$$y(t) = \frac{4}{3}\cos(t) - \frac{1}{3}\cos(2t) + \frac{1}{6}\int_0^t [2\sin(t - \tau) - \sin(2(t - \tau))]g(\tau)\,d\tau.$$

22.

(a)
$$\int_{-\infty}^t \delta(\tau)\,d\tau = \begin{cases} 0 & t < 0 \\ 1 & t > 0 \end{cases}$$

implies

$$\int_{-\infty}^t \delta(\tau)\,d\tau = u(t) \implies \delta(t) = u'(t).$$

(b) Applying the Laplace transform and the initial conditions, we have

$$[s^2 + 2\delta + 1]Y(s) = \frac{1}{s}.$$

Therefore,

$$Y(s) = \frac{1}{(s + \delta)^2 + (1 - \delta^2)} \cdot \frac{1}{s}.$$

Therefore,

$$y(t) = \frac{1}{\sqrt{1 - \delta^2}}\int_0^t e^{-\delta\tau}\sin(\sqrt{1 - \delta^2}\tau)\,d\tau$$

for $0 < \delta < 1$. For $\delta = 1$, it is straightforward to show that

$$y(t) = -te^{-t} - e^{-t} + 1.$$

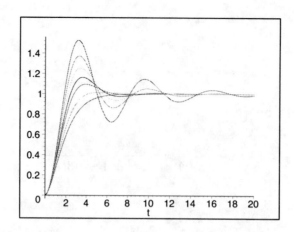

23. Taking the Laplace transform of the integral equation, we have

$$\mathcal{L}[\phi(t)] + \mathcal{L}\left[\int_0^t k(t - \xi)\phi(\xi)\,d\xi\right] = \mathcal{L}[f(t)]$$

which implies

$$\mathcal{L}[\phi(t)] + \mathcal{L}[k(t)] \cdot \mathcal{L}[\phi(t)] = \mathcal{L}[f(t)].$$

Therefore,

$$\mathcal{L}[\phi(t)](1 + \mathcal{L}[k(t)]) = \mathcal{L}[f(t)] \implies \mathcal{L}[\phi(t)] = \frac{\mathcal{L}[f(t)]}{1 + \mathcal{L}[k(t)]}.$$

24.

(a) Applying the result in problem 23 above, we have

$$\mathcal{L}[\phi(t)] = \frac{\mathcal{L}[\sin(2t)]}{1 + \mathcal{L}[t]}$$

$$= \frac{2/(s^2 + 4)}{1 + 1/s^2}$$

$$= \frac{2s^2}{(s^2 + 1)(s^2 + 4)}.$$

Using partial fractions, we can write

$$\frac{2s^2}{(s^2 + 1)(s^2 + 4)} = \frac{2}{3}\left[\frac{4}{s^2 + 4} - \frac{1}{s^2 + 1}\right].$$

Therefore,

$$\phi(t) = \frac{2}{3}\left[2\sin(2t) - \sin(t)\right].$$

(b) Differentiating the equation once, we have

$$\phi'(t) + \int_0^t \phi(\xi)\,d\xi = 2\cos(2t).$$

112

Differentiating the equation again, we have

$$\phi''(t) + \phi(t) = -4\sin(2t).$$

Using equation (i), we see that

$$\phi(0) = \sin(2 \cdot 0) = 0.$$

Further, using the equation for the first derivative of (i), we see that

$$\phi'(0) = 2\cos(2 \cdot 0) = 2.$$

(c) Letting $\Phi(s) = \mathcal{L}[\phi(t)]$ and taking the Laplace transform of the equation in part (ii), we have

$$[s^2\Phi(s) - s\phi(0) - \phi'(0)] + \Phi(s) = -\frac{8}{s^2 + 4}.$$

Applying the initial conditions, we have

$$[s^2 + 1]\Phi(s) = 2 - \frac{8}{s^2 + 4}.$$

Therefore,

$$\Phi(s) = \frac{2}{s^2 + 1} - \frac{8}{(s^2 + 1)(s^2 + 4)}$$

$$= \frac{2}{s^2 + 1} + \frac{8}{3}\left[\frac{1}{s^2 + 4} - \frac{1}{s^2 + 1}\right]$$

$$= \frac{4}{3}\frac{2}{s^2 + 4} - \frac{2}{3}\frac{1}{s^2 + 1}$$

which implies

$$\phi(t) = \frac{4}{3}\sin(2t) - \frac{2}{3}\sin(t).$$

25.

(a) Taking the Laplace transform of the equation, we have

$$\Phi(s) + \frac{1}{s^2}\Phi(s) = \frac{1}{s}.$$

Therefore,

$$\Phi(s) = \frac{s}{s^2 + 1},$$

which implies

$$\phi(t) = \cos(t).$$

(b) Differentiating the equation once, we have

$$\phi'(t) + \int_0^t \phi(\xi)\, d\xi = 0.$$

Differentiating the equation again, we have

$$\phi''(t) + \phi(t) = 0.$$

Using equation (i), we see that
$$\phi(0) = 1$$
Further, using the equation for the first derivative of (i), we see that

$$\phi'(0) = 0.$$

(c) Letting $\Phi(s) = \mathcal{L}[\phi(t)]$ and taking the Laplace transform of the equation in part (ii), we have
$$[s^2\Phi(s) - s\phi(0) - \phi'(0)] + \Phi(s) = 0.$$

Applying the initial conditions, we have

$$[s^2 + 1]\Phi(s) = s.$$

Therefore,

$$\Phi(s) = \frac{s}{s^2 + 1}$$

which implies

$$\phi(t) = \cos(t).$$

26.

(a) Taking the Laplace transform of the equation, we have

$$\Phi(s) - \frac{1}{s^2}\Phi(s) = \frac{1}{s}.$$

Therefore,

$$\Phi(s) = \frac{s}{s^2 - 1},$$

which implies

$$\phi(t) = \cosh(t).$$

(b) Differentiating the equation once, we have

$$\phi'(t) - \int_0^t \phi(\xi)\, d\xi = 0.$$

Differentiating the equation again, we have

$$\phi''(t) - \phi(t) = 0.$$

Using equation (i), we see that

$$\phi(0) = 1$$

Further, using the equation for the first derivative of (i), we see that

$$\phi'(0) = 0.$$

(c) Letting $\Phi(s) = \mathcal{L}[\phi(t)]$ and taking the Laplace transform of the equation in part (ii), we have

$$[s^2\Phi(s) - s\phi(0) - \phi'(0)] - \Phi(s) = 0.$$

Applying the initial conditions, we have

$$[s^2 - 1]\Phi(s) = s.$$

Therefore,

$$\Phi(s) = \frac{s}{s^2 - 1}$$

which implies

$$\phi(t) = \cosh(t).$$

27.

(a) Taking the Laplace transform of the equation, we have

$$\Phi(s) + \frac{2s}{s^2 + 1}\Phi(s) = \frac{1}{s + 1}.$$

Therefore,

$$\Phi(s) = \frac{s^2 + 1}{(s + 1)^3}$$

$$= \frac{1}{s + 1} - \frac{2}{(s + 1)^2} + \frac{2}{(s + 1)^3}.$$

which implies

$$\phi(t) = e^{-t} - 2te^{-t} + t^2e^{-t}.$$

(b) Differentiating the equation once, we have

$$\phi'(t) + 2\phi(t) - 2\int_0^t \sin(t - \xi)\phi(\xi)\,d\xi = -e^{-t}.$$

Differentiating the equation again, we have

$$\phi''(t) + 2\phi'(t) - 2 \int_0^t \cos(t - \xi)\phi(\xi)\, d\xi = e^{-t}.$$

Using equation (i), we see that

$$\phi(0) = 1$$

Further, using the equation for the first derivative of (i), we see that

$$\phi'(0) = -3.$$

(c) Letting $\Phi(s) = \mathcal{L}[\phi(t)]$ and taking the Laplace transform of the equation in part (ii), we have

$$[s^2\Phi(s) - s\phi(0) - \phi'(0)] + 2[s\Phi(s) - \phi(0)] - 2\frac{s}{s^2 + 1}\Phi(s) = \frac{1}{s + 1}.$$

Applying the initial conditions, we have

$$\left[s^2 + 2s - \frac{2s}{s^2 + 1}\right]\Phi(s) = s - 1 + \frac{1}{s + 1}.$$

Therefore,

$$\Phi(s) = \frac{(s - 1)(s^2 + 1)}{s^4 + 2s^3 + s^2} + \frac{s^2 + 1}{(s + 1)(s^4 + 2s^3 + s^2)}$$

$$= -\frac{1}{s^2} - \frac{4}{(s + 1)^2} + \frac{3}{s} - \frac{2}{s + 1} + \frac{1}{s^2} + \frac{2}{(s + 1)^2} - \frac{3}{s} + \frac{3}{s + 1} + \frac{2}{(s + 1)^3}$$

$$= -\frac{2}{(s + 1)^2} + \frac{1}{s + 1} + \frac{2}{(s + 1)^3}$$

which implies

$$\phi(t) = -2te^{-t} + e^{-t} + t^2 e^{-t}.$$

28.

(a) Taking the Laplace transform of the equation, we have

$$s\Phi(s) + \frac{1}{s^2}\Phi(s) = \frac{1}{s^2}.$$

Therefore,

$$\Phi(s) = \frac{1}{s^3 + 1}$$

$$= \frac{2 - s}{3(s^2 - s + 1)} + \frac{1}{3}\frac{1}{s + 1}$$

$$= -\frac{1}{3}\left[\frac{s - 1/2}{(s - 1/2)^2 + 3/4}\right] + \frac{1/2}{(s - 1/2)^2 + 3/4} + \frac{1}{3}\frac{1}{s + 1}$$

which implies

$$\phi(t) = -\frac{1}{3}e^{t/2}\cos(\sqrt{3}t/2) + \frac{1}{\sqrt{3}}e^{t/2}\sin(\sqrt{3}t/2) + \frac{1}{3}e^{-t}.$$

(b) Differentiating the equation once, we have

$$\phi''(t) + \int_0^t \phi(\xi)\, d\xi = 1.$$

Differentiating the equation again, we have

$$\phi'''(t) + \phi(t) = 0.$$

Using the equation for ϕ', we see that

$$\phi'(0) = 0.$$

Further, using the equation for ϕ'', we see that

$$\phi''(0) = 1.$$

(c) Letting $\Phi(s) = \mathcal{L}[\phi(t)]$ and taking the Laplace transform of the equation in part (ii), we have

$$[s^3\Phi(s) - s^2\phi(0) - s\phi'(0) - \phi''(0)] + \Phi(s) = 0.$$

Applying the initial conditions, we have

$$[s^3 + 1]\Phi(s) = 1.$$

Therefore,

$$\Phi(s) = \frac{1}{s^3 + 1}.$$

Using the same analysis as part (a), we see that

$$\phi(t) = -\frac{1}{3}e^{t/2}\cos(\sqrt{3}t/2) + \frac{1}{\sqrt{3}}e^{t/2}\sin(\sqrt{3}t/2) + \frac{1}{3}e^{-t}.$$

29.

(a) Taking the Laplace transform of the equation, we have

$$s\Phi(s) - 1 - \frac{1}{s^3}\Phi(s) = -\frac{1}{s^2}.$$

Therefore,

$$\Phi(s) = \frac{s}{s^2 + 1}$$

which implies

$$\phi(t) = \cos(t).$$

(b) Differentiating the equation once, we have

$$\phi''(t) - \int_0^t (t - \xi)\phi(\xi)\,d\xi = -1.$$

Differentiating the equation again, we have

$$\phi'''(t) - \int_0^t phi(\xi)\,d\xi = 0.$$

Differentiating again, we see that

$$\phi''''(t) - \phi(t) = 0.$$

Using the equations above for the derivatives of ϕ, we see that $\phi'(0) = 0$, $\phi''(0) = -1$ and $\phi'''(0) = 0$.

(c) Letting $\Phi(s) = \mathcal{L}[\phi(t)]$ and taking the Laplace transform of the equation in part (ii), we have

$$[s^4\Phi(s) - s^3\phi(0) - s^2\phi'(0) - s\phi''(0) - \phi'''(0)] - \Phi(s) = 0.$$

Applying the initial conditions, we have

$$[s^4 - 1]\Phi(s) = s^3 - s.$$

Therefore,

$$\Phi(s) = \frac{s^3 - s}{s^4 - 1}$$

$$= \frac{s}{s^2 + 1}.$$

Therefore,

$$\phi(t) = \cos(t).$$

30.

(a) Taking the Laplace transform of the equation, we have

$$s\Phi(s) - 1 + \Phi(s) = \frac{1}{s^2 + 1}\Phi(s).$$

Therefore,

$$\Phi(s) = \frac{s^2 + 1}{s(s^2 + s + 1)}$$

$$= \frac{1}{s} - \frac{1}{s^2 + s + 1}$$

$$= \frac{1}{s} - \frac{2}{\sqrt{3}}\frac{\sqrt{3}/2}{(s + 1/2)^2 + 3/4},$$

which implies

$$\phi(t) = 1 - \frac{2}{\sqrt{3}}e^{-t/2}\sin\left(\frac{\sqrt{3}}{2}t\right).$$

(b) Differentiating the equation once, we have

$$\phi''(t) + \phi'(t) = \int_0^t \cos(t - \xi)\phi(\xi)\, d\xi.$$

Differentiating the equation again, we have

$$\phi'''(t) + \phi''(t) = -\phi(t) - \int_0^t \sin(t - \xi)\phi(\xi)\, d\xi$$
$$= -\phi'(t).$$

Therefore,
$$\phi'''(t) + \phi''(t) + \phi'(t) = 0.$$

Using the equations above for the derivatives of ϕ, we see that $\phi'(0) = -1$ and $\phi''(0) = 1$.

(c) Letting $\Phi(s) = \mathcal{L}[\phi(t)]$ and taking the Laplace transform of the equation in part (ii), we have

$$[s^3\Phi(s) - s^2\phi(0) - s\phi'(0) - \phi''(0)] + [s^2\Phi(s) - s\phi(0) - \phi'(0)] + [s\Phi(s) - \phi(0)] = 0.$$

Applying the initial conditions, we have

$$[s^3 + s^2 + s]\Phi(s) = s^2 + 1.$$

Therefore,

$$\Phi(s) = \frac{s^2 + 1}{s^3 + s^2 + s}.$$

Using the analysis from part (a), we conclude that

$$\phi(t) = 1 - \frac{2}{\sqrt{3}}e^{-t/2}\sin\left(\frac{\sqrt{3}}{2}t\right).$$

31.

(a) We note that

$$\int_0^b \frac{f(y)}{\sqrt{b - y}}\, dy = \left(\frac{1}{\sqrt{y}} * f\right)(b).$$

Taking the Laplace transform of both sides of this equation, we see that

$$\frac{T_0}{s} = \frac{1}{\sqrt{2g}}F(s) \cdot \mathcal{L}\left[\frac{1}{\sqrt{y}}\right].$$

Then using the fact that

$$\mathcal{L}\left[\frac{1}{\sqrt{y}}\right] = \frac{\Gamma(1/2)}{s^{1/2}} = \sqrt{\frac{\pi}{s}},$$

we conclude that

$$\frac{T_0}{s} = \frac{1}{\sqrt{2g}} F(s) \cdot \sqrt{\pi s},$$

which implies

$$F(s) = \sqrt{\frac{2g}{\pi}} \cdot \frac{T_0}{\sqrt{s}}.$$

Taking the inverse transform, we see that

$$f(y) = \frac{T_0}{\pi} \sqrt{\frac{2g}{y}}.$$

(b) Combining equations (i) and (iv), we see that

$$\frac{2gT_0^2}{\pi^2 y} = 1 + \left(\frac{dx}{dy}\right)^2.$$

Solving this equation for dx/dy, we see that

$$\frac{dx}{dy} = \sqrt{\frac{2\alpha - y}{y}},$$

where $\alpha = gT_0^2/\pi^2$.

(c) Consider the change of variables $y = 2\alpha \sin^2(\theta/2)$. Using the chain rule, we have

$$\frac{dy}{dx} = 2\alpha \sin(\theta/2) \cos(\theta/2) \cdot \frac{d\theta}{dx}$$

and

$$\frac{dx}{dy} = \frac{1}{2\alpha \sin(\theta/2) \cos(\theta/2)} \cdot \frac{dx}{d\theta}.$$

It follows that

$$\frac{dx}{d\theta} = 2\alpha \sin(\theta/2) \cos(\theta/2) \sqrt{\frac{\cos^2(\theta/2)}{\sin^2(\theta/2)}}$$

$$= 2\alpha \cos^2(\theta/2)$$

$$= \alpha + \alpha \cos(\theta).$$

Now integrating this equation for $dx/d\theta$, we have

$$x(\theta) = \alpha\theta + \alpha \sin(\theta) + C.$$

Since the curve passes through the origin, we have $x(0) = 0 = y(0)$. Therefore, $C = 0$, and $x(\theta) = \alpha\theta + \alpha \sin(\theta)$. We also have

$$y(\theta) = 2\alpha \sin^2(\theta/2)$$

$$= \alpha - \alpha \cos(\theta).$$

120

Section 5.9

1. Let Y_1 represent the output before the junction point. Let Y_2 represent the output after the junction point. Let U_1 represent the output of controller G_1 and let U_2 represent the output of controller G_2. Then we have the following system of equations

$$Y_1 = H_1[F - U_1]$$
$$U_1 = G_1 Y_1$$
$$Y_2 = Y_1 + U_2$$
$$U_2 = G_2[F - U_1].$$

Solving the equation for Y_1, we have

$$Y_1 = H_1 F - H_1 G_1 Y_1 \implies Y_1 = \frac{H_1 F}{1 + H_1 G_1}.$$

Then

$$Y_2 = Y_1 + U_2$$
$$\implies Y_2 = \frac{H_1 F}{1 + H_1 G_1} + G_2 \left[F - \frac{G_1 H_1 F}{1 + H_1 G_1} \right]$$
$$\implies Y_2 = \frac{H_1 F + G_2 F}{1 + H_1 G_1}.$$

2. Let Y_1 represent the output after H_1, but before the junction point. Let Y_2 represent the output between H_2 and H_3. Let Y_3 represent the output after H_3. Let U_1 represent the output of controller G_1 and let U_2 represent the output of controller G_2. Then we have the following system of equations

$$Y_1 = H_1[F - U_2]$$
$$Y_2 = H_2[Y_1 - U_1]$$
$$Y_3 = H_3 Y_2$$
$$U_1 = G_1 Y_3$$
$$U_2 = G_2 Y_2.$$

Replacing U_1 and U_2 in the first two equations, we have

$$Y_1 = H_1[F - G_2 Y_2]$$
$$Y_2 = H_2[Y_1 - G_1 Y_3].$$

Using the fact that $Y_3 = H_3 Y_2$, we can eliminate Y_2 in the above equations. We conclude that

$$Y_1 = H_1 F - G_2 H_1 \left(\frac{Y_3}{H_3} \right)$$
$$\frac{Y_3}{H_3} = H_2 Y_1 - G_1 H_2 Y_3.$$

Substituting Y_1 into the second of these equations, we conclude that

$$\frac{Y_3}{H_3} = H_2 \left[\frac{H_1 H_3 F - G_2 H_1 Y_3}{H_3} \right] - G_1 H_2 Y_3.$$

Solving this equation for Y_3, we have

$$Y_3 = \frac{H_1 H_2 H_3 F}{1 + G_2 H_1 H_2 + G_1 H_2 H_3}.$$

Therefore, we conclude that the overall transfer function is

$$H(s) = \frac{Y_3}{F} = \frac{H_1 H_2 H_3}{1 + G_2 H_1 H_2 + G_1 H_2 H_3}.$$

3. Consider the function $h(t) = t^k e^{\alpha t}$. We will prove that

$$\int_0^\infty |h(t)|\, dt \le \frac{k!}{|\alpha|^{k+1}}$$

by induction on k. First, for $k = 1$, we have

$$\int_0^\infty t e^{\alpha t}\, dt = \lim_{b \to \infty} \left. \frac{t e^{\alpha t}}{\alpha} \right|_0^b - \int_0^\infty \frac{e^{\alpha t}}{\alpha}\, dt$$

$$= - \lim_{b \to \infty} \left. \frac{e^{\alpha t}}{\alpha^2} \right|_0^b$$

$$= \frac{1}{\alpha^2}.$$

Now assume that

$$\int_0^\infty t^k e^{\alpha t}\, dt \le \frac{k!}{|\alpha|^{k+1}}$$

to prove that

$$\int_0^\infty t^{k+1} e^{\alpha t}\, dt \le \frac{(k+1)!}{|\alpha|^{k+2}}.$$

Integrating by parts, we see that

$$\int_0^\infty t^{k+1} e^{\alpha t}\, dt = \lim_{b \to \infty} \left. \frac{t^{k+1} e^{\alpha t}}{\alpha} \right|_0^b - \int_0^\infty \frac{e^{\alpha t}}{\alpha} (k+1) t^k\, dt$$

$$= - \int_0^\infty \frac{e^{\alpha t}(k+1) t^k}{\alpha}\, dt.$$

From our inductive assumption,

$$\left| \int_0^\infty \frac{e^{\alpha t}(k+1) t^k}{\alpha}\, dt \right| \le \frac{k+1}{\alpha} \cdot \frac{k!}{|\alpha|^{k+1}} = \frac{(k+1)!}{|\alpha|^{k+2}}.$$

Further, $|t^k e^{\alpha t} \cos(\beta t)| \le t^k e^{\alpha t}$ and $|t^k e^{\alpha t} \sin(\beta t)| \le t^k e^{\alpha t}$. Therefore, the same bounds hold on integrals involving those functions.

4. We will prove this statement by induction. Consider the case when $p_n(x)$ is a first-degree polynomial. That is, $p_1(x) = x + a$. In this case, there is exactly one real root, $x = -a$. If this root is negative, then $-a < 0$ which implies that $a > 0$. Now we will proceed inductively. Let

$$p_n(x) = x^n + a_{n-1}x^{n-1} + \ldots + a_1 x + a_0.$$

Let $c_1, \ldots, c_n$ be the n roots of p_n. Our inductive assumption is that if $\text{Re}(c_i) < 0$ for $i = 1, \ldots, n$, then $a_0, \ldots, a_{n-1}$ are all positive. We now consider a polynomial of degree $n+1$. Consider

$$p_{n+1}(x) = x^{n+1} + a_n x^n + \ldots + a_1 x + a_0.$$

Let $c_1, \ldots, c_{n+1}$ denote the n roots of p_n. We assume that $\text{Re}(c_i) < 0$ for $i = 1, \ldots, n+1$. We will show that $a_0, \ldots, a_n > 0$. Since c_1 is one of the roots, then we can write

$$p_{n+1}(x) = (x - c_1)(x^n + b_{n-1}x^{n-1} + \ldots + b_1 x + b_0).$$

Let

$$q_n(x) \equiv x^n + b_{n-1}x^{n-1} + \ldots + b_1 x + b_0.$$

Now q_n is an nth degree polynomial whose roots are given by $c_2, \ldots, c_{n+1}$. Since all of these roots satisfy $\text{Re}(c_i) < 0$, by our inductive assumption, $b_0, \ldots, b_{n-1} > 0$. Therefore, multiplying $(x - c_1)$ by q_n, we have

$$\begin{aligned} p_{n+1}(x) &= (x - c_1)(x^n + b_{n-1}x^{n-1} + \ldots + b_1 x + b_0) \\ &= x^{n+1} + (b_{n-1} - c_1)x^n + (b_{n-2} - b_{n-1}c_1)x^{n-1} + \ldots + (b_0 - b_1 c_1)x - b_0 c_1. \end{aligned}$$

Under the assumption that $b_0, \ldots, b_{n-1} > 0$ and $\text{Re}(c_1) < 0$, we conclude that each of these coefficients is positive.

5. The associated Routh table is given by

s^3	1	a_1	0
s^2	a_2	a_0	0
s	$\frac{a_1 a_2 - a_0}{a_2}$	0	
s^0	a_0	0	

In order for all the roots to have negative real parts, we need all entries in the first column to be positive. Therefore, we need (1) $a_2 > 0$, (2) $(a_1 a_2 - a_0)/a_2 > 0$ and (3) $a_0 > 0$. Therefore, all the roots will lie in the left half plane if and only if (1) $a_2 > 0$, (2) $a_1 a_2 > a_0$ and (3) $a_0 > 0$.

6. The associated Routh table is given by

s^3	1	3	0
s^2	2	7	0
s	$-1/2$	0	
s^0	7	0	

There are two sign changes in the first column. Therefore, there are two roots with positive real parts. Using a computer, we see that the roots are approximately -2.13 and $0.066 \pm 1.81i$. As shown by the Routh table, two of the roots have positive real parts.

7. The associated Routh table is given by

s^3	1	9	0
s^2	5	5	0
s	8	0	
s^0	5	0	

There are no sign changes in the first column. Therefore, there are no roots with positive real parts. Using a computer, we see that the roots are -1 and $-2 \pm i$. As shown by the Routh table, none of the roots have positive real parts.

8. The associated Routh table is given by

s^4	1	23	-34	0
s^3	9	1	0	
s^2	$\frac{206}{9}$	$-\frac{306}{9}$	0	
s	$\frac{2960}{206}$	0		
s^0	$-\frac{306}{9}$	0		

There are one sign change in the first column. Therefore, there is one root with positive real part. Using a computer, we see that the roots are 1, -2 and $-4 \pm i$. As shown by the Routh table, exactly one of the roots has a positive real part.

9. The associated Routh table is given by

s^4	1	-5	34	0
s^3	5	-35	0	
s^2	2	34	0	
s	-120	0		
s^0	34	0		

There are two sign changes in the first column. Therefore, there are two roots with positive real parts. Using a computer, we see that the roots are 1, 2 and $-4 \pm i$. As shown by the Routh table, exactly two of the roots have positive real parts.

10. The associated Routh table is given by

s^4	1	78	221	0
s^3	14	206	0	
s^2	$\frac{443}{7}$	221	0	
s	$\frac{69600}{443}$	0		
s^0	221	0		

There are no sign changes in the first column. Therefore, there are no roots with positive real parts. Using a computer, we see that the roots are $-3 \pm 2i$ and $-4 \pm i$. As shown by the Routh table, none of the roots have positive real parts.

11. The associated Routh table is given by

s^4	1	24	100	0
s^3	8	32	0	
s^2	20	100	0	
s	-8	0		
s^0	100	0		

There are two sign changes in the first column. Therefore, there are two roots with positive real parts. Using a computer, we see that the roots are approximately $0.141 \pm 2.141i$ and $-4.141 \pm 2.141i$. As shown by the Routh table, two of the roots have positive real parts.

12. The associated Routh table is given by

s^3	1	$K - 6$	0
s^2	5	K	0
s	$\frac{4K-30}{5}$	0	
s^0	K	0	

If we want all the poles to lie in the left half plane, we need all the entries in the first column to have the same sign. That is, we need $4K - 30 > 0$ and $K > 0$. We see that these two inequalities will be satisfied if and only if $K > 15/2$. Below we show a plot of the roots of the polynomial $s^3 + 5s^2 + (K - 6)s + K$ for $K = -10, 0, 10, 20, 30$.

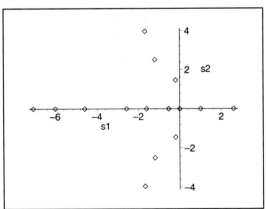

13. The associated Routh table is given by

s^3	1	6	0
s^2	4	$4 + K$	0
s	$\frac{20-K}{4}$	0	
s^0	$4 + K$	0	

125

If we want all the poles to lie in the left half plane, we need all the entries in the first column to have the same sign. That is, we need $20 - K > 0$ and $4 + K > 0$. We see that these two inequalities will be satisfied if and only if $20 > K > -4$. Below we show a plot of the roots of the polynomial $s^3 + 4s^2 + 6s + 4 + K$ for $K = -20, -10, 0, 10, 20, 30$.

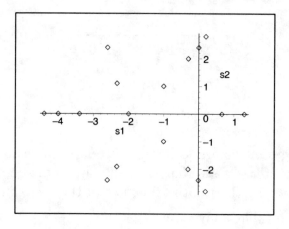

14. The associated Routh table is given by

s^3	1	10	0
s^2	3	$8 + K$	0
s	$\frac{22-K}{3}$	0	
s^0	$8 + K$	0	

If we want all the poles to lie in the left half plane, we need all the entries in the first column to have the same sign. That is, we need $22 - K > 0$ and $8 + K > 0$. We see that these two inequalities will be satisfied if and only if $22 > K > -8$. Below we show a plot of the roots of the polynomial $s^3 + 3s^2 + 10s + 8 + K$ for $K = -20, -10, 0, 10, 20, 30$.

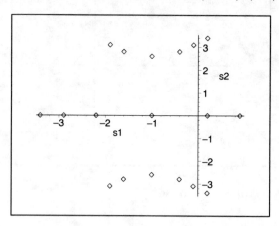

15. The associated Routh table is given by

s^4	1	14	$16+K$	0
s^3	4	20	0	
s^2	9	$16+K$	0	
s	$\frac{116-4K}{9}$	0		
s^0	$16+K$	0		

If we want all the poles to lie in the left half plane, we need all the entries in the first column to have the same sign. That is, we need $116 - 4K > 0$ and $16 + K > 0$. We see that these two inequalities will be satisfied if and only if $29 > K > -16$. Below we show a plot of the roots of the polynomial $s^4 + 4s^3 + 14s^2 + 20s + 16 + K$ for $K = -30, -10, 0, 20, 40$.

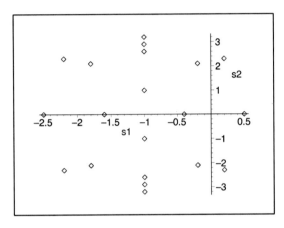

Chapter 6
Section 6.1

1.

(a)

$$\mathbf{A} + 3\mathbf{B} = \begin{pmatrix} e^t & 2e^{-t} & e^{2t} \\ 2e^t & e^{-t} & -e^{2t} \\ -e^t & 3e^{-t} & 2e^{2t} \end{pmatrix} + \begin{pmatrix} 6e^t & 3e^{-t} & 9e^{2t} \\ -3e^t & 6e^{-t} & 3e^{2t} \\ 9e^t & -3e^{-t} & -3e^{2t} \end{pmatrix}$$

$$= \begin{pmatrix} 7e^t & 5e^{-t} & 10e^{2t} \\ -e^t & 7e^{-t} & 2e^{2t} \\ 8e^t & 0 & -e^{2t} \end{pmatrix}.$$

(b)

$$\mathbf{AB} = \begin{pmatrix} 2e^{2t} - 2 + 3e^{3t} & 1 + 4e^{-2t} - e^t & 3e^{3t} + 2e^t - e^{4t} \\ 4e^{2t} - 1 - 3e^{3t} & 2 + 2e^{-2t} + e^t & 6e^{3t} + e^t + e^{4t} \\ -2e^{2t} - 3 + 6e^{3t} & -1 + 6e^{-2t} - 2e^t & -3e^{3t} + 3e^t - 2e^{4t} \end{pmatrix}.$$

(c)

$$\frac{d\mathbf{A}}{dt} = \begin{pmatrix} e^t & -2e^{-t} & 2e^{2t} \\ 2e^t & -e^{-t} & -2e^{2t} \\ -e^t & -3e^{-t} & 4e^{2t} \end{pmatrix}.$$

(d)

$$\int \mathbf{A}(t)\,dt = \begin{pmatrix} e^t & -2e^{-t} & e^{2t}/2 \\ 2e^t & -e^{-t} & -e^{2t}/2 \\ -e^t & -3e^{-t} & e^{2t} \end{pmatrix} + \mathbf{C}.$$

Therefore,

$$\int_0^1 \mathbf{A}(t)\,dt = \begin{pmatrix} e & -2e^{-1} & e^2/2 \\ 2e & -e^{-1} & -e^2/2 \\ -e & -3e^{-1} & e^2 \end{pmatrix} - \begin{pmatrix} 1 & -2 & 1/2 \\ 2 & -1 & -1/2 \\ -1 & -3 & 1 \end{pmatrix}$$

$$= \begin{pmatrix} e - 1 & 2 - 2e^{-1} & e^2/2 - 1/2 \\ 2e - 2 & 1 - e^{-1} & 1/2 - e^2/2 \\ 1 - e & 3 - 3e^{-1} & e^2 - 1 \end{pmatrix}.$$

2. First, we note that

$$\mathbf{x}' = \begin{pmatrix} -6 \\ 8 \\ 4 \end{pmatrix} e^{-t} + \begin{pmatrix} 0 \\ 4 \\ -4 \end{pmatrix} e^{2t}.$$

At the same time, we calculate

$$\begin{pmatrix} 1 & 1 & 1 \\ 2 & 1 & -1 \\ 0 & -1 & 1 \end{pmatrix} \mathbf{x} = \begin{pmatrix} 1 & 1 & 1 \\ 2 & 1 & -1 \\ 0 & -1 & 1 \end{pmatrix} \begin{pmatrix} 6 \\ -8 \\ -4 \end{pmatrix} e^{-t} + \begin{pmatrix} 1 & 1 & 1 \\ 2 & 1 & -1 \\ 0 & -1 & 1 \end{pmatrix} \begin{pmatrix} 0 \\ 2 \\ -2 \end{pmatrix} e^{2t}$$

$$= \begin{pmatrix} -6 \\ 8 \\ 4 \end{pmatrix} e^{-t} + \begin{pmatrix} 0 \\ 4 \\ -4 \end{pmatrix} e^{2t}.$$

3. First, we see that

$$\Psi' = \begin{pmatrix} e^t & -2e^{-2t} & 3e^{3t} \\ -4e^t & 2e^{-2t} & 6e^{3t} \\ -e^t & 2e^{-2t} & 3e^{3t} \end{pmatrix}.$$

1

At the same time,

$$
\begin{pmatrix} 1 & -1 & 4 \\ 3 & 2 & -1 \\ 2 & 1 & -1 \end{pmatrix} \Psi = \begin{pmatrix} 1 & -1 & 4 \\ 3 & 2 & -1 \\ 2 & 1 & -1 \end{pmatrix} \begin{pmatrix} e^t & e^{-2t} & e^{3t} \\ -4e^t & -e^{-2t} & 2e^{3t} \\ -e^t & -e^{-2t} & e^{3t} \end{pmatrix}
$$
$$
= \begin{pmatrix} e^t & -2e^{-2t} & 3e^{3t} \\ -4e^t & 2e^{-2t} & 6e^{3t} \\ -e^t & 2e^{-2t} & 3e^{3t} \end{pmatrix}.
$$

4. Let $x_1 = y$, $x_2 = y'$, $x_3 = y''$ and $x_4 = y'''$. Then

$$
\begin{aligned}
x_1' &= y' = x_2 \\
x_2' &= y'' = x_3 \\
x_3' &= y''' = x_4 \\
x_4' &= y'''' = -4y''' - 3y + t = -4x_4 - 3x_1 + t.
\end{aligned}
$$

Therefore,

$$
\begin{pmatrix} x_1 \\ x_2 \\ x_3 \\ x_4 \end{pmatrix}' = \begin{pmatrix} 0 & 1 & 0 & 0 \\ 0 & 0 & 1 & 0 \\ 0 & 0 & 0 & 1 \\ -3 & 0 & 0 & -4 \end{pmatrix} \begin{pmatrix} x_1 \\ x_2 \\ x_3 \\ x_4 \end{pmatrix} + \begin{pmatrix} 0 \\ 0 \\ 0 \\ t \end{pmatrix}.
$$

5. Let $x_1 = y$, $x_2 = y'$ and $x_3 = y''$. Then

$$
\begin{aligned}
x_1' &= y' = x_2 \\
x_2' &= y'' = x_3 \\
x_3' &= y''' = -\frac{\sin t}{t} y'' - \frac{3}{t} y + \cos t = -\frac{\sin t}{t} x_3 - \frac{3}{t} x_1 + \cos t.
\end{aligned}
$$

Therefore,

$$
\begin{pmatrix} x_1 \\ x_2 \\ x_3 \end{pmatrix}' = \begin{pmatrix} 0 & 1 & 0 \\ 0 & 0 & 1 \\ -\frac{3}{t} & 0 & -\frac{\sin t}{t} \end{pmatrix} \begin{pmatrix} x_1 \\ x_2 \\ x_3 \end{pmatrix} + \begin{pmatrix} 0 \\ 0 \\ \cos t \end{pmatrix}.
$$

6. Let $x_1 = y$, $x_2 = y'$, $x_3 = y''$ and $x_4 = y'''$. Then

$$
\begin{aligned}
x_1' &= y' = x_2 \\
x_2' &= y'' = x_3 \\
x_3' &= y''' = x_4 \\
x_4' &= y'''' = -\frac{e^t}{t(t-1)} y'' - \frac{4t^2}{t(t-1)} y = -\frac{e^t}{t(t-1)} x_3 - \frac{4t^2}{t(t-1)} x_1.
\end{aligned}
$$

Therefore,

$$
\begin{pmatrix} x_1 \\ x_2 \\ x_3 \\ x_4 \end{pmatrix}' = \begin{pmatrix} 0 & 1 & 0 & 0 \\ 0 & 0 & 1 & 0 \\ 0 & 0 & 0 & 1 \\ -\frac{4t^2}{t(t-1)} & 0 & -\frac{e^t}{t(t-1)} & 0 \end{pmatrix} \begin{pmatrix} x_1 \\ x_2 \\ x_3 \\ x_4 \end{pmatrix}.
$$

7. Let $x_1 = y$, $x_2 = y'$, and $x_3 = y''$. Then

$$
\begin{aligned}
x_1' &= y' = x_2 \\
x_2' &= y'' = x_3 \\
x_3' &= y''' = -ty'' - t^2 y' - t^2 y + \ln t = -tx_3 - t^2 x_2 - t^2 x_1 + \ln t.
\end{aligned}
$$

2

Therefore,

$$\begin{pmatrix} x_1 \\ x_2 \\ x_3 \end{pmatrix}' = \begin{pmatrix} 0 & 1 & 0 \\ 0 & 0 & 1 \\ -t^2 & -t^2 & -t \end{pmatrix} \begin{pmatrix} x_1 \\ x_2 \\ x_3 \end{pmatrix} + \begin{pmatrix} 0 \\ 0 \\ \ln t \end{pmatrix}.$$

8. Let $y_1 = y$, $y_2 = y'$, $y_3 = y''$ and $y_4 = y'''$. Then

$$y_1' = y' = y_2$$
$$y_2' = y'' = y_3$$
$$y_3' = y''' = y_4$$
$$y_4' = y'''' = -\frac{(x+1)}{(x-1)}y'' - \frac{\tan x}{(x-1)}y = -\frac{(x+1)}{(x-1)}y_3 - \frac{\tan x}{(x-1)}y_1.$$

Therefore,

$$\begin{pmatrix} y_1 \\ y_2 \\ y_3 \\ y_4 \end{pmatrix}' = \begin{pmatrix} 0 & 1 & 0 & 0 \\ 0 & 0 & 1 & 0 \\ 0 & 0 & 0 & 1 \\ -\frac{\tan x}{x-1} & 0 & -\frac{x+1}{x-1} & 0 \end{pmatrix} \begin{pmatrix} y_1 \\ y_2 \\ y_3 \\ y_4 \end{pmatrix}.$$

9. Let $y_1 = y$, $y_2 = y'$, $y_3 = y''$, $y_4 = y'''$, $y_5 = y^{(4)}$ and $y_6 = y^{(5)}$. Then

$$y_1' = y' = y_2$$
$$y_2' = y'' = y_3$$
$$y_3' = y''' = y_4$$
$$y_4' = y'''' = y_5$$
$$y_5' = y^{(5)} = y_6$$
$$y_6' = -\frac{x^2}{x^2-4}y'' - \frac{9}{x^2-4}y = -\frac{x^2}{x^2-4}y_3 - \frac{9}{x^2-4}y_1.$$

Therefore,

$$\begin{pmatrix} y_1 \\ y_2 \\ y_3 \\ y_4 \\ y_5 \\ y_6 \end{pmatrix}' = \begin{pmatrix} 0 & 1 & 0 & 0 & 0 & 0 \\ 0 & 0 & 1 & 0 & 0 & 0 \\ 0 & 0 & 0 & 1 & 0 & 0 \\ 0 & 0 & 0 & 0 & 1 & 0 \\ 0 & 0 & 0 & 0 & 0 & 1 \\ -\frac{9}{x^2-4} & 0 & -\frac{x^2}{x^2-4} & 0 & 0 & 0 \end{pmatrix} \begin{pmatrix} y_1 \\ y_2 \\ y_3 \\ y_4 \\ y_5 \\ y_6 \end{pmatrix}.$$

10. The mass balance law (22) says that the amount of lead in compartment i satisfies the equation

$$\frac{dx_i}{dt} = \text{input rate} - \text{output rate}.$$

For compartment 1, the lead is coming in from compartment 2 at the input rate $k_{12}x_2$ and in from compartment 3 at the rate $k_{13}x_3$. It is also coming in from an outside source at the rate $L(t)$. it is leaving compartment 1 at the output rate of $k_{31}x_1 + k_{21}x_2 + k_{01}x_1$. Therefore,

$$\frac{dx_1}{dt} = (L(t) + k_{12}x_2 + k_{13}x_3) - (k_{31} + k_{21} + k_{01})x_1.$$

For compartment 2, the lead is coming in from compartment 1 at the input rate $k_{21}x_1$ and leaving at the output rate $k_{12}x_2 + k_{02}x_2$. Therefore,

$$\frac{dx_2}{dt} = k_{21}x_1 - (k_{12} + k_{02})x_2.$$

11.

$$\begin{pmatrix} x_1 \\ x_2 \\ x_3 \end{pmatrix}' = \begin{pmatrix} -(k_{21} + k_{31} + k_{01}) & k_{12} & k_{13} \\ k_{21} & -(k_{02} + k_{12}) & 0 \\ k_{31} & 0 & -k_{13} \end{pmatrix} \begin{pmatrix} x_1 \\ x_2 \\ x_3 \end{pmatrix} + \begin{pmatrix} L(t) \\ 0 \\ 0 \end{pmatrix}.$$

Therefore,

$$\mathbf{K} = \begin{pmatrix} -(k_{21} + k_{31} + k_{01}) & k_{12} & k_{13} \\ k_{21} & -(k_{02} + k_{12}) & 0 \\ k_{31} & 0 & -k_{13} \end{pmatrix}$$

and

$$\mathbf{g}(t) = \begin{pmatrix} L(t) \\ 0 \\ 0 \end{pmatrix}.$$

12. From the left loop, we have the equation

$$L_1 i_1' + v_1 = e(t).$$

From the right loop, we have

$$L_2 i_2' + v_2 + R i_2 - v_1 = 0.$$

The equations for the voltages on the capacitors are given by

$$C_1 v_1' = q_1' = i_3 = i_1 - i_2$$
$$C_2 v_2' = q_2' = i_2.$$

We can rewrite these equations as the system

$$\frac{d}{dt} \begin{pmatrix} v_1 \\ v_2 \\ i_1 \\ i_2 \end{pmatrix} = \begin{pmatrix} 0 & 0 & 1/C_1 & -1/C_1 \\ 0 & 0 & 0 & 1/C_2 \\ -1/L_1 & 0 & 0 & 0 \\ 1/L_2 & -1/L_2 & 0 & -R/L_2 \end{pmatrix} \begin{pmatrix} v_1 \\ v_2 \\ i_1 \\ i_2 \end{pmatrix} + \begin{pmatrix} 0 \\ 0 \\ e(t)/L_1 \\ 0 \end{pmatrix}.$$

13.

(a) By equation (ii),

$$\frac{d}{dt}[x_1(t) + x_2(t)] = -L_{21}x_1 + L_{12}x_2 + L_{21}x_1 - L_{12}x_2$$
$$= (-L_{21} + L_{21})x_1 + (L_{12} - L_{12})x_2 = 0.$$

(b)

$$\begin{pmatrix} x_1 \\ x_2 \end{pmatrix}' = \begin{pmatrix} -L_{21} & L_{12} \\ L_{21} & -L_{12} \end{pmatrix} \begin{pmatrix} x_1 \\ x_2 \end{pmatrix}$$

implies

$$A - \lambda I = \begin{pmatrix} -L_{21} - \lambda & L_{12} \\ L_{21} & -L_{12} - \lambda \end{pmatrix},$$

which implies $\det(A - \lambda I) = \lambda^2 + (L_{12} + L_{21})\lambda$. Therefore, $\det(A - \lambda I) = 0$ implies $\lambda(\lambda + L_{12} + L_{21}) = 0$. Therefore, $\lambda = 0$ or $\lambda = -(L_{12} + L_{21})$. Now $\lambda = 0$ implies

$$A - \lambda I = A = \begin{pmatrix} -L_{21} & L_{12} \\ L_{21} & -L_{12} \end{pmatrix}.$$

Therefore,

$$\mathbf{v_1} = \begin{pmatrix} L_{12} \\ L_{21} \end{pmatrix}$$

4

is an eigenvector for $\lambda_1 = 0$, and

$$\mathbf{x_1}(t) = \begin{pmatrix} L_{12} \\ L_{21} \end{pmatrix}$$

is a solution of this system. Next, $\lambda = -(L_{12} + L_{21})$ implies

$$A - \lambda I = \begin{pmatrix} L_{12} & L_{12} \\ L_{21} & L_{21} \end{pmatrix}.$$

Therefore,

$$\mathbf{v_2} = \begin{pmatrix} 1 \\ -1 \end{pmatrix}$$

is an eigenvector for $\lambda_2 = -(L_{12} + L_{21})$, and

$$\mathbf{x_2}(t) = e^{-(L_{12}+L_{21})t} \begin{pmatrix} 1 \\ -1 \end{pmatrix}$$

is a solution of the system. Therefore, the general solution is given by

$$\begin{pmatrix} x_1 \\ x_2 \end{pmatrix} = c_1 \begin{pmatrix} L_{12} \\ L_{21} \end{pmatrix} + c_2 e^{-(L_{12}+L_{21})t} \begin{pmatrix} 1 \\ -1 \end{pmatrix}.$$

The initial condition $x_1(0) = \alpha$ and $x_2(0) = 0$ implies

$$\begin{pmatrix} x_1(0) \\ x_2(0) \end{pmatrix} = c_1 \begin{pmatrix} L_{12} \\ L_{21} \end{pmatrix} + c_2 \begin{pmatrix} 1 \\ -1 \end{pmatrix} = \begin{pmatrix} \alpha \\ 0 \end{pmatrix}.$$

This equation implies

$$c_1 L_{12} + c_2 = \alpha$$
$$c_1 L_{21} - c_2 = 0.$$

The solution of this system is $c_1 = \alpha/(L_{12} + L_{21})$ and $c_2 = L_{21}\alpha/(L_{12} + L_{21})$. Therefore,

$$\mathbf{x}(t) = \frac{\alpha}{L_{12} + L_{21}} \begin{pmatrix} L_{12} \\ L_{21} \end{pmatrix} + \frac{L_{21}\alpha}{L_{12} + L_{21}} e^{-(L_{12}+L_{21})t} \begin{pmatrix} 1 \\ -1 \end{pmatrix}.$$

(c)

$$x_1(t) = \frac{\alpha}{L_{12} + L_{21}} L_{12} + \frac{L_{21}\alpha}{L_{12} + L_{21}} e^{-(L_{12}+L_{21})t}.$$

Therefore,

$$\overline{x_1} = \lim_{t \to \infty} x_1(t) = \frac{\alpha L_{12}}{L_{12} + L_{21}}.$$

$$x_2(t) = \frac{\alpha}{L_{12} + L_{21}} L_{21} - \frac{L_{21}\alpha}{L_{12} + L_{21}} e^{-(L_{12}+L_{21})t}.$$

Therefore,

$$\overline{x_2} = \lim_{t \to \infty} x_2(t) = \frac{\alpha L_{21}}{L_{12} + L_{21}}.$$

Since the decaying term $e^{-(L_{12}+L_{21})t}$ contains L_{12}, L_{21} in the exponent, the rate of decay depends on L_{12}, L_{21}.

(d)

(e)

As the ratio L_{21}/L_{12} increases, the amount of tracer in compartment 1 decreases and the amount of tracer in compartment 2 increases. As $L_{21}/L_{12} \to \infty$, the amount of tracer in compartment 2 approaches α (the initial amount of tracer injected into the system) while the amount of tracer in compartment 1 approaches zero. Similarly, as $L_{21}/L_{12} \to 0$, the amount of tracer in compartment 1 approaches α while the amount of tracer in compartment 2 approaches zero.

14.

(a)

$$y_1'' = -\frac{(K_1 + K_2)}{m_1}y_1 + \frac{K_2}{m_1}y_2 + \frac{F_1(t)}{m_1}$$
$$y_2'' = \frac{K_2}{m_2}y_1 - \frac{(K_2 + K_3)}{m_2}y_2 + \frac{F_2(t)}{m_2}.$$

Therefore,

$$\begin{pmatrix} y_1 \\ y_2 \end{pmatrix}'' = \begin{pmatrix} -\frac{(K_1+K_2)}{m_1} & \frac{K_2}{m_1} \\ \frac{K_2}{m_2} & -\frac{(K_2+K_3)}{m_2} \end{pmatrix} \begin{pmatrix} y_1 \\ y_2 \end{pmatrix} + \begin{pmatrix} \frac{F_1(t)}{m_1} \\ \frac{F_2(t)}{m_2} \end{pmatrix}.$$

(b)

$$K = \begin{pmatrix} -\frac{(K_1+K_2)}{m_1} & \frac{K_2}{m_1} \\ \frac{K_2}{m_2} & -\frac{(K_2+K_3)}{m_2} \end{pmatrix}$$

will be symmetric if

$$K = K^T = \begin{pmatrix} -\frac{(K_1+K_2)}{m_1} & \frac{K_2}{m_2} \\ \frac{K_2}{m_1} & -\frac{(K_2+K_3)}{m_2} \end{pmatrix}.$$

In particular, K will be symmetric as long as $\frac{K_2}{m_1} = \frac{K_2}{m_2}$. That is, K will be symmetric as long as $m_1 = m_2$.

6

15. If only the displacement of mass m_2 is observable, then we want

$$\mathbf{Cx} = \begin{pmatrix} 0 \\ 1 \\ 0 \\ 0 \end{pmatrix}.$$

Therefore,

$$\mathbf{C} = \begin{pmatrix} 0 & 0 & 0 & 0 \\ 0 & 1 & 0 & 0 \\ 0 & 0 & 0 & 0 \\ 0 & 0 & 0 & 0 \end{pmatrix}.$$

Section 6.2

1. By Corollary 6.2.8, since the functions $p_1(t) = 4$, $p_2(t) = 3$ and $g(t) = t$ are continuous on $(-\infty, \infty)$, the solution is sure to exist for all $t \in (-\infty, \infty)$.

2. We will first rewrite the equation as

$$y''' + \frac{\sin t}{t} y'' + \frac{3}{t} y = \frac{\cos t}{t}.$$

Since the coefficient functions $p_1(t) = \frac{\sin t}{t}$, $p_2(t) = \frac{3}{t}$ and $g(t) = \frac{\cos t}{t}$ are continuous for all $t \neq 0$, the solution is sure to exist in the intervals $(-\infty, 0)$ and $(0, \infty)$.

3. We will rewrite the equation and apply Corollary 6.2.8. We rewrite the equation as

$$y^{(4)} + \frac{e^t}{t(t-1)} y'' + \frac{4t^2}{t(t-1)} y = 0.$$

Since the coefficient functions $p_1(t) = \frac{e^t}{t(t-1)}$ and $p_2(t) = \frac{4t^2}{t(t-1)}$ are continuous for all $t \neq 0, 1$, the solution is sure to exist in the intervals $(-\infty, 0)$, $(0, 1)$ and $(1, \infty)$.

4. The coefficient functions are $p_1(t) = t$, $p_2(t) = t^2$, $p_3(t) = t^3$ and $g(t) = ln(t)$. Since these functions are continuous for all $t > 0$, the solution is sure to exist in the interval $(0, \infty)$.

5. First, we rewrite the equation as

$$y^{(4)} + \frac{x+1}{x-1} y'' + \frac{\tan x}{x-1} y = 0.$$

The coefficient functions are $p_1(x) = \frac{x+1}{x-1}$ and $p_2(x) = \frac{\tan x}{x-1}$. These functions are continuous for all $x \neq 1, \pm \pi/2, \pm 3\pi/2, \dots$. Therefore, the solution is sure to exist in the intervals $\dots, (-3\pi/2, -\pi/2), (-\pi/2, 1), (1, \pi/2), (\pi/2, 3\pi/$

6. First, we rewrite the equation as

$$y^{(6)} + \frac{x^2}{x^2-4} y'' + \frac{9}{x^2-4} y = 0.$$

The coefficient functions are $p_1(x) = \frac{x^2}{x^2-4}$ and $p_2(x) = \frac{9}{x^2-4}$. Since these functions are continuous for all $x \neq \pm 2$, the solution is sure to exist in the intervals $(-\infty, -2)$, $(-2, 2)$ and $(2, \infty)$.

7. First, we let

$$\mathbf{X}(t) = \begin{pmatrix} e^t & e^{-t} & 2e^{4t} \\ 2e^t & -2e^{-t} & 2e^{4t} \\ -e^t & e^{-t} & -8e^{4t} \end{pmatrix}.$$

For the first method of computing $W[\mathbf{x_1}, \mathbf{x_2}, \mathbf{x_3}]$, we begin by computing $|\mathbf{X}(t)|$. We see that

$$|\mathbf{X}(t)| = e^t(16e^{3t} - 2e^{3t}) - e^{-t}(-16e^{5t} + 2e^{5t}) + 2e^{4t}(2 - 2)$$
$$= 14e^{4t} + 14e^{4t} = 28e^{4t}.$$

Then, evaluating $|\mathbf{X}(0)|$, we have $|\mathbf{X}(0)| = 28$.

For the second method, we start by evaluating $\mathbf{X}(0)$. We see that

$$\mathbf{X}(0) = \begin{pmatrix} 1 & 1 & 2 \\ 2 & -2 & 2 \\ -1 & 1 & -8 \end{pmatrix}.$$

Then, $|\mathbf{X}(0)| = 28$.

8. We need to check to see if these functions are linearly independent. To do so, we will compute the Wronskian of these solutions. It suffices to check the Wronskian at one point. Therefore, we will compute $W[\mathbf{x_1}, \mathbf{x_2}, \mathbf{x_3}](0)$. Letting $\mathbf{X}(t)$ denote the matrix whose columns are $\mathbf{x_1}, \mathbf{x_2}$ and $\mathbf{x_3}$, $W[\mathbf{x_1}, \mathbf{x_2}, \mathbf{x_3}](0) = \det \mathbf{X}(0)$. Now

$$\mathbf{X}(t) = \begin{pmatrix} e^{-t} & e^{-t} & 0 \\ 0 & -2e^{-t} & 2e^{-t} \\ -e^{-t} & e^{-t} & -2e^{-t} \end{pmatrix}$$

implies

$$\mathbf{X}(0) = \begin{pmatrix} 1 & 1 & 0 \\ 0 & -2 & 2 \\ -1 & 1 & -2 \end{pmatrix}.$$

Therefore,

$$W[\mathbf{x_1}, \mathbf{x_2}, \mathbf{x_3}](0) = \det \mathbf{X}(0)$$
$$= 1(4 - 2) - 1(0 + 2) = 0.$$

Since $W[\mathbf{x_1}, \mathbf{x_2}, \mathbf{x_3}](0) = 0$, these functions are linearly dependent, and, therefore, do not form a fundamental set of solutions.

9. We proceed as in the solution to problem 8 above, by calculating $W[\mathbf{x_1}, \mathbf{x_2}, \mathbf{x_3}](0)$. Now

$$\mathbf{X}(t) = \begin{pmatrix} e^{-t} & e^{-t} & 2e^{8t} \\ 0 & -4e^{-t} & e^{8t} \\ -e^{-t} & e^{-t} & 2e^{8t} \end{pmatrix}$$

implies

$$\mathbf{X}(0) = \begin{pmatrix} 1 & 1 & 2 \\ 0 & -4 & 1 \\ -1 & 1 & 2 \end{pmatrix}.$$

Therefore,

$$W[\mathbf{x_1}, \mathbf{x_2}, \mathbf{x_3}](0) = \det \mathbf{X}(0)$$
$$= 1(-8 - 1) - 1(1 + 8) = -18.$$

Since $W[\mathbf{x_1}, \mathbf{x_2}, \mathbf{x_3}](0) \neq 0$, these functions are linearly independent, and, therefore, do form a fundamental set of solutions.

10.

(a)

$$W' = (p_{11} + p_{22} + \ldots + p_{nn})W \implies W = Ce^{(p_{11} + p_{22} + \ldots + p_{nn})t}W.$$

Suppose $W(t_0) = 0$ for some t_0. Then $C = 0$ which implies $W \equiv 0$. Therefore, W is either identically zero or else never vanishes on I.

8

(b) Consider the case $n = 3$. Then

$$W[y_1, \ldots, y_n] = \begin{vmatrix} y_1 & y_2 & y_3 \\ y_1' & y_2' & y_3' \\ y_1'' & y_2'' & y_3'' \end{vmatrix}$$

$$= y_1[y_2'y_3'' - y_3'y_2''] - y_2[y_1'y_3'' - y_3'y_1''] + y_3[y_1'y_2'' - y_2'y_1'']$$

Now we are assuming that y_1, y_2, y_3 are solutions of

$$\frac{d^n y}{dt^n} + p_1(t)\frac{d^{n-1}y}{dt^{n-1}} + \ldots + p_{n-1}(t)\frac{dy}{dt} + p_n(t)y = g(t).$$

This equation is written in matrix form in equation (23). By equation (i), we know that if $g(t) = 0$, then W will satisfy the differential equation

$$W' = -p_1 W.$$

Therefore, we know the contributions made by the coefficients $p_1, \ldots, p_{n-1}$. It suffices to look at the contributions made by $g(t)$. Therefore, in the following calculation, we will assume that $p_1, \ldots, p_{n-1} = 0$. When doing so, we see that

$$\begin{aligned} W' &= y_1'[y_2'y_3'' - y_3'y_2''] + y_1[y_2''y_3'' + y_2'y_3''' - y_3''y_2'' - y_3'y_2'''] \\ &\quad - y_2'[y_1'y_3'' - y_3'y_1''] - y_2[y_1''y_3'' + y_1'y_3''' - y_3''y_1'' - y_3'y_1'''] \\ &\quad + y_3'[y_1'y_2'' - y_2'y_1''] - y_3[y_1''y_2'' + y_1'y_2''' - y_2''y_1'' - y_2'y_1'''] \\ &= y_1[y_2'y_3''' - y_3'y_2'''] - y_2[y_1'y_3''' - y_3'y_1'''] + y_3[y_1'y_2''' - y_2'y_1'''] \\ &= y_1[y_2'g - y_3'g] - y_2[y_1'g - y_3'g] + y_3[y_1'g - y_2'g] \\ &= \begin{vmatrix} y_1 & y_2 & y_3 \\ y_1' & y_2' & y_3' \\ g & g & g \end{vmatrix}. \end{aligned}$$

In general, we conclude that

$$W' = -p_1 W + |D|,$$

where

$$D = \begin{pmatrix} y_1 & y_2 & \cdots & y_n \\ y_1' & y_2' & \cdots & y_n' \\ \vdots & \vdots & \cdots & \vdots \\ y_1^{(n-2)} & y_2^{(n-2)} & \cdots & y_n^{(n-2)} \\ g & g & \cdots & g \end{pmatrix}.$$

11.

$$\begin{aligned} y_1 = 1 &\implies y_1''' + y_1' = 0 \\ y_2 = \cos t &\implies y_2''' + y_2' = \sin t - \sin t = 0 \\ y_3 = \sin t &\implies y_3''' + y_3' = -\cos t + \cos t = 0. \end{aligned}$$

Therefore, y_1, y_2, y_3 are all solutions of the differential equation. We now compute their Wronskian. We have

$$W[y_1, y_2, y_3] = \begin{vmatrix} 1 & \cos t & \sin t \\ 0 & -\sin t & \cos t \\ 0 & -\cos t & -\sin t \end{vmatrix}$$

which implies $W[y_1, y_2, y_3] = 1(\sin^2 t + \cos^2 t) = 1$.

12.

$$y_1 = 1 \implies y_1^{(4)} + y_1'' = 0$$

$$y_2 = t \implies y_2^{(4)} + y_2'' = 0$$

$$y_3 = \cos t \implies y_3^{(4)} + y_3'' = \cos t - \cos t = 0$$

$$y_4 = \sin t \implies y_4^{(4)} + y_4'' = \sin t - \sin t = 0.$$

Therefore, y_1, y_2, y_3, y_4 are all solutions of the differential equation. We now compute their Wronskian. We have

$$W[y_1, y_2, y_3, y_4] = \begin{vmatrix} 1 & t & \cos t & \sin t \\ 0 & 1 & -\sin t & \cos t \\ 0 & 0 & -\cos t & -\sin t \\ 0 & 0 & \sin t & -\cos t \end{vmatrix}$$

which implies $W[y_1, y_2, y_3, y_4] = 1(\cos^2 t + \sin^2 t) = 1$.

13.

$$y_1 = e^t \implies y_1''' + 2y_1'' - y_1' - 2y_1 = e^t + 2e^t - e^t - 2e^t = 0$$

$$y_2 = e^{-t} \implies y_2''' + 2y_2'' - y_2' - 2y_2 = -e^{-t} + 2e^{-t} + e^{-t} - 2e^{-t} = 0$$

$$y_3 = e^{-2t} \implies y_3''' + 2y_3'' - y_3' - 2y_3 = -8e^{-2t} + 8e^{-2t} + 2e^{-2t} - 2e^{-2t} = 0$$

Therefore, y_1, y_2, y_3 are all solutions of the differential equation. We now compute their Wronskian. We have

$$W[y_1, y_2, y_3] = \begin{vmatrix} e^t & e^{-t} & e^{-2t} \\ e^t & -e^{-t} & -2e^{-2t} \\ e^t & e^{-t} & 4e^{-2t} \end{vmatrix}$$

which implies

$$W[y_1, y_2, y_3] = e^t(-4e^{-3t} + 2e^{-3t}) - e^{-t}(4e^{-t} + 2e^{-t}) + e^{-2t}(2)$$

$$= -6e^{-2t}.$$

14.

$$y_1 = 1 \implies y_1^{(4)} + 2y_1''' + y_1'' = 0$$

$$y_2 = t \implies y_2^{(4)} + 2y_2''' + y_2'' = 0$$

$$y_3 = e^{-t} \implies y_3^{(4)} + 2y_3''' + y_3'' = e^{-t} - 2e^{-t} + e^{-t} = 0$$

$$y_4 = te^{-t} \implies y_4^{(4)} + 2y_4''' + y_4'' = (-4e^{-t} + te^{-t}) + 2(3e^{-t} - te^{-t}) + (-2e^{-t} + te^{-t}) = 0.$$

Therefore, y_1, y_2, y_3, y_4 are all solutions of the differential equation. We now compute their Wronskian. We have

$$W[y_1, y_2, y_3, y_4] = \begin{vmatrix} 1 & t & e^{-t} & te^{-t} \\ 0 & 1 & -e^{-t} & e^{-t} - te^{-t} \\ 0 & 0 & e^{-t} & -2e^{-t} + te^{-t} \\ 0 & 0 & -e^{-t} & 3e^{-t} - te^{-t} \end{vmatrix}$$

which implies

$$W[y_1, y_2, y_3, y_4] = 1(e^{-t}(3e^{-t} - te^{-t}) + e^{-t}(-2e^{-t} + te^{-t}))$$

$$= e^{-2t}.$$

15.

$$y_1 = 1 \implies xy_1''' - y_1'' = 0$$
$$y_2 = x \implies xy_2''' - y_2'' = 0$$
$$y_3 = x^3 \implies xy_3''' - y_3'' = 6x - 6x = 0.$$

Therefore, y_1, y_2, y_3 are all solutions of the differential equation. We now compute their Wronskian. We have

$$W[y_1, y_2, y_3] = \begin{vmatrix} 1 & x & x^3 \\ 0 & 1 & 3x^2 \\ 0 & 0 & 6x \end{vmatrix}$$

which implies

$$W[y_1, y_2, y_3] = 6x.$$

16.

$$y_1 = x \implies x^3 y_1''' + x^2 y_1'' - 2xy_1' + 2y_1 = -2x + 2x = 0$$
$$y_2 = x^2 \implies x^3 y_2''' + x^2 y_2'' - 2xy_2' + 2y_2 = 2x^2 - 2x(2x) + 2x^2 = 0$$
$$y_3 = \frac{1}{x} \implies x^3 y_3''' + x^2 y_3'' - 2xy_3' + 2y_3 = x^3(-6/x^4) + x^2(2/x^3) - 2x(-1/x^2) + 2(1/x) = 0.$$

Therefore, y_1, y_2, y_3 are all solutions of the differential equation. We now compute their Wronskian. We have

$$W[y_1, y_2, y_3] = \begin{vmatrix} x & x^2 & 1/x \\ 1 & 2x & -1/x^2 \\ 0 & 2 & 2/x^3 \end{vmatrix}$$

which implies

$$W[y_1, y_2, y_3] = \frac{6}{x}.$$

17.

$$L[c_1 y_1 + c_2 y_2] = (c_1 y_1 + c_2 y_2)^{(n)} + p_1(t)(c_1 y_1 + c_2 y_2)^{(n-1)} + \ldots + p_n(t)(c_1 y_1 + c_2 y_2)$$
$$= c_1 y_1^{(n)} + c_2 y_2^{(n)} + c_1 p_1(t) y_1^{(n-1)} + c_2 p_1(t) y_2^{(n-1)} + \ldots + c_1 p_n(t) y_1 + c_2 p_n(t) y_2$$
$$= c_1 L[y_1] + c_2 L[y_2].$$

Section 6.3

1. We will write the system of equations in matrix form as $\mathbf{x}' = A\mathbf{x}$. Here, we have

$$\begin{pmatrix} x_1 \\ x_2 \\ x_3 \end{pmatrix}' = \begin{pmatrix} -4 & 1 & 0 \\ 1 & -5 & 1 \\ 0 & 1 & -4 \end{pmatrix} \begin{pmatrix} x_1 \\ x_2 \\ x_3 \end{pmatrix}.$$

To solve this system, we need to compute the eigenvalues and eigenvectors of A. We have

$$A - \lambda I = \begin{pmatrix} -4 - \lambda & 1 & 0 \\ 1 & -5 - \lambda & 1 \\ 0 & 1 & -4 - \lambda \end{pmatrix}.$$

Therefore,

$$\det(A - \lambda I) = -\lambda^3 - 13\lambda^2 - 54\lambda - 72 = -(\lambda + 3)(\lambda + 6)(\lambda + 4).$$

11

Therefore, the eigenvalues are $\lambda = -3, -4, -6$.

First, $\lambda = -3$ implies

$$A - \lambda I = \begin{pmatrix} -1 & 1 & 0 \\ 1 & -2 & 1 \\ 0 & 1 & -1 \end{pmatrix}$$

$$\rightarrow \begin{pmatrix} 1 & -1 & 0 \\ 0 & 1 & -1 \\ 0 & 0 & 0 \end{pmatrix}$$

after elementary row operations. Therefore,

$$\mathbf{v_1} = \begin{pmatrix} 1 \\ 1 \\ 1 \end{pmatrix}$$

is an eigenvector for $\lambda = -3$, and, consequently,

$$\mathbf{x_1}(t) = e^{-3t} \begin{pmatrix} 1 \\ 1 \\ 1 \end{pmatrix}$$

is a solution of the system.

Next, $\lambda = -4$ implies

$$A - \lambda I = \begin{pmatrix} 0 & 1 & 0 \\ 1 & -1 & 1 \\ 0 & 1 & 0 \end{pmatrix}$$

$$\rightarrow \begin{pmatrix} 1 & -1 & 1 \\ 0 & 1 & 0 \\ 0 & 0 & 0 \end{pmatrix}$$

after elementary row operations. Therefore,

$$\mathbf{v_2} = \begin{pmatrix} 1 \\ 0 \\ -1 \end{pmatrix}$$

is an eigenvector for $\lambda = -4$, and, consequently,

$$\mathbf{x_2}(t) = e^{-4t} \begin{pmatrix} 1 \\ 0 \\ -1 \end{pmatrix}$$

is a solution of the system.

Finally, $\lambda = -6$ implies

$$A - \lambda I = \begin{pmatrix} 2 & 1 & 0 \\ 1 & 1 & 1 \\ 0 & 1 & 2 \end{pmatrix}$$

$$\rightarrow \begin{pmatrix} 1 & 1/2 & 0 \\ 0 & 1 & 2 \\ 0 & 0 & 0 \end{pmatrix}$$

after elementary row operations. Therefore,

$$\mathbf{v_3} = \begin{pmatrix} 1 \\ -2 \\ 1 \end{pmatrix}$$

is an eigenvector for $\lambda = -6$, and, consequently,

$$\mathbf{x_3}(t) = e^{-6t} \begin{pmatrix} 1 \\ -2 \\ 1 \end{pmatrix}$$

is a solution of the system. Therefore, the general solution is

$$\mathbf{x}(t) = c_1 e^{-3t} \begin{pmatrix} 1 \\ 1 \\ 1 \end{pmatrix} + c_2 e^{-4t} \begin{pmatrix} 1 \\ 0 \\ -1 \end{pmatrix} + c_3 e^{-6t} \begin{pmatrix} 1 \\ -2 \\ 1 \end{pmatrix}.$$

2. We will write the system of equations in matrix form as $\mathbf{x}' = A\mathbf{x}$. Here, we have

$$\begin{pmatrix} x_1 \\ x_2 \\ x_3 \end{pmatrix}' = \begin{pmatrix} 1 & 4 & 4 \\ 0 & 3 & 2 \\ 0 & 2 & 3 \end{pmatrix} \begin{pmatrix} x_1 \\ x_2 \\ x_3 \end{pmatrix}.$$

To solve this system, we need to compute the eigenvalues and eigenvectors of A. We have

$$A - \lambda I = \begin{pmatrix} 1 - \lambda & 4 & 4 \\ 0 & 3 - \lambda & 2 \\ 0 & 2 & 3 - \lambda \end{pmatrix}.$$

Therefore,

$$\det(A - \lambda I) = (1 - \lambda)(\lambda^2 - 6\lambda + 5) = (1 - \lambda)(\lambda - 5)(\lambda - 1).$$

Therefore, the eigenvalues are $\lambda = 1, 5$.

First, $\lambda = 1$ implies

$$A - \lambda I = \begin{pmatrix} 0 & 4 & 4 \\ 0 & 2 & 2 \\ 0 & 2 & 2 \end{pmatrix}$$

$$\rightarrow \begin{pmatrix} 0 & 1 & 1 \\ 0 & 0 & 0 \\ 0 & 0 & 0 \end{pmatrix}$$

after elementary row operations. Therefore,

$$\mathbf{v_1} = \begin{pmatrix} 1 \\ 1 \\ -1 \end{pmatrix}$$

and

$$\mathbf{v_2} = \begin{pmatrix} 2 \\ 1 \\ -1 \end{pmatrix}$$

are linearly independent eigenvectors for $\lambda = 1$, and, consequently,

$$\mathbf{x_1}(t) = e^t \begin{pmatrix} 1 \\ 1 \\ -1 \end{pmatrix}$$

and

$$\mathbf{x_2}(t) = e^t \begin{pmatrix} 2 \\ 1 \\ -2 \end{pmatrix}$$

13

are solutions of the system.

Next, $\lambda = 5$ implies

$$A - \lambda I = \begin{pmatrix} -4 & 4 & 4 \\ 0 & -2 & 2 \\ 0 & 2 & -2 \end{pmatrix}$$

$$\rightarrow \begin{pmatrix} 1 & 0 & -2 \\ 0 & 1 & -1 \\ 0 & 0 & 0 \end{pmatrix}$$

after elementary row operations. Therefore,

$$\mathbf{v_3} = \begin{pmatrix} 2 \\ 1 \\ 1 \end{pmatrix}$$

is an eigenvector for $\lambda = 5$, and, consequently,

$$\mathbf{x_3}(t) = e^{5t} \begin{pmatrix} 2 \\ 1 \\ 1 \end{pmatrix}$$

is a solution of the system.

Therefore, the general solution is

$$\mathbf{x}(t) = c_1 e^t \begin{pmatrix} 1 \\ 1 \\ -1 \end{pmatrix} + c_2 e^t \begin{pmatrix} 2 \\ 1 \\ -1 \end{pmatrix} + c_3 e^{5t} \begin{pmatrix} 2 \\ 1 \\ 1 \end{pmatrix}.$$

3. We will write the system of equations in matrix form as $\mathbf{x}' = A\mathbf{x}$. Here, we have

$$\begin{pmatrix} x_1 \\ x_2 \\ x_3 \end{pmatrix}' = \begin{pmatrix} 2 & -4 & 2 \\ -4 & 2 & -2 \\ 2 & -2 & -1 \end{pmatrix} \begin{pmatrix} x_1 \\ x_2 \\ x_3 \end{pmatrix}.$$

To solve this system, we need to compute the eigenvalues and eigenvectors of A. We have

$$A - \lambda I = \begin{pmatrix} 2 - \lambda & -4 & 2 \\ -4 & 2 - \lambda & -2 \\ 2 & -2 & -1 - \lambda \end{pmatrix}.$$

Therefore,

$$\det(A - \lambda I) = -\lambda^3 + 3\lambda^2 + 24\lambda + 28 = -(\lambda + 2)^2(\lambda - 7).$$

Therefore, the eigenvalues are $\lambda = -2, 7$.

First, $\lambda = -2$ implies

$$A - \lambda I = \begin{pmatrix} 4 & -4 & 2 \\ -4 & 4 & -2 \\ 2 & -2 & 1 \end{pmatrix}$$

$$\rightarrow \begin{pmatrix} 2 & -2 & 1 \\ 0 & 0 & 0 \\ 0 & 0 & 0 \end{pmatrix}$$

after elementary row operations. Therefore,

$$\mathbf{v_1} = \begin{pmatrix} 1 \\ 1 \\ 0 \end{pmatrix}$$

14

and

$$\mathbf{v_2} = \begin{pmatrix} 1 \\ 0 \\ -2 \end{pmatrix}$$

are linearly independent eigenvectors for $\lambda = -2$, and, consequently,

$$\mathbf{x_1}(t) = e^{-2t} \begin{pmatrix} 1 \\ 1 \\ 0 \end{pmatrix}$$

and

$$\mathbf{x_2}(t) = e^{-2t} \begin{pmatrix} 1 \\ 0 \\ -2 \end{pmatrix}$$

are solutions of the system.

Next, $\lambda = 7$ implies

$$A - \lambda I = \begin{pmatrix} -5 & -4 & 2 \\ -4 & -5 & -2 \\ 2 & -2 & -8 \end{pmatrix}$$

$$\rightarrow \begin{pmatrix} 1 & -1 & -4 \\ 0 & 1 & 2 \\ 0 & 0 & 0 \end{pmatrix}$$

after elementary row operations. Therefore,

$$\mathbf{v_3} = \begin{pmatrix} 2 \\ -2 \\ 1 \end{pmatrix}$$

is an eigenvector for $\lambda = 7$, and, consequently,

$$\mathbf{x_3}(t) = e^{7t} \begin{pmatrix} 2 \\ -2 \\ 1 \end{pmatrix}$$

is a solution of the system.

Therefore, the general solution is

$$\mathbf{x}(t) = c_1 e^{-2t} \begin{pmatrix} 1 \\ 1 \\ 0 \end{pmatrix} + c_2 e^{-2t} \begin{pmatrix} 1 \\ 0 \\ -2 \end{pmatrix} + c_3 e^{7t} \begin{pmatrix} 2 \\ -2 \\ 1 \end{pmatrix}.$$

4. We will write the system of equations in matrix form as $\mathbf{x'} = A\mathbf{x}$. Here, we have

$$\begin{pmatrix} x_1 \\ x_2 \\ x_3 \end{pmatrix}' = \begin{pmatrix} -2 & 2 & -1 \\ -2 & 3 & -2 \\ -2 & 4 & -3 \end{pmatrix} \begin{pmatrix} x_1 \\ x_2 \\ x_3 \end{pmatrix}.$$

To solve this system, we need to compute the eigenvalues and eigenvectors of A. We have

$$A - \lambda I = \begin{pmatrix} -2-\lambda & 2 & -1 \\ -2 & 3-\lambda & -2 \\ -2 & 4 & -3-\lambda \end{pmatrix}.$$

Therefore,

$$\det(A - \lambda I) = (\lambda + 1)(-\lambda^2 - \lambda) = -\lambda(\lambda + 1)^2.$$

Therefore, the eigenvalues are $\lambda = -1, 0$.

First, $\lambda = -1$ implies

$$A - \lambda I = \begin{pmatrix} -1 & 2 & -1 \\ -2 & 4 & -2 \\ -2 & 4 & -2 \end{pmatrix}$$

$$\rightarrow \begin{pmatrix} 1 & -2 & 1 \\ 0 & 0 & 0 \\ 0 & 0 & 0 \end{pmatrix}$$

after elementary row operations. Therefore,

$$\mathbf{v_1} = \begin{pmatrix} 2 \\ 1 \\ 0 \end{pmatrix}$$

and

$$\mathbf{v_2} = \begin{pmatrix} 1 \\ 0 \\ -1 \end{pmatrix}$$

are linearly independent eigenvectors for $\lambda = -2$, and, consequently,

$$\mathbf{x_1}(t) = e^{-t} \begin{pmatrix} 2 \\ 1 \\ 0 \end{pmatrix}$$

and

$$\mathbf{x_2}(t) = e^{-t} \begin{pmatrix} 1 \\ 0 \\ -1 \end{pmatrix}$$

are solutions of the system.

Next, $\lambda = 0$ implies

$$A - \lambda I = \begin{pmatrix} -2 & 2 & -1 \\ -2 & 3 & -2 \\ -2 & 4 & -3 \end{pmatrix}$$

$$\rightarrow \begin{pmatrix} 1 & -1 & 1/2 \\ 0 & 1 & -1 \\ 0 & 0 & 0 \end{pmatrix}$$

after elementary row operations. Therefore,

$$\mathbf{v_3} = \begin{pmatrix} 1 \\ 2 \\ 2 \end{pmatrix}$$

is an eigenvector for $\lambda = 0$, and, consequently,

$$\mathbf{x_3}(t) = \begin{pmatrix} 1 \\ 2 \\ 2 \end{pmatrix}$$

is a solution of the system.

Therefore, the general solution is

$$\mathbf{x}(t) = c_1 e^{-t} \begin{pmatrix} 2 \\ 1 \\ 0 \end{pmatrix} + c_2 e^{-t} \begin{pmatrix} 1 \\ 0 \\ -1 \end{pmatrix} + c_3 \begin{pmatrix} 1 \\ 2 \\ 2 \end{pmatrix}.$$

16

5. To solve this system, we need to compute the eigenvalues and eigenvectors of A. We have

$$A - \lambda I = \begin{pmatrix} 1 - \lambda & 1 & 2 \\ 1 & 2 - \lambda & 1 \\ 2 & 1 & 1 - \lambda \end{pmatrix}.$$

Therefore,

$$\det(A - \lambda I) = \lambda^3 - 4\lambda^2 - \lambda + 4 = (\lambda - 4)(\lambda - 1)(\lambda + 1).$$

Therefore, the eigenvalues are $\lambda = -1, 1, 4$.

First, $\lambda = -1$ implies

$$A - \lambda I = \begin{pmatrix} 2 & 1 & 2 \\ 1 & 3 & 1 \\ 2 & 1 & 2 \end{pmatrix}$$

$$\rightarrow \begin{pmatrix} 1 & 0 & 1 \\ 0 & 1 & 0 \\ 0 & 0 & 0 \end{pmatrix}$$

after elementary row operations. Therefore,

$$\mathbf{v_1} = \begin{pmatrix} 1 \\ 0 \\ -1 \end{pmatrix}$$

is an eigenvector for $\lambda = -1$, and, consequently,

$$\mathbf{x_1}(t) = e^{-t} \begin{pmatrix} 1 \\ 0 \\ -1 \end{pmatrix}$$

is a solution of the system.

Next, $\lambda = 1$ implies

$$A - \lambda I = \begin{pmatrix} 0 & 1 & 2 \\ 1 & 1 & 1 \\ 2 & 1 & 0 \end{pmatrix}$$

$$\rightarrow \begin{pmatrix} 1 & 0 & -1 \\ 0 & 1 & 2 \\ 0 & 0 & 0 \end{pmatrix}$$

after elementary row operations. Therefore,

$$\mathbf{v_2} = \begin{pmatrix} 1 \\ -2 \\ 1 \end{pmatrix}$$

is an eigenvector for $\lambda = 1$, and, consequently,

$$\mathbf{x_2}(t) = e^{t} \begin{pmatrix} 1 \\ -2 \\ 1 \end{pmatrix}$$

is a solution of the system.

Finally, $\lambda = 4$ implies

$$A - \lambda I = \begin{pmatrix} -3 & 1 & 2 \\ 1 & -2 & 1 \\ 2 & 1 & -3 \end{pmatrix}$$

$$\rightarrow \begin{pmatrix} 1 & 0 & -1 \\ 0 & 1 & -1 \\ 0 & 0 & 0 \end{pmatrix}$$

after elementary row operations. Therefore,

$$\mathbf{v_3} = \begin{pmatrix} 1 \\ 1 \\ 1 \end{pmatrix}$$

is an eigenvector for $\lambda = 4$, and, consequently,

$$\mathbf{x_3}(t) = e^{4t} \begin{pmatrix} 1 \\ 1 \\ 1 \end{pmatrix}$$

is a solution of the system.

Therefore, the general solution is

$$\mathbf{x}(t) = c_1 e^{-t} \begin{pmatrix} 1 \\ 0 \\ -1 \end{pmatrix} + c_2 e^{t} \begin{pmatrix} 1 \\ -2 \\ 1 \end{pmatrix} + c_3 e^{4t} \begin{pmatrix} 1 \\ 1 \\ 1 \end{pmatrix}.$$

6. To solve this system, we need to compute the eigenvalues and eigenvectors of A. We have

$$A - \lambda I = \begin{pmatrix} 3 - \lambda & 2 & 4 \\ 2 & -\lambda & 2 \\ 4 & 2 & 3 - \lambda \end{pmatrix}.$$

Therefore,

$$\det(A - \lambda I) = \lambda^3 - 6\lambda^2 - 15\lambda - 8.$$

Therefore, the eigenvalues are $\lambda = 8, -1$.

First, $\lambda = 8$ implies

$$A - \lambda I = \begin{pmatrix} -5 & 2 & 4 \\ 2 & -8 & 2 \\ 4 & 2 & -5 \end{pmatrix}$$

$$\rightarrow \begin{pmatrix} 1 & 0 & -1 \\ 0 & 2 & -1 \\ 0 & 0 & 0 \end{pmatrix}$$

after elementary row operations. Therefore,

$$\mathbf{v_1} = \begin{pmatrix} 2 \\ 1 \\ 2 \end{pmatrix}$$

is an eigenvector for $\lambda = 8$, and, consequently,

$$\mathbf{x_1}(t) = e^{8t} \begin{pmatrix} 2 \\ 1 \\ 2 \end{pmatrix}$$

is a solution of the system.

Next, $\lambda = -1$ implies

$$A - \lambda I = \begin{pmatrix} 4 & 2 & 4 \\ 2 & 1 & 2 \\ 4 & 2 & 4 \end{pmatrix}$$

$$\rightarrow \begin{pmatrix} 2 & 1 & 2 \\ 0 & 0 & 0 \\ 0 & 0 & 0 \end{pmatrix}$$

18

after elementary row operations. Therefore,

$$\mathbf{v_2} = \begin{pmatrix} 1 \\ -2 \\ 0 \end{pmatrix}$$

and

$$\mathbf{v_3} = \begin{pmatrix} 0 \\ -2 \\ 1 \end{pmatrix}$$

are eigenvectors for $\lambda = -1$. Consequently,

$$\mathbf{x_2}(t) = e^{-t} \begin{pmatrix} 1 \\ -2 \\ 0 \end{pmatrix}$$

and

$$\mathbf{x_3}(t) = e^{-t} \begin{pmatrix} 0 \\ -2 \\ 1 \end{pmatrix}$$

are solutions of the system.

Therefore, the general solution is

$$\mathbf{x}(t) = c_1 e^{8t} \begin{pmatrix} 2 \\ 1 \\ 2 \end{pmatrix} + c_2 e^{-t} \begin{pmatrix} 1 \\ -2 \\ 0 \end{pmatrix} + c_3 e^{-t} \begin{pmatrix} 0 \\ -2 \\ 1 \end{pmatrix}.$$

7. To solve this system, we need to compute the eigenvalues and eigenvectors of A. We have

$$A - \lambda I = \begin{pmatrix} 1 - \lambda & 1 & 1 \\ 2 & 1 - \lambda & -1 \\ -8 & -5 & -3 - \lambda \end{pmatrix}.$$

Therefore,

$$\det(A - \lambda I) = \lambda^3 + \lambda^2 - 4\lambda - 4.$$

Therefore, the eigenvalues are $\lambda = 2, -1, -2$.

First, $\lambda = 2$ implies

$$A - \lambda I = \begin{pmatrix} -1 & 1 & 1 \\ 2 & -1 & -1 \\ -8 & -5 & -5 \end{pmatrix}$$

$$\rightarrow \begin{pmatrix} 1 & 0 & 0 \\ 0 & 1 & 1 \\ 0 & 0 & 0 \end{pmatrix}$$

after elementary row operations. Therefore,

$$\mathbf{v_1} = \begin{pmatrix} 0 \\ 1 \\ -1 \end{pmatrix}$$

is an eigenvector for $\lambda = 2$, and, consequently,

$$\mathbf{x_1}(t) = e^{2t} \begin{pmatrix} 0 \\ 1 \\ -1 \end{pmatrix}$$

19

is a solution of the system.

Next, $\lambda = -1$ implies

$$A - \lambda I = \begin{pmatrix} 2 & 1 & 1 \\ 2 & 2 & -1 \\ -8 & -5 & -2 \end{pmatrix}$$

$$\rightarrow \begin{pmatrix} 2 & 0 & 3 \\ 0 & 1 & -2 \\ 0 & 0 & 0 \end{pmatrix}$$

after elementary row operations. Therefore,

$$\mathbf{v_2} = \begin{pmatrix} 3 \\ -4 \\ -2 \end{pmatrix}$$

is an eigenvector for $\lambda = -1$, and, consequently,

$$\mathbf{x_2}(t) = e^{-t} \begin{pmatrix} 3 \\ -4 \\ -2 \end{pmatrix}$$

is a solution of the system.

Finally, $\lambda = -2$ implies

$$A - \lambda I = \begin{pmatrix} 3 & 1 & 1 \\ 2 & 3 & -1 \\ -8 & -5 & -1 \end{pmatrix}$$

$$\rightarrow \begin{pmatrix} 7 & 0 & 4 \\ 0 & 7 & -5 \\ 0 & 0 & 0 \end{pmatrix}$$

after elementary row operations. Therefore,

$$\mathbf{v_3} = \begin{pmatrix} 4 \\ -5 \\ -7 \end{pmatrix}$$

is an eigenvector for $\lambda = -2$, and, consequently,

$$\mathbf{x_3}(t) = e^{-2t} \begin{pmatrix} 4 \\ -5 \\ -7 \end{pmatrix}$$

is a solution of the system.

Therefore, the general solution is

$$\mathbf{x}(t) = c_1 e^{2t} \begin{pmatrix} 0 \\ 1 \\ -1 \end{pmatrix} + c_2 e^{-t} \begin{pmatrix} 3 \\ -4 \\ -2 \end{pmatrix} + c_3 e^{-2t} \begin{pmatrix} 4 \\ -5 \\ -7 \end{pmatrix}.$$

8. To solve this system, we need to compute the eigenvalues and eigenvectors of A. We have

$$A - \lambda I = \begin{pmatrix} 1 - \lambda & -1 & 4 \\ 3 & 2 - \lambda & -1 \\ 2 & 1 & -1 - \lambda \end{pmatrix}.$$

Therefore,
$$\det(A - \lambda I) = (\lambda - 1)(\lambda + 2)(\lambda - 3).$$

Therefore, the eigenvalues are $\lambda = 1, -2, 3$.

First, $\lambda = 1$ implies

$$A - \lambda I = \begin{pmatrix} 0 & -1 & 4 \\ 3 & 1 & -1 \\ 2 & 1 & -2 \end{pmatrix}$$
$$\rightarrow \begin{pmatrix} 1 & 0 & 1 \\ 0 & 1 & -4 \\ 0 & 0 & 0 \end{pmatrix}$$

after elementary row operations. Therefore,

$$\mathbf{v_1} = \begin{pmatrix} 1 \\ -4 \\ -1 \end{pmatrix}$$

is an eigenvector for $\lambda = 1$, and, consequently,

$$\mathbf{x_1}(t) = e^t \begin{pmatrix} 1 \\ -4 \\ -1 \end{pmatrix}$$

is a solution of the system.

Next, $\lambda = -2$ implies

$$A - \lambda I = \begin{pmatrix} 3 & -1 & 4 \\ 3 & 4 & -1 \\ 2 & 1 & 1 \end{pmatrix}$$
$$\rightarrow \begin{pmatrix} 1 & 0 & 1 \\ 0 & 1 & -1 \\ 0 & 0 & 0 \end{pmatrix}$$

after elementary row operations. Therefore,

$$\mathbf{v_2} = \begin{pmatrix} 1 \\ -1 \\ -1 \end{pmatrix}$$

is an eigenvector for $\lambda = -2$, and, consequently,

$$\mathbf{x_2}(t) = e^{-2t} \begin{pmatrix} 1 \\ -1 \\ -1 \end{pmatrix}$$

is a solution of the system.

Finally, $\lambda = 3$ implies

$$A - \lambda I = \begin{pmatrix} -2 & -1 & 4 \\ 3 & -1 & -1 \\ 2 & 1 & -4 \end{pmatrix}$$
$$\rightarrow \begin{pmatrix} 1 & 0 & -1 \\ 0 & 1 & -2 \\ 0 & 0 & 0 \end{pmatrix}$$

after elementary row operations. Therefore,

$$\mathbf{v_3} = \begin{pmatrix} 1 \\ 2 \\ 1 \end{pmatrix}$$

is an eigenvector for $\lambda = 3$, and, consequently,

$$\mathbf{x_3}(t) = e^{3t} \begin{pmatrix} 1 \\ 2 \\ 1 \end{pmatrix}$$

is a solution of the system.

Therefore, the general solution is

$$\mathbf{x}(t) = c_1 e^t \begin{pmatrix} 1 \\ -4 \\ -1 \end{pmatrix} + c_2 e^{-2t} \begin{pmatrix} 1 \\ -1 \\ -1 \end{pmatrix} + c_3 e^{3t} \begin{pmatrix} 1 \\ 2 \\ 1 \end{pmatrix}.$$

9. We need to find the eigenvalues and eigenvectors.

$$A - \lambda I = \begin{pmatrix} 1 - \lambda & 1 & 2 \\ 0 & 2 - \lambda & 2 \\ -1 & 1 & 3 - \lambda \end{pmatrix}$$

implies

$$\det(A - \lambda I) = \lambda^3 - 6\lambda^2 + 11\lambda - 6.$$

Therefore, the eigenvalues are $\lambda = 1, 2, 3$.

First, for $\lambda = 1$,

$$A - \lambda I = \begin{pmatrix} 0 & 1 & 2 \\ 0 & 1 & 2 \\ -1 & 1 & 2 \end{pmatrix} \rightarrow \begin{pmatrix} 1 & 0 & 0 \\ 0 & 1 & 2 \\ 0 & 0 & 0 \end{pmatrix}$$

after elementary row operations. Therefore,

$$\mathbf{v_1} = \begin{pmatrix} 0 \\ -2 \\ 1 \end{pmatrix}$$

is an eigenvector for $\lambda = 1$ and

$$\mathbf{x_1}(t) = e^t \begin{pmatrix} 0 \\ -2 \\ 1 \end{pmatrix}$$

is a solution of our system.

Second, for $\lambda = 2$,

$$A - \lambda I = \begin{pmatrix} -1 & 1 & 2 \\ 0 & 0 & 2 \\ -1 & 1 & 1 \end{pmatrix} \rightarrow \begin{pmatrix} 1 & -1 & 0 \\ 0 & 0 & 1 \\ 0 & 0 & 0 \end{pmatrix}$$

after elementary row operations. Therefore,

$$\mathbf{v_2} = \begin{pmatrix} 1 \\ 1 \\ 0 \end{pmatrix}$$

is an eigenvector for $\lambda = 2$ and

$$\mathbf{x_2}(t) = e^{2t} \begin{pmatrix} 1 \\ 1 \\ 0 \end{pmatrix}$$

is a solution of our system.

Last, for $\lambda = 3$,

$$A - \lambda I = \begin{pmatrix} -2 & 1 & 2 \\ 0 & -1 & 2 \\ -1 & 1 & 0 \end{pmatrix} \rightarrow \begin{pmatrix} 1 & 0 & -2 \\ 0 & 1 & -2 \\ 0 & 0 & 0 \end{pmatrix}$$

after elementary row operations. Therefore,

$$\mathbf{v_3} = \begin{pmatrix} 2 \\ 2 \\ 1 \end{pmatrix}$$

is an eigenvector for $\lambda = 3$ and

$$\mathbf{x_3}(t) = e^{3t} \begin{pmatrix} 2 \\ 2 \\ 1 \end{pmatrix}$$

is a solution of our system.

Therefore, the general solution is

$$\mathbf{x}(t) = c_1 e^t \begin{pmatrix} 0 \\ -2 \\ 1 \end{pmatrix} + c_2 e^{2t} \begin{pmatrix} 1 \\ 1 \\ 0 \end{pmatrix} + c_3 e^{3t} \begin{pmatrix} 2 \\ 2 \\ 1 \end{pmatrix}.$$

The initial condition implies

$$\mathbf{x}(0) = c_1 \begin{pmatrix} 0 \\ -2 \\ 1 \end{pmatrix} + c_2 \begin{pmatrix} 1 \\ 1 \\ 0 \end{pmatrix} + c_3 \begin{pmatrix} 2 \\ 2 \\ 1 \end{pmatrix} = \begin{pmatrix} 2 \\ 0 \\ 1 \end{pmatrix}.$$

The solution of this equation is $c_1 = 1$, $c_2 = 2$ and $c_3 = 0$. Therefore, the solution is

$$\mathbf{x}(t) = e^t \begin{pmatrix} 0 \\ -2 \\ 1 \end{pmatrix} + 2e^{2t} \begin{pmatrix} 1 \\ 1 \\ 0 \end{pmatrix}.$$

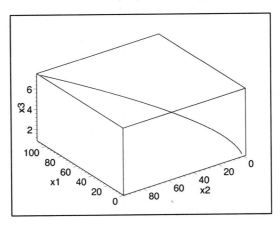

The solution $\mathbf{x}$ satisfies $|\mathbf{x}| \to \infty$ as $t \to \infty$.

10. We need to find the eigenvalues and eigenvectors.

$$A - \lambda I = \begin{pmatrix} -\lambda & 0 & -1 \\ 2 & -\lambda & 0 \\ -1 & 2 & 4 - \lambda \end{pmatrix}$$

implies

$$\det(A - \lambda I) = \lambda^3 - 4\lambda^2 - \lambda + 4.$$

23

Therefore, the eigenvalues are $\lambda = -1, 1, 4$.

First, for $\lambda = -1$,

$$A - \lambda I = \begin{pmatrix} 1 & 0 & -1 \\ 2 & 1 & 0 \\ -1 & 2 & 5 \end{pmatrix} \rightarrow \begin{pmatrix} 1 & 0 & -1 \\ 0 & 1 & 2 \\ 0 & 0 & 0 \end{pmatrix}$$

after elementary row operations. Therefore,

$$\mathbf{v_1} = \begin{pmatrix} 1 \\ -2 \\ 1 \end{pmatrix}$$

is an eigenvector for $\lambda = -1$ and

$$\mathbf{x_1}(t) = e^{-t} \begin{pmatrix} 1 \\ -2 \\ 1 \end{pmatrix}$$

is a solution of our system.

Second, for $\lambda = 1$,

$$A - \lambda I = \begin{pmatrix} -1 & 0 & -1 \\ 2 & -1 & 0 \\ -1 & 2 & 3 \end{pmatrix} \rightarrow \begin{pmatrix} 1 & 0 & 1 \\ 0 & 1 & 2 \\ 0 & 0 & 0 \end{pmatrix}$$

after elementary row operations. Therefore,

$$\mathbf{v_2} = \begin{pmatrix} 1 \\ 2 \\ -1 \end{pmatrix}$$

is an eigenvector for $\lambda = 1$ and

$$\mathbf{x_2}(t) = e^{t} \begin{pmatrix} 1 \\ 2 \\ -1 \end{pmatrix}$$

is a solution of our system.

Last, for $\lambda = 4$,

$$A - \lambda I = \begin{pmatrix} -4 & 0 & -1 \\ 2 & -4 & 0 \\ -1 & 2 & 0 \end{pmatrix} \rightarrow \begin{pmatrix} 4 & 0 & 1 \\ 0 & 8 & 1 \\ 0 & 0 & 0 \end{pmatrix}$$

after elementary row operations. Therefore,

$$\mathbf{v_3} = \begin{pmatrix} 2 \\ 1 \\ -8 \end{pmatrix}$$

is an eigenvector for $\lambda = 4$ and

$$\mathbf{x_3}(t) = e^{4t} \begin{pmatrix} 2 \\ 1 \\ -8 \end{pmatrix}$$

is a solution of our system.

Therefore, the general solution is

$$\mathbf{x}(t) = c_1 e^{-t} \begin{pmatrix} 1 \\ -2 \\ 1 \end{pmatrix} + c_2 e^{t} \begin{pmatrix} 1 \\ 2 \\ -1 \end{pmatrix} + c_3 e^{4t} \begin{pmatrix} 2 \\ 1 \\ -8 \end{pmatrix}.$$

The initial condition implies

$$\mathbf{x}(0) = c_1 \begin{pmatrix} 1 \\ -2 \\ 1 \end{pmatrix} + c_2 \begin{pmatrix} 1 \\ 2 \\ -1 \end{pmatrix} + c_3 \begin{pmatrix} 2 \\ 1 \\ -8 \end{pmatrix} = \begin{pmatrix} 7 \\ 5 \\ 5 \end{pmatrix}.$$

The solution of this equation is $c_1 = 3$, $c_2 = 6$ and $c_3 = -1$. Therefore, the solution is

$$\mathbf{x}(t) = 3e^{-t} \begin{pmatrix} 1 \\ -2 \\ 1 \end{pmatrix} + 6e^t \begin{pmatrix} 1 \\ 2 \\ -1 \end{pmatrix} - e^{4t} \begin{pmatrix} 2 \\ 1 \\ -8 \end{pmatrix}.$$

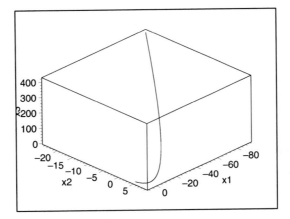

The solution $\mathbf{x}$ satisfies $|\mathbf{x}| \to \infty$ as $t \to \infty$.

11. We need to find the eigenvalues and eigenvectors.

$$A - \lambda I = \begin{pmatrix} -1 - \lambda & 0 & 3 \\ 0 & -2 - \lambda & 0 \\ 3 & 0 & -1 - \lambda \end{pmatrix}$$

implies

$$\det(A - \lambda I) = -(\lambda + 2)(\lambda^2 + 2\lambda - 8) = -(\lambda + 2)(\lambda + 4)(\lambda - 2).$$

Therefore, the eigenvalues are $\lambda = -2, -4, 2$.

First, for $\lambda = -2$,

$$A - \lambda I = \begin{pmatrix} 1 & 0 & 3 \\ 0 & 0 & 0 \\ 3 & 0 & 1 \end{pmatrix} \to \begin{pmatrix} 1 & 0 & 3 \\ 0 & 0 & 1 \\ 0 & 0 & 0 \end{pmatrix}$$

after elementary row operations. Therefore,

$$\mathbf{v_1} = \begin{pmatrix} 0 \\ 1 \\ 0 \end{pmatrix}$$

is an eigenvector for $\lambda = -2$ and

$$\mathbf{x_1}(t) = e^{-2t} \begin{pmatrix} 0 \\ 1 \\ 0 \end{pmatrix}$$

is a solution of our system.

Second, for $\lambda = -4$,

$$A - \lambda I = \begin{pmatrix} 3 & 0 & 3 \\ 0 & 2 & 0 \\ 3 & 0 & 3 \end{pmatrix} \to \begin{pmatrix} 1 & 0 & 1 \\ 0 & 1 & 0 \\ 0 & 0 & 0 \end{pmatrix}$$

25

after elementary row operations. Therefore,

$$\mathbf{v_2} = \begin{pmatrix} 1 \\ 0 \\ -1 \end{pmatrix}$$

is an eigenvector for $\lambda = -4$ and

$$\mathbf{x_2}(t) = e^{-4t} \begin{pmatrix} 1 \\ 0 \\ -1 \end{pmatrix}$$

is a solution of our system.

Last, for $\lambda = 2$,

$$A - \lambda I = \begin{pmatrix} -3 & 0 & 3 \\ 0 & -4 & 0 \\ 3 & 0 & -3 \end{pmatrix} \rightarrow \begin{pmatrix} 1 & 0 & -1 \\ 0 & 1 & 0 \\ 0 & 0 & 0 \end{pmatrix}$$

after elementary row operations. Therefore,

$$\mathbf{v_3} = \begin{pmatrix} 1 \\ 0 \\ 1 \end{pmatrix}$$

is an eigenvector for $\lambda = 2$ and

$$\mathbf{x_3}(t) = e^{2t} \begin{pmatrix} 1 \\ 0 \\ 1 \end{pmatrix}$$

is a solution of our system.

Therefore, the general solution is

$$\mathbf{x}(t) = c_1 e^{-2t} \begin{pmatrix} 0 \\ 1 \\ 0 \end{pmatrix} + c_2 e^{-4t} \begin{pmatrix} 1 \\ 0 \\ -1 \end{pmatrix} + c_3 e^{2t} \begin{pmatrix} 1 \\ 0 \\ 1 \end{pmatrix}.$$

The initial condition implies

$$\mathbf{x}(0) = c_1 \begin{pmatrix} 0 \\ 1 \\ 0 \end{pmatrix} + c_2 \begin{pmatrix} 1 \\ 0 \\ -1 \end{pmatrix} + c_3 \begin{pmatrix} 1 \\ 0 \\ 1 \end{pmatrix} = \begin{pmatrix} 2 \\ -1 \\ -2 \end{pmatrix}.$$

The solution of this equation is $c_1 = -1$, $c_2 = 2$ and $c_3 = 0$. Therefore, the solution is

$$\mathbf{x}(t) = -e^{-2t} \begin{pmatrix} 0 \\ 1 \\ 0 \end{pmatrix} + 2e^{-4t} \begin{pmatrix} 1 \\ 0 \\ -1 \end{pmatrix}.$$

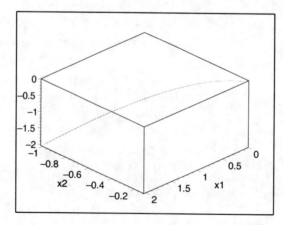

The solution tends to the origin approaching the eigenvector $(0, -1, 0)^T$ as $t \to \infty$.

12. We need to find the eigenvalues and eigenvectors.

$$A - \lambda I = \begin{pmatrix} \frac{1}{2} - \lambda & -1 & -\frac{3}{2} \\ \frac{3}{2} & -2 - \lambda & -\frac{3}{2} \\ -2 & 2 & 1 - \lambda \end{pmatrix}$$

implies

$$\det(A - \lambda I) = -\lambda^3 - frac12\lambda^2 + \lambda + \frac{1}{2} = -\left(\lambda + \frac{1}{2}\right)(\lambda - 1)(\lambda + 1).$$

Therefore, the eigenvalues are $\lambda = -1/2, 1, -1$.

First, for $\lambda = -1/2$,

$$A - \lambda I = \begin{pmatrix} 1 & -1 & -\frac{3}{2} \\ \frac{3}{2} & -\frac{3}{2} & -\frac{3}{2} \\ -2 & 2 & \frac{3}{2} \end{pmatrix} \to \begin{pmatrix} 1 & -1 & -\frac{3}{2} \\ 0 & 0 & 1 \\ 0 & 0 & 0 \end{pmatrix}$$

after elementary row operations. Therefore,

$$\mathbf{v_1} = \begin{pmatrix} 1 \\ 1 \\ 0 \end{pmatrix}$$

is an eigenvector for $\lambda = -1/2$ and

$$\mathbf{x_1}(t) = e^{-t/2} \begin{pmatrix} 1 \\ 1 \\ 0 \end{pmatrix}$$

is a solution of our system.

Second, for $\lambda = 1$,

$$A - \lambda I = \begin{pmatrix} -\frac{1}{2} & -1 & -\frac{3}{2} \\ \frac{3}{2} & -3 & -\frac{3}{2} \\ -2 & 2 & 0 \end{pmatrix} \to \begin{pmatrix} 1 & 2 & 3 \\ 0 & 1 & 1 \\ 0 & 0 & 0 \end{pmatrix}$$

after elementary row operations. Therefore,

$$\mathbf{v_2} = \begin{pmatrix} 1 \\ 1 \\ -1 \end{pmatrix}$$

is an eigenvector for $\lambda = 1$ and

$$\mathbf{x_2}(t) = e^t \begin{pmatrix} 1 \\ 1 \\ -1 \end{pmatrix}$$

is a solution of our system.

Last, for $\lambda = -1$,

$$A - \lambda I = \begin{pmatrix} \frac{3}{2} & -1 & -\frac{3}{2} \\ \frac{3}{2} & -1 & -\frac{3}{2} \\ -2 & 2 & 2 \end{pmatrix} \to \begin{pmatrix} 1 & -1 & -1 \\ 0 & 1 & 0 \\ 0 & 0 & 0 \end{pmatrix}$$

after elementary row operations. Therefore,

$$\mathbf{v_3} = \begin{pmatrix} 1 \\ 0 \\ 1 \end{pmatrix}$$

is an eigenvector for $\lambda = -1$ and

$$\mathbf{x_3}(t) = e^{-t} \begin{pmatrix} 1 \\ 0 \\ 1 \end{pmatrix}$$

is a solution of our system.

Therefore, the general solution is

$$\mathbf{x}(t) = c_1 e^{-t/2} \begin{pmatrix} 1 \\ 1 \\ 0 \end{pmatrix} + c_2 e^t \begin{pmatrix} 1 \\ 1 \\ -1 \end{pmatrix} + c_3 e^{-t} \begin{pmatrix} 1 \\ 0 \\ 1 \end{pmatrix}.$$

The initial condition implies

$$\mathbf{x}(0) = c_1 \begin{pmatrix} 1 \\ 1 \\ 0 \end{pmatrix} + c_2 \begin{pmatrix} 1 \\ 1 \\ -1 \end{pmatrix} + c_3 \begin{pmatrix} 1 \\ 0 \\ 1 \end{pmatrix} = \begin{pmatrix} 2 \\ 1 \\ 1 \end{pmatrix}.$$

The solution of this equation is $c_1 = 1$, $c_2 = 0$ and $c_3 = 1$. Therefore, the solution is

$$\mathbf{x}(t) = -e^{-t/2} \begin{pmatrix} 1 \\ 1 \\ 0 \end{pmatrix} + e^{-t} \begin{pmatrix} 1 \\ 0 \\ 1 \end{pmatrix}.$$

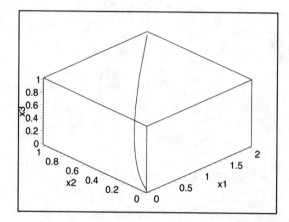

The solution tends to the origin approaching the eigenvector $(1,1,0)^T$ as $t \to \infty$.

13.

$$\mathbf{x}' = k A \mathbf{x}$$

where

$$\mathbf{x} = \begin{pmatrix} x_1 \\ x_2 \\ x_3 \\ x_4 \end{pmatrix}$$

and

$$A = \begin{pmatrix} -1 & 1 & 0 & 0 \\ 1 & -2 & 1 & 0 \\ 0 & 1 & -2 & 1 \\ 0 & 0 & 1 & -1 \end{pmatrix}.$$

To find the general solution, we look for the eigenvalues of A. We see that

$$A - \lambda I = \begin{pmatrix} -1-\lambda & 1 & 0 & 0 \\ 1 & -2-\lambda & 1 & 0 \\ 0 & 1 & -2-\lambda & 1 \\ 0 & 0 & 1 & -1-\lambda \end{pmatrix}.$$

Therefore,

$$\det(A - \lambda I) = \lambda^4 + 6\lambda^3 + 10\lambda^2 + 4\lambda = \lambda(\lambda+2)(\lambda^2 + 4\lambda + 2).$$

Therefore, the eigenvalues are $\lambda = 0, -2, -2 \pm \sqrt{2}$.

First, $\lambda = 0$ implies

$$A - \lambda I = \begin{pmatrix} -1 & 1 & 0 & 0 \\ 1 & -2 & 1 & 0 \\ 0 & 1 & -2 & 1 \\ 0 & 0 & 1 & -1 \end{pmatrix} \rightarrow \begin{pmatrix} 1 & -1 & 0 & 0 \\ 0 & 1 & -1 & 0 \\ 0 & 0 & 1 & -1 \\ 0 & 0 & 0 & 0 \end{pmatrix}$$

after elementary row operations. Therefore,

$$\mathbf{v_1} = \begin{pmatrix} 1 \\ 1 \\ 1 \\ 1 \end{pmatrix}$$

is an eigenvector for $\lambda = 0$ and

$$\mathbf{x_1}(t) = \begin{pmatrix} 1 \\ 1 \\ 1 \\ 1 \end{pmatrix}$$

is a solution of the system.

Second, $\lambda = -2$ implies

$$A - \lambda I = \begin{pmatrix} 1 & 1 & 0 & 0 \\ 1 & 0 & 1 & 0 \\ 0 & 1 & 0 & 1 \\ 0 & 0 & 1 & 1 \end{pmatrix} \rightarrow \begin{pmatrix} 1 & 1 & 0 & 0 \\ 0 & 1 & -1 & 0 \\ 0 & 0 & 1 & 1 \\ 0 & 0 & 0 & 0 \end{pmatrix}.$$

Therefore,

$$\mathbf{v_2} = \begin{pmatrix} 1 \\ -1 \\ -1 \\ 1 \end{pmatrix}$$

is an eigenvector for $\lambda = -2$ and

$$\mathbf{x_2}(t) = e^{-2t} \begin{pmatrix} 1 \\ -1 \\ -1 \\ 1 \end{pmatrix}$$

is a solution of the system.

Third, $\lambda = -2 + \sqrt{2}$ implies

$$A - \lambda I = \begin{pmatrix} 1-\sqrt{2} & 1 & 0 & 0 \\ 1 & -\sqrt{2} & 1 & 0 \\ 0 & 1 & -\sqrt{2} & 1 \\ 0 & 0 & 1 & 1-\sqrt{2} \end{pmatrix} \rightarrow \begin{pmatrix} 1 & -\sqrt{2} & 1 & 0 \\ 0 & 1 & 1 & 0 \\ 0 & 0 & 1 & 1-\sqrt{2} \\ 0 & 0 & 0 & 0 \end{pmatrix}.$$

Therefore,

$$\mathbf{v_3} = \begin{pmatrix} 1 \\ -1+\sqrt{2} \\ 1-\sqrt{2} \\ -1 \end{pmatrix}$$

is an eigenvector for $\lambda = -2 + \sqrt{2}$ and

$$\mathbf{x_3}(t) = e^{(-2+\sqrt{2})t} \begin{pmatrix} 1 \\ -1+\sqrt{2} \\ 1-\sqrt{2} \\ -1 \end{pmatrix}$$

is a solution of the system.

Fourth, $\lambda = -2 - \sqrt{2}$ implies

$$A - \lambda I = \begin{pmatrix} 1+\sqrt{2} & 1 & 0 & 0 \\ 1 & \sqrt{2} & 1 & 0 \\ 0 & 1 & \sqrt{2} & 1 \\ 0 & 0 & 1 & 1+\sqrt{2} \end{pmatrix} \rightarrow \begin{pmatrix} 1 & \sqrt{2} & 1 & 0 \\ 0 & 1 & 1 & 0 \\ 0 & 0 & 1 & 1+\sqrt{2} \\ 0 & 0 & 0 & 0 \end{pmatrix}.$$

Therefore,

$$\mathbf{v_4} = \begin{pmatrix} 1 \\ -1-\sqrt{2} \\ 1+\sqrt{2} \\ -1 \end{pmatrix}$$

is an eigenvector for $\lambda = -2 - \sqrt{2}$ and

$$\mathbf{x_4}(t) = e^{(-2-\sqrt{2})t} \begin{pmatrix} 1 \\ -1-\sqrt{2} \\ 1+\sqrt{2} \\ -1 \end{pmatrix}$$

is a solution of the system.

Therefore, the general solution is

$$\mathbf{x}(t) = c_1 \begin{pmatrix} 1 \\ 1 \\ 1 \\ 1 \end{pmatrix} + c_2 e^{-2t} \begin{pmatrix} 1 \\ -1 \\ -1 \\ 1 \end{pmatrix} + c_3 e^{(-2+\sqrt{2})t} \begin{pmatrix} 1 \\ -1+\sqrt{2} \\ 1-\sqrt{2} \\ -1 \end{pmatrix} + c_4 e^{(-2-\sqrt{2})t} \begin{pmatrix} 1 \\ -1-\sqrt{2} \\ 1+\sqrt{2} \\ -1 \end{pmatrix}$$

Below is a graph of the components of the eigenvectors of A.

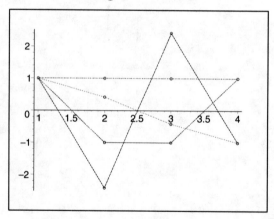

As $t \to \infty$, the solution approaches the equilibrium state $(1,1,1,1)^T$. The eigenvalue $\lambda_3 = -2 + \sqrt{2}$ controls the long term decay rate towards equilibrium.

14.

(a)

$$x_1' = k(x_2 - x_1)$$
$$x_2' = k(x_1 - 2x_2 + x_3)$$
$$x_3' = k(x_2 - 2x_3 + x_4) \quad \text{but } x_4(t) = 0 \implies x_3' = k(x_2 - 2x_3)$$

Therefore,

$$\begin{pmatrix} x_1 \\ x_2 \\ x_3 \end{pmatrix}' = k \begin{pmatrix} -1 & 1 & 0 \\ 1 & -2 & 1 \\ 0 & 1 & -2 \end{pmatrix} \begin{pmatrix} x_1 \\ x_2 \\ x_3 \end{pmatrix}.$$

30

(b)

$$\frac{d}{dt}(x_1 + x_2 + x_3) = k(x_2 - x_1) + k(x_1 - 2x_2 + x_3) + k(x_2 - 2x_3)$$
$$= (-k + k)x_1 + (k - 2k + k)x_2 + (k - 2k)x_3$$
$$= -kx_3.$$

Therefore, the rate of change of particles leaving the first three lattice points is proportional to the rate at which they leave lattice point three.

(c)

$$A - \lambda I = \begin{pmatrix} -1 - \lambda & 1 & 0 \\ 1 & -2 - \lambda & 1 \\ 0 & 1 & -2 - \lambda \end{pmatrix}$$

implies

$$\det(A - \lambda I) = -(\lambda^3 + 5\lambda^2 + 6\lambda + 1).$$

The eigenvalues of A are $\lambda_1 \approx -.1981$, $\lambda_2 \approx -1.555$ and $\lambda_3 \approx -3.247$. Corresponding eigenvectors are given by

$$\mathbf{v_1} \approx \begin{pmatrix} 2.25 \\ 1.8 \\ 1 \end{pmatrix} \quad \mathbf{v_2} \approx \begin{pmatrix} 0.55 \\ -1.25 \\ 1 \end{pmatrix} \quad \mathbf{v_3} \approx \begin{pmatrix} -0.8 \\ 0.45 \\ 1 \end{pmatrix}.$$

Below is a plot of the components of the eigenvectors above.

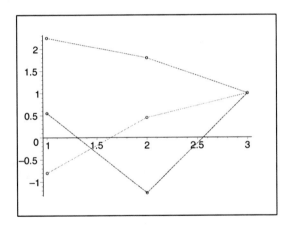

(d) The general solution is given by

$$\mathbf{x}(t) = c_1 e^{\lambda_1 t}\mathbf{v_1} + c_2 e^{\lambda_2 t}\mathbf{v_2} + c_2 e^{\lambda_3 t}\mathbf{v_3},$$

where $\lambda_1, \lambda_2, \lambda_3$ and $\mathbf{v_1}, \mathbf{v_2}, \mathbf{v_3}$ are listed above. Since all the eigenvalues are negative, as $t \to \infty$,

$$\mathbf{x}(t) \to \begin{pmatrix} 0 \\ 0 \\ 0 \end{pmatrix}.$$

15.

$$A - \lambda I = \begin{pmatrix} 2 - \lambda & -4 & -3 \\ 3 & -5 - \lambda & -3 \\ -2 & 2 & 1 - \lambda \end{pmatrix}.$$

Therefore,

$$\det(A - \lambda I) = -\lambda^3 - 2\lambda^2 + \lambda + 2 = -(\lambda + 2)(\lambda + 1)(\lambda - 1).$$

Therefore, the eigenvalues are given by $\lambda = -2, -1, 1$. The corresponding eigenvectors are given as follows. For $\lambda_1 = -2$,

$$A - \lambda I = \begin{pmatrix} 4 & -4 & -3 \\ 3 & -3 & -3 \\ -2 & 2 & 3 \end{pmatrix} \rightarrow \begin{pmatrix} 1 & -1 & -1 \\ 0 & 0 & 1 \\ 0 & 0 & 0 \end{pmatrix}$$

after elementary row operations. Therefore,

$$\mathbf{v_1} = \begin{pmatrix} 1 \\ 1 \\ 0 \end{pmatrix}.$$

For $\lambda_2 = -1$,

$$A - \lambda I = \begin{pmatrix} 3 & -4 & -3 \\ 3 & -4 & -3 \\ -2 & 2 & 2 \end{pmatrix} \rightarrow \begin{pmatrix} 1 & -1 & -1 \\ 0 & 1 & 0 \\ 0 & 0 & 0 \end{pmatrix}$$

after elementary row operations. Therefore,

$$\mathbf{v_2} = \begin{pmatrix} 1 \\ 0 \\ 1 \end{pmatrix}.$$

For $\lambda_3 = 1$,

$$A - \lambda I = \begin{pmatrix} 1 & -4 & -3 \\ 3 & -6 & -3 \\ -2 & 2 & 0 \end{pmatrix} \rightarrow \begin{pmatrix} 1 & -4 & -3 \\ 0 & 1 & 1 \\ 0 & 0 & 0 \end{pmatrix}$$

after elementary row operations. Therefore,

$$\mathbf{v_3} = \begin{pmatrix} 1 \\ 1 \\ -1 \end{pmatrix}.$$

Therefore, the general solution is

$$\mathbf{x}(t) = c_1 e^{-2t} \begin{pmatrix} 1 \\ 1 \\ 0 \end{pmatrix} + c_2 e^{-t} \begin{pmatrix} 1 \\ 0 \\ 1 \end{pmatrix} + c_3 e^{t} \begin{pmatrix} 1 \\ 1 \\ -1 \end{pmatrix}.$$

If we want the solution to tend to $(0,0,0)^T$ as $t \to \infty$, we need $c_3 = 0$. That is, we need the initial condition $\mathbf{x_0}$ to satisfy

$$\mathbf{x_0} = c_1 \begin{pmatrix} 1 \\ 1 \\ 0 \end{pmatrix} + c_2 \begin{pmatrix} 1 \\ 0 \\ 1 \end{pmatrix}.$$

Therefore, letting

$$\mathbf{u_1} = \begin{pmatrix} 1 \\ 1 \\ 0 \end{pmatrix} \quad \text{and} \quad \mathbf{u_2} = \begin{pmatrix} 1 \\ 0 \\ 1 \end{pmatrix},$$

and letting

$$S = \{\mathbf{u} : \mathbf{u} = a_1 \mathbf{u_1} + a_2 \mathbf{u_2}, -\infty < a_1, a_2 < \infty\},$$

then for any $\mathbf{x_0} \in S$, the solution $\mathbf{x}(t) \to (0,0,0)^T$ as $t \to \infty$.

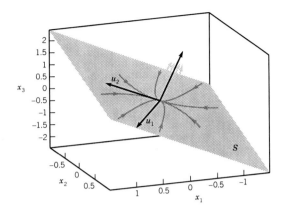

If $\mathbf{x_0} \notin S$, then $\mathbf{x}(t)$ approaches the line determined by

$$\mathbf{v_3} = \begin{pmatrix} 1 \\ 1 \\ -1 \end{pmatrix}$$

as $t \to \infty$.

16. The eigenvalues of this matrix are given by $\lambda = -2, -1, 1$. In particular, $\lambda = -1$ has multiplicity two. The corresponding eigenvectors are given as follows. For $\lambda_1 = -2$,

$$\mathbf{v_1} = \begin{pmatrix} 1 \\ 0 \\ -1 \\ 0 \end{pmatrix}$$

is an associated eigenvector. For $\lambda_2 = -1$, two linearly independent eigenvectors are given by

$$\mathbf{v_2} = \begin{pmatrix} 1 \\ -1 \\ 1 \\ 0 \end{pmatrix} \quad \text{and} \quad \begin{pmatrix} 0 \\ 1 \\ 0 \\ 1 \end{pmatrix}.$$

For $\lambda_3 = 1$,

$$\mathbf{v_4} = \begin{pmatrix} 1 \\ 1 \\ 0 \\ 1 \end{pmatrix}$$

is an associated eigenvector. Therefore, the general solution is given by

$$\mathbf{x}(t) = c_1 e^{-2t} \begin{pmatrix} 1 \\ 0 \\ -1 \\ 0 \end{pmatrix} + c_2 e^{-t} \begin{pmatrix} 1 \\ -1 \\ 1 \\ 0 \end{pmatrix} + c_3 e^{-t} \begin{pmatrix} 0 \\ 1 \\ 0 \\ 1 \end{pmatrix} + c_4 e^{t} \begin{pmatrix} 1 \\ 1 \\ 0 \\ 1 \end{pmatrix}.$$

As long as $c_4 = 0$, the solution will tend to $(0, 0, 0)^T$ as $t \to \infty$. In order to guarantee that $c_4 = 0$, we need

$$\mathbf{x_0} = c_1 \begin{pmatrix} 1 \\ 0 \\ -1 \\ 0 \end{pmatrix} + c_2 \begin{pmatrix} 1 \\ -1 \\ 1 \\ 0 \end{pmatrix} + c_3 \begin{pmatrix} 0 \\ 1 \\ 0 \\ 1 \end{pmatrix}.$$

33

Therefore, letting

$$\mathbf{u_1} = \begin{pmatrix} 1 \\ 0 \\ -1 \\ 0 \end{pmatrix} \quad \mathbf{u_2} = \begin{pmatrix} 1 \\ -1 \\ 1 \\ 0 \end{pmatrix} \quad \mathbf{u_3} = \begin{pmatrix} 0 \\ 1 \\ 0 \\ 1 \end{pmatrix},$$

and defining S as indicated, for $\mathbf{x_0} \in S$, $\mathbf{x}(t) \to (0,0,0,0)^T$ as $t \to \infty$.

17.

Writing our system in matrix form, we have

$$\begin{pmatrix} m_1 \\ m_2 \\ m_3 \end{pmatrix}' = \begin{pmatrix} -k_1 & 0 & 0 \\ k_1 & -k_2 & 0 \\ 0 & k_2 & 0 \end{pmatrix} \begin{pmatrix} m_1 \\ m_2 \\ m_3 \end{pmatrix}.$$

Then

$$A - \lambda I = \begin{pmatrix} -k_1 - \lambda & 0 & 0 \\ k_1 & -k_2 - \lambda & 0 \\ 0 & k_2 & -\lambda \end{pmatrix}$$

implies

$$\det(A - \lambda I) = -\lambda(\lambda + k_1)(\lambda + k_2).$$

Therefore, $\lambda = 0, -k_1, -k_2$. We will first consider the case when $k_1 \neq k_2$. First, for $\lambda_1 = 0$,

$$A - \lambda I = \begin{pmatrix} = -k_1 & 0 & 0 \\ k_1 & -k_2 & 0 \\ 0 & k_2 & 0 \end{pmatrix} \to \begin{pmatrix} 1 & 0 & 0 \\ 0 & 1 & 0 \\ 0 & 0 & 0 \end{pmatrix}$$

after elementary row operations. Therefore,

$$\mathbf{v_1} = \begin{pmatrix} 0 \\ 0 \\ 1 \end{pmatrix}.$$

Next, for $\lambda_2 = -k_1$,

$$A - \lambda I = \begin{pmatrix} 0 & 0 & 0 \\ k_1 & k_1 - k_2 & 0 \\ 0 & k_2 & k_1 \end{pmatrix} \to \begin{pmatrix} k_1 & k_1 - k_2 & 0 \\ 0 & k_2 & k_1 \\ 0 & 0 & 0 \end{pmatrix}$$

after elementary row operations. Therefore,

$$\mathbf{v_2} = \begin{pmatrix} k_2 - k_1 \\ k_1 \\ -k_2 \end{pmatrix}.$$

Third, for $\lambda_3 = -k_2$,

$$A - \lambda I = \begin{pmatrix} k_2 - k_1 & 0 & 0 \\ k_1 & 0 & 0 \\ 0 & k_2 & k_2 \end{pmatrix} \to \begin{pmatrix} 1 & 0 & 0 \\ 0 & 1 & 1 \\ 0 & 0 & 0 \end{pmatrix}$$

after elementary row operations. Therefore,

$$\mathbf{v_3} = \begin{pmatrix} 0 \\ 1 \\ -1 \end{pmatrix}.$$

34

Therefore, the general solution is given by

$$\mathbf{x}(t) = c_1 \begin{pmatrix} 0 \\ 0 \\ 1 \end{pmatrix} + c_2 e^{-k_1 t} \begin{pmatrix} k_2 - k_1 \\ k_1 \\ -k_2 \end{pmatrix} + c_3 e^{-k_2 t} \begin{pmatrix} 0 \\ 1 \\ -1 \end{pmatrix}.$$

Using the initial conditions, we have

$$\mathbf{x}(0) = c_1 \begin{pmatrix} 0 \\ 0 \\ 1 \end{pmatrix} + c_2 \begin{pmatrix} k_2 - k_1 \\ k_1 \\ -k_2 \end{pmatrix} + c_3 \begin{pmatrix} 0 \\ 1 \\ -1 \end{pmatrix} = \begin{pmatrix} m_0 \\ 0 \\ 0 \end{pmatrix}$$

Solving this system, we have $c_1 = m_0$, $c_2 = m_0/(k_2 - k - 1)$ and $c_3 = -m_0 k_1/(k_2 - k_1)$. Therefore, for $k_1 \neq k_2$, the solution is given by

$$\mathbf{x}(t) = m_0 \begin{pmatrix} 0 \\ 0 \\ 1 \end{pmatrix} + \frac{m_0}{k_2 - k_1} e^{-k_1 t} \begin{pmatrix} k_2 - k_1 \\ k_1 \\ -k_2 \end{pmatrix} - \frac{m_0 k_1}{k_2 - k_1} e^{-k_2 t} \begin{pmatrix} 0 \\ 1 \\ -1 \end{pmatrix}.$$

Now, let us consider the case when $k_1 = k_2$. As above, for $\lambda_1 = 0$, we have

$$\begin{pmatrix} 0 \\ 0 \\ 1 \end{pmatrix}$$

and, therefore,

$$\mathbf{x_1}(t) = \begin{pmatrix} 0 \\ 0 \\ 1 \end{pmatrix}$$

is one solution of the system. Now in this case, $\lambda_2 = -k_1$ is an eigenvalue with multiplicity 2. As above, for $\lambda_2 = -k_1$,

$$A - \lambda I = \begin{pmatrix} 0 & 0 & 0 \\ k_1 & k_1 - k_2 & 0 \\ 0 & k_2 & k_1 \end{pmatrix} = \begin{pmatrix} 0 & 0 & 0 \\ k_1 & 0 & 0 \\ 0 & k_1 & k_1 \end{pmatrix} \rightarrow \begin{pmatrix} 1 & 0 & 0 \\ 0 & 1 & 1 \\ 0 & 0 & 0 \end{pmatrix}$$

Therefore,

$$\mathbf{v_2} = \begin{pmatrix} 0 \\ 1 \\ -1 \end{pmatrix}$$

is an associated eigenvector and

$$\mathbf{x_2}(t) = e^{-k_1 t} \begin{pmatrix} 0 \\ 1 \\ -1 \end{pmatrix}.$$

To find another solution associated with λ_2, we need to find $\mathbf{w}$ satisfying $(A - \lambda I)\mathbf{w} = \mathbf{v_2}$. That is, we need to find $\mathbf{w}$ satisfying

$$\begin{pmatrix} 0 & 0 & 0 \\ k_1 & 0 & 0 \\ 0 & k_1 & k_1 \end{pmatrix} \mathbf{w} = \begin{pmatrix} 0 \\ 1 \\ -1 \end{pmatrix}.$$

We see that

$$\mathbf{w} = \begin{pmatrix} 1/k_1 \\ 0 \\ -1/k_1 \end{pmatrix}$$

is such a vector. Therefore,

$$\mathbf{x_3}(t) = t e^{-k_1 t} \begin{pmatrix} 0 \\ 1 \\ -1 \end{pmatrix} + e^{-k_1 t} \begin{pmatrix} 1/k_1 \\ 0 \\ -1/k_1 \end{pmatrix}$$

35

is another solution of our system. Therefore, for $k_1 = k_2$, the general solution is given by

$$\mathbf{x}(t) = c_1 \begin{pmatrix} 0 \\ 0 \\ 1 \end{pmatrix} + c_2 e^{-k_1 t} \begin{pmatrix} 0 \\ 1 \\ -1 \end{pmatrix} + c_3 e^{-k_1 t} \begin{pmatrix} 1 \\ k_1 t \\ -k_1 t - 1 \end{pmatrix}.$$

Using our initial conditions, we have

$$\mathbf{x}(0) = c_1 \begin{pmatrix} 0 \\ 0 \\ 1 \end{pmatrix} + c_2 \begin{pmatrix} 0 \\ 1 \\ -1 \end{pmatrix} + c_3 \begin{pmatrix} 1 \\ 0 \\ -1 \end{pmatrix} = \begin{pmatrix} m_0 \\ 0 \\ 0 \end{pmatrix}.$$

The solution of this system is $c_1 = m_0$, $c_2 = 0$ and $c_3 = m_0$. Therefore, for $k_1 = k_2$, the solution of our equation is

$$\mathbf{x}(t) = m_0 \begin{pmatrix} 0 \\ 0 \\ 1 \end{pmatrix} + m_0 e^{-k_1 t} \begin{pmatrix} 1 \\ k_1 t \\ -k_1 t - 1 \end{pmatrix}.$$

18. The eigenvalues are given by $\lambda = -1, -2, -3, -4$. Their associated eigenvectors are given by

$$\mathbf{v_1} = \begin{pmatrix} -1 \\ 0 \\ 1 \\ 0 \end{pmatrix}, \quad \mathbf{v_2} = \begin{pmatrix} 1 \\ 0 \\ -1 \\ 1 \end{pmatrix}, \quad \mathbf{v_3} = \begin{pmatrix} 0 \\ 1 \\ 0 \\ 1 \end{pmatrix}, \quad \mathbf{v_4} = \begin{pmatrix} -1 \\ 0 \\ 0 \\ 1 \end{pmatrix}.$$

Therefore, a fundamental set of solutions is given by

$$e^{-t} \begin{pmatrix} -1 \\ 0 \\ 1 \\ 0 \end{pmatrix}, \quad e^{-2t} \begin{pmatrix} 1 \\ 0 \\ -1 \\ 1 \end{pmatrix}, \quad e^{-3t} \begin{pmatrix} 0 \\ 1 \\ 0 \\ 1 \end{pmatrix}, \quad e^{-4t} \begin{pmatrix} -1 \\ 0 \\ 0 \\ 1 \end{pmatrix}.$$

19. The eigenvalues are given by $\lambda = -3, 3$. First, $\lambda_1 = -3$ has multiplicity 1 and an associated eigenvector

$$\mathbf{v_1} = \begin{pmatrix} 1 \\ -2 \\ 0 \\ 1 \end{pmatrix}.$$

Second, $\lambda_2 = 3$ has multiplicity 3 with 3 associated eigenvectors,

$$\mathbf{v_2} = \begin{pmatrix} -1 \\ 0 \\ 0 \\ 1 \end{pmatrix}, \quad \mathbf{v_3} = \begin{pmatrix} 0 \\ 0 \\ 1 \\ 0 \end{pmatrix}, \quad \mathbf{v_4} = \begin{pmatrix} 2 \\ 1 \\ 0 \\ 0 \end{pmatrix}.$$

Therefore, a fundamental set of solutions is given by

$$e^{-3t} \begin{pmatrix} 1 \\ -2 \\ 0 \\ 1 \end{pmatrix}, \quad e^{3t} \begin{pmatrix} -1 \\ 0 \\ 0 \\ 1 \end{pmatrix}, \quad e^{3t} \begin{pmatrix} 0 \\ 0 \\ 1 \\ 0 \end{pmatrix}, \quad e^{3t} \begin{pmatrix} 2 \\ 1 \\ 0 \\ 0 \end{pmatrix}.$$

20. The eigenvalues are given by $\lambda = 0, 9, -6$. First, $\lambda_1 = 0$ has multiplicity 2 and two associated eigenvectors

$$\mathbf{v_1} = \begin{pmatrix} 1 \\ 1 \\ 0 \\ -3 \end{pmatrix}, \mathbf{v_2} = \begin{pmatrix} 1 \\ 0 \\ 1 \\ -2 \end{pmatrix}$$

Second, $\lambda_2 = 9$ and $\lambda_3 = -6$ each have multiplicity 1 with the following associated eigenvectors

$$\mathbf{v_3} = \begin{pmatrix} 1 \\ 2 \\ -1 \\ -1 \end{pmatrix} \quad \mathbf{v_4} = \begin{pmatrix} -1 \\ -2 \\ 4 \\ 1 \end{pmatrix},$$

respectively. Therefore, a fundamental set of solutions is given by

$$\begin{pmatrix} 1 \\ 1 \\ 0 \\ -3 \end{pmatrix}, \quad \begin{pmatrix} 1 \\ 0 \\ 1 \\ -2 \end{pmatrix}, \quad e^{9t} \begin{pmatrix} 1 \\ 2 \\ -1 \\ -1 \end{pmatrix}, \quad e^{-6t} \begin{pmatrix} -1 \\ -2 \\ 4 \\ 1 \end{pmatrix}.$$

21. The eigenvalues are given by $\lambda = -2, 2, -4, 4$. The associated eigenvectors are given by

$$\mathbf{v_1} = \begin{pmatrix} -1 \\ 0 \\ 1 \\ 0 \end{pmatrix}, \mathbf{v_2} = \begin{pmatrix} -1 \\ -1 \\ 1 \\ 1 \end{pmatrix}, \quad \mathbf{v_3} = \begin{pmatrix} 1 \\ 0 \\ 0 \\ 1 \end{pmatrix}, \quad \mathbf{v_4} = \begin{pmatrix} 0 \\ 1 \\ 0 \\ 1 \end{pmatrix},$$

respectively. Therefore, a fundamental set of solutions is given by

$$e^{-2t} \begin{pmatrix} -1 \\ 0 \\ 1 \\ 0 \end{pmatrix}, \quad e^{2t} \begin{pmatrix} -1 \\ -1 \\ 1 \\ 1 \end{pmatrix}, \quad e^{-4t} \begin{pmatrix} 1 \\ 0 \\ 0 \\ 1 \end{pmatrix}, \quad e^{4t} \begin{pmatrix} 0 \\ 1 \\ 0 \\ 1 \end{pmatrix}.$$

22. The eigenvalues are given by $\lambda = -4, -3, -2, -1$. First, the eigenvalues $\lambda = -2$ has multiplicity 2 and 2 associated eigenvectors

$$\mathbf{v_1} = \begin{pmatrix} 1 \\ 0 \\ -1 \\ 1 \\ 0 \end{pmatrix}, \quad \mathbf{v_2} = \begin{pmatrix} 1 \\ 0 \\ 0 \\ 0 \\ 1 \end{pmatrix}.$$

All other eigenvalues have multiplicity 1. Eigenvectors for $\lambda = -4, -3, -1$ are given by

$$\mathbf{v_3} = \begin{pmatrix} 1 \\ 0 \\ 0 \\ -1 \\ 1 \end{pmatrix}, \mathbf{v_4} = \begin{pmatrix} 0 \\ 1 \\ 0 \\ 1 \\ 0 \end{pmatrix}, \quad \mathbf{v_5} = \begin{pmatrix} 1 \\ 0 \\ -1 \\ 0 \\ 1 \end{pmatrix},$$

respectively. Therefore, a fundamental set of solutions is given by

$$e^{-4t} \begin{pmatrix} 1 \\ 0 \\ 0 \\ -1 \\ 1 \end{pmatrix}, \quad e^{-3t} \begin{pmatrix} 0 \\ 1 \\ 0 \\ 1 \\ 0 \end{pmatrix}, \quad e^{-2t} \begin{pmatrix} 1 \\ 0 \\ -1 \\ 1 \\ 0 \end{pmatrix}, \quad e^{-2t} \begin{pmatrix} 1 \\ 0 \\ 0 \\ 0 \\ 1 \end{pmatrix}, \quad e^{-t} \begin{pmatrix} 1 \\ 0 \\ -1 \\ 0 \\ 1 \end{pmatrix}.$$

23. The eigenvalues are given by $\lambda = -2, -1, 1, 2$. First, the eigenvalues $\lambda = -2$ has multiplicity 2 and 2 associated eigenvectors

$$\mathbf{v_1} = \begin{pmatrix} 0 \\ 1 \\ 0 \\ 1 \\ 0 \end{pmatrix}, \quad \mathbf{v_2} = \begin{pmatrix} 1 \\ 0 \\ 2 \\ 0 \\ 1 \end{pmatrix}.$$

All other eigenvalues have multiplicity 1. Eigenvectors for $\lambda = -1, 1, 2$ are given by

$$\mathbf{v_3} = \begin{pmatrix} 1 \\ 0 \\ 0 \\ -1 \\ 1 \end{pmatrix}, \mathbf{v_4} = \begin{pmatrix} 2 \\ 0 \\ 0 \\ 0 \\ 1 \end{pmatrix}, \quad \mathbf{v_5} = \begin{pmatrix} 1 \\ 1 \\ -1 \\ 1 \\ 0 \end{pmatrix},$$

respectively. Therefore, a fundamental set of solutions is given by

$$e^{-2t} \begin{pmatrix} 0 \\ 1 \\ 0 \\ 1 \\ 0 \end{pmatrix}, \quad e^{-2t} \begin{pmatrix} 1 \\ 0 \\ 2 \\ 0 \\ 1 \end{pmatrix}, \quad e^{-t} \begin{pmatrix} 1 \\ 0 \\ 0 \\ -1 \\ 1 \end{pmatrix}, \quad e^{t} \begin{pmatrix} 2 \\ 0 \\ 0 \\ 0 \\ 1 \end{pmatrix}, \quad e^{2t} \begin{pmatrix} 1 \\ 1 \\ -1 \\ 1 \\ 0 \end{pmatrix}.$$

Section 6.4

1.

$$A - \lambda I = \begin{pmatrix} -2 - \lambda & 2 & 1 \\ -2 & 2 - \lambda & 2 \\ 2 & -3 & -3 - \lambda \end{pmatrix}$$

implies

$$\det(A - \lambda I) = -\lambda^3 - 3\lambda^2 - 4\lambda - 2 = -(\lambda + 1)(\lambda^2 + 2\lambda + 2).$$

Therefore, the eigenvalues are given by $\lambda = -1$ and $\lambda = -1 \pm i$. First, for $\lambda = -1$, we have

$$A - \lambda I = \begin{pmatrix} -1 & 2 & 1 \\ -2 & 3 & 2 \\ 2 & -3 & -2 \end{pmatrix} \rightarrow \begin{pmatrix} 1 & -2 & -1 \\ 0 & 1 & 0 \\ 0 & 0 & 0 \end{pmatrix}$$

after elementary row operations. Therefore,

$$\mathbf{v_1} = \begin{pmatrix} 1 \\ 0 \\ 1 \end{pmatrix}$$

is an associated eigenvector, and

$$\mathbf{x_1}(t) = e^{-t} \begin{pmatrix} 1 \\ 0 \\ 1 \end{pmatrix}$$

is a solution of our system.

Next, for $\lambda = -1 + i$,

$$A - \lambda I = \begin{pmatrix} -1 - i & 2 & 1 \\ -2 & 3 - i & 2 \\ 2 & -3 & -2 - i \end{pmatrix} \rightarrow \begin{pmatrix} 1 & 0 & \frac{1}{2} - \frac{i}{2} \\ 0 & 1 & 1 \\ 0 & 0 & 0 \end{pmatrix}$$

after elementary row operations. Therefore,

$$\mathbf{v_2} = \begin{pmatrix} -\frac{1}{2} + \frac{i}{2} \\ 1 \\ -1 \end{pmatrix}$$

is an associated eigenvector. Further,

$$\mathbf{u}(t) = e^{(-1+i)t} \left[\begin{pmatrix} -\frac{1}{2} \\ 1 \\ -1 \end{pmatrix} + i \begin{pmatrix} \frac{1}{2} \\ 0 \\ 0 \end{pmatrix} \right]$$

38

is a solution of our system. We know that if $\mathbf{u}(t)$ is a solution, then $\mathrm{Re}(\mathbf{u})$ and $\mathrm{Im}(\mathbf{u})$ are also solutions. Consequently, we get the following two linearly independent solutions.

$$\mathbf{x_2}(t) = \mathrm{Re}(\mathbf{u}) = e^{-t}\begin{pmatrix} -\frac{1}{2}\cos t - \frac{1}{2}\sin t \\ \cos t \\ -\cos t \end{pmatrix}$$

and

$$\mathbf{x_3}(t) = \mathrm{Im}(\mathbf{u}) = e^{-t}\begin{pmatrix} -\frac{1}{2}\sin t + \frac{1}{2}\cos t \\ \sin t \\ -\sin t \end{pmatrix}.$$

We conclude that the general solution is given by

$$\mathbf{x}(t) = c_1 e^{-t}\begin{pmatrix} 1 \\ 0 \\ 1 \end{pmatrix} + c_2 e^{-t}\begin{pmatrix} -\frac{1}{2}\cos t - \frac{1}{2}\sin t \\ \cos t \\ -\cos t \end{pmatrix} + c_3 e^{-t}\begin{pmatrix} -\frac{1}{2}\sin t + \frac{1}{2}\cos t \\ \sin t \\ -\sin t \end{pmatrix}.$$

2.

$$A - \lambda I = \begin{pmatrix} 2-\lambda & -4 & -1 \\ 1 & 1-\lambda & 3 \\ 3 & -4 & -2-\lambda \end{pmatrix}$$

implies

$$\det(A - \lambda I) = -\lambda^3 + \lambda^2 - 15\lambda - 17 = -(\lambda+1)(\lambda^2 - 2\lambda + 17).$$

Therefore, the eigenvalues are given by $\lambda = -1$ and $\lambda = 1 \pm 4i$. First, for $\lambda = -1$, we have

$$A - \lambda I = \begin{pmatrix} 3 & -4 & -1 \\ 1 & 2 & 3 \\ 3 & -4 & -1 \end{pmatrix} \rightarrow \begin{pmatrix} 1 & 2 & 3 \\ 0 & 1 & 1 \\ 0 & 0 & 0 \end{pmatrix}$$

after elementary row operations. Therefore,

$$\mathbf{v_1} = \begin{pmatrix} 1 \\ 1 \\ -1 \end{pmatrix}$$

is an associated eigenvector, and

$$\mathbf{x_1}(t) = e^{-t}\begin{pmatrix} 1 \\ 1 \\ -1 \end{pmatrix}$$

is a solution of our system.

Next, for $\lambda = 1 + 4i$,

$$A - \lambda I = \begin{pmatrix} 1-4i & -4 & -1 \\ 1 & -4i & 3 \\ 3 & -4 & -3-4i \end{pmatrix} \rightarrow \begin{pmatrix} 1 & -4i & 3 \\ 0 & 1 & i \\ 0 & 0 & 0 \end{pmatrix}$$

after elementary row operations. Therefore,

$$\mathbf{v_2} = \begin{pmatrix} 1 \\ -i \\ 1 \end{pmatrix}$$

is an associated eigenvector. Further,

$$\mathbf{u}(t) = e^{(1+4i)t}\left[\begin{pmatrix} 1 \\ 0 \\ 1 \end{pmatrix} + i\begin{pmatrix} 0 \\ -1 \\ 0 \end{pmatrix} \right]$$

is a solution of our system. We know that if $\mathbf{u}(t)$ is a solution, then $\text{Re}(\mathbf{u})$ and $\text{Im}(\mathbf{u})$ are also solutions. Consequently, we get the following two linearly independent solutions.

$$\mathbf{x_2}(t) = \text{Re}(\mathbf{u}) = e^t \begin{pmatrix} \cos(4t) \\ \sin(4t) \\ \cos(4t) \end{pmatrix}$$

and

$$\mathbf{x_3}(t) = \text{Im}(\mathbf{u}) = e^t \begin{pmatrix} \sin(4t) \\ -\cos(4t) \\ \sin(4t) \end{pmatrix}.$$

We conclude that the general solution is given by

$$\mathbf{x}(t) = c_1 e^{-t} \begin{pmatrix} 1 \\ 1 \\ -1 \end{pmatrix} + c_2 e^t \begin{pmatrix} \cos(4t) \\ \sin(4t) \\ \cos(4t) \end{pmatrix} + c_3 e^t \begin{pmatrix} \sin(4t) \\ -\cos(4t) \\ \sin(4t) \end{pmatrix}.$$

3.
$$A - \lambda I = \begin{pmatrix} -\lambda & -2 & -1 \\ 1 & -1-\lambda & 1 \\ 1 & -2 & -2-\lambda \end{pmatrix}$$

implies

$$\det(A - \lambda I) = -\lambda^3 - 3\lambda^2 - 7\lambda - 5 = -(\lambda+1)(\lambda^2 + 2\lambda + 5).$$

Therefore, the eigenvalues are given by $\lambda = -1$ and $\lambda = -1 \pm 2i$. First, for $\lambda = -1$, we have

$$A - \lambda I = \begin{pmatrix} 1 & -2 & -1 \\ 1 & 0 & 1 \\ 1 & -2 & -1 \end{pmatrix} \rightarrow \begin{pmatrix} 1 & -2 & -1 \\ 0 & 1 & 1 \\ 0 & 0 & 0 \end{pmatrix}$$

after elementary row operations. Therefore,

$$\mathbf{v_1} = \begin{pmatrix} 1 \\ 1 \\ -1 \end{pmatrix}$$

is an associated eigenvector, and

$$\mathbf{x_1}(t) = e^{-t} \begin{pmatrix} 1 \\ 1 \\ -1 \end{pmatrix}$$

is a solution of our system.

Next, for $\lambda = -1 + 2i$,

$$A - \lambda I = \begin{pmatrix} 1-2i & -2 & -1 \\ 1 & -2i & 1 \\ 1 & -2 & -1-2i \end{pmatrix} \rightarrow \begin{pmatrix} 1 & -2i & 1 \\ 0 & 1 & i \\ 0 & 0 & 0 \end{pmatrix}$$

after elementary row operations. Therefore,

$$\mathbf{v_2} = \begin{pmatrix} 1 \\ -i \\ 1 \end{pmatrix}$$

is an associated eigenvector. Further,

$$\mathbf{u}(t) = e^{(-1+2i)t} \left[\begin{pmatrix} 1 \\ 0 \\ 1 \end{pmatrix} + i \begin{pmatrix} 0 \\ -1 \\ 0 \end{pmatrix} \right]$$

40

is a solution of our system. We know that if $\mathbf{u}(t)$ is a solution, then $\text{Re}(\mathbf{u})$ and $\text{Im}(\mathbf{u})$ are also solutions. Consequently, we get the following two linearly independent solutions.

$$\mathbf{x_2}(t) = \text{Re}(\mathbf{u}) = e^{-t} \begin{pmatrix} \cos(2t) \\ \sin(2t) \\ \cos(2t) \end{pmatrix}$$

and

$$\mathbf{x_3}(t) = \text{Im}(\mathbf{u}) = e^{-t} \begin{pmatrix} \sin(2t) \\ -\cos(2t) \\ \sin(2t) \end{pmatrix}.$$

We conclude that the general solution is given by

$$\mathbf{x}(t) = c_1 e^{-t} \begin{pmatrix} 1 \\ 1 \\ -1 \end{pmatrix} + c_2 e^{-t} \begin{pmatrix} \cos(2t) \\ \sin(2t) \\ \cos(2t) \end{pmatrix} + c_3 e^{-t} \begin{pmatrix} \sin(2t) \\ -\cos(2t) \\ \sin(2t) \end{pmatrix}.$$

4.

$$A - \lambda I = \begin{pmatrix} -4 - \lambda & 2 & -1 \\ -6 & -\lambda & -3 \\ 0 & 8/3 & -2 - \lambda \end{pmatrix}$$

implies

$$\det(A - \lambda I) = -\lambda^3 - 6\lambda^2 - 28\lambda - 40 = -(\lambda + 2)(\lambda^2 + 4\lambda + 20).$$

Therefore, the eigenvalues are given by $\lambda = -2$ and $\lambda = -2 \pm 4i$. First, for $\lambda = -2$, we have

$$A - \lambda I = \begin{pmatrix} -2 & 2 & -1 \\ -6 & 2 & -3 \\ 0 & 8/3 & 0 \end{pmatrix} \rightarrow \begin{pmatrix} 1 & -1 & 1/2 \\ 0 & 1 & 0 \\ 0 & 0 & 0 \end{pmatrix}$$

after elementary row operations. Therefore,

$$\mathbf{v_1} = \begin{pmatrix} 1 \\ 0 \\ -2 \end{pmatrix}$$

is an associated eigenvector, and

$$\mathbf{x_1}(t) = e^{-2t} \begin{pmatrix} 1 \\ 0 \\ -2 \end{pmatrix}$$

is a solution of our system.

Next, for $\lambda = -2 + 4i$,

$$A - \lambda I = \begin{pmatrix} -2 - 4i & 2 & -1 \\ -6 & 2 - 4i & -3 \\ 0 & 8/3 & -4i \end{pmatrix} \rightarrow \begin{pmatrix} 1 & \frac{-1+2i}{3} & \frac{1}{2} \\ 0 & 1 & -\frac{3}{2}i \\ 0 & 0 & 0 \end{pmatrix}$$

after elementary row operations. Therefore,

$$\mathbf{v_2} = \begin{pmatrix} \frac{1+i}{2} \\ \frac{3}{2}i \\ 1 \end{pmatrix}$$

is an associated eigenvector. Further,

$$\mathbf{u}(t) = e^{(-2+4i)t} \left[\begin{pmatrix} \frac{1}{2} \\ 0 \\ 1 \end{pmatrix} + i \begin{pmatrix} \frac{1}{2} \\ \frac{3}{2} \\ 0 \end{pmatrix} \right]$$

is a solution of our system. We know that if $\mathbf{u}(t)$ is a solution, then $\text{Re}(\mathbf{u})$ and $\text{Im}(\mathbf{u})$ are also solutions. Consequently, we get the following two linearly independent solutions.

$$\mathbf{x_2}(t) = \text{Re}(\mathbf{u}) = e^{-2t} \begin{pmatrix} \frac{1}{2}\cos(4t) - \frac{1}{2}\sin(4t) \\ -\frac{3}{2}\sin(4t) \\ \cos(4t) \end{pmatrix}$$

and

$$\mathbf{x_3}(t) = \text{Im}(\mathbf{u}) = e^{-2t} \begin{pmatrix} \frac{1}{2}\sin(4t) + \frac{1}{2}\cos(4t) \\ \frac{3}{2}\cos(4t) \\ \sin(4t) \end{pmatrix}.$$

We conclude that the general solution is given by

$$\mathbf{x}(t) = c_1 e^{-2t} \begin{pmatrix} 1 \\ 0 \\ -2 \end{pmatrix} + c_2 e^{-2t} \begin{pmatrix} \frac{1}{2}\cos(4t) - \frac{1}{2}\sin(4t) \\ -\frac{3}{2}\sin(4t) \\ \cos(4t) \end{pmatrix} + c_3 e^{-2t} \begin{pmatrix} \frac{1}{2}\sin(4t) + \frac{1}{2}\cos(4t) \\ \frac{3}{2}\cos(4t) \\ \sin(4t) \end{pmatrix}.$$

5.

$$A - \lambda I = \begin{pmatrix} -7 - \lambda & 6 & -6 \\ -9 & 5 - \lambda & -9 \\ 0 & -1 & -1 - \lambda \end{pmatrix}$$

implies

$$\det(A - \lambda I) = -\lambda^3 - 3\lambda^2 - 12\lambda - 10$$

Therefore, the eigenvalues are given by $\lambda = -1$ and $\lambda = -1 \pm 3i$. First, for $\lambda = -1$, we have

$$A - \lambda I = \begin{pmatrix} -6 & 6 & -6 \\ -9 & 6 & -9 \\ 0 & -1 & 0 \end{pmatrix} \rightarrow \begin{pmatrix} 1 & 0 & 1 \\ 0 & 1 & 0 \\ 0 & 0 & 0 \end{pmatrix}$$

after elementary row operations. Therefore,

$$\mathbf{v_1} = \begin{pmatrix} 1 \\ 0 \\ -1 \end{pmatrix}$$

is an associated eigenvector, and

$$\mathbf{x_1}(t) = e^{-t} \begin{pmatrix} 1 \\ 0 \\ -1 \end{pmatrix}$$

is a solution of our system.

Next, for $\lambda = -1 + 3i$,

$$A - \lambda I = \begin{pmatrix} -6 - 3i & 6 & -6 \\ -9 & 6 - 3i & -9 \\ 0 & -1 & -3i \end{pmatrix} \rightarrow \begin{pmatrix} 1 & 0 & 2 + 2i \\ 0 & 1 & 3i \\ 0 & 0 & 0 \end{pmatrix}$$

after elementary row operations. Therefore,

$$\mathbf{v_2} = \begin{pmatrix} 2 + 2i \\ 3i \\ -1 \end{pmatrix}$$

is an associated eigenvector. Further,

$$\mathbf{u}(t) = e^{(-1+3i)t} \left[\begin{pmatrix} 2 \\ 0 \\ -1 \end{pmatrix} + i \begin{pmatrix} 2 \\ 3 \\ 0 \end{pmatrix} \right]$$

is a solution of our system. We know that if $\mathbf{u}(t)$ is a solution, then $\text{Re}(\mathbf{u})$ and $\text{Im}(\mathbf{u})$ are also solutions. Consequently, we get the following two linearly independent solutions.

$$\mathbf{x_2}(t) = \text{Re}(\mathbf{u}) = e^{-t} \begin{pmatrix} 2\cos(3t) - 2\sin(3t) \\ -3\sin(3t) \\ -\cos(3t) \end{pmatrix}$$

and

$$\mathbf{x_3}(t) = \text{Im}(\mathbf{u}) = \begin{pmatrix} 2\sin(3t) + 2\cos(3t) \\ 3\cos(3t) \\ -\sin(3t) \end{pmatrix}.$$

We conclude that the general solution is given by

$$\mathbf{x}(t) = c_1 e^{-t} \begin{pmatrix} 1 \\ 0 \\ -1 \end{pmatrix} + c_2 e^{-t} \begin{pmatrix} 2\cos(3t) - 2\sin(3t) \\ -3\sin(3t) \\ -\cos(3t) \end{pmatrix} + c_3 \begin{pmatrix} 2\sin(3t) + 2\cos(3t) \\ 3\cos(3t) \\ -\sin(3t) \end{pmatrix}.$$

6.

$$A - \lambda I = \begin{pmatrix} \frac{1}{3} - \lambda & \frac{1}{3} & -\frac{5}{3} \\ -\frac{7}{3} & -\frac{1}{3} - \lambda & \frac{5}{3} \\ 1 & -1 & -1 - \lambda \end{pmatrix}$$

implies

$$\det(A - \lambda I) = -\lambda^3 - \lambda^2 - 4\lambda - 4 = -(\lambda + 1)(\lambda^2 + 4).$$

Therefore, the eigenvalues are given by $\lambda = -1$ and $\lambda = \pm 2i$. First, for $\lambda = -1$, we have

$$A - \lambda I = \begin{pmatrix} \frac{4}{3} & \frac{1}{3} & -\frac{5}{3} \\ -\frac{7}{3} & \frac{2}{3} & \frac{5}{3} \\ 1 & -1 & 0 \end{pmatrix} \rightarrow \begin{pmatrix} 1 & -1 & 0 \\ 0 & 1 & -1 \\ 0 & 0 & 0 \end{pmatrix}$$

after elementary row operations. Therefore,

$$\mathbf{v_1} = \begin{pmatrix} 1 \\ 1 \\ 1 \end{pmatrix}$$

is an associated eigenvector, and

$$\mathbf{x_1}(t) = e^{-t} \begin{pmatrix} 1 \\ 1 \\ 1 \end{pmatrix}$$

is a solution of our system.
  Next, for $\lambda = 2i$,

$$A - \lambda I = \begin{pmatrix} \frac{1}{3} - 2i & \frac{1}{3} & -\frac{5}{3} \\ -\frac{7}{3} & -\frac{1}{3} - 2i & \frac{5}{3} \\ 1 & -1 & -1 - 2i \end{pmatrix} \rightarrow \begin{pmatrix} 1 & -1 & -1 - 2i \\ 0 & 1 & 1 + i \\ 0 & 0 & 0 \end{pmatrix}$$

after elementary row operations. Therefore,

$$\mathbf{v_2} = \begin{pmatrix} i \\ -1 - i \\ 1 \end{pmatrix}$$

is an associated eigenvector. Further,

$$\mathbf{u}(t) = e^{2it} \left[\begin{pmatrix} 0 \\ -1 \\ 1 \end{pmatrix} + i \begin{pmatrix} 1 \\ -1 \\ 0 \end{pmatrix} \right]$$

is a solution of our system. We know that if $\mathbf{u}(t)$ is a solution, then $\mathrm{Re}(\mathbf{u})$ and $\mathrm{Im}(\mathbf{u})$ are also solutions. Consequently, we get the following two linearly independent solutions.

$$\mathbf{x_2}(t) = \mathrm{Re}(\mathbf{u}) = \begin{pmatrix} -\sin(2t) \\ -\cos(2t) + \sin(2t) \\ \cos(2t) \end{pmatrix}$$

and

$$\mathbf{x_3}(t) = \mathrm{Im}(\mathbf{u}) = \begin{pmatrix} \cos(2t) \\ -\sin(2t) - \cos(2t) \\ \sin(2t) \end{pmatrix}.$$

We conclude that the general solution is given by

$$\mathbf{x}(t) = c_1 e^{-t} \begin{pmatrix} 1 \\ 1 \\ 1 \end{pmatrix} + c_2 \begin{pmatrix} -\sin(2t) \\ -\cos(2t) + \sin(2t) \\ \cos(2t) \end{pmatrix} + c_3 \begin{pmatrix} \cos(2t) \\ -\sin(2t) - \cos(2t) \\ \sin(2t) \end{pmatrix}.$$

7.

$$A - \lambda I = \begin{pmatrix} 1 - \lambda & 1 & 1 \\ 2 & 1 - \lambda & -1 \\ -8 & -5 & -3 - \lambda \end{pmatrix}$$

implies

$$\det(A - \lambda I) = -\lambda^3 - \lambda^2 + 4\lambda + 4 = -(\lambda + 1)(\lambda^2 - 4).$$

Therefore, the eigenvalues are given by $\lambda = -1, 2, -2$. First, for $\lambda = -1$, we have

$$A - \lambda I = \begin{pmatrix} 2 & 1 & 1 \\ 2 & 2 & -1 \\ -8 & -5 & -2 \end{pmatrix} \rightarrow \begin{pmatrix} 2 & 1 & 1 \\ 0 & 1 & -2 \\ 0 & 0 & 0 \end{pmatrix}$$

after elementary row operations. Therefore,

$$\mathbf{v_1} = \begin{pmatrix} -\frac{3}{2} \\ 2 \\ 1 \end{pmatrix}$$

is an associated eigenvector, and

$$\mathbf{x_1}(t) = e^{-t} \begin{pmatrix} -\frac{3}{2} \\ 2 \\ 1 \end{pmatrix}$$

is a solution of our system.

Next, for $\lambda = 2$,

$$A - \lambda I = \begin{pmatrix} -1 & 1 & 1 \\ 2 & -1 & -1 \\ -8 & -5 & -5 \end{pmatrix} \rightarrow \begin{pmatrix} 1 & -1 & -1 \\ 0 & 1 & 1 \\ 0 & 0 & 0 \end{pmatrix}$$

after elementary row operations. Therefore,

$$\mathbf{v_2} = \begin{pmatrix} 0 \\ 1 \\ -1 \end{pmatrix}$$

is an associated eigenvector.

Then, for $\lambda = -2$,

$$A - \lambda I = \begin{pmatrix} 3 & 1 & 1 \\ 2 & 3 & -1 \\ -8 & -5 & -1 \end{pmatrix} \rightarrow \begin{pmatrix} 1 & \frac{1}{3} & \frac{1}{3} \\ 0 & 1 & -\frac{5}{7} \\ 0 & 0 & 0 \end{pmatrix}$$

44

after elementary row operations. Therefore,

$$\mathbf{v_3} = \begin{pmatrix} -\frac{4}{7} \\ \frac{5}{7} \\ 1 \end{pmatrix}$$

is an associated eigenvector. We conclude that the general solution is given by

$$\mathbf{x}(t) = c_1 e^{-t} \begin{pmatrix} -\frac{3}{2} \\ 2 \\ 1 \end{pmatrix} + c_2 e^{2t} \begin{pmatrix} 0 \\ 1 \\ -1 \end{pmatrix} + c_3 e^{-2t} \begin{pmatrix} -\frac{4}{7} \\ \frac{5}{7} \\ 1 \end{pmatrix}.$$

8.

$$A - \lambda I = \begin{pmatrix} 1 - \lambda & -1 & 4 \\ 3 & 2 - \lambda & -1 \\ 2 & 1 & -1 - \lambda \end{pmatrix}$$

implies

$$\det(A - \lambda I) = -\lambda^3 + 2\lambda^2 + 5\lambda + 6 = -(\lambda - 1)(\lambda - 3)(\lambda + 2).$$

Therefore, the eigenvalues are given by $\lambda = 1, 3, -2$. First, for $\lambda = 1$, we have

$$A - \lambda I = \begin{pmatrix} 0 & -1 & 4 \\ 3 & 1 & -1 \\ 2 & 1 & -2 \end{pmatrix} \rightarrow \begin{pmatrix} 1 & \frac{1}{2} & -1 \\ 0 & 1 & -4 \\ 0 & 0 & 0 \end{pmatrix}$$

after elementary row operations. Therefore,

$$\mathbf{v_1} = \begin{pmatrix} -1 \\ 4 \\ 1 \end{pmatrix}$$

is an associated eigenvector, and

$$\mathbf{x_1}(t) = e^t \begin{pmatrix} -1 \\ 4 \\ 1 \end{pmatrix}$$

is a solution of our system.
Next, for $\lambda = 3$,

$$A - \lambda I = \begin{pmatrix} -2 & -1 & 4 \\ 3 & -1 & -1 \\ 2 & 1 & -4 \end{pmatrix} \rightarrow \begin{pmatrix} 1 & \frac{1}{2} & -2 \\ 0 & 1 & -2 \\ 0 & 0 & 0 \end{pmatrix}$$

after elementary row operations. Therefore,

$$\mathbf{v_2} = \begin{pmatrix} 1 \\ 2 \\ 1 \end{pmatrix}$$

is an associated eigenvector, and

$$\mathbf{x_2}(t) = e^{3t} \begin{pmatrix} 1 \\ 2 \\ 1 \end{pmatrix}$$

is a solution of our system.
Then, for $\lambda = -2$,

$$A - \lambda I = \begin{pmatrix} 3 & -1 & 4 \\ 3 & 4 & -1 \\ 2 & 1 & 1 \end{pmatrix} \rightarrow \begin{pmatrix} 1 & \frac{1}{2} & \frac{1}{2} \\ 0 & 1 & -1 \\ 0 & 0 & 0 \end{pmatrix}$$

after elementary row operations. Therefore,

$$\mathbf{v_3} = \begin{pmatrix} -1 \\ 1 \\ 1 \end{pmatrix}$$

is an associated eigenvector, and

$$\mathbf{x_3}(t) = e^{-2t} \begin{pmatrix} -1 \\ 1 \\ 1 \end{pmatrix}$$

is a solution of our system. We conclude that the general solution is given by

$$\mathbf{x}(t) = c_1 e^t \begin{pmatrix} -1 \\ 4 \\ 1 \end{pmatrix} + c_2 e^{3t} \begin{pmatrix} 1 \\ 2 \\ 1 \end{pmatrix} + c_3 e^{-2t} \begin{pmatrix} -1 \\ 1 \\ 1 \end{pmatrix}.$$

9.

(a)

$$A - \lambda I = \begin{pmatrix} \frac{3}{4} - \lambda & \frac{29}{4} & -\frac{11}{2} \\ -\frac{3}{4} & \frac{3}{4} - \lambda & -\frac{5}{2} \\ \frac{5}{4} & \frac{11}{4} & -\frac{5}{2} - \lambda \end{pmatrix}$$

implies

$$\det(A - \lambda I) = -\lambda^3 - \lambda^2 - 16\lambda - 16 = -(\lambda + 1)(\lambda^2 + 16).$$

Therefore, the eigenvalues are given by $\lambda = -1, \pm 4i$.

If $\lambda = -1$, then

$$A - \lambda I = \begin{pmatrix} \frac{3}{4} + 1 & \frac{29}{4} & -\frac{11}{2} \\ -\frac{3}{4} & \frac{3}{4} + 1 & -\frac{5}{2} \\ \frac{5}{4} & \frac{11}{4} & -\frac{5}{2} + 1 \end{pmatrix} \rightarrow \begin{pmatrix} 1 & 0 & 1 \\ 0 & 1 & -1 \\ 0 & 0 & 0 \end{pmatrix}.$$

Therefore,

$$\mathbf{v_1} = \begin{pmatrix} 1 \\ -1 \\ -1 \end{pmatrix}$$

is an associated eigenvector. If $\lambda = 4i$, then

$$A - \lambda I = \begin{pmatrix} \frac{3}{4} - 4i & \frac{29}{4} & -\frac{11}{2} \\ -\frac{3}{4} & \frac{3}{4} - 4i & -\frac{5}{2} \\ \frac{5}{4} & \frac{11}{4} & -\frac{5}{2} - 4i \end{pmatrix} \rightarrow \begin{pmatrix} 1 & 0 & -2 - i \\ 0 & 1 & -i \\ 0 & 0 & 0 \end{pmatrix}.$$

Therefore,

$$\mathbf{v} = \begin{pmatrix} 2 + i \\ i \\ 1 \end{pmatrix}$$

is an associated eigenvector, and

$$\mathbf{u}(t) = e^{4it} \left[\begin{pmatrix} 2 \\ 0 \\ 1 \end{pmatrix} + i \begin{pmatrix} 1 \\ 1 \\ 0 \end{pmatrix} \right]$$

is a solution of our system. We can write this solution as two linearly independent real-valued solutions by looking at the real and imaginary parts of $\mathbf{u}$. In particular, we see that

$$\mathbf{x_1}(t) = \text{Re}(\mathbf{u}) = \begin{pmatrix} 2 \\ 0 \\ 1 \end{pmatrix} \cos(4t) - \begin{pmatrix} 1 \\ 1 \\ 0 \end{pmatrix} \sin(4t)$$

46

and

$$\mathbf{x_2}(t) = \mathrm{Im}(\mathbf{u}) = \begin{pmatrix} 2 \\ 0 \\ 1 \end{pmatrix} \sin(4t) + \begin{pmatrix} 1 \\ 1 \\ 0 \end{pmatrix} \cos(4t)$$

are two linearly independent solutions. We conclude that the general solution is given by

$$\mathbf{x}(t) = c_1 e^{-t} \begin{pmatrix} 1 \\ -1 \\ -1 \end{pmatrix} + c_2 \left[\begin{pmatrix} 2 \\ 0 \\ 1 \end{pmatrix} \cos(4t) - \begin{pmatrix} 1 \\ 1 \\ 0 \end{pmatrix} \sin(4t) \right] + c_3 \left[\begin{pmatrix} 2 \\ 0 \\ 1 \end{pmatrix} \sin(4t) + \begin{pmatrix} 1 \\ 1 \\ 0 \end{pmatrix} \cos(4t) \right]$$

If $c_1 = 0$, then the solution will lie on a closed curve in the plane spanned by

$$\mathbf{a} = \begin{pmatrix} 2 \\ 0 \\ 1 \end{pmatrix} \qquad \mathbf{b} = \begin{pmatrix} 1 \\ 1 \\ 0 \end{pmatrix}.$$

(b)

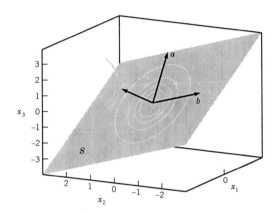

(c) For $\mathbf{x_0} \notin S$, as $t \to \infty$, the solution will tend towards one of the closed curves in S, spiraling about the line spanned by the eigenvector $(1, -1, -1)^T$.

10.

(a) The eigenvalues of this matrix are given by $\lambda = \pm 3i, -1 \pm 2i$. As in the solution to #9 above, the vectors $\mathbf{a}$ and $\mathbf{b}$ satisfying the requested condition will be given by looking at the eigenvectors for the purely imaginary eigenvalues. Here, we see that the purely imaginary eigenvalues are $\pm 3i$. For $\lambda = 3i$,

$$\mathbf{v} = \begin{pmatrix} 1 \\ i \\ i \\ 1 \end{pmatrix}$$

is an associated eigenvector. As in the solution to #9 above, the real and imaginary parts of $\mathbf{v}$ will give us the desired vectors $\mathbf{a}$ and $\mathbf{b}$. In particular, we have

$$\mathbf{a} = \begin{pmatrix} 1 \\ 0 \\ 0 \\ 1 \end{pmatrix} \qquad \mathbf{b} = \begin{pmatrix} 0 \\ 1 \\ 1 \\ 0 \end{pmatrix}.$$

For $\mathbf{x_0} \in S$, where S is the plane spanned by $\mathbf{a}$ and $\mathbf{b}$, the solution will be given by

$$\mathbf{x}(t) = c_1 \left[\begin{pmatrix} 1 \\ 0 \\ 0 \\ 1 \end{pmatrix} \cos(3t) - \begin{pmatrix} 0 \\ 1 \\ 1 \\ 0 \end{pmatrix} \sin(3t) \right] + c_2 \left[\begin{pmatrix} 1 \\ 0 \\ 0 \\ 1 \end{pmatrix} \sin(3t) + \begin{pmatrix} 0 \\ 1 \\ 1 \\ 0 \end{pmatrix} \cos(3t) \right],$$

and, therefore, solutions will stay on a closed curve in S.

(b) We consider the other eigenvalues $-1 \pm 2i$. For $\lambda = -1 + 2i$, an associated eigenvector is

$$\mathbf{v} = \begin{pmatrix} -1 - i \\ -1 + i \\ 1 \\ -2 - i \end{pmatrix}.$$

Therefore,

$$\mathbf{u}(t) = e^{(-1+2i)t} \left[\begin{pmatrix} -1 \\ -1 \\ 1 \\ -2 \end{pmatrix} + i \begin{pmatrix} -1 \\ 1 \\ 0 \\ -1 \end{pmatrix} \right]$$

is a solution associated with this eigenvector. Looking at the real and imaginary parts of $\mathbf{u}$, we get the following two solutions of our system. We have

$$\mathbf{x_1}(t) = e^{-t} \left[\begin{pmatrix} -1 \\ -1 \\ 1 \\ -2 \end{pmatrix} \cos(2t) - \begin{pmatrix} -1 \\ 1 \\ 0 \\ -1 \end{pmatrix} \sin(2t) \right]$$

and

$$\mathbf{x_2}(t) = e^{-t} \left[\begin{pmatrix} -1 \\ -1 \\ 1 \\ -2 \end{pmatrix} \sin(2t) + \begin{pmatrix} -1 \\ 1 \\ 0 \\ -1 \end{pmatrix} \cos(2t) \right].$$

Using those solutions given by the eigenvalues $\pm 3i$, found in part (a) above, we conclude that the general solution of our system is

$$\mathbf{x}(t) = c_1 \left[\begin{pmatrix} 1 \\ 0 \\ 0 \\ 1 \end{pmatrix} \cos(3t) - \begin{pmatrix} 0 \\ 1 \\ 1 \\ 0 \end{pmatrix} \sin(3t) \right] + c_2 \left[\begin{pmatrix} 1 \\ 0 \\ 0 \\ 1 \end{pmatrix} \sin(3t) + \begin{pmatrix} 0 \\ 1 \\ 1 \\ 0 \end{pmatrix} \cos(3t) \right]$$

$$+ c_3 e^{-t} \left[\begin{pmatrix} -1 \\ -1 \\ 1 \\ -2 \end{pmatrix} \cos(2t) - \begin{pmatrix} -1 \\ 1 \\ 0 \\ -1 \end{pmatrix} \sin(2t) \right] + c_4 e^{-t} \left[\begin{pmatrix} -1 \\ -1 \\ 1 \\ -2 \end{pmatrix} \sin(2t) + \begin{pmatrix} -1 \\ 1 \\ 0 \\ -1 \end{pmatrix} \cos(2t) \right].$$

Since the components associated with e^{-t} will decay to zero as $t \to \infty$, we see that if $\mathbf{x_0} \notin S$, then $\mathbf{x}(t) \to S$ as $t \to \infty$.

11.

(a) Suppose that $c_1 \mathbf{a} + c_2 \mathbf{b} = 0$. Since $\mathbf{a}$ and $\mathbf{b}$ are the real and imaginary parts of the vector $\mathbf{v_1}$, $\mathbf{a} = (\mathbf{v_1} + \overline{\mathbf{v}}_1)/2$ and $\mathbf{b} = (\mathbf{v_1} - \overline{\mathbf{v}}_1)/2i$. Therefore,

$$c_1(\mathbf{v_1} + \overline{\mathbf{v}}_1) - ic_2(\mathbf{v_1} - \overline{\mathbf{v}}_1) = 0,$$

which leads to

$$(c_1 - ic_2)\mathbf{v_1} + (c_1 + ic_2)\overline{\mathbf{v}}_1 = 0.$$

(b) Since $\mathbf{v_1}$ and $\overline{\mathbf{v}}_1$ are linearly independent, we must have

$$c_1 - ic_2 = 0$$
$$c_1 + ic_2 = 0.$$

It follows that $c_1 = c_2 = 0$.

(c) Consider the equation $c_1\mathbf{x_1}(t_0) + c_2\mathbf{x_2}(t_0) = 0$. Using equation (4), we can then write

$$c_1 e^{\mu t_0}(\mathbf{a}\cos\nu t_0 - \mathbf{b}\sin\nu t_0) + c_2 e^{\mu t_0}(\mathbf{a}\sin\nu t_0 + \mathbf{b}\cos\nu t_0) = 0.$$

Rearranging the terms and dividing by the exponential,

$$(c_1 + c_2)\cos(\nu t_0)\mathbf{a} + (c_2 - c_1)\sin(\nu t_0)\mathbf{b} = 0.$$

From part (b), since $\mathbf{a}$ and $\mathbf{b}$ are linearly independent, it follows that

$$(c_1 + c_2)\cos(\nu t_0) = (c_2 - c_1)\sin(\nu t_0) = 0.$$

Without loss of generality, we may assume that the trigonometric factors are nonzero. We then conclude that $c_1 + c_2 = 0$ and $c_2 - c_1 = 0$, which leads to $c_1 = c_2 = 0$. Therefore, $\mathbf{x_1}(t)$ and $\mathbf{x_2}(t)$ are linearly independent at the point t_0 and therefore at every point.

12.

(a) Set $\mathbf{y} = (y_1, y_2)^T$. We can rewrite Eqns. (14) in the form

$$\begin{pmatrix} 2 & 0 \\ 0 & 9/4 \end{pmatrix}\begin{pmatrix} y_1'' \\ y_2'' \end{pmatrix} = \begin{pmatrix} -4 & 3 \\ 3 & -27/4 \end{pmatrix}\begin{pmatrix} y_1 \\ y_2 \end{pmatrix}.$$

Multiplying both sides of this equation by the inverse of the diagonal matrix, we have

$$\begin{pmatrix} y_1'' \\ y_2'' \end{pmatrix} = \begin{pmatrix} -2 & 3/2 \\ 4/3 & -3 \end{pmatrix}\begin{pmatrix} y_1 \\ y_2 \end{pmatrix}.$$

(b) Substituting $\mathbf{y} = e^{\omega t}\mathbf{u}$, we have

$$\omega^2\begin{pmatrix} u_1 \\ u_2 \end{pmatrix}e^{\omega t} = \begin{pmatrix} -2 & 3/2 \\ 4/3 & -3 \end{pmatrix}\begin{pmatrix} u_1 \\ u_2 \end{pmatrix}e^{\omega t}.$$

We can rewrite this equation as

$$(\mathbf{B} - \omega^2 I)\mathbf{u} = 0.$$

(c) The eigenvalues of B are $\omega_1^2 = -1$ and $\omega_2^2 = -4$ with corresponding eigenvectors

$$\mathbf{v_1} = \begin{pmatrix} 3 \\ 2 \end{pmatrix}, \qquad \mathbf{v_2} = \begin{pmatrix} 3 \\ -4 \end{pmatrix},$$

respectively.

(d) The linearly independent solutions are

$$\mathbf{y_1}(t) = C_1\begin{pmatrix} 3 \\ 2 \end{pmatrix}e^{it}$$

$$\mathbf{y_2}(t) = C_2\begin{pmatrix} 3 \\ -4 \end{pmatrix}e^{2it}.$$

Writing the solutions in terms of real-valued functions, we find that the component functions are given by

$$y_1 = 3c_1\cos(t) + 3c_2\sin(t) + 3c_3\cos(2t) + 3c_4\sin(2t)$$
$$y_2 = 2c_1\cos(t) + 2c_2\sin(t) - 4c_3\cos(2t) - 4c_4\sin(2t).$$

(e) Differentiating these expressions, we see that

$$y_1' = -3c_1\sin(t) + 3c_2\cos(t) - 6c_3\sin(2t) + 6c_4\cos(2t)$$
$$y_2' = -2c_1\sin(t) + 2c_2\cos(t) + 8c_3\sin(2t) - 8c_4\cos(2t).$$

13.

(a)

$$A - \lambda I = \begin{pmatrix} -\lambda & 0 & 1 & 0 \\ 0 & -\lambda & 0 & 1 \\ -2 & 1 & -\lambda & 0 \\ 1 & -2 & 0 & -\lambda \end{pmatrix}$$

implies

$$\det(A - \lambda I) = \lambda^4 + 4\lambda^2 + 3 = (\lambda^2 + 3)(\lambda^2 + 1).$$

Therefore, the eigenvalues of A are $\pm\sqrt{3}i$ and $\pm i$.

First, for $\lambda = i$,

$$A - \lambda I = \begin{pmatrix} -i & 0 & 1 & 0 \\ 0 & -i & 0 & 1 \\ -2 & 1 & -i & 0 \\ 1 & -2 & 0 & -i \end{pmatrix} \rightarrow \begin{pmatrix} 1 & -2 & 0 & -i \\ 0 & 1 & 0 & i \\ 0 & 0 & 1 & -1 \\ 0 & 0 & 0 & 0 \end{pmatrix},$$

after elementary row operations. Therefore,

$$\mathbf{v_1} = \begin{pmatrix} -i \\ -i \\ 1 \\ 1 \end{pmatrix}$$

is an associated eigenvector for $\lambda = i$. Since $A\mathbf{v_1} = i\mathbf{v_1}$, then $A\overline{\mathbf{v_1}} = -i\overline{\mathbf{v_1}}$. Therefore,

$$\mathbf{v_2} = \begin{pmatrix} i \\ i \\ 1 \\ 1 \end{pmatrix}$$

is an eigenvector associated with $\lambda = -i$.

Next, for $\lambda = \sqrt{3}i$,

$$A - \lambda I = \begin{pmatrix} -\sqrt{3}i & 0 & 1 & 0 \\ 0 & -\sqrt{3}i & 0 & 1 \\ -2 & 1 & -\sqrt{3}i & 0 \\ 1 & -2 & 0 & -\sqrt{3}i \end{pmatrix} \rightarrow \begin{pmatrix} 1 & -2 & 0 & -\sqrt{3}i \\ 0 & 1 & 0 & \frac{1}{\sqrt{3}}i \\ 0 & 0 & 1 & 1 \\ 0 & 0 & 0 & 0 \end{pmatrix},$$

after elementary row operations. Therefore,

$$\mathbf{v_3} = \begin{pmatrix} -\frac{1}{\sqrt{3}}i \\ \frac{1}{\sqrt{3}}i \\ 1 \\ -1 \end{pmatrix}$$

is an associated eigenvector for $\lambda = \sqrt{3}i$. Since $A\mathbf{v_3} = i\mathbf{v_3}$, then $A\overline{\mathbf{v_1}} = -\sqrt{3}i\overline{\mathbf{v_1}}$. Therefore,

$$\mathbf{v_4} = \begin{pmatrix} \frac{1}{\sqrt{3}}i \\ -\frac{1}{\sqrt{3}}i \\ 1 \\ -1 \end{pmatrix}$$

is an eigenvector associated with $\lambda = -\sqrt{3}i$.

(b) Let $\mathbf{u_1}(t)$ be the solution associated with $\lambda = i$. From part (a), we see that

$$\mathbf{u_1}(t) = e^{it} \begin{pmatrix} -i \\ -i \\ 1 \\ 1 \end{pmatrix}.$$

Taking the real and imaginary parts of $\mathbf{u_1}$, we have the following two linearly independent real-valued solutions,

$$\mathbf{x_1}(t) = \mathrm{Re}(\mathbf{u_1}) = \begin{pmatrix} 0 \\ 0 \\ 1 \\ 1 \end{pmatrix} \cos(t) - \begin{pmatrix} -1 \\ -1 \\ 0 \\ 0 \end{pmatrix} \sin t$$

and

$$\mathbf{x_2}(t) = \mathrm{Im}(\mathbf{u_1}) = \begin{pmatrix} 0 \\ 0 \\ 1 \\ 1 \end{pmatrix} \sin t + \begin{pmatrix} -1 \\ -1 \\ 0 \\ 0 \end{pmatrix} \cos t.$$

Then, let $\mathbf{u_2}(t)$ be the solution associated with $\lambda = \sqrt{3}i$. From part (a), we see that

$$\mathbf{u_2}(t) = e^{\sqrt{3}it} \begin{pmatrix} -\frac{1}{\sqrt{3}}i \\ \frac{1}{\sqrt{3}}i \\ 1 \\ -1 \end{pmatrix}.$$

Taking the real and imaginary parts of $\mathbf{u_2}$, we have two more solutions of our system,

$$\mathbf{x_3}(t) = \mathrm{Re}(\mathbf{u_2}) = \begin{pmatrix} 0 \\ 0 \\ 1 \\ -1 \end{pmatrix} \cos(\sqrt{3}t) - \begin{pmatrix} -\frac{1}{\sqrt{3}} \\ \frac{1}{\sqrt{3}} \\ 0 \\ 0 \end{pmatrix} \sin(\sqrt{3}t)$$

and

$$\mathbf{x_4}(t) = \mathrm{Im}(\mathbf{u_2}) = \begin{pmatrix} 0 \\ 0 \\ 1 \\ -1 \end{pmatrix} \sin(\sqrt{3}t) + \begin{pmatrix} -\frac{1}{\sqrt{3}} \\ \frac{1}{\sqrt{3}} \\ 0 \\ 0 \end{pmatrix} \cos(\sqrt{3}t).$$

Therefore, the general solution of our system is

$$\mathbf{x}(t) = c_1 \begin{pmatrix} \sin t \\ \sin t \\ \cos t \\ \cos t \end{pmatrix} + c_2 \begin{pmatrix} -\cos t \\ -\cos t \\ \sin t \\ \sin t \end{pmatrix} + c_3 \begin{pmatrix} \sin(\sqrt{3}t) \\ -\sin(\sqrt{3}t) \\ \sqrt{3}\cos(\sqrt{3}t) \\ -\sqrt{3}\cos(\sqrt{3}t) \end{pmatrix} + c_4 \begin{pmatrix} -\cos(\sqrt{3}t) \\ \cos(\sqrt{3}t) \\ \sqrt{3}\sin(\sqrt{3}t) \\ -\sqrt{3}\sin(\sqrt{3}t) \end{pmatrix}$$

(c) Both masses will vibrate at the lowest frequency $\omega_1 = 1$ if we choose $c_3 = c_4 = 0$. In this case, the first mode of vibration occurs for any c_1 and c_2 not both zero and

$$x_1(t) = c_1 \sin(t) - c_2 \cos(t) = x_2(t).$$

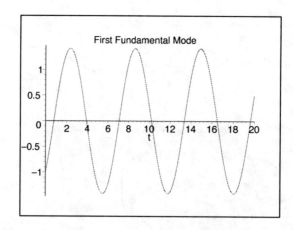

Both masses will vibrate at the second frequency $\omega_2 = \sqrt{3}$ if we choose $c_1 = c_2 = 0$. In this case, the first mode of vibration occurs for any c_3 and c_4 not both zero and

$$x_1(t) = c_3 \sin(\sqrt{3}t) - c_4 \cos(\sqrt{3}t)$$
$$x_2(t) = -c_3 \sin(\sqrt{3}t) + c_4 \cos(\sqrt{3}t).$$

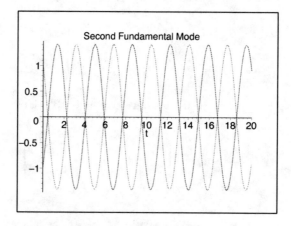

(d) Using the initial conditions and the general solution found in part (b), we have

$$\mathbf{x}(0) = c_1 \begin{pmatrix} 0 \\ 0 \\ 1 \\ 1 \end{pmatrix} + c_2 \begin{pmatrix} -1 \\ -1 \\ 0 \\ 0 \end{pmatrix} + c_3 \begin{pmatrix} 0 \\ 0 \\ \sqrt{3} \\ -\sqrt{3} \end{pmatrix} + c_4 \begin{pmatrix} -1 \\ 1 \\ 0 \\ 0 \end{pmatrix} = \begin{pmatrix} -1 \\ 3 \\ 0 \\ 0 \end{pmatrix}.$$

The solution of this system is $c_1 = 0$, $c_2 = -1$, $c_3 = 0$ and $c_4 = 2$. Therefore, the solution satisfying the given initial condition is

$$\mathbf{x}(t) = \begin{pmatrix} \cos t \\ \cos t \\ -\sin t \\ -\sin t \end{pmatrix} + 2 \begin{pmatrix} -\cos(\sqrt{3}t) \\ \cos(\sqrt{3}t) \\ \sqrt{3}\sin(\sqrt{3}t) \\ -\sqrt{3}\sin(\sqrt{3}t) \end{pmatrix}$$

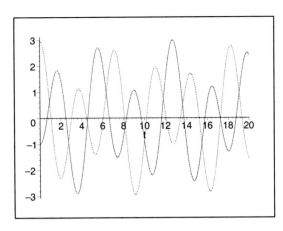

(e) In the case when the initial condition is $\mathbf{x}(0) = (1, 1, 0, 1)^T$, the solution is

$$\mathbf{x}(t) = \frac{1}{2}\begin{pmatrix} \sin t \\ \sin t \\ \cos t \\ \cos t \end{pmatrix} - \begin{pmatrix} -\cos t \\ -\cos t \\ \sin t \\ \sin t \end{pmatrix} - \frac{1}{2\sqrt{3}}\begin{pmatrix} \sin(\sqrt{3}t) \\ -\sin(\sqrt{3}t) \\ \sqrt{3}\cos(\sqrt{3}t) \\ -\sqrt{3}\cos(\sqrt{3}t) \end{pmatrix}$$

Plots of x_1 and x_2 are shown below:

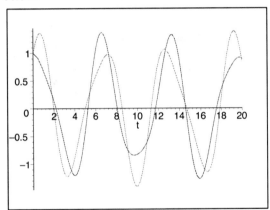

14.

$$A - \lambda I = \begin{pmatrix} -\lambda & 0 & 1 & 0 \\ 0 & -\lambda & 0 & 1 \\ -2k/m & k/m & -\lambda & 0 \\ k/m & -2k/m & 0 & -\lambda \end{pmatrix}$$

Therefore,

$$\det(A - \lambda I) = \lambda^4 + 4\frac{k}{m}\lambda^2 + 3\frac{k^2}{m^2} = \lambda^4 + 4\omega^2\lambda^2 + 3\omega^4 = (\lambda^2 + 3\omega^2)(\lambda^2 + \omega^2).$$

Therefore, the eigenvalues are given by $\lambda = \pm\omega i, \pm\sqrt{3}\omega i$. The eigenvector for $\lambda = \omega i$ is

$$\mathbf{v_1} = \begin{pmatrix} -\omega i \\ -\omega i \\ 1 \\ 1 \end{pmatrix}$$

Therefore,

$$\mathbf{u_1}(t) = e^{i\omega t}\begin{pmatrix} -\omega i \\ -\omega i \\ 1 \\ 1 \end{pmatrix}$$

53

is one solution. Taking the real and imaginary parts, we have

$$\mathbf{x_1}(t) = \text{Re}(\mathbf{u_1}(t)) = \begin{pmatrix} 0 \\ 0 \\ 1 \\ 1 \end{pmatrix} \cos(\omega t) - \begin{pmatrix} -\omega \\ -\omega \\ 0 \\ 0 \end{pmatrix} \sin(\omega t)$$

and

$$\mathbf{x_2}(t) = \text{Im}(\mathbf{u_1}(t)) = \begin{pmatrix} 0 \\ 0 \\ 1 \\ 1 \end{pmatrix} \sin(\omega t) + \begin{pmatrix} -\omega \\ -\omega \\ 0 \\ 0 \end{pmatrix} \cos(\omega t).$$

The eigenvector for $\lambda = \sqrt{3}\omega i$ is

$$\mathbf{v_2} = \begin{pmatrix} -\frac{1}{\sqrt{3}}\omega i \\ \frac{1}{\sqrt{3}}\omega i \\ 1 \\ -1 \end{pmatrix}.$$

Therefore,

$$\mathbf{u_2}(t) = e^{i\sqrt{3}\omega t} \begin{pmatrix} -\frac{1}{\sqrt{3}}\omega i \\ \frac{1}{\sqrt{3}}\omega i \\ 1 \\ -1 \end{pmatrix}$$

is another solution. Taking the real and imaginary parts, we have

$$\mathbf{x_3}(t) = \text{Re}(\mathbf{u_2})(t) = \begin{pmatrix} 0 \\ 0 \\ 1 \\ -1 \end{pmatrix} \cos(\sqrt{3}\omega t) - \begin{pmatrix} -\frac{1}{\sqrt{3}}\omega \\ \frac{1}{\sqrt{3}}\omega \\ 0 \\ 0 \end{pmatrix} \sin(\sqrt{3}\omega t)$$

and

$$\mathbf{x_4}(t) = \text{Im}(\mathbf{u_2})(t) = \begin{pmatrix} 0 \\ 0 \\ 1 \\ -1 \end{pmatrix} \sin(\sqrt{3}\omega t) + \begin{pmatrix} -\frac{1}{\sqrt{3}}\omega \\ \frac{1}{\sqrt{3}}\omega \\ 0 \\ 0 \end{pmatrix} \cos(\sqrt{3}\omega t).$$

Therefore, the general solution is

$$\mathbf{x}(t) = c_1 \begin{pmatrix} \omega\sin(\omega t) \\ \omega\sin(\omega t) \\ \cos(\omega t) \\ \cos(\omega t) \end{pmatrix} + c_2 \begin{pmatrix} -\omega\cos(\omega t) \\ -\omega\cos(\omega t) \\ \sin(\omega t) \\ \sin(\omega t) \end{pmatrix} + c_3 \begin{pmatrix} \omega\sin(\sqrt{3}\omega t) \\ -\omega\sin(\sqrt{3}\omega t) \\ \sqrt{3}\cos(\sqrt{3}\omega t) \\ -\sqrt{3}\cos(\sqrt{3}\omega t) \end{pmatrix} + c_4 \begin{pmatrix} -\omega\cos(\sqrt{3}\omega t) \\ \omega\cos(\sqrt{3}\omega t) \\ \sqrt{3}\sin(\sqrt{3}\omega t) \\ -\sqrt{3}\sin(\sqrt{3}\omega t) \end{pmatrix}.$$

The two frequencies of vibration are ω and $\sqrt{3}\omega$. The first mode corresponds to ω and the second mode corresponds to $\sqrt{3}\omega$.

15. The eigenvalues are given by $-1, 1, -1 \pm 4i$. The corresponding eigenvectors are

$$\mathbf{v_1} = \begin{pmatrix} -1 \\ 0 \\ 0 \\ 1 \end{pmatrix}, \quad \mathbf{v_2} = \begin{pmatrix} 1 \\ 1 \\ 0 \\ 1 \end{pmatrix}, \quad \mathbf{v_3} = \begin{pmatrix} 1 \\ i \\ 1 \\ i \end{pmatrix}, \quad \mathbf{v_4} = \begin{pmatrix} 1 \\ -i \\ 1 \\ -i \end{pmatrix}.$$

Therefore, a fundamental set of solutions is given by

$$e^{-t} \begin{pmatrix} -1 \\ 0 \\ 0 \\ 1 \end{pmatrix}, \quad e^{t} \begin{pmatrix} 1 \\ 1 \\ 0 \\ 1 \end{pmatrix}, \quad e^{(-1+4i)t} \begin{pmatrix} 1 \\ i \\ 1 \\ i \end{pmatrix}, \quad e^{(-1-4i)t} \begin{pmatrix} 1 \\ -i \\ 1 \\ -i \end{pmatrix}.$$

In order to write the solutions as real-valued solutions, we look at the solution given from the eigenvalue $-1 + 4i$, and take the real and imaginary parts of that solution. In particular, for

$$\mathbf{u}(t) = e^{(-1+4i)t} \begin{pmatrix} 1 \\ i \\ 1 \\ i \end{pmatrix},$$

we have the two linearly independent real-valued solutions

$$\mathbf{x_1}(t) = \mathrm{Re}(\mathbf{u}(t)) = e^{-t} \left[\begin{pmatrix} 1 \\ 0 \\ 1 \\ 0 \end{pmatrix} \cos(4t) - \begin{pmatrix} 0 \\ 1 \\ 0 \\ 1 \end{pmatrix} \sin(4t) \right]$$

and

$$\mathbf{x_2}(t) = \mathrm{Im}(\mathbf{u}(t)) = e^{-t} \left[\begin{pmatrix} 1 \\ 0 \\ 1 \\ 0 \end{pmatrix} \sin(4t) + \begin{pmatrix} 0 \\ 1 \\ 0 \\ 1 \end{pmatrix} \cos(4t) \right].$$

Therefore, we conclude that a fundamental set of real-valued solutions is given by

$$e^{-t} \begin{pmatrix} -1 \\ 0 \\ 0 \\ 1 \end{pmatrix}, \quad e^{t} \begin{pmatrix} 1 \\ 1 \\ 0 \\ 1 \end{pmatrix}, \quad e^{-t} \begin{pmatrix} \cos(4t) \\ -\sin(4t) \\ \cos(4t) \\ -\sin(4t) \end{pmatrix}, \quad e^{-t} \begin{pmatrix} \sin(4t) \\ \cos(4t) \\ \sin(4t) \\ \cos(4t) \end{pmatrix}.$$

16. The eigenvalues are approximately $-2 \pm i, -1 \pm 2i$. The corresponding eigenvectors are approximately

$$\mathbf{v_1} = \begin{pmatrix} \frac{1}{2} - \frac{1}{2}i \\ 1 \\ 1 \\ 0 \end{pmatrix}, \quad \mathbf{v_2} = \begin{pmatrix} \frac{1}{2} + \frac{1}{2}i \\ 1 \\ 1 \\ 0 \end{pmatrix}, \quad \mathbf{v_3} = \begin{pmatrix} -1 - i \\ -1 + i \\ 1 \\ -1 \end{pmatrix}, \quad \mathbf{v_4} = \begin{pmatrix} -1 + i \\ -1 - i \\ 1 \\ -1 \end{pmatrix}.$$

Therefore, a fundamental set of solutions is given by Looking at $\mathbf{v}_1$ and $\mathbf{v}_3$, we have the following two solutions:

$$\mathbf{u_1}(t) = e^{(-2+i)t} \left[\begin{pmatrix} \frac{1}{2} \\ 1 \\ 1 \\ 0 \end{pmatrix} + i \begin{pmatrix} -\frac{1}{2} \\ 0 \\ 0 \\ 0 \end{pmatrix} \right]$$

$$\mathbf{u_2}(t) = e^{(-1+2i)t} \left[\begin{pmatrix} -1 \\ -1 \\ 1 \\ -1 \end{pmatrix} + i \begin{pmatrix} -1 \\ 1 \\ 0 \\ 0 \end{pmatrix} \right].$$

Looking at the real and imaginary parts of $\mathbf{u_1}(t)$, we have the following two linearly independent real-valued solutions,

$$\mathbf{x_1}(t) = e^{-2t} \begin{pmatrix} \frac{1}{2}\cos t + \frac{1}{2}\sin t \\ \cos t \\ \cos t \\ 0 \end{pmatrix}$$

$$\mathbf{x_2}(t) = e^{-2t} \begin{pmatrix} \frac{1}{2}\sin t - \frac{1}{2}\cos t \\ \sin t \\ \sin t \\ 0 \end{pmatrix}.$$

Looking at the real and imaginary parts of $\mathbf{u_2}(t)$, we have the following two linearly independent real-valued solutions,

$$\mathbf{x_3}(t) = e^{-t} \begin{pmatrix} -\cos(2t) + \sin(2t) \\ -\cos(2t) - \sin(2t) \\ \cos(2t) \\ -\cos(2t) \end{pmatrix}$$

$$\mathbf{x_4}(t) = e^{-t} \begin{pmatrix} -\sin(2t) - \cos(2t) \\ -\sin(2t) + \cos(2t) \\ \sin(2t) \\ -\sin(2t) \end{pmatrix}.$$

17. The eigenvalues are given by $\pm 2i, \pm 3i$. An eigenvector for $2i$ is given by

$$\mathbf{v_1} = \begin{pmatrix} 1 + i \\ 1 - i \\ -1 \\ 1 \end{pmatrix}.$$

Therefore, one solution of the system is given by

$$\mathbf{u_1}(t) = e^{2it} \begin{pmatrix} 1 + i \\ 1 - i \\ -1 \\ 1 \end{pmatrix}.$$

Taking the real and imaginary parts of $\mathbf{u_1}$, we have the two linearly independent real-valued solutions

$$\mathbf{x_1}(t) = \begin{pmatrix} 1 \\ 1 \\ -1 \\ 1 \end{pmatrix} \cos(2t) - \begin{pmatrix} 1 \\ -1 \\ 0 \\ 0 \end{pmatrix} \sin(2t)$$

and

$$\mathbf{x_2}(t) = \begin{pmatrix} 1 \\ 1 \\ -1 \\ 1 \end{pmatrix} \sin(2t) + \begin{pmatrix} 1 \\ -1 \\ 0 \\ 0 \end{pmatrix} \cos(2t).$$

Next, we look at an eigenvector for $3i$. An associated eigenvector is

$$\mathbf{v_2} = \begin{pmatrix} 1 \\ 1 + i \\ 1 + i \\ 0 \end{pmatrix}.$$

Taking the real and imaginary parts, we have two more real-valued solutions,

$$\mathbf{x_3}(t) = \begin{pmatrix} 1 \\ 1 \\ 1 \\ 0 \end{pmatrix} \cos(3t) - \begin{pmatrix} 0 \\ 1 \\ 1 \\ 0 \end{pmatrix} \sin(3t)$$

and

$$\mathbf{x_4}(t) = \begin{pmatrix} 1 \\ 1 \\ 1 \\ 0 \end{pmatrix} \sin(3t) + \begin{pmatrix} 0 \\ 1 \\ 1 \\ 0 \end{pmatrix} \cos(3t).$$

56

Therefore, a fundamental set of solutions is given by

$$\begin{pmatrix} \cos(2t) - \sin(2t) \\ \cos(2t) + \sin(2t) \\ -\cos(2t) \\ \cos(2t) \end{pmatrix}, \begin{pmatrix} \sin(2t) + \cos(2t) \\ \sin(2t) - \cos(2t) \\ -\sin(2t) \\ \sin(2t) \end{pmatrix}, \begin{pmatrix} \cos(3t) \\ \cos(3t) - \sin(3t) \\ \cos(3t) - \sin(3t) \\ 0 \end{pmatrix}, \begin{pmatrix} \sin(3t) \\ \sin(3t) + \cos(3t) \\ \sin(3t) + \cos(3t) \\ 0 \end{pmatrix}.$$

18. The eigenvalues are given by $\lambda = -1, -2, -2 \pm 3i$. Eigenvectors associated with $\lambda = -1, -2$ are given by

$$\mathbf{v_1} = \begin{pmatrix} -1 \\ 1 \\ 0 \\ 1 \end{pmatrix}, \quad \mathbf{v_2} = \begin{pmatrix} 1 \\ 0 \\ -1 \\ 1 \end{pmatrix},$$

respectively. Therefore, solutions associated with these eigenvalues are given by

$$e^{-t} \begin{pmatrix} -1 \\ 1 \\ 0 \\ 1 \end{pmatrix}, \quad e^{-2t} \begin{pmatrix} 1 \\ 0 \\ -1 \\ 1 \end{pmatrix}.$$

An eigenvector for $\lambda = -2 + 3i$ is given by

$$\mathbf{v_3} = \begin{pmatrix} 1 \\ i \\ i \\ 1 \end{pmatrix}.$$

Therefore,

$$\mathbf{u}(t) = e^{(-2+3i)t} \begin{pmatrix} 1 \\ i \\ i \\ 1 \end{pmatrix}.$$

is another solution of the system. Taking the real and imaginary parts of $\mathbf{u}$, we have the following two linearly independent real-valued solutions,

$$\mathbf{x_3}(t) = e^{-2t} \left[\begin{pmatrix} 1 \\ 0 \\ 0 \\ 1 \end{pmatrix} \cos(3t) - \begin{pmatrix} 0 \\ 1 \\ 1 \\ 0 \end{pmatrix} \sin(3t) \right]$$

and

$$\mathbf{x_4}(t) = e^{-2t} \left[\begin{pmatrix} 1 \\ 0 \\ 0 \\ 1 \end{pmatrix} \sin(3t) + \begin{pmatrix} 0 \\ 1 \\ 1 \\ 0 \end{pmatrix} \cos(3t) \right].$$

Therefore, a fundamental set of solutions is given by

$$e^t \begin{pmatrix} -1 \\ 1 \\ 0 \\ 1 \end{pmatrix}, \quad e^{-2t} \begin{pmatrix} 1 \\ 0 \\ -1 \\ 1 \end{pmatrix}, e^{-2t} \begin{pmatrix} \cos(3t) \\ -\sin(3t) \\ -\sin(3t) \\ \cos(3t) \end{pmatrix}, e^{-2t} \begin{pmatrix} \sin(3t) \\ \cos(3t) \\ \cos(3t) \\ \sin(3t) \end{pmatrix}.$$

Section 6.5

1.

$$A - \lambda I = \begin{pmatrix} 3 - \lambda & -2 \\ 2 & -2 - \lambda \end{pmatrix}$$

implies
$$\det(A - \lambda I) = \lambda^2 - \lambda - 2.$$
Therefore, the eigenvalues are $\lambda = 2, -1$. For $\lambda = 2$,
$$A - \lambda I = \begin{pmatrix} 1 & -2 \\ 2 & -4 \end{pmatrix} \to \begin{pmatrix} 1 & -2 \\ 0 & 0 \end{pmatrix}.$$
therefore,
$$\mathbf{v_1} = \begin{pmatrix} 2 \\ 1 \end{pmatrix}$$
is an eigenvector for $\lambda = 2$. Therefore,
$$\mathbf{x_1}(t) = \begin{pmatrix} 2 \\ 1 \end{pmatrix} e^{2t}$$
is a solution of the system. For $\lambda = -1$,
$$A - \lambda I = \begin{pmatrix} 4 & -2 \\ 2 & -1 \end{pmatrix} \to \begin{pmatrix} 2 & -1 \\ 0 & 0 \end{pmatrix}.$$
therefore,
$$\mathbf{v_2} = \begin{pmatrix} 1 \\ 2 \end{pmatrix}$$
is an eigenvector for $\lambda = -1$. Therefore,
$$\mathbf{x_2}(t) = \begin{pmatrix} 1 \\ 2 \end{pmatrix} e^{-t}$$
is a solution of the system. We conclude that
$$\mathbf{X}(t) = \begin{pmatrix} e^{-t} & 2e^{2t} \\ 2e^{-t} & e^{2t} \end{pmatrix}$$
is a fundamental matrix for this system. The fundamental matrix e^{At} is given by $e^{At} = \mathbf{X}(t)\mathbf{X}^{-1}(0)$. Therefore,
$$e^{At} = \frac{1}{3} \begin{pmatrix} e^{-t} & 2e^{2t} \\ 2e^{-t} & e^{2t} \end{pmatrix} \begin{pmatrix} -1 & 2 \\ 2 & -1 \end{pmatrix} = \frac{1}{3} \begin{pmatrix} -e^{-t} + 4e^{2t} & 2e^{-t} - 2e^{2t} \\ -2e^{-t} + 2e^{2t} & 4e^{-t} - e^{2t} \end{pmatrix}.$$

2.
$$A - \lambda I = \begin{pmatrix} -\frac{3}{4} - \lambda & \frac{1}{2} \\ \frac{1}{8} & -\frac{3}{4} - \lambda \end{pmatrix}$$
implies
$$\det(A - \lambda I) = \lambda^2 + \frac{3}{2}\lambda + \frac{1}{2}.$$
Therefore, the eigenvalues are $\lambda = -1/2, -1$. For $\lambda = -1/2$,
$$A - \lambda I = \begin{pmatrix} -\frac{1}{4} & \frac{1}{2} \\ \frac{1}{8} & -\frac{1}{4} \end{pmatrix} \to \begin{pmatrix} 1 & -2 \\ 0 & 0 \end{pmatrix}.$$
therefore,
$$\mathbf{v_1} = \begin{pmatrix} 2 \\ 1 \end{pmatrix}$$
is an eigenvector for $\lambda = -1/2$. Therefore,
$$\mathbf{x_1}(t) = \begin{pmatrix} 2 \\ 1 \end{pmatrix} e^{-t/2}$$

58

is a solution of the system. For $\lambda = -1$,

$$A - \lambda I = \begin{pmatrix} \frac{1}{4} & \frac{1}{2} \\ \frac{1}{8} & \frac{1}{4} \end{pmatrix} \rightarrow \begin{pmatrix} 1 & 2 \\ 0 & 0 \end{pmatrix}.$$

Therefore,

$$\mathbf{v_2} = \begin{pmatrix} 2 \\ -1 \end{pmatrix}$$

is an eigenvector for $\lambda = -1$. Therefore,

$$\mathbf{x_2}(t) = \begin{pmatrix} 2 \\ -1 \end{pmatrix} e^{-t}$$

is a solution of the system. We conclude that

$$\mathbf{X}(t) = \begin{pmatrix} 2e^{-t/2} & 2e^{-t} \\ e^{-t/2} & -e^{-t} \end{pmatrix}$$

is a fundamental matrix for this system. The fundamental matrix e^{At} is given by $e^{At} = \mathbf{X}(t)\mathbf{X}^{-1}(0)$. Therefore,

$$e^{At} = \frac{1}{4} \begin{pmatrix} 2e^{-t/2} & 2e^{-t} \\ e^{-t/2} & -e^{-t} \end{pmatrix} \begin{pmatrix} 1 & 2 \\ 1 & -2 \end{pmatrix} = \frac{1}{4} \begin{pmatrix} 2e^{-t/2} + 2e^{-t} & 4e^{-t/2} - 4e^{-t} \\ e^{-t/2} - e^{-t} & 2e^{-t/2} + 2e^{-t} \end{pmatrix}.$$

3.

$$A - \lambda I = \begin{pmatrix} 3 - \lambda & -4 \\ 1 & -1 - \lambda \end{pmatrix}$$

implies

$$\det(A - \lambda I) = \lambda^2 - 2\lambda + 1 = (\lambda - 1)^2.$$

Therefore, $\lambda = 1$ is the only eigenvalue. For $\lambda = 1$,

$$A - \lambda I = \begin{pmatrix} 2 & -4 \\ 1 & -2 \end{pmatrix} \rightarrow \begin{pmatrix} 1 & -2 \\ 0 & 0 \end{pmatrix}.$$

therefore,

$$\mathbf{v} = \begin{pmatrix} 2 \\ 1 \end{pmatrix}$$

is an eigenvector for $\lambda = 1$. Therefore,

$$\mathbf{x_1}(t) = \begin{pmatrix} 2 \\ 1 \end{pmatrix} e^t$$

is a solution of the system. To find another solution, we need to look for a vector $\mathbf{w}$ satisfying $(A - I)\mathbf{w} = \mathbf{v}$. That is, we need to find $\mathbf{w}$ such that

$$\begin{pmatrix} 2 & -4 \\ 1 & -2 \end{pmatrix} \mathbf{w} = \begin{pmatrix} 2 \\ 1 \end{pmatrix}.$$

The vector

$$\mathbf{w} = \begin{pmatrix} 1 \\ 0 \end{pmatrix}$$

satisfies this equation. Therefore,

$$\mathbf{x_2}(t) = \begin{pmatrix} 2 \\ 1 \end{pmatrix} te^t + \begin{pmatrix} 1 \\ 0 \end{pmatrix} e^t$$

is a second solution. We conclude that

$$\mathbf{X}(t) = \begin{pmatrix} 2e^t & 2te^t + e^t \\ e^t & te^t \end{pmatrix}$$

is a fundamental matrix for this system. The fundamental matrix e^{At} is given by $e^{At} = \mathbf{X}(t)\mathbf{X}^{-1}(0)$. Therefore,

$$e^{At} = \begin{pmatrix} 2e^t & 2te^t + e^t \\ e^t & te^t \end{pmatrix} \begin{pmatrix} 0 & 1 \\ 1 & -2 \end{pmatrix} = \begin{pmatrix} 2te^t + e^t & -4te^t \\ te^t & e^t - 2te^t \end{pmatrix}$$

4.

$$A - \lambda I = \begin{pmatrix} 4 - \lambda & -2 \\ 8 & -4 - \lambda \end{pmatrix}$$

implies

$$\det(A - \lambda I) = \lambda^2.$$

Therefore, $\lambda = 0$ is the only eigenvalue. For $\lambda = 0$,

$$A - \lambda I = \begin{pmatrix} 4 & -2 \\ 8 & -4 \end{pmatrix} \rightarrow \begin{pmatrix} 2 & -1 \\ 0 & 0 \end{pmatrix}.$$

therefore,

$$\mathbf{v} = \begin{pmatrix} 1 \\ 2 \end{pmatrix}$$

is an eigenvector for $\lambda = 0$. Therefore,

$$\mathbf{x_1}(t) = \begin{pmatrix} 1 \\ 2 \end{pmatrix}$$

is a solution of the system. To find another solution, we need to look for a vector $\mathbf{w}$ satisfying $A\mathbf{w} = \mathbf{v}$. That is, we need to find $\mathbf{w}$ such that

$$\begin{pmatrix} 4 & -2 \\ 8 & -4 \end{pmatrix} \mathbf{w} = \begin{pmatrix} 1 \\ 2 \end{pmatrix}.$$

The vector

$$\mathbf{w} = \begin{pmatrix} \frac{1}{4} \\ 0 \end{pmatrix}$$

satisfies this equation. Therefore,

$$\mathbf{x_2}(t) = t \begin{pmatrix} 1 \\ 2 \end{pmatrix} + \begin{pmatrix} \frac{1}{4} \\ 0 \end{pmatrix}$$

is a second solution. We conclude that

$$\mathbf{X}(t) = \begin{pmatrix} 1 & t + \frac{1}{4} \\ 2 & 2t \end{pmatrix}$$

is a fundamental matrix for this system. The fundamental matrix e^{At} is given by $e^{At} = \mathbf{X}(t)\mathbf{X}^{-1}(0)$. Therefore,

$$e^{At} = \begin{pmatrix} 1 & t + \frac{1}{4} \\ 2 & 2t \end{pmatrix} \begin{pmatrix} 0 & \frac{1}{2} \\ 4 & -2 \end{pmatrix} = \begin{pmatrix} 1 + 4t & -2t \\ 8t & 1 - 4t \end{pmatrix}$$

5.

$$A - \lambda I = \begin{pmatrix} 2 - \lambda & -5 \\ 1 & -2 - \lambda \end{pmatrix}$$

implies

$$\det(A - \lambda I) = \lambda^2 + 1.$$

Therefore, the eigenvalues are $\lambda = \pm i$. For $\lambda = i$,

$$A - \lambda I = \begin{pmatrix} 2 - i & -5 \\ 1 & -2 - i \end{pmatrix} \rightarrow \begin{pmatrix} 1 & -2 - i \\ 0 & 0 \end{pmatrix}.$$

therefore,

$$\mathbf{v_1} = \begin{pmatrix} 2 + i \\ 1 \end{pmatrix}$$

is an eigenvector for $\lambda = i$. Therefore,

$$\mathbf{x_1}(t) = e^{it} \begin{pmatrix} 2+i \\ 1 \end{pmatrix}$$

is a solution of the system. Taking the real and imaginary parts of $\mathbf{u}(t)$, we have the two linearly independent real-valued solutions

$$\mathbf{x_1}(t) = \begin{pmatrix} 2\cos t - \sin t \\ \cos t \end{pmatrix}$$

$$\mathbf{x_2}(t) = \begin{pmatrix} 2\sin t + \cos t \\ \sin t \end{pmatrix}.$$

We conclude that

$$\mathbf{X}(t) = \begin{pmatrix} 2\cos t - sint & 2\sin t + \cos t \\ \cos t & \sin t \end{pmatrix}$$

is a fundamental matrix for this system. The fundamental matrix e^{At} is given by $e^{At} = \mathbf{X}(t)\mathbf{X}^{-1}(0)$. Therefore,

$$e^{At} = \begin{pmatrix} 2\cos t - sint & 2\sin t + \cos t \\ \cos t & \sin t \end{pmatrix} \begin{pmatrix} 0 & 1 \\ 1 & -2 \end{pmatrix} = \begin{pmatrix} 2\sin t + \cos t & -5\sin t \\ \sin t & \cos t - 2\sin t \end{pmatrix}.$$

6.

$$A - \lambda I = \begin{pmatrix} -1 - \lambda & -4 \\ 1 & -1 - \lambda \end{pmatrix}$$

implies

$$\det(A - \lambda I) = \lambda^2 + 2\lambda + 5.$$

Therefore, the eigenvalues are $\lambda = -1 \pm 2i$. For $\lambda = -1 + 2i$,

$$A - \lambda I = \begin{pmatrix} -2i & -4 \\ 1 & -2i \end{pmatrix} \rightarrow \begin{pmatrix} 1 & -2i \\ 0 & 0 \end{pmatrix}.$$

therefore,

$$\mathbf{v_1} = \begin{pmatrix} 2i \\ 1 \end{pmatrix}$$

is an eigenvector for $\lambda = -1 + 2i$. Therefore,

$$\mathbf{x_1}(t) = e^{(-1+2i)t} \begin{pmatrix} 2i \\ 1 \end{pmatrix}$$

is a solution of the system. Taking the real and imaginary parts of $\mathbf{u}(t)$, we have the two linearly independent real-valued solutions

$$\mathbf{x_1}(t) = e^{-t} \begin{pmatrix} -2\sin(2t) \\ \cos(2t) \end{pmatrix}$$

$$\mathbf{x_2}(t) = e^{-t} \begin{pmatrix} 2\cos(2t) \\ \sin(2t) \end{pmatrix}.$$

We conclude that

$$\mathbf{X}(t) = e^{-t} \begin{pmatrix} -2sin(2t) & 2\cos(2t) \\ \cos(2t) & \sin(2t) \end{pmatrix}$$

is a fundamental matrix for this system. The fundamental matrix e^{At} is given by $e^{At} = \mathbf{X}(t)\mathbf{X}^{-1}(0)$. Therefore,

$$e^{At} = \frac{1}{2}e^{-t} \begin{pmatrix} -2sin(2t) & 2\cos(2t) \\ \cos(2t) & \sin(2t) \end{pmatrix} \begin{pmatrix} 0 & 2 \\ 1 & 0 \end{pmatrix} = \frac{1}{2}e^{-t} \begin{pmatrix} 2\cos(2t) & -4\sin(2t) \\ \sin(2t) & 2\cos(2t) \end{pmatrix}.$$

7.

$$A - \lambda I = \begin{pmatrix} 5 - \lambda & -1 \\ 3 & 1 - \lambda \end{pmatrix}$$

implies

$$\det(A - \lambda I) = \lambda^2 - 6\lambda + 8.$$

Therefore, the eigenvalues are $\lambda = 2, 4$. For $\lambda = 2$,

$$A - \lambda I = \begin{pmatrix} 3 & -1 \\ 3 & -1 \end{pmatrix} \rightarrow \begin{pmatrix} 3 & -1 \\ 0 & 0 \end{pmatrix}.$$

therefore,

$$\mathbf{v_1} = \begin{pmatrix} 1 \\ 3 \end{pmatrix}$$

is an eigenvector for $\lambda = 2$. Therefore,

$$\mathbf{x_1}(t) = \begin{pmatrix} 1 \\ 3 \end{pmatrix} e^{2t}$$

is a solution of the system. For $\lambda = 4$,

$$A - \lambda I = \begin{pmatrix} 1 & -1 \\ 3 & -3 \end{pmatrix} \rightarrow \begin{pmatrix} 1 & -1 \\ 0 & 0 \end{pmatrix}.$$

Therefore,

$$\mathbf{v_2} = \begin{pmatrix} 1 \\ 1 \end{pmatrix}$$

is an eigenvector for $\lambda = 4$. Therefore,

$$\mathbf{x_2}(t) = \begin{pmatrix} 1 \\ 1 \end{pmatrix} e^{4t}$$

is a solution of the system. We conclude that

$$\mathbf{X}(t) = \begin{pmatrix} e^{2t} & e^{4t} \\ 3e^{2t} & e^{4t} \end{pmatrix}$$

is a fundamental matrix for this system. The fundamental matrix e^{At} is given by $e^{At} = \mathbf{X}(t)\mathbf{X}^{-1}(0)$. Therefore,

$$e^{At} = \frac{1}{2} \begin{pmatrix} e^{2t} & e^{4t} \\ 3e^{2t} & e^{4t} \end{pmatrix} \begin{pmatrix} -1 & 1 \\ 3 & -1 \end{pmatrix} = \frac{1}{2} \begin{pmatrix} -e^{2t} + 3e^{4t} & e^{2t} - e^{4t} \\ -3e^{2t} + 3e^{4t} & 3e^{2t} - e^{4t} \end{pmatrix}.$$

8.

$$A - \lambda I = \begin{pmatrix} 1 - \lambda & -1 \\ 5 & -3 - \lambda \end{pmatrix}$$

implies

$$\det(A - \lambda I) = \lambda^2 + 2\lambda + 2.$$

Therefore, the eigenvalues are $\lambda = -1 \pm i$. For $\lambda = -1 + i$,

$$A - \lambda I = \begin{pmatrix} 2 - i & -1 \\ 5 & -2 - i \end{pmatrix} \rightarrow \begin{pmatrix} 5 & -2 - i \\ 0 & 0 \end{pmatrix}.$$

therefore,

$$\mathbf{v_1} = \begin{pmatrix} 2 + i \\ 5 \end{pmatrix}$$

is an eigenvector for $\lambda = -1 + i$. Therefore,

$$\mathbf{x_1}(t) = e^{(-1+i)t} \begin{pmatrix} 2 + i \\ 5 \end{pmatrix}$$

is a solution of the system. Taking the real and imaginary parts of $\mathbf{u}(t)$, we have the two linearly independent real-valued solutions

$$\mathbf{x_1}(t) = e^{-t}\begin{pmatrix} 2\cos(t) - \sin(t) \\ 5\cos(t) \end{pmatrix}$$

$$\mathbf{x_2}(t) = e^{-t}\begin{pmatrix} 2\sin(t) + \cos(t) \\ 5\sin(t) \end{pmatrix}.$$

We conclude that

$$\mathbf{X}(t) = e^{-t}\begin{pmatrix} 2\cos(t) - \sin(t) & 2\sin(t) + \cos(t) \\ 5\cos(t) & 5\sin(t) \end{pmatrix}$$

is a fundamental matrix for this system. The fundamental matrix e^{At} is given by $e^{At} = \mathbf{X}(t)\mathbf{X}^{-1}(0)$. Therefore,

$$e^{At} = \frac{1}{5}e^{-t}\begin{pmatrix} 2\cos(t) - \sin(t) & 2\sin(t) + \cos(t) \\ 5\cos(t) & 5\sin(t) \end{pmatrix}\begin{pmatrix} 0 & 1 \\ 5 & -2 \end{pmatrix} = e^{-t}\begin{pmatrix} 2\sin(t) + \cos(t) & -\sin(t) \\ 5\sin(t) & \cos(t) - 2\sin(t) \end{pmatrix}.$$

9.

$$A - \lambda I = \begin{pmatrix} 2-\lambda & -1 \\ 3 & -2-\lambda \end{pmatrix}$$

implies

$$\det(A - \lambda I) = \lambda^2 - 1.$$

Therefore, the eigenvalues are $\lambda = 1, -1$. For $\lambda = 1$,

$$A - \lambda I = \begin{pmatrix} 1 & -1 \\ 3 & -3 \end{pmatrix} \to \begin{pmatrix} 1 & -1 \\ 0 & 0 \end{pmatrix}.$$

Therefore,

$$\mathbf{v_1} = \begin{pmatrix} 1 \\ 1 \end{pmatrix}$$

is an eigenvector for $\lambda = 1$. Therefore,

$$\mathbf{x_1}(t) = \begin{pmatrix} 1 \\ 1 \end{pmatrix}e^t$$

is a solution of the system. For $\lambda = -1$,

$$A - \lambda I = \begin{pmatrix} 3 & -1 \\ 3 & -1 \end{pmatrix} \to \begin{pmatrix} 3 & -1 \\ 0 & 0 \end{pmatrix}.$$

Therefore,

$$\mathbf{v_2} = \begin{pmatrix} 1 \\ 3 \end{pmatrix}$$

is an eigenvector for $\lambda = -1$. Therefore,

$$\mathbf{x_2}(t) = \begin{pmatrix} 1 \\ 3 \end{pmatrix}e^{-t}$$

is a solution of the system. We conclude that

$$\mathbf{X}(t) = \begin{pmatrix} e^t & e^{-t} \\ e^t & 3e^{-t} \end{pmatrix}$$

is a fundamental matrix for this system. The fundamental matrix e^{At} is given by $e^{At} = \mathbf{X}(t)\mathbf{X}^{-1}(0)$. Therefore,

$$e^{At} = \frac{1}{2}\begin{pmatrix} e^t & e^{-t} \\ e^t & 3e^{-t} \end{pmatrix}\begin{pmatrix} 3 & -1 \\ -1 & 1 \end{pmatrix} = \frac{1}{2}\begin{pmatrix} 3e^t - e^{-t} & -e^t + e^{-t} \\ 3e^t - 3e^{-t} & -e^t + 3e^{-t} \end{pmatrix}.$$

10.

$$A - \lambda I = \begin{pmatrix} 1 - \lambda & 1 \\ 4 & -2 - \lambda \end{pmatrix}$$

implies

$$\det(A - \lambda I) = \lambda^2 + \lambda - 6.$$

Therefore, the eigenvalues are $\lambda = 2, -3$. For $\lambda = 2$,

$$A - \lambda I = \begin{pmatrix} -1 & 1 \\ 4 & -4 \end{pmatrix} \to \begin{pmatrix} 1 & -1 \\ 0 & 0 \end{pmatrix}.$$

Therefore,

$$\mathbf{v_1} = \begin{pmatrix} 1 \\ 1 \end{pmatrix}$$

is an eigenvector for $\lambda = 2$. Therefore,

$$\mathbf{x_1}(t) = \begin{pmatrix} 1 \\ 1 \end{pmatrix} e^{2t}$$

is a solution of the system. For $\lambda = -3$,

$$A - \lambda I = \begin{pmatrix} 4 & 1 \\ 4 & 1 \end{pmatrix} \to \begin{pmatrix} 4 & 1 \\ 0 & 0 \end{pmatrix}.$$

Therefore,

$$\mathbf{v_2} = \begin{pmatrix} 1 \\ -4 \end{pmatrix}$$

is an eigenvector for $\lambda = -3$. Therefore,

$$\mathbf{x_2}(t) = \begin{pmatrix} 1 \\ -4 \end{pmatrix} e^{-3t}$$

is a solution of the system. We conclude that

$$\mathbf{X}(t) = \begin{pmatrix} e^{2t} & e^{-3t} \\ e^{2t} & -4e^{-3t} \end{pmatrix}$$

is a fundamental matrix for this system. The fundamental matrix e^{At} is given by $e^{At} = \mathbf{X}(t)\mathbf{X}^{-1}(0)$. Therefore,

$$e^{At} = \frac{1}{5} \begin{pmatrix} e^{2t} & e^{-3t} \\ e^{2t} & -4e^{-3t} \end{pmatrix} \begin{pmatrix} 4 & 1 \\ 1 & -1 \end{pmatrix} = \frac{1}{5} \begin{pmatrix} 4e^{2t} + e^{-3t} & e^{2t} - e^{-3t} \\ 4e^{2t} - 4e^{-3t} & e^{2t} + 4e^{-3t} \end{pmatrix}.$$

11.

$$A - \lambda I = \begin{pmatrix} -\frac{3}{2} - \lambda & 1 \\ -\frac{1}{4} & -\frac{1}{2} - \lambda \end{pmatrix}$$

implies

$$\det(A - \lambda I) = \lambda^2 + 2\lambda + 1 = (\lambda + 1)^2.$$

Therefore, $\lambda = -1$ is the only eigenvalue. For $\lambda = -1$,

$$A - \lambda I = \begin{pmatrix} -\frac{1}{2} & 1 \\ -\frac{1}{4} & \frac{1}{2} \end{pmatrix} \to \begin{pmatrix} 1 & -2 \\ 0 & 0 \end{pmatrix}.$$

Therefore,

$$\mathbf{v} = \begin{pmatrix} 2 \\ 1 \end{pmatrix}$$

is an eigenvector for $\lambda = -1$. Therefore,

$$\mathbf{x_1}(t) = \begin{pmatrix} 2 \\ 1 \end{pmatrix} e^{-t}$$

is a solution of the system. To find another solution, we need to look for a vector $\mathbf{w}$ satisfying $(A+I)\mathbf{w} = \mathbf{v}$. That is, we need to find $\mathbf{w}$ such that

$$\begin{pmatrix} -\frac{1}{2} & 1 \\ -\frac{1}{4} & \frac{1}{2} \end{pmatrix} \mathbf{w} = \begin{pmatrix} 2 \\ 1 \end{pmatrix}.$$

The vector

$$\mathbf{w} = \begin{pmatrix} -4 \\ 0 \end{pmatrix}$$

satisfies this equation. Therefore,

$$\mathbf{x_2}(t) = \begin{pmatrix} 2 \\ 1 \end{pmatrix} te^{-t} + \begin{pmatrix} -4 \\ 0 \end{pmatrix} e^{-t}$$

is a second solution. We conclude that

$$X(t) = \begin{pmatrix} 2e^{-t} & 2te^{-t} - 4e^{-t} \\ e^{-t} & te^{-t} \end{pmatrix}$$

is a fundamental matrix for this system. The fundamental matrix e^{At} is given by $e^{At} = X(t)X^{-1}(0)$. Therefore,

$$e^{At} = \begin{pmatrix} 2e^{-t} & 2te^{-t} - 4e^{-t} \\ e^{-t} & te^{-t} \end{pmatrix} \begin{pmatrix} 0 & 1 \\ -\frac{1}{4} & \frac{1}{2} \end{pmatrix} = \begin{pmatrix} e^{-t} - \frac{1}{2}te^{-t} & te^{-t} \\ -\frac{1}{4}te^{-t} & e^{-t} + \frac{1}{2}te^{-t} \end{pmatrix}.$$

12.

$$A - \lambda I = \begin{pmatrix} -3 - \lambda & \frac{5}{2} \\ -\frac{5}{2} & 2 - \lambda \end{pmatrix}$$

implies

$$\det(A - \lambda I) = \lambda^2 + \lambda + \frac{1}{4} = \left(\lambda + \frac{1}{2}\right)^2.$$

Therefore, $\lambda = -1/2$ is the only eigenvalue. For $\lambda = -1/2$,

$$A - \lambda I = \begin{pmatrix} -\frac{5}{2} & \frac{5}{2} \\ -\frac{5}{2} & \frac{5}{2} \end{pmatrix} \rightarrow \begin{pmatrix} 1 & -1 \\ 0 & 0 \end{pmatrix}.$$

Therefore,

$$\mathbf{v} = \begin{pmatrix} 1 \\ 1 \end{pmatrix}$$

is an eigenvector for $\lambda = -1/2$. Therefore,

$$\mathbf{x_1}(t) = \begin{pmatrix} 1 \\ 1 \end{pmatrix} e^{-t/2}$$

is a solution of the system. To find another solution, we need to look for a vector $\mathbf{w}$ satisfying $(A+I)\mathbf{w} = \mathbf{v}$. That is, we need to find $\mathbf{w}$ such that

$$\begin{pmatrix} -\frac{5}{2} & \frac{5}{2} \\ -\frac{5}{2} & \frac{5}{2} \end{pmatrix} \mathbf{w} = \begin{pmatrix} 1 \\ 1 \end{pmatrix}.$$

The vector

$$\mathbf{w} = \begin{pmatrix} -\frac{2}{5} \\ 0 \end{pmatrix}$$

satisfies this equation. Therefore,

$$\mathbf{x_2}(t) = \begin{pmatrix} 1 \\ 1 \end{pmatrix} te^{-t/2} + \begin{pmatrix} -\frac{2}{5} \\ 0 \end{pmatrix} e^{-t/2}$$

is a second solution. We conclude that

$$X(t) = \begin{pmatrix} e^{-t/2} & te^{-t/2} - \frac{2}{5}e^{-t/2} \\ e^{-t/2} & te^{-t/2} \end{pmatrix}$$

is a fundamental matrix for this system. The fundamental matrix e^{At} is given by $e^{At} = \mathbf{X}(t)\mathbf{X}^{-1}(0)$. Therefore,

$$e^{At} = \begin{pmatrix} e^{-t/2} & te^{-t/2} - \frac{2}{5}e^{-t/2} \\ e^{-t/2} & te^{-t/2} \end{pmatrix} \begin{pmatrix} 0 & 1 \\ -\frac{5}{2} & \frac{5}{2} \end{pmatrix} = \begin{pmatrix} e^{-t/2} - \frac{5}{2}te^{-t/2} & \frac{5}{2}te^{-t/2} \\ -\frac{5}{2}te^{-t/2} & e^{-t/2} + \frac{5}{2}te^{-t/2} \end{pmatrix}.$$

13. The eigenvalues of this matrix are $\lambda = -1, -2, 2$. For $\lambda = -1$,

$$A - \lambda I = \begin{pmatrix} 2 & 1 & 1 \\ 2 & 2 & -1 \\ -8 & -5 & -2 \end{pmatrix} \rightarrow \begin{pmatrix} 2 & 0 & 3 \\ 0 & 1 & -2 \\ 0 & 0 & 0 \end{pmatrix}.$$

Therefore,

$$\mathbf{v_1} = \begin{pmatrix} 3 \\ -4 \\ -2 \end{pmatrix}$$

is an eigenvector for $\lambda = -1$. Therefore,

$$\mathbf{x_1}(t) = \begin{pmatrix} 3 \\ -4 \\ -2 \end{pmatrix} e^{-t}$$

is a solution of the system. For $\lambda = -2$,

$$A - \lambda I = \begin{pmatrix} 3 & 1 & 1 \\ 2 & 3 & -1 \\ -8 & -5 & -1 \end{pmatrix} \rightarrow \begin{pmatrix} 7 & 0 & 4 \\ 0 & 7 & -5 \\ 0 & 0 & 0 \end{pmatrix}.$$

Therefore,

$$\mathbf{v_2} = \begin{pmatrix} 4 \\ -5 \\ -7 \end{pmatrix}$$

is an eigenvector for $\lambda = -2$. Therefore,

$$\mathbf{x_2}(t) = \begin{pmatrix} 4 \\ -5 \\ -7 \end{pmatrix} e^{-2t}$$

is a solution of the system. For $\lambda = 2$,

$$A - \lambda I = \begin{pmatrix} -1 & 1 & 1 \\ 2 & -1 & -1 \\ -8 & -5 & -5 \end{pmatrix} \rightarrow \begin{pmatrix} 1 & 0 & 0 \\ 0 & 1 & 1 \\ 0 & 0 & 0 \end{pmatrix}.$$

Therefore,

$$\mathbf{v_3} = \begin{pmatrix} 0 \\ 1 \\ -1 \end{pmatrix}$$

is an eigenvector for $\lambda = 2$. Therefore,

$$\mathbf{x_3}(t) = \begin{pmatrix} 0 \\ 1 \\ -1 \end{pmatrix} e^{2t}$$

is a solution of the system. We conclude that

$$\mathbf{X}(t) = \begin{pmatrix} 3e^{-t} & 4e^{-2t} & 0 \\ -4e^{-t} & -5e^{-2t} & e^{2t} \\ -2e^{-t} & -7e^{-2t} & -e^{2t} \end{pmatrix}$$

is a fundamental matrix for this system. The fundamental matrix e^{At} is given by $e^{At} = \mathbf{X}(t)\mathbf{X}^{-1}(0)$. Therefore,

$$e^{At} = \begin{pmatrix} 3e^{-t} & 4e^{-2t} & 0 \\ -4e^{-t} & -5e^{-2t} & e^{2t} \\ -2e^{-t} & -7e^{-2t} & -e^{2t} \end{pmatrix} \begin{pmatrix} 1 & 1/3 & 1/3 \\ -1/2 & -1/4 & -1/4 \\ 3/2 & 13/12 & 1/12 \end{pmatrix}$$

$$= \begin{pmatrix} -2e^{-2t} + 3e^{-t} & -e^{-2t} + e^{-t} & -e^{-2t} + e^{-t} \\ \frac{5}{2}e^{-2t} - 4e^{-t} + \frac{3}{2}e^{2t} & \frac{5}{4}e^{-2t} - \frac{4}{3}e^{-t} + \frac{13}{12}e^{2t} & \frac{5}{4}e^{-2t} - \frac{4}{3}e^{-t} + \frac{1}{12}e^{2t} \\ \frac{7}{2}e^{-2t} - 2e^{-t} - \frac{3}{2}e^{2t} & \frac{7}{4}e^{-2t} - \frac{2}{3}e^{-t} - \frac{13}{12}e^{2t} & \frac{7}{4}e^{-2t} - \frac{2}{3}e^{-t} - \frac{1}{12}e^{2t} \end{pmatrix}.$$

14. The eigenvalues of this matrix are $\lambda = 1, -2, 3$. For $\lambda = 1$,

$$A - \lambda I = \begin{pmatrix} 0 & -1 & 4 \\ 3 & 1 & -1 \\ 2 & 1 & -2 \end{pmatrix} \rightarrow \begin{pmatrix} 1 & 0 & 1 \\ 0 & 1 & -4 \\ 0 & 0 & 0 \end{pmatrix}.$$

Therefore,

$$\mathbf{v_1} = \begin{pmatrix} 1 \\ -4 \\ -1 \end{pmatrix}$$

is an eigenvector for $\lambda = 1$. Therefore,

$$\mathbf{x_1}(t) = \begin{pmatrix} 1 \\ -4 \\ -1 \end{pmatrix} e^t$$

is a solution of the system. For $\lambda = -2$,

$$A - \lambda I = \begin{pmatrix} 3 & -1 & 4 \\ 3 & 4 & -1 \\ 2 & 1 & 1 \end{pmatrix} \rightarrow \begin{pmatrix} 1 & 0 & 1 \\ 0 & 1 & -1 \\ 0 & 0 & 0 \end{pmatrix}.$$

Therefore,

$$\mathbf{v_2} = \begin{pmatrix} 1 \\ -1 \\ -1 \end{pmatrix}$$

is an eigenvector for $\lambda = -2$. Therefore,

$$\mathbf{x_2}(t) = \begin{pmatrix} 1 \\ -1 \\ -1 \end{pmatrix} e^{-2t}$$

is a solution of the system. For $\lambda = 3$,

$$A - \lambda I = \begin{pmatrix} -2 & -1 & 4 \\ 3 & -1 & -1 \\ 2 & 1 & -4 \end{pmatrix} \rightarrow \begin{pmatrix} 1 & 0 & -1 \\ 0 & 1 & -2 \\ 0 & 0 & 0 \end{pmatrix}.$$

Therefore,

$$\mathbf{v_3} = \begin{pmatrix} 1 \\ 2 \\ 1 \end{pmatrix}$$

is an eigenvector for $\lambda = 3$. Therefore,

$$\mathbf{x_3}(t) = \begin{pmatrix} 1 \\ 2 \\ 1 \end{pmatrix} e^{3t}$$

is a solution of the system. We conclude that

$$\mathbf{X}(t) = \begin{pmatrix} e^t & e^{-2t} & e^{3t} \\ -4e^t & -e^{-2t} & 2e^{3t} \\ -e^t & -e^{-2t} & e^{3t} \end{pmatrix}$$

is a fundamental matrix for this system. The fundamental matrix e^{At} is given by $e^{At} = \mathbf{X}(t)\mathbf{X}^{-1}(0)$. Therefore,

$$e^{At} = \begin{pmatrix} e^t & e^{-2t} & e^{3t} \\ -4e^t & -e^{-2t} & 2e^{3t} \\ -e^t & -e^{-2t} & e^{3t} \end{pmatrix} \begin{pmatrix} 1/6 & -1/3 & 1/2 \\ 1/3 & 1/3 & -1 \\ 1/2 & 0 & 1/2 \end{pmatrix}$$

$$= \begin{pmatrix} \frac{1}{6}e^t + \frac{1}{3}e^{-2t} + \frac{1}{2}e^{3t} & -\frac{1}{3}e^t + \frac{1}{3}e^{-2t} & \frac{1}{2}e^t - e^{-2t} + \frac{1}{2}e^{3t} \\ -\frac{2}{3}e^t - \frac{1}{3}e^{-2t} + e^{3t} & \frac{4}{3}e^t - \frac{1}{3}e^{-2t} & -2e^t + e^{-2t} + e^{3t} \\ -\frac{1}{6}e^t - \frac{1}{3}e^{-2t} + \frac{1}{2}e^{3t} & \frac{1}{3}e^t - \frac{1}{3}e^{-2t} & -\frac{1}{2}e^t + e^{-2t} + \frac{1}{2}e^{3t} \end{pmatrix}.$$

15. The fundamental matrix found in problem 6 was

$$e^{At} = \frac{1}{2}e^{-t}\begin{pmatrix} 2\cos(2t) & -4\sin(2t) \\ \sin(2t) & 2\cos(2t) \end{pmatrix}.$$

Therefore, the solution of the IVP is given by

$$\mathbf{x} = e^{At}\mathbf{x_0}$$

$$= \frac{1}{2}e^{-t}\begin{pmatrix} 2\cos(2t) & -4\sin(2t) \\ \sin(2t) & 2\cos(2t) \end{pmatrix}\begin{pmatrix} 3 \\ 1 \end{pmatrix}$$

$$= e^{-t}\begin{pmatrix} 3\cos(2t) - 2\sin(2t) \\ \frac{3}{2}\sin(2t) + \cos(2t) \end{pmatrix}.$$

16. The fundamental matrix found in problem 9 was

$$e^{At} = \frac{1}{2}\begin{pmatrix} 3e^t - e^{-t} & -e^t + e^{-t} \\ 3e^t - 3e^{-t} & -e^t + 3e^{-t} \end{pmatrix}.$$

Therefore, the solution of the IVP is given by

$$\mathbf{x} = e^{At}\mathbf{x_0}$$

$$= \frac{1}{2}\begin{pmatrix} 3e^t - e^{-t} & -e^t + e^{-t} \\ 3e^t - 3e^{-t} & -e^t + 3e^{-t} \end{pmatrix}\begin{pmatrix} 2 \\ -1 \end{pmatrix}$$

$$= \frac{1}{2}\begin{pmatrix} 7e^t - 3e^{-t} \\ 7e^t - 9e^{-t} \end{pmatrix}.$$

17. First,

$$sI_2 - A = \begin{pmatrix} s+4 & 1 \\ -1 & s+2 \end{pmatrix}$$

implies

$$(sI_2 - A)^{-1} = \frac{1}{(s+3)^2}\begin{pmatrix} s+2 & -1 \\ 1 & s+4 \end{pmatrix}.$$

Therefore,

$$e^{At} = \mathcal{L}^{-1}\left((sI_2 - A)^{-1}\right)$$

$$= \begin{pmatrix} \mathcal{L}^{-1}\left(\frac{s+2}{(s+3)^2}\right) & \mathcal{L}^{-1}\left(-\frac{1}{(s+3)^2}\right) \\ \mathcal{L}^{-1}\left(\frac{1}{(s+3)^2}\right) & \mathcal{L}^{-1}\left(\frac{s+4}{(s+3)^2}\right) \end{pmatrix}$$

68

Using the fact that

$$\frac{s+2}{(s+3)^2} = \frac{1}{s+3} - \frac{1}{(s+3)^2}$$

$$\frac{s+4}{(s+3)^2} = \frac{1}{s+3} + \frac{1}{(s+3)^2}$$

combined with the table of Laplace transforms, we conclude that

$$e^{At} = \begin{pmatrix} e^{-3t} - te^{-3t} & -te^{-3t} \\ te^{-3t} & e^{-3t} + te^{-3t} \end{pmatrix}.$$

18. First,

$$sI_2 - A = \begin{pmatrix} s-5 & 1 \\ -1 & s-3 \end{pmatrix}$$

implies

$$(sI_2 - A)^{-1} = \frac{1}{(s-4)^2} \begin{pmatrix} s-3 & -1 \\ 1 & s-5 \end{pmatrix}.$$

Therefore,

$$e^{At} = \mathcal{L}^{-1}\left((sI_2 - A)^{-1} \right)$$

$$= \begin{pmatrix} \mathcal{L}^{-1}\left(\frac{s-3}{(s-4)^2} \right) & \mathcal{L}^{-1}\left(-\frac{1}{(s-4)^2} \right) \\ \mathcal{L}^{-1}\left(\frac{1}{(s-4)^2} \right) & \mathcal{L}^{-1}\left(\frac{s-5}{(s-4)^2} \right) \end{pmatrix}.$$

Using the fact that

$$\frac{s-3}{(s-4)^2} = \frac{1}{s-4} + \frac{1}{(s-4)^2}$$

$$\frac{s-5}{(s-4)^2} = \frac{1}{s-4} - \frac{1}{(s-4)^2}$$

combined with the table of Laplace transforms, we conclude that

$$e^{At} = \begin{pmatrix} e^{4t} + te^{4t} & -te^{4t} \\ te^{4t} & e^{4t} - te^{4t} \end{pmatrix}.$$

19. First,

$$sI_2 - A = \begin{pmatrix} s+1 & 5 \\ -1 & s-3 \end{pmatrix}$$

implies

$$(sI_2 - A)^{-1} = \frac{1}{(s-1)^2 + 1} \begin{pmatrix} s-3 & -5 \\ 1 & s+1 \end{pmatrix}.$$

Therefore,

$$e^{At} = \mathcal{L}^{-1}\left((sI_2 - A)^{-1} \right)$$

$$= \begin{pmatrix} \mathcal{L}^{-1}\left(\frac{s-3}{(s-1)^2+1} \right) & \mathcal{L}^{-1}\left(-\frac{5}{(s-1)^2+1} \right) \\ \mathcal{L}^{-1}\left(\frac{1}{(s-1)^2+1} \right) & \mathcal{L}^{-1}\left(\frac{s+1}{(s-1)^2+1} \right) \end{pmatrix}.$$

Using the fact that

$$\frac{s-3}{(s-1)^2 + 1} = \frac{s-1}{(s-1)^2 + 1} - \frac{2}{(s-1)^2 + 1}$$

$$\frac{s+1}{(s-1)^2 + 1} = \frac{s-1}{(s-1)^2 + 1} + \frac{2}{(s-1)^2 + 1}$$

combined with the table of Laplace transforms, we conclude that

$$e^{At} = \begin{pmatrix} e^t \cos t - 2e^t \sin t & -5e^t \sin t \\ e^t \sin t & e^t \cos t + 2e^t \sin t \end{pmatrix}.$$

20. First,

$$sI_3 - A = \begin{pmatrix} s & -1 & 1 \\ -1 & s & -1 \\ -1 & -1 & s \end{pmatrix}$$

implies

$$(sI_3 - A)^{-1} = \begin{pmatrix} \frac{1}{s} & \frac{1}{(s+1)s} & -\frac{1}{(s+1)s} \\ \frac{1}{(s-1)s} & \frac{s^2+1}{s(s^2-1)} & \frac{1}{(s+1)s} \\ \frac{1}{(s-1)s} & \frac{1}{(s-1)s} & \frac{1}{s} \end{pmatrix}.$$

Now $e^{At} = \mathcal{L}^{-1}\left((sI_3 - A)^{-1}\right)$ Using the following facts that

$$\frac{1}{(s+1)s} = -\frac{1}{s+1} + \frac{1}{s}$$

$$\frac{1}{(s-1)s} = \frac{1}{s-1} - \frac{1}{s}$$

$$\frac{s^2+1}{s(s^2-1)} = -\frac{1}{s} + \frac{1}{s-1} + \frac{1}{s+1}$$

combined with the table of Laplace transforms, we conclude that

$$e^{At} = \begin{pmatrix} 1 & 1 - e^{-t} & e^{-t} - 1 \\ e^t - 1 & e^t + e^{-t} - 1 & 1 - e^{-t} \\ e^t - 1 & e^t - 1 & 1 \end{pmatrix}.$$

21.

(a) Letting $x_1 = u$ and $x_2 = u'$, then $u'' = x_2'$. Therefore, in terms of the new variables, we have

$$x_2' + \omega^2 x_1 = 0$$

with the initial conditions $x_1(0) = u_0$ and $x_2(0) = v_0$. The equivalent first order system is

$$x_1' = x_2$$
$$x_2' = -\omega^2 x_1$$

which can be expressed in the form

$$\begin{pmatrix} x_1 \\ x_2 \end{pmatrix}' = \begin{pmatrix} 0 & 1 \\ -\omega^2 & 0 \end{pmatrix} \begin{pmatrix} x_1 \\ x_2 \end{pmatrix}$$

with initial conditions

$$\begin{pmatrix} x_1(0) \\ x_2(0) \end{pmatrix} = \begin{pmatrix} u_0 \\ v_0 \end{pmatrix}.$$

(b) Setting

$$\mathbf{A} = \begin{pmatrix} 0 & 1 \\ -\omega^2 & 0 \end{pmatrix},$$

we see that

$$\mathbf{A}^2 = -\omega^2 \mathbf{I}, \mathbf{A}^3 = -\omega^2 \mathbf{A}, \mathbf{A}^4 = \omega^4 \mathbf{I}.$$

By induction, it follows that

$$\mathbf{A}^{2k} = (-1)^k \omega^{2k} \mathbf{I}$$
$$\mathbf{A}^{2k+1} = (-1)^k \omega^{2k} \mathbf{A}.$$

Therefore,

$$
\begin{aligned}
e^{\mathbf{A}t} &= \sum_{k=0}^{\infty} \left[(-1)^k \frac{\omega^{2k} t^{2k}}{(2k)!} \mathbf{I} + (-1)^k \frac{\omega^{2k} t^{2k+1}}{(2k+1)!} \mathbf{A} \right] \\
&= \left[\sum_{k=0}^{\infty} (-1)^k \frac{\omega^{2k} t^{2k}}{(2k)!} \right] \mathbf{I} + \frac{1}{\omega} \left[\sum_{k=0}^{\infty} (-1)^k \frac{\omega^{2k+1} t^{2k+1}}{(2k+1)!} \right] \mathbf{A} \\
&= \mathbf{I} \cos(\omega t) + \mathbf{A} \frac{\sin(\omega t)}{\omega}.
\end{aligned}
$$

(c) Using the formula for $e^{\mathbf{A}t}$ found in part (b), we can conclude that the solution of the IVP is given by

$$
\begin{aligned}
\mathbf{x} &= e^{\mathbf{A}t} \mathbf{x_0} \\
&= \left[\cos(\omega t) \mathbf{I} + \frac{1}{\omega} \sin(\omega t) \mathbf{A} \right] \begin{pmatrix} u_0 \\ v_0 \end{pmatrix} \\
&= \cos(\omega t) \begin{pmatrix} u_0 \\ v_0 \end{pmatrix} + \frac{1}{\omega} \sin(\omega t) \begin{pmatrix} v_0 \\ -\omega^2 u_0 \end{pmatrix}.
\end{aligned}
$$

22. The eigenvalues of A are $-k_1, -k_2, 0$. The associated eigenvectors are

$$
\mathbf{v_1} = \begin{pmatrix} \frac{-k_1+k_2}{k_2} \\ \frac{k_1}{k_2} \\ 1 \end{pmatrix}, \quad
\mathbf{v_2} = \begin{pmatrix} 0 \\ 1 \\ 1 \end{pmatrix}, \quad
\mathbf{v_3} = \begin{pmatrix} 0 \\ 0 \\ 1 \end{pmatrix}.
$$

Therefore,

$$
\mathbf{X}(t) = \begin{pmatrix}
\frac{-k_1+k_2}{k_2} e^{-k_1 t} & 0 & 0 \\
\frac{k_1}{k_2} e^{-k_1 t} & e^{-k_2 t} & 0 \\
e^{-k_1 t} & e^{-k_2 t} & 1
\end{pmatrix}
$$

is a fundamental matrix. Then

$$
\begin{aligned}
e^{At} &= \mathbf{X}(t)\mathbf{X}^{-1}(0) \\
&= \begin{pmatrix}
\frac{-k_1+k_2}{k_2} e^{-k_1 t} & 0 & 0 \\
\frac{k_1}{k_2} e^{-k_1 t} & e^{-k_2 t} & 0 \\
e^{-k_1 t} & e^{-k_2 t} & 1
\end{pmatrix}
\begin{pmatrix}
\frac{k_2}{-k_1+k_2} & 0 & 0 \\
-\frac{k_1}{-k_1+k_2} & 1 & 0 \\
-1 & -1 & 1
\end{pmatrix} \\
&= \begin{pmatrix}
e^{-k_1 t} & 0 & 0 \\
\frac{k_1}{-k_1+k_2} e^{-k_1 t} - \frac{k_1}{-k_1+k_2} e^{-k_2 t} & e^{-k_2 t} & 0 \\
\frac{k_2}{-k_1+k_2} e^{-k_1 t} - \frac{k_1}{-k_1+k_2} e^{-k_2 t} - 1 & e^{-k_2 t} - 1 & 1
\end{pmatrix}.
\end{aligned}
$$

Therefore, the solution of $\mathbf{m}' = A\mathbf{m}$ satisfying the initial condition $\mathbf{m}(0) = (m_0, 0, 0)^T$ is given by

$$
\begin{aligned}
\mathbf{m}(t) = e^{At}\mathbf{m}(0) &= \begin{pmatrix}
e^{-k_1 t} & 0 & 0 \\
\frac{k_1}{-k_1+k_2} e^{-k_1 t} - \frac{k_1}{-k_1+k_2} e^{-k_2 t} & e^{-k_2 t} & 0 \\
\frac{k_2}{-k_1+k_2} e^{-k_1 t} - \frac{k_1}{-k_1+k_2} e^{-k_2 t} - 1 & e^{-k_2 t} - 1 & 1
\end{pmatrix}
\begin{pmatrix} m_0 \\ 0 \\ 0 \end{pmatrix} \\
&= m_0 \begin{pmatrix}
e^{-k_1 t} \\
\frac{k_1}{-k_1+k_2} e^{-k_1 t} - \frac{k_1}{-k_1+k_2} e^{-k_2 t} \\
\frac{k_2}{-k_1+k_2} e^{-k_1 t} - \frac{k_1}{-k_1+k_2} e^{-k_2 t} - 1
\end{pmatrix}.
\end{aligned}
$$

23.

(a)

$$T^{-1}AT = D \implies TT^{-1}AT = TD \implies AT = TD \implies ATT^{-1} = TDT^{-1} \implies A = TDT^{-1}.$$

Therefore, $A^2 = (TDT^{-1})(TDT^{-1}) = TD^2T^{-1}$. Now, we will prove the general result by induction. Assume $A^n = TD^nT^{-1}$ to show that $A^{n+1} = TD^{n+1}T^{-1}$.

$$A^{n+1} = AA^n = ATD^nT^{-1} = (TDT^{-1})(TD^nT^{-1}) = TDD^nT^{-1} = TD^{n+1}T^{-1}.$$

(b)

$$\begin{aligned}
e^{At} &= I_n + At + \frac{1}{2!}A^2t^2 + \frac{1}{3!}A^3t^3 + \dots \\
&= TT^{-1} + TDT^{-t}t + \frac{1}{2!}TD^2T^{-1}t^2 + \frac{1}{3!}TD^3T^{-1}t^3 + \dots \\
&= T\left[T^{-1} + DT^{-1}t + \frac{1}{2!}D^2T^{-1}t^2 + \frac{1}{3!}D^3T^{-1}t^3 + \dots\right] \\
&= T\left[I_n + Dt + \frac{1}{2!}D^2t^2 + \frac{1}{3!}D^3t^3 + \dots\right]T^{-1} \\
&= Te^{Dt}T^{-1}.
\end{aligned}$$

24.

(a) Assuming that $\mathbf{x} = \phi(t)$ is a solution, then $\phi' = \mathbf{A}\phi$ with $\phi(0) = \mathbf{x}^0$. Integrating both sides of this equation, we have

$$\phi(t) - \phi(0) = \int_0^t \mathbf{A}\phi(s)\,ds,$$

which implies

$$\phi(t) = \mathbf{x_0} + \int_0^t \mathbf{A}\phi(s)\,ds.$$

(b)

$$\begin{aligned}
\phi^{(1)}(t) &= \mathbf{x_0} + \int_0^t \mathbf{A}\mathbf{x_0}\,ds \\
&= \mathbf{x_0} + \mathbf{A}\mathbf{x_0}t \\
&= (\mathbf{I} + \mathbf{A}t)\mathbf{x_0}.
\end{aligned}$$

(c) Repeating this process, we see that

$$\begin{aligned}
\phi^{(2)}(t) &= \mathbf{x_0} + \int_0^t \mathbf{A}(\mathbf{I} + \mathbf{A}t)\mathbf{x_0}\,ds \\
&= \mathbf{x_0} + \mathbf{A}\mathbf{x_0}t + \mathbf{A}^2\mathbf{x_0}\frac{t^2}{2} \\
&= \left(\mathbf{I} + \mathbf{A}t + \mathbf{A}^2\frac{t^2}{2}\right)\mathbf{x_0}.
\end{aligned}$$

By induction, if we assume that

$$\phi^{(n)}(t) = \left(\mathbf{I} + \mathbf{A}t + \mathbf{A}^2\frac{t^2}{2} + \dots + \mathbf{A}^n\frac{t^n}{n!}\right)\mathbf{x_0},$$

it follows that

$$\phi^{(n+1)}(t) = \int_0^t \mathbf{A}\left(\mathbf{I} + \mathbf{A}t + \mathbf{A}^2\frac{t^2}{2} + \ldots + \mathbf{A}^n\frac{t^n}{n!}\right)\mathbf{x_0}\,ds$$

$$= \mathbf{A}\left(\mathbf{I}t + \mathbf{A}\frac{t^2}{2} + \mathbf{A}^2\frac{t^3}{3!} + \ldots + \mathbf{A}^n\frac{t^{n+1}}{(n+1)!}\right)\mathbf{x_0}$$

$$= \left(\mathbf{A}t + \mathbf{A}^2\frac{t^2}{2} + \mathbf{A}^3\frac{t^3}{3!} + \ldots + \mathbf{A}^{n+1}\frac{t^{n+1}}{(n+1)!}\right)\mathbf{x_0}$$

(d) Define $\phi^{(\infty)}(t) = \lim_{n\to\infty}\phi^{(n)}(t)$. It can be shown that the limit exists. In fact,

$$\phi^{(\infty)}(t) = e^{\mathbf{A}t}\mathbf{x_0}.$$

Term-by-term differentiation results in

$$\frac{d}{dt}\phi^{(\infty)}(t) = \frac{d}{dt}\left(\mathbf{I} + \mathbf{A}t + \mathbf{A}^2\frac{t^2}{2} + \ldots + \mathbf{A}^n\frac{t^n}{n!} + \ldots\right)\mathbf{x_0}$$

$$= \left(\mathbf{A} + \mathbf{A}^2 t + \ldots + \mathbf{A}^n\frac{t^{n-1}}{(n-1)!} + \ldots\right)\mathbf{x_0}$$

$$= \mathbf{A}\left(\mathbf{I} + \mathbf{A}t + \mathbf{A}^2\frac{t^2}{2} + \ldots + \mathbf{A}^n\frac{t^n}{n!} + \ldots\right)\mathbf{x_0}.$$

That is,

$$\frac{d}{dt}\phi^{(\infty)}(t) = \mathbf{A}\phi^{(\infty)}(t).$$

Furthermore, $\phi^{(\infty)}(0) = \mathbf{x_0}$. Based on uniqueness of solutions, we conclude that $\phi(t) = \phi^{(\infty)}(t) = e^{\mathbf{A}t}\mathbf{x_0}$.

Section 6.6

1. By the product rule and the Fundamental Theorem of Calculus,

$$\mathbf{x}'_p = \mathbf{X}'(t)\int_{t_1}^t \mathbf{X}^{-1}(s)\mathbf{g}(s)\,ds + \mathbf{X}(t)\mathbf{X}^{-1}(t)\mathbf{g}(t)$$

$$= \mathbf{P}(t)\mathbf{X}(t)\int_{t_1}^t \mathbf{X}^{-1}(s)\mathbf{g}(s)\,ds + \mathbf{g}(t)$$

$$= \mathbf{P}(t)\mathbf{x}_p + \mathbf{g}(t).$$

2. The eigenvalues of

$$\begin{pmatrix} 2 & -1 \\ 3 & -2 \end{pmatrix}$$

are given by $\lambda_1 = 1$ and $\lambda_2 = -1$. Corresponding eigenvectors are given by

$$\mathbf{v_1} = \begin{pmatrix} 1 \\ 1 \end{pmatrix}, \quad \mathbf{v_2} = \begin{pmatrix} 1 \\ 3 \end{pmatrix}.$$

Therefore, two linearly independent solutions are given by

$$\mathbf{x_1}(t) = \begin{pmatrix} 1 \\ 1 \end{pmatrix}e^t, \quad \mathbf{x_2}(t) = \begin{pmatrix} 1 \\ 3 \end{pmatrix}e^{-t}$$

and

$$\mathbf{X}(t) = \begin{pmatrix} e^t & e^{-t} \\ e^t & 3e^{-t} \end{pmatrix}$$

73

is a fundamental matrix. In order to calculate the general solution, we need to calculate $\int_{t_1}^{t} \mathbf{X}^{-1}(s)\mathbf{g}(s)\,ds$. We see that

$$\mathbf{X}^{-1}(s) = \frac{1}{2}\begin{pmatrix} 3e^{-s} & -e^{-s} \\ -e^{s} & e^{s} \end{pmatrix}.$$

Therefore,

$$\begin{aligned}
\int_{t_1}^{t} \mathbf{X}^{-1}(s)\mathbf{g}(s)\,ds &= \frac{1}{2}\int_{t_1}^{t}\begin{pmatrix} 3e^{-s} & -e^{-s} \\ -e^{s} & e^{s} \end{pmatrix}\begin{pmatrix} e^{s} \\ s \end{pmatrix}\,ds \\
&= \frac{1}{2}\int_{t_1}^{t}\begin{pmatrix} 3 - se^{-s} \\ -e^{2s} + se^{s} \end{pmatrix}\,ds \\
&= \frac{1}{2}\begin{pmatrix} 3t + te^{-t} + e^{-t} \\ -\frac{1}{2}e^{2t} + te^{t} - e^{t} \end{pmatrix} + \mathbf{c}.
\end{aligned}$$

Then the general solution will be given by

$$\begin{aligned}
\mathbf{x}(t) &= \mathbf{X}(t)\mathbf{c} + \mathbf{X}(t)\int_{t_1}^{t}\mathbf{X}^{-1}(s)\mathbf{g}(s)\,ds \\
&= \begin{pmatrix} e^{t} & e^{-t} \\ e^{t} & 3e^{-t} \end{pmatrix}\mathbf{c} + \begin{pmatrix} e^{t} & e^{-t} \\ e^{t} & 3e^{-t} \end{pmatrix}\left[\frac{1}{2}\begin{pmatrix} 3t + te^{-t} + e^{-t} \\ -\frac{1}{2}e^{2t} + te^{t} - e^{t} \end{pmatrix} + \mathbf{c}\right] \\
&= c_1 e^{t}\begin{pmatrix} 1 \\ 1 \end{pmatrix} + c_2 e^{-t}\begin{pmatrix} 1 \\ 3 \end{pmatrix} + \begin{pmatrix} \left(\frac{3}{2}t - \frac{1}{4}\right)e^{t} + t \\ \left(\frac{3}{2}t - \frac{3}{4}\right)e^{t} + 2t - 1 \end{pmatrix}
\end{aligned}$$

3. The eigenvalues of

$$\begin{pmatrix} 1 & \sqrt{3} \\ \sqrt{3} & -1 \end{pmatrix}$$

are given by $\lambda_1 = 2$ and $\lambda_2 = -2$. Corresponding eigenvectors are given by

$$\mathbf{v_1} = \begin{pmatrix} \sqrt{3} \\ 1 \end{pmatrix}, \quad \mathbf{v_2} = \begin{pmatrix} 1 \\ -\sqrt{3} \end{pmatrix}.$$

Therefore, two linearly independent solutions of the homogeneous equation are given by

$$\mathbf{x_1}(t) = \begin{pmatrix} \sqrt{3} \\ 1 \end{pmatrix}e^{2t}, \quad \mathbf{x_2}(t) = \begin{pmatrix} 1 \\ -\sqrt{3} \end{pmatrix}e^{-2t}$$

and

$$\mathbf{X}(t) = \begin{pmatrix} \sqrt{3}e^{2t} & e^{-2t} \\ e^{2t} & -\sqrt{3}e^{-2t} \end{pmatrix}$$

is a fundamental matrix. In order to calculate the general solution, we need to calculate $\int_{t_1}^{t} \mathbf{X}^{-1}(s)\mathbf{g}(s)\,ds$. We see that

$$\mathbf{X}^{-1}(s) = \frac{1}{4}\begin{pmatrix} \sqrt{3}e^{-2s} & e^{-2s} \\ e^{2s} & -\sqrt{3}e^{2s} \end{pmatrix}.$$

Therefore,

$$\begin{aligned}
\int_{t_1}^{t} \mathbf{X}^{-1}(s)\mathbf{g}(s)\,ds &= \frac{1}{4}\int_{t_1}^{t}\begin{pmatrix} \sqrt{3}e^{-2s} & e^{-2s} \\ e^{2s} & -\sqrt{3}e^{2s} \end{pmatrix}\begin{pmatrix} e^{s} \\ \sqrt{3}e^{-s} \end{pmatrix}\,ds \\
&= \frac{1}{4}\int_{t_1}^{t}\begin{pmatrix} \sqrt{3}e^{-s} + \sqrt{3}e^{-3s} \\ e^{3s} - 3e^{s} \end{pmatrix}\,ds \\
&= \frac{1}{4}\begin{pmatrix} -\sqrt{3}e^{-t} - \frac{1}{\sqrt{3}}e^{-3t} \\ \frac{1}{3}e^{3t} - 3e^{t} \end{pmatrix} + \mathbf{c}.
\end{aligned}$$

Then the general solution will be given by

$$\mathbf{x}(t) = \mathbf{X}(t)\mathbf{c} + \mathbf{X}(t)\int_{t_1}^{t}\mathbf{X}^{-1}(s)\mathbf{g}(s)\,ds$$

$$= \begin{pmatrix} \sqrt{3}e^{2t} & e^{-2t} \\ e^{2t} & -\sqrt{3}e^{-2t} \end{pmatrix}\mathbf{c} + \begin{pmatrix} \sqrt{3}e^{2t} & e^{-2t} \\ e^{2t} & -\sqrt{3}e^{-2t} \end{pmatrix}\left[\frac{1}{4}\begin{pmatrix} -\sqrt{3}e^{-t} - \frac{1}{\sqrt{3}}e^{-3t} \\ \frac{1}{3}e^{3t} - 3e^{t} \end{pmatrix} + \mathbf{c}\right]$$

$$= c_1 e^{2t}\begin{pmatrix} \sqrt{3} \\ 1 \end{pmatrix} + c_2 e^{-2t}\begin{pmatrix} 1 \\ -\sqrt{3} \end{pmatrix} + \begin{pmatrix} -\frac{2}{3}e^{t} - e^{-t} \\ -\frac{1}{\sqrt{3}}e^{t} + \frac{2}{\sqrt{3}}e^{-t} \end{pmatrix}$$

4. The eigenvalues of

$$\begin{pmatrix} 2 & -5 \\ 1 & -2 \end{pmatrix}$$

are given by $\lambda_1 = i$ and $\lambda_2 = -i$. For $\lambda = i$,

$$\mathbf{v} = \begin{pmatrix} 2 + i \\ 1 \end{pmatrix}$$

is a corresponding eigenvector, and

$$\mathbf{u}(t) = e^{it}\begin{pmatrix} 2 + i \\ 1 \end{pmatrix}$$

is a solution of the homogeneous equation. Looking at the real and imaginary parts of $\mathbf{u}$, we have the following two linearly independent, real-valued solutions of the homogeneous equation:

$$\mathbf{x_1}(t) = \begin{pmatrix} 2 \\ 1 \end{pmatrix}\cos t - \begin{pmatrix} 1 \\ 0 \end{pmatrix}\sin t$$

and

$$\mathbf{x_2}(t) = \begin{pmatrix} 2 \\ 1 \end{pmatrix}\sin t + \begin{pmatrix} 1 \\ 0 \end{pmatrix}\cos t.$$

Therefore,

$$\mathbf{X}(t) = \begin{pmatrix} 2\cos t - \sin t & 2\sin t + \cos t \\ \cos t & \sin t \end{pmatrix}$$

is a fundamental matrix for this equation. In order to calculate the general solution, we need to calculate $\int_{t_1}^{t}\mathbf{X}^{-1}(s)\mathbf{g}(s)\,ds$. We see that

$$\mathbf{X}^{-1}(s) = \begin{pmatrix} -\sin s & 2\sin s + \cos s \\ \cos s & -2\cos s + \sin s \end{pmatrix}.$$

Therefore,

$$\int_{t_1}^{t}\mathbf{X}^{-1}(s)\mathbf{g}(s)\,ds = \int_{t_1}^{t}\begin{pmatrix} -\sin s & 2\sin s + \cos s \\ \cos s & -2\cos s + \sin s \end{pmatrix}\begin{pmatrix} -\cos s \\ \sin s \end{pmatrix}ds$$

$$= \int_{t_1}^{t}\begin{pmatrix} 2\sin s\cos s + 2\sin^2 s \\ -2\cos s\sin s + \sin^2 s - \cos^2 s \end{pmatrix}ds$$

$$= \int_{t_1}^{t}\begin{pmatrix} \sin(2s) + 1 - \cos(2s) \\ -\sin(2s) - \cos(2s) \end{pmatrix}ds$$

$$= \frac{1}{2}\begin{pmatrix} -\cos(2t) + 2t - \sin(2t) \\ \cos(2t) - \sin(2t) \end{pmatrix} + \mathbf{c}.$$

Then the general solution will be given by

$$\mathbf{x}(t) = \mathbf{X}(t)\mathbf{c} + \mathbf{X}(t)\int_{t_1}^{t} \mathbf{X}^{-1}(s)\mathbf{g}(s)\,ds$$

$$= \begin{pmatrix} 2\cos t - \sin t & 2\sin t + \cos t \\ \cos t & \sin t \end{pmatrix}\mathbf{c} + \frac{1}{2}\begin{pmatrix} 2\cos t - \sin t & 2\sin t + \cos t \\ \cos t & \sin t \end{pmatrix}\begin{pmatrix} -\cos(2t) + 2t - \sin(2t) \\ \cos(2t) - \sin(2t) \end{pmatrix}$$

$$= c_1\begin{pmatrix} 2\cos t - \sin t \\ \cos t \end{pmatrix} + c_2\begin{pmatrix} 2\sin t + \cos t \\ \sin t \end{pmatrix} + \frac{1}{2}\begin{pmatrix} 4t\cos t - 2t\sin t - \cos t - 3\sin t \\ 2t\cos t - \cos t - \sin t \end{pmatrix}.$$

We note that in multiplying the last two matrices, we made use of the trig identities, $\cos(2t) = 2\cos^2 t - 1$ and $\cos(2t) = 1 - \sin^2 t$.

5. The eigenvalues of

$$\begin{pmatrix} 1 & 1 \\ 4 & -2 \end{pmatrix}$$

are given by $\lambda_1 = -3$ and $\lambda_2 = 2$. Corresponding eigenvectors are given by

$$\mathbf{v_1} = \begin{pmatrix} 1 \\ -4 \end{pmatrix}, \quad \mathbf{v_2} = \begin{pmatrix} 1 \\ 1 \end{pmatrix}.$$

Therefore, two linearly independent solutions of the homogeneous equation are given by

$$\mathbf{x_1}(t) = \begin{pmatrix} 1 \\ -4 \end{pmatrix}e^{-3t}, \quad \mathbf{x_2}(t) = \begin{pmatrix} 1 \\ 1 \end{pmatrix}e^{2t}$$

and

$$\mathbf{X}(t) = \begin{pmatrix} e^{-3t} & e^{2t} \\ -4e^{-3t} & e^{2t} \end{pmatrix}$$

is a fundamental matrix for this equation. In order to calculate the general solution, we need to calculate $\int_{t_1}^{t} \mathbf{X}^{-1}(s)\mathbf{g}(s)\,ds$. We see that

$$\mathbf{X}^{-1}(s) = \frac{1}{5}\begin{pmatrix} e^{3s} & -e^{3s} \\ 4e^{-2s} & e^{-2s} \end{pmatrix}.$$

Therefore,

$$\int_{t_1}^{t} \mathbf{X}^{-1}(s)\mathbf{g}(s)\,ds = \frac{1}{5}\int_{t_1}^{t}\begin{pmatrix} e^{3s} & -e^{3s} \\ 4e^{-2s} & e^{-2s} \end{pmatrix}\begin{pmatrix} e^{-2s} \\ -2e^s \end{pmatrix}ds$$

$$= \frac{1}{5}\int_{t_1}^{t}\begin{pmatrix} e^{s} + 2e^{4s} \\ 4e^{-4s} - 2e^{-s} \end{pmatrix}ds$$

$$= \frac{1}{5}\begin{pmatrix} e^{t} + \frac{1}{2}e^{4t} \\ -e^{-4t} + 2e^{-t} \end{pmatrix} + \mathbf{c}.$$

Then the general solution will be given by

$$\mathbf{x}(t) = \mathbf{X}(t)\mathbf{c} + \mathbf{X}(t)\int_{t_1}^{t}\mathbf{X}^{-1}(s)\mathbf{g}(s)\,ds$$

$$= \begin{pmatrix} e^{-3t} & e^{2t} \\ -4e^{-3t} & e^{2t} \end{pmatrix}\mathbf{c} + \frac{1}{5}\begin{pmatrix} e^{-3t} & e^{2t} \\ -4e^{-3t} & e^{2t} \end{pmatrix}\begin{pmatrix} e^{t} + \frac{1}{2}e^{4t} \\ -e^{-4t} + 2e^{-t} \end{pmatrix}$$

$$= c_1 e^{-3t}\begin{pmatrix} 1 \\ -4 \end{pmatrix} + c_2 e^{2t}\begin{pmatrix} 1 \\ 1 \end{pmatrix} + \begin{pmatrix} \frac{1}{2}e^{t} \\ -e^{-2t} \end{pmatrix}.$$

6. The eigenvalues of A are given by $\lambda_1 = 1$ and $\lambda_2 = 2$ with multiplicity 2. An eigenvector for λ_1 is given by

$$\mathbf{v_1} = \begin{pmatrix} 1 \\ 0 \\ 1 \end{pmatrix}.$$

Two eigenvectors for λ_2 are given by

$$\mathbf{v_2} = \begin{pmatrix} 0 \\ 1 \\ 1 \end{pmatrix}, \quad \mathbf{v_3} = \begin{pmatrix} 1 \\ -2 \\ 0 \end{pmatrix}.$$

Therefore, three solutions of the homogeneous equation are given by

$$\mathbf{x_1}(t) = e^t \begin{pmatrix} 1 \\ 0 \\ 1 \end{pmatrix}, \quad \mathbf{x_2}(t) = e^{2t} \begin{pmatrix} 0 \\ 1 \\ 1 \end{pmatrix}, \quad \mathbf{x_3}(t) = e^{2t} \begin{pmatrix} 1 \\ -2 \\ 0 \end{pmatrix},$$

and

$$\mathbf{X}(t) = \begin{pmatrix} e^t & 0 & e^{2t} \\ 0 & e^{2t} & -2e^{2t} \\ e^t & e^{2t} & 0 \end{pmatrix}$$

is a fundamental matrix for this equation. Further,

$$\mathbf{X}^{-1}(s) = \begin{pmatrix} 2e^{-s} & e^{-s} & -e^{-s} \\ -2e^{-2s} & -e^{-2s} & 2e^{-2s} \\ -e^{-2s} & -e^{-2s} & e^{-2s} \end{pmatrix}.$$

Therefore,

$$\int_{t_1}^{t} \mathbf{X}^{-1}(s)\mathbf{g}(s)\,ds = \int_{t_1}^{t} \begin{pmatrix} 2e^{-s} & e^{-s} & -e^{-s} \\ -2e^{-2s} & -e^{-2s} & 2e^{-2s} \\ -e^{-2s} & -e^{-2s} & e^{-2s} \end{pmatrix} \begin{pmatrix} 1 \\ s \\ e^{-s} \end{pmatrix} ds$$

$$= \int_{t_1}^{t} \begin{pmatrix} 2e^{-s} + se^{-s} - e^{-2s} \\ -2e^{-2s} - se^{-2s} + 2e^{-3s} \\ -e^{-2s} - se^{-2s} + e^{-3s} \end{pmatrix} ds$$

$$= \begin{pmatrix} -3e^{-t} - te^{-t} + \frac{1}{2}e^{-2t} \\ \frac{5}{4}e^{-2t} + \frac{1}{2}te^{-2t} - \frac{2}{3}e^{-3t} \\ \frac{3}{4}e^{-2t} + \frac{1}{2}te^{-2t} - \frac{1}{3}e^{-3t} \end{pmatrix} + \mathbf{c}.$$

Therefore, the general solution is given by

$$\mathbf{x}(t) = \mathbf{X}(t)\mathbf{c} + \mathbf{X}(t) \int_{t_1}^{t} \mathbf{X}^{-1}(s)\mathbf{g}(s)\,ds$$

$$= \begin{pmatrix} e^t & 0 & e^{2t} \\ 0 & e^{2t} & -2e^{2t} \\ e^t & e^{2t} & 0 \end{pmatrix} \mathbf{c} + \begin{pmatrix} e^t & 0 & e^{2t} \\ 0 & e^{2t} & -2e^{2t} \\ e^t & e^{2t} & 0 \end{pmatrix} \begin{pmatrix} -3e^{-t} - te^{-t} + \frac{1}{2}e^{-2t} \\ \frac{5}{4}e^{-2t} + \frac{1}{2}te^{-2t} - \frac{2}{3}e^{-3t} \\ \frac{3}{4}e^{-2t} + \frac{1}{2}te^{-2t} - \frac{1}{3}e^{-3t} \end{pmatrix}$$

$$= c_1 e^t \begin{pmatrix} 1 \\ 0 \\ 1 \end{pmatrix} + c_2 e^{2t} \begin{pmatrix} 0 \\ 1 \\ 1 \end{pmatrix} + c_3 e^{2t} \begin{pmatrix} 1 \\ -2 \\ 0 \end{pmatrix} + \begin{pmatrix} -\frac{9}{4} - \frac{1}{2}t + \frac{1}{6}e^{-t} \\ -\frac{1}{4} - \frac{1}{2}t \\ -\frac{7}{4} - \frac{1}{2}t - \frac{1}{6}e^{-t} \end{pmatrix}.$$

7. The eigenvalues of A are given by $\lambda_1 = -2$ and $\lambda_2 = -1$ with multiplicity 2. An eigenvector for λ_1 is given by

$$\mathbf{v_1} = \begin{pmatrix} 1 \\ -2 \\ 1 \end{pmatrix}.$$

Two eigenvectors for λ_2 are given by

$$\mathbf{v_2} = \begin{pmatrix} -1 \\ 1 \\ 0 \end{pmatrix}, \quad \mathbf{v_3} = \begin{pmatrix} 1 \\ 0 \\ 1 \end{pmatrix}.$$

Therefore, three solutions of the homogeneous equation are given by

$$\mathbf{x_1}(t) = e^{-2t}\begin{pmatrix} 1 \\ -2 \\ 1 \end{pmatrix}, \quad \mathbf{x_2}(t) = e^{-t}\begin{pmatrix} -1 \\ 1 \\ 0 \end{pmatrix}, \quad \mathbf{x_3}(t) = e^{-t}\begin{pmatrix} 1 \\ 0 \\ 1 \end{pmatrix},$$

and

$$\mathbf{X}(t) = \begin{pmatrix} e^{-2t} & -e^{-t} & e^{-t} \\ -2e^{-2t} & e^{-t} & 0 \\ e^{-2t} & 0 & e^{-t} \end{pmatrix}$$

is a fundamental matrix for this equation. Further,

$$\mathbf{X}^{-1}(s) = \frac{1}{2}\begin{pmatrix} -e^{2s} & -e^{2s} & e^{2s} \\ -2e^{s} & 0 & 2e^{s} \\ e^{s} & e^{s} & e^{s} \end{pmatrix}.$$

Therefore,

$$\int_{t_1}^{t} \mathbf{X}^{-1}(s)\mathbf{g}(s)\,ds = \frac{1}{2}\int_{t_1}^{t}\begin{pmatrix} -e^{2s} & -e^{2s} & e^{2s} \\ -2e^{s} & 0 & 2e^{s} \\ e^{s} & e^{s} & e^{s} \end{pmatrix}\begin{pmatrix} 1 \\ s \\ e^{-3s} \end{pmatrix}\,ds$$

$$= \frac{1}{2}\int_{t_1}^{t}\begin{pmatrix} -e^{2s} - se^{2s} + e^{-s} \\ -2e^{s} + 2e^{-2s} \\ e^{s} + se^{s} + e^{-2s} \end{pmatrix}\,ds$$

$$= \frac{1}{2}\begin{pmatrix} -\frac{1}{4}e^{2t} - \frac{1}{2}te^{2t} - e^{-t} \\ -2e^{t} - e^{-2t} \\ te^{t} - \frac{1}{2}e^{-2t} \end{pmatrix} + \mathbf{c}.$$

Therefore, the general solution is given by

$$\mathbf{x}(t) = \mathbf{X}(t)\mathbf{c} + \mathbf{X}(t)\int_{t_1}^{t}\mathbf{X}^{-1}(s)\mathbf{g}(s)\,ds$$

$$= \begin{pmatrix} e^{-2t} & -e^{-t} & e^{-t} \\ -2e^{-2t} & e^{-t} & 0 \\ e^{-2t} & 0 & e^{-t} \end{pmatrix}\mathbf{c} + \frac{1}{2}\begin{pmatrix} e^{-2t} & -e^{-t} & e^{-t} \\ -2e^{-2t} & e^{-t} & 0 \\ e^{-2t} & 0 & e^{-t} \end{pmatrix}\begin{pmatrix} -\frac{1}{4}e^{2t} - \frac{1}{2}te^{2t} - e^{-t} \\ -2e^{t} - e^{-2t} \\ te^{t} - \frac{1}{2}e^{-2t} \end{pmatrix}$$

$$= c_1 e^{-2t}\begin{pmatrix} 1 \\ -2 \\ 1 \end{pmatrix} + c_2 e^{-t}\begin{pmatrix} -1 \\ 1 \\ 0 \end{pmatrix} + c_3 e^{-t}\begin{pmatrix} 1 \\ 0 \\ 1 \end{pmatrix} + \begin{pmatrix} \frac{7}{8} + \frac{1}{4}t - \frac{1}{4}e^{-3t} \\ -\frac{3}{4} + \frac{1}{2}t + \frac{1}{2}e^{-3t} \\ -\frac{1}{8} + \frac{1}{4}t - \frac{3}{4}e^{-3t} \end{pmatrix}.$$

8. The eigenvalues of A are given by $\lambda_1 = -1$ and $\lambda_2 = -2$ with multiplicity 2. An eigenvector for λ_1 is given by

$$\mathbf{v_1} = \begin{pmatrix} 1 \\ 0 \\ 1 \end{pmatrix}.$$

Two eigenvectors for λ_2 are given by

$$\mathbf{v_2} = \begin{pmatrix} 0 \\ -3 \\ 1 \end{pmatrix}, \quad \mathbf{v_3} = \begin{pmatrix} 1 \\ 2 \\ 0 \end{pmatrix}.$$

Therefore, three solutions of the homogeneous equation are given by

$$\mathbf{x_1}(t) = e^{-t}\begin{pmatrix} 1 \\ 0 \\ 1 \end{pmatrix}, \quad \mathbf{x_2}(t) = e^{-2t}\begin{pmatrix} 0 \\ -3 \\ 1 \end{pmatrix}, \quad \mathbf{x_3}(t) = e^{-2t}\begin{pmatrix} 1 \\ 2 \\ 0 \end{pmatrix},$$

and

$$\mathbf{X}(t) = \begin{pmatrix} e^{-t} & 0 & e^{-2t} \\ 0 & -3e^{-2t} & 2e^{-2t} \\ e^{-t} & e^{-2t} & 0 \end{pmatrix}$$

is a fundamental matrix for this equation. Further,

$$\mathbf{X}^{-1}(s) = \begin{pmatrix} -2e^s & e^s & 3e^s \\ 2e^{2s} & -e^{2s} & -2e^{2s} \\ 3e^{2s} & -e^{2s} & -3e^{2s} \end{pmatrix}.$$

Therefore,

$$\int_{t_1}^t \mathbf{X}^{-1}(s)\mathbf{g}(s)\,ds = \int_{t_1}^t \begin{pmatrix} -2e^s & e^s & 3e^s \\ 2e^{2s} & -e^{2s} & -2e^{2s} \\ 3e^{2s} & -e^{2s} & -3e^{2s} \end{pmatrix} \begin{pmatrix} s \\ 0 \\ \cos s \end{pmatrix} ds$$

$$= \int_{t_1}^t \begin{pmatrix} -2se^s + 3e^s\cos s \\ 2se^{2s} - 2e^{2s}\cos s \\ 3se^{2s} - 3e^{2s}\cos s \end{pmatrix} ds$$

$$= \begin{pmatrix} -2te^t + 2e^t + \frac{3}{2}e^t\cos t + \frac{3}{2}e^t\sin t \\ te^{2t} - \frac{1}{2}e^{2t} - \frac{4}{5}e^{2t}\cos t - \frac{2}{5}e^{2t}\sin t \\ \frac{3}{2}te^{2t} - \frac{3}{4}e^{2t} - \frac{6}{5}e^{2t}\cos t - \frac{3}{5}e^{2t}\sin t \end{pmatrix} + \mathbf{c}.$$

Therefore, the general solution is given by

$$\mathbf{x}(t) = \mathbf{X}(t)\mathbf{c} + \mathbf{X}(t)\int_{t_1}^t \mathbf{X}^{-1}(s)\mathbf{g}(s)\,ds$$

$$= \begin{pmatrix} e^{-t} & 0 & e^{-2t} \\ 0 & -3e^{-2t} & 2e^{-2t} \\ e^{-t} & e^{-2t} & 0 \end{pmatrix} \mathbf{c} + \begin{pmatrix} e^{-t} & 0 & e^{-2t} \\ 0 & -3e^{-2t} & 2e^{-2t} \\ e^{-t} & e^{-2t} & 0 \end{pmatrix} \begin{pmatrix} -2te^t + 2e^t + \frac{3}{2}e^t\cos t + \frac{3}{2}e^t\sin t \\ te^{2t} - \frac{1}{2}e^{2t} - \frac{4}{5}e^{2t}\cos t - \frac{2}{5}e^{2t}\sin t \\ \frac{3}{2}te^{2t} - \frac{3}{4}e^{2t} - \frac{6}{5}e^{2t}\cos t - \frac{3}{5}e^{2t}\sin t \end{pmatrix}$$

$$= c_1 e^{-t}\begin{pmatrix} 1 \\ 0 \\ 1 \end{pmatrix} + c_2 e^{-2t}\begin{pmatrix} 0 \\ -3 \\ 1 \end{pmatrix} + c_3 e^{-2t}\begin{pmatrix} 1 \\ 2 \\ 0 \end{pmatrix} + \begin{pmatrix} \frac{5}{4} - \frac{1}{2}t + \frac{3}{10}\cos t + \frac{9}{10}\sin t \\ 0 \\ \frac{3}{2} - t + \frac{7}{10}\cos t + \frac{11}{10}\sin t \end{pmatrix}.$$

9. The eigenvalues of A are given by $\lambda_1 = -2$, $\lambda_2 = -1$ and $\lambda_3 = 1$ with eigenvectors

$$\mathbf{v_1} = \begin{pmatrix} 1 \\ -2 \\ 1 \end{pmatrix}, \quad \mathbf{v_2} = \begin{pmatrix} -1 \\ 0 \\ 1 \end{pmatrix}, \quad \mathbf{v_3} = \begin{pmatrix} 1 \\ 1 \\ 1 \end{pmatrix}.$$

Therefore, three solutions of the homogeneous equation are given by

$$\mathbf{x_1}(t) = e^{-2t}\begin{pmatrix} 1 \\ -2 \\ 1 \end{pmatrix}, \quad \mathbf{x_2}(t) = e^{-t}\begin{pmatrix} -1 \\ 0 \\ 1 \end{pmatrix}, \quad \mathbf{x_3}(t) = e^t\begin{pmatrix} 1 \\ 1 \\ 1 \end{pmatrix},$$

and

$$\mathbf{X}(t) = \begin{pmatrix} e^{-2t} & -e^{-t} & e^t \\ -2e^{-2t} & 0 & e^t \\ e^{-2t} & e^{-t} & e^t \end{pmatrix}$$

is a fundamental matrix for this equation. Further,

$$\mathbf{X}^{-1}(s) = \frac{1}{6}\begin{pmatrix} e^{2s} & -2e^{2s} & e^{2s} \\ -3e^s & 0 & 3e^s \\ 2e^{-s} & 2e^{-s} & 2e^{-s} \end{pmatrix}.$$

79

Therefore,

$$\int_{t_1}^{t} \mathbf{X}^{-1}(s)\mathbf{g}(s)\,ds = \frac{1}{6}\int_{t_1}^{t}\begin{pmatrix} e^{2s} & -2e^{2s} & e^{2s} \\ -3e^{s} & 0 & 3e^{s} \\ 2e^{-s} & 2e^{-s} & 2e^{-s} \end{pmatrix}\begin{pmatrix} 0 \\ \sin s \\ 0 \end{pmatrix}ds$$

$$= \frac{1}{6}\int_{t_1}^{t}\begin{pmatrix} -2e^{2s}\sin s \\ 0 \\ 2e^{-s}\sin s \end{pmatrix}ds$$

$$= \frac{1}{6}\begin{pmatrix} \frac{2}{5}e^{2t}\cos t - \frac{4}{5}e^{2t}\sin t \\ 0 \\ -e^{-t}\cos t - e^{-t}\sin t \end{pmatrix} + \mathbf{c}.$$

Therefore, the general solution is given by

$$\mathbf{x}(t) = \mathbf{X}(t)\mathbf{c} + \mathbf{X}(t)\int_{t_1}^{t}\mathbf{X}^{-1}(s)\mathbf{g}(s)\,ds$$

$$= \begin{pmatrix} e^{-2t} & -e^{-t} & e^{t} \\ -2e^{-2t} & 0 & e^{t} \\ e^{-2t} & e^{-t} & e^{t} \end{pmatrix}\mathbf{c} + \frac{1}{6}\begin{pmatrix} e^{-2t} & -e^{-t} & e^{t} \\ -2e^{-2t} & 0 & e^{t} \\ e^{-2t} & e^{-t} & e^{t} \end{pmatrix}\begin{pmatrix} \frac{2}{5}e^{2t}\cos t - \frac{4}{5}e^{2t}\sin t \\ 0 \\ -e^{-t}\cos t - e^{-t}\sin t \end{pmatrix}$$

$$= c_1 e^{-2t}\begin{pmatrix} 1 \\ -2 \\ 1 \end{pmatrix} + c_2 e^{-t}\begin{pmatrix} -1 \\ 0 \\ 1 \end{pmatrix} + c_3 e^{t}\begin{pmatrix} 1 \\ 1 \\ 1 \end{pmatrix} + \begin{pmatrix} -\frac{1}{10}\cos t - \frac{3}{10}\sin t \\ \frac{1}{10}\sin t - \frac{3}{10}\cos t \\ -\frac{1}{10}\cos t - \frac{3}{10}\sin t \end{pmatrix}.$$

10.

(a) The eigenvalues of A are approximately $\lambda_1 \approx -0.198$, $\lambda_2 \approx -3.25$ and $\lambda_3 \approx -1.55$. The eigenvectors are approximately

$$\mathbf{v_1} = \begin{pmatrix} 0.445 \\ 0.802 \\ 1 \end{pmatrix}, \quad \mathbf{v_2} = \begin{pmatrix} 1.802 \\ -2.247 \\ 1 \end{pmatrix}, \quad \mathbf{v_3} = \begin{pmatrix} -1.247 \\ -0.555 \\ 1 \end{pmatrix},$$

Therefore,

$$\mathbf{X}(t) = \begin{pmatrix} 0.445e^{-0.198t} & 1.802e^{-3.25t} & -1.247e^{-1.55t} \\ 0.802e^{-0.198t} & -2.247e^{-3.25t} & -0.555e^{-1.55t} \\ e^{-0.198t} & e^{-3.25t} & e^{-1.55t} \end{pmatrix}$$

is a fundamental matrix. Then

$$\mathbf{X}^{-1}(s) = \begin{pmatrix} 0.242e^{0.198s} & 0.436e^{0.198s} & 0.543e^{0.198s} \\ 0.194e^{3.25s} & -0.242e^{3.25s} & 0.107e^{3.25s} \\ -0.436e^{1.55s} & -0.194e^{1.55s} & 0.349e^{1.55s} \end{pmatrix}$$

implies

$$\int_{t_1}^{t}\mathbf{X}^{-1}(s)\mathbf{g}(s)\,ds = \int_{t_1}^{t}\begin{pmatrix} 0.242e^{0.198s} & 0.436e^{0.198s} & 0.543e^{0.198s} \\ 0.194e^{3.25s} & -0.242e^{3.25s} & 0.107e^{3.25s} \\ -0.436e^{1.55s} & -0.194e^{1.55s} & 0.349e^{1.55s} \end{pmatrix}\begin{pmatrix} 1 \\ 0 \\ 0 \end{pmatrix}ds$$

$$= \int_{t_1}^{t}\begin{pmatrix} 0.242e^{0.198s} \\ 0.194e^{3.25s} \\ -0.436e^{1.55s} \end{pmatrix}$$

$$= \begin{pmatrix} 1.22e^{0.198t} \\ .0597e^{3.25t} \\ -0.281e^{1.55t} \end{pmatrix} + \mathbf{c}.$$

80

Therefore, the general solution is

$$\mathbf{x}(t) = \mathbf{X}(t)\mathbf{c} + \mathbf{X}(t) \int_{t_1}^{t} \mathbf{X}^{-1}(s)\mathbf{g}(s)\,ds$$

$$= \begin{pmatrix} 0.445e^{-0.198t} & 1.802e^{-3.25t} & -1.247e^{-1.55t} \\ 0.802e^{-0.198t} & -2.247e^{-3.25t} & -0.555e^{-1.55t} \\ e^{-0.198t} & e^{-3.25t} & e^{-1.55t} \end{pmatrix} \mathbf{c}$$

$$+ \begin{pmatrix} 0.445e^{-0.198t} & 1.802e^{-3.25t} & -1.247e^{-1.55t} \\ 0.802e^{-0.198t} & -2.247e^{-3.25t} & -0.555e^{-1.55t} \\ e^{-0.198t} & e^{-3.25t} & e^{-1.55t} \end{pmatrix} \begin{pmatrix} 1.22e^{0.198t} \\ .0597e^{3.25t} \\ -0.281e^{1.55t} \end{pmatrix}$$

$$= \begin{pmatrix} 0.445e^{-0.198t} & 1.802e^{-3.25t} & -1.247e^{-1.55t} \\ 0.802e^{-0.198t} & -2.247e^{-3.25t} & -0.555e^{-1.55t} \\ e^{-0.198t} & e^{-3.25t} & e^{-1.55t} \end{pmatrix} \mathbf{c} + \begin{pmatrix} 1 \\ 1 \\ 1 \end{pmatrix}$$

Now, the initial condition implies

$$\mathbf{x}(0) = \begin{pmatrix} 0.445 & 1.802 & -1.247 \\ 0.802 & -2.247 & -0.555 \\ 1 & 1 & 1 \end{pmatrix} \mathbf{c} + \begin{pmatrix} 1 \\ 1 \\ 1 \end{pmatrix} = \begin{pmatrix} 0 \\ 0 \\ 0 \end{pmatrix}.$$

That is,

$$\mathbf{X}(0)\mathbf{c} = \begin{pmatrix} -1 \\ -1 \\ -1 \end{pmatrix}.$$

Therefore,

$$\mathbf{c} = \mathbf{X}^{-1}(0) \begin{pmatrix} -1 \\ -1 \\ -1 \end{pmatrix}$$

$$= \begin{pmatrix} 0.242 & 0.436 & 0.543 \\ 0.194 & -0.242 & 0.107 \\ -0.436 & -0.194 & 0.349 \end{pmatrix} \begin{pmatrix} -1 \\ -1 \\ -1 \end{pmatrix}$$

$$= \begin{pmatrix} -1.221 \\ -.059 \\ .281 \end{pmatrix}$$

Therefore, the solution is given by

$$\mathbf{x}(t) = -1.221e^{-0.198t} \begin{pmatrix} 0.445 \\ 0.802 \\ 1 \end{pmatrix} - .059e^{-3.25t} \begin{pmatrix} 1.802 \\ -2.247 \\ 1 \end{pmatrix} + .281e^{-1.55t} \begin{pmatrix} -1.247 \\ -0.555 \\ 1 \end{pmatrix} + \begin{pmatrix} 1 \\ 1 \\ 1 \end{pmatrix}.$$

As $t \to \infty$, $\mathbf{x}(t) \to (1,1,1)^T$. Below are the graphs of x_1, x_2 and x_3 respectively.

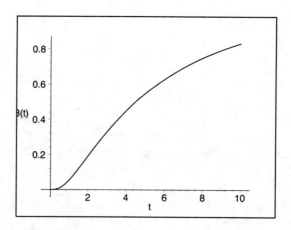

(b) The calculation proceeds as in part (a). We need to recalculate the term involving **g** because of a new inhomogeneous term. In particular, we have

$$\int_{t_1}^{t} \mathbf{X}^{-1}(s)\mathbf{g}(s)\,ds = \int_{t_1}^{t} \begin{pmatrix} 0.242e^{0.198s} & 0.436e^{0.198s} & 0.543e^{0.198s} \\ 0.194e^{3.25s} & -0.242e^{3.25s} & 0.107e^{3.25s} \\ -0.436e^{1.55s} & -0.194e^{1.55s} & 0.349e^{1.55s} \end{pmatrix} \begin{pmatrix} 1-\cos(\omega s) \\ 0 \\ 0 \end{pmatrix} ds$$

$$= \int_{t_1}^{t} \left[\begin{pmatrix} 0.242e^{0.198s} \\ 0.194e^{3.25s} \\ -0.436e^{1.55s} \end{pmatrix} - \begin{pmatrix} 0.242e^{0.198s}\cos(\omega s) \\ 0.194e^{3.25s}\cos(\omega s) \\ -0.436e^{1.55s}\cos(\omega s) \end{pmatrix} \right] ds$$

$$= \begin{pmatrix} 1.22e^{.198t} - (0.048e^{.198t}\cos(\omega t))/(0.039+\omega^2) - (0.242\omega e^{.198t}\sin(\omega t))/(0.039+\omega^2) \\ 0.059e^{3.25t} - (.644e^{3.25t}\cos(\omega t))/(10.56+\omega^2) - (.198\omega e^{3.25t}\sin(\omega t))/(10.56+\omega^2) \\ -0.281e^{1.55t} + (0.676e^{1.55t}\cos(\omega t))/(2.4+\omega^2) + (0.436\omega e^{1.55t}\sin(\omega t))/(2.4+\omega^2) \end{pmatrix} + \mathbf{c}$$

Using our answer from part (a) and the calculation above, we see that the general solution is given by

$$\mathbf{x}(t) = \begin{pmatrix} 0.445e^{-0.198t} & 1.802e^{-3.25t} & -1.247e^{-1.55t} \\ 0.802e^{-0.198t} & -2.247e^{-3.25t} & -0.555e^{-1.55t} \\ e^{-0.198t} & e^{-3.25t} & e^{-1.55t} \end{pmatrix} \mathbf{c} + \begin{pmatrix} 1 \\ 1 \\ 1 \end{pmatrix}$$

$$+ \cos(\omega t)\left[\frac{1}{0.039+\omega^2}\begin{pmatrix} -0.0214 \\ -0.038 \\ -0.048 \end{pmatrix} + \frac{1}{10.56+\omega^2}\begin{pmatrix} -1.16 \\ 1.45 \\ -0.644 \end{pmatrix} + \frac{1}{2.4+\omega^2}\begin{pmatrix} -0.843 \\ -0.375 \\ 0.676 \end{pmatrix} \right]$$

$$+ \omega\sin(\omega t)\left[\frac{1}{0.039+\omega^2}\begin{pmatrix} -0.108 \\ -0.194 \\ -0.242 \end{pmatrix} + \frac{1}{10.56+\omega^2}\begin{pmatrix} -.357 \\ 0.445 \\ -0.198 \end{pmatrix} + \frac{1}{2.4+\omega^2}\begin{pmatrix} -0.544 \\ -0.242 \\ 0.435 \end{pmatrix} \right]$$

Using the initial condition, we have

$$\mathbf{x}(0) = \begin{pmatrix} 0.445 & 1.802 & -1.247 \\ 0.802 & -2.247 & -0.555 \\ 1 & 1 & 1 \end{pmatrix} \mathbf{c} + \begin{pmatrix} 1 \\ 1 \\ 1 \end{pmatrix}$$

$$+ \left[\frac{1}{0.039+\omega^2}\begin{pmatrix} -0.0214 \\ -0.038 \\ -0.048 \end{pmatrix} + \frac{1}{10.56+\omega^2}\begin{pmatrix} -1.16 \\ 1.45 \\ -0.644 \end{pmatrix} + \frac{1}{2.4+\omega^2}\begin{pmatrix} -0.843 \\ -0.375 \\ 0.676 \end{pmatrix} \right] = \begin{pmatrix} 0 \\ 0 \\ 0 \end{pmatrix}.$$

For $\omega = 1$, this equation reduces to

$$\begin{pmatrix} 0.445 & 1.802 & -1.247 \\ 0.802 & -2.247 & -0.555 \\ 1 & 1 & 1 \end{pmatrix} \mathbf{c} = \begin{pmatrix} -0.611 \\ -0.728 \\ -1.097 \end{pmatrix}$$

82

Using $\mathbf{X}^{-1}(0)$, we have

$$
\mathbf{c} = \begin{pmatrix} 0.242 & 0.436 & 0.543 \\ 0.194 & -0.242 & 0.107 \\ -0.436 & -0.194 & 0.349 \end{pmatrix} \begin{pmatrix} -0.611 \\ -0.728 \\ -1.097 \end{pmatrix}
$$

$$
= \begin{pmatrix} -1.06 \\ -0.0597 \\ 0.0248 \end{pmatrix}
$$

Therefore, the solution is given by $\mathbf{x}(t)$ above with this vector $\mathbf{c}$. Below we show component plots of x_1, x_2, x_3 respectively.

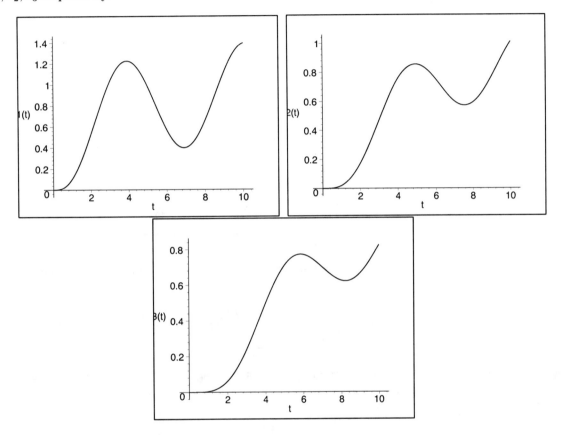

If $\omega = 4$, we proceed similarly. The equation for $\mathbf{x}(0)$ reduces to

$$
\begin{pmatrix} 0.445 & 1.802 & -1.247 \\ 0.802 & -2.247 & -0.555 \\ 1 & 1 & 1 \end{pmatrix} \mathbf{c} = \begin{pmatrix} -0.909 \\ -1.03 \\ -1.01 \end{pmatrix}
$$

Using $\mathbf{X}^{-1}(0)$, we have

$$
\mathbf{c} = \begin{pmatrix} 0.242 & 0.436 & 0.543 \\ 0.194 & -0.242 & 0.107 \\ -0.436 & -0.194 & 0.349 \end{pmatrix} \begin{pmatrix} -0.909 \\ -1.03 \\ -1.01 \end{pmatrix}
$$

$$
= \begin{pmatrix} -1.22 \\ -0.035 \\ 0.0244 \end{pmatrix}.
$$

In this case the solution is given by $\mathbf{x}(t)$ above with $\omega = 4$ and $\mathbf{c}$ as defined here. Below we show component plots of x_1, x_2, x_3 respectively.

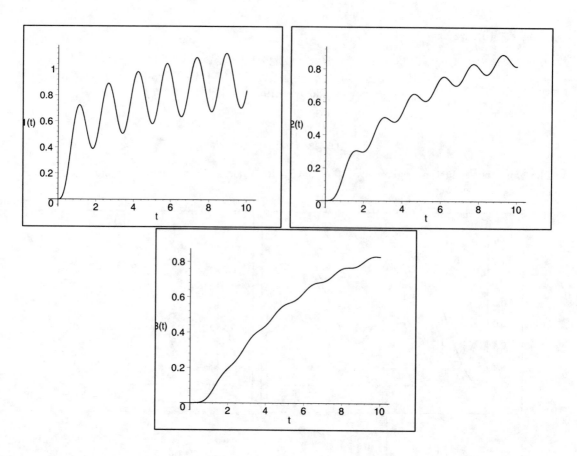

11. Assuming another lattice point to the right of x_3 which is an absorbing boundary, our system of equations can be written as

$$\begin{pmatrix} x_1 \\ x_2 \\ x_3 \end{pmatrix}' = k \begin{pmatrix} -2 & 1 & 0 \\ 1 & -2 & 1 \\ 0 & 1 & -2 \end{pmatrix} \begin{pmatrix} x_1 \\ x_2 \\ x_3 \end{pmatrix} + k \begin{pmatrix} f(t) \\ 0 \\ 0 \end{pmatrix}.$$

For $k = 1$, $f(t) = 1$ with the given initial conditions, we show component plots of x_1, x_2, x_3 below.

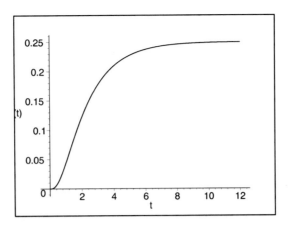

As $t \to \infty$, we see that $\mathbf{x}(t) \to (3/4, 1/2, 1/4)^T$. For $k = 1$, $f(t) = 1 - \cos(t)$, we show component plots of x_1, x_2, x_3 below:

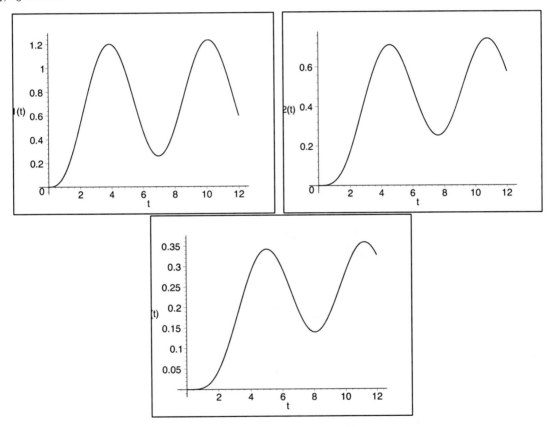

For $k = 1$, $f(t) = 1 - \cos(4t)$, we show component plots of x_1, x_2, x_3 below

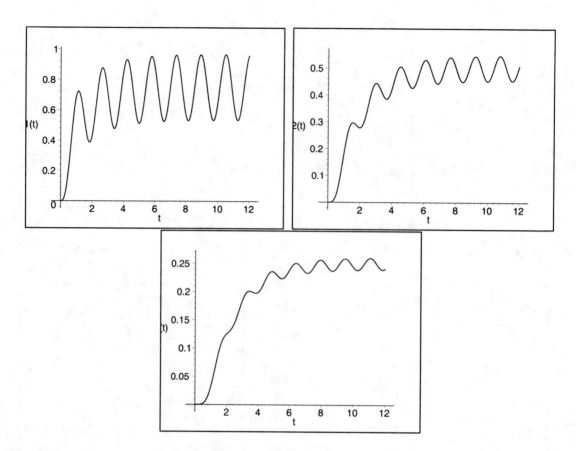

12. Plugging in the values for all the k_{ij} terms and L, we have

$$\begin{pmatrix} x_1 \\ x_2 \\ x_3 \end{pmatrix}' = \begin{pmatrix} -.0359 & .012 & .000035 \\ .011 & -.028 & 0 \\ .0039 & 0 & -.000035 \end{pmatrix} \begin{pmatrix} x_1 \\ x_2 \\ x_3 \end{pmatrix} + \begin{pmatrix} 35 \\ 0 \\ 0 \end{pmatrix}$$

The solution for $0 \le t \le 35$ will be given by

$$\mathbf{x}(t) = e^{At}\mathbf{x_0} + e^{At} \int_0^t e^{-As}\mathbf{g}(s)\, ds.$$

Since $\mathbf{x_0} = (0,0,0)^T$, the solution will just be given by

$$\mathbf{x}(t) = e^{At} \int_0^t e^{-As}\mathbf{g}(s)\, ds.$$

Using the fundamental matrix e^{At}, we have

$$e^{A(t-s)}\mathbf{g}(s) = 35 \begin{pmatrix} .66e^{-.044(t-s)} + .34e^{-.02(t-s)} + .00014e^{-.000031(t-s)} \\ .45e^{-.02(t-s)} - .45e^{-.044(t-s)} + .000055e^{-.000031(t-s)} \\ -.067e^{-.02(t-s)} - .059e^{-.044(t-s)} + .13e^{-.000031(t-s)} \end{pmatrix}.$$

Therefore, for $0 \le t \le 365$,

$$\mathbf{x}(t) = \int_0^t e^{A(t-s)}\mathbf{g}(s)\, ds$$

$$= 35[1 - e^{-.044t}] \begin{pmatrix} 15 \\ -10.2 \\ -1.34 \end{pmatrix} + 35[1 - e^{-.02t}] \begin{pmatrix} 17 \\ 22.5 \\ -3.35 \end{pmatrix} + 35[1 - e^{-.000031t}] \begin{pmatrix} 4.5 \\ 1.77 \\ 4193 \end{pmatrix}.$$

86

At $t = 365$,
$$\mathbf{x}(365) \approx \begin{pmatrix} 1121 \\ 431 \\ 1487 \end{pmatrix}.$$

Then, the inhomogeneous term vanishes for $t \geq 365$. Therefore, for $t \geq 365$, the solution is given by

$$\mathbf{x}(t) = e^{A(t-365)}\mathbf{x}(365)$$

$$= e^{-.044(t-365)} \begin{pmatrix} 528 \\ -357 \\ -47 \end{pmatrix} + e^{-.02(t-365)} \begin{pmatrix} 591 \\ 788 \\ -117 \end{pmatrix} + e^{-.000031(t-365)} \begin{pmatrix} 1.8 \\ 1.3 \\ 1609 \end{pmatrix}.$$

13. Using the values specified for m_1, m_2, k_1, k_2, the gain function is given by

$$\mathbf{G}(i\omega) = -(\mathbf{A} - i\omega\mathbf{I})^{-1}\mathbf{B}$$

where

$$\mathbf{A} = \begin{pmatrix} 0 & 0 & 1 & 0 \\ 0 & 0 & 0 & 1 \\ -2 & 3/2 & -\gamma/2 & 0 \\ 3/2 & -16/9 & 0 & -4\gamma/9 \end{pmatrix}$$

and

$$\mathbf{B} = \begin{pmatrix} 0 \\ 0 \\ e^{i\omega t} \\ 0 \end{pmatrix}.$$

We can compute $|G_1(i\omega)|$, $|G_2(i\omega)|$ by looking at the first and second components of the function $G(i\omega)$ above. For $\gamma = 1$, we have

$$|G_1(i\omega)| = \left| \frac{2(-4i\omega + 9\omega^2 - 16)}{-17\omega^2 - 17i\omega^3 + 18\omega^4 + 32i\omega + 28} \right|$$

$$|G_2(i\omega)| = \left| \frac{24}{-72\omega^2 - 17i\omega^3 + 18\omega^4 + 32i\omega + 28} \right|$$

The gain functions corresponding to $\gamma = 0.5$ and $\gamma = 0.1$ are shown below:

14. For

$$\mathbf{x}_p(t) = t \begin{pmatrix} a_1 \\ a_2 \end{pmatrix} + \begin{pmatrix} b_1 \\ b_2 \end{pmatrix},$$

$$\mathbf{x_p}' = \begin{pmatrix} a_1 \\ a_2 \end{pmatrix}.$$

Therefore,

$$\begin{pmatrix} a_1 \\ a_2 \end{pmatrix} = \begin{pmatrix} (-2a_1 + a_2)t - 2b_1 + b_2 \\ (a_1 - 2a_2 + 3)t + b_1 - 2b_2 \end{pmatrix}.$$

Equating like coefficients, we have $-2a_1 + a_2 = 0$, $-2b_1 + b_2 = a_1$, $a_1 - 2a_2 + 3 = 0$, $b_1 - 2b_2 = a_2$. Solving this system of equations, we have $a_1 = 1$, $a_2 = 2$, $b_1 = -12/9$, $b_2 = -15/9$. Therefore,

$$\mathbf{x_p}(t) = t \begin{pmatrix} 1 \\ 2 \end{pmatrix} + \begin{pmatrix} -\frac{12}{9} \\ -\frac{15}{9} \end{pmatrix}.$$

15. For

$$\mathbf{x}_p(t) = \cos t \begin{pmatrix} a_1 \\ a_2 \end{pmatrix} + \sin t \begin{pmatrix} b_1 \\ b_2 \end{pmatrix},$$

$$\mathbf{x_p}' = -\sin t \begin{pmatrix} a_1 \\ a_2 \end{pmatrix} + \cos t \begin{pmatrix} b_1 \\ b_2 \end{pmatrix}.$$

Therefore,

$$\begin{pmatrix} -a_1 \sin t + b_1 \cos t \\ -a_2 \sin t + b_2 \cos t \end{pmatrix} = \begin{pmatrix} (-2a_1 + a_2)\cos t + (-2b_1 + b_2 + 1)\sin t \\ (a_1 - 2a_2)\cos t + (b_1 - 2b_2)\sin t \end{pmatrix}.$$

Equating like coefficients, we have $-2a_1 + a_2 - b_1 = 0$, $-2b_1 + b_2 + a_1 + 1 = 0$, $a_1 - 2a_2 - b_2 = 0$, $b_1 - 2b_2 + a_2 = 0$. Solving this system of equations, we have $a_1 = -3/10$, $a_2 = -1/5$, $b_1 = 2/5$, $b_2 = 1/10$. Therefore,

$$\mathbf{x_p}(t) = \cos t \begin{pmatrix} -\frac{3}{10} \\ -\frac{1}{5} \end{pmatrix} + \sin t \begin{pmatrix} \frac{2}{5} \\ \frac{1}{10} \end{pmatrix}.$$

16. For

$$\mathbf{x}_p(t) = e^{-t} \begin{pmatrix} b_1 \\ b_2 \end{pmatrix} + te^{-t} \begin{pmatrix} a_1 \\ a_2 \end{pmatrix},$$

$$\mathbf{x_p}' = -e^{-t} \begin{pmatrix} b_1 \\ b_2 \end{pmatrix} + e^{-t} \begin{pmatrix} a_1 \\ a_2 \end{pmatrix} - te^{-t} \begin{pmatrix} a_1 \\ a_2 \end{pmatrix}.$$

Therefore,

$$\begin{pmatrix} -b_1 e^{-t} + a_1 e^{-t} - a_1 te^{-t} \\ -b_2 e^{-t} + a_2 e^{-t} - a_2 te^{-t} \end{pmatrix} = \begin{pmatrix} (-2b_1 + b_2 + 2)e^{-t} + (-2a_1 + a_2)te^{-t} \\ (b_1 - 2b_2)e^{-t} + (a_1 - 2a_2)te^{-t} \end{pmatrix}.$$

Equating like coefficients, we have $-b_1 + b_2 + 2 - a_1 = 0$, $-a_1 + a_2 = 0$, $b_1 - b_2 - a_2 = 0$. Using the fact that $a_1 = a_2$, we have $-b_1 + b_2 + 2 = b_1 - b_2$ which implies $2b_1 - 2b_2 = 2$, or $b_1 - b_2 = 1$. Therefore, $a_2 = b_1 - b_2 = 1$, and, since $a_1 = a_2$, we must also have $a_1 = 1$. Then $b_1 - b_2 = 1$ implies b_1, b_2 are arbitrary as long as $b_1 = 1 + b_2$. In particular, take $b_1 = 1$, $b_2 = 0$. We conclude that

$$\mathbf{x_p}(t) = e^{-t} \begin{pmatrix} 1 \\ 0 \end{pmatrix} + te^{-t} \begin{pmatrix} 1 \\ 1 \end{pmatrix}$$

is a particular solution.

Section 6.7

1. The only eigenvalue of A is $\lambda = 1$ with multiplicity 2. We see that

$$A - \lambda I = \begin{pmatrix} 2 & -4 \\ 1 & -2 \end{pmatrix}$$
$$(A - \lambda I)^2 = \begin{pmatrix} 0 & 0 \\ 0 & 0 \end{pmatrix}.$$

Therefore, two linearly independent solutions of $(A - \lambda I)^2 \mathbf{v} = 0$ are

$$\mathbf{v_1} = \begin{pmatrix} 1 \\ 0 \end{pmatrix}, \quad \mathbf{v_2} = \begin{pmatrix} 0 \\ 1 \end{pmatrix}.$$

Then by Theorem 6.7.1, two linearly independent solutions of our system will be given by the following equation:

$$\mathbf{x_k}(t) = e^t[\mathbf{v_k} + t(A - I)\mathbf{v_k}]$$

for $k = 1, 2$. Using the values for $\mathbf{v_k}$ and $A - \lambda I$ above, we have

$$\mathbf{x_1}(t) = e^t \left[\begin{pmatrix} 1 \\ 0 \end{pmatrix} + t \begin{pmatrix} 2 \\ 1 \end{pmatrix} \right]$$
$$\mathbf{x_2}(t) = e^t \left[\begin{pmatrix} 0 \\ 1 \end{pmatrix} + t \begin{pmatrix} -4 \\ -2 \end{pmatrix} \right].$$

Therefore,

$$\mathbf{X}(t) = e^t \begin{pmatrix} 1 + 2t & -4t \\ t & 1 - 2t \end{pmatrix}$$

is a fundamental matrix for this system.

2. The only eigenvalue of A is $\lambda = 0$ with multiplicity 2. We see that

$$A - \lambda I = \begin{pmatrix} 4 & -2 \\ 8 & -4 \end{pmatrix}$$
$$(A - \lambda I)^2 = \begin{pmatrix} 0 & 0 \\ 0 & 0 \end{pmatrix}.$$

Therefore, two linearly independent solutions of $(A - \lambda I)^2 \mathbf{v} = 0$ are

$$\mathbf{v_1} = \begin{pmatrix} 1 \\ 0 \end{pmatrix}, \quad \mathbf{v_2} = \begin{pmatrix} 0 \\ 1 \end{pmatrix}.$$

Then by Theorem 6.7.1, two linearly independent solutions of our system will be given by the following equation:

$$\mathbf{x_k}(t) = [\mathbf{v_k} + tA\mathbf{v_k}]$$

for $k = 1, 2$. Using the values for $\mathbf{v_k}$ and A above, we have

$$\mathbf{x_1}(t) = \left[\begin{pmatrix} 1 \\ 0 \end{pmatrix} + t \begin{pmatrix} 4 \\ 8 \end{pmatrix}\right]$$

$$\mathbf{x_2}(t) = \left[\begin{pmatrix} 0 \\ 1 \end{pmatrix} + t \begin{pmatrix} -2 \\ -4 \end{pmatrix}\right].$$

Therefore,

$$\mathbf{X}(t) = \begin{pmatrix} 1 + 4t & -2t \\ 8t & 1 - 4t \end{pmatrix}$$

is a fundamental matrix for this system.

3. The only eigenvalue of A is $\lambda = 2$ with multiplicity 3. We see that

$$A - \lambda I = \begin{pmatrix} -1 & 1 & 1 \\ 2 & -1 & -1 \\ -3 & 2 & 2 \end{pmatrix}$$

$$(A - \lambda I)^2 = \begin{pmatrix} 0 & 0 & 0 \\ -1 & 1 & 1 \\ 1 & -1 & -1 \end{pmatrix}$$

$$(A - \lambda I)^3 = \begin{pmatrix} 0 & 0 & 0 \\ 0 & 0 & 0 \\ 0 & 0 & 0 \end{pmatrix}.$$

Therefore, three linearly independent solutions of $(A - \lambda I)^3 \mathbf{v} = 0$ are

$$\mathbf{v_1} = \begin{pmatrix} 1 \\ 0 \\ 0 \end{pmatrix}, \quad \mathbf{v_2} = \begin{pmatrix} 0 \\ 1 \\ 0 \end{pmatrix}, \quad \mathbf{v_3} = \begin{pmatrix} 0 \\ 0 \\ 1 \end{pmatrix}.$$

Then by Theorem 6.7.1, three linearly independent solutions of our system will be given by the following equation:

$$\mathbf{x_k}(t) = e^{2t}\left[\mathbf{v_k} + t(A - 2I)\mathbf{v_k} + \frac{t^2}{2}(A - 2I)^2\mathbf{v_k}\right]$$

for $k = 1, 2, 3$. Using the values for $\mathbf{v_k}$ and $A - \lambda I$ above, we have

$$\mathbf{x_1}(t) = e^{2t}\left[\begin{pmatrix} 1 \\ 0 \\ 0 \end{pmatrix} + t \begin{pmatrix} -1 \\ 2 \\ -3 \end{pmatrix} + \frac{t^2}{2}\begin{pmatrix} 0 \\ -1 \\ 1 \end{pmatrix}\right]$$

$$\mathbf{x_2}(t) = e^{2t}\left[\begin{pmatrix} 0 \\ 1 \\ 0 \end{pmatrix} + t \begin{pmatrix} 1 \\ -1 \\ 2 \end{pmatrix} + \frac{t^2}{2}\begin{pmatrix} 0 \\ 1 \\ -1 \end{pmatrix}\right]$$

$$\mathbf{x_3}(t) = e^{2t}\left[\begin{pmatrix} 0 \\ 0 \\ 1 \end{pmatrix} + t \begin{pmatrix} 1 \\ -1 \\ 2 \end{pmatrix} + \frac{t^2}{2}\begin{pmatrix} 0 \\ 1 \\ -1 \end{pmatrix}\right].$$

Therefore,

$$\mathbf{X}(t) = e^{2t}\begin{pmatrix} 1 - t & t & t \\ 2t - \frac{t^2}{2} & 1 - t + \frac{t^2}{2} & -t + \frac{t^2}{2} \\ -3t + \frac{t^2}{2} & 2t - \frac{t^2}{2} & 1 + 2t - \frac{t^2}{2} \end{pmatrix}$$

is a fundamental matrix for this system.

4. The only eigenvalue of A is $\lambda = 1$ with multiplicity 3. We see that

$$A - \lambda I = \begin{pmatrix} 4 & -3 & -2 \\ 8 & -6 & -4 \\ -4 & 3 & 2 \end{pmatrix}$$

$$(A - \lambda I)^2 = \begin{pmatrix} 0 & 0 & 0 \\ 0 & 0 & 0 \\ 0 & 0 & 0 \end{pmatrix}$$

$$(A - \lambda I)^3 = \begin{pmatrix} 0 & 0 & 0 \\ 0 & 0 & 0 \\ 0 & 0 & 0 \end{pmatrix}.$$

Therefore, three linearly independent solutions of $(A - \lambda I)^3 \mathbf{v} = 0$ are

$$\mathbf{v_1} = \begin{pmatrix} 1 \\ 0 \\ 0 \end{pmatrix}, \quad \mathbf{v_2} = \begin{pmatrix} 0 \\ 1 \\ 0 \end{pmatrix}, \quad \mathbf{v_3} = \begin{pmatrix} 0 \\ 0 \\ 1 \end{pmatrix}.$$

Then by Theorem 6.7.1, three linearly independent solutions of our system will be given by the following equation:

$$\mathbf{x_k}(t) = e^t [\mathbf{v_k} + t(A - I)\mathbf{v_k}]$$

for $k = 1, 2, 3$. Using the values for $\mathbf{v_k}$ and $A - \lambda I$ above, we have

$$\mathbf{x_1}(t) = e^t \left[\begin{pmatrix} 1 \\ 0 \\ 0 \end{pmatrix} + t \begin{pmatrix} 4 \\ 8 \\ -4 \end{pmatrix} \right]$$

$$\mathbf{x_2}(t) = e^t \left[\begin{pmatrix} 0 \\ 1 \\ 0 \end{pmatrix} + t \begin{pmatrix} -3 \\ -6 \\ 3 \end{pmatrix} \right]$$

$$\mathbf{x_3}(t) = e^t \left[\begin{pmatrix} 0 \\ 0 \\ 1 \end{pmatrix} + t \begin{pmatrix} -2 \\ -4 \\ 2 \end{pmatrix} \right].$$

Therefore,

$$\mathbf{X}(t) = e^t \begin{pmatrix} 1 + 4t & -3t & -2t \\ 8t & 1 - 6t & -4t \\ -4t & 3t & 1 + 2t \end{pmatrix}$$

is a fundamental matrix for this system.

5. The eigenvalues of A are $\lambda = -1$ with multiplicity 1 and $\lambda = 2$ with multiplicity 2. For $\lambda = -1$,

$$A - \lambda I = \begin{pmatrix} -6 & 9 & -6 \\ -8 & 12 & -7 \\ -2 & 3 & 0 \end{pmatrix} \rightarrow \begin{pmatrix} 1 & -\frac{3}{2} & 0 \\ 0 & 0 & 1 \\ 0 & 0 & 0 \end{pmatrix}.$$

Therefore,

$$\mathbf{v_1} = \begin{pmatrix} \frac{3}{2} \\ 1 \\ 0 \end{pmatrix}$$

is an eigenvector associated with $\lambda = -1$ and

$$\mathbf{x_1}(t) = e^{-t} \begin{pmatrix} \frac{3}{2} \\ 1 \\ 0 \end{pmatrix}$$

is a solution of our system.

For $\lambda = 2$,

$$A - \lambda I = \begin{pmatrix} -9 & 9 & -6 \\ -8 & 9 & -7 \\ -2 & 3 & -3 \end{pmatrix} \rightarrow \begin{pmatrix} 1 & -1 & \frac{2}{3} \\ 0 & 1 & -\frac{5}{3} \\ 0 & 0 & 0 \end{pmatrix}.$$

As this eigenspace is only one-dimensional, we look at $(A - \lambda I)^2$. We see that for $\lambda = 2$,

$$(A - \lambda I)^2 = \begin{pmatrix} 21 & -18 & 9 \\ 14 & -12 & 6 \\ 0 & 0 & 0 \end{pmatrix}.$$

To find two solutions associated with the eigenvalues $\lambda = 2$, we look for two linearly independent solutions of $(A - \lambda I)^2 \mathbf{v} = 0$. We see that

$$(A - \lambda I)^2 \rightarrow \begin{pmatrix} 7 & -6 & 3 \\ 0 & 0 & 0 \\ 0 & 0 & 0 \end{pmatrix}$$

after elementary row operations. Therefore, any vectors $\mathbf{v}$ lying in the plane $7x - 6y + 3z = 0$ will satisfy the equation $(A - \lambda I)^2 \mathbf{v} = 0$. We see that

$$\mathbf{v_2} = \begin{pmatrix} 6 \\ 7 \\ 0 \end{pmatrix}, \quad \mathbf{v_3} = \begin{pmatrix} 3 \\ 0 \\ -7 \end{pmatrix}$$

are two linearly independent vectors satisfying this condition. Then, using Theorem 6.7.1, two solutions of our system will be given by the equation

$$\mathbf{x_k}(t) = e^t[\mathbf{v_k} + t(A - I)\mathbf{v_k}]$$

for $k = 2, 3$. Using the values for $\mathbf{v_k}$ and $A - \lambda I$ above, we have

$$\mathbf{x_2}(t) = e^{2t}\left[\begin{pmatrix} 6 \\ 7 \\ 0 \end{pmatrix} + t\begin{pmatrix} 9 \\ 15 \\ 9 \end{pmatrix}\right]$$

$$\mathbf{x_3}(t) = e^{2t}\left[\begin{pmatrix} 3 \\ 0 \\ -7 \end{pmatrix} + t\begin{pmatrix} 15 \\ 25 \\ 15 \end{pmatrix}\right].$$

Therefore,

$$\mathbf{X}(t) = \begin{pmatrix} \frac{3}{2}e^{-t} & e^{2t}(6 + 9t) & e^{2t}(3 + 15t) \\ e^{-t} & e^{2t}(7 + 15t) & 25te^{2t} \\ 0 & 9te^{2t} & e^{2t}(-7 + 15t) \end{pmatrix}$$

is a fundamental matrix for this system.

6. The only eigenvalue of A is $\lambda = 1$ with multiplicity 3. For $\lambda = 1$,

$$A - \lambda I = \begin{pmatrix} 4 & 6 & 2 \\ -2 & -3 & -1 \\ -2 & -3 & -1 \end{pmatrix}$$

which implies

$$(A - \lambda I)^2 = \begin{pmatrix} 0 & 0 & 0 \\ 0 & 0 & 0 \\ 0 & 0 & 0 \end{pmatrix}$$

$$(A - \lambda I)^2 = \begin{pmatrix} 0 & 0 & 0 \\ 0 & 0 & 0 \\ 0 & 0 & 0 \end{pmatrix}.$$

Therefore,

$$\mathbf{v_1} = \begin{pmatrix} 1 \\ 0 \\ 0 \end{pmatrix}, \quad \mathbf{v_2} = \begin{pmatrix} 0 \\ 1 \\ 0 \end{pmatrix}, \quad \mathbf{v_3} = \begin{pmatrix} 0 \\ 0 \\ 1 \end{pmatrix}$$

are linearly independent solutions of $(A - \lambda I)^2 \mathbf{v} = 0$ for $\lambda = 1$. Then by Theorem 6.7.1,

$$\mathbf{x_k} = e^t \left[\mathbf{v_k} + t \begin{pmatrix} 4 & 6 & 2 \\ -2 & -3 & -1 \\ -2 & -3 & -1 \end{pmatrix} \mathbf{v_k} \right]$$

is a solution for $k = 1, 2, 3$.

Therefore,

$$\mathbf{x_1}(t) = e^t \begin{pmatrix} 1 \\ 0 \\ 0 \end{pmatrix} + te^t \begin{pmatrix} 4 \\ -2 \\ -2 \end{pmatrix}, \quad \mathbf{x_2}(t) = e^t \begin{pmatrix} 0 \\ 1 \\ 0 \end{pmatrix} + te^t \begin{pmatrix} 6 \\ -3 \\ -3 \end{pmatrix}, \quad \mathbf{x_3}(t) = e^t \begin{pmatrix} 0 \\ 0 \\ 1 \end{pmatrix} + te^t \begin{pmatrix} 2 \\ -1 \\ -1 \end{pmatrix}$$

are three linearly independent solutions. Therefore,

$$\mathbf{X}(t) = \begin{pmatrix} e^t + 4te^t & 6te^t & 2te^t \\ -2te^t & e^t - 3te^t & -te^t \\ -2te^t & -3te^t & e^t - te^t \end{pmatrix}$$

is a fundamental matrix for this system.

7. The eigenvalues of A are $\lambda = -1 \pm i$ and $\lambda = -1$ with multiplicity 2. First, $\lambda = -1 + i$, has an eigenvector

$$\mathbf{v_1} = \begin{pmatrix} \frac{7}{5} - \frac{4}{5}i \\ 1 \\ -\frac{3}{5} - \frac{4}{5}i \\ -1 + i \end{pmatrix}.$$

Therefore,

$$\mathbf{u}(t) = e^{(-1+i)t} \begin{pmatrix} \frac{7}{5} - \frac{4}{5}i \\ 1 \\ -\frac{3}{5} - \frac{4}{5}i \\ -1 + i \end{pmatrix}$$

is a solution of this system. Taking the real and imaginary parts of $\mathbf{u}$, we have the following two linearly independent real-valued solutions,

$$\mathbf{x_1}(t) = e^{-t} \left[\begin{pmatrix} \frac{7}{5} \\ 1 \\ -\frac{3}{5} \\ -1 \end{pmatrix} \cos t - \begin{pmatrix} -\frac{4}{5} \\ 0 \\ -\frac{4}{5} \\ 1 \end{pmatrix} \sin t \right], \quad \mathbf{x_2}(t) = e^{-t} \left[\begin{pmatrix} \frac{7}{5} \\ 1 \\ -\frac{3}{5} \\ -1 \end{pmatrix} \sin t + \begin{pmatrix} -\frac{4}{5} \\ 0 \\ -\frac{4}{5} \\ 1 \end{pmatrix} \cos t \right].$$

Next, we look at the eigenvalue $\lambda = -1$. For $\lambda = -1$,

$$A - \lambda I = \begin{pmatrix} -1 & -7 & -7 & -5 \\ -3 & -7 & -7 & -7 \\ 1 & 1 & 1 & 1 \\ 2 & 8 & 8 & 7 \end{pmatrix}.$$

Therefore,

$$(A - \lambda I)^2 = \begin{pmatrix} 5 & 9 & 9 & 12 \\ 3 & 7 & 7 & 8 \\ -1 & -5 & -5 & -4 \\ -4 & -6 & -6 & -9 \end{pmatrix} \rightarrow \begin{pmatrix} 1 & 0 & 0 & \frac{3}{2} \\ 0 & 1 & 1 & \frac{1}{2} \\ 0 & 0 & 0 & 0 \\ 0 & 0 & 0 & 0 \end{pmatrix}.$$

93

Therefore,

$$\mathbf{v_3} = \begin{pmatrix} 3 \\ 1 \\ 0 \\ -2 \end{pmatrix}, \quad \mathbf{v_4} = \begin{pmatrix} 3 \\ 0 \\ 1 \\ -2 \end{pmatrix}$$

are two linearly independent solutions of $(A - \lambda I)^2 \mathbf{v} = 0$. Then by Theorem 6.7.1, for $\lambda = -1$, $k = 3, 4$,

$$\mathbf{x_k}(t) = e^{-t} \left[\mathbf{v_k} + t(A - \lambda I)\mathbf{v_k} \right]$$

are two more linearly independent solutions of our system. Plugging in the values for $\mathbf{v_3}, \mathbf{v_4}$ and $A + I$, we have

$$\mathbf{x_3}(t) = e^{-t} \left[\begin{pmatrix} 3 \\ 1 \\ 0 \\ -2 \end{pmatrix} + t \begin{pmatrix} 0 \\ -2 \\ 2 \\ 0 \end{pmatrix} \right], \quad \mathbf{x_4}(t) = e^{-t} \left[\begin{pmatrix} 3 \\ 0 \\ 1 \\ -2 \end{pmatrix} + t \begin{pmatrix} 0 \\ -2 \\ 2 \\ 0 \end{pmatrix} \right].$$

Therefore,

$$\mathbf{X}(t) = \begin{pmatrix} \frac{7}{5}e^{-t}\cos t + \frac{4}{5}e^{-t}\sin t & \frac{7}{5}e^{-t} - \frac{4}{5}e^{-t}\cos t & 3e^{-t} & 3e^{-t} \\ e^{-t}\cos t & e^{-t}\sin t & e^{-t} - 2te^{-t} & -2te^{-t} \\ -\frac{3}{5}e^{-t}\cos t + \frac{4}{5}e^{-t}\sin t & -\frac{3}{5}e^{-t}\sin t - \frac{4}{5}e^{-t}\cos t & 2te^{-t} & e^{-t} + 2te^{-t} \\ -e^{-t}\cos t - e^{-t}\sin t & -e^{-t}\sin t + e^{-t}\cos t & -2e^{-t} & -2e^{-t} \end{pmatrix}$$

is a fundamental matrix for this system.

8. The only eigenvalue of A is $\lambda = -1$ with multiplicity 4. Then for $\lambda = -1$,

$$(A - \lambda I)^2 = \begin{pmatrix} -2 & 2 & 4 & -2 \\ -3 & 3 & 6 & -3 \\ 1 & -1 & -2 & 1 \\ 1 & -1 & -2 & 1 \end{pmatrix}$$

$$(A - \lambda I)^3 = \begin{pmatrix} 0 & 0 & 0 & 0 \\ 0 & 0 & 0 & 0 \\ 0 & 0 & 0 & 0 \\ 0 & 0 & 0 & 0 \end{pmatrix}$$

$$(A - \lambda I)^4 = \begin{pmatrix} 0 & 0 & 0 & 0 \\ 0 & 0 & 0 & 0 \\ 0 & 0 & 0 & 0 \\ 0 & 0 & 0 & 0 \end{pmatrix}.$$

Therefore,

$$\mathbf{v_1} = \begin{pmatrix} 1 \\ 0 \\ 0 \\ 0 \end{pmatrix}, \mathbf{v_2} = \begin{pmatrix} 0 \\ 1 \\ 0 \\ 0 \end{pmatrix}, \mathbf{v_3} = \begin{pmatrix} 0 \\ 0 \\ 1 \\ 0 \end{pmatrix}, \mathbf{v_4} = \begin{pmatrix} 0 \\ 0 \\ 0 \\ 1 \end{pmatrix}$$

are four linearly independent solutions of $(A - \lambda I)^4 \mathbf{v} = 0$. Then, by Theorem 6.7.1, for $\lambda = -1$, $k = 1, 2, 3, 4$, $\mathbf{v_k}$ as defined above,

$$\mathbf{x_k}(t) = e^{-t} \left[\mathbf{v_k} + t(A - \lambda I)\mathbf{v_k} + \frac{t^2}{2}(A - \lambda I)^2 \mathbf{v_k} \right]$$

are linearly independent solutions of our system. Then, plugging in the values for $\mathbf{v_k}$, $A - \lambda I$ and $(A - \lambda I)^2$

above, we have the following four solutions,

$$\mathbf{x_1}(t) = e^{-t}\left[\begin{pmatrix}1\\0\\0\\0\end{pmatrix} + t\begin{pmatrix}2\\2\\-1\\-2\end{pmatrix} + \frac{t^2}{2}\begin{pmatrix}-2\\-3\\1\\1\end{pmatrix}\right]$$

$$\mathbf{x_2}(t) = e^{-t}\left[\begin{pmatrix}0\\1\\0\\0\end{pmatrix} + t\begin{pmatrix}-1\\-1/2\\1/2\\3/2\end{pmatrix} + \frac{t^2}{2}\begin{pmatrix}2\\3\\-1\\-1\end{pmatrix}\right]$$

$$\mathbf{x_3}(t) = e^{-t}\left[\begin{pmatrix}0\\0\\1\\0\end{pmatrix} + t\begin{pmatrix}-2\\-1\\1\\3\end{pmatrix} + \frac{t^2}{2}\begin{pmatrix}4\\6\\-2\\-2\end{pmatrix}\right]$$

$$\mathbf{x_4}(t) = e^{-t}\left[\begin{pmatrix}0\\0\\0\\1\end{pmatrix} + t\begin{pmatrix}3\\7/2\\-3/2\\-5/2\end{pmatrix} + \frac{t^2}{2}\begin{pmatrix}-2\\-3\\1\\1\end{pmatrix}\right]$$

Therefore,

$$\mathbf{X}(t) = e^{-t}\begin{pmatrix} 1+2t-t^2 & -t+t^2 & -2t+2t^2 & 3t-t^2 \\ 2t-\frac{3}{2}t^2 & 1-\frac{1}{2}t+\frac{3}{2}t^2 & -t+3t^2 & \frac{7}{2}t-\frac{3}{2}t^2 \\ -t+\frac{1}{2}t^2 & \frac{1}{2}t-\frac{1}{2}t^2 & 1+t-t^2 & -\frac{3}{2}t+\frac{1}{2}t^2 \\ -2t+\frac{1}{2}t^2 & \frac{3}{2}t-\frac{1}{2}t^2 & 3t-t^2 & 1-\frac{5}{2}t+\frac{1}{2}t^2 \end{pmatrix}$$

is a fundamental matrix for this system.

9.

$$A - \lambda I = \begin{pmatrix} 1-\lambda & -4 \\ 4 & -7-\lambda \end{pmatrix}.$$

Therefore, $\det(A-\lambda I) = \lambda^2 + 6\lambda + 9 = (\lambda+3)^2$. Therefore, $\lambda = -3$ is the only eigenvalue of A. For $\lambda = -3$,

$$(A-\lambda I)^2 = \begin{pmatrix} 0 & 0 \\ 0 & 0 \end{pmatrix}.$$

Therefore,

$$\mathbf{v_1} = \begin{pmatrix}1\\0\end{pmatrix}, \quad \mathbf{v_2} = \begin{pmatrix}0\\1\end{pmatrix}$$

are linearly independent solutions of $(A-\lambda I)^2 \mathbf{v} = 0$. Then, by Theorem 6.7.1, for $\lambda = 1$, $k = 1, 2$,

$$\mathbf{x_k}(t) = e^t[\mathbf{v_k} + t(A-\lambda I)\mathbf{v_k}]$$

are linearly independent solutions of our system. Plugging in for $\mathbf{v_k}$ and $A-\lambda I$, we have

$$\mathbf{x_1}(t) = e^{-3t}\left[\begin{pmatrix}1\\0\end{pmatrix} + t\begin{pmatrix}4\\4\end{pmatrix}\right]$$

$$\mathbf{x_2}(t) = e^{-3t}\left[\begin{pmatrix}0\\1\end{pmatrix} + t\begin{pmatrix}-4\\-4\end{pmatrix}\right].$$

Therefore,

$$\mathbf{X}(t) = e^{-3t}\begin{pmatrix} 1+4t & -4t \\ 4t & 1-4t \end{pmatrix}$$

is a fundamental matrix.

$$\mathbf{X}(0) = \begin{pmatrix}1 & 0\\0 & 1\end{pmatrix} \implies \mathbf{X}^{-1}(0) = \begin{pmatrix}1 & 0\\0 & 1\end{pmatrix}.$$

95

Therefore,

$$e^{At} = \mathbf{X}(t)\mathbf{X}^{-1}(0) = \mathbf{X}(t) = e^{-3t} \begin{pmatrix} 1 + 4t & -4t \\ 4t & 1 - 4t \end{pmatrix}.$$

Therefore, the solution of this initial value problem is given by

$$\mathbf{x}(t) = e^{At}\mathbf{x_0}$$

$$= e^{-3t} \begin{pmatrix} 1 + 4t & -4t \\ 4t & 1 - 4t \end{pmatrix} \begin{pmatrix} 3 \\ 2 \end{pmatrix}$$

$$= e^{-3t} \begin{pmatrix} 3 + 4t \\ 2 + 4t \end{pmatrix}.$$

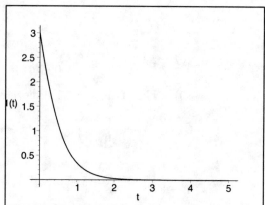

10.

$$A - \lambda I = \begin{pmatrix} 3 - \lambda & -4 \\ 1 & -1 - \lambda \end{pmatrix}.$$

Therefore, $\det(A - \lambda I) = \lambda^2 - 2\lambda + 1 = (\lambda - 1)^2$. Therefore, $\lambda = 1$ is the only eigenvalue of A. For $\lambda = 1$,

$$(A - \lambda I)^2 = \begin{pmatrix} 0 & 0 \\ 0 & 0 \end{pmatrix}.$$

Therefore,

$$\mathbf{v_1} = \begin{pmatrix} 1 \\ 0 \end{pmatrix}, \quad \mathbf{v_2} = \begin{pmatrix} 0 \\ 1 \end{pmatrix}$$

are linearly independent solutions of $(A - \lambda I)^2 \mathbf{v} = 0$. Then, by Theorem 6.7.1, for $\lambda = 1$, $k = 1, 2$,

$$\mathbf{x_k}(t) = e^t [\mathbf{v_k} + t(A - \lambda I)\mathbf{v_k}]$$

are linearly independent solutions of our system. Plugging in for $\mathbf{v_k}$ and $A - \lambda I$, we have

$$\mathbf{x_1}(t) = e^t \left[\begin{pmatrix} 1 \\ 0 \end{pmatrix} + t \begin{pmatrix} 2 \\ 1 \end{pmatrix} \right]$$

$$\mathbf{x_2}(t) = e^t \left[\begin{pmatrix} 0 \\ 1 \end{pmatrix} + t \begin{pmatrix} -4 \\ -2 \end{pmatrix} \right].$$

Therefore,

$$\mathbf{X}(t) = \begin{pmatrix} e^t + 2te^t & -4te^t \\ te^t & e^t - 2te^t \end{pmatrix}$$

is a fundamental matrix.

$$\mathbf{X}(0) = \begin{pmatrix} 1 & 0 \\ 0 & 1 \end{pmatrix} \implies \mathbf{X}^{-1}(0) = \begin{pmatrix} 1 & 0 \\ 0 & 1 \end{pmatrix}.$$

Therefore,

$$e^{At} = \mathbf{X}(t)\mathbf{X}^{-1}(0) = \mathbf{X}(t) = \begin{pmatrix} e^t + 2te^t & -4te^t \\ te^t & e^t - 2te^t \end{pmatrix}.$$

Therefore, the solution of this initial value problem is given by

$$\mathbf{x}(t) = e^{At}\mathbf{x_0}$$
$$= \begin{pmatrix} e^t + 2te^t & -4te^t \\ te^t & e^t - 2te^t \end{pmatrix} \begin{pmatrix} -2 \\ 4 \end{pmatrix}$$
$$= \begin{pmatrix} -2e^t - 20te^t \\ 4e^t - 10te^t \end{pmatrix}.$$

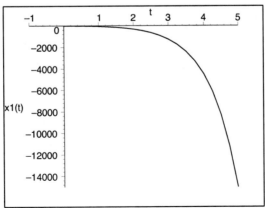

11. The eigenvalues of A are $\lambda = 2$ which has multiplicity 1 and $\lambda = 1$ which has multiplicity 2. First, for $\lambda = 2$,

$$\mathbf{v_1} = \begin{pmatrix} 0 \\ -3 \\ 1 \end{pmatrix}$$

is a corresponding eigenvector, and, therefore,

$$\mathbf{x_1}(t) = e^{2t} \begin{pmatrix} 0 \\ -3 \\ 1 \end{pmatrix}$$

is one solution of the system.

Next, for $\lambda = 1$,

$$A - \lambda I = \begin{pmatrix} 3 & 1 & 3 \\ 6 & 3 & 6 \\ -5 & -2 & -5 \end{pmatrix}$$

$$(A - \lambda I)^2 = \begin{pmatrix} 0 & 0 & 0 \\ 6 & 3 & 6 \\ -2 & -1 & -2 \end{pmatrix}.$$

Therefore,

$$\mathbf{v_2} = \begin{pmatrix} 0 \\ -2 \\ 1 \end{pmatrix}, \quad \mathbf{v_3} = \begin{pmatrix} 1 \\ -2 \\ 0 \end{pmatrix}$$

are two linearly independent solutions of $(A - \lambda I)^2 \mathbf{v} = 0$. Therefore, by Theorem 6.7.1, for $\lambda = 1$, $k = 2, 3$,

$$\mathbf{x_k(t)} = e^t \left[\mathbf{v_k} + t(A - \lambda I)\mathbf{v_k} \right]$$

are two linearly independent solutions of our system. Plugging in for $\mathbf{v_k}$ and $A - \lambda I$, we have the following two solutions,

$$\mathbf{x_2(t)} = e^t \left[\begin{pmatrix} 0 \\ -2 \\ 1 \end{pmatrix} + t \begin{pmatrix} 1 \\ 0 \\ -1 \end{pmatrix} \right]$$

$$\mathbf{x_3(t)} = e^t \left[\begin{pmatrix} 1 \\ -2 \\ 0 \end{pmatrix} + t \begin{pmatrix} 1 \\ 0 \\ -1 \end{pmatrix} \right].$$

Therefore,

$$\mathbf{X}(t) = \begin{pmatrix} 0 & te^t & e^t + te^t \\ -3e^{2t} & -2e^t & -2e^t \\ e^{2t} & e^t - te^t & -te^t \end{pmatrix}$$

is a fundamental matrix.

$$\mathbf{X}(0) = \begin{pmatrix} 0 & 0 & 1 \\ -3 & -2 & -2 \\ 1 & 1 & 0 \end{pmatrix} \implies \mathbf{X}^{-1}(0) = \begin{pmatrix} -2 & -1 & -2 \\ 2 & 1 & 3 \\ 1 & 0 & 0 \end{pmatrix}.$$

Therefore,

$$e^{At} = \mathbf{X}(t)\mathbf{X}^{-1}(0) = \begin{pmatrix} 3te^t + e^t & te^t & 3te^t \\ 6e^{2t} - 6e^t & 3e^{2t} - 2e^t & 6e^{2t} - 6e^t \\ -2e^{2t} + 2e^t - 3te^t & -e^{2t} + e^t - te^t & -2e^{2t} + 3e^t - 3te^t \end{pmatrix}.$$

Therefore, the solution of this initial value problem is given by

$$\mathbf{x}(t) = e^{At}\mathbf{x}_0$$

$$= \begin{pmatrix} 3te^t + e^t & te^t & 3te^t \\ 6e^{2t} - 6e^t & 3e^{2t} - 2e^t & 6e^{2t} - 6e^t \\ -2e^{2t} + 2e^t - 3te^t & -e^{2t} + e^t - te^t & -2e^{2t} + 3e^t - 3te^t \end{pmatrix} \begin{pmatrix} 1 \\ -1 \\ 1 \end{pmatrix}$$

$$= \begin{pmatrix} 5te^t + e^t \\ 9e^{2t} - 10e^t \\ -3e^{2t} + 4e^t - 5te^t \end{pmatrix}.$$

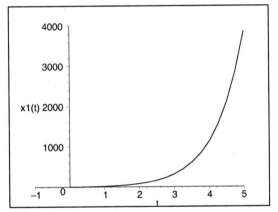

12. The only eigenvalue of A is $\lambda = -2$ which has multiplicity 3. For $\lambda = -2$,

$$A - \lambda I = \begin{pmatrix} 3 & 1 & 0 \\ -14 & -3 & 1 \\ 15 & 5 & 0 \end{pmatrix}$$

$$(A - \lambda I)^2 = \begin{pmatrix} -5 & 0 & 1 \\ 15 & 0 & -3 \\ -25 & 0 & 5 \end{pmatrix}$$

$$(A - \lambda I)^3 = \begin{pmatrix} 0 & 0 & 0 \\ 0 & 0 & 0 \\ 0 & 0 & 0 \end{pmatrix}.$$

Therefore,

$$\mathbf{v}_1 = \begin{pmatrix} 1 \\ 0 \\ 0 \end{pmatrix}, \quad \mathbf{v}_2 = \begin{pmatrix} 0 \\ 1 \\ 0 \end{pmatrix}, \quad \mathbf{v}_3 = \begin{pmatrix} 0 \\ 0 \\ 1 \end{pmatrix}$$

are three linearly independent solutions of $(A-\lambda I)^3\mathbf{v}=0$. Therefore, by Theorem 6.7.1, for $\lambda=1$, $k=1,2,3$,

$$\mathbf{x_k(t)}=e^{-2t}\left[\mathbf{v_k}+t(A-\lambda I)\mathbf{v_k}+\frac{t^2}{2}(A-\lambda I)\mathbf{v_k}\right]$$

are three linearly independent solutions of our system. Plugging in for $\mathbf{v_k}$ and $A-\lambda I$, we have the following three solutions,

$$\mathbf{x_1(t)}=e^{-2t}\left[\begin{pmatrix}1\\0\\0\end{pmatrix}+t\begin{pmatrix}3\\-14\\15\end{pmatrix}+\frac{t^2}{2}\begin{pmatrix}-5\\15\\-25\end{pmatrix}\right]$$

$$\mathbf{x_2(t)}=e^{-2t}\left[\begin{pmatrix}0\\1\\0\end{pmatrix}+t\begin{pmatrix}1\\-3\\5\end{pmatrix}\right]$$

$$\mathbf{x_3(t)}=e^{-2t}\left[\begin{pmatrix}0\\0\\1\end{pmatrix}+t\begin{pmatrix}0\\1\\0\end{pmatrix}+\frac{t^2}{2}\begin{pmatrix}1\\-3\\5\end{pmatrix}\right].$$

Therefore,

$$\mathbf{X}(t)=e^{-2t}\begin{pmatrix}1+3t-\frac{5}{2}t^2 & t & \frac{1}{2}t^2\\-14t+\frac{15}{2}t^2 & 1-3t & t-\frac{3}{2}t^2\\15t-\frac{25}{2}t^2 & 5t & 1+\frac{5}{2}t^2\end{pmatrix}$$

is a fundamental matrix.

$$\mathbf{X}(0)=\begin{pmatrix}1&0&0\\0&1&0\\0&0&1\end{pmatrix}\implies \mathbf{X}^{-1}(0)=\begin{pmatrix}1&0&0\\0&1&0\\0&0&1\end{pmatrix}.$$

Therefore,

$$e^{At}=\mathbf{X}(t)\mathbf{X}^{-1}(0)=e^{-2t}\begin{pmatrix}1+3t-\frac{5}{2}t^2 & t & \frac{1}{2}t^2\\-14t+\frac{15}{2}t^2 & 1-3t & t-\frac{3}{2}t^2\\15t-\frac{25}{2}t^2 & 5t & 1+\frac{5}{2}t^2\end{pmatrix}.$$

Therefore, the solution of this initial value problem is given by

$$\mathbf{x}(t)=e^{At}\mathbf{x_0}$$

$$=e^{-2t}\begin{pmatrix}1+3t-\frac{5}{2}t^2 & t & \frac{1}{2}t^2\\-14t+\frac{15}{2}t^2 & 1-3t & t-\frac{3}{2}t^2\\15t-\frac{25}{2}t^2 & 5t & 1+\frac{5}{2}t^2\end{pmatrix}\begin{pmatrix}5\\7\\-4\end{pmatrix}$$

$$=e^{-2t}\begin{pmatrix}5+22t-\frac{29}{2}t^2\\-95t+\frac{87}{2}t^2+7\\110t-\frac{145}{2}t^2-4\end{pmatrix}.$$

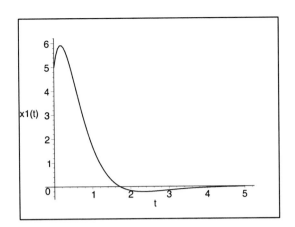

Chapter 7
Section 7.1

1.

(a) $-y + xy = 0$ implies $y(-1 + x) = 0$ implies $x = 1$ or $y = 0$. Then, $x + 2xy = 0$ implies $x(1 + 2y) = 0$ implies $x = 0$ or $y = -1/2$. Therefore, the critical points are $(1, -1/2)$ and $(0, 0)$.

(b)

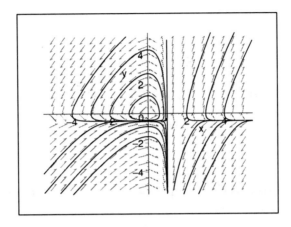

(c) The critical point $(0, 0)$ is a center, therefore, stable. The critical point $(1, -1/2)$ is a saddle point, therefore, unstable.

2.

(a) $1 + 2y = 0$ implies $y = -1/2$. Then, $1 - 3x^2 = 0$ implies $x = \pm 1/\sqrt{3}$. Therefore, the critical points are $(-1/\sqrt{3}, -1/2)$ and $(1/\sqrt{3}, -1/2)$.

(b)

(c) The critical point $(-1/\sqrt{3}, -1/2)$ is a saddle point, therefore, unstable. The critical point $(1/\sqrt{3}, -1/2)$ is a center, therefore, stable.

3.

(a) The equation $2x - x^2 - xy = 0$ implies $x = 0$ or $x + y = 2$. The equation $3y - 2y^2 - 3xy = 0$ implies $y = 0$ or $3x + 2y = 3$. Solving these equations, we have the critical points $(0, 0)$, $(0, 3/2)$, $(2, 0)$ and $(-1, 3)$.

(b)

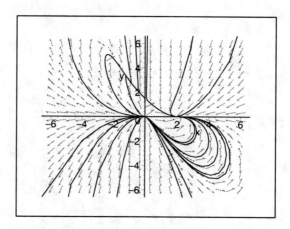

(c) The critical point $(0, 0)$ is an unstable node. The critical point $(0, 3/2)$ is a saddle point, therefore, unstable. The critical point $(2, 0)$ is an asymptotically stable node. The critical point $(-1, 3)$ is an asymptotically stable node.

(d) For $(2, 0)$, the basin of attraction is the first quadrant plus the region in the fourth quadrant bounded by the trajectories heading away from $(0, 0)$ but looping back towards $(2, 0)$. For $(-1, 3)$, the basin of attraction is bounded to the right by the y-axis and to the left by those trajectories leaving $(0, 0)$ but looping back towards $(-1, 3)$.

4.

(a) The equation $-(x - y)(1 - x - y) = 0$ implies $x - y = 0$ or $x + y = 1$. The equation $-x(2 + y) = 0$ implies $x = 0$ or $y = -2$. Solving these equations, we have the critical points $(0, 0)$, $(0, 1)$, $(-2, -2)$ and $(3, -2)$.

(b)

(c) The critical point $(0, 0)$ is an asymptotically stable spiral point. The critical point $(0, 1)$ is a saddle point, therefore, unstable. The critical point $(-2, -2)$ is a saddle point, therefore, unstable. The critical point $(3, -2)$ is a saddle point, therefore, unstable.

(d) For $(0, 0)$, the basin of attraction is bounded below by the line $y = -2$, to the right by a trajectory passing near the point $(2, 0)$, to the left by a trajectory heading towards (and then away from) the unstable critical point $(0, 1)$, and above by a trajectory heading towards (and then away from) the unstable critical point $(0, 1)$.

5.

(a) The equation $x(2 - x - y) = 0$ implies $x = 0$ or $x + y = 2$. If $x = 0$, the equation $-x + 3y - 2xy = 0$ implies $y = 0$. If $x + y = 2$, then the equation $-x + 3y - 2xy = 0$ can be reduced to $y^2 - 1 = 0$. Therefore, $y = \pm 1$. Now if $y = 1$, then $x = 1$. If $y = -1$, then $x = 3$. Therefore, the critical points are $(0, 0)$, $(1, 1)$ and $(3, -1)$.

(b)

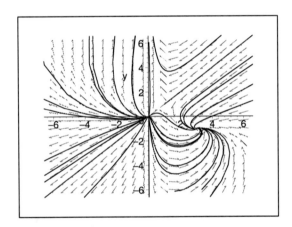

(c) The critical point $(0, 0)$ is an unstable node. The critical point $(1, 1)$ is a saddle point, therefore, unstable. The critical point $(3, -1)$ is an asymptotically stable spiral point.

(d) For $(3, -1)$, the basin of attraction is bounded to the left by the $y-$axis and above by a trajectory heading into (and away from) the unstable critical point $(1, 1)$.

6.

(a) The equation $(2 - x)(y - x) = 0$ implies $x = 2$ or $x = y$. The equation $y(2 - x - x^2) = 0$ implies $y = 0$ or $x = -2$ or $x = 1$. The solutions of those two equations are the critical points $(0, 0)$, $(2, 0)$, $(-2, -2)$ and $(1, 1)$.

(b)

(c) The critical point $(0,0)$ is a saddle point, therefore, unstable. The critical point $(2,0)$ is also a saddle point, therefore, unstable. The critical point $(-2,-2)$ is an asymptotically stable spiral point. The critical point $(1,1)$ is also an asymptotically stable spiral point.

(d) For $(-2,-2)$, the basin of attraction is the region bounded above by the $x-$axis and to the right by the line $x = 2$. For $(1,1)$, the basin of attraction is the region bounded below by the $x-$axis and to the right by the line $x = 2$.

7.

(a) The equation $(2-y)(x-y) = 0$ implies $y = 2$ or $x = y$. The equation $(1+x)(x+y) = 0$ implies $x = -1$ or $x = -y$. The solutions of those two equations are the critical points $(0,0)$, $(-1,2)$, $(-2,2)$ and $(-1,-1)$.

(b)

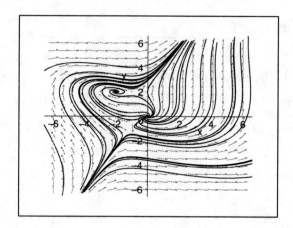

(c) The critical point $(0,0)$ is an unstable spiral point. The critical point $(-1,2)$ is a saddle point, therefore, unstable. The critical point $(-2,2)$ is an asymptotically stable spiral point. The critical point $(-1,-1)$ is a saddle point, therefore, unstable.

(d) For $(-2,2)$, the basin of attraction is a circular region lying in the second quadrant, close to the $x-$ and $y-$axes, bounded by trajectories leaving the unstable spiral point $(0,0)$.

4

8.

(a) The equation $x(2-x-y) = 0$ implies $x = 0$ or $x+y = 2$. The equation $(1-y)(2+x) = 0$ implies $y = 1$ or $x = -2$. The solutions of those two equations are the critical points $(0,1)$, $(1,1)$ and $(-2,4)$.

(b)

(c) The critical point $(0,1)$ is a saddle point, therefore, unstable. The critical point $(1,1)$ is an asymptotically stable node. The critical point $(-2,4)$ is an unstable spiral point.

(d) For $(1,1)$, the basin of attraction is the right half plane.

9.

(a) The equation $(2+x)(3y-x) = 0$ implies $x = -2$ or $x = 3y$. The equation $(3-x)(y-x) = 0$ implies $x = 3$ or $x = y$. The solutions of these two equations are the critical points $(0,0)$, $(3,1)$ and $(-2,-2)$.

(b)

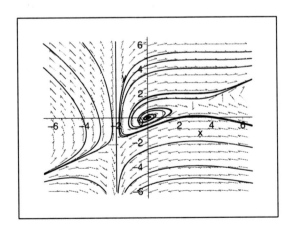

(c) The critical point $(0,0)$ is an unstable spiral point. The critical point $(3,1)$ is a saddle point, therefore, unstable. The critical point $(-2,-2)$ is also a saddle point, therefore, unstable.

5

10.

(a) The equation $(1 + y)(1 - x) = 0$ implies $x = 1$ or $y = -1$. If $x = 1$, then $x - 2y - x^2 = 0 \implies y = 0$. If $y = -1$, then $x - 2y - x^2 = 0 \implies x^2 - x - 2 = 0 \implies x = -1, 2$. Therefore, the critical points are $(1, 0), (-1, -1), (2, -1)$.

(b)

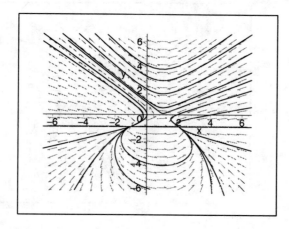

(c) The critical point $(1, 0)$ is a saddle point, therefore, unstable. The critical point $(-1, -1)$ is an unstable node. The critical point $(2, -1)$ is an asymptotically stable node.

(d) The basin of attraction for $(2, -1)$ is bounded above by the trajectory heading towards and then away from the unstable critical point $(1, 0)$. The basin is bounded on the left by a trajectory travelling away from the unstable critical point $(-1, -1)$.

11.

(a) The equation $-y = 0$ implies $y = 0$. Then if $y = 0$, the equation $-3y - x(x-1)(x-2) = 0$ reduces to $x = 0, 1, 2$. Therefore, the critical points are $(0, 0), (1, 0), (2, 0)$.

(b)

(c) The critical point $(0,0)$ is a saddle point, therefore, unstable. The critical point $(1,0)$ is an asymptotically stable node. The critical point $(2,0)$ is a saddle point, therefore, unstable.

(d) The basin of attraction for $(1,0)$ is bounded on the left by trajectories heading towards and then away from the unstable critical point $(0,0)$. The basin is bounded on the right by trajectories travelling towards and then away from the unstable critical point $(2,0)$.

12.

(a) The critical points are $(-2,1), (-2,1/2), (2,1), (1/3,-2/3)$.

(b)

(c) The critical point $(-2,1)$ is an unstable node. The critical point $(-2,1/2)$ is a saddle point, therefore, unstable. The critical point $(2,1)$ is a saddle point, therefore, unstable. The critical point $(1/3,-2/3)$ is an asymptotically stable spiral point.

(d) The basin of attraction for $(1/3,-2/3)$ is bounded on the left by the line $x=-2$ and bounded above by the line $y=1$.

13.

(a) The equation $dx/dt = 0$ implies $x = 0$ or $x + 3y = 2$. The equation $dy/dt = 0$ implies $y = 0$, $x = 3$, or $x = -2$. Solving these two equations simultaneously, we see that the critical points are $(0,0), (2,0), (3,-1/3), (-2,4/3)$.

(b)

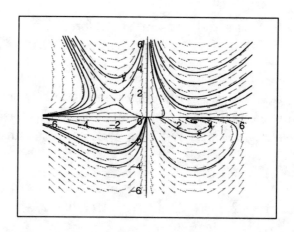

(c) The critical point $(0,0)$ is an unstable node. The critical point $(2,0)$ is a saddle point, therefore, unstable. The critical point $(3,-1/3)$ is an asymptotically stable spiral point. The critical point $(-2,4/3)$ is a saddle point, therefore, unstable.

(d) The basin of attraction for $(3,-1/3)$ is the fourth quadrant.

14.

(a) The equation $dx/dt = 0$ implies $y = -1$ or $x - 2y = -1$. The equation $dy/dt = 0$ implies $y = 0$ or $2x + y = 3$. Solving these two equations simultaneously, we see that the critical points are $(2,-1),(-1,0),(1,1)$.

(b)

(c) The critical point $(2,-1)$ is a stable spiral point. The critical point $(-1,0)$ is a saddle point, therefore, unstable. The critical point $(1,1)$ is an unstable spiral point.

15.

(a) The equation $dx/dt = 0$ implies $y = 1$, $x = 1$ or $x = -2$. The equation $dy/dt = 0$ implies $x = 0$ or $y = 2$. Solving these two equations simultaneously, we see that the critical points are $(0,1),(1,2),(-2,2)$.

(b)

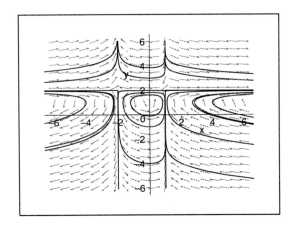

(c) The critical point $(0, 1)$ is a center, therefore, stable. The critical point $(1, 2)$ is a saddle point, therefore, unstable. The critical point $(-2, 2)$ is a saddle point, therefore, unstable.

16.

(a) The equation $dx/dt = 0$ implies $y = \pm 3$. The equation $dy/dt = 0$ implies $x = 1$ or $x = -y$. Solving these two equations simultaneously, we see that the critical points are $(1, 3), (-3, 3), (1, -3), (3, -3)$.

(b)

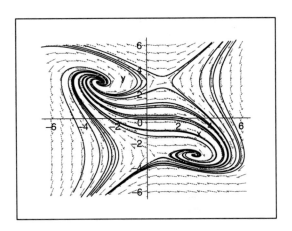

(c) The critical point $(1, 3)$ is a saddle point, therefore, unstable. The critical point $(-3, 3)$ is an unstable spiral point. The critical point $(1, -3)$ is a saddle point, therefore, unstable. The critical point $(3, -3)$ is an asymptotically stable spiral point.

(d) The basin of attraction for $(3, -3)$ is bounded below and to the right by trajectories heading towards and then away from the unstable critical point $(1, -3)$. The basin is bounded above by trajectories heading towards and then away (to the left) from the unstable critical point $(1, 3)$.

9

17.

(a) The equation $dx/dt = 0$ implies $y = 0$. The equation $dy/dt = 0$ implies $x = 0$ or $x = \pm\sqrt{6}$. Solving these two equations simultaneously, we see that the critical points are $(0,0), (\sqrt{6},0), (-\sqrt{6},0)$.

(b)

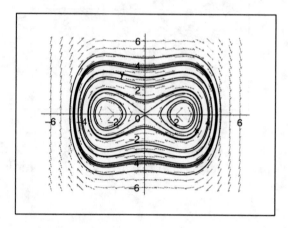

(c) The critical point $(0,0)$ is a saddle point, therefore, unstable. The critical point $(\sqrt{6},0)$ is a center, therefore, stable. The critical point $(-\sqrt{6},0)$ is a center, therefore, stable.

18.

(a) The equation $dy/dt = 0$ implies $y = 0$ or $x = 1/2$. Using these values in the equation for $dx/dt = 0$, we see that the critical points are $(0,0), (1,0), (1/2, 1/4)$.

(b)

(c) The critical point $(0,0)$ is a saddle point, therefore, unstable. The critical point $(1,0)$ is a saddle point, therefore, unstable. The critical point $(1/2, 1/4)$ is a center, therefore, stable.

19.

(a) The trajectories are solutions of the differential equation

$$\frac{dy}{dx} = \frac{-\omega^2 \sin x}{y}.$$

Rewriting this equation as $\omega^2 \sin x\, dx + y\, dy = 0$, we see that this ODE is exact with

$$\frac{\partial H}{\partial x} = \omega^2 \sin x \quad \text{and} \quad \frac{\partial H}{\partial y} = y.$$

Integrating the first equation, we find that $H(x, y) = -\omega^2 \cos x + f(y)$. Differentiating this equation with respect to y, we find that $H_y = f'(y) = y$ which implies that $f(y) = y^2/2 + C$. Therefore, the solutions of the ODE are level curves of

$$H(x, y) = -\omega^2 \cos x + y^2/2.$$

Adding an arbitrary constant does not affect the trajectories. Therefore, the trajectories can be written as

$$\frac{1}{2}y^2 + \omega^2(1 - \cos x) = C,$$

where C is an arbitrary constant.

(b) Multiplying by mL^2 and reverting to the original physical variables, we obtain

$$\frac{1}{2}mL^2 \left(\frac{d\theta}{dt}\right)^2 + mL^2\omega^2(1 - \cos\theta) = mL^2 c.$$

Since $\omega^2 = g/L$, this equation can be written as

$$\frac{1}{2}mL^2 \left(\frac{d\theta}{dt}\right)^2 + mgL(1 - \cos\theta) = E,$$

where $E = mL^2 c$.

(c) The absolute velocity of the point mass is given by $v = L d\theta/dt$. The kinetic energy of the mass is $T = mv^2/2$. Choosing the rest position as the datum, that is, the level of zero potential energy, the gravitational potential energy of the point mass is $V = mgL(1 - \cos\theta)$. It follows that the total energy, $T + V$, is constant along the trajectories.

21.

(a)

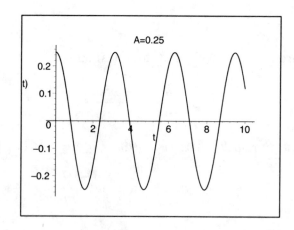

Since the system is undamped, and $y(0) = 0$, the amplitude is 0.25. The period is estimated at $\tau \approx 3.16$.

(b)

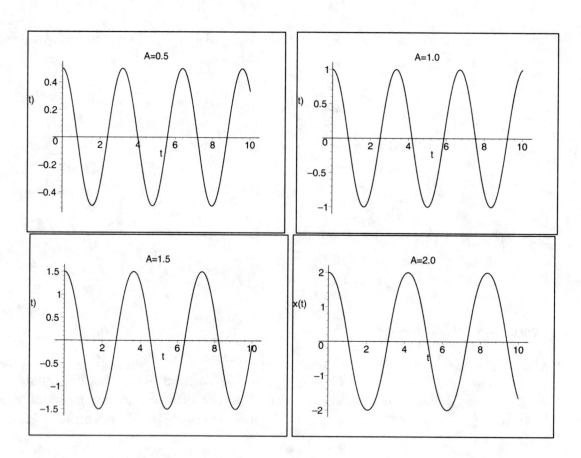

In each case, the amplitudes are given by $R = 0.5, 1.0, 1.5, 2.0$, respectively. The periods T are estimated at $3.20, 3.35, 3.63, 4.17$, respectively.

(c) Since the system is conservative, the amplitude is equal to the initial amplitude. The period of motion is an increasing function of the initial position A.

12

It appears that as $A \to 0$, the period approaches π, the period of the corresponding linear pendulum.

(d)

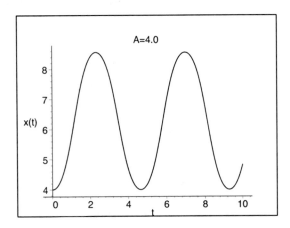

The pendulum is released from rest at an inclination of $4 - \pi$ radians from vertical. The pendulum will swing past the lower equilibrium position $(\theta = 2\pi)$ and come to rest, momentarily, at a maximum rotational displacement of $\theta_{max} = 4\pi - 4$. The transition between the two dynamics occurs at $A = \pi$.

21.

(a)

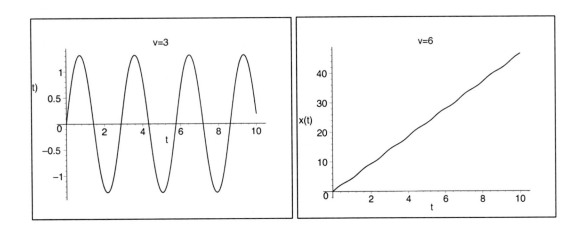

For initial velocity $v = 3$, the pendulum will swing back and forth about its equilibrium position. For initial velocity $v = 6$, the pendulum will not swing back and forth, but instead will continue to rotate about the origin indefinitely.

(b) By plotting the graphs of the solutions for different values of v, we conclude that the transition takes place at $v_c \approx 4.90$.

22.

(a) We see that

$$\frac{d}{d\theta}\left[\left(\frac{d\theta}{dt}\right)^2\right] = 2\frac{d\theta}{dt}\left[\frac{d^2\theta}{dt\,d\theta} + \frac{d^2\theta}{dt^2}\frac{dt}{d\theta}\right]$$

$$= 2\frac{d\theta}{dt}\left[0 + \frac{d^2\theta}{dt^2}\frac{dt}{d\theta}\right]$$

$$= 2\frac{d^2\theta}{dt^2}.$$

Therefore,

$$mL^2\frac{d^2\theta}{dt^2} + mgL\sin\theta = 0 \implies \frac{1}{2}mL^2\frac{d}{d\theta}\left[\left(\frac{d\theta}{dt}\right)^2\right] = -mgL\sin\theta.$$

Integrating this equation with respect to θ, we have

$$\frac{1}{2}m\left(L\frac{d\theta}{dt}\right)^2 = mgL(\cos\theta - \cos\alpha).$$

Separating variables and taking the square root of both sides of the equation, we have

$$dt = \pm\sqrt{\frac{L}{2g}}\frac{d\theta}{\sqrt{\cos\theta - \cos\alpha}}.$$

The negative square root is chosen because the bob is being released from a positive angle α. Therefore, $d\theta/dt$ will be negative.

(b) Integrating the last equation above from α to 0, we have

$$\int_\alpha^0 dt = -\sqrt{\frac{L}{2g}}\int_\alpha^0 \frac{d\theta}{\sqrt{\cos\theta - \cos\alpha}}$$

$$\implies t(0) - t(\alpha) = -\sqrt{\frac{L}{2g}}\int_\alpha^0 \frac{d\theta}{\sqrt{\cos\theta - \cos\alpha}}.$$

First, $t(\alpha) = 0$, since the bob starts at $\theta = \alpha$. Further, if the period of the pendulum is T, then the bob will first go by the equilibrium position $\theta = 0$ at time $T/4$. Therefore, we conclude that

$$\frac{T}{4} = -\sqrt{\frac{L}{2g}}\int_\alpha^0 \frac{d\theta}{\sqrt{\cos\theta - \cos\alpha}}$$

(c) By using the identities $\cos\theta = 1 - 2\sin^2(\theta/2)$ and $\cos\theta = 1 - 2\sin^2(\alpha/2)$, we can rewrite the equation in part (b) as

$$T = -4\sqrt{\frac{L}{2g}}\int_\alpha^0 \frac{d\theta}{\sqrt{2(\sin^2(\alpha/2) - \sin^2(\theta/2))}}.$$

Next, making the change of variables $\sin(\theta/2) = k\sin\phi$ with $k = \sin(\alpha/2)$, our equation becomes

$$T = -4\sqrt{\frac{L}{2g}}\int_{\pi/2}^0 \frac{(2k\cos\phi\, d\phi/\cos(\theta/2))}{\sqrt{2(k^2 - k^2\sin^2\phi)}}$$

$$= 4\sqrt{\frac{L}{g}}\int_0^{\pi/2} \frac{\cos\phi\, d\phi}{\cos(\theta/2)\sqrt{1 - \sin^2\phi}}$$

$$= 4\sqrt{\frac{L}{g}}\int_0^{\pi/2} \frac{\sqrt{1 - \sin^2\phi}\, d\phi}{\sqrt{1 - \sin^2(\theta/2)}\sqrt{1 - \sin^2\phi}}$$

$$= 4\sqrt{\frac{L}{g}}\int_0^{\pi/2} \frac{d\phi}{\sqrt{1 - k^2\sin^2\phi}}$$

(d) Set $k = \sin(\alpha/2) = \sin(A/2)$ and $g/L = 4$.

23.

$$\frac{d\Phi}{dt} = \frac{d\phi}{dt}(t - s) = F(\phi(t - s), \psi(t - s)) = F(\Phi, \Psi) = F(x, y)$$

and

$$\frac{d\Psi}{dt} = \frac{d\psi}{dt}(t - s) = G(\phi(t - s), \psi(t - s)) = F(\Phi, \Psi) = G(x, y).$$

Therefore, $\Phi(t), \Psi(t)$ is a solution for $\alpha + s < t < \beta + s$.

24. Let C_0 be the trajectory generated by the solution $x = \phi_0(t)$, $y = \psi_0(t)$ with $\phi_0(t_0) = x_0$, $\psi_0(t_0) = y_0$ and let C_1 be the trajectory generated by the solution $x = \phi_1(t), y = \psi_1(t)$

with $\phi_1(t_1) = x_0$, $\psi_1(t_1) = y_0$. From problem 23, we know that $\Phi_1(t) = \phi_1(t - (t_0 - t_1))$, $\Psi_1(t) = \psi_1(t - (t_0 - t_1))$ is a solution. Further, $\Phi_1(t_0) = \phi_1(t_1) = x_0$ and $\Psi_1(t_0) = y_0$. Then, by uniqueness, $\phi_0(t) = \Phi_1(t)$ and $\psi_0(t) = \Phi_1(t)$. Therefore, the trajectories must be the same.

25. If we assume that a trajectory can reach a critical point (x_0, y_0) in a finite length of time, then we would have two trajectories passing through the same point. This contradicts the result in problem 24.

26. Since the trajectory is closed, there is at least one point (x_0, y_0) such that $\phi(t_0) = x_0$, $\psi(t_0) = y_0$ and a number $T > 0$ such that $\phi(t_0 + T) = x_0$, $\psi(t_0 + T) = y_0$. From problem 23, we know that $\Phi(t) = \psi(t + T)$, $\Psi(t) = \psi(t + T)$ will also be a solution. But, then by uniqueness $\Phi(t) = \phi(t)$ and $\Psi(t) = \psi(t)$ for all t. Therefore, $\phi(t + T) = \phi(t)$ and $\psi(t + T) = \psi(t)$ for all t. Therefore, the solution is periodic with period T.

Section 7.2

1.

(a) The equation $dx/dt = 0$ implies $y = 2x$. The equation $dy/dt = 0$ implies $y = x^2$. Therefore, for these equations to both be satisfied, we need $2x = x^2$ which means $x = 0$ or $x = 2$. Therefore, the two critical points are $(0, 0)$ and $(2, 4)$.

(b) Here, we have $F(x, y) = -2x + y$ and $G(x, y) = x^2 - y$. Therefore, the Jacobian matrix for this system is

$$\begin{pmatrix} F_x & F_y \\ G_x & G_y \end{pmatrix} = \begin{pmatrix} -2 & 1 \\ 2x & -1 \end{pmatrix}.$$

Therefore, near the critical point $(0, 0)$, the Jacobian matrix is

$$\begin{pmatrix} F_x(0,0) & F_y(0,0) \\ G_x(0,0) & G_y(0,0) \end{pmatrix} = \begin{pmatrix} -2 & 1 \\ 0 & -1 \end{pmatrix}$$

and the corresponding linear system near $(0, 0)$ is

$$\frac{d}{dt}\begin{pmatrix} x \\ y \end{pmatrix} = \begin{pmatrix} -2 & 1 \\ 0 & -1 \end{pmatrix}\begin{pmatrix} x \\ y \end{pmatrix}.$$

Near the critical point $(2, 4)$, the Jacobian matrix is

$$\begin{pmatrix} F_x(2,4) & F_y(2,4) \\ G_x(2,4) & G_y(2,4) \end{pmatrix} = \begin{pmatrix} -2 & 1 \\ 4 & -1 \end{pmatrix}$$

and the corresponding linear system near $(2, 4)$ is

$$\frac{d}{dt}\begin{pmatrix} u \\ v \end{pmatrix} = \begin{pmatrix} -2 & 1 \\ 4 & -1 \end{pmatrix}\begin{pmatrix} u \\ v \end{pmatrix}$$

where $u = x - 2$ and $v = y - 4$.

16

(c) The eigenvalues of the linear system near $(0,0)$ are $\lambda = -1, -2$. From this, we can conclude that $(0,0)$ is an asymptotically stable node for the nonlinear system. The eigenvalues of the linear system near $(2,4)$ are $(-3 \pm \sqrt{17})/2$. Since one of these eigenvalues is positive and one is negative, the critical point $(2,4)$ is an unstable saddle point for the nonlinear system.

(d)

(e) The basin of attraction for the asymptotically stable critical point $(0,0)$ is bounded on the right by trajectories heading towards and then away and up to the right from the critical point $(2,4)$.

2.

(a) To find the critical points, we need to solve the equations $y = x$ and $x - 2y + xy - 2 = 0$. Plugging x in for y in the second equation, we see that we have $x = 2$ and $x = -1$. Therefore, the two critical points are $(2,2)$ and $(-1,-1)$.

(b) Here, we have $F(x,y) = x - y$ and $G(x,y) = x - 2y + xy - 2$. Therefore, the Jacobian matrix for this system is

$$\begin{pmatrix} F_x & F_y \\ G_x & G_y \end{pmatrix} = \begin{pmatrix} 1 & -1 \\ 1+y & -2+x \end{pmatrix}.$$

Therefore, near the critical point $(2,2)$, the Jacobian matrix is

$$\begin{pmatrix} F_x(2,2) & F_y(2,2) \\ G_x(2,2) & G_y(2,2) \end{pmatrix} = \begin{pmatrix} 1 & -1 \\ 3 & 0 \end{pmatrix}$$

and the corresponding linear system near $(2,2)$ is

$$\frac{d}{dt}\begin{pmatrix} u \\ v \end{pmatrix} = \begin{pmatrix} 1 & -1 \\ 3 & 0 \end{pmatrix}\begin{pmatrix} u \\ v \end{pmatrix}$$

where $u = x - 2$ and $v = y - 2$. Near the critical point $(-1,-1)$, the Jacobian matrix is

$$\begin{pmatrix} F_x(-1,-1) & F_y(-1,-1) \\ G_x(-1,-1) & G_y(-1,-1) \end{pmatrix} = \begin{pmatrix} 1 & -1 \\ 0 & -3 \end{pmatrix}$$

17

and the corresponding linear system near $(-1, -1)$ is

$$\frac{d}{dt}\begin{pmatrix} u \\ v \end{pmatrix} = \begin{pmatrix} 1 & -1 \\ 0 & -3 \end{pmatrix}\begin{pmatrix} u \\ v \end{pmatrix}$$

where $u = x + 1$ and $v = y + 1$.

(c) The eigenvalues of the linear system near $(2, 2)$ are $\lambda = (1 \pm \sqrt{11}i)/2$. From this, we can conclude that $(2, 2)$ is an unstable spiral point for the nonlinear system. The eigenvalues of the linear system near $(-1, -1)$ are $\lambda = 1, -3$. Since one of these eigenvalues is positive and one is negative, the critical point $(-1, -1)$ is an unstable saddle point for the nonlinear system.

(d)

3.

(a) To find the critical points, we need to solve the equations $x = -y^2$ and $x = -2y$. In order for these two equations to be satisfied simultaneously, we need $y^2 = 2y$. Therefore, $y = 0$ or $y = 2$. Therefore, the two critical points are $(0, 0)$ and $(-4, 2)$.

(b) Here, we have $F(x, y) = x + y^2$ and $G(x, y) = x + 2y$. Therefore, the Jacobian matrix for this system is

$$\begin{pmatrix} F_x & F_y \\ G_x & G_y \end{pmatrix} = \begin{pmatrix} 1 & 2y \\ 1 & 2 \end{pmatrix}.$$

Therefore, near the critical point $(0, 0)$, the Jacobian matrix is

$$\begin{pmatrix} F_x(0, 0) & F_y(0, 0) \\ G_x(0, 0) & G_y(0, 0) \end{pmatrix} = \begin{pmatrix} 1 & 0 \\ 1 & 2 \end{pmatrix}$$

and the corresponding linear system near $(0, 0)$ is

$$\frac{d}{dt}\begin{pmatrix} x \\ y \end{pmatrix} = \begin{pmatrix} 1 & 0 \\ 1 & 2 \end{pmatrix}\begin{pmatrix} x \\ y \end{pmatrix}.$$

18

Near the critical point $(-4, 2)$, the Jacobian matrix is

$$\begin{pmatrix} F_x(-4, 2) & F_y(-4, 2) \\ G_x(-4, 2) & G_y(-4, 2) \end{pmatrix} = \begin{pmatrix} 1 & 4 \\ 1 & 2 \end{pmatrix}$$

and the corresponding linear system near $(-4, 2)$ is

$$\frac{d}{dt}\begin{pmatrix} u \\ v \end{pmatrix} = \begin{pmatrix} 1 & 4 \\ 1 & 2 \end{pmatrix}\begin{pmatrix} u \\ v \end{pmatrix}$$

where $u = x + 4$ and $v = y - 2$.

(c) The eigenvalues of the linear system near $(0, 0)$ are $\lambda = 1, 2$. From this, we can conclude that $(0, 0)$ is an unstable node for the nonlinear system. The eigenvalues of the linear system near $(-4, 2)$ are $\lambda = (3 \pm \sqrt{17})/2$. Since one of these eigenvalues is positive and one is negative, the critical point $(-4, 2)$ is an unstable saddle point for the nonlinear system.

(d)

4.

(a) To find the critical points, we need to solve the equations $x = y^2$ and $x - 2y + x^2 = 0$. Plugging $x = y^2$ into the second equation, we have $y^4 + y^2 - 2y = 0$. The real solutions of this equation are $y = 0, 1$. Therefore, the two critical points are $(0, 0)$ and $(1, 1)$.

(b) Here, we have $F(x, y) = x - y^2$ and $G(x, y) = x - 2y + x^2$. Therefore, the Jacobian matrix for this system is

$$\begin{pmatrix} F_x & F_y \\ G_x & G_y \end{pmatrix} = \begin{pmatrix} 1 & -2y \\ 1 + 2x & -2 \end{pmatrix}.$$

Therefore, near the critical point $(0, 0)$, the Jacobian matrix is

$$\begin{pmatrix} F_x(0, 0) & F_y(0, 0) \\ G_x(0, 0) & G_y(0, 0) \end{pmatrix} = \begin{pmatrix} 1 & 0 \\ 1 & -2 \end{pmatrix}$$

19

and the corresponding linear system near $(0,0)$ is

$$\frac{d}{dt}\begin{pmatrix} x \\ y \end{pmatrix} = \begin{pmatrix} 1 & 0 \\ 1 & -2 \end{pmatrix}\begin{pmatrix} x \\ y \end{pmatrix}.$$

Near the critical point $(1,1)$, the Jacobian matrix is

$$\begin{pmatrix} F_x(1,1) & F_y(1,1) \\ G_x(1,1) & G_y(1,1) \end{pmatrix} = \begin{pmatrix} 1 & -2 \\ 3 & -2 \end{pmatrix}$$

and the corresponding linear system near $(1,1)$ is

$$\frac{d}{dt}\begin{pmatrix} u \\ v \end{pmatrix} = \begin{pmatrix} 1 & -2 \\ 3 & -2 \end{pmatrix}\begin{pmatrix} u \\ v \end{pmatrix}$$

where $u = x - 1$ and $v = y - 1$.

(c) The eigenvalues of the linear system near $(0,0)$ are $\lambda = 1, -2$. From this, we can conclude that $(0,0)$ is an unstable saddle point for the nonlinear system. The eigenvalues of the linear system near $(1,1)$ are $\lambda = (-1 \pm \sqrt{15}i)/2$. From this, we can conclude that $(1,1)$ is an asymptotically stable spiral point for the nonlinear system.

(d)

(e) The basin of attraction for the asymptotically stable point $(1,1)$ is bounded above by a trajectory heading over the spiral region, towards $(0,0)$ and then away from $(0,0)$ up and to the left. The basin of attraction is bounded below by a trajectory heading towards $(0,0)$ from the right and then heading away from $(0,0)$ up and to the left.

5.

(a) To find the critical points, we need to solve the equations $(2 + x)(y - x) = 0$ and $(4 - x)(y + x) = 0$. Solving this system of equations, we see that the critical points are given by $(0,0)$, $(4,4)$ and $(-2,2)$.

(b) Here, we have $F(x, y) = (2 + x)(y - x)$ and $G(x, y) = (4 - x)(y + x)$. Therefore, the Jacobian matrix for this system is

$$\begin{pmatrix} F_x & F_y \\ G_x & G_y \end{pmatrix} = \begin{pmatrix} -2 - 2x + y & 2 + x \\ 4 - y - 2x & 4 - x \end{pmatrix}.$$

Therefore, near the critical point $(0, 0)$, the Jacobian matrix is

$$\begin{pmatrix} F_x(0, 0) & F_y(0, 0) \\ G_x(0, 0) & G_y(0, 0) \end{pmatrix} = \begin{pmatrix} -2 & 2 \\ 4 & 4 \end{pmatrix}$$

and the corresponding linear system near $(0, 0)$ is

$$\frac{d}{dt} \begin{pmatrix} x \\ y \end{pmatrix} = \begin{pmatrix} -2 & 2 \\ 4 & 4 \end{pmatrix} \begin{pmatrix} x \\ y \end{pmatrix}.$$

Near the critical point $(-2, 2)$, the Jacobian matrix is

$$\begin{pmatrix} F_x(-2, 2) & F_y(-2, 2) \\ G_x(-2, 2) & G_y(-2, 2) \end{pmatrix} = \begin{pmatrix} 4 & 0 \\ 6 & 6 \end{pmatrix}$$

and the corresponding linear system near $(-2, 2)$ is

$$\frac{d}{dt} \begin{pmatrix} u \\ v \end{pmatrix} = \begin{pmatrix} 4 & 0 \\ 6 & 6 \end{pmatrix} \begin{pmatrix} u \\ v \end{pmatrix}$$

where $u = x + 2$ and $v = y - 2$. Near the critical point $(4, 4)$, the Jacobian matrix is

$$\begin{pmatrix} F_x(4, 4) & F_y(4, 4) \\ G_x(4, 4) & G_y(4, 4) \end{pmatrix} = \begin{pmatrix} -6 & 6 \\ -8 & 0 \end{pmatrix}$$

and the corresponding linear system near $(4, 4)$ is

$$\frac{d}{dt} \begin{pmatrix} u \\ v \end{pmatrix} = \begin{pmatrix} -6 & 6 \\ -8 & 0 \end{pmatrix} \begin{pmatrix} u \\ v \end{pmatrix}$$

where $u = x - 4$ and $v = y - 4$.

(c) The eigenvalues of the linear system near $(0, 0)$ are $\lambda = 1 \pm \sqrt{17}$. From this, we can conclude that $(0, 0)$ is an unstable saddle point for the nonlinear system. The eigenvalues of the linear system near $(-2, 2)$ are $\lambda = 4, 6$. From this, we can conclude that $(-2, 2)$ is an unstable node for the nonlinear system. The eigenvalues of the linear system near $(4, 4)$ are $\lambda = -3 \pm i\sqrt{39}$. From this, we can conclude that $(4, 4)$ is an asymptotically stable spiral point the nonlinear system.

(d)

(e) The basin of attraction for the asymptotically stable point $(4, 4)$ is bounded below by trajectories heading in towards the origin and bounded on the left by trajectories heading away from $(-2, 2)$.

6.

(a) To find the critical points, we need to solve the equations $x - x^2 - xy = 0$ and $3y - xy - 2y^2 = 0$. Solving this system of equations, we see that the critical points are given by $(0, 0)$, $(0, 3/2)$, $(1, 0)$, $(-1, 2)$.

(b) Here, we have $F(x, y) = x - x^2 - xy$ and $G(x, y) = 3y - xy - 2y^2$. Therefore, the Jacobian matrix for this system is

$$\begin{pmatrix} F_x & F_y \\ G_x & G_y \end{pmatrix} = \begin{pmatrix} 1 - 2x - y & -x \\ -y & 3 - x - 4y \end{pmatrix}.$$

Therefore, near the critical point $(0, 0)$, the Jacobian matrix is

$$\begin{pmatrix} 1 & 0 \\ 0 & 3 \end{pmatrix}$$

and the corresponding linear system near $(0, 0)$ is

$$\frac{d}{dt} \begin{pmatrix} x \\ y \end{pmatrix} = \begin{pmatrix} 1 & 0 \\ 0 & 3 \end{pmatrix} \begin{pmatrix} x \\ y \end{pmatrix}.$$

Near the critical point $(0, 3/2)$, the Jacobian matrix is

$$\begin{pmatrix} -1/2 & 0 \\ -3/2 & -3 \end{pmatrix}$$

and the corresponding linear system near $(0, 3/2)$ is

$$\frac{d}{dt} \begin{pmatrix} u \\ v \end{pmatrix} = \begin{pmatrix} -1/2 & 0 \\ -3/2 & -3 \end{pmatrix} \begin{pmatrix} u \\ v \end{pmatrix}$$

where $u = x$ and $v = y - 3/2$. Near the critical point $(1, 0)$, the Jacobian matrix is

$$\begin{pmatrix} -1 & -1 \\ 0 & 2 \end{pmatrix}$$

and the corresponding linear system near $(1, 0)$ is

$$\frac{d}{dt}\begin{pmatrix} u \\ v \end{pmatrix} = \begin{pmatrix} -1 & -1 \\ 0 & 2 \end{pmatrix}\begin{pmatrix} u \\ v \end{pmatrix}$$

where $u = x - 1$ and $v = y$. Near the critical point $(-1, 2)$, the Jacobian matrix is

$$\begin{pmatrix} 1 & 1 \\ -2 & -4 \end{pmatrix}$$

and the corresponding linear system near $(-1, 2)$ is

$$\frac{d}{dt}\begin{pmatrix} u \\ v \end{pmatrix} = \begin{pmatrix} 1 & 1 \\ -2 & -4 \end{pmatrix}\begin{pmatrix} u \\ v \end{pmatrix}$$

where $u = x + 1$ and $v = y - 2$.

(c) The eigenvalues of the linear system near $(0, 0)$ are $\lambda = 1, 3$. From this, we can conclude that $(0, 0)$ is an unstable node for the nonlinear system. The eigenvalues of the linear system near $(0, 3/2)$ are $\lambda = -1/2, -3$. From this, we can conclude that $(0, 3/2)$ is an asymptotically stable node for the nonlinear system. The eigenvalues of the linear system near $(1, 0)$ are $\lambda = -1, 2$. From this, we can conclude that $(1, 0)$ is a saddle point for the nonlinear system. The eigenvalues of the linear system near $(-1, 2)$ are $\lambda = (-3 \pm \sqrt{17})/2$. From this, we can conclude that $(-1, 2)$ is a saddle point for the nonlinear system.

(d)

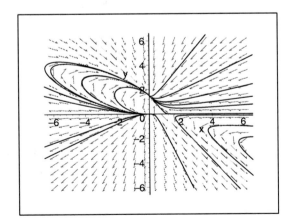

(e) The basin of attraction for the asymptotically stable point $(0, 3/2)$ consists of the first quadrant combined with trajectories heading into the second quadrant from $(0, 0)$ and towards $(0, 3/2)$.

7.

(a) To find the critical points, we need to solve the equations $1 - y = 0$ and $x^2 - y^2 = 0$. Solving this system of equations, we see that the critical points are given by $(-1, 1)$ and $(1, 1)$.

(b) Here, we have $F(x, y) = 1 - y$ and $G(x, y) = x^2 - y^2$. Therefore, the Jacobian matrix for this system is
$$\begin{pmatrix} F_x & F_y \\ G_x & G_y \end{pmatrix} = \begin{pmatrix} 0 & -1 \\ 2x & -2y \end{pmatrix}.$$
Therefore, near the critical point $(-1, 1)$, the Jacobian matrix is
$$\begin{pmatrix} 0 & -1 \\ -2 & -2 \end{pmatrix}.$$
Near the critical point $(1, 1)$, the Jacobian matrix is
$$\begin{pmatrix} 0 & -1 \\ 2 & -2 \end{pmatrix}.$$

(c) The eigenvalues of the linear system near $(-1, 1)$ are $\lambda = -1 \pm \sqrt{3}$. From this, we can conclude that $(-1, 1)$ is a saddle point for the nonlinear system. The eigenvalues of the linear system near $(1, 1)$ are $\lambda = -1 \pm i$. From this, we can conclude that $(1, 1)$ is an asymptotically stable spiral point for the nonlinear system.

(d)

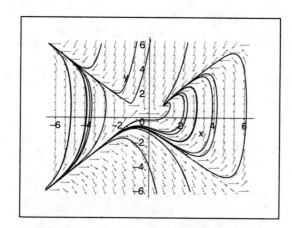

(e) The basin of attraction for the asymptotically stable point $(1, 1)$ is bounded on the left by trajectories just to the left of the $y-$axis and below by trajectories heading away from the saddle point $(-1, 1)$ and back towards $(1, 1)$.

8.

(a) To find the critical points, we need to solve the equations $x - x^2 - 2xy = 0$ and $-y(x+1) = 0$. The second equation implies $x = -1$ or $y = 0$. Plugging these values into the first equation, we see that the three critical points are $(0, 0)$, $(1, 0)$ and $(-1, 1)$.

(b) Here, we have $F(x, y) = x - x^2 - 2xy$ and $G(x, y) = -y(x + 1)$. Therefore, the Jacobian matrix for this system is

$$\begin{pmatrix} F_x & F_y \\ G_x & G_y \end{pmatrix} = \begin{pmatrix} 1 - 2x - 2y & -2x \\ -y & -x - 1 \end{pmatrix}.$$

Therefore, near the critical point $(0, 0)$, the Jacobian matrix is

$$\begin{pmatrix} F_x(0,0) & F_y(0,0) \\ G_x(0,0) & G_y(0,0) \end{pmatrix} = \begin{pmatrix} 1 & 0 \\ 0 & -1 \end{pmatrix}$$

and the corresponding linear system near $(0, 0)$ is

$$\frac{d}{dt}\begin{pmatrix} x \\ y \end{pmatrix} = \begin{pmatrix} 1 & 0 \\ 0 & -1 \end{pmatrix}\begin{pmatrix} x \\ y \end{pmatrix}.$$

Near the critical point $(1, 0)$, the Jacobian matrix is

$$\begin{pmatrix} F_x(1,0) & F_y(1,0) \\ G_x(1,0) & G_y(1,0) \end{pmatrix} = \begin{pmatrix} -1 & -2 \\ 0 & -2 \end{pmatrix}$$

and the corresponding linear system near $(1, 0)$ is

$$\frac{d}{dt}\begin{pmatrix} u \\ v \end{pmatrix} = \begin{pmatrix} -1 & -2 \\ 0 & -2 \end{pmatrix}\begin{pmatrix} u \\ v \end{pmatrix}$$

where $u = x - 1$ and $v = y$. Near the critical point $(-1, 1)$, the Jacobian matrix is

$$\begin{pmatrix} F_x(-1,1) & F_y(-1,1) \\ G_x(-1,1) & G_y(-1,1) \end{pmatrix} = \begin{pmatrix} 1 & 2 \\ -1 & 0 \end{pmatrix}$$

and the corresponding linear system near $(-1, 1)$ is

$$\frac{d}{dt}\begin{pmatrix} u \\ v \end{pmatrix} = \begin{pmatrix} 1 & 2 \\ -1 & 0 \end{pmatrix}\begin{pmatrix} u \\ v \end{pmatrix}$$

where $u = x + 1$ and $v = y - 1$.

(c) The eigenvalues of the linear system near $(0, 0)$ are $\lambda = 1, -1$. From this, we can conclude that $(0, 0)$ is an unstable saddle point for the nonlinear system. The eigenvalues of the linear system near $(1, 0)$ are $\lambda = -1, -2$. From this, we can conclude that $(1, 0)$ is an asymptotically stable node for the nonlinear system. The eigenvalues of the linear system near $(-1, 1)$ are $\lambda = (1 \pm \sqrt{7}i)/2$. From this, we can conclude that $(-1, 1)$ is an unstable spiral point.

(d)

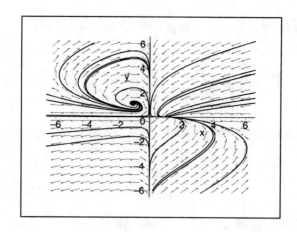

(e) The basin of attraction for the asymptotically stable point $(1, 0)$ is the right half plane.

9.

(a) To find the critical points, we need to solve the equations $-(x - y)(1 - x - y) = 0$ and $x(2 + y) = 0$. Solving these equations, we find that the critical points are $(0, 0)$, $(-2, -2)$, $(0, 1)$ and $(3, -2)$.

(b) Here, we have $F(x, y) = -(x - y)(1 - x - y)$ and $G(x, y) = x(2 + y)$. Therefore, the Jacobian matrix for this system is

$$\begin{pmatrix} F_x & F_y \\ G_x & G_y \end{pmatrix} = \begin{pmatrix} 2x - 1 & 1 - 2y \\ 2 + y & x \end{pmatrix}.$$

Therefore, near the critical point $(0, 0)$, the Jacobian matrix is

$$\begin{pmatrix} -1 & 1 \\ 2 & 0 \end{pmatrix}.$$

Near the critical point $(0, 1)$, the Jacobian matrix is

$$\begin{pmatrix} -1 & -1 \\ 3 & 0 \end{pmatrix}.$$

Near the critical point $(-2, -2)$, the Jacobian matrix is

$$\begin{pmatrix} -5 & 5 \\ 0 & -2 \end{pmatrix}.$$

Near the critical point $(3, -2)$, the Jacobian matrix is

$$\begin{pmatrix} 5 & 5 \\ 0 & 3 \end{pmatrix}.$$

(c) The eigenvalues of the linear system near $(0,0)$ are $\lambda = 1, -2$. From this, we can conclude that $(0,0)$ is an unstable saddle point for the nonlinear system. The eigenvalues of the linear system near $(0,1)$ are $\lambda = (-1 \pm i\sqrt{11})/2$. From this, we can conclude that $(0,1)$ is an asymptotically stable spiral for the nonlinear system. The eigenvalues of the linear system near $(-2,-2)$ are $\lambda = -2, -5$. From this, we can conclude that $(-2,-2)$ is an asymptotically stable node. The eigenvalues of the linear system near $(3,-2)$ are $\lambda = 3, 5$. From this, we can conclude that $(3,-2)$ is an unstable node.

(d)

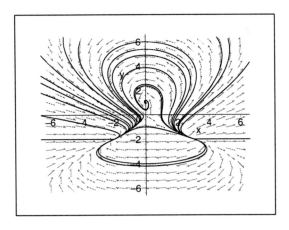

10.

(a) To find the critical points, we need to solve the equations $x + x^2 + y^2 = 0$ and $y - xy = 0$. Solving these equations, we find that the critical points are $(0,0)$ and $(-1,0)$.

(b) Here, we have $F(x,y) = x + x^2 + y^2$ and $G(x,y) = y - xy$. Therefore, the Jacobian matrix for this system is

$$\begin{pmatrix} F_x & F_y \\ G_x & G_y \end{pmatrix} = \begin{pmatrix} 1 + 2x & 2y \\ -y & 1 - x \end{pmatrix}.$$

Therefore, near the critical point $(0,0)$, the Jacobian matrix is

$$\begin{pmatrix} 1 & 0 \\ 0 & 1 \end{pmatrix}.$$

Near the critical point $(-1,0)$, the Jacobian matrix is

$$\begin{pmatrix} -1 & 0 \\ 0 & 3 \end{pmatrix}.$$

(c) The eigenvalues of the linear system near $(0,0)$ are $\lambda = 1$. From this, we can conclude that $(0,0)$ is an unstable node or spiral point for the nonlinear system. The eigenvalues of the linear system near $(-1,0)$ are $\lambda = -1, 3$. From this, we can conclude that $(-1,0)$ is a saddle point for the nonlinear system.

27

(d)

11.

(a) To find the critical points, we need to solve the equations $2x+y+xy^3 = 0$ and $x-2y-xy = 0$. Substituting $y = x/(x+2)$ into the first equation results in

$$3x^4 + 13x^3 + 28x^2 + 20x = 0.$$

One root of this equation is $x = 0$. The only other real root is

$$x = \frac{1}{9}\left[\left(287 + 18\sqrt{2019}\right)^{1/3} - 83\left(287 + 18\sqrt{2019}\right)^{-1/3} - 13\right].$$

Therefore, the critical points are $(0,0)$ and $(-1.19345, 1.4797)$.

(b) Here, we have $F(x,y) = 2x+y+xy^3$ and $G(x,y) = x-2y-xy$. Therefore, the Jacobian matrix for this system is

$$\begin{pmatrix} F_x & F_y \\ G_x & G_y \end{pmatrix} = \begin{pmatrix} 2+y^3 & 1+3xy^2 \\ 1-y & -2-x \end{pmatrix}.$$

Therefore, near the critical point $(0,0)$, the Jacobian matrix is

$$\begin{pmatrix} 2 & 1 \\ 1 & -2 \end{pmatrix}.$$

Near the critical point $(-1.19345, 1.4797)$, the Jacobian matrix is

$$\begin{pmatrix} -1.2399 & -6.8393 \\ 2.4797 & -0.8065 \end{pmatrix}.$$

(c) The eigenvalues of the linear system near $(0,0)$ are $\lambda = \pm\sqrt{5}$. From this, we can conclude that $(0,0)$ is an unstable saddle point for the nonlinear system. The eigenvalues of the linear system near $(-1.19345, 1.4797)$ are $\lambda = -1.0232 \pm 4.1125i$. From this, we can conclude that $(-1.19345, 1.4797)$ is an asymptotically stable spiral point for the nonlinear system.

(d)

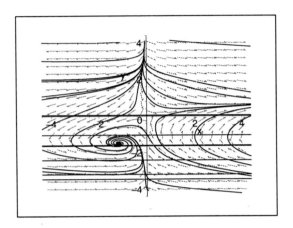

12.

(a) To find the critical points, we need to solve the equations $(1+x)\sin y = 0$ and $1 - x - \cos y = 0$. If $x = -1$, then we must have $\cos y = 2$, which is impossible. Therefore, $\sin y = 0$, which implies that $y = n\pi$, $n = 0, \pm 1, \pm 2, \ldots$ Based on the second equation, $x = 1 - \cos n\pi$. It follows that the critical points are located at $(0, 2k\pi)$ and $(2, (2k+1)\pi)$ where $k = 0, \pm 1, \pm 2, \ldots$

(b) Here, we have $F(x, y) = (1+x)\sin y$ and $G(x, y) = 1 - x - \cos y$. Therefore, the Jacobian matrix for this system is

$$\begin{pmatrix} F_x & F_y \\ G_x & G_y \end{pmatrix} = \begin{pmatrix} \sin y & (1+x)\cos y \\ -1 & \sin y \end{pmatrix}.$$

Therefore, near the critical points $(0, 2k\pi)$, the Jacobian matrix is

$$\begin{pmatrix} 0 & 1 \\ -1 & 0 \end{pmatrix}.$$

Near the critical point $(2, (2k+1)\pi)$, the Jacobian matrix is

$$\begin{pmatrix} 0 & -3 \\ -1 & 0 \end{pmatrix}.$$

(c) The eigenvalues of the linear system near $(0, 2k\pi)$ are $\lambda = \pm i$. Based on this information, we cannot make a conclusion about the nature of the critical points near $(0, 2k\pi)$ for the nonlinear system. The eigenvalues of the linear system near $(2, (2k+1)\pi)$ are $\lambda = \pm\sqrt{3}$. From this, we can conclude that the critical points $(2, (2k+1)\pi)$ are saddles.

(d)

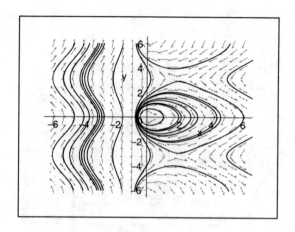

Upon looking at the phase portrait, we see that the critical points $(0, 2k\pi)$ are centers.

13.

(a) To find the critical points, we need to solve the equations $x - y^2 = 0$ and $y - x^2 = 0$. Substituting $y = x^2$ into the first equation, results in $x - x^4 = 0$ which has real roots $x = 0, 1$. Therefore, the critical points are $(0, 0)$ and $(1, 1)$.

(b) Here, we have $F(x, y) = x - y^2$ and $G(x, y) = y - x^2$. Therefore, the Jacobian matrix for this system is

$$\begin{pmatrix} F_x & F_y \\ G_x & G_y \end{pmatrix} = \begin{pmatrix} 1 & -2y \\ -2x & 1 \end{pmatrix}.$$

Therefore, near the critical point $(0, 0)$, the Jacobian matrix is

$$\begin{pmatrix} 1 & 0 \\ 0 & 1 \end{pmatrix}.$$

Near the critical point $(1, 1)$, the Jacobian matrix is

$$\begin{pmatrix} 1 & -2 \\ -2 & 1 \end{pmatrix}.$$

(c) There is a repeated eigenvalue $\lambda = 1$ for the linear system near $(0, 0)$. Based on this information, we cannot make a conclusion about the nature of the critical point near $(0, 0)$ for the nonlinear system. The eigenvalues of the linear system near $(1, 1)$ are $\lambda = 3, -1$. From this, we can conclude that the critical point $(1, 1)$ is a saddle.

(d)

30

Upon looking at the phase portrait, we see that the critical point $(0, 0)$ is an unstable node.

14.

(a) To find the critical points, we need to solve the equations $1 - xy = 0$ and $x - y^3 = 0$. After multiplying the second equation by y, it follows that $y = \pm 1$. Therefore, the critical points of the system are $(1, 1)$ and $(-1, -1)$.

(b) Here, we have $F(x, y) = 1 - xy$ and $G(x, y) = x - y^3$. Therefore, the Jacobian matrix for this system is

$$\begin{pmatrix} F_x & F_y \\ G_x & G_y \end{pmatrix} = \begin{pmatrix} -y & -x \\ 1 & -3y^2 \end{pmatrix}.$$

Therefore, near the critical point $(1, 1)$, the Jacobian matrix is

$$\begin{pmatrix} -1 & -1 \\ 1 & -3 \end{pmatrix}.$$

Near the critical point $(-1, -1)$, the Jacobian matrix is

$$\begin{pmatrix} 1 & 1 \\ 1 & -3 \end{pmatrix}.$$

(c) The eigenvalues of the linear system near $(1, 1)$ are $\lambda = -2, -2$; that is, there is one repeated eigenvalue. Therefore, we cannot draw a conclusion about the nature of the critical point $(1, 1)$ for the nonlinear system. The eigenvalues of the linear system near $(-1, -1)$ are $-1 \pm \sqrt{5}$. From this, we can conclude that the critical point $(1, 1)$ is a saddle.

(d)

Upon looking at the phase portrait, we see that the critical point $(1, 1)$ is an asymptotically stable node.

15.

(a) To find the critical points, we need to solve the equations

$$-2x - y - x(x^2 + y^2) = 0$$
$$x - y + y(x^2 + y^2) = 0.$$

It is clear that $(0, 0)$ is a critical point. Solving the first equation for y, we find that

$$y = \frac{-1 \pm \sqrt{1 - 8x^2 - 4x^4}}{2x}.$$

Substitution of these relations into the second equation results in two equations of the form $f_1(x) = 0$ and $f_2(x) = 0$. Plotting these functions, we note that only $f_1(x) = 0$ has real roots given by $x \approx \pm 0.33076$. It follows that the additional critical points are $(-0.33076, 1.0924)$ and $(0.33076, -1.0924)$.

(b) Here, we have $F(x, y) = -2x - y - x(x^2 + y^2)$ and $G(x, y) = x - y + y(x^2 + y^2)$. Therefore, the Jacobian matrix for this system is

$$\begin{pmatrix} F_x & F_y \\ G_x & G_y \end{pmatrix} = \begin{pmatrix} -2 - 3x^2 - y^2 & -1 - 2xy \\ 1 + 2xy & -1 + x^2 + 3y^2 \end{pmatrix}.$$

Therefore, near the critical point $(0, 0)$, the Jacobian matrix is

$$\begin{pmatrix} -2 & -1 \\ 1 & -1 \end{pmatrix}.$$

Near the critical points $(-0.33076, 1.0924)$ and $(0.33076, -1.0924)$, the Jacobian matrix is

$$\begin{pmatrix} -3.5216 & -0.27735 \\ 0.27735 & 2.6895 \end{pmatrix}.$$

32

(c) The eigenvalues of the linear system near $(0,0)$ are $\lambda = (-3 \pm i\sqrt{3})/2$ Therefore, $(0,0)$ is an asymptotically stable spiral point for the nonlinear system. The eigenvalues of the linear system near $(-0.33076, 1.0924)$ and $(0.33076, -1.0924)$ are $-3.5092, 2.6771$. From this, we can conclude that these two critical points are saddles.

(d)

16.

(a) To find the critical points, we need to solve the equations

$$y + x(1 - x^2 - y^2) = 0$$
$$-x + y(1 - x^2 - y^2) = 0.$$

Multiplying the first equation by y, the second equation by x and subtracting the second from the first, we conclude that $x^2 + y^2 = 0$. Therefore, the only critical point is $(0,0)$.

(b) Here, we have $F(x,y) = y + x(1 - x^2 - y^2)$ and $G(x,y) = -x + y(1 - x^2 - y^2)$. Therefore, the Jacobian matrix for this system is

$$\begin{pmatrix} F_x & F_y \\ G_x & G_y \end{pmatrix} = \begin{pmatrix} 1 - 3x^2 - y^2 & 1 - 2xy \\ -1 - 2xy & 1 - x^2 - 3y^2 \end{pmatrix}.$$

Therefore, near the critical point $(0,0)$, the Jacobian matrix is

$$\begin{pmatrix} 1 & 1 \\ -1 & 1 \end{pmatrix}.$$

(c) The eigenvalues of the linear system near $(0,0)$ are $\lambda = 1 \pm i$. Therefore, $(0,0)$ is an unstable spiral point for the nonlinear system.

(d)

17.

(a) The critical points occur when either $y = -2$ or $y = 0.5x$ and either $x = 2$ or $y = -0.5x$. Solving these equations simultaneously, we have the critical points $(0,0)$, $(2,-2)$, $(2,1)$, and $(4,-2)$.

(b) Here, we have $F(x,y) = (2+y)(y-0.5x)$ and $G(x,y) = (2-x)(y+0.5x)$. Therefore, the Jacobian matrix for this system is

$$\begin{pmatrix} F_x & F_y \\ G_x & G_y \end{pmatrix} = \begin{pmatrix} -1-0.5y & 2+2y-0.5x \\ -y+1-x & 2-x \end{pmatrix}.$$

Therefore, near the critical point $(0,0)$, the Jacobian matrix is

$$\begin{pmatrix} F_x(0,0) & F_y(0,0) \\ G_x(0,0) & G_y(0,0) \end{pmatrix} = \begin{pmatrix} -1 & 2 \\ 1 & 2 \end{pmatrix}$$

and the corresponding linear system near $(0,0)$ is

$$\frac{d}{dt}\begin{pmatrix} x \\ y \end{pmatrix} = \begin{pmatrix} -1 & 2 \\ 1 & 2 \end{pmatrix}\begin{pmatrix} x \\ y \end{pmatrix}.$$

Near the critical point $(2,-2)$, the Jacobian matrix is

$$\begin{pmatrix} F_x(2,-2) & F_y(2,-2) \\ G_x(2,-2) & G_y(2,-2) \end{pmatrix} = \begin{pmatrix} 0 & -3 \\ 1 & 0 \end{pmatrix}$$

and the corresponding linear system near $(2,-2)$ is

$$\frac{d}{dt}\begin{pmatrix} u \\ v \end{pmatrix} = \begin{pmatrix} 0 & -3 \\ 1 & 0 \end{pmatrix}\begin{pmatrix} u \\ v \end{pmatrix}$$

where $u = x - 2$ and $v = y + 2$. Near the critical point $(2,1)$, the Jacobian matrix is

$$\begin{pmatrix} F_x(2,1) & F_y(2,1) \\ G_x(2,1) & G_y(2,1) \end{pmatrix} = \begin{pmatrix} -\frac{3}{2} & 3 \\ -2 & 0 \end{pmatrix}$$

and the corresponding linear system near $(2, 1)$ is

$$\frac{d}{dt}\begin{pmatrix} u \\ v \end{pmatrix} = \begin{pmatrix} -\frac{3}{2} & 3 \\ -2 & 0 \end{pmatrix}\begin{pmatrix} u \\ v \end{pmatrix}$$

where $u = x - 2$ and $v = y - 1$. Near the critical point $(4, -2)$, the Jacobian matrix is

$$\begin{pmatrix} F_x(4, -2) & F_y(4, -2) \\ G_x(4, -2) & G_y(4, -2) \end{pmatrix} = \begin{pmatrix} 0 & -4 \\ -1 & -2 \end{pmatrix}$$

and the corresponding linear system near $(4, -2)$ is

$$\frac{d}{dt}\begin{pmatrix} u \\ v \end{pmatrix} = \begin{pmatrix} 0 & -4 \\ -1 & -2 \end{pmatrix}\begin{pmatrix} u \\ v \end{pmatrix}$$

where $u = x - 4$ and $v = y + 2$.

(c) The eigenvalues of the linear system near $(0, 0)$ are $\lambda = (1 \pm \sqrt{17})/2$. From this, we can conclude that $(0, 0)$ is an unstable saddle point for the nonlinear system. The eigenvalues of the linear system near $(2, -2)$ are $\lambda = \pm\sqrt{3}i$. From this, we can only say that $(2, -2)$ is a center or spiral point for the nonlinear system, and we do not determine the stability. The eigenvalues of the linear system near $(2, 1)$ are $\lambda = (-3 \pm \sqrt{87}i)/4$. From this, we can conclude that $(2, 1)$ is an asymptotically stable spiral point. The eigenvalues of the linear system near $(4, -2)$ are $\lambda = -1 \pm \sqrt{5}$. From this, we can conclude that $(4, -2)$ is an unstable saddle point.

(d)

From the phase portrait, we can see that $(2, -2)$ is a stable center for the nonlinear system.

(e) The basin of attraction for the asymptotically stable point $(2, 1)$ is bounded to the left and below by trajectories heading towards and then away from the unstable saddle point $(0, 0)$. The basin of attraction is bounded to the right by trajectories heading around the left side of the center $(2, -2)$.

35

18.

(a) The critical points occur when either $y = \pm 2$ and either $x = -1.5$ or $x = y$. Therefore, we see that the critical points are given by $(-1.5, 2)$, $(2, 2)$, $(-1.5, -2)$ and $(-2, -2)$.

(b) Here, we have $F(x, y) = 4 - y^2$ and $G(x, y) = (1.5 + x)(y - x)$. Therefore, the Jacobian matrix for this system is

$$\begin{pmatrix} F_x & F_y \\ G_x & G_y \end{pmatrix} = \begin{pmatrix} 0 & -2y \\ y - 2x - 1.5 & 1.5 + x \end{pmatrix}.$$

Therefore, near the critical point $(-1.5, 2)$, the Jacobian matrix is

$$\begin{pmatrix} F_x(-1.5, 2) & F_y(-1.5, 2) \\ G_x(-1.5, 2) & G_y(-1.5, 2) \end{pmatrix} = \begin{pmatrix} 0 & -4 \\ \frac{7}{2} & 0 \end{pmatrix}$$

and the corresponding linear system near $(-1.5, 2)$ is

$$\frac{d}{dt} \begin{pmatrix} u \\ v \end{pmatrix} = \begin{pmatrix} 0 & -4 \\ \frac{7}{2} & 0 \end{pmatrix} \begin{pmatrix} u \\ v \end{pmatrix}$$

where $u = x + 1.5$ and $v = y - 2$. Near the critical point $(2, 2)$, the Jacobian matrix is

$$\begin{pmatrix} F_x(2, 2) & F_y(2, 2) \\ G_x(2, 2) & G_y(2, 2) \end{pmatrix} = \begin{pmatrix} 0 & -4 \\ -\frac{7}{2} & \frac{7}{2} \end{pmatrix}$$

and the corresponding linear system near $(2, 2)$ is

$$\frac{d}{dt} \begin{pmatrix} u \\ v \end{pmatrix} = \begin{pmatrix} 0 & -4 \\ -\frac{7}{2} & \frac{7}{2} \end{pmatrix} \begin{pmatrix} u \\ v \end{pmatrix}$$

where $u = x - 2$ and $v = y - 2$. Near the critical point $(-1.5, -2)$, the Jacobian matrix is

$$\begin{pmatrix} F_x(-1.5, -2) & F_y(-1.5, -2) \\ G_x(-1.5, -2) & G_y(-1.5, -2) \end{pmatrix} = \begin{pmatrix} 0 & 4 \\ -\frac{1}{2} & 0 \end{pmatrix}$$

and the corresponding linear system near $(-1.5, -2)$ is

$$\frac{d}{dt} \begin{pmatrix} u \\ v \end{pmatrix} = \begin{pmatrix} 0 & 4 \\ -\frac{1}{2} & 0 \end{pmatrix} \begin{pmatrix} u \\ v \end{pmatrix}$$

where $u = x + 1.5$ and $v = y + 2$. Near the critical point $(-2, -2)$, the Jacobian matrix is

$$\begin{pmatrix} F_x(-2, -2) & F_y(-2, -2) \\ G_x(-2, -2) & G_y(-2, -2) \end{pmatrix} = \begin{pmatrix} 0 & 4 \\ \frac{1}{2} & -\frac{1}{2} \end{pmatrix}$$

and the corresponding linear system near $(-2, -2)$ is

$$\frac{d}{dt} \begin{pmatrix} u \\ v \end{pmatrix} = \begin{pmatrix} 0 & 4 \\ \frac{1}{2} & -\frac{1}{2} \end{pmatrix} \begin{pmatrix} u \\ v \end{pmatrix}$$

where $u = x + 2$ and $v = y + 2$.

(c) The eigenvalues of the linear system near $(-1.5, 2)$ are $\lambda = \pm\sqrt{14}i$. From this, we can only conclude that $(-1.5, 2)$ is a center or a spiral point, and we cannot determine its stability. The eigenvalues of the linear system near $(2, 2)$ are $\lambda = (7 \pm \sqrt{273})/4$. From this, we can conclude that $(2, 2)$ is an unstable saddle point for the nonlinear system. The eigenvalues of the linear system near $(-1.5, -2)$ are $\lambda = \pm\sqrt{2}i$. From this, we can only conclude that $(-1.5, -2)$ is a either a center or a spiral point, and we cannot determine its stability. The eigenvalues of the linear system near $(-2, -2)$ are $\lambda = (-1 \pm \sqrt{33})/4$. From this, we can conclude that $(-2, -2)$ is an unstable saddle point.

(d)

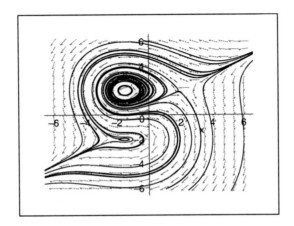

From the phase portrait above, we can see that $(-1.5, 2)$ and $(-1.5, -2)$ are stable centers for the nonlinear system.

19.

(a) The critical points occur when either $y = 1$ or $y = 2x$ and either $x = -2$ or $x = 2y$. Therefore, we see that the critical points are $(0, 0)$, $(2, 1)$, $(-2, 1)$ and $(-2, -4)$.

(b) Here, we have $F(x, y) = (1 - y)(2x - y)$ and $G(x, y) = (2 + x)(x - 2y)$. Therefore, the Jacobian matrix for this system is

$$
\begin{pmatrix} F_x & F_y \\ G_x & G_y \end{pmatrix} = \begin{pmatrix} 2 - 2y & -2x - 1 + 2y \\ 2 + 2x - 2y & -4 - 2x \end{pmatrix}.
$$

Therefore, near the critical point $(0, 0)$, the Jacobian matrix is

$$
\begin{pmatrix} F_x(0, 0) & F_y(0, 0) \\ G_x(0, 0) & G_y(0, 0) \end{pmatrix} = \begin{pmatrix} 2 & -1 \\ 2 & -4 \end{pmatrix}
$$

and the corresponding linear system near $(0, 0)$ is

$$
\frac{d}{dt} \begin{pmatrix} x \\ y \end{pmatrix} = \begin{pmatrix} 2 & -1 \\ 2 & -4 \end{pmatrix} \begin{pmatrix} x \\ y \end{pmatrix}.
$$

Near the critical point $(2, 1)$, the Jacobian matrix is

$$\begin{pmatrix} F_x(2,1) & F_y(2,1) \\ G_x(2,1) & G_y(2,1) \end{pmatrix} = \begin{pmatrix} 0 & -3 \\ 4 & -8 \end{pmatrix}$$

and the corresponding linear system near $(2, 1)$ is

$$\frac{d}{dt}\begin{pmatrix} u \\ v \end{pmatrix} = \begin{pmatrix} 0 & -3 \\ 4 & -8 \end{pmatrix}\begin{pmatrix} u \\ v \end{pmatrix}$$

where $u = x - 2$ and $v = y - 1$. Near the critical point $(-2, 1)$, the Jacobian matrix is

$$\begin{pmatrix} F_x(-2,1) & F_y(-2,1) \\ G_x(-2,1) & G_y(-2,1) \end{pmatrix} = \begin{pmatrix} 0 & 5 \\ -4 & 0 \end{pmatrix}$$

and the corresponding linear system near $(-2, 1)$ is

$$\frac{d}{dt}\begin{pmatrix} u \\ v \end{pmatrix} = \begin{pmatrix} 0 & 5 \\ -4 & 0 \end{pmatrix}\begin{pmatrix} u \\ v \end{pmatrix}$$

where $u = x - 2$ and $v = y - 1$. Near the critical point $(-2, -4)$, the Jacobian matrix is

$$\begin{pmatrix} F_x(-2,-4) & F_y(-2,-4) \\ G_x(-2,-4) & G_y(-2,-4) \end{pmatrix} = \begin{pmatrix} 10 & -5 \\ 6 & 0 \end{pmatrix}$$

and the corresponding linear system near $(-2, -4)$ is

$$\frac{d}{dt}\begin{pmatrix} u \\ v \end{pmatrix} = \begin{pmatrix} 10 & -5 \\ 6 & 0 \end{pmatrix}\begin{pmatrix} u \\ v \end{pmatrix}$$

where $u = x + 2$ and $v = y + 4$.

(c) The eigenvalues of the linear system near $(0,0)$ are $\lambda = -1 \pm \sqrt{7}$. From this, we can conclude that $(0,0)$ is an unstable saddle point for the nonlinear system. The eigenvalues of the linear system near $(2, 1)$ are $\lambda = -2, -6$. From this, we can conclude that $(2, 1)$ is an asymptotically stable node for the nonlinear system. The eigenvalues of the linear system near $(-2, 1)$ are $\lambda = \pm 2\sqrt{5}i$. From this, we can only conclude that $(-2, 1)$ is a either a center or a spiral point, and we cannot determine its stability. The eigenvalues of the linear system near $(-2, -4)$ are $\lambda = 5 \pm \sqrt{5}i$. From this, we can conclude that $(-2, -4)$ is an unstable spiral point.

(d)

38

From the phase portrait above, we can see that $(-2, 1)$ is a stable center for the nonlinear system.

(e) The basin of attraction for the asymptotically stable node $(2, 1)$ is bounded on the left by trajectories approaching the stable center $(-2, 1)$ and also by those trajectories heading to the left away from the unstable spiral point $(-2, -4)$.

20.

(a) Looking at the equation $dy/dt = 0$, we see that we need $x = 0$, $y = 0$ or $y = 3$. Plugging each of these values into the equation $dx/dt = 0$, we conclude that the critical points are $(0, 0)$, $(2, 3)$ and $(-2, 3)$.

(b) Here, we have $F(x, y) = 2x^2y - 3x^2 - 4y$ and $G(x, y) = -2xy^2 + 6xy$. Therefore, the Jacobian matrix for this system is

$$\begin{pmatrix} F_x & F_y \\ G_x & G_y \end{pmatrix} = \begin{pmatrix} 4xy - 6x & 2x^2 - 4 \\ -2y^2 + 6y & -4xy + 6x \end{pmatrix}.$$

Therefore, near the critical point $(0, 0)$, the Jacobian matrix is

$$\begin{pmatrix} F_x(0, 0) & F_y(0, 0) \\ G_x(0, 0) & G_y(0, 0) \end{pmatrix} = \begin{pmatrix} 0 & -4 \\ 0 & 0 \end{pmatrix}$$

and the corresponding linear system near $(0, 0)$ is

$$\frac{d}{dt} \begin{pmatrix} x \\ y \end{pmatrix} = \begin{pmatrix} 0 & -4 \\ 0 & 0 \end{pmatrix} \begin{pmatrix} x \\ y \end{pmatrix}.$$

Near the critical point $(2, 3)$, the Jacobian matrix is

$$\begin{pmatrix} F_x(2, 3) & F_y(2, 3) \\ G_x(2, 3) & G_y(2, 3) \end{pmatrix} = \begin{pmatrix} 12 & 4 \\ 0 & -12 \end{pmatrix}$$

and the corresponding linear system near $(2, 3)$ is

$$\frac{d}{dt} \begin{pmatrix} u \\ v \end{pmatrix} = \begin{pmatrix} 12 & 4 \\ 0 & -12 \end{pmatrix} \begin{pmatrix} u \\ v \end{pmatrix}$$

where $u = x - 2$ and $v = y - 3$. Near the critical point $(-2, 3)$, the Jacobian matrix is

$$\begin{pmatrix} F_x(-2, 3) & F_y(-2, 3) \\ G_x(-2, 3) & G_y(-2, 3) \end{pmatrix} = \begin{pmatrix} -12 & 4 \\ 0 & 12 \end{pmatrix}$$

and the corresponding linear system near $(-2, 3)$ is

$$\frac{d}{dt} \begin{pmatrix} u \\ v \end{pmatrix} = \begin{pmatrix} -12 & 4 \\ 0 & 12 \end{pmatrix} \begin{pmatrix} u \\ v \end{pmatrix}$$

where $u = x + 2$ and $v = y - 3$.

(c) The eigenvalue of the linear system near $(0, 0)$ is $\lambda = 0$. Therefore, we can make no conclusion about the critical point $(0, 0)$ for the nonlinear system. The eigenvalues of the linear system near $(2, 3)$ are $\lambda = 12, -12$. From this, we can conclude that $(2, 3)$ is an unstable saddle point for the nonlinear system. The eigenvalues of the linear system near $(-2, 3)$ are $\lambda = 12, -12$. From this, we can conclude that $(-2, 3)$ is an unstable saddle point for the nonlinear system.

(d)

21.

(a) The critical points occur when $y = 0$ and $x + 2x^3 = 0$. From the second equation, we need $x = 0$ or $1 + 2x^2 = 0$. The only real solution of those equations is $x = 0$. Therefore, the only critical point is $(0, 0)$.

(b) The Jacobian matrix is given by

$$\begin{pmatrix} F_x & F_y \\ G_x & G_y \end{pmatrix} = \begin{pmatrix} 0 & 1 \\ 1 + 6x^2 & 0 \end{pmatrix}.$$

Therefore, coefficient of the linearized system near $(0, 0)$ is

$$\begin{pmatrix} 0 & 1 \\ 1 & 0 \end{pmatrix}.$$

That is,

$$\frac{dx}{dt} = y$$
$$\frac{dy}{dt} = x.$$

Therefore, $dy/dx = x/y$. Separating variables, this equation can be written as $y\,dy = x\,dx$. Integrating this equation, we have $y^2 = x^2 + C$. Therefore, the trajectories consist of a family of hyperbolas.

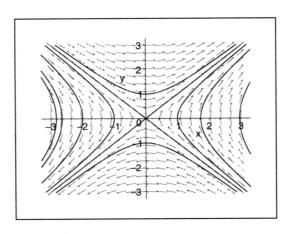

(c) For the nonlinear system, the equation for dy/dx is given by $dy/dx = (x + 2x^3)/y$. Separating variables, this equation can be written as $y\,dy = (x + 2x^3)\,dx$. Integrating this equation, we have $y^2 = x^2 + x^4 + C$. Therefore, the trajectories are level curves of $H(x, y) = x^2 + x^4 - y^2$. The trajectories that enter and leave the origin are no longer straight lines.

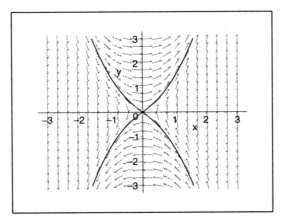

22.

(a) The critical points occur when $x = 0$ and $-2y + x^3 = 0$. Plugging the first equation into the second, we see that the only critical point is $(0, 0)$.

(b) The Jacobian matrix is given by

$$\begin{pmatrix} F_x & F_y \\ G_x & G_y \end{pmatrix} = \begin{pmatrix} 1 & 0 \\ 3x^2 & -2 \end{pmatrix}.$$

Therefore, coefficient of the linearized system near $(0, 0)$ is

$$\begin{pmatrix} 1 & 0 \\ 0 & -2 \end{pmatrix}.$$

That is,

$$\frac{dx}{dt} = x$$
$$\frac{dy}{dt} = -2y.$$

41

Therefore, $dy/dx = -2y/x$. Separating variables, this equation can be written as $dy/-2y = dx/x$. Integrating this equation, we have $x^2 y = C$. Therefore, the trajectories consist of the level curves of $H(x, y) = x^2 y$.

(c) For the nonlinear system, the equation for dy/dx is given by $dy/dx = (-2y + x^3)/x$. This equation can be rewritten as $(2y - x^3)\,dx + x\,dy = 0$. Multiplying this equation by x, we can rewrite the equation as $(2xy - x^4)\,dx + x^2\,dy = 0$. We notice that this equation is exact. Integrating $2xy - x^4$ with respect to x, we have $H(x, y) = x^2 y - x^5/5 + h(y)$. Then differentiating with respect to y, we have $H_y = x^2 + h'(y) = x^2$. Therefore, the trajectories are level curves of $H(x, y) = x^2 y - x^5/5$. The trajectories that enter and leave the origin are shown below:

23.

(a)

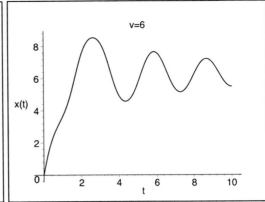

The first graph represents a pendulum which will swing back and forth about the equilibrium position $\theta = 0$. The second graph represents a pendulum which will swing completely around the origin before swinging back and forth about the equilibrium position $\theta = 2\pi$.

(b) Considering the graph of x versus t for various values of t, we see that the critical value of v occurs at $v_c \approx 5.41$.

(c) Below we show the graphs of x versus t for $\gamma = 1/2$.

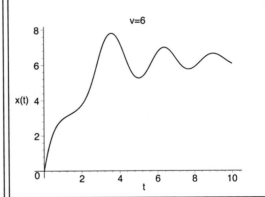

For $\gamma = 1/2$, the critical value of v will be $v_c \approx 5.94$. As γ increases, the critical value v_c increases.

24.

(a) It is clear to see that the origin is a critical point of each system. For both systems, the linearized system is

$$\frac{dx}{dt} = y$$
$$\frac{dy}{dt} = -x.$$

43

The associated Jacobian matrix is

$$\begin{pmatrix} 0 & 1 \\ -1 & 0 \end{pmatrix}.$$

The eigenvalues of this system are $\lambda = \pm i$. Therefore, $(0,0)$ is a center for each system.

(b) For $g_1(x,y) = x(x^2+y^2)$, we see that $g_1(x,y)/r \to 0$ as $r \to 0$. Also, $g_2(x,y) = y(x^2+y^2)$ implies $g_2(x,y)/r \to 0$ as $r \to 0$. In addition, both g_1 and g_2 have continuous first partial derivatives. Therefore, we conclude that each system is almost linear.

(c) For system (ii), we notice that

$$r\frac{dr}{dt} = x\frac{dx}{dt} + y\frac{dy}{dt} = xy - x^2(x^2+y^2) - xy - y^2(x^2+y^2) = -(x^2+y^2)^2 = -r^4.$$

Therefore, $r' = -r^3$. Consequently, $r^2 = 1/(2t+c)$ where $c = 1/r_0^2$. Since $r \to 0$ as $t \to 0$, regardless of the value of r_0, the origin is an asymptotically stable equilibrium point.

On the other hand, for system (i),

$$r\frac{dr}{dt} = x\frac{dx}{dt} + y\frac{dy}{dt} = xy + x^2(x^2+y^2) - xy + y^2(x^2+y^2) = (x^2+y^2)^2 = r^4.$$

Therefore, $r' = r^3$. Therefore, $r^2 = (c-2t)/(2t-c)^2$. Imposing the initial condition $r(0) = r_0$, we obtain a specific solution

$$r^2 = -\frac{r_0^2}{2r_0^2 t - 1}.$$

Since the solution becomes unbounded as $t \to 1/2r_0^2$, the critical point is unstable.

25. The characteristic equation for the coefficient matrix is $\lambda^2 + 1 = 0$, which has roots $\lambda = \pm i$. Therefore, the critical point at the origin is a center. For the perturbed matrix, the characteristic equation is $\lambda^2 - 2\epsilon\lambda + 1 + \epsilon^2 = 0$. This equation has roots $\lambda = \epsilon \pm i$. Therefore, as long as $\epsilon \neq 0$, the critical point of the perturbed system is a spiral point. Its stability depends on the sign of ϵ.

26. The characteristic equation for the coefficient matrix is $(\lambda + 1)^2 = 0$, which has the repeated root $\lambda = -1$. Therefore, the critical point is an asymptotically stable node. For the perturbed matrix, the characteristic equation is $\lambda^2 + 2\lambda + 1 + \epsilon = 0$. This equation has roots $\lambda = -1 \pm \sqrt{-\epsilon}$. If $\epsilon > 0$, then the roots are complex and the critical point is a stable spiral. If $\epsilon < 0$, then the roots are real and both negative, in which case the critical point remains a stable node.

27.

(a) Let $y = x'$. Then $y' = x'' = -c(x)x' - g(x) = -c(x)y - g(x)$. Therefore, we can write the system as

$$\frac{dx}{dt} = y$$

$$\frac{dy}{dt} = -c(x)y - g(x).$$

44

(b) Clearly, we see that $(0,0)$ is a critical point of this system. The Jacobian matrix for the system is given by

$$\begin{pmatrix} 0 & 1 \\ -g'(0) & -c(0) \end{pmatrix}.$$

Therefore, we can rewrite the system as

$$\begin{pmatrix} x \\ y \end{pmatrix}' = \begin{pmatrix} 0 & 1 \\ -g'(0) & -c(0) \end{pmatrix} \begin{pmatrix} x \\ y \end{pmatrix} + \begin{pmatrix} 0 \\ g'(0)x + c(0)y - c(x)y - g(x) \end{pmatrix}.$$

We see that

$$\frac{g_2(x,y)}{r} = \frac{g'(0)x + c(0)y - c(x)y - g(x)}{r}$$

$$= \frac{g'(0)r\cos\theta + c(0)r\sin\theta - c(r\cos\theta)r\sin\theta - g(r\cos\theta)}{r}$$

$$= g'(0)\cos\theta + c(0)\sin\theta - c(r\cos\theta)\sin\theta - \frac{g(r\cos\theta)}{r}.$$

Then, since $c(r\cos\theta) \to c(0)$ as $r \to 0$ and $g(r\cos\theta)/r\cos\theta \to g'(0)$ as $r \to 0$, we conclude that $g_2(x,y)/r \to 0$ as $r \to 0$.

(c) The eigenvalues satisfy $\lambda^2 + c(0)\lambda + g'(0) = 0$. Therefore, the eigenvalues are given by

$$\lambda = \frac{-c(0) \pm \sqrt{c(0)^2 - 4g'(0)}}{2}.$$

Consider the case when $c(0) > 0$ and $g'(0) > 0$. If $c(0)^2 - 4g'(0) \geq 0$, then we have real, negative roots. If $c(0)^2 - 4g'(0) < 0$, then we have complex roots with negative real part. In the first of these cases, the origin is a stable node. In the second case, the origin is a stable spiral. In either case, the critical point $(0,0)$ is asymptotically stable.

Next, consider the case when $c(0) < 0$ and $g'(0) < 0$. If $c(0)^2 - 4g'(0) > 0$, then we have real roots of opposite sign. If $c(0)^2 - 4g'(0) < 0$, then we have complex roots with a negative real part. In either of these cases, the critical point $(0,0)$ is unstable.

Section 7.3

1.

(a)

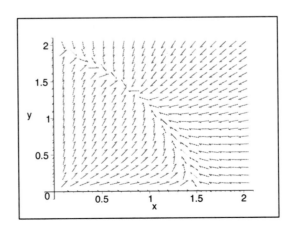

(b) The critical points are solutions of the system

$$x(1.5 - x - 0.5y) = 0$$
$$y(2 - y - 0.75x) = 0.$$

The four critical points are $(0,0)$, $(0,2)$, $(1.5,0)$ and $(0.8,1.4)$.

(c) The Jacobian matrix is

$$\mathbf{J} = \begin{pmatrix} 3/2 - 2x - y/2 & -x/2 \\ -3y/4 & 2 - 3x/4 - 2y \end{pmatrix}.$$

At $(0,0)$,

$$\mathbf{J}(0,0) = \begin{pmatrix} 3/2 & 0 \\ 0 & 2 \end{pmatrix}.$$

The associated eigenvalues and eigenvectors are $\lambda_1 = 3/2$, $\mathbf{v}_1 = (1,0)^T$ and $\lambda_2 = 2$, $\mathbf{v}_2 = (0,1)^T$. The eigenvalues are positive. Therefore, the origin is an unstable node.

At $(0,2)$,

$$\mathbf{J}(0,2) = \begin{pmatrix} 1/2 & 0 \\ -3/2 & -2 \end{pmatrix}.$$

The associated eigenvalues and eigenvectors are $\lambda_1 = 1/2$, $\mathbf{v}_1 = (1, -0.6)^T$ and $\lambda_2 = -2$, $\mathbf{v}_2 = (0,1)^T$. The eigenvalues have opposite sign. Therefore, $(0,2)$ is a saddle, which is unstable.

At $(1.5, 0)$,

$$\mathbf{J}(1.5, 0) = \begin{pmatrix} -1.5 & -0.75 \\ 0 & 0.875 \end{pmatrix}.$$

The associated eigenvalues and eigenvectors are $\lambda_1 = -1.5$, $\mathbf{v}_1 = (1,0)^T$ and $\lambda_2 = 0.875$, $\mathbf{v}_2 = (-0.31579, 1)^T$. The eigenvalues are opposite sign. Therefore, $(1.5, 0)$ is a saddle, which is unstable.

At $(0.8, 1.4)$,

$$\mathbf{J}(0.8, 1.4) = \begin{pmatrix} -0.8 & -0.4 \\ -1.05 & -1.4 \end{pmatrix}.$$

The associated eigenvalues and eigenvectors are $\lambda_1 = (-11 + \sqrt{51})/10$, $\mathbf{v}_1 = (1, (3 - \sqrt{51})/4)^T$ and $\lambda_2 = (-11 - \sqrt{51})/10$, $\mathbf{v}_2 = (1, (3 + \sqrt{51})/4)^T$. The eigenvalues are negative. Therefore, $(0.8, 1.4)$ is a stable node, which is asymptotically stable.

(d,e)

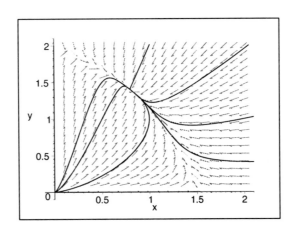

(f) Except for initial conditions lying on the coordinate axes, almost all trajectories converge to the stable node $(0.8, 1.4)$.

2.

(a)

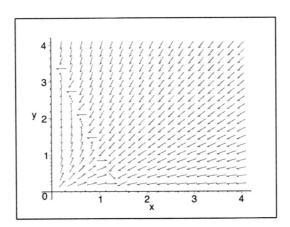

(b) The critical points are solutions of the system

$$x(1.5 - x - 0.5y) = 0$$
$$y(2 - 0.5y - 1.5x) = 0.$$

The four critical points are $(0,0)$, $(0,4)$, $(1.5,0)$ and $(1,1)$.

(c) The Jacobian matrix is

$$\mathbf{J} = \begin{pmatrix} 3/2 - 2x - y/2 & -x/2 \\ -3y/2 & 2 - 3x/2 - y \end{pmatrix}.$$

At $(0,0)$,

$$\mathbf{J}(0,0) = \begin{pmatrix} 3/2 & 0 \\ 0 & 2 \end{pmatrix}.$$

The associated eigenvalues and eigenvectors are $\lambda_1 = 3/2$, $\mathbf{v}_1 = (1,0)^T$ and $\lambda_2 = 2$, $\mathbf{v}_2 = (0,1)^T$. The eigenvalues are positive. Therefore, the origin is an unstable node.

At $(0,4)$,

$$\mathbf{J}(0,4) = \begin{pmatrix} -1/2 & 0 \\ -6 & -2 \end{pmatrix}.$$

The associated eigenvalues and eigenvectors are $\lambda_1 = -1/2$, $\mathbf{v}_1 = (1,-4)^T$ and $\lambda_2 = -2$, $\mathbf{v}_2 = (0,1)^T$. The eigenvalues are both negative. Therefore, $(0,4)$ is a a stable node, which is asymptotically stable.

At $(1.5,0)$,

$$\mathbf{J}(1.5,0) = \begin{pmatrix} -1.5 & -0.75 \\ 0 & -0.25 \end{pmatrix}.$$

The associated eigenvalues and eigenvectors are $\lambda_1 = -1.5$, $\mathbf{v}_1 = (1,0)^T$ and $\lambda_2 = -0.25$, $\mathbf{v}_2 = (3,-5)^T$. The eigenvalues are both negative. Therefore, this critical point is a stable node, which is asymptotically stable.

At $(1,1)$,

$$\mathbf{J}(1,1) = \begin{pmatrix} -1 & -1/2 \\ -3/2 & -1/2 \end{pmatrix}.$$

The associated eigenvalues and eigenvectors are $\lambda_1 = (-3 + \sqrt{13})/4$, $\mathbf{v}_1 = (1, -(1 + \sqrt{13})/2)^T$ and $\lambda_2 = -(3 + \sqrt{13})/4$, $\mathbf{v}_2 = (0, (-1 + \sqrt{13})/2)^T$. The eigenvalues are of opposite sign. Therefore, $(1,1)$ is a saddle, which is unstable.

(d,e)

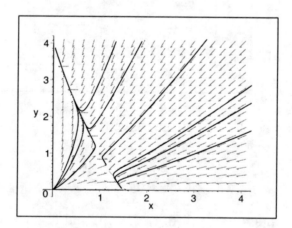

(f) Trajectories approaching the critical point $(1,1)$ form a separatrix. Solutions on either side of the separatrix approach either $(0,4)$ or $(1.5,0)$.

3.

(a)

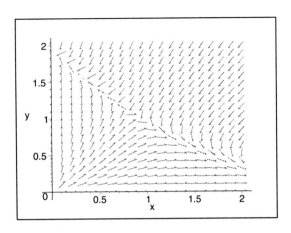

(b) The critical points are solutions of the system

$$x(1.5 - 0.5x - y) = 0$$
$$y(2 - y - 1.125x) = 0.$$

The four critical points are $(0, 0)$, $(0, 2)$, $(3, 0)$ and $(4/5, 11/10)$.

(c) The Jacobian matrix is

$$\mathbf{J} = \begin{pmatrix} 3/2 - x - y & -x \\ -1.125y & 2 - 2y - 1.125x \end{pmatrix}.$$

At $(0, 0)$,

$$\mathbf{J}(0, 0) = \begin{pmatrix} 3/2 & 0 \\ 0 & 2 \end{pmatrix}.$$

The associated eigenvalues and eigenvectors are $\lambda_1 = 3/2$, $\mathbf{v}_1 = (1, 0)^T$ and $\lambda_2 = 2$, $\mathbf{v}_2 = (0, 1)^T$. The eigenvalues are positive. Therefore, the origin is an unstable node.

At $(0, 2)$,

$$\mathbf{J}(0, 2) = \begin{pmatrix} -1/2 & 0 \\ -9/4 & -2 \end{pmatrix}.$$

The associated eigenvalues and eigenvectors are $\lambda_1 = -1/2$, $\mathbf{v}_1 = (1, -3/2)^T$ and $\lambda_2 = -2$, $\mathbf{v}_2 = (0, 1)^T$. The eigenvalues are both negative. Therefore, $(0, 2)$ is a a stable node, which is asymptotically stable.

At $(3, 0)$,

$$\mathbf{J}(3, 0) = \begin{pmatrix} -3/2 & -3 \\ 0 & -11/8 \end{pmatrix}.$$

The associated eigenvalues and eigenvectors are $\lambda_1 = -3/2$, $\mathbf{v}_1 = (1, 0)^T$ and $\lambda_2 = -11/8$, $\mathbf{v}_2 = (-24, 1)^T$. The eigenvalues are both negative. Therefore, this critical point is a stable node, which is asymptotically stable.

At $(4/5, 11/10)$,

$$\mathbf{J}(4/5, 11/10) = \begin{pmatrix} -2/5 & -4/5 \\ -99/80 & -11/10 \end{pmatrix}.$$

49

The associated eigenvalues and eigenvectors are $\lambda_1 = -3/4 + \sqrt{445}/20$, $\mathbf{v}_1 = (1, (7 - \sqrt{445})/16)^T$ and $\lambda_2 = -3/4 - \sqrt{445}/20$, $\mathbf{v}_2 = (0, (7 + \sqrt{445})/16)^T$. The eigenvalues are of opposite sign. Therefore, $(4/5, 11/10)$ is a saddle, which is unstable.

(d,e)

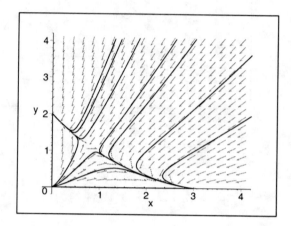

(f) Trajectories approaching the critical point $(4/5, 11/10)$ form a separatrix. Solutions on either side of the separatrix approach either $(3, 0)$ or $(0, 2)$.

4.

(a)

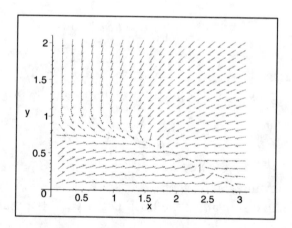

(b) The critical points are solutions of the system

$$x(1.5 - 0.5x - y) = 0$$
$$y(0.75 - y - 0.125x) = 0.$$

The four critical points are $(0, 0)$, $(0, 3/4)$, $(3, 0)$ and $(2, 1/2)$.

(c) The Jacobian matrix is

$$\mathbf{J} = \begin{pmatrix} 3/2 - x - y & -x \\ -y/8 & 3/4 - x/8 - 2y \end{pmatrix}.$$

At $(0, 0)$,
$$\mathbf{J}(0,0) = \begin{pmatrix} 3/2 & 0 \\ 0 & 3/4 \end{pmatrix}.$$

The associated eigenvalues and eigenvectors are $\lambda_1 = 3/2$, $\mathbf{v}_1 = (1,0)^T$ and $\lambda_2 = 3/4$, $\mathbf{v}_2 = (0,1)^T$. The eigenvalues are positive. Therefore, the origin is an unstable node.

At $(0, 3/4)$,
$$\mathbf{J}(0,3/4) = \begin{pmatrix} 3/4 & 0 \\ -3/32 & -3/4 \end{pmatrix}.$$

The associated eigenvalues and eigenvectors are $\lambda_1 = 3/4$, $\mathbf{v}_1 = (-16,1)^T$ and $\lambda_2 = -3/4$, $\mathbf{v}_2 = (0,1)^T$. The eigenvalues are of opposite sign. Therefore, $(0, 3/4)$ is a a saddle, which is unstable.

At $(3, 0)$,
$$\mathbf{J}(3,0) = \begin{pmatrix} -3/2 & -3 \\ 0 & 3/8 \end{pmatrix}.$$

The associated eigenvalues and eigenvectors are $\lambda_1 = -3/2$, $\mathbf{v}_1 = (1,0)^T$ and $\lambda_2 = 3/8$, $\mathbf{v}_2 = (-8,5)^T$. The eigenvalues are of opposite sign. Therefore, this critical point is a saddle, which is unstable.

At $(2, 1/2)$,
$$\mathbf{J}(2,1/2) = \begin{pmatrix} -1 & -2 \\ -1/16 & -1/2 \end{pmatrix}.$$

The associated eigenvalues and eigenvectors are $\lambda_1 = (-3 + \sqrt{3})/4$, $\mathbf{v}_1 = (1, -(1 + \sqrt{3})/8)^T$ and $\lambda_2 = -(3 + \sqrt{3})/4$, $\mathbf{v}_2 = (0, (-1 + \sqrt{3})/8)^T$. The eigenvalues are both negative. Therefore, $(2, 1/2)$ is a stable node, which is asymptotically stable.

(d,e)

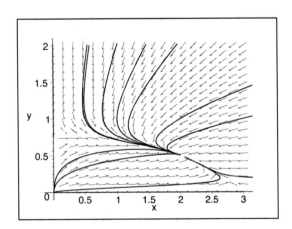

(f) Except for initial conditions lying along the coordinate axes, the other solutions converge to the stable node $(2, 1/2)$.

5.

(a)

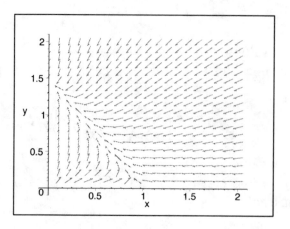

(b) The critical points are solutions of the system

$$x(1 - x - y) = 0$$
$$y(1.5 - y - x) = 0.$$

The three critical points are $(0,0)$, $(0,3/2)$ and $(1,0)$.

(c) The Jacobian matrix is

$$\mathbf{J} = \begin{pmatrix} 1 - 2x - y & -x \\ -y & 1.5 - 2y - x \end{pmatrix}.$$

At $(0,0)$,

$$\mathbf{J}(0,0) = \begin{pmatrix} 1 & 0 \\ 0 & 1.5 \end{pmatrix}.$$

The associated eigenvalues and eigenvectors are $\lambda_1 = 1$, $\mathbf{v}_1 = (1,0)^T$ and $\lambda_2 = 1.5$, $\mathbf{v}_2 = (0,1)^T$. The eigenvalues are positive. Therefore, the origin is an unstable node.

At $(0,3/2)$,

$$\mathbf{J}(0,3/2) = \begin{pmatrix} -1/2 & 0 \\ -3/2 & -3/2 \end{pmatrix}.$$

The associated eigenvalues and eigenvectors are $\lambda_1 = -1/2$, $\mathbf{v}_1 = (1,-3/2)^T$ and $\lambda_2 = -3/2$, $\mathbf{v}_2 = (0,1)^T$. The eigenvalues are both negative. Therefore, $(0,3/2)$ is a stable node, which is asymptotically stable.

At $(1,0)$,

$$\mathbf{J}(1,0) = \begin{pmatrix} -1 & -1 \\ 0 & 1/2 \end{pmatrix}.$$

The associated eigenvalues and eigenvectors are $\lambda_1 = -1$, $\mathbf{v}_1 = (1,0)^T$ and $\lambda_2 = 1/2$, $\mathbf{v}_2 = (1,-3/2)^T$. The eigenvalues are of opposite sign. Therefore, this critical point is a saddle, which is unstable.

(d,e)

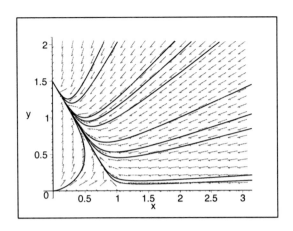

(f) All trajectories converge to the stable node $(0, 1.5)$.

6.

(a)

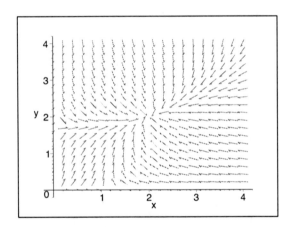

(b) The critical points are solutions of the system

$$x(1 - x + 0.5y) = 0$$
$$y(2.5 - 1.5y + 0.25x) = 0.$$

The four critical points are $(0, 0)$, $(0, 5/3)$, $(1, 0)$ and $(2, 2)$.

(c) The Jacobian matrix is

$$\mathbf{J} = \begin{pmatrix} 1 - 2x + 0.5y & 0.5x \\ 0.25y & 2.5 - 3y + 0.25x \end{pmatrix}.$$

At $(0, 0)$,

$$\mathbf{J}(0, 0) = \begin{pmatrix} 1 & 0 \\ 0 & 2.5 \end{pmatrix}.$$

The associated eigenvalues and eigenvectors are $\lambda_1 = 1$, $\mathbf{v}_1 = (1, 0)^T$ and $\lambda_2 = 2.5$, $\mathbf{v}_2 = (0, 1)^T$. The eigenvalues are positive. Therefore, the origin is an unstable node.

At $(0, 5/3)$,

$$\mathbf{J}(0, 5/3) = \begin{pmatrix} 11/6 & 0 \\ 5/12 & -5/2 \end{pmatrix}.$$

The associated eigenvalues and eigenvectors are $\lambda_1 = 11/6$, $\mathbf{v}_1 = (52/5, 1)^T$ and $\lambda_2 = -5/2$, $\mathbf{v}_2 = (0, 1)^T$. The eigenvalues are of opposite sign. Therefore, $(0, 5/3)$ is a saddle, which is unstable.

At $(1, 0)$,

$$\mathbf{J}(1, 0) = \begin{pmatrix} -1 & 1/2 \\ 0 & 11/4 \end{pmatrix}.$$

The associated eigenvalues and eigenvectors are $\lambda_1 = -1$, $\mathbf{v}_1 = (1, 0)^T$ and $\lambda_2 = 11/4$, $\mathbf{v}_2 = (1, 15/2)^T$. The eigenvalues are of opposite sign. Therefore, this critical point is a saddle, which is unstable.

At $(2, 2)$,

$$\mathbf{J}(2, 2) = \begin{pmatrix} -2 & 1 \\ 1/2 & -3 \end{pmatrix}.$$

The associated eigenvalues and eigenvectors are $\lambda_1 = (-5 + \sqrt{3})/2$, $\mathbf{v}_1 = (1, (-1 + \sqrt{3})/2)^T$ and $\lambda_2 = -(5 + \sqrt{3})/2$, $\mathbf{v}_2 = (1, -(1 + \sqrt{3})/2)^T$. The eigenvalues are both negative. Therefore, this critical point is a stable node, which is asymptotically stable.

(d,e)

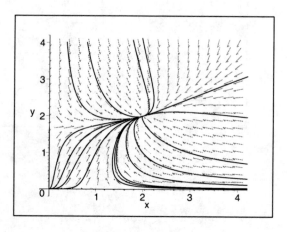

(f) All trajectories converge to the stable node $(0, 1.5)$.

7. We see that

$$(\sigma_1 X + \sigma_2 Y)^2 - 4\sigma_1\sigma_2 XY = \sigma_1^2 X^2 + 2\sigma_1\sigma_2 XY + \sigma_2^2 Y^2 - 4\sigma_1\sigma_2 XY$$
$$= (\sigma_1 X - \sigma_2 Y)^2.$$

The stated identity follows from the identity above. Since all parameters and variables are positive,

$$(\sigma_1 X + \sigma_2 Y)^2 - 4(\sigma_1\sigma_2 - \alpha_1\alpha_2)XY \geq 0.$$

Therefore, the eigenvalues can never be complex.

8.

(a) The critical points are solutions of

$$x(\epsilon_1 - \sigma_1 x - \alpha_1 y) = 0$$
$$y(\epsilon_2 - \sigma_2 y - \alpha_2 x) = 0.$$

If $x = 0$, then either $y = 0$ or $y = \epsilon_2/\sigma_2$. If $\epsilon_1 - \sigma_1 x - \alpha_1 y = 0$, then solving for x, we see that $x = (\epsilon_1 - \alpha_1 y)/\sigma_1$. Substituting x into the second equation, we have

$$(\sigma_1 \sigma_2 - \alpha_1 \alpha_2)y^2 - (\sigma_1 \epsilon_2 - \epsilon_1 \alpha_2)y = 0.$$

Based on the assumption that $\sigma_1 \sigma_2 - \alpha_1 \alpha_2 = 0$, it follows that $(\sigma_1 \epsilon_2 - \epsilon_1 \alpha_2)y = 0$. So if $\sigma_1 \epsilon_2 - \epsilon_1 \alpha_2 \neq 0$, then $y = 0$, and the critical points are located at $(0,0)$, $(0, \epsilon_2, \sigma_2)$ and $(\epsilon_1/\sigma_1, 0)$.

For the case when $\sigma_1 \epsilon_2 - \epsilon_1 \alpha_2 = 0$, y can be arbitrary. Using the fact that $x = (\epsilon_1 - \alpha_1 y)/\sigma_1$, we conclude that all points on the line $\sigma_1 x + \alpha_1 y = \epsilon_1$ are critical points.

(b) The Jacobian matrix is

$$\mathbf{J} = \begin{pmatrix} \epsilon_1 - 2\sigma_1 x - \alpha_1 y & -\alpha_1 x \\ -\alpha_2 y & \epsilon_2 - 2\sigma_2 y - \alpha_2 x \end{pmatrix}.$$

At $(0,0)$,

$$\mathbf{J}(0,0) = \begin{pmatrix} \epsilon_1 & 0 \\ 0 & \epsilon_2 \end{pmatrix}.$$

The associated eigenvalues are $\lambda = epsilon_1, \epsilon_2$. Since both eigenvalues are positive, the origin is an unstable node.

At $(0, \epsilon_2/\sigma_2)$,

$$\mathbf{J}(0, \epsilon_2/\sigma_2) = \begin{pmatrix} (\epsilon_1 \alpha_2 - \sigma_1 \epsilon_2)/\alpha_2 & 0 \\ \epsilon_2 \alpha_2/\sigma_2 & -\epsilon_2 \end{pmatrix}.$$

The associated eigenvalues are $\lambda = (\epsilon_1 \alpha_2 - \sigma_1 \epsilon_2)/\alpha_2, -\epsilon_2$. If $\sigma_1 \epsilon_2 - \epsilon_1 \alpha_2 > 0$, then both eigenvalues are negative, and the point $(0, \epsilon_2/\sigma_2)$ is a stable node, which is asymptotically stable. If $\sigma_1 \epsilon_2 - \epsilon_1 \alpha_2 < 0$, then the eigenvalues are of opposite sign, and the point $(0, \epsilon_2/\sigma_2)$ is a saddle, which is unstable.

At $(\epsilon_1/\sigma_1, 0)$,

$$\mathbf{J}(\epsilon_1/\sigma_1) = \begin{pmatrix} -\epsilon_1 & -\epsilon_1 \alpha_1/\sigma_1 \\ 0 & (\sigma_1 \epsilon_2 - \epsilon_1 \alpha_2)/\sigma_1 \end{pmatrix}.$$

The associated eigenvalues are $\lambda = (\sigma_1 \epsilon_2 - \epsilon_1 \alpha_2)/\sigma_1, -\epsilon_1$. If $\sigma_1 \epsilon_2 - \epsilon_1 \alpha_2 > 0$, then the eigenvalues are of opposite sign, and the point $(\epsilon_1/\sigma_1, 0)$ is a saddle, which is unstable. If $\sigma_1 \epsilon_2 - \epsilon_1 \alpha_2 < 0$, then both eigenvalues are negative, and the point $(\epsilon_1/\sigma_1, 0)$ is a stable nod, which is asymptotically stable.

(c) When $\sigma_1 \epsilon_2 - \epsilon_1 \alpha_2 = 0$, the set of critical points consists of $(0,0)$ and all the points on the line $\sigma_1 x + \alpha_1 y = \epsilon_1$. Based on part (b), the origin is still an unstable node. Setting $y = (\epsilon_1 - \sigma_1 x)/\alpha_1$, the Jacobian matrix along the line stated above is

$$\mathbf{J} = \begin{pmatrix} -\sigma_1 x & -\alpha_1 x \\ -\alpha_2(\epsilon_1 - \sigma_1 x)/\alpha_1 & \alpha_2 x - \epsilon_1 \alpha_2/\sigma_1 \end{pmatrix}.$$

the characteristic equation of that matrix is $\lambda^2 + (\epsilon_1\alpha_2 - \alpha_2\sigma_1 x + \sigma_1^2 x)\lambda/\sigma_1 = 0$. Using the hypothesis that $\sigma_1\sigma_2 - \alpha_1\alpha_2 = 0$, we can rewrite the characteristic equation as

$$\lambda^2 + [\epsilon_2 - \alpha_2 x + \sigma_1 x]\lambda = 0.$$

We note that $0 \leq x \leq \epsilon_1/\sigma_1$. Since the coefficient in the quadratic equation is linear, and

$$\epsilon_2 - \alpha_2 x + \sigma_1 = \begin{cases} \epsilon_2 & x = 0 \\ \epsilon_1 & x = \epsilon_1/\sigma_1, \end{cases}$$

it follows that the coefficient is positive for $0 \leq x \leq \epsilon_1/\sigma_1$. Therefore, along the line $\sigma_1 x + \alpha_1 y = \epsilon_1$, one eigenvalue is zero and the other is negative. Therefore, these critical points are stable nodes, which are asymptotically stable.

9.

(a) The critical points are solutions of

$$x(1 - x - y) + \delta a = 0$$
$$y(0.75 - y - 0.5x) + \delta b = 0.$$

We assume the solutions can be written in the form

$$x = x_0 + x_1\delta + x_2\delta^2 + \ldots$$
$$y = y_0 + y_1\delta + y_2\delta^2 + \ldots$$

Substituting these series expansions results in

$$x_0(1 - x_0 - y_0) + (x_1 - 2x_1x_0 - x_0y_1 - x_1y_0 + a)\delta + \ldots = 0$$
$$y_0(0.75 - y_0 - 0.5x_0) + (0.75y_1 - 2y_0y_1 - x_1y_0/2 - x_0y_1/2 + b)\delta + \ldots = 0.$$

(b) As $\delta \to 0$, the equations reduce to the original system of equations in (3). In that case, the critical points are $x_0 = y_0 = 0.5$.

(c) Setting the coefficients of the linear terms equal to zero, we find that

$$- y_1/2 - x_1/2 + a = 0$$
$$- x_1/4 - y_1/2 + b = 0,$$

which has solutions $x_1 = 4a - 4b$ and $y_1 = -2a + 4b$.

(d) If $b < a$, then there will be an increase in the level of species 1. On the other hand, at points where $b > a$, $x_1\delta < 0$. Similarly, if $2b > a$, then there will be an increase in the level of species 2. On the other hand, at points where $2b < a$, $y_1\delta < 0$. It follows that if $b < a < 2b$, the level of both species will increase. Otherwise, the level of one species will increase while the level of the other species decreases.

10.

(a) The critical points are solutions to the system:

$$-y = 0$$
$$-\gamma y - x(x - 0.15)(x - 2) = 0.$$

Setting $y = 0$, the second equation becomes $x(x - 0.15)(x - 2) = 0$. Therefore, the critical points are $(0,0)$, $(0.15, 0)$ and $(2, 0)$. The Jacobian matrix is

$$\mathbf{J} = \begin{pmatrix} 0 & -1 \\ -3x^2 + 4.3x - 0.3 & -\gamma \end{pmatrix}.$$

At $(0, 0)$,

$$\mathbf{J}(0, 0) = \begin{pmatrix} 0 & -1 \\ -0.3 & -\gamma \end{pmatrix}.$$

The associated eigenvalues are $\lambda = -\gamma/2 \pm \sqrt{25\gamma^2 + 30}/10$. Since the eigenvalues have opposite sign, the origin is a saddle, which is unstable.

At $(0.15, 0)$,

$$\mathbf{J}(0.15, 0) = \begin{pmatrix} 0 & -1 \\ 0.2775 & -\gamma \end{pmatrix}.$$

The associated eigenvalues are $\lambda = -\gamma/2 \pm \sqrt{100\gamma^2 - 111}/20$. If $100\gamma^2 - 111 \geq 0$, then the eigenvalues are real. Since $\lambda_1\lambda_2 = 0.2775$, both eigenvalues will have the same sign. Therefore, the critical point is a node which its stability dependent on the sign of γ. If $100\gamma^2 - 111 < 0$, then the eigenvalues are complex conjugates. In that case, the critical point $(0.15, 0)$ is a spiral, with its stability dependent on the sign of γ.

At $(2, 0)$,

$$\mathbf{J}(2, 0) = \begin{pmatrix} 0 & -1 \\ -3.7 & -\gamma \end{pmatrix}.$$

The associated eigenvalues are $\lambda = -\gamma/2 \pm \sqrt{25\gamma^2 + 370}/10$. Since the eigenvalues have opposite sign, $(2, 0)$ is a saddle, which is unstable.

(b)

(c) Based on the phase portraits above, it is clear that the required value of γ satisfies $0.8 < \gamma < 1.5$. Numerical experiments show that the required value of $\gamma \approx 1.20$.

11.

(a) Below we sketch some nullclines:

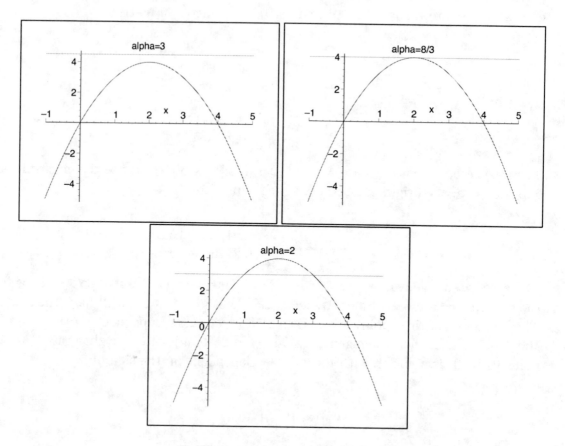

(b) The critical points are solutions of

$$-4x + y + x^2 = 0$$
$$\frac{3}{2}\alpha - y = 0.$$

The solutions of these equations are

$$\left(2 \pm \sqrt{4 - \frac{3}{2}\alpha}, \frac{3}{2}\alpha\right)$$

and exist for $\alpha \leq 8/3$.

(c) For $\alpha = 2$, the critical points are $(1, 3)$ and $(3, 3)$. The Jacobian matrix is

$$\mathbf{J} = \begin{pmatrix} -4 + 2x & 1 \\ 0 & -1 \end{pmatrix}.$$

At $(1, 3)$,
$$\mathbf{J}(1, 3) = \begin{pmatrix} -2 & 1 \\ 0 & -1 \end{pmatrix}.$$

The eigenvalues are $\lambda = -2, -1$. Since they are both negative, $(1, 3)$ is a stable node, which is asymptotically stable.

At $(3, 3)$,
$$\mathbf{J}(3, 3) = \begin{pmatrix} 2 & 1 \\ 0 & -1 \end{pmatrix}.$$

The eigenvalues are $\lambda = 2, -1$. Since they are of opposite sign $(3, 3)$ is a saddle, which is unstable.

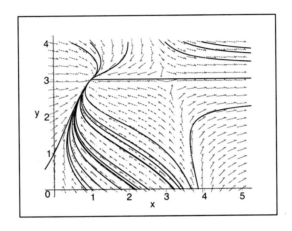

(d) The bifurcation value is $\alpha_0 = 8/3$. At this value α_0, the critical point is $(2, 4)$. The Jacobian matrix is
$$\mathbf{J}(2, 4) = \begin{pmatrix} 0 & 1 \\ 0 & -1 \end{pmatrix}.$$

The eigenvalues are $\lambda = 0, 1$.

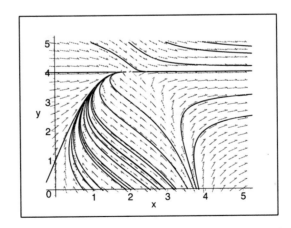

(e) Below we show the phase portrait for $\alpha = 3$.

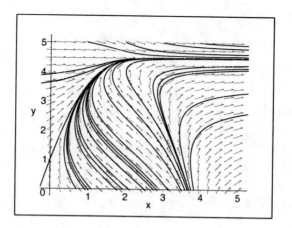

12.

(a) Below we sketch some nullclines:

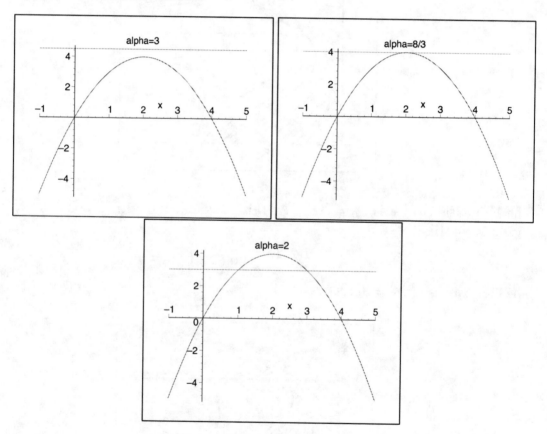

(b) The critical points are solutions of

$$\frac{3}{2}\alpha - y = 0. \qquad\qquad -4x + y + x^2 = 0$$

The solutions of these equations are

$$\left(2 \pm \sqrt{4 - \frac{3}{2}\alpha}, \frac{3}{2}\alpha\right)$$

60

and exist for $\alpha \le 8/3$.

(c) For $\alpha = 2$, the critical points are $(1,3)$ and $(3,3)$. The Jacobian matrix is

$$\mathbf{J} = \begin{pmatrix} 0 & -1 \\ -4 + 2x & 1 \end{pmatrix}.$$

At $(1,3)$,

$$\mathbf{J}(1,3) = \begin{pmatrix} 0 & -1 \\ -2 & 1 \end{pmatrix}.$$

The eigenvalues are $\lambda = -1, 2$. Since they are of opposite sign $(1,3)$ is a saddle, which is unstable.

At $(3,3)$,

$$\mathbf{J}(3,3) = \begin{pmatrix} 0 & -1 \\ 2 & 1 \end{pmatrix}.$$

The eigenvalues are $\lambda = (1 \pm i\sqrt{7})/2$. Therefore, $(3,3)$ is an unstable spiral.

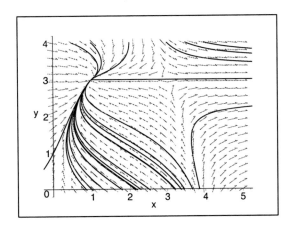

(d) The bifurcation value is $\alpha_0 = 8/3$. At this value α_0, the critical point is $(2,4)$. The Jacobian matrix is

$$\mathbf{J}(2,4) = \begin{pmatrix} 0 & -1 \\ 0 & 1 \end{pmatrix}.$$

The eigenvalues are $\lambda = 0, 1$.

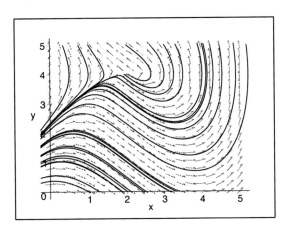

61

(e) Below we show the phase portrait for $\alpha = 3$.

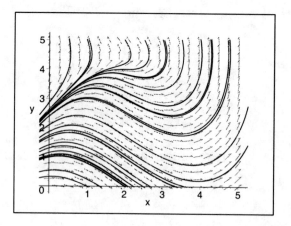

13.

(a) Below we sketch some nullclines:

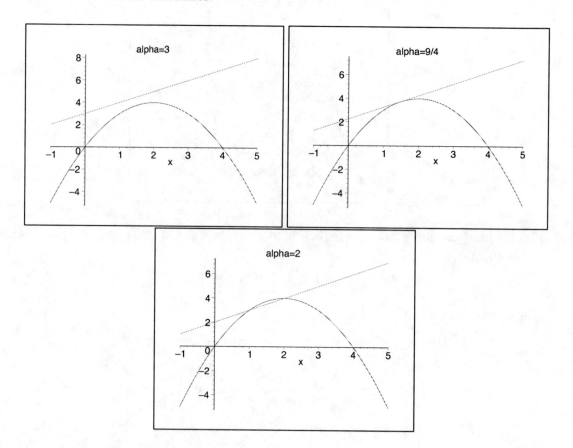

(b) The critical points are solutions of

$$-4x + y + x^2 = 0$$
$$\alpha - x + y = 0.$$

The solutions of these equations are

$$x_0 = \frac{3 + \sqrt{9 - 4\alpha}}{2}, y_0 = \alpha + \frac{3 + \sqrt{9 - 4\alpha}}{2}$$

and

$$x_0 = \frac{3 - \sqrt{9 - 4\alpha}}{2}, y_0 = \alpha + \frac{3 - \sqrt{9 - 4\alpha}}{2}$$

and exist for $\alpha \leq 9/4$.

(c) For $\alpha = 2$, the critical points are $(1, 3)$ and $(2, 4)$. The Jacobian matrix is

$$\mathbf{J} = \begin{pmatrix} 2x - 4 & 1 \\ -1 & 1 \end{pmatrix}.$$

At $(1, 3)$,

$$\mathbf{J}(1, 3) = \begin{pmatrix} -2 & 1 \\ -1 & 1 \end{pmatrix}.$$

The eigenvalues are $\lambda = (-1 \pm \sqrt{5})/2$. Since they are of opposite sign $(1, 3)$ is a saddle, which is unstable.

At $(2, 4)$,

$$\mathbf{J}(2, 4) = \begin{pmatrix} 0 & 1 \\ -1 & 1 \end{pmatrix}.$$

The eigenvalues are $\lambda = (1 \pm i\sqrt{3})/2$. Therefore, $(2, 4)$ is an unstable spiral.

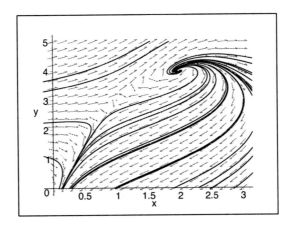

(d) The bifurcation value is $\alpha_0 = 9/4$. At this value α_0, the critical point is $(3/2, 15/4)$. The Jacobian matrix is

$$\mathbf{J}(3/2, 15/4) = \begin{pmatrix} -1 & 1 \\ -1 & 1 \end{pmatrix}.$$

The eigenvalues are both $\lambda = 0$.

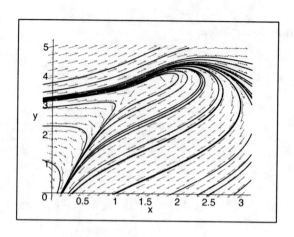

(e) Below we show the phase portrait for $\alpha = 3$.

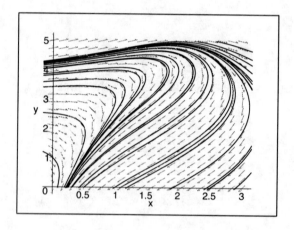

14.

(a) Below we sketch some nullclines:

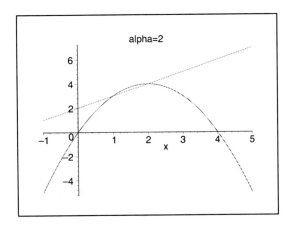

(b) The critical points are solutions of

$$\alpha - x + y = 0$$
$$-4x + y + x^2 = 0.$$

The solutions of these equations are

$$x_0 = \frac{3 + \sqrt{9 - 4\alpha}}{2}, y_0 = \alpha + \frac{3 + \sqrt{9 - 4\alpha}}{2}$$

and

$$x_0 = \frac{3 - \sqrt{9 - 4\alpha}}{2}, y_0 = \alpha + \frac{3 - \sqrt{9 - 4\alpha}}{2}$$

and exist for $\alpha \leq 9/4$.

(c) For $\alpha = 2$, the critical points are $(1, 3)$ and $(2, 4)$. The Jacobian matrix is

$$\mathbf{J} = \begin{pmatrix} -1 & 1 \\ 2x - 4 & 1 \end{pmatrix}.$$

At $(1, 3)$,

$$\mathbf{J}(1, 3) = \begin{pmatrix} -1 & 1 \\ -2 & 1 \end{pmatrix}.$$

The eigenvalues are $\lambda = \pm i$. Therefore, $(1, 3)$ is a center, which is stable.

At $(2, 4)$,

$$\mathbf{J}(2, 4) = \begin{pmatrix} -1 & 1 \\ 0 & 1 \end{pmatrix}.$$

The eigenvalues are $\lambda = 1, -1$. Therefore, $(2, 4)$ is a saddle point, which is unstable.

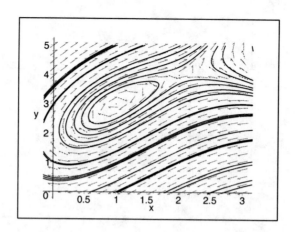

(d) The bifurcation value is $\alpha_0 = 9/4$. At this value α_0, the critical point is $(3/2, 15/4)$. The Jacobian matrix is

$$\mathbf{J}(3/2, 15/4) = \begin{pmatrix} -1 & 1 \\ -1 & 1 \end{pmatrix}.$$

The eigenvalues are both $\lambda = 0$.

(e) Below we show the phase portrait for $\alpha = 3$.

15.

(a) The equation $x' = 0$ implies $x = -2$ or $x - y = 1$. The equation $y' = 0$ implies $y = 1$ or $x + \alpha y = -1$. Solving these equations, we see that the critical points are $(2, 1)$, $(-2, 1)$, $(-2, 1/\alpha)$ and $((-1 + \alpha)/(1 + \alpha), -2/(1 + \alpha))$.

(b) When $\alpha_0 = 1$, the second and third critical points listed above coincide. When $\alpha_0 = -3$, the first and fourth critical points listed above coincide, but here we are only considering $\alpha > 0$. Therefore, $\alpha_0 = 1$.

(c) Here, we have $F(x, y) = (2 + x)(1 - x + y)$ and $G(x, y) = (y - 1)(1 + x + \alpha y)$. Therefore, the Jacobian matrix for this system is

$$\begin{pmatrix} F_x & F_y \\ G_x & G_y \end{pmatrix} = \begin{pmatrix} -1 - 2x + y & 2 + x \\ y - 1 & 1 + x + 2\alpha y - \alpha \end{pmatrix}.$$

We will look at the linear systems near the second and third critical points above, namely $(-2, 1)$ and $(-2, 1/\alpha)$. Near the critical point $(-2, 1)$, the Jacobian matrix is

$$\begin{pmatrix} F_x(-2, 1) & F_y(-2, 1) \\ G_x(-2, 1) & G_y(-2, 1) \end{pmatrix} = \begin{pmatrix} 4 & 0 \\ 0 & -1 + \alpha \end{pmatrix}$$

and the corresponding linear system near $(-2, 1)$ is

$$\frac{d}{dt} \begin{pmatrix} u \\ v \end{pmatrix} = \begin{pmatrix} 4 & 0 \\ 0 & -1 + \alpha \end{pmatrix} \begin{pmatrix} u \\ v \end{pmatrix}$$

where $u = x + 2$ and $v = y - 1$. Near the critical point $(-2, 1/\alpha)$, the Jacobian matrix is

$$\begin{pmatrix} F_x(-2, 1/\alpha) & F_y(-2, 1/\alpha) \\ G_x(-2, 1/\alpha) & G_y(-2, 1/\alpha) \end{pmatrix} = \begin{pmatrix} 3 + \frac{1}{\alpha} & 0 \\ -1 + \frac{1}{\alpha} & 1 - \alpha \end{pmatrix}$$

and the corresponding linear system near $(-2, 1/\alpha)$ is

$$\frac{d}{dt} \begin{pmatrix} u \\ v \end{pmatrix} = \begin{pmatrix} 3 + \frac{1}{\alpha} & 0 \\ -1 + \frac{1}{\alpha} & 1 - \alpha \end{pmatrix} \begin{pmatrix} u \\ v \end{pmatrix}$$

where $u = x + 2$ and $v = y - 1/\alpha$.

The eigenvalues for the linearized system near $(-2, 1)$ are given by $\lambda = 4, -1 + \alpha$. The eigenvalues for the linearized system near $(-2, 1\alpha)$ are given by $\lambda = 3 + \frac{1}{\alpha}, 1 - \alpha$. In part (b), we determined that the bifurcation point was $\alpha_0 = 1$. Here, we see that if $\alpha > 1$, then $(-2, 1)$ will have two positive eigenvalues associated with it, and, therefore, be an unstable node, while $(-2, 1/\alpha)$ will have eigenvalues of opposite signs, and, therefore, by an unstable saddle point. If $\alpha < 1$, then $(-2, 1)$ will have eigenvalues of the opposite sign, and, therefore, be an unstable saddle point, while $(-2, 1/\alpha)$ will have two positive eigenvalues, and, therefore, be an unstable node.

(d) The phase portraits below are for $\alpha = 2$ and $\alpha = 1/2$, respectively.

16.

(a) The equation $x' = 0$ implies $y = 0$ or $2x - 3y = \alpha$. The equation $y' = 0$ implies $x = 2$ or $y = -3$. Solving these equations, we see that the critical points are $(2,0)$, $(2, (4-\alpha)/3)$ and $((\alpha-9)/2, -3)$.

(b) When $\alpha_0 = 4$, the first and second critical points listed above coincide. When $\alpha_0 = 13$, the second and third critical points listed above coincide.

(c) Here, we have $F(x, y) = y(\alpha - 2x + 3y)$ and $G(x, y) = (2 - x)(3 + y)$. Therefore, the Jacobian matrix for this system is

$$\begin{pmatrix} F_x & F_y \\ G_x & G_y \end{pmatrix} = \begin{pmatrix} -2y & \alpha - 2x + 6y \\ -3 - y & 2 - x \end{pmatrix}.$$

We will look at the linear systems near the critical points above. Near the critical point $(2, 0)$, the Jacobian matrix is

$$\begin{pmatrix} F_x(2,0) & F_y(2,0) \\ G_x(2,0) & G_y(2,0) \end{pmatrix} = \begin{pmatrix} 0 & \alpha - 4 \\ -3 & 0 \end{pmatrix}$$

and the corresponding linear system near $(2, 0)$ is

$$\frac{d}{dt} \begin{pmatrix} u \\ v \end{pmatrix} = \begin{pmatrix} 0 & \alpha - 4 \\ -3 & 0 \end{pmatrix} \begin{pmatrix} u \\ v \end{pmatrix}$$

68

where $u = x - 2$ and $v = y$. Near the critical point $(2, (4 - \alpha)/3)$, the Jacobian matrix is

$$\begin{pmatrix} F_x(2, (4-\alpha)/3) & F_y(2, (4-\alpha)/3) \\ G_x(2, (4-\alpha)/3) & G_y(2, (4-\alpha)/3) \end{pmatrix} = \begin{pmatrix} \frac{2\alpha-8}{3} & 4-\alpha \\ \frac{\alpha-13}{3} & 0 \end{pmatrix}$$

and the corresponding linear system near $(2, (4 - \alpha)/3)$ is

$$\frac{d}{dt}\begin{pmatrix} u \\ v \end{pmatrix} = \begin{pmatrix} \frac{2\alpha-8}{3} & 4-\alpha \\ \frac{\alpha-13}{3} & 0 \end{pmatrix}\begin{pmatrix} u \\ v \end{pmatrix}$$

where $u = x - 2$ and $v = y - (4 - \alpha)/3$. Near the critical point $((\alpha - 9)/2, -3)$, the Jacobian matrix is

$$\begin{pmatrix} F_x((\alpha-9)/2, -3) & F_y((\alpha-9)/2, -3) \\ G_x((\alpha-9)/2, -3) & G_y((\alpha-9)/2, -3) \end{pmatrix} = \begin{pmatrix} 6 & -9 \\ 0 & \frac{13-\alpha}{2} \end{pmatrix}$$

and the corresponding linear system near $((\alpha - 9)/2, -3)$ is

$$\frac{d}{dt}\begin{pmatrix} u \\ v \end{pmatrix} = \begin{pmatrix} 6 & -9 \\ 0 & \frac{13-\alpha}{2} \end{pmatrix}\begin{pmatrix} u \\ v \end{pmatrix}$$

where $u = x - (\alpha - 9)/2$ and $v = y + 3$.

The eigenvalues for the linearized system near $(2, 0)$ are given by $\lambda = \pm\sqrt{12 - 3\alpha}$. The eigenvalues for the linearized system near $(2, (4 - \alpha)/3)$ are given by $\lambda = [(\alpha - 4) \pm \sqrt{-2\alpha^2 + 43\alpha - 140}]/3$. The eigenvalues for the linearized system near $((\alpha-9)/2, 3)$ are given by $\lambda = 6, (13 - \alpha)/2$. In part (b), we determined that the bifurcation points were $\alpha_0 = 4, 13$. First, we consider $\alpha_0 = 4$. At this value for α, the first and second critical points coincide. For α near 4, but $\alpha > 4$, the eigenvalues for the linearized system about the critical point $(2, 0)$ will be purely imaginary. Therefore, we cannot conclude from this test the behavior of the nonlinear system about $(2, 0)$. At the same time, for α near 4, but $\alpha > 4$, the eigenvalues for the linearized system about the critical point $(2, (4-\alpha)/3)$ will be of opposite sign, and, therefore, $(2, (4 - \alpha)/3)$ will be an unstable saddle point. On the other hand, for α near 4, but $\alpha < 4$, the eigenvalues associated with $(2, 0)$ will be real-valued, of opposite sign, and, therefore, $(2, 0)$ will be an unstable saddle point. For α near 4, but $\alpha < 4$, the eigenvalues associated with $(2, (4 - \alpha)/3$ will have negative real part and non-zero imaginary part. Therefore, in this case, $(2, (4 - \alpha)/3)$ will be an asymptotically stable spiral point.

Next, we consider the second and third critical points near $\alpha_0 = 13$. For α near 13, but $\alpha > 13$, the eigenvalues associated with $(2, (4 - \alpha)/3)$ will be both be positive, and, therefore, $(2, (4 - \alpha)/3)$ will be an unstable node. At the same time, for $\alpha > 13$, the eigenvalues associated with $((\alpha - 9)/2, -3)$ will be real-valued with opposite signs, and, therefore, $((\alpha - 9)/2, -3)$ will be an unstable saddle point. For α near 13 but $\alpha < 13$, the eigenvalues associated with $(2, (4 - \alpha)/3)$ will be real-valued of opposite signs, and, therefore, $(2, (4 - \alpha)/3)$ will be an unstable saddle point. For $\alpha < 13$, the eigenvalues associated with $((\alpha - 9)/2, -3)$ will both be real-valued and positive, and, therefore, $((\alpha - 9)/2, -3)$ will be an unstable node.

(d) In the first two phase portraits, we take $\alpha = 5$ and $\alpha = 3$, respectively.

In these next two phase portraits, we take $\alpha = 14$ and $\alpha = 12$, respectively.

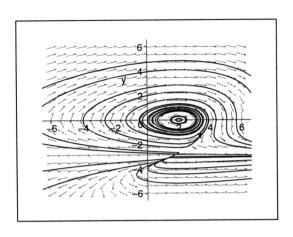

17.

(a) The critical points need to satisfy the system of equations

$$x(4 - x - y) = 0$$
$$y(2 + 2\alpha - y - \alpha x) = 0.$$

The four critical points are $(0,0)$, $(0, 2 + 2\alpha)$, $(4, 0)$, $(2, 2)$.

(b) The Jacobian matrix is given by

$$\mathbf{J} = \begin{pmatrix} 4 - 2x - y & -x \\ -\alpha y & 2 + 2\alpha - 2y - \alpha x \end{pmatrix}.$$

Therefore, at $(2, 2)$,

$$\mathbf{J}(2, 2) = \begin{pmatrix} -2 & -2 \\ -2\alpha & -2 \end{pmatrix}.$$

For $\alpha = 0.75$,

$$\mathbf{J}(2, 2) = \begin{pmatrix} -2 & -2 \\ -3/2 & -2 \end{pmatrix}.$$

The eigenvalues of this matrix are $\lambda = -2 \pm \sqrt{3}$. Since both of these eigenvalues are negative, for $\alpha = 0.75$, $(2, 2)$ is a stable node, which is asymptotically stable. For $\alpha = 1.25$,

$$\mathbf{J}(2, 2) = \begin{pmatrix} -2 & -2 \\ -5/2 & -2 \end{pmatrix}.$$

The eigenvalues of this matrix are $\lambda = -2 \pm \sqrt{5}$. Since these eigenvalues are of opposite sign, for $\alpha = 1.25$, $(2, 2)$ is a saddle point which is unstable.

In general, the eigenvalues at $(2, 2)$ are given by $-2 \pm 2\sqrt{\alpha}$. The nature of the critical point will change when $2\sqrt{\alpha} = 2$; that is, at $\alpha_0 = 1$. At this value of α, the number of negative eigenvalues changes from two to one.

(c) From the Jacobian matrix in part (b), we see that the approximate linear system is

$$\begin{pmatrix} u \\ v \end{pmatrix}' = \begin{pmatrix} -2 & -2 \\ -2\alpha & -2 \end{pmatrix} \begin{pmatrix} u \\ v \end{pmatrix}.$$

71

(d) As shown in part (b), the value $\alpha_0 = 1$.

(e)

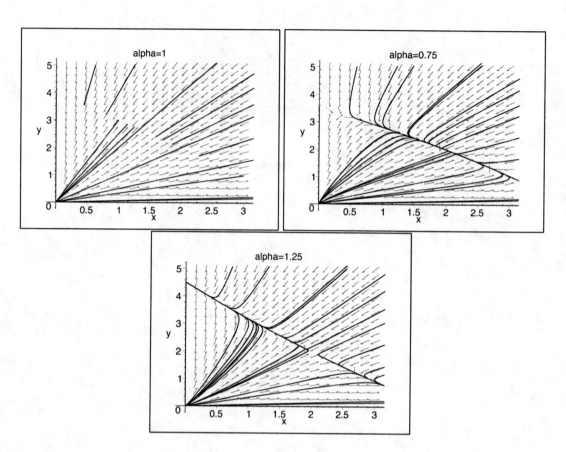

Section 7.4

1.

(a)

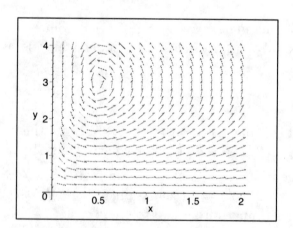

(b) The critical points are solutions of the system

$$x(1.5 - 0.5y) = 0$$
$$y(-0.5 + x) = 0.$$

The two critical points are $(0, 0)$ and $(0.5, 3)$.

(c) The Jacobian matrix is

$$\mathbf{J} = \begin{pmatrix} 3/2 - y/2 & -x/2 \\ y & -1/2 + x \end{pmatrix}.$$

At $(0, 0)$,

$$\mathbf{J}(0, 0) = \begin{pmatrix} 3/2 & 0 \\ 0 & -1/2 \end{pmatrix}.$$

The eigenvalues and eigenvectors are $\lambda_1 = 3/2$, $\mathbf{v}_1 = (1, 0)^T$ and $\lambda_2 = -1/2$, $\mathbf{v}_2 = (0, 1)^T$. The eigenvalues are of opposite sign. Therefore, $(0, 0)$ is a saddle point, which is unstable.

At $(0.5, 3)$,

$$\mathbf{J}(0.5, 3) = \begin{pmatrix} 0 & -1/4 \\ 3 & 0 \end{pmatrix}.$$

The eigenvalues and eigenvectors are $\lambda_1 = i\sqrt{3}/2$, $\mathbf{v}_1 = (1, -2i\sqrt{3})^T$ and $\lambda_2 = -i\sqrt{3}/2$, $\mathbf{v}_2 = (1, 2i\sqrt{3})^T$. The eigenvalues are purely imaginary. Therefore, $(0.5, 3)$ is a center, which is stable.

(d,e)

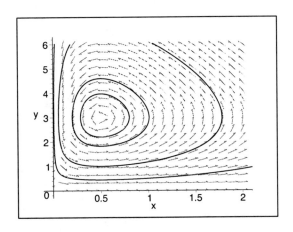

(f) Except for solutions along the coordinate axes, the other trajectories are closed curves about the critical point $(0.5, 3)$.

2.

(a)

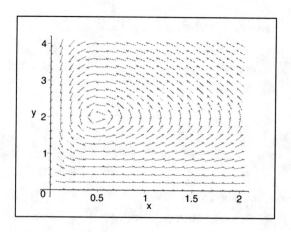

(b) The critical points are solutions of the system

$$x(1 - 0.5y) = 0$$
$$y(-0.25 + 0.5x) = 0.$$

The two critical points are $(0, 0)$ and $(0.5, 2)$.

(c) The Jacobian matrix is

$$\mathbf{J} = \begin{pmatrix} 1 - y/2 & -x/2 \\ y/2 & -1/4 + x/2 \end{pmatrix}.$$

At $(0, 0)$,

$$\mathbf{J}(0, 0) = \begin{pmatrix} 1 & 0 \\ 0 & -1/4 \end{pmatrix}.$$

The eigenvalues and eigenvectors are $\lambda_1 = 1$, $\mathbf{v}_1 = (1, 0)^T$ and $\lambda_2 = -1/4$, $\mathbf{v}_2 = (0, 1)^T$. The eigenvalues are of opposite sign. Therefore, $(0, 0)$ is a saddle point, which is unstable.

At $(0.5, 2)$,

$$\mathbf{J}(0.5, 2) = \begin{pmatrix} 0 & -1/4 \\ 1 & 0 \end{pmatrix}.$$

The eigenvalues and eigenvectors are $\lambda_1 = i/2$, $\mathbf{v}_1 = (1, -2i)^T$ and $\lambda_2 = -i/2$, $\mathbf{v}_2 = (1, 2i)^T$. The eigenvalues are purely imaginary. Therefore, $(0.5, 2)$ is a center, which is stable.

(d,e)

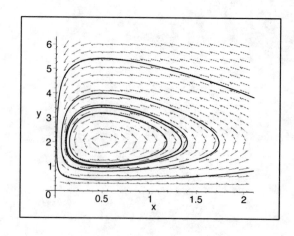

(f) Except for solutions along the coordinate axes, the other trajectories are closed curves about the critical point $(0.5, 2)$.

3.

(a)

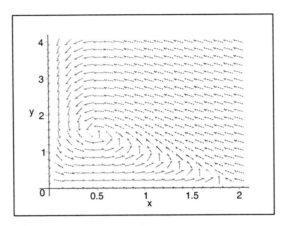

(b) The critical points are solutions of the system

$$x(1 - 0.5x - 0.5y) = 0$$
$$y(-0.25 + 0.5x) = 0.$$

The three critical points are $(0, 0)$, $(2, 0)$ and $(1/2, 3/2)$.

(c) The Jacobian matrix is

$$\mathbf{J} = \begin{pmatrix} 1 - x - y/2 & -x/2 \\ y/2 & -1/4 + x/2 \end{pmatrix}.$$

At $(0, 0)$,

$$\mathbf{J}(0, 0) = \begin{pmatrix} 1 & 0 \\ 0 & -1/4 \end{pmatrix}.$$

The eigenvalues and eigenvectors are $\lambda_1 = 1$, $\mathbf{v}_1 = (1, 0)^T$ and $\lambda_2 = -1/4$, $\mathbf{v}_2 = (0, 1)^T$. The eigenvalues are of opposite sign. Therefore, $(0, 0)$ is a saddle point, which is unstable.

At $(2, 0)$,

$$\mathbf{J}(2, 0) = \begin{pmatrix} -1 & -1 \\ 0 & 3/4 \end{pmatrix}.$$

The eigenvalues and eigenvectors are $\lambda_1 = -1$, $\mathbf{v}_1 = (1, 0)^T$ and $\lambda_2 = 3/4$, $\mathbf{v}_2 = (1, -7/4)^T$. The eigenvalues have opposite sign. Therefore, $(2, 0)$ is a saddle point, which is unstable.

At $(1/2, 3/2)$,

$$\mathbf{J}(1/2, 3/2) = \begin{pmatrix} -1/4 & -1/4 \\ 3/4 & 0 \end{pmatrix}.$$

The eigenvalues and eigenvectors are $\lambda_1 = (-1 + i\sqrt{11})/8$, $\mathbf{v}_1 = ((-1 + i\sqrt{11})/6, 1)^T$ and $\lambda_2 = (-1 - i\sqrt{11})/8$, $\mathbf{v}_2 = ((-1 - i\sqrt{11})/6, 1)^T$. The eigenvalues have negative real part. Therefore, $(1/2, 3/2)$ is a stable spiral, which is asymptotically stable.

(d,e)

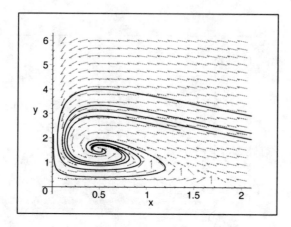

(f) Except for solutions along the coordinate axes, the other trajectories spiral towards the critical point $(1/2, 3/2)$.

4.

(a)

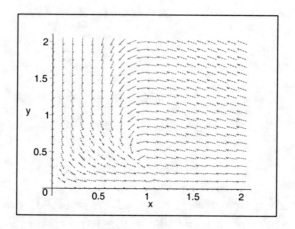

(b) The critical points are solutions of the system

$$x(1.125 - x - 0.5y) = 0$$
$$y(-1 + x) = 0.$$

The three critical points are $(0, 0)$, $(9/8, 0)$ and $(1, 1/4)$.

(c) The Jacobian matrix is

$$\mathbf{J} = \begin{pmatrix} 9/8 - 2x - y/2 & -x/2 \\ y & -1 + x \end{pmatrix}.$$

At $(0, 0)$,

$$\mathbf{J}(0,0) = \begin{pmatrix} 9/8 & 0 \\ 0 & -1 \end{pmatrix}.$$

The eigenvalues and eigenvectors are $\lambda_1 = 9/8$, $\mathbf{v}_1 = (1,0)^T$ and $\lambda_2 = -1$, $\mathbf{v}_2 = (0,1)^T$. The eigenvalues are of opposite sign. Therefore, $(0,0)$ is a saddle point, which is unstable.

At $(9/8, 0)$,

$$\mathbf{J}(9/8,0) = \begin{pmatrix} -9/8 & -9/16 \\ 0 & 1/8 \end{pmatrix}.$$

The eigenvalues and eigenvectors are $\lambda_1 = -9/8$, $\mathbf{v}_1 = (1,0)^T$ and $\lambda_2 = 1/8$, $\mathbf{v}_2 = (9, -20)^T$. The eigenvalues have opposite sign. Therefore, $(9/8, 0)$ is a saddle point, which is unstable.

At $(1, 1/4)$,

$$\mathbf{J}(1,1/4) = \begin{pmatrix} -1 & -1/2 \\ 1/4 & 0 \end{pmatrix}.$$

The eigenvalues and eigenvectors are $\lambda_1 = (-2 + \sqrt{2})/4$, $\mathbf{v}_1 = (-2 + \sqrt{2}, 1)^T$ and $\lambda_2 = (-2 - \sqrt{2})/4$, $\mathbf{v}_2 = (-2 - \sqrt{2}, 1)^T$. The eigenvalues are both negative. Therefore, $(1, 1/4)$ is a stable node, which is asymptotically stable.

(d,e)

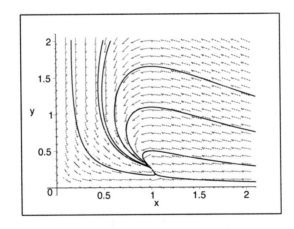

(f) Except for solutions along the coordinate axes, the other trajectories converge to the critical point $(1, 1/4)$.

5.

(a)

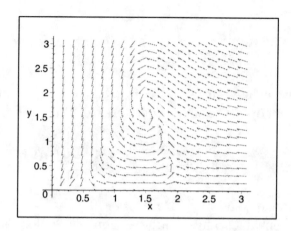

(b) The critical points are solutions of the system

$$x(-1 + 2.5x - 0.3y - x^2) = 0$$
$$y(-1.5 + x) = 0.$$

The four critical points are $(0, 0)$, $(1/2, 0)$, $(2, 0)$ and $(3/2, 5/3)$.

(c) The Jacobian matrix is

$$\mathbf{J} = \begin{pmatrix} -1 + 5x - 3x^2 - 3y/10 & -3x/10 \\ y & -3/2 + x \end{pmatrix}.$$

At $(0, 0)$,

$$\mathbf{J}(0, 0) = \begin{pmatrix} -1 & 0 \\ 0 & -3/2 \end{pmatrix}.$$

The eigenvalues and eigenvectors are $\lambda_1 = -1$, $\mathbf{v}_1 = (1, 0)^T$ and $\lambda_2 = -3/2$, $\mathbf{v}_2 = (0, 1)^T$. The eigenvalues are both negative. Therefore, $(0, 0)$ is a stable node, which is asymptotically stable.

At $(1/2, 0)$,

$$\mathbf{J}(1/2, 0) = \begin{pmatrix} 3/4 & -3/20 \\ 0 & -1 \end{pmatrix}.$$

The eigenvalues and eigenvectors are $\lambda_1 = 3/4$, $\mathbf{v}_1 = (1, 0)^T$ and $\lambda_2 = -1$, $\mathbf{v}_2 = (3, 35)^T$. The eigenvalues have opposite sign. Therefore, $(1/2, 0)$ is a saddle point, which is unstable.

At $(2, 0)$,

$$\mathbf{J}(2, 0) = \begin{pmatrix} -3 & -3/5 \\ 0 & 1/2 \end{pmatrix}.$$

The eigenvalues and eigenvectors are $\lambda_1 = -3$, $\mathbf{v}_1 = (1, 0)^T$ and $\lambda_2 = 1/2$, $\mathbf{v}_2 = (6, -35)^T$. The eigenvalues have opposite sign. Therefore, $(2, 0)$ is a saddle, which is unstable.

At $(3/2, 5/3)$,

$$\mathbf{J}(3/2, 5/3) = \begin{pmatrix} -3/4 & -9/20 \\ 5/3 & 0 \end{pmatrix}.$$

The eigenvalues and eigenvectors are $\lambda_1 = (-3 + i\sqrt{39})/8$, $\mathbf{v}_1 = ((-9 + i3\sqrt{39})/40, 1)^T$ and $\lambda_2 = (-3 - i\sqrt{39})/8$, $\mathbf{v}_2 = ((-9 - i3\sqrt{39})/40, 1)^T$. The eigenvalues have negative real part. Therefore, $(3/2, 5/3)$ is a stable spiral, which is asymptotically stable.

(d,e)

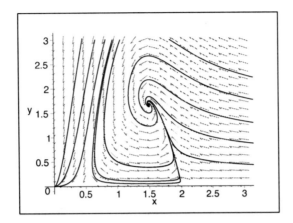

(f) The solution curve which is heading towards $(1/2, 0)$ is a separatrix. Trajectories on either side of that solution curve converge to the node $(0, 0)$ or the stable spiral at $(3/2, 5/3)$.

6. Given that t is measured from the time x is a maximum, we have

$$x = \frac{c}{\gamma} + \frac{cK}{\gamma} \cos(\sqrt{ac}\, t) \qquad y = \frac{a}{\alpha} + K\frac{a}{\alpha}\sqrt{\frac{c}{\alpha}} \sin(\sqrt{ac}\, t).$$

The period of oscillation is $T = 2\pi/\sqrt{ac}$. Based on the properties of $\cos(x)$ and $\sin(x)$, we see that the prey population, x is at maximum at $t = 0$ and $t = T$. It reaches a minimum at $t = T/2$. Its rate of increase is greatest at $t = 3T/4$. The rate of the decrease of the prey population is greatest at $t = T/4$. The predator population, y, is maximum at $t = T/4$. It is a minimum at $t = 3T/4$. The rate of increase of the predator population is greatest at $t = 0$ and $t = T$. The rate of decrease of the predator population is greatest at $t = T/2$.

In what follows, we consider problem 2 with an initial condition of $x(0) = 0.7$ and $y(0) = 2$. The critical point of interest is $(0.5, 2)$. Since $a = 1$ and $c = 1/4$, the period of oscillation is $T = 4\pi$.

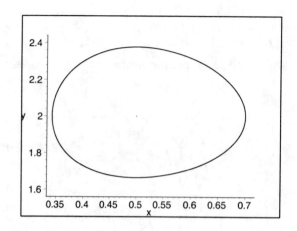

7.

(a) Looking at the coefficient of the trigonometric functions in equations (24), we see that the ratio will be given by

$$\frac{(c\gamma)}{(a/\alpha)\sqrt{c/a}} = \frac{\alpha\sqrt{c}}{\gamma\sqrt{a}}.$$

(b) For system (2), $a = 1$, $\alpha = 0.5$, $c = 0.75$ and $\gamma = 0.25$. Therefore, the ratio is $0.5\sqrt{0.75}/0.25 = \sqrt{3}$.

(c) The amplitude for the prey function in Figure 7.4.3 is approximately 2.5 and the amplitude for the predator function in Figure 7.4.3 is approximately 1.5. Using these approximations, we say that the ratio is approximately 5/3 which is close to $\sqrt{3}$.

(d) The we show graphs of the prey and predators for initial conditions $x(0) = 1$, $y(0) = 1/2$. In this case, we emphasize that the ratio of the amplitudes is approximately $5.3/3 \approx 1.77$. In particular, we note that the ratio of the amplitudes slowly increases as the initial condition (x_0, y_0) moves away from the equilibrium point $(3, 2)$.

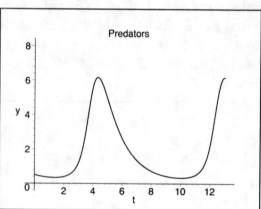

8.

(a) The period of oscillation for the linear system is $T = 2\pi/\sqrt{ac}$. In system (2), $a = 1$ and $c = 0.75$. Therefore, the period is estimated as $T = 2\pi/\sqrt{0.75} \approx 7.2552$.

(b) For the nonlinear system, the period is approximately 7.6. Therefore, the estimate is close to the actual period.

(c) For initial conditions $x(0) = 3.5, 4.0, 4.5, 5.0$ with $y(0) = 2$, the periods are $7.26, 7.29, 7.34, 7.42$, respectively. We see the period increases as the initial point moves away from the critical point $(3, 2)$.

9.

(a)

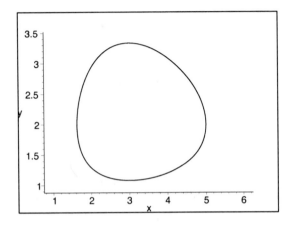

The period of oscillation is determined by observing when the trajectory becomes a closed curve. In this case, $T \approx 6.45$.

(b)

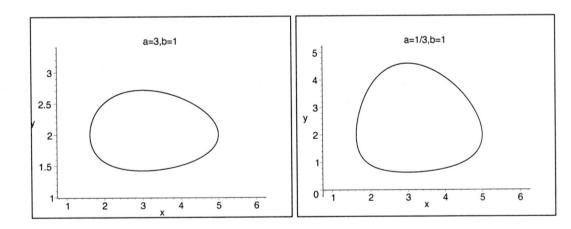

For $a = 3$, $T \approx 3.69$. For $a = 1/3$, $T \approx 11.44$.

(c)

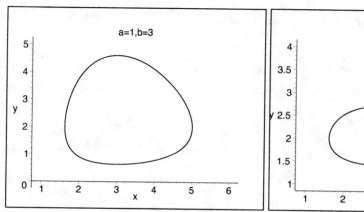

For $b = 3$, $T \approx 3.82$. For $b = 1/3$, $T \approx 11.06$.

(d) If one parameter is fixed, the period varies inversely with the other parameter.

10.

(a) First, we notice that

$$\int_A^{A+T} \cos(\sqrt{ac}\, t + \phi)\, dt = \int_A^{A+T} \sin(\sqrt{ac}\, t + \phi)\, dt.$$

Therefore,

$$\bar{x} = \frac{1}{T} \int_A^{A+T} \frac{c}{\gamma}\, dt = \frac{c}{\gamma}$$

$$\bar{y} = \frac{1}{T} \int_A^{A+T} \frac{a}{\alpha}\, dt = \frac{a}{\alpha}.$$

(b) One way to estimate the mean values is to find a horizontal line such that the area above the line is approximately equal to the area under the line. From Figure 7.4.3, it appears that $\bar{x} \approx 3.25$ and $\bar{y} \approx 2.0$. Here, $c = 0.75$ and $\gamma = 0.25$. Therefore, $c/\gamma = 3$. Further, $a = 1$ and $\alpha = 0.5$. Therefore, $a/\alpha = 2$.

(c) Consider the system

$$\frac{dx}{dt} = x\left(1 - \frac{y}{2}\right)$$

$$\frac{dy}{dt} = y\left(-\frac{3}{4} + \frac{x}{4}\right).$$

Here, we plot the solutions for initial conditions $x(0) = 3$ and $y(0) = 2.5$.

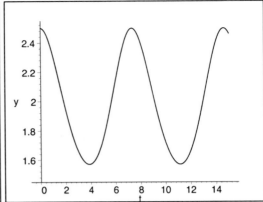

We estimate $\overline{x} = 3.05$ and $\overline{y} = 2$. Here, we plot the solutions for initial conditions $x(0) = 3$ and $y(0) = 3$.

 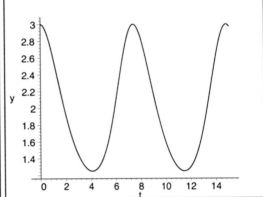

We estimate $\overline{x} = 3.25$ and $\overline{y} = 2.125$. The mean values increase as the amplitude increases. Therefore, in particular, the mean values increase as the initial conditions move farther from the critical point.

11.

(a) Looking at the equation for $x' = 0$, we need $x = 0$ or $\sigma x + 0.5y = 0$ Looking at the equation for $y' = 0$, we need $y = 0$ or $x = 3$. Therefore, the critical points are given by $(0,0)$, $(1/\sigma, 0)$ and $(3, 2 - 6\sigma)$. As σ increases from zero, the critical point $(1/\sigma, 0)$ approaches the origin, and the critical point $(3, 2 - 6\sigma)$ will eventually leave the first quadrant and enter the fourth quadrant.

(b) Here, we have $F(x, y) = x(1 - \sigma x - 0.5y)$ and $G(x, y) = y(-0.75 + 0.25x)$. Therefore, the Jacobian matrix for this system is

$$\begin{pmatrix} F_x & F_y \\ G_x & G_y \end{pmatrix} = \begin{pmatrix} 1 - 2\sigma x - 0.5y & -0.5x \\ 0.25y & -0.75 + 0.25x \end{pmatrix}.$$

We will look at the linear systems near the critical points above. Near the critical point $(0, 0)$, the Jacobian matrix is

$$\begin{pmatrix} F_x(0,0) & F_y(0,0) \\ G_x(0,0) & G_y(0,0) \end{pmatrix} = \begin{pmatrix} 1 & 0 \\ 0 & -\frac{3}{4} \end{pmatrix}$$

83

and the corresponding linear system near $(0,0)$ is

$$\frac{d}{dt}\begin{pmatrix} x \\ y \end{pmatrix} = \begin{pmatrix} 1 & 0 \\ 0 & -\frac{3}{4} \end{pmatrix}\begin{pmatrix} x \\ y \end{pmatrix}.$$

Near the critical point $(1/\sigma, 0)$, the Jacobian matrix is

$$\begin{pmatrix} F_x(1/\sigma,0) & F_y(1/\sigma,0) \\ G_x(1/\sigma,0) & G_y(1/\sigma,0) \end{pmatrix} = \begin{pmatrix} -1 & -\frac{1}{2\sigma} \\ 0 & \frac{1-3\sigma}{4\sigma} \end{pmatrix}$$

and the corresponding linear system near $(1/\sigma, 0)$ is

$$\frac{d}{dt}\begin{pmatrix} u \\ v \end{pmatrix} = \begin{pmatrix} -1 & -\frac{1}{2\sigma} \\ 0 & \frac{1-3\sigma}{4\sigma} \end{pmatrix}\begin{pmatrix} u \\ v \end{pmatrix}$$

where $u = x - 1/\sigma$ and $v = y$. Near the critical point $(3, 2 - 6\sigma)$, the Jacobian matrix is

$$\begin{pmatrix} F_x(3,2-6\sigma) & F_y(3,2-6\sigma) \\ G_x(3,2-6\sigma) & G_y(3,2-6\sigma) \end{pmatrix} = \begin{pmatrix} -3\sigma & -\frac{3}{2} \\ \frac{1-3\sigma}{2} & 0 \end{pmatrix}$$

and the corresponding linear system near $(3, 2 - 6\sigma)$ is

$$\frac{d}{dt}\begin{pmatrix} u \\ v \end{pmatrix} = \begin{pmatrix} -3\sigma & -\frac{3}{2} \\ \frac{1-3\sigma}{2} & 0 \end{pmatrix}\begin{pmatrix} u \\ v \end{pmatrix}$$

where $u = x - 3$ and $v = y - 2 + 6\sigma$.

The eigenvalues for the linearized system near $(0,0)$ are given by $\lambda = 1, -3/4$. Therefore, $(0,0)$ is a saddle point. The eigenvalues for the linearized system near $(1/\sigma, 0)$ are given by $\lambda = -1, (1 - 3\sigma)/(4\sigma)$. For $\sigma < 1/3$, there will be one positive eigenvalue and one negative eigenvalue. In this case, $(1/\sigma, 0)$ will be a saddle point. For $\sigma > 1/3$, both eigenvalues will be negative, in which case $(1/\sigma, 0)$ will be an asymptotically stable node. The eigenvalues for the linearized system near $(3, 2 - 6\sigma)$ are $\lambda = (-3\sigma \pm \sqrt{9\sigma^2 + 9\sigma - 3})/2$. Solving the polynomial equation $9\sigma^2 + 9\sigma - 3 = 0$, we see that the eigenvalues will have non-zero imaginary part if $0 < \sigma < (\sqrt{21} - 3)/6$. In this case, since the real part, -3σ will be negative, the point $(3, 2 - 6\sigma)$ will be an asymptotically stable spiral point. If $\sigma > (\sqrt{21} - 3)/6$, then the eigenvalues will both be real. We just need to determine whether they will have the same sign or opposite signs. Solving the equation $-3\sigma + \sqrt{9\sigma^2 + 9\sigma - 3} = 0$, we see that the cut-off is $\sigma = 1/3$. In particular, we conclude that if $(\sqrt{21} - 3)/6 < \sigma < 1/3$, then this critical point will have two real-valued eigenvalues which are negative, in which case this critical point will be an asymptotically stable node. If $\sigma > 1/3$, however, the eigenvalues will be real-valued, but with opposite signs, in which case $(3, 2 - 6\sigma)$ will be a saddle point.

We see that the critical point $(3, 2 - 6\sigma)$ is the critical point in the first quadrant if $0 < \sigma < 1/3$. From the analysis above, we see that the nature of the critical point changes at $\sigma_1 = (\sqrt{21} - 3)/6$. In particular, at this value of σ, the critical point switches from an asymptotically stable spiral point to an asymptotically stable node.

(c) The two phase portraits below are shown $\sigma = 0.1$ and $\sigma = 0.3$, respectively.

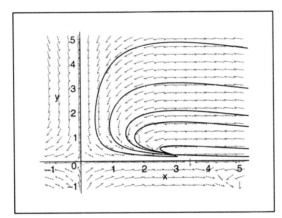

(d) As σ increases, the spiral behavior disappears. For smaller values of σ, the number of prey will decrease, causing a decrease in the number of predators, but then triggering an increase in the number of prey and eventually an increase in the number of predators. This cycle will continue to repeat as the system approaches the equilibrium point. As the value for σ increases, the cycling behavior between the predators and prey goes away.

12.

(a) The critical points are given by $(0, 0)$, $(a/\sigma, 0)$ and $(c/\gamma, (a\gamma - \sigma c)/(\alpha\gamma))$. We need to assume that $a/\sigma > c/\gamma$ in order to guarantee that we have an equilibrium solution in which y (the number of predators) is positive. As σ increases from zero, the critical point $(a/\sigma, 0)$ approaches the origin, and the critical point $(c/\gamma, (a\gamma - \sigma c)/(\alpha\gamma))$ will eventually leave the first quadrant and enter the fourth quadrant.

(b) Here, we have $F(x, y) = x(a - \sigma x - \alpha y)$ and $G(x, y) = y(-c + \gamma x)$. Therefore, the Jacobian matrix for this system is

$$\begin{pmatrix} F_x & F_y \\ G_x & G_y \end{pmatrix} = \begin{pmatrix} a - 2\sigma x - \alpha y & -\alpha x \\ \gamma y & -c + \gamma x \end{pmatrix}.$$

85

We will look at the linear systems near the critical points above. Near the critical point $(0,0)$, the Jacobian matrix is

$$\begin{pmatrix} F_x(0,0) & F_y(0,0) \\ G_x(0,0) & G_y(0,0) \end{pmatrix} = \begin{pmatrix} a & 0 \\ 0 & -c \end{pmatrix}$$

and the corresponding linear system near $(0,0)$ is

$$\frac{d}{dt}\begin{pmatrix} x \\ y \end{pmatrix} = \begin{pmatrix} a & 0 \\ 0 & -c \end{pmatrix}\begin{pmatrix} x \\ y \end{pmatrix}.$$

Near the critical point $(a/\sigma, 0)$, the Jacobian matrix is

$$\begin{pmatrix} F_x(a/\sigma,0) & F_y(a/\sigma,0) \\ G_x(a/\sigma,0) & G_y(a/\sigma,0) \end{pmatrix} = \begin{pmatrix} -a & -\frac{a\alpha}{\sigma} \\ 0 & \frac{a\gamma-c\sigma}{\sigma} \end{pmatrix}$$

and the corresponding linear system near $(a/\sigma, 0)$ is

$$\frac{d}{dt}\begin{pmatrix} u \\ v \end{pmatrix} = \begin{pmatrix} -a & -\frac{a\alpha}{\sigma} \\ 0 & \frac{a\gamma-c\sigma}{\sigma} \end{pmatrix}\begin{pmatrix} u \\ v \end{pmatrix}$$

where $u = x - a/\sigma$ and $v = y$. Near the critical point $(c/\gamma, (a\gamma - \sigma c)/(\alpha\gamma))$, the Jacobian matrix is

$$\begin{pmatrix} F_x(c/\gamma,(a\gamma-\sigma c)/(\alpha\gamma)) & F_y(c/\gamma,(a\gamma-\sigma c)/(\alpha\gamma)) \\ G_x(c/\gamma,(a\gamma-\sigma c)/(\alpha\gamma)) & G_y(c/\gamma,(a\gamma-\sigma c)/(\alpha\gamma)) \end{pmatrix} = \begin{pmatrix} -\frac{\sigma c}{\gamma} & -\frac{\alpha c}{\gamma} \\ \frac{a\gamma-c\sigma}{\gamma} & 0 \end{pmatrix}$$

and the corresponding linear system near $(c/\gamma, (a\gamma - \sigma c)/(\alpha\gamma))$ is

$$\frac{d}{dt}\begin{pmatrix} u \\ v \end{pmatrix} = \begin{pmatrix} -\frac{\sigma c}{\gamma} & -\frac{\alpha c}{\gamma} \\ \frac{a\gamma-c\sigma}{\gamma} & 0 \end{pmatrix}\begin{pmatrix} u \\ v \end{pmatrix}$$

where $u = x - c/\gamma$ and $v = y - (a\gamma - \sigma c)/(\alpha\gamma)$.

The eigenvalues for the linearized system near $(0,0)$ are given by $\lambda = a, -c$. Therefore, $(0,0)$ is a saddle point. The eigenvalues for the linearized system near $(a/\sigma, 0)$ are given by $\lambda = -a, (a\gamma - c\sigma)/\sigma$. For $\sigma < a\gamma/c$, there will be one positive eigenvalue and one negative eigenvalue. In this case, $(a/\sigma, 0)$ will be a saddle point. For $\sigma > a\gamma/c$, both eigenvalues will be negative, in which case $(a/\sigma, 0)$ will be an asymptotically stable node. The eigenvalues for the linearized system near $(c/\gamma, (a\gamma - \sigma c)/(\alpha\gamma))$ are $\lambda = [-c\sigma \pm \sqrt{c^2\sigma^2 + 4c^2 a\sigma - 4ac a\gamma}]/(2\gamma)$. Solving the quadratic equation $c^2\sigma^2 + 4c^2 a\sigma - 4ac a\gamma = 0$, we see that the eigenvalues will have non-zero imaginary part if $0 < \sigma < (-2c\alpha + 2\sqrt{c^2\alpha^2 + ac\alpha\gamma})/c$. We notice that $(-2c\alpha + 2\sqrt{c^2\alpha^2 + ac\alpha\gamma})/c < a\gamma/c$, as claimed. In this case, since the real part, $-c\sigma$ will be negative, the point $(c/\gamma, (a\gamma - \sigma c)/(\alpha\gamma))$ will be an asymptotically stable spiral point. If $\sigma > (-2c\alpha + 2\sqrt{c^2\alpha^2 + ac\alpha\gamma})/c$, then the eigenvalues will both be real. We just need to determine whether they will have the same sign or opposite signs. Solving the equation $-c\sigma + \sqrt{c^2\sigma^2 + 4c^2 a\sigma - 4ac a\gamma} = 0$, we see that the cut-off is $\sigma = a\gamma/c$. In particular, we conclude that if $(-2c\alpha + 2\sqrt{c^2\alpha^2 + ac\alpha\gamma})/c < \sigma < a\gamma/c$, then this critical point will have two real-valued eigenvalues which are negative, in which case this critical point will be an asymptotically stable node. If $\sigma > a\gamma/c$, however, the eigenvalues will be real-valued, but with opposite signs, in which case this critical point will be a saddle point.

(c) As shown in part (b), at $\sigma = (-2c\alpha + 2\sqrt{c^2\alpha^2 + ac\alpha\gamma})/c$, the nature of the critical point will change from an asymptotically stable spiral point to an asymptotically stable node.

(d) The behavior is similar to the behavior in problem 11. See the solution to 11(d) above.

13.

(a) Solving the equations $x' = 0$ and $y' = 0$ simultaneously, we arrive at the critical points $(0,0)$, $(5,0)$ and $(2,2.4)$.

(b) Here, we have $F(x,y) = x(1 - 0.2x - \frac{2y}{x+6})$ and $G(x,y) = y(-0.25 + \frac{x}{x+6})$. Therefore, the Jacobian matrix for this system is

$$\begin{pmatrix} F_x & F_y \\ G_x & G_y \end{pmatrix} = \begin{pmatrix} 1 - 0.4x - \frac{12y}{(x+6)^2} & -\frac{2x}{x+6} \\ \frac{6y}{(x+6)^2} & -\frac{1}{4} + \frac{x}{x+6} \end{pmatrix}.$$

We will look at the linear systems near the critical points above. Near the critical point $(0,0)$, the Jacobian matrix is

$$\begin{pmatrix} F_x(0,0) & F_y(0,0) \\ G_x(0,0) & G_y(0,0) \end{pmatrix} = \begin{pmatrix} 1 & 0 \\ 0 & -\frac{1}{4} \end{pmatrix}$$

and the corresponding linear system near $(0,0)$ is

$$\frac{d}{dt}\begin{pmatrix} x \\ y \end{pmatrix} = \begin{pmatrix} 1 & 0 \\ 0 & -\frac{1}{4} \end{pmatrix}\begin{pmatrix} x \\ y \end{pmatrix}.$$

Near the critical point $(5,0)$, the Jacobian matrix is

$$\begin{pmatrix} F_x(5,0) & F_y(5,0) \\ G_x(5,0) & G_y(5,0) \end{pmatrix} = \begin{pmatrix} -1 & -\frac{10}{11} \\ 0 & \frac{9}{44} \end{pmatrix}$$

and the corresponding linear system near $(5,0)$ is

$$\frac{d}{dt}\begin{pmatrix} u \\ v \end{pmatrix} = \begin{pmatrix} -1 & -\frac{10}{11} \\ 0 & \frac{9}{44} \end{pmatrix}\begin{pmatrix} u \\ v \end{pmatrix}$$

where $u = x - 5$ and $v = y$. Near the critical point $(2,2.4)$, the Jacobian matrix is

$$\begin{pmatrix} F_x(2,2.4) & F_y(2,2.4) \\ G_x(2,2.4) & G_y(2,2.4) \end{pmatrix} = \begin{pmatrix} -\frac{1}{4} & -\frac{1}{2} \\ \frac{9}{40} & 0 \end{pmatrix}$$

and the corresponding linear system near $(2,2.4)$ is

$$\frac{d}{dt}\begin{pmatrix} u \\ v \end{pmatrix} = \begin{pmatrix} -\frac{1}{4} & -\frac{1}{2} \\ \frac{9}{40} & 0 \end{pmatrix}\begin{pmatrix} u \\ v \end{pmatrix}$$

where $u = x - 2$ and $v = y - 2.4$.

The eigenvalues for the linearized system near $(0,0)$ are $\lambda = 1, -1/4$. Therefore, $(0,0)$ is a saddle point. The eigenvalues for the linearized system near $(5,0)$ are $\lambda = -1, 9/44$. Therefore, $(5,0)$ is a saddle point. The eigenvalues for the linearized system near $(2,2.4)$ are $\lambda = (-5 \pm i\sqrt{155})/40$. Since the real part of these eigenvalues is negative, $(2,2.4)$ is an asymptotically stable spiral point.

87

(c)

14.

(a) If the prey are harvested, then there will be less prey available for the predators, thus causing a decrease in the number of predators and allowing more of the prey that are not harvested to survive. (As we will see below, the number of prey will not change.) If the predators are harvested, then there will be less predators to eat the prey, thus, allowing the number of prey to increase. As a result, with more prey available, a larger percentage of the predators which are not harvested will be able to survive. (As we will see below the total number of predators will remain the same.) If they are both harvested, then initially there will be less prey available for the predators, causing a decrease in the number of predators, thus leading to an increase in the number of prey which survive.

(b) The equilibrium solution will occur when $x' = 0$, $y' = 0$. Solving these equations, we see that the equilibrium solution (with non-zero amounts of predators and prey) is given by $((c + E_2)/\gamma, (a - E_1)/\alpha)$. Therefore, if $E_1 > 0$ and $E_2 = 0$, then the number of prey stay the same, but the number of predators decreases.

(c) Using the equilibrium solution from part (b), we see that if $E_1 = 0$, $E_2 > 0$, then the number of prey increases, while the number of predators stays the same.

(d) If both $E_1 > 0$ and $E_2 > 0$, then the number of prey increases and the number of predators decreases.

15.

(a) Solving the equations $x' = 0$ and $y' = 0$, we see that the equilibrium solution (in which we have a non-zero number of predators and prey) is given by $((c + E_2)/gamma, (a\gamma - E_1\gamma - \sigma c - \sigma E_2)/(\alpha\gamma))$. Therefore, if $E_1 > 0$, but $E_2 = 0$, then the number of prey stays the same, while the number of predators decreases.

(b) If $E_1 = 0$, but $E_2 > 0$, then the number of prey increases, while the number of predators decreases.

(c) If both $E_1 > 0$ and $E_2 > 0$, then the number of prey increases, while the number of predators decreases.

16.

(a) Similarly analysis as the solution to 14(a).

(b) If $H_2 = 0$, then solving the predator equation $y' = 0$, we see that the number of prey in the equilibrium solution will be $x = 3$. Then, plugging $x = 3$ into the equation for $x' = 0$, we conclude that the number of predators in the equilibrium solution will be $y = 2 - 2H_1/3$. Therefore, if $H_1 > 0$, but $H_2 = 0$, then the number of prey will remain the same, while the number of predators will decrease.

(c) If $H_1 = 0$, then solving the prey equation $x' = 0$, we see that the number of predators in the equilibrium solution will be $y = 2$. Plugging $y = 2$ into the predators equation $y' = 0$, we conclude that the number of prey in the equilibrium solution will be $x = 2H_2 + 3$. Therefore, if $H_1 = 0$ but $H_2 > 0$, then the number of prey will increase, while the number of predators will stay the same.

(d) Now consider $H_1, H_2 > 0$. The equilibrium solution must satisfy $x(1 - 0.5y) - H_1 = 0$ and $y(-0.75 + 0.25x) - H_2 = 0$. In other words, it must satisfy $x(1 - 0.5y) = H_1 > 0$ and $y(-0.75 + 0.25x) = H_2 > 0$. Therefore, from the first equation, we need $x > 0$ and $1 - 0.5y > 0$ or $x < 0$ and $1 - 0.5y < 0$, but we are assuming the number of prey is positive, therefore, we must have $1 - 0.5y > 0$. In other words, $2 > y$. Therefore, the number of predators will decrease from the number of predators in the presence of no harvesting. Similarly, we must have $-0.75 + 0.25x > 0$ which implies $x > 3$. Therefore, the number of prey will increase from the number of prey in the presence of no harvesting.

Section 7.5

1. The equilibrium solutions of the ODE are $r = 0$ and $r = 1$. We notice that for $0 < r < 1$, $dr/dt > 0$, while for $r > 1$, $dr/dt < 0$. Therefore, $r = 0$ is an unstable critical point, while $r = 1$ is an asymptotically stable critical point. A limit cycle is given by $r = 1$, $\theta = t + t_0$, which is asymptotically stable.

2. The equilibrium solutions are given by $r = 0$ and $r = 1$. We notice that for $0 < r < 1$, $dr/dt > 0$, while for $r > 1$, $dr/dt > 0$, as well. Therefore, $r = 0$ is an unstable critical point, while $r = 1$ is a semistable critical point. A limit cycle is given by $r = 1$, $\theta = -t + t_0$, which is semistable.

3. The equilibrium solutions are given by $r = 0$, $r = 1$ and $r = 3$. We notice that $dr/dt > 0$ for $0 < r < 1$ and $r > 3$, while $dr/dt < 0$ for $1 < r < 3$. Therefore, $r = 0$ is an unstable critical point, $r = 1$ is an asymptotically stable critical point, and $r = 3$ is an unstable critical point. A limit cycle is given by $r = 1$, $\theta = t + t_0$, which is asymptotically stable. Another limit cycle is given by $r = 3$, $\theta = t + t_0$. This limit cycle is unstable.

4. The equilibrium solutions are given by $r = 0$, $r = 1$ and $r = 2$. We notice that $dr/dt < 0$ for $0 < r < 1$ and $r > 2$, while $dr/dt > 0$ for $1 < r < 2$. Therefore, $r = 0$ is a stable critical

point, $r = 1$ is an unstable critical point, and $r = 2$ is an asymptotically stable critical point. A limit cycle is given by $r = 1$, $\theta = -t + t_0$, which is unstable. Another limit cycle is given by $r = 2$, $\theta = -t + t_0$, which is asymptotically stable.

5. The equilibrium solutions of the ODE are given by $r = n$, $n = 0, 1, 2, \ldots$ Based on the sign of dr/dt in the neighborhood of each critical value, we see that the equilibrium solutions corresponding to $r = 2k$ for $k = 1, 2, \ldots$ are unstable periodic solutions with $\theta = t + t_0$. The equilibrium solutions $r = 2k + 1$ for $k = 0, 1, 2, \ldots$ correspond to stable limit cycles with $\theta = t + t_0$. The solution $r = 0$ corresponds to an unstable critical point.

6. The equilibrium solutions are given by $r = 0$, $r = 2$ and $r = 3$. We notice that $dr/dt < 0$ for $0 < r < 2$ and $2 < r < 3$, while $dr/dt > 0$ for $r > 3$. Therefore, $r = 0$ is a stable critical point, $r = 2$ is a semistable critical point, and $r = 3$ is an unstable critical point. A semistable limit cycle is given by $r = 2$, $\theta = -t + t_0$. An unstable limit cycle is given by $r = 3$, $\theta = -t + t_0$.

7.

$$y\frac{dx}{dt} - x\frac{dy}{dt} = r\sin\theta\left[r\cos\theta - r\sin\theta\frac{d\theta}{dt}\right] - r\cos\theta\left[r_t\sin\theta + r\cos\theta\frac{d\theta}{dt}\right]$$

$$= -r^2\sin^2\theta\frac{d\theta}{dt} - r^2\cos^2\theta\frac{d\theta}{dt}$$

$$= -r^2\frac{d\theta}{dt}.$$

8.

(a) Using the fact that

$$r\frac{dr}{dt} = x\frac{dx}{dt} + y\frac{dy}{dt},$$

here we have

$$r\frac{dr}{dt} = \frac{x^2 f(r)}{r} + \frac{y^2 f(r)}{r}$$

$$= \frac{r^2 f(r)}{r}$$

$$= rf(r).$$

Therefore, $rdr/dt = rf(r)$, which implies $dr/dt = f(r)$. Therefore, we have periodic solutions corresponding to the zeros of $f(r)$. To find the direction of motion on the closed trajectories, we use the fact that

$$-r^2\frac{d\theta}{dt} = y\frac{dx}{dt} - x\frac{dy}{dt}.$$

Therefore, for this system, we have

$$-r^2\frac{d\theta}{dt} = -y^2 + \frac{xyf(r)}{r} - x^2 - \frac{xyf(r)}{r}$$

$$= -x^2 - y^2$$

$$= -r^2.$$

90

Therefore, $d\theta\,dt = 1$, which implies $\theta = t + t_0$. Therefore, the closed trajectories will move in the counter-clockwise direction.

(b) By part (a), we know the periodic solutions will be given by the zeros of f. The zeros are $r = 0, 1, 2, 3$. Using the fact that $dr/dt = f(r)$, we see that $dr/dt > 0$ if $0 < r < 1$ and $r > 3$ and $dr/dt < 0$ if $1 < r < 2$ and $2 < r < 3$. Therefore, $r = 0$ is unstable, $r = 1$ is asymptotically stable, $r = 2$ is semistable, and $r = 3$ is unstable. We conclude that there is an asymptotically stable limit cycle at $r = 1$ with $\theta = t + t_0$, a semistable limit cycle at $r = 2$ with $\theta = t + t_0$ and an unstable periodic solution at $r = 3$ with $\theta = t + t_0$.

9. Using the fact that
$$r\frac{dr}{dt} = x\frac{dx}{dt} + y\frac{dy}{dt},$$
here we have
$$r\frac{dr}{dt} = xy + \frac{x^2}{\sqrt{x^2 + y^2}}(x^2 + y^2 - 2) - xy + \frac{y^2}{\sqrt{x^2 + y^2}}(x^2 + y^2 - 2)$$
$$= r(r^2 - 2).$$

Therefore, $dr/dt = r^2 - 2$. Therefore, we have one critical point at $r = \sqrt{2}$. We see that $dr/dt > 0$ if $r > \sqrt{2}$ and $dr/dt < 0$ if $r < \sqrt{2}$. Therefore, $r = \sqrt{2}$ is unstable. To find the direction of motion on the closed trajectories, we use the fact that
$$-r^2\frac{d\theta}{dt} = y\frac{dx}{dt} - x\frac{dy}{dt}.$$

Therefore, here we have
$$-r^2\frac{d\theta}{dt} = y^2 + x^2 = r^2.$$

Therefore, $d\theta/dt = -1$, which implies $\theta = -t + t_0$.

10. Given $F(x, y) = a_{11}x + a_{12}y$ and $G(x, y) = a_{21}x + a_{22}y$, it follows that
$$F_x + G_y = a_{11} + a_{22}.$$

Based on the hypothesis, $F_x + G_y$ is either always positive or always negative on the entire plane. By Theorem 7.5.2, the system cannot have a nontrivial periodic solution.

11. Given that $F(x, y) = x + y + x^3 - y^2$ and $G(x, y) = -x + 2y + x^2y + y^3/3$,
$$F_x + G_y = 3 + 4x^2 + y^2$$

is positive for all (x, y). Therefore, by Theorem 7.5.2, the system cannot have a nontrivial periodic solution.

12. Given that $F(x, y) = -2x - 3y - xy^2$ and $G(x, y) = y + x^3 - x^2y$,
$$F_x + G_y = -1 - x^2 - y^2$$

is negative for all (x, y). Therefore, by Theorem 7.5.2, the system cannot have a nontrivial periodic solution.

13. We parametrize the curve C by t. Therefore, we can rewrite the line integral as

$$\int_C [F(x, y)\, dy - G(x, y)\, dx] = \int_t^{t+T} [F(\phi(t), \psi(t))\psi'(t) - G(\phi(t), \psi(t))\phi'(t)]\, dt$$

$$= \int_t^{t+T} [\phi'(t)\psi'(t) - \psi'(t)\phi'(t)]\, dt = 0.$$

Then, using Green's Theorem, we must have

$$\iint_R [F_x(x, y) + G_y(x, y)]\, dA = 0.$$

If this integral is zero, there must be at least one point in A for which $F_x(x, y) + G_y(x, y) = 0$. Therefore, if we have a nontrivial periodic solution, we must have $F_x + G_y = 0$ somewhere in A.

14.

(a) Based on the given graphs, we estimate the periods as follows: for $\mu = 0.2$, the period $T \approx 6.29$, for $\mu = 1$, the period $T \approx 6.66$, for $\mu = 5$, the period $T \approx 11.6$.

(b) We choose initial conditions $x(0) = 2$, $y(0) = 0$.

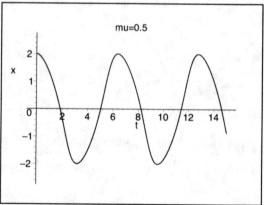

For $\mu = 0.5$, $T \approx 6.38$.

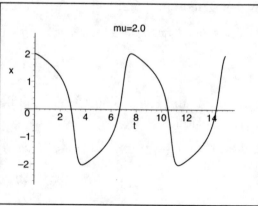

For $\mu = 2.0$, $T \approx 7.65$.

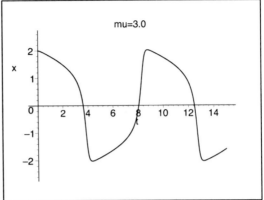

For $\mu = 3.0$, $T \approx 8.86$.

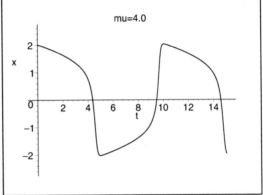

For $\mu = 4.0$, $T \approx 10.25$.

(c)

The period, T, appears to be a quadratic function of μ.

15.

(a) Letting $x = u$ and $y = u'$, we obtain the system

$$\frac{dx}{dt} = y$$

$$\frac{dy}{dt} = -x + \mu \left(1 - \frac{1}{3}y^2\right) y.$$

(b) To find the critical points, we need to solve the system

$$y = 0$$

$$-x + \mu \left(1 - \frac{1}{3}y^2\right) y = 0.$$

We see that the only solution of this system is $(0, 0)$. Therefore, the only critical point is $(0, 0)$. We notice that this system is almost linear. Therefore, we look at the Jacobian matrix. We see that

$$\mathbf{J} = \begin{pmatrix} 0 & 1 \\ -1 & \mu - \mu y^2 \end{pmatrix}.$$

Therefore,

$$\mathbf{J}(0, 0) = \begin{pmatrix} 0 & 1 \\ -1 & \mu \end{pmatrix}.$$

The eigenvalues are $\lambda = (\mu \pm \sqrt{\mu^2 - 4})/2$. If $\mu = 0$, the equation reduces to the ODE for the simple harmonic oscillator. In that case, the eigenvalues are purely imaginary and $(0, 0)$ is a center, which is stable. If $0 < \mu < 2$, the eigenvalues have non-zero imaginary part with positive real part. In that case, the critical point $(0, 0)$ is an unstable spiral. If $\mu > 0$, the eigenvalues are real and both positive. In that case, the origin is an unstable node.

(c) We will consider initial conditions $x(0) = 2$, $y(0) = 0$.

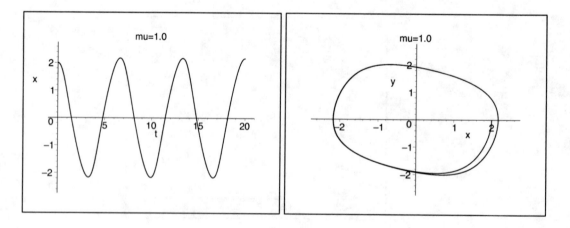

For $\mu = 1.0$, $A \approx 2.16$ and $T \approx 6.65$.

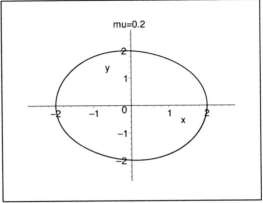

(d)

For $\mu = 0.2$, $A \approx 2.00$ and $T \approx 6.30$.

 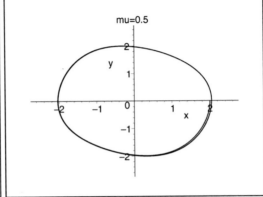

For $\mu = 0.5$, $A \approx 2.04$ and $T \approx 6.38$.

 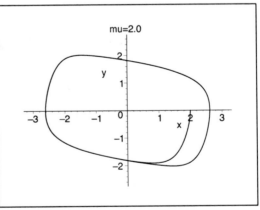

For $\mu = 2.0$, $A \approx 2.6$ and $T \approx 7.62$.

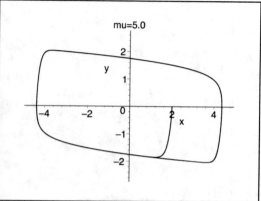

For $\mu = 5.0$, $A \approx 4.37$ and $T \approx 11.61$.

(e)

	A	T
$\mu = 0.2$	2.00	6.30
$\mu = 0.5$	2.04	6.38
$\mu = 1.0$	2.16	6.65
$\mu = 2.0$	2.6	7.62
$\mu = 5.0$	4.37	11.61

16.

(a) The critical points are solutions of

$$\mu x + y - x(x^2 + y^2) = 0$$
$$-x + \mu y - y(x^2 + y^2) = 0.$$

Multiplying the first equation by y, the second equation by x and subtracting the second equation from the first, we have

$$x^2 + y^2 = 0.$$

Therefore, the only critical point is the origin.

(b) The Jacobian matrix is given by

$$\mathbf{J} = \begin{pmatrix} \mu - 3x^2 - y^2 & 1 - 2xy \\ -1 - 2xy & \mu - x^2 - 3y^2 \end{pmatrix}.$$

Therefore, at $(0,0)$,

$$\mathbf{J}(0,0) = \begin{pmatrix} \mu & 1 \\ -1 & \mu \end{pmatrix}.$$

Therefore, the linear system near the origin is given by

$$x' = \mu x + y$$
$$y' = -x + \mu y.$$

The eigenvalues for this system are $\lambda = \mu \pm i$. For $\mu < 0$, the origin is a stable spiral. For $\mu = 0$, the origin is a center. For $\mu > 0$, the origin is an unstable spiral.

(c) As usual, let $x = r\cos\theta$ and $y = r\sin\theta$. Then multiplying the first equation of our system by x and the second equation by y, we have

$$xx' = \mu x^2 + xy - x^2(x^2 + y^2)$$
$$yy' = -xy + \mu y^2 - y^2(x^2 + y^2).$$

Adding these two equations results in

$$xx' + yy' = \mu(x^2 + y^2) - (x^2 + y^2)^2 = \mu r^2 - r^4.$$

Then, using the fact that

$$xx' + yy' = rr',$$

we conclude that $rr' = \mu r^2 - r^4$, and, therefore,

$$\frac{dr}{dt} = \mu r - r^3.$$

To find an equation for $d\theta/dt$, we multiply the first equation by y, the second equation by x and subtract the second equation from the first. We conclude that

$$yx' - xy' = x^2 + y^2 = r^2.$$

Using the fact that

$$-r^2\frac{d\theta}{dt} = yx' - xy',$$

we conclude that

$$-r^2\frac{d\theta}{dt} = r^2 \implies \frac{d\theta}{dt} = -1.$$

(d) Using the equation for dr/dt found in part (c), we see that there is a critical point at $r = 0$ and $r = \sqrt{\mu}$. Further, we note that $dr/dt > 0$ for $0 < r < \sqrt{\mu}$ and $dr/dt < 0$ for $r > \sqrt{\mu}$. Therefore, $r = \sqrt{\mu}$ is an asymptotically stable limit cycle which will attract all nonzero solutions.

17.

(a) At a critical point, we need $x' = y = 0$. Then plugging $y = 0$ into the equation for $y' = 0$, we have $-x = 0$. Therefore, $(0,0)$ is the only critical point. To determine its type and stability, we need to look at the Jacobian matrix,

$$\begin{pmatrix} F_x & F_y \\ G_x & G_y \end{pmatrix} = \begin{pmatrix} 0 & 1 \\ -1 - 2\mu xy & \mu - \mu x^2 \end{pmatrix}.$$

Therefore, the Jacobian matrix near $(0,0)$ is

$$\begin{pmatrix} 0 & 1 \\ -1 & \mu \end{pmatrix}.$$

97

The eigenvalues of this matrix are $\lambda = (\mu \pm \sqrt{\mu^2 - 4})/2$. If $\mu \geq 2$, then the eigenvalues are real and positive. In this case, the critical point is an unstable node. If $0 < \mu < 2$, then the eigenvalues have a non-zero imaginary part with a positive real part. In this case, the critical point is an unstable spiral point. If $\mu = 0$, then the eigenvalues are purely imaginary. In this case, we can only say that the critical point is either a center or spiral point, and we cannot determine the stability. If $-2 < \mu < 0$, then the eigenvalues have a non-zero imaginary part with a negative real part. In this case, the critical point is an asymptotically stable spiral point. If $\mu \leq -2$, then the eigenvalues are both real and negative. In this case, the critical point is an asymptotically stable node.

(b)

(c) Below we show the phase portraits in the case $\mu = -2, -3, -4$.

98

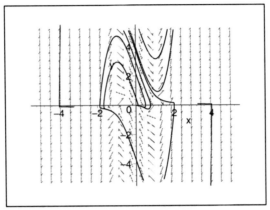

(d) Below we show the phase portraits in the cases $\mu = -0.5, -0.25, 0.25, 0.5$, respectively.

18.

(a) The critical points must satisfy $x = 0$ or $0.2x + \dfrac{2y}{x+6} = 2.4$ and $y = 0$ or $\dfrac{x}{x+6} = 0.25$. Solving these equations, we see that the critical points are $(0,0)$, $(12,0)$ and $(2,8)$.

(b) To determine the type and stability of each critical point, we need to look at the Jacobian matrix

$$\begin{pmatrix} F_x & F_y \\ G_x & G_y \end{pmatrix} = \begin{pmatrix} 2.4 - 0.4x - \frac{12y}{(x+6)^2} & -\frac{2x}{x+6} \\ \frac{6y}{(x+6)^2} & -0.25 + \frac{x}{x+6} \end{pmatrix}.$$

Therefore, the Jacobian near $(0, 0)$ is

$$\begin{pmatrix} 2.4 & 0 \\ 0 & -0.25 \end{pmatrix}.$$

The Jacobian matrix near $(12, 0)$ is

$$\begin{pmatrix} -\frac{12}{5} & \frac{4}{3} \\ 0 & \frac{5}{12} \end{pmatrix}.$$

The Jacobian matrix near $(2, 8)$ is

$$\begin{pmatrix} \frac{1}{10} & -\frac{1}{2} \\ \frac{3}{4} & 0 \end{pmatrix}.$$

The eigenvalues for $(0, 0)$ are $\lambda = 2.4, -0.25$. Therefore, $(0, 0)$ is a saddle point. The eigenvalues for $(12, 0)$ are $\lambda = -12/5, 5/12$. Therefore, $(12, 0)$ is a saddle point. The eigenvalues for $(2, 8)$ are $\lambda = .05 \pm 0.61i$. Therefore, $(2, 8)$ is an unstable spiral point.

(c)

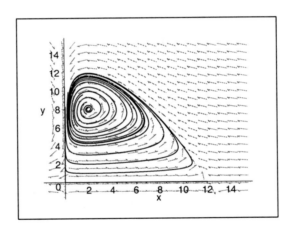

19.

(a) The critical points are $(0, 0)$, $(5a, 0)$ and $(2, 4a - 1.6)$.

(b) To determine the type and stability of the critical point in the interior of the first quadrant, we calculate the Jacobian matrix

$$\begin{pmatrix} F_x & F_y \\ G_x & G_y \end{pmatrix} = \begin{pmatrix} a - 0.4x - \frac{12y}{(x+6)^2} & -\frac{2x}{x+6} \\ \frac{6y}{(x+6)^2} & -0.25 + \frac{x}{x+6} \end{pmatrix}.$$

Therefore, the Jacobian matrix near $(2, 4a - 1.6)$ is

$$\begin{pmatrix} \frac{1}{4}a - \frac{1}{2} & -\frac{1}{2} \\ \frac{3}{8}a - \frac{3}{20} & 0 \end{pmatrix}.$$

101

The eigenvalues of this matrix are given by $\lambda = -0.25 + 0.125a \pm 0.025\sqrt{220 - 400a + 25a^2}$. We see that the term inside the square root will be negative for $8 - 2\sqrt{345}/5 < a < 8 + 2\sqrt{345}/5$. Therefore, for a in this range, the critical point will be a spiral point. We see that at $a = 2$, the real part of the eigenvalues will be zero. Therefore, at $a_0 = 2$, the critical point switches from an asymptotically stable spiral point to an unstable spiral point.

(c)

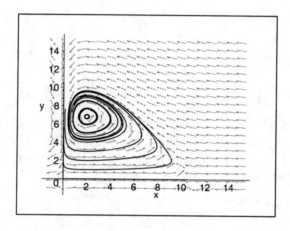

The limit cycle gets bigger as a increases.

20.

(a) The critical points are solutions of

$$1 - (b+1)x + x^2y/4 = 0$$
$$bx - x^2y/4 = 0.$$

Adding the two equations, we conclude that $x = 1$. Substituting $x = 1$ into either equation, we conclude that $y = 4b$. Therefore, the only critical point is $(1, 4b)$.

(b) The Jacobian matrix for the linearized system is

$$\mathbf{J} = \frac{1}{4}\begin{pmatrix} -4(b+1) + 2xy & x^2 \\ 4b - 2xy & -x^2 \end{pmatrix}.$$

At the critical point $(1, 4b)$,

$$\mathbf{J}(1, 4b) = \begin{pmatrix} b - 1 & 1/4 \\ -b & -1/4 \end{pmatrix}.$$

The eigenvalues of this system are

$$\lambda = -\frac{5 - 4b \pm \sqrt{9 - 40b + 16b^2}}{8}$$
$$= -\frac{(5/4 - b) \pm \sqrt{(5/4 - b)^2 - 1}}{2}.$$

102

(c) Letting $L = 5/4 - b$, the eigenvalues can be expressed as

$$\lambda = -\frac{L \pm \sqrt{L^2 - 1}}{2}.$$

The eigenvalues will be real if $L^2 \geq 1$ and have non-zero imaginary part if $L^2 < 1$.

For the complex case $(-1 < L < 1)$, the equilibrium point is a stable spiral if $0 < L < 1$. It is an unstable spiral if $-1 < L < 0$. Using the fact that $L = 5/4 - b$, we conclude that the equilibrium point is an asymptotically stable spiral if $1/4 < b < 5/4$ and an unstable spiral if $5/4 < b < 9/4$.

For the real case, $L^2 > 1$, the equilibrium point is a node. The equilibrium is a stable node if $L > 1$ and an unstable node if $L < -1$. Using the fact that $L = 5/4 - b$, we conclude that the equilibrium point is an asymptotically stable node if $0 < b < 1/4$ and an unstable node if $b > 9/4$.

(d) From part (c), we see that the equilibrium point changes from an asymptotically stable spiral to an unstable spiral when $b_0 = 5/4$.

(e)

(f)

 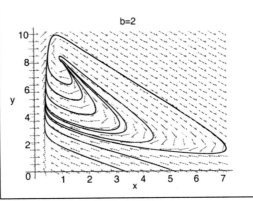

21.

(a) The critical points are solutions of

$$3\left(x + y - \frac{1}{3}x^3 - k\right) = 0$$

$$-\frac{1}{3}(x + 0.8y - 0.7) = 0.$$

Multiplying the second equation by 11.25 and adding to the first equation, we have

$$x^3 + 0.75x - 2.625 + 3k = 0.$$

Letting $f(x) = x^3 + 0.75x - 2.625 + 3k$, we see that $f'(x) = 3x^2 + 0.75 > 0$ for all x. Therefore, there is only one value of x which will satisfy the above system. For that value of x, we can calculate y and will find one critical point.

(b) Using the equation in part (a) for x and setting $k = 0$, we see that the x−coordinate of the critical point is $x = 1.1994$. Substituting that value for x into the second equation, we conclude that $y = -0.6246$. The Jacobian matrix is given by

$$\mathbf{J} = \begin{pmatrix} 3 - 3x^2 & 3 \\ -1/3 & -4/15 \end{pmatrix}.$$

Therefore, at $(1.1994, -0.6246)$,

$$\mathbf{J}(1.1994, -0.6246) = \begin{pmatrix} -1.316 & 3 \\ -1/3 & -4/15 \end{pmatrix}.$$

The eigenvalues of this matrix are $\lambda = -0.791 \pm 0.851i$. Therefore, the critical point $(1.1994, -0.6246)$ is an asymptotically stable spiral point.

Now using the equation in part (a) and setting $k = 0.5$, we see that the x−coordinate of the critical point is $x = 0.80485$. Substituting that value for x into the second equation, we conclude that $y = -0.13106$. Therefore, at $(0.80485, -0.13106)$,

$$\mathbf{J}(0.80485, -0.13106) = \begin{pmatrix} 1.05665 & 3 \\ -1/3 & -4/15 \end{pmatrix}.$$

The eigenvalues of this matrix are $\lambda = 0.395 \pm 0.7498i$. Therefore, the critical point $(0.80485, -0.13106)$ is an unstable spiral point.

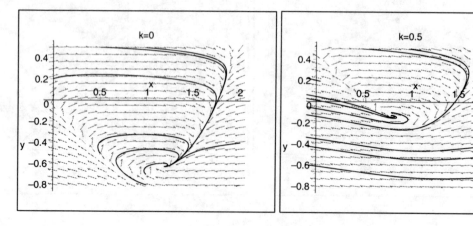

(c) By numerical calculation, we see that $k_0 \approx 0.3465$. For this value of k_0, we calculate the critical point as in part (b). In particular, we find that the critical point is $(0.9545, -0.31813)$.

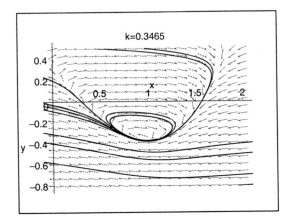

(d) In what follows, we consider initial conditions $x(0) = 0.5$, $y(0) = 0.5$.

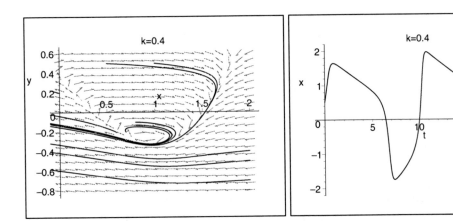

For $k = 0.4$, $T \approx 11.23$.

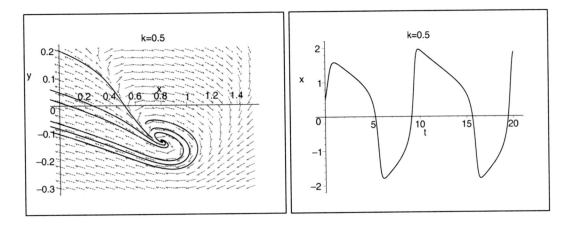

For $k = 0.5$, $T \approx 10.37$.

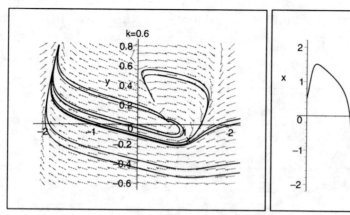

For $k = 0.6$, $T \approx 9.93$.

(e) By numerical analysis, we see that $k_1 \approx 1.4035$.

Section 7.6

1.

(a) Since the eigenvalues must be solutions of $-(8/3 + \lambda)(\lambda^2 + 11\lambda - 10(r - 1)) = 0$, we see that the eigenvalues are given by $\lambda = -8/3$ and the roots of $\lambda^2 + 11\lambda - 10(r - 1) = 0$. The roots of the quadratic equation are $\lambda = (-11 \pm sqrt81 + 40r)/2$. Therefore, the eigenvalues are as stated in equation (7).

(b) The corresponding eigenvectors are given by $\mathbf{v}_1 = (0, 0, 1)^T$ for $\lambda_1 = -8/3$, $\mathbf{v}_2 = ((-9 + \sqrt{81 + 40r})/2r, 1, 0)^T$ for $\lambda_2 = (-11 + \sqrt{81 + 40r})/2$ and $\mathbf{v}_3 = ((-9 - \sqrt{81 + 40r})/2r, 1, 0)^T$ for $\lambda_3 = (-11 - \sqrt{81 + 40r})/2$

(c) For $r = 28$, using the formulas for the eigenvalues and eigenvectors from part (b), we see that $\lambda_1 = -8/3$ with eigenvector $\mathbf{v}_1 = (0, 0, 1)^T$, $\lambda_2 \approx 11.8277$ with eigenvector $\mathbf{v}_2 \approx (20, 43.6554, 0)^T$, and $\lambda_3 \approx -22.8277$ with eigenvector $\mathbf{v}_3 \approx (20, -25.6554, 0)^T$.

2.

(a) Using the values $\sigma = 10$ and $b = 8/3$, the Jacobian matrix associated with our nonlinear system is

$$\mathbf{J} = \begin{pmatrix} -10 & 10 & 0 \\ r - z & -1 & -x \\ y & x & -8/3 \end{pmatrix}.$$

Therefore, at the critical point $P_2 = (\sqrt{8(r-1)/3}, \sqrt{8(r-1)/3}, r - 1)$,

$$\mathbf{J}(P_2) = \begin{pmatrix} -10 & 10 & 0 \\ 1 & -1 & -\sqrt{8(r-1)/3} \\ \sqrt{8(r-1)/3} & \sqrt{8(r-1)/3} & -8/3 \end{pmatrix},$$

as stated in equation (8).

106

(b) The eigenvalues of the system must satisfy $\det(A - \lambda I) = 0$ where $A = \mathbf{J}(P_2)$. We see that

$$A - \lambda I = \begin{pmatrix} -10 - \lambda & 10 & 0 \\ 1 & -1 - \lambda & -\sqrt{8(r-1)/3} \\ \sqrt{8(r-1)/3} & \sqrt{8(r-1)/3} & -8/3 - \lambda \end{pmatrix}.$$

Therefore,

$$\begin{aligned}
\det(A - \lambda I) &= (-10 - \lambda)[(1 + \lambda)(8/3 + \lambda) + (8(r-1)/3)] \\
&\quad - 10[(-8/3 - \lambda) + 8(r-1)/3] \\
&= (-10 - \lambda)[\lambda^2 + 11\lambda/3 + 8/3 + 8(r-1)/3] + 10(8/3 + \lambda) - 80(r-1)/3 \\
&= -10\lambda^2 - 110\lambda/3 - 80/3 - 80(r-1)/3 - \lambda^3 - 11\lambda^2/3 - 8\lambda/3 \\
&\quad - 8(r-1)\lambda/3 + 80/3 + 10\lambda - 80(r-1)/3 \\
&= -\lambda^3 - 41\lambda^2/3 - 8(r+10)\lambda/3 - 160(r-1)/3.
\end{aligned}$$

Setting this equation equal to zero and multiplying by -3, we arrive at equation (9).

(c) For $r = 28$, equation (9) becomes

$$3\lambda^3 + 41\lambda^2 + 304\lambda + 4320 = 0.$$

The solutions of this equation are $\lambda = -13.8546, 0.093956 \pm 10.1945i$.

3.

(a) For $r = 28$, in problem 2, we saw that the real part of the complex roots was positive. By numerical investigation, we see that the real part changes sign at $r \approx 24.737$.

(b) Suppose a cubic polynomial $x^3 + Ax^2 + Bx + C$ has one real zero and two pure imaginary zeros. Then the polynomial can be factored as $(x - \lambda_1)(x^2 + \lambda_2)$ where $\lambda_2 > 0$. Therefore,

$$x^3 + Ax^2 + Bx + C = (x - \lambda_1)(x^2 + \lambda_2) \implies A = -\lambda_1, B = \lambda_2, C = -\lambda_1\lambda_2 = AB.$$

Therefore, if $AB \neq C$, the cubic polynomial will not have the specified type of roots.

(c) First, we rewrite the equation as

$$x^3 + \frac{41}{3}\lambda^2 + \frac{8(r+10)}{3} + \frac{160(r-1)}{3} = 0.$$

Using the result from part (b), we need to find when $AB = C$, where $A = 41/3$, $B = 8(r + 10)/3$ and $C = 160(r - 1)/3$. Setting $AB = C$, we have the equation $328(r + 10)/9 = 160(r - 1)/3$. Solving this equation, we see that the real part of the complex roots changes sign when $r = 470/19$.

4.

(a)

$$\frac{dV}{dt} = \frac{\partial V}{\partial x}\frac{dx}{dt} + \frac{\partial V}{\partial y}\frac{dy}{dt} + \frac{\partial V}{\partial z}\frac{dz}{dt}$$

$$= 2rx\frac{dx}{dt} + 2\sigma y\frac{dy}{dt} + 2\sigma(z-2r)\frac{dz}{dt}$$

$$= 2\sigma rx(-x+y) + 2\sigma y(rx - y - xz) + 2\sigma(z-2r)(-bz+xy)$$

$$== -2\sigma rx^2 - 2\sigma y^2 - 2\sigma bz^2 + 4\sigma rbz$$

$$= -2\sigma[rx^2 + y^2 + bz^2 - 2rbz]$$

$$= -2\sigma[rx^2 + y^2 + b(z-r)^2 - br^2].$$

(b) We need $dV/dt < 0$ when $V(x,y,z) = c$. From part (a), we know that

$$\frac{dV}{dt} = -2\sigma[rx^2 + y^2 + b(z-r)^2 - br^2].$$

Therefore, we need to find a sufficient condition on c such that if $V(x,y,z) = c$, then $rx^2 + y^2 + b(z-r)^2 - br^2 > 0$. Let

$$f(x,y,z) = rx^2 + y^2 + b(z-r)^2 - br^2.$$

We want to insure that $f(x,y,z) > 0$ for all (x,y,z) satisfying $V(x,y,z) = c$. Therefore, we will minimize f subject to the constraint that $V = c$. Then, we will put a condition on c which guarantees that the minimum value of f is positive. Therefore, we need to minimize f subject to the constraint $V = c$. We have a Lagrange multiplier problem. We need to consider $\Delta f = \lambda \Delta g$. Setting up that system of equations, we have

$$2rx = \lambda 2rx$$

$$2y = \lambda 2ry$$

$$2b(z-r) = \lambda 2\sigma(z-2r).$$

The solutions of the first equation are $x = 0$ or $\lambda = 1$. The solutions of the second equation are $y = 0$ or $\lambda = 1/\sigma$. The solution of the last equation is $z = r(b - 2\lambda\sigma)/(b - \lambda\sigma)$. We see that if $x = 0 = y$, then $z = c/\sigma + 2r$. In that case $f(x,y,z) = b(c/\sigma + r)^2 - br^2 > 0$. If $x = 0$, $\lambda = 1/\sigma$, then $z = r(b-2)/(b-1)$. Then using the equation $V(x,y,z) = c$, we conclude that

$$y^2 = \frac{c}{\sigma} - r^2\left(\frac{b-2}{b-1} - 2\right)^2.$$

Therefore,

$$f(x,y,z) = \frac{c}{\sigma} - r^2\left(\frac{b-2}{b-1} - 2\right)^2 + br^2\left(\frac{b-2}{b-1} - 1\right)^2 - br^2.$$

Therefore, in order to guarantee that $f(x,y,z) > 0$ for all (x,y,z) satisfying $V(x,y,z) = c$, we choose c such that

$$c > \sigma r^2\left(\frac{b-2}{b-1} - 2\right)^2 - \sigma br^2\left(\frac{b-2}{b-1} - 1\right)^2 + \sigma br^2.$$

Finally, if $y = 0$ and $\lambda = 1$, then $z = r(b - 2\sigma)/(b - \sigma)$. Then using the equation $V(x, y, z) = c$, we see that

$$x^2 = \frac{c}{r} - \sigma r \left(\frac{b - 2\sigma}{b - \sigma} - 2 \right)^2.$$

Therefore,

$$f(x, y, z) = c - \sigma r^2 \left(\frac{b - 2\sigma}{b - \sigma} - 2 \right)^2 + br^2 \left(\frac{b - 2\sigma}{b - \sigma} - 1 \right)^2 - br^2.$$

Therefore, in order to guarantee that $f(x, y, z) > 0$ for all (x, y, z) satisfying $V(x, y, z) = c$, we choose c such that

$$c > \sigma r^2 \left(\frac{b - 2\sigma}{b - \sigma} - 2 \right)^2 - br^2 \left(\frac{b - 2\sigma}{b - \sigma} - 1 \right)^2 + br^2.$$

Therefore, if c satisfies both inequalities above, then we have a sufficient condition to guarantee that $f(x, y, z) > 0$ for all (x, y, z) satisfying $V(x, y, z) = c$.

5. First, we plot x versus t corresponding to $r = 28$ with an initial condition of $(5, 5, 5)$:

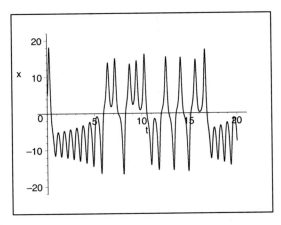

Next, we plot x versus t corresponding to $r = 28$ with an initial condition of $(5.01, 5, 5)$:

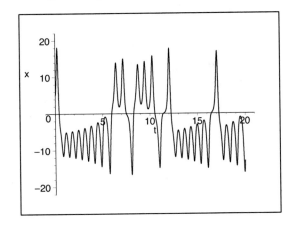

6. First, we plot the projections in the $xy-$plane:

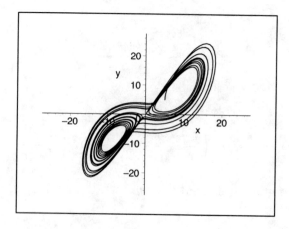

Next, we plot the projections in the $xz-$plane:

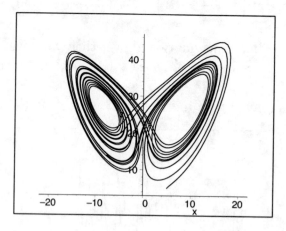

7.

(a) First, we plot x versus t corresponding to $r = 21$ with an initial condition of $(3, 8, 0)$.

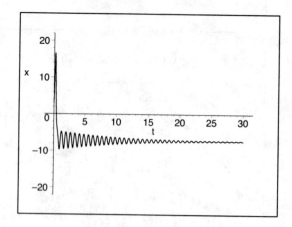

Here is a plot of x versus t corresponding to $r = 21$ with an initial condition of $(5, 5, 5)$.

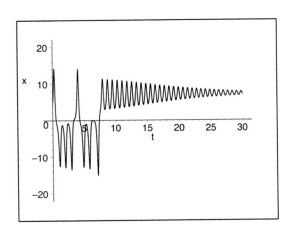

Below is a plot of x versus t corresponding to $r = 21$ with an initial condition of $(5, 5, 10)$.

(b) For $r = 22$,

111

For $r = 23$,

For $r = 24$,

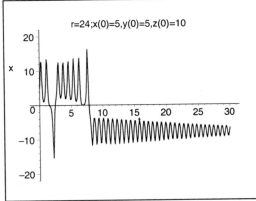

(c) Below, we consider $r = 24.5$,

8.

(a)

The period appears to be $T \approx 1.12$.

(b)

(c)

9.

(a)

(b)

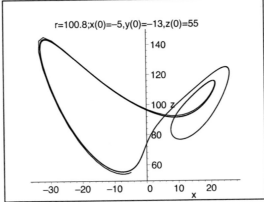

Printed in the United States
95443LV00002B/1-14/A